FUNCTIONAL AND CLINICAL NEUROANATOMY

FUNCTIONAL AND CLINICAL NEUROANATOMY
A Guide for Health Care Professionals

JAHANGIR MOINI
Professor of Science and Health (RET), Eastern Florida State College, Palm Bay, FL, United States

PIROUZ PIRAN
Neurologist, Clinical and Research, Johns Hopkins University, Baltimore, MD, United States

Academic Press is an imprint of Elsevier
125 London Wall, London EC2Y 5AS, United Kingdom
525 B Street, Suite 1650, San Diego, CA 92101, United States
50 Hampshire Street, 5th Floor, Cambridge, MA 02139, United States
The Boulevard, Langford Lane, Kidlington, Oxford OX5 1GB, United Kingdom

© 2020 Elsevier Inc. All rights reserved.

No part of this publication may be reproduced or transmitted in any form or by any means, electronic or mechanical, including photocopying, recording, or any information storage and retrieval system, without permission in writing from the publisher. Details on how to seek permission, further information about the Publisher's permissions policies and our arrangements with organizations such as the Copyright Clearance Center and the Copyright Licensing Agency, can be found at our website: www.elsevier.com/permissions.

This book and the individual contributions contained in it are protected under copyright by the Publisher (other than as may be noted herein).

Notices

Knowledge and best practice in this field are constantly changing. As new research and experience broaden our understanding, changes in research methods, professional practices, or medical treatment may become necessary.

Practitioners and researchers must always rely on their own experience and knowledge in evaluating and using any information, methods, compounds, or experiments described herein. In using such information or methods they should be mindful of their own safety and the safety of others, including parties for whom they have a professional responsibility.

To the fullest extent of the law, neither the Publisher nor the authors, contributors, or editors, assume any liability for any injury and/or damage to persons or property as a matter of products liability, negligence or otherwise, or from any use or operation of any methods, products, instructions, or ideas contained in the material herein.

Library of Congress Cataloging-in-Publication Data
A catalog record for this book is available from the Library of Congress

British Library Cataloguing-in-Publication Data
A catalogue record for this book is available from the British Library

ISBN: 978-0-12-817424-1

For information on all Academic Press publications
visit our website at https://www.elsevier.com/books-and-journals

Publisher: Nikki Levy
Acquisitions Editor: Natalie Farra
Editorial Project Manager: Timothy Bennett
Production Project Manager: Bharatwaj Varatharajan
Cover Designer: Greg Harris

Typeset by SPi Global, India

Contents

About the Authors	xv
Preface	xvii
Acknowledgments	xix

1. Histophysiology — 1

Anatomical divisions of the nervous system	1
Functional divisions of the nervous system	2
Neuroglia	3
Neurons	9
Membrane potential	16
Synapses	26
Neuronal integration of excitatory and inhibitory stimuli	30
Degeneration and regeneration of nerve cells	34
Clinical considerations	36
Clinical cases	45
Key terms	48
Suggested readings	49

2. Embryology — 51

Development of the neural tube	51
Neural crest	55
Cranial placodes	56
Medulla spinalis (spinal cord)	56
Myelencephalon (medulla oblongata)	58
Metencephalon	59
Mesencephalon (midbrain)	59
Diencephalon development	60
Telencephalon development	60
Central nervous system malformation	61
Clinical considerations	63
Clinical cases	71
Key terms	74
Suggested readings	75

3. Gross anatomy of the brain — 77

Brain divisions	77
Clinical considerations	84

Clinical cases	90
Key terms	92
Suggested readings	93

4. Meninges and ventricles — 95

Meninges	95
Dural nerve supply	98
Dural arterial supply	100
Dural venous sinuses	101
Meninges of the spinal cord	103
Ventricles	104
Cerebrospinal fluid	106
Clinical considerations	112
Clinical cases	125
Key terms	128
Suggested readings	129

5. Blood supply of the CNS — 131

Arteries of the brain	131
Cerebral arterial circle	135
Meningeal arteries	136
Veins of the brain	136
Venous dural sinuses	139
Cavernous sinus	140
Arteries of the spinal cord	140
Venous drainage of the spinal cord	141
Clinical considerations	142
Clinical cases	172
Key terms	175
Suggested readings	176

6. Cerebral cortex — 177

Cerebrum	177
Gross anatomy	177
Cerebral cortex	179
Neocortex	179
Functional cerebral cortex areas	179
Electroencephalogram	184
Split-brain syndrome	186
White matter of the cerebrum	186

Higher cortical functions	187
Clinical considerations	193
Clinical cases	235
Key terms	238
Suggested readings	239

7. Basal nuclei — 241

Structure of the basal nuclei	241
Functional considerations	242
Clinical considerations	244
Clinical cases	262
Key terms	265
Suggested readings	265

8. Diencephalon: Thalamus and hypothalamus — 267

Epithalamus	267
Thalamus	270
Subthalamus	272
Hypothalamus	273
Blood supply to the diencephalon	276
Internal capsule	277
Functions of the thalamic nuclei and their major connections	278
Functional considerations of the hypothalamus	279
Regulation of the autonomic nervous system	280
Clinical considerations	281
Clinical cases	288
Key terms	291
Suggested readings	291

9. Brainstem — 293

The rule of 4 of the brainstem	293
Midbrain	294
Pons	296
Medulla oblongata	297
Corticobulbar tracts	301
Functional considerations	301
Clinical considerations	302
Clinical cases	312
Key terms	315
Suggested readings	316

Contents

10. Cranial nerves — **319**
- Olfactory nerves (I) — 319
- Optic nerves (II) — 321
- Oculomotor nerves (III) — 322
- Trochlear nerves (IV) — 322
- Trigeminal nerves (V) — 323
- Abducens nerves (VI) — 324
- Facial nerves (VII) — 326
- Vestibulocochlear nerves (VIII) — 326
- Glossopharyngeal nerves (IX) — 327
- Vagus nerves (X) — 329
- Accessory nerves (XI) — 330
- Hypoglossal nerves (XII) — 331
- Clinical considerations — 331
- Clinical cases — 339
- Key terms — 342
- Suggested readings — 343

11. Trigeminal and facial nerves — **345**
- Trigeminal nerve — 345
- Facial nerve — 348
- Clinical considerations — 350
- Clinical cases — 357
- Key terms — 360
- Suggested readings — 360

12. Auditory system — **363**
- Anatomical structures — 363
- Hearing process — 367
- Vestibule — 371
- Semicircular canals — 373
- Cochlea and cochlear duct — 374
- Perceiving sound — 382
- Pathway of sound waves — 383
- Neural pathway of hearing — 383
- Hearing tests — 385
- Equilibrium — 386
- Clinical considerations — 390
- Key terms — 390
- Suggested readings — 391

13. Auditory system lesions and disorders — 393

Tinnitus	393
Hearing loss	394
Sudden deafness	403
Vertigo	404
Meniere's disease	405
Treatment	406
Benign paroxysmal positional vertigo	406
Ramsay Hunt syndrome	407
Acute vestibular neuronitis	408
Drug-induced ototoxicity	408
Acoustic neuroma	409
Labyrinthitis	411
Clinical cases	412
Key terms	415
Suggested readings	415

14. Visual system — 417

Anatomy of the eyeball	417
Visual acuity	427
Pupillary light reflexes and pathway	428
Visual pathways	428
Visual pigments	431
Clinical considerations	432
Clinical cases	461
Key terms	464
Suggested readings	465

15. Limbic, olfactory, and gustatory systems — 467

Limbic system	467
Olfactory system	476
Gustatory system	481
Clinical considerations	485
Clinical cases	492
Key terms	493
Suggested readings	494

16. Cerebellum — 497

Anatomy of the cerebellum	497
Cerebellar cortex	501

	Major functions of the cerebellum	508
	Clinical considerations	508
	Clinical cases	513
	Key terms	516
	Suggested readings	516
17.	**Autonomic nervous system**	**519**
	Autonomic outflow	522
	Sympathetic division	524
	Parasympathetic division	528
	Autonomic plexuses	529
	Autonomic innervation of the head	530
	Visceral afferent pathways	532
	Hierarchy of the autonomic nervous system	534
	Functional considerations	534
	Clinical considerations	537
	Clinical cases	543
	Key terms	545
	Suggested readings	546
18.	**Neurotransmitters**	**549**
	Classifications of neurotransmitters	549
	Types of neurotransmitters	549
	Opioid peptides	559
	Nonopioid neuropeptides	560
	Neuromodulators	560
	Functional considerations	561
	Clinical considerations	563
	Clinical cases	579
	Key terms	581
	Suggested readings	582
19.	**Spinal cord**	**585**
	Gross anatomy and protection	585
	Cross-sectional anatomy	588
	Innervation of specific body regions	591
	Ascending and descending pathways	595
	Integrative pathways	604
	Reflex activity	604
	Spinal reflexes	605
	Blood supply of the spinal cord	613

Clinical considerations	614
Key terms	614
Suggested readings	616

20. Spinal cord lesions and disorders — 617

Myelopathies	617
Spina bifida	635
Syringomyelia	637
Spinal cord trauma and compression	639
Spinal cord tumors	642
Clinical cases	644
Key terms	645
Suggested readings	646

21. Complete neurological exam — 647

History	647
Neurologic examination	648
Key terms	666
Suggested readings	666

22. Neurologic diagnostic procedures — 669

Lumbar puncture	669
CT scan	671
Magnetic resonance imaging	675
Cerebral catheter angiography	676
Duplex Doppler ultrasonography	676
Myelography	678
Electroencephalography	678
Electromyography	679
Key terms	680
Suggested readings	680

Glossary	*683*
Index	*719*

About the Authors

Dr. Moini was Assistant Professor at Tehran University for 9 years. He was Department Chair for Health and Science at Everest University for 15 years, and was a professor at that college for a total of 24 years. For the past 6 years, he has been a professor of Science and Health at Eastern Florida State College. Dr. Moini has taught anatomy and physiology for 30 years. He is now retired. He is also an internationally published author of 39 books over the past 20 years.

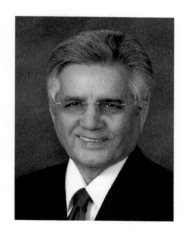

Dr. Pirouz Piran is currently serving as a clinical and research neuro-critical care fellow physician at the Johns Hopkins University Hospital. He finished his vascular neurology fellowship training at the Weill Cornell Medical Center in New York City. His neurology residency training was at the Cleveland Clinic in Florida where he served as the chief resident during his final year. He received his medical degree from Beheshti University of Medical Sciences in Tehran, Iran.

Preface

The understanding of extreme complexity of the nervous system requires significant thought and study. The study of neuroanatomy helps us to understand normal as well as abnormal behaviors. This book offers an easy-to-understand format detailing the structures and functions of the nervous system. It presents important diseases and conditions related to all parts of neuroanatomy. This book also emphasizes the methods and treatments used in practice today. It also covers updated forms of technology used for the diagnosis of neurological disorders. The book contains "Focus On" boxes that highlight key subjects, section reviews throughout the chapters, chapter objectives, summaries, and clinical considerations. Unique clinical cases, several per chapter, are included and feature critical thinking questions with answers.

The authors have focused on specifically targeted health professionals when writing this book, including medical students, nurses, physician assistants, nurse practitioners, and medical assistants who work in the field of neurology. It is also an invaluable resource for graduate and postgraduate students in neuroscience. The 22 chapters of this book begin with histophysiology, embryology, gross anatomy of the brain, meninges, ventricles, blood circulation, and the cranial and spinal nerves. It also emphasizes various sensory systems, the autonomic nervous system, and neurotransmitters. At the end of the book, there are chapters about complete neurological examinations and diagnostic procedures.

Acknowledgments

The authors appreciate the contributions of everyone who assisted in the creation of this book, especially Nikki Levy, Natalie Farra, Timothy Bennett, Bharatwaj Varatharajan, Greg Harris, and Greg Vadimsky. Also, they would like to thank Dr. Morvarid Moini for contributing artwork.

CHAPTER 1

Histophysiology

The human *nervous system* is the primary controlling and communicating system of the body. The nervous system includes multiple nervous organs, including the *brain*; *spinal cord*; *receptors* in complex sense organs such as the eye and ear; and *nerves* that link with other body systems. The complex sense organs receive collected information from external and internal stimuli, process it, and send signals to initiate required responses. **Nervous tissue** makes up the organs of the nervous system and supports blood vessels as well as connective tissues.

Neurons are the basic structural and functional units of the nervous system. They are specialized cells used for intercellular communication and serve as the transmitting cells of the nervous system. Nervous tissue contains another group of cells besides neurons. These supporting cells are called *neuroglia*, or *glial cells*, which are essential for neurons to survive and function. Neuroglia preserve physical and biochemical structures of nervous tissue. Glial cells are divided into microglia and macroglia (astrocytes, oligodendrocytes, and ependymal cells).

Anatomical divisions of the nervous system

The two divisions of the nervous system include the *central nervous system* (CNS) and *peripheral nervous system* (PNS) (see Fig. 1.1). The CNS consists of the brain and the spinal cord.

The CNS integrates, processes, and coordinates sensory data and motor commands. Sensory data concerns conditions inside or outside of the body. Motor commands regulate skeletal muscles and glands. The brain controls higher functions such as intelligence, memory, learning, cognition, and emotion.

The PNS includes the nervous tissue that is not the part of the CNS. The two PNS subdivisions are the *somatic nervous system* (SNS) and *autonomic nervous system* (ANS). The PNS brings sensory information to the CNS. It also uses motor commands to communicate with peripheral tissues and organs. *Nerve fibers* are made up by bundles of axons. These bundles carry sensory information and motor commands in the PNS. They are known as *peripheral nerves*, or more simply *nerves*. The nerves connected to the brain are referred to as *cranial nerves*, while the nerves attached to the spinal cord are called *spinal nerves*.

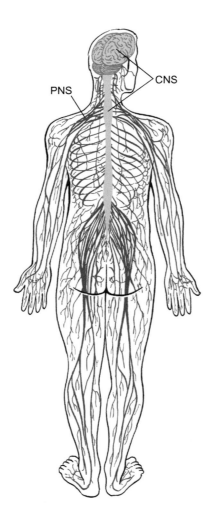

Fig. 1.1 The central nervous system and peripheral nervous system.

Functional divisions of the nervous system

The PNS is divided into afferent and efferent divisions, unlike the CNS, which does not have such divisions. The *afferent* division of the PNS carries sensory information to the CNS from peripheral tissue and organ *receptors*. These sensory structures detect changes in the internal or external environment and respond to certain stimuli. Receptors consist of simple and complex types. Single cells have thin extensions of cytoplasm as receptors. Complex receptor organs include the eye and ear. Neurons as well as specialized tissue cells may also serve as sensory receptors.

Motor commands from the CNS (brain and spinal cord) are carried by the *efferent* division of the PNS to the muscles, glands, and adipose tissues. These target *effectors* respond to these stimuli. The efferent division has two subdivisions:
- The SNS regulates skeletal muscle contractions. *Voluntary* muscle contractions are consciously controlled. *Involuntary* contractions are controlled subconsciously. The automatic responses involved in involuntary contractions are called *reflexes*.
- The *ANS* is also known as the *visceral motor system*. It subconsciously regulates cardiac muscle, smooth muscles, adipose tissues, and glandular secretions. The ANS includes two divisions: the *parasympathetic nervous system* and the *sympathetic nervous system*. These subdivisions usually have opposite actions. For example, sympathetic activity increases the heart rate while parasympathetic activity slows it.

The *ANS* within the gastrointestinal system is also called the *enteric nervous system* (ENS), a network of neurons and nerves in the walls of the digestive tract. The sympathetic and parasympathetic divisions affect ENS activities. However, the ENS is able to begin and regulate many local visceral reflexes with no input from the CNS. The ENS and the brain use identical neurotransmitters.

Section review
1. What are the subdivisions of the nervous system?
2. What are the functions of the ANS?
3. What are the differences between the afferent and efferent divisions?

Neuroglia

The functions of **neuroglia** include the separation and the protection of neurons, providing support for nervous tissue, and assisting in the regulation of interstitial fluid composition (ependymal cells and astrocytes). There are different types of neuroglia with different functions as mentioned above. Approximately 90% of the nervous system consists of neuroglia. Therefore, there are more neuroglia than neurons. In the CNS, there four types of neuroglia, while in the PNS there are only two types. Table 1.1 summarizes various types of neuroglia and their functions.

Neuroglia of the CNS

The CNS contains four types of neuroglia: *microglia, astrocytes, oligodendrocytes,* and *ependymal cells*.

Microglia
Microglia are the least common and smallest CNS neuroglia. They are phagocytic cells with thin, finely branched processes. They migrate through nervous tissue, engulfing

Table 1.1 Types of neuroglia.

Central nervous system	1. Microglia	Small, stationary phagocytes that remove debris, wastes, and pathogens from cells. They can enlarge and move due to stimulation, and also eliminate synapses (synaptic pruning)
	2. Astrocytes	Have a central body with many radiating processes. They maintain the blood-brain barrier; provide structural support; control amounts of ions, nutrients, and dissolved gases; absorb and recycle neurotransmitters; form scar tissue after injuries
	3. Oligodendrocytes	Have a central body with processes that wrap around their neuron processes. They myelinate CNS axons; provide framework
	4. Ependymal cells	They form sheets with the motile cilia, and line ventricles of the brain and the central spinal cord canal; produce cerebrospinal fluid, and also propel it
Peripheral nervous system	1. Schwann cells	Each cell entirely wraps around the neuron processes, with an outer portion called the neurilemma. They provide myelination in the PNS; surround all PNS axons; assist in repairs following injuries
	2. Satellite cells	Also called satellite glial cells, they cover the surface of nerve cell bodies in sensory, sympathetic, and parasympathetic ganglia. In the neurons around ganglia, they control levels of neurotransmitters, oxygen, carbon dioxide, and nutrients; surround neuron cell bodies

cellular debris, wastes, and pathogens. Microglia originate in early embryonic development from the mesodermal cell layer. Microglia migrate into the CNS while it is forming.

Astrocytes

Astrocytes are the most numerous and the largest neuroglia in the CNS. They are so named because of their star-like shape (see Fig. 1.2). Astrocytes have many thin cytoplasmic processes, which end in *vascular feet*, connecting them to the outside of nearby capillary walls. Astrocytes are filled with microfilaments that extend across their entire width as well as their processes. They are connected together by **tight junctions**. Astrocytes have the following functions:
- *Maintaining the blood-brain barrier (BBB)*. The **BBB** is a selectively semipermeable membrane between the capillary flow and the brain and spinal cord. This barrier allows some substances to pass through, but blocks the passage of others. The BBB is created by astrocytes, and isolates the CNS from the general circulation. Astrocytes release chemicals that help maintain the permeability of the capillary endothelial cells.

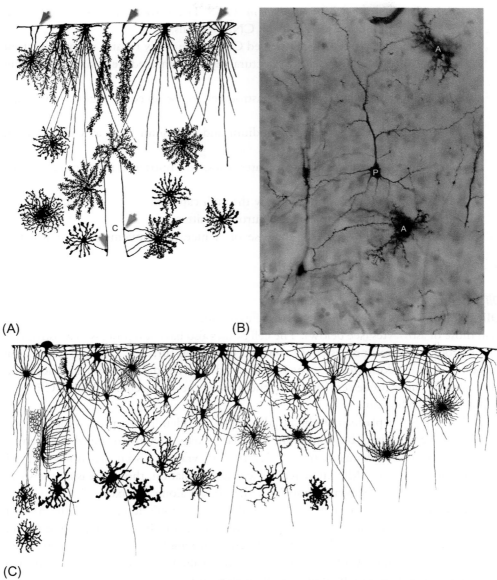

Fig. 1.2 Protoplasmic astrocytes in the cerebral cortex, stained by the Golgi method. (A) Astrocytes have enlarged end-feet, covering the surface of the CNS (*blue arrows*), contacting capillaries (letter "C", *green arrows*) and contacting neurons (not shown). (B) Golgi-stained astrocytes (letter "A") and pyramidalneurons (letter "P"). (C) Golgi-stained astrocytes from the frontal cortex of a 42-year-old woman, demonstrating their morphological diversity. *(Part A from Ramóny Cajal, S. Histologie du système nerveux de l'homme et des vertébrés; Paris: Maloine, 1909–1911. Part B courtesy of McMullen, N. T. University of Arizona, College of Medicine. Part C from Retzius, G. Biologische Untersuchungen. Die Neuroglia des Gehirns beim Menschen und bei Saeugethieren, Vol. 6; Jena: Verlagvon Gustav Fischer, 1894.)*

- *Creating a three-dimensional framework for the CNS.* The astrocyte cytoskeleton helps to provide a structural framework for the CNS neurons.
- *Repairing damaged nervous tissue.* Damaged CNS tissue rarely regains normal function. Astrocytes have the ability to make structural repairs to injured areas, stabilizing tissues and preventing additional injury.
- *Controlling the interstitial environment.* Astrocytes are believed to regulate interstitial fluid as follows:
  1. They regulate concentration of sodium and potassium ions, as well as carbon dioxide.
  2. They quickly transport dissolved gases, ions, and nutrients, ions between capillaries and neurons.
  3. They control volume of blood flow through capillaries.
  4. They absorb and recycle certain neurotransmitters.
  5. They release chemicals that increase or reduce the movement of information across axon terminals.

Oligodendrocytes and myelination

Oligodendrocytes have thin cytoplasmic extensions, but their cell bodies are smaller than those of astrocytes (see below). The cell bodies also have less process than astrocytes. The oligodendrocytes' processes are attached to neuron surfaces. The functions of the processes are still unknown. The processes that end on the surfaces of *axons* are more fully understood. These processes, insulating them from the extracellular fluid (ECF), surround many CNS axons.

As it matures, the plasma membrane adjacent to oligodendrocyte process forms a large pad, while the cytoplasm becomes thin. This flat structure wraps around the axolemma and forms concentric plasma membrane layers. The wrapping is known as *myelin* and is made up of membranes. Myelin provides electrical insulation, thus increasing the speed of electrical transition (action potentials) as they travel along the axon.

Oligodendrocytes form a *myelin sheath* along the axon's length, and the axon is referred to as being *myelinated*. The larger areas of the axon that are wrapped in myelin are known as *internodes*, which are about 1–2 mm in length. Separating adjacent internodes are small gaps, only a few micrometers long, called *nodes*, or **nodes of Ranvier**. The collateral branches of an axon's begin at these nodes.

Myelinated axons have a glossy white appearance, mostly due to the lipids in the myelin. The areas consisting mostly of myelinated axons are known as the *white matter* of the CNS. *Unmyelinated axons* do not have a complete covering of oligodendrocytes, and are common where collaterals and short axons synapse with neuron cell bodies. Regions with dendrites, neuron cell bodies, and unmyelinated axons are gray in color. Therefore, they are called the *gray matter* of the CNS. Oligodendrocytes are structurally important because they join clusters of axons.

Ependymal cells

The longitudinal axis of the brain and the spinal cord contains a central passageway filled with protective *cerebrospinal fluid* (CSF). The CSF also surrounds the brain and the spinal cord. Our brain and spinal cord float in the CSF. The central passageway of the spinal cord is referred to as the *central canal*. In several brain areas, the passageway creates enlarged cavities called *ventricles*. The central canal and ventricles are lined with **ependymal cells**, which help to produce and control CSF. Also, ependymal cells have *cilia* that assist in the circulation of CSF inside the brain ventricles.

Neuroglia of the PNS

Cell bodies of PNS are clustered together forming ganglia. Neuroglial processes insulate cell bodies of neurons, and most PNS axons, from their surroundings. The two types of neuroglia in the PNS are satellite cells and Schwann cells.

Schwann cells

Schwann cells are also known as *neurolemmocytes*, and have two types of formations. They may form a thick sheath of myelin or create indented plasma membrane folds around peripheral axons throughout the PNS. Where a Schwann cell covers an axon, the outer Schwann cell surface is known as the **neurilemma**. This shields the axon from interstitial fluids. *Nodes* are the gaps between Schwann cells. Schwann cells are basically the oligodendrocytes of our PNS. Only peripheral nerves have neurilemma (sheath of Schwann) in addition to the myelin sheath. The term neurilemma is also spelled *neurolemma*.

A myelinating Schwann cell myelinates a single axon. However, a CNS oligodendrocyte may myelinate several axons. Nonmyelinating Schwann cells may *enclose* parts of several unmyelinated axons. A series of Schwann cells encloses an axon along its entire length.

Satellite cells

Satellite cells surround neuron cell bodies in ganglia. They control the interstitial fluid around neurons, similar to the function of astrocytes in the CNS.

> **Focus on demyelination**
> *Demyelination* is progressive destruction of myelin sheaths. It can occur in the CNS and in the PNS. It causes a loss of sensation and motor control, resulting in numbness and paralysis. Several unrelated conditions can damage to myelin, and present symptoms similar to those of demyelination. Chronic exposure to arsenic, lead, mercury, or other *heavy-metal ions* can cause heavy metal poisoning. This leads to neuroglial damage and demyelination.

Diphtheria is a bacterial infection that produces a toxin that damages Schwann cells and also destroys PNS myelin sheaths. Demyelination occurs leading to sensory and motor problems. A fatal paralysis often results. Today, diphtheria is very rare because of immunization.

Multiple sclerosis (MS) is the one of the most common demyelinating conditions. It is distinguished by repeated demyelinating episodes that affect axons of the brain, spinal cord, and optic nerve. Therefore MS only affects the CNS. Recall that the optic nerve is the only nerve in the PNS in which axons are myelinated by oligodendrocytes. The actual cause of MS is not known. However, myelin is attacked because of an autoimmune reaction. Inflammation develops at various sites where white matter is located. The disease is named after the multiple "scars" that develop in the brain. Signs and symptoms include partial vision loss, speech and balance problems, generalized motor incoordination as well as the loss of bowel and bladder control. Most frequently, the patient has cyclical attacks and periods of recovery, which are called "relapsing-remitting." The first episode occurs between 20 and 40 years of age. It is nearly twice as common in women. MS is more prevalent in cold regions, and in people of Northern European descent. In young adults, MS is the most common disabling neurological disease.

Guillain-Barre syndrome (GBS) is a demyelinating disease of the PNS. In contrast to MS, GBS usually involves a single attack that affects the myelin of the axons of the PNS. Recall that Schwann cells make the myelin of the PNS. GBS is also thought to be an autoimmune disease. Patients with GBS usually present with a progressive ascending paralysis. GBS can be fatal as it can paralyze the respiratory muscles.

Focus on abnormalities of neuroglia

When there are excessive or insufficient amounts of neuroglial cells, health can be affected. While neuroglia may divide quickly causing tumors, neurons do not divide. Damage to the spinal cord may result in the destruction of neuroglia. The axons will no longer be able to produce myelin. Due to overgrowth of neuroglia, scars form that slow the ability to recover lost function.

Neural responses to injuries

Neurons have only limited responses to injury. Generally, more repair is possible in PNS neurons than in those of the CNS. Nissl bodies scatter, and an increased rate of proteins synthesis causes the nucleus to move away from its centralized location. If it recovers and begins to function normally again, the neuron will regain its normal appearance.

Recovery is based on events inside the axon. A crushing injury causes a localized decrease in blood flow and oxygen, making the affected axolemma unexcitable. When the pressure is relieved in <72 h, the neuron has the ability to recover in some cases.

However, more prolonged or severe pressure causes effects similar to those caused when an axon is severed.

Schwann cells of the PNS help to repair damaged nerves. When the axons are damaged, a degeneration process begins in axons distal to the injury site. This degeneration process is called **Wallerian degeneration**. Macrophages move into the area, cleaning up debris. The Schwann cells proliferate, forming a cellular cord that resembles the original axon's path. The neuronal axon grows in the injury site. Schwann cells then wrap around the axon. An axon that grows along the correct cord of Schwann cells may reestablish its normal synaptic contacts over time. However, if this does not occur, or the axon stops growing, normal function will not return. In the majority of cases, the growing axon will arrive at the correct location if the edges of the original nerve bundle that were cut still remain in contact. Axons usually regenerate at the rate of 1 mm per day.

Regeneration is more difficult in the CNS because: (1) there is much more axon involvement, (2) axon growth across the injured area is prevented by scar tissue produced from astrocytes, and (3) axon growth is blocked by the release of chemicals from the astrocytes.

Section review
1. What are the functional differences between glial cells in the CNS and PNS?
2. Which neuroglia are the largest and the most numerous in the CNS?
3. What is the role of the Schwann cells in the PNS?
4. What is a neurilemma?

Neurons

The **neuron** is the basic structural and functional unit of the nervous system. They are generally large, highly specialized cells. Neurons are able to provide electrical impulses along their axons, and their structures are different than the structures of glial cells.

Functional characteristics of neurons

Most neurons live much longer than glial cells and have a high metabolic rate. These cells have excitable plasma membranes, similar to skeletal muscle cells. A large amount of energy in the neuron is required due to the generation and **propagation** of action potentials (see below). During the functions of neurons, much energy is used to synthesize and secrete chemical compounds needed to send information between neurons. In active neurons, mitochondria produce a large amount of energy.

Most neurons do not have centrioles, which are organelles that function in mitosis. The centrioles help organize the microtubules that move chromosomes, and the cytoskeleton. Therefore, the lack of centrioles in the neurons of the CNS prevents them from dividing. Since they cannot divide, they cannot be replaced if lost to injury or disease.

The structure of neurons

Neurons have many shapes and are mostly located in the CNS. A neuron has four basic regions: a large *cell body*; multiple short, branched *dendrites*; one long *axon*; and terminal axon branches called *telodendria*.

The cell body

The *cell body* is also called the *soma*. It contains a large, round nucleus with an obvious *nucleolus*. The nucleus is usually centrally located within the cell body and contains genes. The **perikaryon** is the cytoplasm that surrounds the nucleus. The nucleus is surrounded by a plasma membrane. Mature neurons do not duplicate their chromosomes, but only function in gene expression. The chromosomes are uncoiled, instead of being compacted. The nucleolus assists in synthesizing ribosomal ribonucleic acid (rRNA) and in assembling ribosome subunits. The size of the nucleolus is large due to high rates of protein synthesis.

The *nuclear envelope* is a specialized region of the rough endoplasmic reticulum (RER) in the cytoplasm. The envelope has two layers, fine **nuclear pores**, and allows materials to diffuse in and out of the nucleus. Ribosomal subunits, once formed, easily pass through the nuclear pores into the cytoplasm.

The cytoplasm has both granular and agranular endoplasmic reticulum. The cytoplasm also includes Nissl bodies (also called *Nissl substance*), the Golgi apparatus (or complex), mitochondria, microfilaments, microtubules, lysosomes, centrioles, lipofuscin, melanin, glycogen, and lipid. The Nissl bodies (named for German neurologist Franz Nissl) synthesize proteins and give a gray color to the areas that contain neuron cell bodies—the *gray matter*. Nissl bodies are distributed through the cytoplasm, except for near the axon, which is called the **axon hillock** ("little hill").

The **Golgi apparatus** in the cytoplasm is a network of irregular threads around the nucleus. It has flattened cisternae and small vesicles of smooth endoplasmic reticulum. The Nissl bodies' protein is transferred inside of the Golgi complex via transport vesicles, and is stored there. Carbohydrates may be added to the protein to form *glycoproteins*. The Golgi complex also assists in producing lysosomes and in synthesizing cell membranes. The synthesis of cell membranes is very important in the formation of synaptic vesicles at axon terminals.

The spherical- or rod-shaped *mitochondria* are scattered through the cell body, dendrites, and axons. They have a double membrane, with the inner membrane having folds or *cristae* that project into their centers. Mitochondria are important in the production of energy. The numerous mitochondria, free and fixed ribosomes, and membranes give the perikaryon a coarse, grainy appearance. Mitochondria generate adenosine triphosphate (ATP) to meet neuronal energy demand. The ribosomes and RER synthesize proteins. Some areas of the perikaryon contain clusters of RER and free ribosomes. These regions, which stain darkly with cresyl violet, are the *Nissl bodies*.

The cytoskeleton of the perikaryon contains **neurofilaments** and **neurotubules**. These are similar to intermediate filaments and microtubules found in other types of cells. The main component of the cytoskeleton is formed by neurofilaments. **Neurofibrils** are bundles of neurofilaments that extend into the dendrites and axon, and provide internal support to them. Neurofibrils are numerous and run parallel to each other. The perikaryon contains organelles that provide energy and synthesize organic materials, especially the chemical *neurotransmitters* that are important in cell-to-cell communication.

Actin **microfilaments** are located beneath the plasma membrane of neurons and form a dense network. Along with the microtubules, the microfilaments assist in forming new cell processes and removing old ones. The **microtubules** extend through the cell body and its processes. In an axon, microtubules are parallel to each other, with the proximal end pointing to the cell body, and the distal end pointing away from it.

Lysosomes are vesicles bound to the plasma membrane. They contain hydrolytic enzymes, and are formed by budding off of the Golgi complex. The function of lysosomes is to digest myelin, pigments, and lipid. **Centrioles** inside immature, dividing nerve cells are small and exist in pairs. The pigment material known as **lipofuscin** consists of yellow-brown granules in the cytoplasm. It accumulates with aging, but is a harmless by-product of metabolism. *Melanin granules* exist in cell cytoplasm as well as in the brain, such as in the substantia nigra of the midbrain. They are related to the neurons' ability to synthesize catecholamine. These neurons' transmitter is *dopamine*. Table 1.2 summarizes the structures and functions of nerve cell bodies.

Neuron processes

Around the cell body are variable processes that include **dendrites**, **axons**, and **telodendria**. Dendrites are thin, sensitive, and branched extensions. They are important for intercellular communication. Typical dendrites have many branches, some of which have fine, long, and spiked projections called *dendritic spines*, which function in synapses. Neurons of the CNS receive information from other neurons mostly at the dendritic spines, which may make up 80%–90% of the neuron's total surface area.

The dendrites of motor neurons are short and tapered. They usually have hundreds of stick-like dendrites that are near the cell body. Almost all organelles in the cell body are also found in dendrites. The dendrites are the primary *input regions* creating a large surface area that receives signals from other neurons. In many parts of the brain, thinner dendrites are highly specialized to collect information. Dendrites bring incoming messages toward cell bodies in the form of short-distance electrical signals called *graded potentials*.

Table 1.2 Structures and functions of nerve cell bodies.

Structures	Functions
Nucleus	Controls cell activities
Perikaryon (cytoplasm)	Surrounds the nucleus and occupies the entire cell body
Ribosomes	Synthesis of proteins
Nissl bodies	Synthesis of proteins
Golgi complex	Adds carbohydrates to protein; forms products to be transported to nerve terminals; forms cell membranes
Mitochondria	Form chemical energy
Neurofibrils	Determine shape of neuron
Microfilaments	Help to form and retract cell processes; assist in cellular transport
Microtubules	Aid in cellular transport
Lysosomes	Digest melanin, pigment, and lipid
Centrioles	Aid in cell division and maintenance of microtubules
Lipofuscin	A harmless metabolic byproduct
Melanin	Related to formation of *dopa*, which is the precursor of the neurotransmitter *dopamine*

An axon is a single, long cytoplasmic process that may be as much as a meter in length. A good example is the axon that extends from the lumbar spine to the big toe, which is up to 4 ft. in length. Therefore, axons are some of the longest cells in the human body. A long axon is also called a *nerve fiber*. There is only one axon per neuron. A nerve is made of several axons. Axons are able to propagate an electrical impulse known as an *action potential* away from the cell body. The **axoplasm** contains lysosomes, mitochondria, neurofibrils, neurotubules, small vesicles, and various enzymes. The **axolemma** surrounds the axoplasm. In the CNS, the axolemma may be exposed to interstitial fluid or covered by neuroglial cellular processes. The *initial segment* of an axon in a common neuron joins its cell body at the thick, cone-shaped *axon hillock*. The axon narrows, forming a thin process of the same diameter over the remainder of its length. Some neurons have axons that are either extremely short or completely absent. Other neurons have axons that make up nearly all of the length of the neuron.

Axons may be branched along their lengths. This produces side branches called **collaterals**, which allow a single neuron to share information with several other cells. The primary axon trunk and collaterals terminate in fine extensions called telodendria, or *terminal branches*. Neurons may have >10,000 terminal branches, which are also called *terminal arborizations*. The actual end point of an axon is called its *terminus*. The telodendria end at knob-like **axon terminals** (*synaptic terminals*), which assist in communication with another cell. A **synapse** is the location in which a neuron communicates with another cell, which may be another neuron.

Axonal (axoplasmic) transport is the movement of materials between the cell body and axon terminals. Materials travel along the axon on neurotubules in the axoplasm. They are pulled by proteins that act as molecular motors, which are called *kinesin* and *dynein*, and use ATP. Some materials move slowly, only a few millimeters per day. This type of transport is called "slow stream." Some vesicles move quicker, traveling as a "fast stream," up to 1000 mm/day.

Axonal transport occurs in both directions at the same time. The flow of materials from the cell body to the axon terminal is carried by **kinesin**, and called **anterograde flow**. Simultaneously, various substances are moved by dynein from the axon terminal toward the cell body. This process is called **retrograde flow**. Materials flow in both directions along the axon by the anterograde flow and retrograde flow. However, the action potential only travels away from the cell body in one direction. Debris or unusual substances appearing in the axon terminal are quickly delivered to the cell body by retrograde flow. Once in the cell body, they may change cell activity by activating or inactivating specific genes.

Focus on deprivation of oxygen in neurons
When neurons are deprived of oxygen, their nuclei shrink. Affected neurons change shape and then disintegrate. *Ischemia, hypoxemia,* or toxins may cause oxygen deficiency. Toxins may block aerobic respiration, preventing neurons from using oxygen.

The classification of neurons

Neurons are classified either by their structures or by their functions. Therefore, understanding these classifications helps in easily remembering the concepts of neurons.

Structural classification
Neurons are classified based on relationships between their dendrites, cell body, and axons, as follows: *anaxonic, bipolar, unipolar,* and *multipolar* (Fig. 1.3):
- **Anaxonic neurons** are small, with numerous dendrites and no obvious axons. Their axons are not easily seen, even under a microscope. Anaxonic neurons are found in the brain and special sense organs. Their functions are not completely understood.
- **Multipolar neurons** have two or more dendrites and one axon. They are the most common CNS neurons. All motor neurons that control skeletal muscles are multipolar neurons. Their longest axons carry motor commands from the spinal cord to small muscles of the toes.
- **Bipolar neurons** have one dendrite and one axon—with the cell body between these two structures. Bipolar neurons are not common. They are located only in

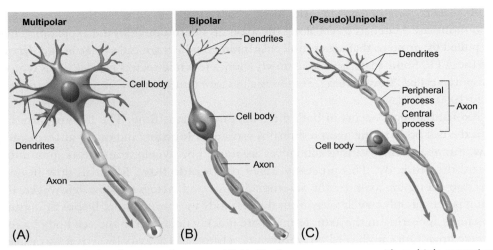

Fig. 1.3 Structural classifications of neurons. (A) Multipolar neuron: neuron with multiple extensions from the cell body. (B) Bipolar neuron: neuron with exactly two extensions from the cell body. (C) (Pseudo-) unipolar neuron: neuron with only one extension from the cell body. The central process is an axon; the peripheral process is a modified axon with branched dendrites at its extremity. (The *red arrows* show the direction of impulse travel.)

the special sense organs. They transmit information about sight, smell, or hearing. Bipolar neurons are relatively small.
- In a **unipolar neuron**, or *pseudounipolar neuron*, the dendrites and axon are continuous with each other, with the cell body located on one side. The base of the neuron lies where the dendrites meet. The remaining process, considered an axon, carries action potentials. The majority of the sensory neurons of the PNS are unipolar. Their axons are very long, 1 m or more, and end at synapses in the CNS. The longest unipolar axons carry sensations from the toes to the spinal cord.

Functional classification
Neurons can also be classified by function as: (1) *sensory neurons*, (2) *motor neurons*, and (3) *interneurons*. These classifications are based on the direction that nerve impulses of neurons travel, to and from the CNS.

Sensory neurons
Sensory neurons form the *afferent* division of the PNS. They transmit impulses from sensory receptors in the skin or internal organs toward the CNS. A **ganglion** is a group of neuron cell bodies located in the PNS. Cell bodies in the brain and spinal cord (CNS) form *nuclei*. Examples include the *trigeminal nuclei*. Sensory neurons are unipolar. Virtually all of their cell bodies are located in peripheral *sensory ganglia*. The processes of sensory neurons are known as *afferent fibers*. They extend between sensory receptors and the CNS. Information moves from sensory receptors to the spinal cord or brain. There are about 10 million

sensory neurons in the body, each collecting information about the external and internal environments. *Somatic sensory neurons* monitor the external environment. **Visceral sensory neurons** monitor the internal environment and organ systems. In the peripheral nerves the somatic fibers innervate skin, muscle, joints, and body walls. Similarly, the visceral fibers innervate the blood vessels and internal organs.

Sensory fibers are called *afferent* and *motor fibers* are called *efferent*. Sensory receptors may be classified into three groups:

- **Interoceptors** monitor the cardiovascular, digestive, reproductive, respiratory, and urinary systems. They provide signals to contract or distend visceral structures. General visceral afferent fibers carry interoceptive data from the receptors of visceral organs.
- **Exteroceptors** provide pressure, temperature, and touch information, and the senses of equilibrium (balance), hearing, sight, smell, and taste.
- **Proprioceptors** monitor skeletal muscle and joint movement and positioning.

Somatic afferent fibers carry data from proprioceptors and exteroceptors.

Motor neurons

Motor neurons form the *efferent* division of the PNS. There are approximately 500,000 motor neurons carrying information from the CNS to peripheral effectors in peripheral tissues and organ systems. *Efferent fibers* are the axons of motor neurons that carry information away from the CNS. The two primary efferent systems are the SNS and the autonomic (visceral) nervous system (ANS).

The SNS includes the *somatic motor neurons*, which innervate skeletal muscles. The SNS is under conscious control. The cell bodies of somatic motor neurons are located within the CNS. Their axons run within peripheral nerves, innervating skeletal muscle fibers at neuromuscular junctions.

The ANS is not consciously controlled. **Visceral motor neurons** stimulate all peripheral effectors except for skeletal muscles. They innervate the cardiac muscle, smooth muscle, adipose tissue, and glands. Visceral motor axons of the CNS innervate additional visceral motor neurons in peripheral *autonomic ganglia*. Here, the neuronal cell bodies innervate and control peripheral effectors. *Preganglionic fibers* are the axons that extend from the CNS to an autonomic ganglion. *Postganglionic fibers* are the axons that connect the ganglion cells with the peripheral effectors.

Interneurons

Interneurons are situated between sensory and motor neurons. There are approximately 20 billion interneurons, or *association neurons*. Most are found in the brain and the spinal cord, and others are within the autonomic ganglia. Interneurons make up >99% of all the neurons in the body. The primary function of interneurons is integration. They carry sensory information and regulate motor activity. More interneurons are activated when a response to stimuli is required to be complex. Interneurons are utilized in all higher functions, including learning, memory, cognition, and planning. Nearly all interneurons are multipolar. However, they are of many different sizes and have different patterns of fiber branches.

Section review
1. How are neurons classified?
2. What are the differences between afferent and efferent fibers?
3. What are the functions of interneurons?

Membrane potential

Membrane potential and resting membrane potential are two unique physiological features of cells. Here we focus on neuron membranes, but the same concepts apply to many other types of cells. Living cells have a membrane potential that changes continuously based on cellular activities. This is further defined as a difference in electrical charges across the plasma membrane. It is called a *potential* since it is a type of stored energy that is known as *potential energy*. When opposite electrical charges, including opposite ions, are separated by a membrane, they have a potential to move toward each other—if the membrane allows them to cross it. A membrane that has this potential is called *polarized*. It has a negative pole and a positive pole. The negative pole is the side where there are excessive amounts of negative ions. The positive pole is the side where there are excessive amounts of positive ions. The magnitude of the potential difference, between the two sides, is measured in millivolts (mV) or volts (V).

This voltage can be measured by using a *voltmeter*. The sign of a membrane's voltage will indicate the charge that exists on the inner surface of a polarized membrane. A negative value, such as $-60\,mV$, indicates that the potential difference has a 60-mV magnitude and that the inner membrane is negative compared to its outer surface. If the measured voltage is positive, it means that inner membrane is positive while the outer membrane is negative.

The **resting membrane potential** is the membrane potential of a cell that is at rest. All neural activities are initiated from a change in the neuron's resting membrane potential. This change is temporary and localized. This effect, known as a **graded potential**, decreases over distance from the stimulus. When the graded potential is sufficiently large, an action potential in the membrane of the axon is triggered. An action potential is simply an electrical signal of the nerve cells.

The resting membrane potential

A neuron that is not conducting electrical signals is described as resting—usually at about $-70\,mV$, though this can vary to some degree. There are three important factors concerning resting membrane potential, which are as follows:

- *There are large differences in the ionic composition of the ECF and intracellular fluid (cytoplasm).* There are high concentrations of sodium (Na^+) and chloride ions (Cl^-) in the ECF. The cytoplasm contains high concentrations of potassium ions (K^+) and negatively charged proteins.
- *Selective permeability of cell membranes.* Since cells have selective permeable membranes, even distribution of ions does not occur. The lipid portions of the plasma membrane keep ions from freely crossing. The ions can only enter or leave the cell through one of many different membrane channels. At the resting membrane potential, ions move through *leak channels*, which are membrane channels that remain open. Certain ions are moved in or out of the cell via active transport mechanisms, which include the sodium–potassium exchange pump.
- *Specific ions have different membrane permeability.* As a result of this factor, a cell's passive and active transport mechanisms do not result in an equal distribution of charges across the plasma membrane. Negatively charged proteins in the cell cannot cross the membrane due to their large size. Therefore, the inner membrane surface has excessive negative charges compared to the outer surface.

Passive as well as active forces determine the membrane potential across the plasma membrane.

The sodium-potassium exchange pump

At normal resting membrane potential, cells push out sodium and bring in potassium (see Fig. 1.4). This process occurs via the sodium–potassium exchange pump, which requires ATP. Three intracellular sodium ions are exchanged for two extracellular potassium ions. At normal resting membrane potential, sodium ions are ejected as quickly as they enter the cell. Therefore, the exchange pump accurately balances the passive forces of diffusion. Since the ionic concentration gradients are kept balanced, the resting membrane potential remains stable.

Graded potentials

Graded potentials are produced by any stimulus that opens a gated channel. Graded potentials are also called **local potentials**. They are membrane potential changes that are not able to spread over long distances away from the stimulation.

The following events occur:
A. Sodium ions enter the cell attracted to negative charges on the inner surface of the membrane. As positive charges spread out, membrane potential moves toward 0 mV. **Depolarization** is the result of any change from resting membrane potential to a less negative potential. Depolarization indicates changes in potential from −70 mV to lesser negative values (−65, −45, −10 mV), and also to membrane potentials above 0 mV (+10 mV, +30 mV). For all of these changes, the membrane potential becomes more positive.

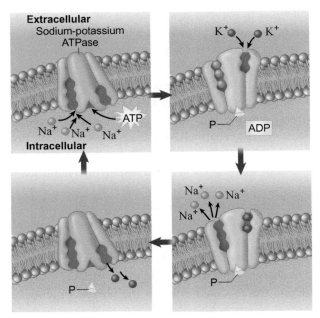

Fig. 1.4 The processes of the sodium-potassium exchange pump.

B. When the plasma membrane depolarizes, its outer surface releases sodium ions. Along with other extracellular sodium ions, these ions move to the open channels and replace ions already inside the cell. The term *local current* describes the movement of positive charges, parallel to the inside and outside of a depolarizing membrane. In graded potentials, the amount of depolarization is reduced over distance away from the stimuli. Local current is greatly reduced since the cytoplasm has significant resistance to the movement of ions. Another factor is that some sodium ions entering cells move out, across the membrane, via sodium leak channels. At a far enough distance away from the entry point, effects on membrane potential cannot be detected. The largest change in membrane potential is related to the size of the stimulus. This size determines how many sodium ion channels are open. More open channels mean that more sodium ions enter, a larger area of membrane is affected, and there is more depolarization.

The membrane potential quickly returns to resting level once the chemical stimulus is removed and normal membrane permeability is restored. *Repolarization* is the method of restoring normal resting membrane potential following depolarization. Repolarization usually requires a combination of ion movement through membrane channels. However, in general, the efflux of potassium is mainly responsible for repolarization. It requires the activities of ion pumps, primarily the sodium-potassium exchange pump.

When a gated potassium ion channel opens because of stimuli, the effects are opposite. Potassium ion outflow increases and the inside of the cell loses positive ions, becoming more negative. **Hyperpolarization** is produced by the loss of positive ions. This is an increase in negativity of the resting membrane potential, such as from -70 to -80 mV, or even more. A local current distributes the effect to adjacent areas of the plasma membrane. The effect decreases over distance from the channels that are open.

Graded potentials occur in the membranes of nerve and muscle cells, fat cells, epithelial cells, gland cells, and many sensory receptors. These potentials often initiate various cell functions, such as when a graded potential at a gland cell's surface triggers exocytosis of secretory vesicles. Another example is when a neuromuscular junction's motor end plate is stimulated by a graded potential due to the actions of ACh. This can then trigger an action potential in nearby areas of the sarcolemma. While graded potentials are supported by the motor end plate, the remainder of the sarcolemma is made up of excitable membrane. These areas are different since they contain voltage-gated ion channels.

If a graded potential causes hyperpolarization in a neuron, an action potential is less likely to occur. If the graded potential causes depolarization, the action potential is more likely to occur.

Passive processes: The electrochemical gradient

Across the plasma membrane, passive processes involve chemical and electrical gradients. There is an attraction between positive and negative charges. When not separated, ions with opposite charges move together and eliminate their potential difference. This movement of charges that eliminates the potential difference is called a *current*. If a plasma membrane or other barrier separates oppositely charged ions, the current's strength relies on how easily the ions cross the membrane. The *resistance* of the membrane measures how much it restricts ion movement. When resistance is high, the current is extremely small, since less ions can cross the membrane. When resistance is low, the current is very large, since more ions can cross the membrane. As ion channels open or close, the resistance of the plasma membrane changes. This alteration can cause differences in the amount of ions being carried in or out of the cytoplasm.

Electrical gradients may oppose or reinforce each ion's chemical gradient. An ion's **electrochemical gradient** is the total of the chemical and electrical forces that act on the ion, across the plasma membrane. The resting membrane potential of most cells, including neurons, is primarily affected by the electrochemical gradients for K^+ and Na^+.

Potassium ions have higher concentrations inside the cells, while their concentrations are very low outside of the cells. Therefore, potassium easily moves out of the cells due to the chemical gradient. This movement is opposed by the electrical gradient. Potassium ions inside and outside the cell are attracted to negative charges inside the plasma membrane. These ions are pushed away by the positive charges outside the plasma membrane. Though the chemical gradient can overpower the electrical gradient, the electrical gradient lessens the force that drives K^+ out of the cell.

The **equilibrium potential** is the membrane potential for an ion at which there is no movement of the ion across the plasma membrane. This equilibrium, for potassium, occurs at a membrane potential of about −90 mV. For neurons, the resting membrane potential is usually −70 mV, which is close to the equilibrium potential for potassium ions. The difference is mostly because of continuous sodium ion leakage into the cell. The equilibrium potential shows ion contribution to the resting membrane potential.

Sodium ions are relatively highly concentrated outside of the cells, but very low in concentrations inside the cell. Therefore, a strong chemical gradient forces sodium ions into the cell. Also, excessive negative charges inside the plasma membrane attract extracellular sodium ions. Electrical forces and chemical forces drive sodium ions into the cell. The equilibrium potential for Na^+ is about +66 mV. The resting membrane potential is very different, since resting membrane permeability to sodium ions is extremely low. This is also true since ion pumps the plasma membrane force sodium ions out as quickly as they cross the membrane.

An electrochemical gradient is an example of *potential energy,* or *stored energy*. A fully charged battery exemplifies this energy. Electrochemical gradients would be eliminated by diffusion if there were no plasma membrane. Each stimulus that increases plasma membrane permeability to sodium or potassium ions creates intense and quick movement of ions. The type of stimulus does not regulate ion movement. This is accomplished by the electrochemical gradient.

Membrane channels change resting membrane potential

The resting membrane potential exists since: (a) cytoplasm has different chemical and ionic compositions compared to ECF and (b) the plasma membrane has selectively permeability. The membrane potential rises and falls because of temporary changes in permeability. The opening or closing of certain membrane channels causes changes in membrane potential.

Membrane channels control the movement of ions across the plasma membrane. Sodium and potassium ions are the critical determinants of the membrane potential of neurons and many other cell types. Sodium and potassium ion channels may be passive or active.

Passive ion channels

Leak channels are passive ion channels that are always open. Their permeability changes as proteins forming these channels change shape due to local conditions. Leak channels are essential for establishing the normal resting membrane potential of cells.

Active ion channels

Gated ion channels are active channels in the plasma membranes. They open or close in response to certain stimuli. There are three types of gated ion channels:

- *Chemically gated ion channels*, or *ligand-gated ion channels*, open or close as they bind certain chemicals or *ligands*. Receptors that bind acetylcholine (ACh) at the neuromuscular junction are examples of chemically gated ion channels. These channels are most abundant on the dendrites and cell bodies of neurons, which is where most synaptic communication occurs.
- *Voltage-gated ion channels* open or close because of changes in the membrane potential. They are special areas of *excitable cell membrane*, which is capable of generating and propagating action potentials. Examples include the axons of multipolar and unipolar neurons, as well as the sarcolemma and T tubules of skeletal and cardiac muscle cells. Sodium, potassium, and calcium ion channels are the most important types of voltage-gated ion channels. Sodium ion channels have two independent gates: an *activation gate* that needs stimulation in order to open and allow sodium ions to enter the cell, and an *inactivation gate* that closes, stopping the entry of sodium ions. There are three different functions of these channels:
    - closed, but able to open
    - activated—open
    - inactivated—closed, and unable to open
- *Mechanically gated ion channels* open or close because of changes in the membrane surface, such as when pressure is applied. These channels are essential for sensory receptors that respond to pressure, touch, or vibration. The distribution of membrane channels is different between the areas of the plasma membrane. This affects the way the cells respond to stimuli, and the parts of the cell that are involved. In a neuron, chemically gated ion channels are present on the dendrites and cell body. Along the axon are voltage-gated sodium ion and potassium ion channels. Voltage-gated calcium ion channels are located at axon terminals.

All gated channels are closed at the resting membrane potential. As the channels open, ion movement across the plasma membrane increases and the membrane potential is changed.

Section review
1. What are a membrane potential and a graded potential?
2. What is an electrochemical gradient?
3. How does a sodium-potassium exchange pump function?

Action potentials

Action potentials are *nerve impulses*, and not graded potentials. They are further defined as changes in membrane potential. Once they begin, they affect the entire excitable membrane. Action potentials are spread (*propagated*) along the surface of an axon. They are not reduced as they move away from their source. Impulses travel along the axon to the axon terminals. Generation and propagation are related. Action potentials must be generated at one location before they can be propagated away from that location.

Threshold and the all-or-none principle

An action potential is stimulated only when a graded potential depolarizes the axolemma to a specific level. The **threshold** is the membrane potential at which an action potential begins. An axon's threshold is usually between -60 and -55 mV. This corresponds to a depolarization of 10–15 mV. Any stimulus that changes resting membrane potential from -70 to -62 mV produces only a graded depolarization and not an action potential. When the stimulus is removed, the membrane potential returns to its resting level. Local currents are created by the graded depolarization of the axon hillock. They cause depolarization of the initial axon segment.

For excitable membranes, including axons, a graded depolarization is like pressure on a gun's trigger. The action potential is similar to that when the gun fires. Every stimulus bringing the membrane to threshold creates identical action potentials. As long as a stimulus exceeds threshold, the action potential is independent of the strength of the depolarizing stimulus. This is known as the *all-or-none principle*. It applies to all excitable membranes. The stimulus either triggers a typical action potential or none at all.

Generation of action potentials

The steps involved in the generation of an action potential from the resting state are shown in Fig. 1.5. The activation gates of the voltage-gated sodium ion channels are closed at resting membrane potential. When the membrane reaches threshold, voltage-gated sodium ion channels open at one location, generally the initial axon segment. Then the following occur:
o Depolarization to threshold.
o Activation of voltage-gated sodium ion channels and rapid depolarization.
o Inactivation of voltage-gated sodium ion channels and activation of voltage-gated potassium ion channels; this begins repolarization.
o Closing of voltage-gated potassium ion channels; this produces a short hyperpolarization, then a return to the resting membrane potential. There are several characteristics that distinguish local potentials from action potentials (Table 1.3).

The refractory period

When additional depolarization occurs, the plasma membrane does not respond normally. This occurs when action potentials begin until resting membrane potential is reestablished. This is the **refractory period** of the membrane. Once voltage-gated sodium ion channels open at threshold, until their inactivation ends, the membrane cannot respond to more stimulation. This is because all of the voltage-gated sodium ion channels

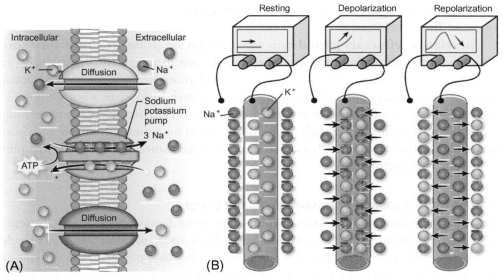

Fig. 1.5 Sodium-potassium pump and propagation of an action potential. (A) Concentration difference of sodium (Na$^+$) and potassium (K$^+$) intracellularly and extracellularly. The direction of active transport by the sodium-potassium pump is also shown. (B) The left diagram represents the polarized state of a neuronal membrane when at rest. The middle and right diagrams represent changes in sodium and potassium membrane permeabilities with depolarization and repolarization.

Table 1.3 A comparison of graded potentials and action potentials.

Graded potentials	Action potentials
No threshold level	Depolarization to threshold must occur before action potential begins
Produced by gated channels on soma and dendrites	Produced by voltage-gated channels on axon and trigger zone
Passive spread outward from site of stimulation	Action potential at one location depolarizes adjacent locations to threshold level
Depolarizing (positive) or hyperpolarizing (negative)	Always depolarizing
Reversible, returning to resting membrane potential if stimulation stops before threshold is reached	Irreversible, continuing to completion once it starts
Local, with effect on membrane potential decreasing over distance from stimulation site	Self-propagated along entire membrane surface; no decrease in strength
Amount of depolarization or hyperpolarization based on intensity of stimulus; summation occurs	All-or-none event; all stimuli exceeding threshold produce identical action potentials; no summation occurs
No refractory period	Refractory period occurs
Occur in most plasma membranes	Occur only in excitable membranes of specialized cells (neurons, muscle cells)
Decremental; signals weaken over distance	Nondecremental; signals have the same strength regardless of distance traveled

are either open or are inactivated. This **absolute refractory period** is the first occurrence and lasts 0.4–1.0 ms.

The **relative refractory period** starts when sodium ion channels return to their normal resting condition. This period continues until the membrane potential is stabilized, at the resting level. If the membrane has enough depolarization, another action potential can occur during this period. This depolarization requires a larger-than-normal stimulus. This is because the local current must bring in sufficient sodium ions to oppose the exiting of positive potassium ions through open voltage-gated potassium ion channels. It also occurs because the membrane is hyperpolarized to a certain amount through the majority of the relative refractory period.

Functions of the sodium-potassium exchange pump

For action potentials, depolarization occurs from inward movement of sodium ions. Repolarization requires loss of potassium ions. Over time, prestimulation ion levels are reestablished by the actions of this pump. For each action potential, the amount of ions is small. Many thousands of action potentials must occur before there is a significant change in intracellular ion concentrations. Therefore, the exchange pump is not important to any single action potential.

However, 1000 action potentials per second can be generated by a completely stimulated neuron. At this point, the exchange pump is required to maintain ion concentrations within the acceptable limits over time. One molecule of ATP is broken down every time the pump exchanges two extracellular potassium ions for three intracellular sodium ions. The membrane protein of the pump is Na^+/K^+ $ATPase$. Energy is supplied to the pump's ions by dividing a phosphate group from one molecule of ATP to form ADP. The function of the neuron can be stopped if the cell uses all of its ATP or if a metabolic poison inactivates Na^+/K^+ ATPase.

Action potential propagation

Events generating action potentials occur in a small part of the plasma membrane surface. Differently from graded potentials, action potentials spread along the excitable membrane. An action potential is relayed from one site to another in a series of steps, with the message being repeated at each step. Since the same events occur repeatedly, *propagation* is the preferred term—instead of *conduction*, which suggests a flow of charges.

Types of propagation

The opening of additional voltage-gated sodium ion channels is triggered when sodium ions move into an axon and depolarize adjacent areas. The reaction spreads across the membrane surface and the action potential is propagated along the length of the axon, eventually reaching the axon terminals. Action potentials travel along axons either by *continuous propagation* (if they are unmyelinated axons) or by *saltatory propagation* (if they are myelinated axons).

Continuous propagation

Action potentials move via **continuous propagation** in unmyelinated axons. The axolemma is organized in adjacent segments. Continuous propagation occurs as follows:
- The membrane potential briefly becomes positive at the peak of the action potential.
- A local current develops as sodium ions begin moving in the cytoplasm and ECF.
- The local current spreads out in all directions, depolarizing the nearby membrane areas. The axon hillock could not to respond with an action potential since it has no voltage-gated sodium ion channels.
- The process continues as if in a chain reaction.

Each time there is development of a local current develops, the action potential moves in one direction: *forward*. This is because the previous axon segment is still in the absolute refractory period. Therefore, action potentials move away from their generation site and do not reverse direction. Over time, the furthest parts of the plasma membrane are affected.

Messages are relayed from one location to another. Distance does not affect this process. The action potential that reaches the axon terminal is exactly the same as the one generated at the initial axon segment. Though the events at each location take about a millisecond, each event must be repeated at every step along the way. For another action potential to occur at the same location, another stimulus must be applied.

Saltatory propagation

In myelinated axons, action potentials move via **saltatory propagation**. In both the CNS and PNS, saltatory propagation carries each action potential along an axon much faster than via continuous propagation. Continuous propagation is not able to occur along a myelinated axon. This is because myelin increases resistance to the flow of ions across the membrane. Ions easily cross the axolemma, but only at its *nodes*. Only nodes with voltage-gated ion channels are able to respond to depolarizing stimuli.

When an action potential arrives at the initial segment of a myelinated axon, the local current jumps over the internodes. It then depolarizes the nearest node to threshold. Since nodes may be 1–2 mm apart in a large myelinated axon, the action potential "skips" from node to node instead of moving along the axon in many tiny steps. Saltatory propagation is not only quicker but also uses lower amounts of energy, since less surface area is involved. Also, fewer sodium ions must be pumped out of the cytoplasm.

Axon diameter and propagation speed

An axon's diameter also affects propagation speed, but not as significantly. Axon diameter is important since ions must move through the cytoplasm to depolarize nearby areas of

the plasma membrane. Cytoplasm offers lower resistance to ion movement than does the axon membrane. An axon with a larger diameter has a lower resistance. Axons may be categorized by their diameter, myelination, and propagation speed:
1. **Type A fibers**—the largest myelinated axons; diameters range from 4 to 20 µm. Action potentials move as fast as 120 m/s or 268 mph.
2. **Type B fibers**—smaller myelinated axons; diameters of 2–4 µm. Propagation speeds average around 18 m/s (about 40 mph).
3. **Type C fibers**—unmyelinated; <2 µm in diameter. They propagate action potentials slowly, at 1 m/s (only 2 mph).

The advantage of myelin is clear when comparing Type C with Type A fibers. The diameter increases by 10 times, and the propagation speed increases by 120 times.

Type A fibers can be motor or sensory. They carry information to the CNS about position, balance, and delicate touch and pressure sensations from the skin surfaces. Motor neurons controlling skeletal muscles send commands via Type A axons. Type B and Type C fibers carry information to and from the CNS, including pain, temperature, general touch, and pressure sensations. Additionally, they carry signals to cardiac muscle, smooth muscles, glands, and other peripheral effectors. Only about one-third of all axons carrying sensory information are myelinated. Most sensory information utilizes the thin Type C fibers. Sensory information related to survival and motor commands that avoid injury use Type A fibers. Less critical information is relayed by Type B or Type C fibers.

Section review
1. What is an action potential?
2. What are the descriptions of threshold and the all-or-none principle?
3. What are the differences between the three types of axon fibers?

Synapses

Effective messages propagated along an axon must also be transferred to another cell. This transfer occurs at a *synapse*, a specialized area where a neuron communicates with another cell. Between two neurons and at a synapse, information passes from the *presynaptic neuron* to the *postsynaptic neuron*. Synapses may involve various types of postsynaptic cells. One example is the neuromuscular junction, which is a synapse where the postsynaptic cell is a skeletal muscle fiber. There are >100 trillion synapses in the human brain alone.

Types of synapses

There are two types of synapses, *electrical* and *chemical*, which function in unique ways. It is important to understand the differences between these two types.

Electrical synapses

At *electrical synapses*, the cells have direct physical contact. The presynaptic and postsynaptic membranes of the two cells are joined at gap junctions (see Fig. 1.6A). The lipid parts of adjacent membranes are separated by only 2 nm. They are positioned by essential membrane proteins called **connexons**. Pores, formed by these proteins, allow ions to pass between the cells. Changes in membrane potential of one cell result in local currents affecting the other cell, much as if they shared a common membrane. Therefore, an electrical synapse propagates action potentials between the cells efficiently and quickly.

In adults, electrical synapses are not common in both the CNS and PNS. They occur in certain brain areas such as the vestibular nuclei, involved in balance, in the eyes, and in one or more pairs of PNS ganglia (ciliary ganglia). They also occur in certain embryonic structures.

Chemical synapses

In a **chemical synapse** (Fig. 1.6B), one neuron sends signals to another neuron. This involves the axon terminal of the **presynaptic neuron**, which sends the message, and the **postsynaptic neuron**, which receives the message. The two cells are separated by the narrow **synaptic cleft**.

The presynaptic cell is commonly a neuron. Specialized receptor cells may create synaptic connections with dendrites. Postsynaptic cells can be neurons or other types of cells. Neurons communicate with other neurons at synapses on dendrites, on cell bodies, or along axons of receiving cells. *Axoaxonic synapses* occur between axons of two neurons.

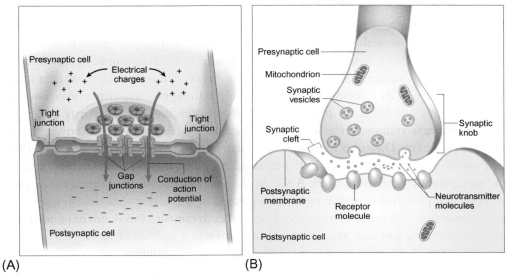

Fig. 1.6 (A) Electrical synapse. (B) Chemical synapse.

Axosomatic synapses have junctions at an axon terminal of one neuron and the cell body of another. In an *axodendritic synapse*, synaptic contact occurs between an axon terminal of one neuron and a dendrite of another neuron.

A **neuromuscular junction** is a synapse between a neuron and a skeletal muscle cell. At a **neuroglandular junction**, one neuron controls or regulates activity of a secretory cell. Neurons also *innervate* many other cell types, including *adipocytes*. The axon terminal of a presynaptic cell, into the synaptic cleft, releases chemicals called neurotransmitters. Neurotransmitters are contained in *synaptic vesicles* inside the axon terminal.

Each postsynaptic cell has its own type of axon terminal. A simple, rounded axon terminal is present when the postsynaptic cell is another neuron. At a synapse, the presynaptic membrane is separated from the postsynaptic membrane by the synaptic cleft. Neurotransmitters are released at the presynaptic membrane, while the postsynaptic membrane has receptors for the neurotransmitters. At a neuromuscular junction, the axon terminal is much more structurally complex.

Axon terminal structures include mitochondria and thousands of vesicles that contain neurotransmitter molecules. Axon terminals reabsorb breakdown molecules of neurotransmitters formed at the synapse, then reproduce the neurotransmitters. The axon terminals also receive continual amounts of neurotransmitters that were created in the cell body, plus enzymes and lysosomes, via anterograde flow. Chemical synapses are much more common than electric synapses. Communications across chemical synapses occur only from presynaptic membranes to postsynaptic membranes, and not in the reverse direction.

Function of chemical synapses

Electrical events trigger the release of neurotransmitters, which flood the synaptic cleft, binding to receptors on the postsynaptic plasma membrane. This changes its permeability and produces graded potentials. This process is similar to the function of neuromuscular junctions. Since chemical synapses do not involve direct cellular joining, there is far more variation in results. At a chemical synapse, arriving action potentials may release enough neurotransmitter to bring the postsynaptic neuron to threshold, or they may not.

Cholinergic synapses release ACh at all neuromuscular junctions that involve skeletal muscle fibers. They also release ACh at many CNS synapses in the CNS, all PNS neuron-to-neuron synapses, and all neuromuscular and neuroglandular junctions in the parasympathetic division of the ANS. At cholinergic synapses between neurons, presynaptic and postsynaptic membranes are separated by a synaptic cleft. Most ACh in an axon terminal is collected in synaptic vesicles that each contains thousands of neurotransmitter molecules. Just one axon terminal may contain a million of these vesicles. The functions of a cholinergic synapse are further explained as follows:

1. An action potential arrives at the presynaptic axon terminal. This depolarizes the membrane and opens its voltage-gated calcium ion channels for a short time.

2. Extracellular calcium ions enter the axon terminal via the voltage-gated calcium channels. These calcium ions then attach to the vesicles that contain ACh. The attachment of the calcium ions to the vesicles causes the release of ACh in the synaptic cleft. The ACh is released in groups of about 3000 molecules, which is the average number of molecules in just one vesicle. The release of ACh stops quickly because active transport activity removes calcium ions rapidly from the cytoplasm in the axon terminal back to the extracellular space. The ions are pumped out of the cell or moved to the mitochondria waiting for another action potential to arrive.
3. ACh binds to receptors on the postsynaptic membrane, depolarizing the membrane. Across the synaptic cleft, ACh diffuses toward the receptors of the postsynaptic membrane. The ACh receptors consist of chemically gated sodium and potassium ion channels. The main response is the increased permeability to sodium ions, causing a depolarization in the postsynaptic membrane of about 20 ms. These cation channels move potassium ions out of the cell. Since sodium ions are driven by a stronger electrochemical gradient, there is slight depolarization of the postsynaptic membrane, which is a graded potential. The more ACh released at the presynaptic membrane, the more there are open cation channels in the postsynaptic membrane, and therefore, more depolarization. If the depolarization brings a nearby section of excitable membrane (such as the initial axon segment) to threshold, an action potential occurs in the postsynaptic neuron.
4. ACh is removed from the synaptic cleft by *acetylcholinesterase* (AChE). The effects of ACh on the postsynaptic membrane are temporary. This is because the enzyme AChE (also called AChE or *cholinesterase*) is contained in the synaptic cleft and postsynaptic membrane. Approximately 50% of ACh released at the presynaptic membrane is degraded prior to reaching the postsynaptic membrane receptors. It only takes about 20 ms for ACh molecules that bind to receptor sites to be broken down.

Via hydrolysis, AChE breaks down molecules of ACh into *acetate* and *choline*. Choline is actively absorbed by axon terminals and used to synthesize more ACh, via acetate provided by *coenzyme A (CoA)*. Coenzymes from vitamins are needed in many enzymatic reactions. Acetate moving away from the synapse can be absorbed and metabolized by postsynaptic cells or by various cells and tissues.

Synaptic delay

A **synaptic delay** is the time required for a signal to cross a synapse between two neurons. There is only 0.2–0.5 ms between the arrival of an action potential at the axon terminal and its effect on the postsynaptic membrane. The majority of this delay is caused by the time needed for calcium ion influx and release of the neurotransmitter release. The delay is not due to neurotransmitter diffusion. The synaptic cleft is thin and neurotransmitters diffuse across it quickly.

If there is a delay of 0.5 ms, an action potential may travel >7 cm (about 3 in) along a myelinated axon. When information is passed along a chain of CNS interneurons, the cumulative synaptic delay may exceed the propagation time along the axons. With less synapses involved, the total synaptic delay is shorter and the response is faster. The fastest reflexes have only one synapse, and a sensory neuron directly controls a motor neuron.

Synaptic fatigue

Synaptic fatigue is also called *short-term synaptic depression* and is defined as a temporary inability of neurons to fire and transmit input signals. It is a form of *synaptic plasticity*, a type of negative feedback. It is mostly a presynaptic phenomenon. Since ACh molecules are recycled, axon terminals are not completely dependent on the ACh from the cell body that is delivered by axonal transport. When stimulation is intense, it may not be possible for resynthesis and transport mechanisms to keep up with neurotransmitter demand. Synaptic fatigue then occurs. The response of the synapse is weakened until ACh is replenished.

> **Focus on synaptic transmission**
> Several drugs, such as diphenylhydantoin, antidepressants classified as selective serotonin reuptake inhibitors (SSRIs), and caffeine may affect synaptic transmission. Diphenylhydantoin limits frequency of action potentials that reach the axon terminal. The SSRIs block serotonin transport into the presynaptic cell, increasing the stimulation of postsynaptic cells. Caffeine stimulates the activity of the nervous system by lowering synaptic thresholds, resulting in the postsynaptic neurons being excited more easily.

Section review
1. What is structure of the axon terminal?
2. What is the role of ACh in the synapses?
3. What is the difference between synaptic delay and fatigue?

Neuronal integration of excitatory and inhibitory stimuli

Each neuron may receive information across thousands of synapses. The effect on membrane potential of the cell body—mostly near the axon hillock—determines how the neuron responds over time. If depolarization occurs at the axon hillock, the membrane potential is affected at the initial segment. When the initial segment reaches threshold, there is generation and propagation of an action potential. The axon hillock actually integrates the stimuli affecting the cell body and dendrites. This determines the rate of action potential generation at the initial segment. It is the most basic level of *information processing* in the nervous system. When the signal reaches the postsynaptic cell, its response depends on the stimulated receptors' actions, and stimuli which influence the cell simultaneously. Excitatory and inhibitory stimuli are regulated via interactions between *postsynaptic potentials*.

Postsynaptic potentials

Postsynaptic potentials are graded potentials that develop in the postsynaptic membrane due to effects of a neurotransmitter. Two major types include: excitatory postsynaptic potentials (EPSP) and inhibitory postsynaptic potentials (IPSPs). An **EPSP** is a graded depolarization caused by the arrival of a neurotransmitter at the postsynaptic membrane. An EPSP develops from chemically gated ion channels that open in the plasma membrane. This leads to membrane depolarization. An example of an EPSP is the graded depolarization caused by the binding of ACh. Since it is a graded potential, an EPSP affects only the area that closely surrounds the synapse.

An **IPSP** develops when a postsynaptic membrane experiences a graded hyperpolarization. The opening of chemically gated potassium ion channels may cause an IPSP to occur. As hyperpolarization continues, the neuron is *inhibited*, since a greater-than-usual depolarizing stimulus is required to bring the membrane potential to threshold. An action potential is usually produced by a stimulus that changes the membrane potential by 10 mV (from -70 to -60 mV). However, if the membrane potential was changed to -85 mV by an IPSP, the same stimulus would depolarize it to just -75 mV, which is below threshold.

Integrating postsynaptic potentials: Summation

Membrane potential only experiences a slight effect from a single EPSP. It usually produces depolarization of nearly 0.5 mV at the postsynaptic membrane. Before an action potential develops in the initial segment, local currents, requiring at least 10 mV, must depolarize the region. A single EPSP will not cause an action potential, even when a synapse is on an axon hillock. Single EPSPs combine through **summation**. This integrates the effects of the graded potentials that affect a single part of the plasma membrane. The graded potentials can be EPSPs, IPSPs, or both. Two forms of summation exist: spatial summation and temporal summation.

Spatial summation

Spatial summation occurs as stimuli are applied simultaneously but in different areas. This has a cumulative effect on the membrane potential. Spatial summation utilizes *multiple synapses* acting at the same time. Each synapse brings sodium ions over the postsynaptic membrane. This causes a graded potential with localized effects. At every active synapse, sodium ions producing the EPSP spread along the inner membrane surface. They mix with those entering at other synapses. The results upon the initial segment are cumulative. The amount of depolarization is based on the number of synapses that are active at each moment, along with the distance from the initial segment. Similar to temporal summation, an action potential occurs when the membrane potential at the initial segment reaches threshold.

Temporal summation

Temporal summation is the addition of stimuli occurring quickly at a *single synapse* that is active *repeatedly*. A usual EPSP lasts about 20 ms. However, with maximum stimulation, action potentials can reach axon terminals every millisecond. The two effects are combined. Each time an action potential arrives, a vesicle group discharges ACh into the synaptic cleft. Additional chemically gated ion channels open every time more ACh molecules arrive at the postsynaptic membrane. The degree of depolarization increases. Several small steps will bring the initial segment to threshold.

Summation of EPSPs and IPSPs

IPSPs also combine in a similar way. The activation of different chemically gated ion channels is involved with EPSPs and IPSPs. This produces opposite effects upon membrane potential. Antagonism between IPSPs and EPSPs is essential in cellular information processing. Similar interactions between EPSPs and IPSPs determine membrane potential where the axon hillock and initial segment meet. Neuromodulators, hormones, or both can alter the postsynaptic membrane's sensitivity to excitatory or inhibitory neurotransmitters. By shifting balance between EPSPs and IPSPs, the compounds promote facilitation or inhibition of CNS and PNS neurons.

Presynaptic regulation: Inhibition and facilitation

Inhibitory or excitatory responses may occur at postsynaptic neurons as well as presynaptic neurons. An axoaxonic (axon-to-axon) synapse at the axon terminal can either decrease or increase the rate of neurotransmitter release at the presynaptic membrane. In some *presynaptic inhibition*, the release of gamma-aminobutyric acid (GABA) restricts the opening of voltage-gated calcium ion channels in axon terminals (Fig. 1.7A). This inhibition reduces the quantity of neurotransmitter that is released as an action potential arrives. It also reduces the effects of synaptic activity on the postsynaptic membrane. Neurotransmitters will be explained in depth in Chapter 19.

In *presynaptic facilitation*, axoaxonic synapse activity increases the quantities of neurotransmitter released as action potentials reach the axon terminal (Fig. 1.7B). This increase strengthens and extends the effects of the neurotransmitter on the postsynaptic membrane. The neurotransmitter *serotonin* aids in presynaptic facilitation. At an axoaxonic synapse, serotonin is released, causing voltage-gated calcium ion channels to remain open longer.

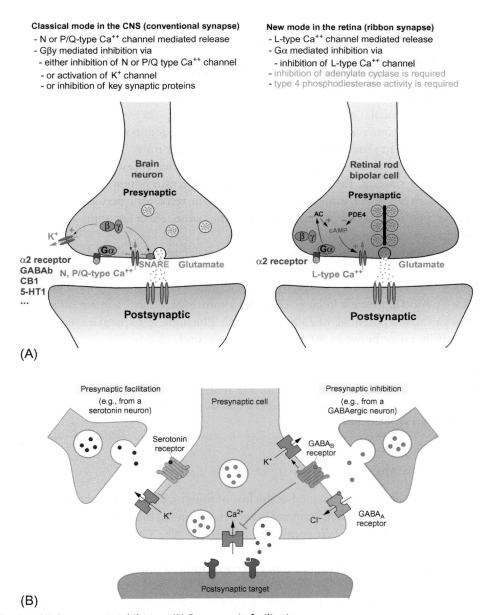

Fig. 1.7 (A) Presynaptic inhibition. (B) Presynaptic facilitation.

The rate of action potential generation

Generally, nervous system messages are processed due to the frequency of action potentials. When action potentials arrive at neuromuscular junctions at the rate of one every second, there may be isolated, grouped twitches in associated skeletal muscle fibers.

A sustained tetanic contraction occurs at the rate of 100 action potentials arriving per second. While only a few action potentials per second, along sensory fiber, may be felt as a very light touch, hundreds of action potentials per second may be felt along that same axon as intense pressure. Generally, the frequency of action potentials relates to the level of sensory stimulation or the strength of motor responses.

An action potential is generated if a graded potential depolarizes the axon hillock for a short time, and the initial segment reaches threshold. More action potentials are produced if the initial segment remains above threshold for a longer time. The *frequency* of action potentials depends on the amount of depolarization above threshold. The higher the degree of depolarization, the greater the frequency of action potentials. The membrane can respond to a second stimulus once the absolute refractory period ends. Holding membrane potential above threshold causes the same effect as applying a second, greater-than-normal stimulus.

Action potentials can be generated at a maximal rate when the relative refractory period is completely eliminated. Therefore, the maximum theoretical action potential frequency is established by the length of the absolute refractory period. The absolute refractory period is shortest in large-diameter axons, which have a *theoretical* maximum frequency of action potentials of 2500 per second. The highest frequencies recorded from axons actually range between 500 and 1000 per second.

Section review
1. What is the difference between EPSPs and IPSPs?
2. What is the difference between temporal and spatial summation?
3. What is a facilitated neuron?

Degeneration and regeneration of nerve cells

Nerve cell degeneration is explained as the loss of functional activity and trophic degeneration of nerve axons and their terminal branches, following the destruction of their cells of origin or interruption of their continuity with these cells. The pathology is characteristic of neurodegenerative diseases. Often the process of nerve degeneration is studied in research on neuroanatomical localization and correlation of the neurophysiology of neural pathways.

Neurodegenerative disease is an umbrella term for a range of conditions that primarily affect the neurons in the human brain. Since neurons don't normally reproduce or replace themselves when they become damaged or die, they cannot be replaced by the body. Neurodegenerative diseases are incurable, debilitating conditions that result in progressive degeneration and/or death of the nerve cells. This causes problems with movement (ataxias) or mental functioning (dementias). Dementias are responsible for the greatest burden of neurodegenerative diseases.

The brain, spinal cord, and peripheral nerves are susceptible to various types of injury ranging from trauma to neurodegenerative diseases that cause progressive deterioration. Because of the complexity of the brain and spinal cord, little spontaneous regeneration, repair, or healing occurs. Therefore, brain damage, paralysis from spinal cord injury, and peripheral nerve damage are often permanent and incapacitating. Patients with serious nervous system injuries or strokes often require lifelong assistance.

In the CNS, regeneration is limited by the inhibitory influences of the glial and extracellular environment. This is partially created by the migration of myelin-associated inhibitors, astrocytes, oligodendrocytes and their precursors, and microglia. Growth factors are not expressed or reexpressed. Glial scars rapidly form, and the glia actually produce factors that inhibit remyelination and axon repair. Axons also lose potential for growth due to aging. Once glial scars form, axons cannot grow across them. It should be stated that CNS axons have been shown to regrow in permissive environments. Therefore, research into CNS nerve regeneration focuses on crossing or eliminating inhibitory lesion sites.

However, neuroregeneration in the PNS occurs to a significant degree. Axonal sprouts form at the proximal stump and grow until they enter the distal stump. The growth of the sprouts is governed by chemotactic factors secreted from Schwann cells. Injury to the PNS immediately elicits the migration of phagocytes, Schwann cells, and macrophages to the lesion site in order to clear away debris such as damaged tissue. When a nerve axon is severed, the end still attached to the cell body is labeled the proximal segment, while the other end is called the distal segment. After injury, the proximal end swells and experiences some retrograde degeneration, but once the debris is cleared, it begins to sprout axons and the presence of growth cones can be detected. The proximal axons are able to regrow as long as the cell body is intact, and they have made contact with the Schwann cells in the endoneurial channel or tube. Human axon growth rates can reach 1 mm per day in small nerves, and 5 mm per day in large nerves. The distal segment, however, experiences Wallerian degeneration within hours of the injury. The axons and myelin degenerate, but the endoneurium remains. In the later stages of regeneration, the remaining endoneurial tube directs axon growth back to the correct targets. During Wallerian degeneration, Schwann cells grow in ordered columns along the endoneurial tube, creating a band of Bungner (boB) that protects and preserves the endoneurial channel. Also, macrophages and Schwann cells release neurotrophic factors that enhance regrowth.

Today's clinicians, scientists, and other specialists are taking a multidisciplinary integrative approach to neuroregeneration for many devastating neurological conditions. Researchers are investigating the effects of restoring cerebrovascular function through the transplantation of induced pluripotent stem (iPS) cell-derived progenitor cells on amyloid pathology and cognitive function in amyloid *Alzheimer's disease* model mice. The iPS cells converted from skin fibroblasts, by transducing various transcription factors, have the potential to generate all tissues in the body, including vascular cells. For *amyotrophic lateral sclerosis (ALS)*, researchers are using adipose-derived mesenchymal stem cells from the patient's own body. These cells are modified in the laboratory and delivered back into the patient's nervous system to promote neuron regeneration.

For *MS*, certain patients experience spontaneous repair of the myelin and nerves—and as a result, therapies are being developed that simulate this repair, promoting recovery of lost function. The regeneration of the myelin sheath can be stimulated by small, folded DNA molecules known as **aptamers**. For *Parkinson's disease*, skin and iPS cell lines are being studied regarding the abilities to generate cells that die as a result of the disease.

The actual *regrowth* of nerve cells, called *axogenesis*, is being studied using *zebrafish*. This research is providing understanding into how nerve cells grow during the development of the nervous system, and how nerve regeneration might be improved after injury. For stroke neuroregeneration, it is believed that mesenchymal stem cells (MSC) can rescue damaged neurons after exposure to oxygen-glucose deprivation (OGD) stress.

Clinical considerations

There are many histological disorders that affect the neurons and glial cells. Demyelination of axons can result in several conditions, and the glial cells may produce various types of tumors.

Myasthenia gravis

Myasthenia gravis is a condition signified by episodic muscle weakness and the ability to become easily fatigued. It is caused by autoantibody- and cell-mediated destruction of the acetylcholine receptors (AChRs), which disrupts neuromuscular transmission. The condition is most common in women between the ages of 20 and 40. Myasthenia gravis is associated with abnormalities of the thymus, thyrotoxicosis, and other autoimmune disorders. Approximately 65% of patients have thymic hyperplasia, while 10% have a thymoma. About 50% of thymomas are malignant. Risk factors include infections, surgery, and drugs such as aminoglycosides, magnesium sulfate, quinine, calcium channel blockers, and procainamide.

Some patients with the generalized form of myasthenia gravis have no antibodies to AChRs in their serum. About 50% have antibodies to muscle-specific receptor tyrosine kinase or *MuSK*. This is a surface membrane enzyme that helps AChR molecules aggregate when the neuromuscular junction is developing. Anti-MuSK antibodies are not present in patients with AChR antibodies or with isolated ocular myasthenia. Patients with anti-MuSK antibodies may have less response to anticholinesterase drugs and need more aggressive early immunotherapy. Ocular myasthenia gravis only affects that extraocular muscles and makes up about 15% of cases.

Congenital myasthenia is a rare autosomal recessive disorder starting in childhood. It is due to structural abnormalities in the postsynaptic receptors and is not an autoimmune disorder. Ophthalmoplegia is usually present. Neonatal myasthenia affects 12% of infants born to mothers who have myasthenia gravis. It is caused by IgG antibodies passively crossing the placenta. Generalized muscle weakness develops, resolving in days to weeks when antibody titers decline. Treatment is usually supportive.

Clinical manifestations
Myasthenia gravis usually causes ptosis, diplopia, and muscle weakness after an affected muscle is used. The weakness resolves after resting the affected muscles, but recurs when they are used again. The ocular muscles are initially affected in 40% of patients, but eventually in 85% of patients. In just 15% of patients, it is only the ocular muscles that are affected. The ability of the hands to grip may be weak to normal, known as *milkmaid's grip*. There may be weakening of the neck muscles. When generalized myasthenia develops following ocular symptoms, it usually does this within the first 3 years. Proximal limb weakness commonly occurs. In some patients, bulbar symptoms develop. These include altered voice, choking, nasal regurgitation, and dysphagia. Sensation and deep tendon reflexes remain normal. The manifestations are of altered intensity, over hours to days.

Myasthenic crisis is a severe, generalized condition. It may involve *quadriparesis* or life-threatening respiratory muscle weakness. However, this form only occurs in about 10% of patients. It is often caused by a supervening infection that reactivates the immune system. Once respiratory insufficiency begins, respiratory failure can develop very quickly.

Diagnosis
The diagnosis of the present form of myasthenia gravis is suggested by the signs and symptoms. Confirmation is by various tests. These include a bedside anticholinesterase test using the short-acting drug *edrophonium*. This will be positive in most patients who have myasthenia with overt weakness. An obviously weakened muscle is used in the test. The patient exercises the affected muscle until fatigue occurs, then 2 mg of edrophonium is administered intravenously. If bradycardia, atrioventricular block, or other adverse reactions do not occur within 30 s, another 8 mg is administered. If the muscle recovers within 2 min, this signifies a positive result. However, this is not definitive for myasthenia gravis since similar improvement occurs in other neuromuscular disorders. During this test, weakness caused by cholinergic crisis may become worse. Therefore, the antidote *atropine* as well as resuscitation equipment must be available during the test.

Confirmation is by serum AChR antibody levels, electromyography, or both. AChR antibodies will be present in 80%–90% of patients with generalized myasthenia, but in only 50% with the ocular form. Antibody levels do not relate to the severity of the disease. About 50% of patients without AChR antibodies will test positive for anti-MuSK antibodies. Electromyography using repetitive stimuli, 2–3 per second, shows a large decrease in amplitude of the compound muscle action potential response in about 60% of patients. Single-fiber electromyography can detect this decrease in >95% of patients.

When myasthenia gravis is diagnosed, a CT (computerized tomography) scan or an MRI (magnetic resonance imaging) of the thorax must be performed to detect a thymoma. Other tests are done to screen for autoimmune disorders often associated with this disease, such as hyperthyroidism, rheumatoid arthritis, systemic lupus erythematosus, and vitamin B12 deficiency. Patients in myasthenic crisis must be evaluated for an infectious trigger. To detect impending respiratory failure, bedside pulmonary function tests such as *forced vital capacity* are performed.

Treatment

For congenital myasthenia gravis, anticholinesterase drugs and immunomodulators are not beneficial. Any patient with respiratory failure must be intubated and placed on a mechanical ventilator. Symptomatic treatment involves anticholinesterase drugs, but these do not change the underlying disease process. These drugs rarely relieve all of the symptoms, and the disease can even become **refractory** to them. The preferred anticholinesterase drugs include pyridostigmine and neostigmine. These drugs can cause abdominal cramps and diarrhea, which are treated with oral atropine or propantheline.

Cholinergic crisis may be caused by doses of pyridostigmine or neostigmine that are too high, resulting in muscle weakness. A mild crisis is difficult to differentiate from worsening myasthenia gravis. Severe cholinergic crisis can usually be differentiated, however, because it causes increased lacrimation and salivation, diarrhea, and tachycardia. An edrophonium test may be useful for patients who were initially responding well to treatment, but then began deteriorating. However, some physicians prefer to initiate respiratory support and stop anticholinesterase drugs for several days in these scenarios.

Regarding immunomodulators, the use of immunosuppressants interrupt the disease process but do not quickly relieve the symptoms. The administration of intravenous immune globulins once daily for 5 days has been shown to improve the condition of 70% of patients in 1–2 weeks, with effects lasting for up to 2 months. Corticosteroids are commonly used as maintenance therapy, yet have only slight immediate effects for myasthenic crisis. >50% of patients actually have acute worsening of their condition after the beginning of high-dose corticosteroids, primarily prednisone. Improvement may occur over several months, and then doses should be reduced to a minimum amount that will control symptoms.

Nearly as effective as corticosteroids is the drug called azathioprine, though benefits may not be significant for many months. Cyclosporine may allow corticosteroid doses to be reduced. Both these drugs require standard precautions. Additional helpful drugs include methotrexate, cyclophosphamide, and mycophenolate mofetil.

For most patients with generalized myasthenia who are under the age of 60, thymectomy is an option. This procedure should be performed for all patients who have a thymoma. After thymectomy, 80% of patients will experience remission or maintenance drug doses can be lowered. Plasmapheresis may be helpful during myasthenic crisis. For patients unresponsive to drugs, plasmapheresis may be useful prior to thymectomy.

> **Focus on tumors of the central nervous system**
> Tumors of the CNS and its associated membranes result in approximately 90,000 deaths in the United States each year. *Primary CNS tumors* originate in the CNS. Almost 50% of CNS tumors are primary tumors. *Secondary CNS tumors* arise from metastasis of cancer cells that originate elsewhere. In adults, primary CNS tumors occur due to the division of abnormal neuroglia instead of the division of abnormal neurons, since typical neurons in adults cannot divide. Through the division of stem cells, neurons increase in amount until children reach the age of 4. This is why primary CNS tumors involving abnormal neurons can occur in young children.

Neuroblastoma

Neuroblastoma is a highly malignant tumor in infants. However, it may also occur in children under 5 years of age. Neuroblastoma is composed of neuroblasts, usually in the medulla of an adrenal gland. However, it is possible for the tumor to form in the nerve tissue of the spinal cord, chest, or neck. According to the National Cancer Institute, neuroblastoma is the first most common cancer in infants, and the third most common cancer in children. >600 cases are diagnosed annually in the United States. It accounts for approximately 15% of all pediatric cancer fatalities. Incidence is higher in non-African-American children, and it is slightly more common in males than females. Neuroblastoma sometimes forms before birth and may be discovered during a fetal ultrasound. Usually, by the time of diagnosis, the cancer has metastasized, most commonly to the lymph nodes, bones, bone marrow, and liver. In infants, metastasis also occurs to the skin.

Neuroblastoma may be caused by a gene mutation, passed from a parent to a child. When this occurs, the child is usually at a younger age, and more than one tumor forms in the adrenal glands. The most common symptoms are abdominal pain, discomfort, and a sense of fullness caused by an abdominal mass. Metastasis-related signs and symptoms may include bone pain due to widespread bone metastasis, a lump where the tumor or tumors exist, **proptosis** of the eyes due to retrobulbar metastasis, *periorbital ecchymosis*, and weakness or paralysis. Respiratory problems, due to liver metastases may occur, especially in infants. Diagnosis may involve physical examination, history, neurological examination, urine catecholamine studies, blood chemistry studies, X-rays, CT scans, biopsy, and ultrasonography. Treatments include surgery followed by observation, chemotherapy, radiation therapy, monoclonal antibody therapy (dinutuximab) with interleukin-2, granulocyte-macrophage colony-stimulating factor, isotretinoin, and iodine 131-MIBG. For patients with changes in the anaplastic lymphoma kinase (ALK) gene, tyrosine kinase inhibitor therapy is indicated.

Astrocytosis

Astrocytosis is an increase in the number of astrocytes, the star-shaped cells that make up the supportive tissues of the brain. It is often observed in irregular zones close to degenerative lesions, abscesses, or certain brain neoplasms. Sometimes, astrocytosis may be diffused in a relatively large region. The condition represents an ongoing repair process.

Gliomas

Gliomas are tumors arising from the supportive tissues of the brain. There are three types of normal glial cells that may produce tumors. These include astrocytes, oligodendrocytes, and ependymal cells. Tumors that have a mixture of these cells are known as *mixed gliomas* or *oligoastrocytomas*. The tumors called *optic nerve gliomas* or *brainstem gliomas*, named after their locations, and not for the types of tissues from which they originate. Other types of gliomas include *ependymomas, oligodendrogliomas,* and *gliomatosis cerebri*.

According to the National Institutes of Health (NIH), gliomas affect <200,000 people in the United States. Gliomas represent about 81% of malignant primary brain tumors. Though they are relatively rare, they cause significant mortality and morbidity. The most common gliomas histology is *glioblastoma*, which makes up approximately 45% of all gliomas. This type has a 5-year relative survival rate of about 5%.

Symptoms are varied, based on the type of gliomas. General symptoms include headache, nausea, visual problems, behavior and personality changes, hormonal disturbances, memory disturbances, increased intracranial pressure, and seizures. The exact cause of gliomas is unknown. Increased intracranial pressure manifests as headache, nausea, and vomiting in early stages, and can progress to coma and even death as the pressure continues to rise.

Treatments are based on tumor type, but generally include surgery, radiation therapy (especially if the tumor is high grade), chemotherapy (also mostly for high-grade tumors), observation, and clinical trials of new medications.

Astrocytoma

Astrocytomas are tumors that arise from astrocytes. The tumors may be benign or malignant. The tumors are graded, from I to IV, based on the normal or abnormal qualities of their cells. *Low-grade* astrocytomas are more common in children and *high-grade* astrocytomas are more common in adults. Low-grade astrocytomas are usually localized and grow slowly. High-grade astrocytomas grow quickly (3–6 months from the first clinical symptoms).

The grades of astrocytomas are as follows:
- *Pilocytic (juvenile) astrocytoma*—Grade I—usually do not spread; the most benign type.
- *Diffuse (low-grade) astrocytoma*—Grade II—usually invade surrounding tissue but grows slowly.
- *Anaplastic astrocytoma*—Grade III—rare type that requires more aggressive treatment.
- *Glioblastoma*—Grade IV—primary tumors are very aggressive, while secondary tumors begin as low-grade tumor and evolve into grade IV. Grade III and IV astrocytomas are usually located in the frontal lobes and cerebral hemispheres (see Figs. 1.8 and 1.9).
- *Subependymal giant cell astrocytoma*—ventricular tumors associated with tuberous sclerosis.

According to Harvard Medical School in 2014, pilocytic astrocytoma accounted for 0.6%–5.1% of all intracranial neoplasms and was the most common primary brain tumor in children. It also accounted for 70%–85% of all cerebellar astrocytomas. Astrocytoma is more common in the Caucasian race, and slightly more common in males than in females.

Astrocytomas may appear anywhere in the CNS. Early symptoms include headaches, seizures, loss of memory, and behavioral changes. The cause is unknown. Treatments include surgery, radiation therapy, observation, follow-up scans, and chemotherapy.

Though most common in the frontal lobes and cerebral hemispheres, grade III and IV astrocytomas may also be located in the brainstem, and are twice as common in men as in

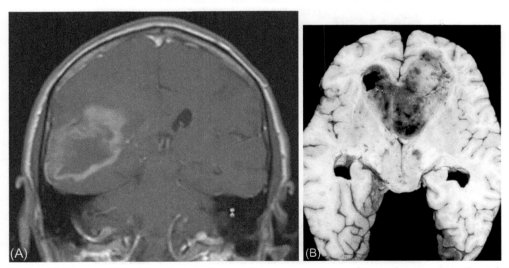

Fig. 1.8 (A) Contrast T1-weighted coronal magnetic resonance image shows a large mass in the right parietal lobe with "ring" enhancement. (B) Glioblastoma appearing as a necrotic, hemorrhagic, and infiltrating mass.

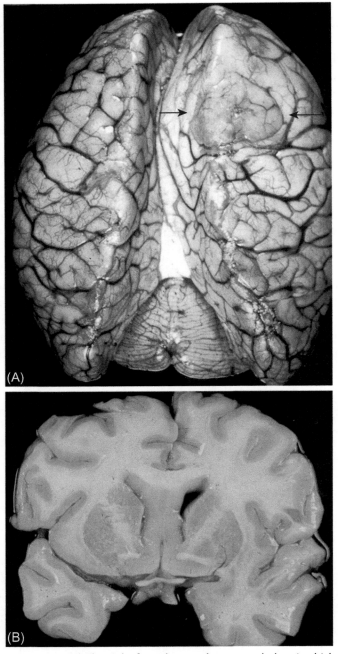

Fig. 1.9 Diffuse astrocytoma. (A) The right frontal tumor has expanded gyri, which led to flattening (*arrows*). (B) There is bilateral expansion of the septum pellucidum by gray, glassy tumor.

women. They are the third most common cancer in people between ages 15 and 34, and fourth most common in people aged 35–54. These astrocytomas are usually large, well circumscribed, and have a variegated pattern. Their peripheral rims are pink-gray and solid. They have a yellow, soft necrotic center and hemorrhaged points. When viewed under a microscope, they show an increase in cellularity, pleomorphic qualities, vascular proliferation, and necrosis. The main histologic difference between anaplastic grade III tumors and grade IV glioblastoma multiforme tumors is necrosis and vascular proliferation. An astrocytoma with hemorrhage and necrosis is grade IV by definition.

Grade IV astrocytomas (or glioblastoma multiforme) make up about 55% of all gliomas. Their molecular pathology shows high vascularization and extensive heterogenic infiltration. Sometimes, they grow large enough to extend from the meningeal surface and push through the ventricular wall. At the time of the patient's death about 50% of these tumors are bilateral, or occupy more than one lobe. In some cases, grade IV astrocytomas have been found outside of the brain and spinal cord.

The grade IV tumors are usually diffuse, with nonspecific clinical manifestations. These include irritability, headache, and personality changes. There is progression to symptoms of increased intracranial pressure, including headache when changing positions, papilledema, and vomiting. About 30%–40% of patients have seizures. Symptoms may further progress to definite focal signs. These include dysphasia, hemiparesis, cranial nerve palsies, dyspraxia, and visual field defects, as well as generalized signs of intracranial pressure. Diagnosis usually takes 3–6 months after the onset of first clinical manifestations. This is because the patient often does not realize the cause of symptoms. For grades III and IV astrocytomas, survival over 5 years is only between 5% and 10%.

Ependymomas

Ependymomas represent 10% of pediatric brain tumors, making them the third most common CNS tumor in children. The mean age at diagnosis is 6 years. However, approximately 30% of ependymomas occur in children under 3 years of age. Ependymomas are primary tumors that develop from the ependymal lining of the ventricular system. As much as 70% of cases occur in the posterior fossa, usually spreading locally to the brainstem. However, ependymomas may develop anywhere in the brain or spinal cord. Tumor cells may spread in the CSF. Ependymomas also affect adults and are slightly more common in males and Caucasians. In the United States, approximately 1340 cases of ependymomas are diagnosed annually. They are not encapsulated (see Fig. 1.10).

The first symptoms are pain and dysesthesias, usually because of increased intracranial pressure. An affected infant may have developmental delays and irritability. There may be changes in concentration, mood, or personality. Additional symptoms include balance and gait disturbances, seizures, and symptoms of spinal cord compression such as back pain, and loss of bladder or bowel control. Diagnosis is based on MRI and biopsy.

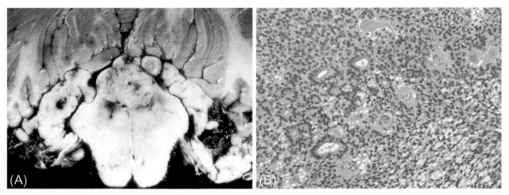

Fig. 1.10 Ependymoma. (A) Tumor growing into the fourth ventricle, distorting, compressing, and infiltrating surrounding structures. (B) Microscopic appearance of ependymoma.

Treatments include surgical removal of the tumor followed by MRI to assess residual tumor. Chemotherapy, radiation therapy, or both follow this. The overall 5-year survival rate is about 50%, but this partially depends on the age of the patient, with children over 5 years having the best prognosis.

Pesticide toxicity

Pesticides are agents that are toxic to insects and other species. Therefore, they are potentially hazardous to humans and other animals. People who use pesticides or those that may come into regular contact with them must understand their relative toxicities and potential health effects. >95% of all pesticide exposures come from dermal exposure, mostly to the hands and forearms. The wearing of protective chemical-resistant gloves can nearly eliminate this type of exposure. The other primary routes of pesticide exposure are via inhalation into the lungs, or by entering the mouth or eyes. Acute affects include dermatitis, rash, blisters, coughing, sneezing, wheezing, rawness of the mucous membranes, reddening, burning, and itching. Systemic effects may include nausea, vomiting, fatigue, headache, intestinal disorders, changes in heart rate, difficulty in breathing, convulsions, and coma. If severe, acute symptoms can be fatal. Aside from the acute effects, there are also long-term chronic effects.

The chronic effects of pesticide toxicity are many. These include birth defects, fetal toxicity, benign or malignant tumors, blood disorders, genetic changes, endocrine disruption, nerve disorders, and effects on reproduction. Chronic toxicity is more difficult to determine through laboratory analysis. Regardless of the type of pesticide, it is important to seek medical attention right away for any symptoms that become serious, and always follow the instructions on the pesticide container for accidental exposures.

Pesticide toxicity inhibits the enzyme *AChE* and causes action potentials to be stimulated. When AChE is inhibited, normal functioning of the nervous system is disrupted, potentially leading to death. The pesticides most harmful to AChE are *carbamate* and *organophosphate* pesticides. Exposure results in continued overexcitation of nerve-to-nerve and nerve-to-muscle communication. Voltage-gated ion channels are opened by pesticides allowing ions to flow into axons, depolarizing their neurons. The neurons become nonfunctional, which is characterized by paralysis. In between the normal state and paralysis, partial depolarization exists. This leaves neurons susceptible to *false triggering*. A small stimulus that normally does not trigger an action potential now produces one easily, since the resting potential slowly climbs to the threshold needed to launch an action potential. Therefore, the pesticide causes the body to exhibit twitching and uncontrolled movements, as nerve impulses trigger muscle movements. It is not always obvious that symptoms are from an acute exposure or due to delayed effects from repeated exposures. People who regularly work with such pesticides should be enrolled in a cholinesterase-monitoring program, which involves regular testing of the blood.

Testing for cholinesterase levels requires two tests, performed at least 72 h apart, but not >14 days apart. Each person has his or her own unique baseline level, to which additional testing is compared. Adverse effects of carbamates last only for a relatively short period of time, but it takes much longer for the body to recover from organophosphate exposures.

Clinical cases

Clinical case 1
1. Which type of neuroglial cells is responsible for myelination of axons in the CNS?
2. In the clinical case below, what is the reason that the patient's vision became abnormal?
3. What drugs can reduce the inflammation caused by MS and delay progression of the disease?

A 25-year-old female has been experiencing muscle fatigue, numbness and tingling in her extremities and blurred vision. Her symptoms were mild and only lasted for a day or two. After several days, her symptoms increased, and she went to see her physician. She complained of severe fatigue and her vision had become much worse in her left eye. The physician ordered an MRI and spinal puncture for a sample of her CSF. The MRI revealed demyelinated axons (plaques) along her optic nerve, cerebellum, brainstem, and spinal cord. Her CSF had elevated levels of proteins and **oligoclonal bands**. The diagnosis of MS was made.

Answers:
1. Oligodendrocytes are responsible for CNS axon myelination.
2. The plaques on this patient's optic nerve affected nerve transmission from the retina to the visual cortex. The optic nerve is the only peripheral nerve in the body that is myelinated by oligodendrocytes and therefore is affected in MS. Recall that MS is a demyelinating disease of the CNS.
3. Corticosteroids reduce the related inflammation but do not change the disease course.

Clinical case 2

1. Based on the clinical case below, what you think the diagnosis will be?
2. What other steps can the pediatrician take to confirm diagnosis?
3. What are the treatments for the patient's condition?

A 3-year-old boy, over the past month, has had trouble sleeping, been more irritable than usual, and complained of stomach pain. He has not had a fever, vomiting, diarrhea, or constipation. His appetite has been much less than normal. The pediatrician palpates an abdominal mass above the umbilicus. The lungs are clear, and the rest of the examination is normal. However, CT scan shows a large mass in the central abdomen that extends to the mid-thorax. Bone marrow biopsy shows small round blue cells. Urine catecholamines are slightly elevated.

Answers:
1. The diagnosis is neuroblastoma, which is a highly malignant cancer that is common in children under the age of 5. Recall that the chromaffin cells of the adrenal medulla are derived from neuro crest cells.
2. The pediatrician can assess if there is bone pain, proptosis of the eyes, periorbital ecchymosis, and weakness or paralysis. Additional diagnostic measures include neurological examination, blood chemistry studies, biopsy, ultrasonography, X-rays, and CT scan.
3. Treatments include surgery followed by observation, chemotherapy, radiation therapy, stem cell rescue, monoclonal antibody therapy (dinutuximab) with interleukin-2, granulocyte-macrophage colony-stimulating factor, isotretinoin, and iodine 131-MIBG.

Clinical case 3

1. In the clinical case below, which type of cells in the brain is related to the immune system?
2. Besides a virus, what are the other causes of encephalitis?
3. What are the risk factors for encephalitis?

A 60-year-old man was diagnosed with encephalitis, which is an acute infection and inflammation of the brain. He previously had a viral infection after being bitten by a mosquito. The patient's symptoms of encephalitis were mild, and he recovered in a few weeks.

Answers:
1. Microglia, or microglial cells, are the immune cells of the brain. They act as macrophages, patrolling for plaques, damaged or unnecessary neurons and synapses, and infectious agents.
2. Encephalitis may also be caused by bacteria, fungi, and autoimmune disorders. By definition, encephalitis alters mental status.
3. The highest risk factor groups for encephalitis are older adults, children under the age of 1 year, and people with weak immune systems.

Clinical case 4
1. Based on the clinical case below, what is the most common site for an ependymoma to occur?
2. What are the initial symptoms of this tumor in infants?
3. What is the diagnosis and treatment based on?

A 4-year-old child was taken to his pediatrician because of irritability, changes in his mood and personality, and poorer than normal concentration. His mother reported that he had two episodes of seizures in the past 2 weeks. The pediatrician ordered an MRI and biopsy, which revealed a tumor in the brain. The diagnosis was ependymoma.

Answers:
1. The most common site for an ependymoma to occur is the posterior fossa of the brain.
2. The initial symptoms of the tumor, usually due to increased intracranial pressure, include developmental delays and irritability. There may be changes in concentration, mood, or personality.
3. Diagnosis is based on an MRI and biopsy. Treatments include surgical removal of the tumor followed by MRI to assess residual tumor. Chemotherapy, radiation therapy, or both follow this.

Clinical case 5
1. What is the prognosis for anaplastic astrocytoma?
2. What are the treatment options?
3. Based on the location of the tumor, what outcomes can this patient expect following treatment?

A 65-year-old man had several episodes of losing consciousness, and his family took him to his physician. The physician ordered a CT scan, which showed a left temporal mass. An MRI confirmed this. After biopsy, he was diagnosed with anaplastic astrocytoma. Surgery was performed followed by radiation therapy.

Answers:
1. Anaplastic astrocytoma, a grade III tumor, is a rare type that requires more aggressive treatment and has a poor prognosis. These tumors lack vascular proliferation and necrosis, but compared to grade II tumors, are more cellular, with more atypical appearances, and have more mitoses.
2. Usually, initial treatment for anaplastic astrocytoma is to surgically remove as much of the tumor as possible, followed by radiation therapy. Chemotherapy is generally not helpful.
3. Based on the area of the brain involved, the patient may experience various types of paralysis, speech difficulties, and cognitive or sensory perception deficits.

Key terms

Absolute refractory period	Motor neurons
Action potentials	Multipolar neurons
Anaxonic neurons	Myasthenia gravis
Anterograde flow	Nervous tissue
Aptamers	Neuroblastoma
Astrocytes	Neurofibrils
Astrocytomas	Neurofilaments
Astrocytosis	Neuroglandular junction
Axolemma	Neurilemma
Axon hillock	Neuroglia
Axon terminals	Neuromuscular junction
Axons	Neuron
Axoplasm	Neurotransmitters
Bipolar neurons	Neurotubules
Blood brain barrier (BBB)	Nodes of Ranvier
Centrioles	Nuclear pores
Chemical synapse	Oligoclonal bands
Cholinergic synapses	Oligodendrocytes
Collaterals	Perikaryon
Connexons	Postsynaptic neuron
Continuous propagation	Postsynaptic potentials
Dendrites	Presynaptic neuron
Depolarization	Propagation
Electrochemical gradient	Proprioceptors
Ependymal cells	Proptosis
Ependymomas	Refractory
Equilibrium potential	Refractory period
Excitatory postsynaptic potential (EPSP)	Relative refractory period
Exteroceptors	Resting membrane potential
Ganglion	Retrograde flow
Gated ion channels	Saltatory propagation
Gliomas	Satellite cells
Golgi apparatus	Schwann cells
Graded potential	Spatial summation
Hyperpolarization	Summation
Inhibitory postsynaptic potential (IPSP)	Synapse
	Synaptic cleft
Interneurons	Synaptic delay
Interoceptors	Synaptic fatigue
Kinesin	Telodendria
Leak channels	Temporal summation
Lipofuscin	Threshold
Local potentials	Tight junctions
Membrane potential	Unipolar neuron
Microfilaments	Visceral motor neurons
Microglia	Visceral sensory neurons
Microtubules	Wallerian degeneration

Suggested readings

1. www.atlasbrain.com/enx/atlas_main.html.
2. Berkowitz, A. *Lange Clinical Neurology and Neuroanatomy—A Localization-Based Approach*. McGraw-Hill Education, 2016.
3. Blumenfeld, H. *Neuroanatomy Through Clinical Cases*, 2nd ed.; Sinauer Associates/Oxford University Press, 2010.
4. www.cern-foundation.org/education/ependymoma-basics/ependymoma-overview.
5. www.columbia/edu/itc/hs/medical/neuroanatomy/neuroanat.
6. https://www.frontiersin.org/journals/neuroanatomy.
7. Gould, D. J.; Brueckner-Collins, J. K. *Sidman's Neuroanatomy—A Programmed Learning Tool*, 2nd ed.; LWW, 2007.
8. Haines, D. E. *Neuroanatomy in Clinical Context—An Atlas of Structures, Sections, Systems, and Syndromes*, 9th ed.; LWW, 2014.
9. https://www.kenhub.com/en/start/neuroanatomy/atlas.
10. https://library.med.utah.edu/webpath/histhtml/neuranat/neuranca.html.
11. https://librepathology.org/wiki/neurohistology.
12. Marieb, E. N.; Hoehn, K. N. *Anatomy & Physiology*, 6th ed.; Pearson, 2016.
13. Marieb, E. N.; Hoehn, K. N. *Human Anatomy & Physiology*, 10th ed.; Pearson, 2015.
14. Martini, F. H.; Nath, J. L.; Bartholomew, E. F. *Fundamentals of Anatomy & Physiology*, 10th ed.; Pearson, 2014.
15. McKinley, M.; O'Loughlin, V.; Bidle, T. *Anatomy & Physiology—An Integrative Approach*, 2nd ed.; McGraw-Hill Education, 2015.
16. http://nba.uth.tmc.edu/neuroanatomy.
17. www.neuroanatomy.org.
18. https://www.neuroscienceassociates.com/technologies/staining.
19. www.neurosurgery.pitt.edu/research/surgical-neuroanatomy-lab.
20. https://pathology.duke/edu/files/neuroanat/neurohistology.html.
21. https://radiopaedia.org/articles/neuroanatomy.
22. Saladin, K. S. *Anatomy & Physiology: The Unity of Form and Function*, 8th ed.; McGraw-Hill Education, 2017.
23. Schuenke, M.; Schulte, E.; Schumacher, U.; MacPherson, B.; Stefan, C. *Thieme Atlas of Anatomy—Head, Neck, and Neuroanatomy*, 2nd ed.; Thieme, 2016.
24. ten Donkelaar, H. J.; Kachlik, D.; Tubbs, R. S. *An Illustrated Terminologia Neuroanatomica: A Concise Encyclopedia of Human Neuroanatomy*. Springer, 2018.
25. Vanderah, T. W.; Gould, D. J. *Nolte's the Human Brain—An Introduction to Its Functional Anatomy*, 7th ed.; Elsevier, 2016.

CHAPTER 2

Embryology

The human nervous system begins development in the embryonic period. It starts as a simple ectodermal tissue. By understanding how the nervous system develops, it is easier to comprehend its ultimate adult structure and function. Our nervous system consists of two systems: the central nervous system (CNS) and the peripheral nervous system (PNS). Basically, the brain and spinal cord are considered as the CNS, and the remaining neuronal tissues are considered as the PNS. CNS congenital malformations are also easier to understand when we are aware of the embryological development of the CNS. As the nervous system develops, neurons proliferate in large amounts. They move from the areas in which they developed to reach their final locations, establishing normal connections with other neurons. Neural development begins early in the embryonic period, but is the last body system to be completed after birth. If something harms a developing embryo, nervous system development may become abnormal.

Early CNS development forms on a simple plate, called the neural plate which folds and forms a groove, later becoming a tube that is open at both ends; the neural tube. If either opening fails to close, a neural tube defect will occur. Most of the nervous system is made up of two different types of cells: neurons and glial cells, which are generated within the neural tube. These cells differentiate into many different, highly specialized cell types, which were discussed in Chapter 1.

Development of the neural tube

The neural tube has three cell layers: the ectoderm, mesoderm, and endoderm. Eventually, these three cell layers form our various body tissues. *Neuroectoderm* derives from ectoderm and forms the skin and neural tissues. The mesoderm forms bone and muscles. The endoderm forms the gastrointestinal and respiratory cell linings. By the 18th gestational day, chemicals are released by the notochord (derived from the mesoderm), which signals the ectoderm to form the neural plate. Soon, this plate folds inward and forms a longitudinal midline neural groove. A parallel neural fold on either side surrounds this groove. The neural groove becomes deeper, while the neural folds move nearer to each other in the dorsal midline. By the end of the third week, neural folds grow over the midline and form the neural tube.

The open ends of the neural tube are the cranial and caudal neuropores. Fusion continues in both directions. The entire neural tube will be closed by the end of the fourth week, a process known as primary neurulation. The rostral neuropore closes at day 25 and the caudal neuropore closes at day 27. The closing neural tube continually separates from the neuroectodermal surface. It leaves groups of cells from the crest of the neural folds behind. These neural crest cells, which are derived from the neuroectoderm, develop into many cell types, including those of the PNS. The neural tube will develop into almost all of the CNS. The neural cavity becomes the brain's ventricular system, as well as the spinal cord's central canal. The sacral spinal cord forms differently. When the neural tube has closed, a second cavity extends into the solid cell mass at its caudal end. This occurs during weeks 5 and 6, and is known as secondary neurulation.

Sensory and motor areas defined by the sulcus limitans

Dorsal-ventral patterns of the spinal cord and brainstem influence development of sensory or motor functions. The ectoderm near the future dorsal neural tube surface and the mesodermal notochord near the ventral surface produce various types of signaling molecules early in embryonic development. The nucleus pulposus, which is the inner core of the vertebral disk, also derives from the notochord. These signaling molecules have opposing concentration gradients, which cause distinct patterns of further development in these neural tube regions.

In the fourth week, a longitudinal groove called the sulcus limitans develops in the lateral wall of the neural tube. This separates the neural tube into dorsal half or alar plate, and a ventral half or basal plate. These plates are important since derivatives of the alar plate are mostly involved in sensory processing, and derivatives of the basal plate are mostly involved in motor processing.

The sulcus limitans will not be present in the future adult spinal cord. However, the central gray matter of the spinal cord is divided into a posterior (dorsal) horn and an anterior (ventral) horn on either side, forming an H-shaped gray matter structure. Sensory and motor neurons travel by different pathways in the spinal cord. Central processes of sensory neurons are derived from neural crest cells. They primarily end in the posterior horn, which has cells with axons forming ascending sensory pathways. These sensory pathways are called *ascending* since they deliver information from the periphery to the brain. The motor pathways are called *descending* as they deliver information from the brain to the skeletal muscles. The anterior (ventral) horn has cell bodies of autonomic and somatic motor neurons. Their axons emerge from the spinal cord to innervate autonomic ganglia and skeletal muscles. The same analogy exists between the sensory alar plate derivatives and the motor basal plate derivatives in the brainstem.

Bulges and flexures of the neural tube

Longitudinal development of the neural tube is also related to different signaling molecules. The neural tube, even before completely closing, has bulges in its rostral end, near the location where the brain will develop. Bends in the tube also begin to appear.

Primary vesicles

In the fourth week, three bulging *vesicles* appear in the neural tube. These are known as the primary vesicles, and are described, from rostral to caudal locations, as follows:
- **Prosencephalon**—which will become the forebrain
- **Mesencephalon**—which will become the midbrain
- **Rhombencephalon**—a rhomboid-shaped structure that merges with the caudal (spinal) part of the neural tube. It is also known as the *hindbrain* and will become pons, medulla, and cerebellum.

Derivatives of the neural tube vesicles are explained in Table 2.1.

Secondary vesicles

By the fifth week, two of the primary vesicles (the prosencephalon and rhombencephalon) subdivide, and form secondary vesicles (see Fig. 2.1). These are as follows:
- **Telencephalon**—the *endbrain*, forming from the prosencephalon. Then, the telencephalon becomes the cerebral hemispheres of an adult brain.
- **Diencephalon**—the *in-between-brain*, also from the prosencephalon
  - *Thalamus/hypothalamus*—form from the diencephalon, along with the retina and several small structures. The thalamus is a large gray matter mass between the cerebral cortex and other structures. The hypothalamus is a control center for autonomic functions.
- **Mesencephalon**—does not divide into secondary vesicles, and will develop into the midbrain, as mentioned above.
- **Metencephalon**—formed from the rhombencephalon, will become the *pons* and *cerebellum*. Between the metencephalon and myelencephalon, the pontine flexure will appear in the dorsal brainstem surface. It is important for the caudal brainstem

Table 2.1 Derivatives of the neural tube vesicles.

Primary vesicle	Secondary vesicle	Cavity	Neural derivatives
Prosencephalon (forebrain)	Telencephalon	Lateral ventricles	Cerebral hemispheres
	Diencephalon	Third ventricle	Thalamus, hypothalamus, retina, other structures
Mesencephalon (midbrain)	Mesencephalon	Cerebral aqueduct	Midbrain
Rhombencephalon (hindbrain)	Metencephalon	Part of fourth ventricle	Pons, cerebellum
	Myelencephalon	Part of fourth ventricle, part of central canal	Medulla oblongata

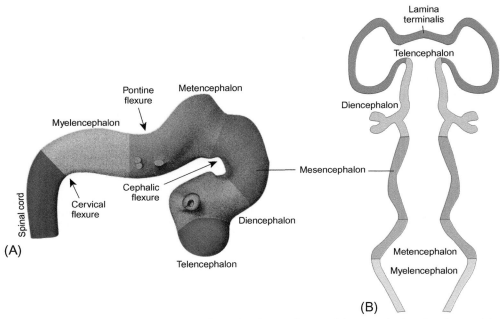

Fig. 2.1 Secondary vesicles during the sixth week. (A) Lateral view of the neural tube, showing vesicles and flexures. (B) Schematic longitudinal section, as though the flexures are straightened out.

structure. The walls of the neural tube spread, forming a diamond-shaped cavity (*rhombencephalon*). Only a thin "roof" of membranes remains over the future fourth ventricle.
- **Myelencephalon**—also from the rhombencephalon, it will become the medulla oblongata, which is the area of the brainstem that merges with the spinal cord.

The alar and basal plates, separated by the sulcus limitans, eventually are located in the floor of the fourth ventricle. The future corresponding adult brainstem (rostral medulla and caudal pons) will have sensory nuclei that are lateral and not posterior to the motor nuclei.

Lateral areas of the alar plate, inside the rostral metencephalon, will become much thicker and form the rhombic lips. Parts of these lips move into the brainstem, forming the cerebellum-related nuclei. Other parts enlarge and fuse at midline, forming a transverse ridge. This is the future *cerebellum* (see Fig. 2.2).

Section review
1. What are the three layers of the neural tube?
2. What is the process of primary neurulation?
3. What structure separates the neural tube into the alar and basal plates?
4. In which horn are the cell bodies of the autonomic motor neurons located?

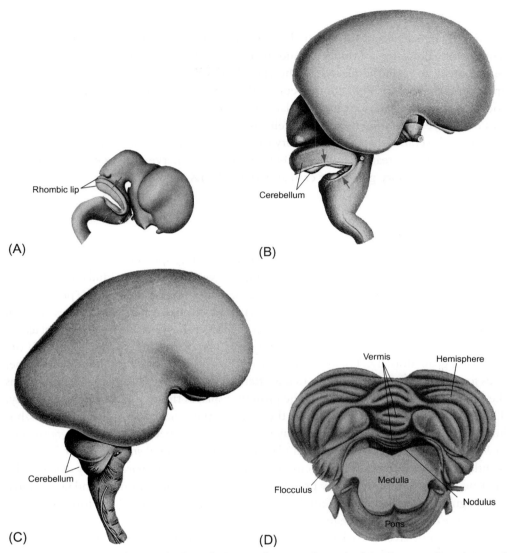

Fig. 2.2 Development of the cerebellum. (A) During the second month of development lateral parts of the alar plate in the rostral metencephalon thicken to form rhombic lips. These continue to enlarge during the third and fourth months (B) and (C), forming the cerebellum. (D) By about 5 months, a series of deep fissures develops in the cerebellar surface, and both midline (vermis) and lateral (hemisphere) zones are apparent.

Neural crest

As the neural tube closes, as far as the future diencephalon, the neural crest becomes pinched off, as continuing bilateral cell strands. Neural crest cells give rise to most of the PNS. This includes sensory cells, autonomic ganglia, Schwann cells, microglia, adrenal medulla secretory cells, melanocytes (cutaneous pigment cells), arachnoid cells, and microglial cells.

Cranial placodes

A series of cranial placodes are formed by a connection of ectoderm near the brainstem's neural crest. This strip continues around the rostral end of the neural plate, thickening in certain areas, to form the placodes. The cranial placodes, along with areas of the neural crest, form the sensory PNS components of the head. The single adenohypophyseal placode will form the anterior pituitary gland. On each side, the lens placode will form the lens of the eye. Migrating neural crest cells, as mentioned before, forms most, if not all of the glia in the cranial ganglia and sensory organs. These cells work with placodes, in different amounts, to create the olfactory epithelium and its nerve, the trigeminal ganglion, inner ear, and sensory ganglia of cranial nerves VII, IX, and X.

Medulla spinalis (spinal cord)

The rest of the neural tube becomes the spinal cord. The spinal cord extends from the medulla oblongata all the way down to the lumbar vertebrae. The medulla spinalis is the clinical name for the *spinal cord*. As it develops, there is symmetric neural tube progression, in layers. As a part of the neural tube, the spinal cord also has alar and basal plates. Ventral growth of the two basal layers and related marginal layers beyond the *floor plate* level leaves a separation known as the ventral median fissure. The mantle and marginal layers of the alar plates grow dorsally. The dorsal marginal layers become fused on the median plane. This forms a dorsal median septum that is often poorly defined. Its external margin forms the dorsal median sulcus. Such midline growth displaces the roof plate region ventrally. This reduces the neural canal, forming the small central canal of the spinal cord, which is lined with *ependymal cells*. The mantle layer of the alar plate becomes the *dorsal gray column* or *horn*. The mantle layer of the basal plate becomes the *ventral gray column* or *horn*. The mantle zone at the plane of the sulcus limitans becomes the *intermediate gray column*. The intermediate gray column is only present in the thoracic, cranial lumbar, and sacral spinal cord segments (see Fig. 2.3). Neural crest cells provide the neurons that form the spinal ganglia at each spinal cord segment.

In the mantle layer of each plate, neurons are arranged in functional columns. General visceral *afferent* and *efferent* neurons are near each other, in their respective gray columns on each side of the dorsal plane, through the sulcus limitans. General somatic afferent and general proprioceptive neuronal columns are dorsal, in the alar plate of the mantle layer. The general somatic efferent column is ventral in the basal plate of the mantle layer. At the levels of the limbs, spinal cord segments responsible for their innervation are enlarged, which forms the cervical and lumbosacral intumescences. There are dorsal, lateral, and ventral funiculi, which are white matter processes, on each side of the spinal cord.

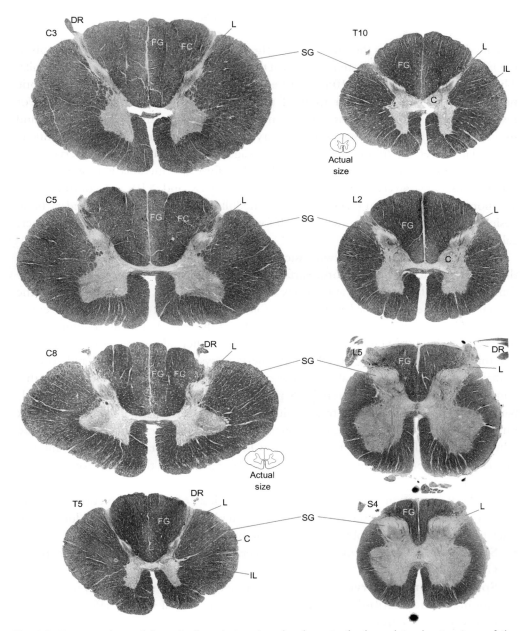

Fig. 2.3 Cross sections of the spinal cord at various levels; note the large lateral extensions of the anterior horns in C5, C8, and L5.

Myelencephalon (medulla oblongata)

The medulla oblongata is the lowest part of the brainstem and is also referred to simply as the *medulla*. Recall that brainstem consists of three parts: the midbrain, pons, and medulla. The three structures contain 10 cranial nerve pairs (except for the olfactory and optic nerves) and are responsible for most of the unconscious activities of the brain. For instance, the medulla's neurons generate nerve centers responsible for pain relay, and movements of the cardiovascular, respiratory, and gastrointestinal systems. The nuclei of cranial nerves IX (glossopharyngeal), X (vagus nerve), XI (accessory nerve), and XII (hypoglossal nerve) are in the medulla. Therefore, the medulla is responsible for tongue movements (via CN XII), swallowing (CN IX and XII) and control of the autonomic nervous system (via CN X). The medulla develops from the secondary brain vesicle—the myelencephalon, which itself formed from the earlier primary brain vesicle called the rhombencephalon.

The medulla oblongata has two basic parts. Its upper, open, or *superior* area is where the fourth ventricle forms the dorsal surface. Its lower, closed, or *inferior* area is where the fourth ventricle has narrowed at the obex of the caudal medulla, surrounding part of the central canal. The *anterior median fissure* has a fold of *pia mater*, and extends along the entire length of the medulla, ending at the lower border of the *pons*, at a small, triangular foramen cecum. There are raised medullary pyramids on either side of this fissure. Inside the pyramids are *pyramidal tracts*, which are the nervous system's corticospinal and corticobulbar tracts. The corticospinal tract (descending pathway) is the main motor tract that controls the movement of different body parts below the neck. Similarly, the corticobulbar pathway is also a descending tract, which controls the movement of the head and neck muscles. At the caudal medulla, these tracts cross over as the *decussation of the pyramids*, which obscures the fissure at this location. Other fibers originating from the anterior median fissure above the decussation, running laterally across the surface of the pons, are called *anterior external arcuate fibers*.

Final differentiation of the medulla occurs at week 20 of gestation. Neuroblasts from the alar plate will produce sensory nuclei, and basal plate neuroblasts will produce motor nuclei. The alar plate neuroblasts form the following:
- **Solitary nucleus**—containing general visceral afferent fibers for taste; and the special visceral afferent column
- **Spinal trigeminal nerve nuclei**—containing the general somatic afferent column
- **Cochlear and vestibular nuclei**—containing the special somatic afferent column
- **Inferior olivary nucleus**—relaying to the cerebellum
- **Dorsal column nuclei**—containing the gracile and cuneate nuclei

The basal plate neuroblasts form the following:
- **Hypoglossal nucleus**—containing general somatic efferent fibers
- **Nucleus ambiguous**—forming the special visceral efferent fibers

- **Dorsal nucleus of the vagus nerve**—forming general visceral efferent fibers
- **Inferior salivatory nucleus**—also forming general visceral efferent fibers

Details regarding these nucleuses will be discussed in detail in the following chapters.

Focus on injury to the medulla oblongata
Injury to the medulla oblongata may cause many different sensory-related problems. These include numbness, difficulty swallowing, acid reflux, paralysis, and lack of movement control. Drugs and other chemicals impact the medulla's functions. For example, an opiate overdose may be fatal because these drugs inhibit medulla activity, and the body becomes unable to perform its vital functions.

Section review
1. How does the neural crest develop?
2. Which structures form the sensory PNS components of the head?
3. What is the clinical name for the spinal cord?
4. What is the primary structure that originates the myelencephalon and medulla oblongata?

Metencephalon

The metencephalon is the embryonic part of the hindbrain. It differentiates into the pons and cerebellum, containing part of the fourth ventricle. The nuclei of the trigeminal nerve (CN V), abducens nerve (CN VI), facial nerve (CN VII), and vestibulocochlear nerve (CN VIII) are in the pons. The metencephalon develops from the higher, or rostral half of the embryonic rhombencephalon. It is differentiated from the myelencephalon by about 5 weeks of development. By the third month, the metencephalon has differentiated into the pons and cerebellum.

Mesencephalon (midbrain)

The mesencephalon, or *midbrain*, is associated with vision, hearing, motor control, sleeping, waking, alertness, and the regulation of temperature. The nuclei of cranial nerves III (ocular nerve) and IV (trochlear nerve) are in the midbrain. The midbrain makes up the tectum, tegmentum, cerebral peduncles, and several nuclei and fasciculi. The midbrain adjoins the metencephalon caudally. It adjoins the diencephalon rostrally. The *substantia nigra* is located in the midbrain and has left and right regions. Parts of the substantia nigra appear darker than nearby areas because of high levels of *neuromelanin* in dopaminergic neurons. The substantia nigra plays an important role in movement coordination. The *tectum* is the dorsal part of the metencephalon. It includes superior and inferior colliculi, which aid in the processing of audio and visual information. The *isthmus* is the main control center for the tectum.

The mesencephalon arises from the second vesicle of the neural tube and remains undivided for the remainder of neural development.Cells in the midbrain continue to multiply and compress the cerebral aqueduct while it is forming. Partial or complete obstruction of the cerebral aqueduct during development can lead to *congenital hydrocephalus*.

Diencephalon development

The diencephalon is made up of structures on either side of the third ventricle. These include the thalamus, hypothalamus and posterior pituitary, epithalamus and pineal body, and subthalamus. The hypothalamus performs many vital functions, mostly related to the regulation of visceral activities. The diencephalon is a primary brain vesicle. Recall that the neural tube forms the mesencephalon, rhombencephalon, and prosencephalon, from which the telencephalon and diencephalon form.

Telencephalon development

Significant growth of the telencephalon dominates future events. This part of the neural tube forms from two swelled areas that are connected at the midline by the thin, membranous lamina terminalis. The basal wall of the telencephalon is near the diencephalon. The basal wall becomes thicker, forming the *primordia* of gray masses. These are known as the basal ganglia or *basal nuclei*. Simultaneously, the diencephalon walls become thicker, forming the thalamus and hypothalamus. These are separated by the hypothalamic sulcus. Over time, the telencephalon folds down along the diencephalon, and they eventually fuse together.

The area of the telencephalon's surface that overlies the area of fusion forms part of the cerebral cortex, known as the insula. Over the next months, the cortex that adjoins the insula becomes greatly expanded until it is totally hidden under the other structures. Both cerebral hemispheres eventually have an arched "C" shape and encircle the insular cortex. Areas of each hemisphere that began as dorsal to the insula are pushed into the temporal lobe.

The cortical area expansion, beginning with the vesicles of the telencephalon, continues as the "C" shape develops, and ends with many folds being created on the hemispheres' surfaces. Each cerebral hemisphere begins with a smooth surface. However, the cerebral hemispheres become more convoluted with continued development.

It is important to understand that growth of the cerebral cortex, cerebellum, and other areas of the CNS involves extensive proliferation and migration of neurons as well as glial cells. This mostly occurs during the third, fourth, and fifth months of development.

However, the formation of neuronal connections continues long after the infant is born. The production of myelin sheaths mostly occurs following birth.

Ventricular system

The neural tube cavity becomes the ventricular system of the adult brain, as well as the central spinal cord canal. The third ventricle forms from the cavity of the diencephalon. The fourth ventricle is formed by the cavity of the pons and rostral medulla.

Each cerebral hemisphere contains a large, C-shaped *lateral ventricle* that communicates with the third ventricle via an interventricular foramen. The third ventricle communicates with the fourth ventricle via the cerebral aqueduct of the midbrain. The roof of the fourth ventricle is thin where the rhombencephalon walls spread apart to form this ventricle. Also, an area that covers the third ventricle's roof and extends onto the telencephalon's surface becomes thin. At both locations, groups of small blood vessels penetrate the ventricular roof. They form the choroid plexus, which produces most of the *cerebrospinal fluid* (*CSF*) that will fill the ventricles. When the cerebral hemispheres become C-shaped, so does the choroid plexus, which protrudes into the lateral ventricles.

Section review
1. What does the metencephalon differentiate into?
2. What are the functions of the mesencephalon?
3. What are the structural components of the diencephalon?
4. From which structure do the basal ganglia originate?
5. Which structure forms the roof of the fourth ventricle?

Central nervous system malformation

When the extremely sensitive development processes of the nervous system are interrupted, a congenital malformation may occur. Table 2.2 summarizes malformations and the times during development that are interrelated. However, the actual causes of many malformations are still not fully understood. Continued research is revealing more and more about environmental, genetic, and molecular factors. Also, normal PNS development is based on very similar genes and molecules to those needed for normal CNS development. In the PNS, certain mutations that affect migration or differentiation of neural crest cells can result in abnormalities of the autonomic nervous system, hearing, skin pigment, and cardiac function.

When there is a defective closure of the neural tube, a variety of congenital malformations of the nervous system can occur. About one per 1000 live births has one of these malformations. If the neural tube fails to completely close, a fatal deformity called

Table 2.2 CNS malformations and related developmental periods.

Primary developments	Week	Malformations
Neural folds and groove Visibility of three primary vesicles Cephalic and cervical flexures Appearance of motor neurons	3	Neural tube defects
Neural tube begins to close (day 22) Rostral end of neural tube closes (day 24) Caudal end of neural tube closes (day 26) Neural crest cells start to migrate Secondary neurulation begins Emergence of motor nerves	4	Neural tube defects and holoprosencephaly
Optic vesicle and pontine flexure develop Visibility of five secondary vesicles Sulcus limitans and sensory ganglia develop Sensory nerves grow into the CNS Rhombic lips develop Basal nuclei begin developing Thalamus and hypothalamus begin developing Autonomic ganglia, lenses of eyes, cochlea begin developing	5	Holoprosencephaly and sacral cord abnormalities
Enlargement of telencephalon Prominence of basal nuclei Completion of secondary neurulation Cerebellum and optic nerve develop Choroid plexus and insula develop	6 and 7	None
Neuronal proliferation occurs Cerebral and cerebellar cortices develop Anterior commissure and optic chiasm develop Internal capsule develops Appearance of reflexes	8–12	Migration and/or proliferation problems such as abnormal cortex or gyri
Neuronal migration and proliferation Glial differentiation Corpus callosum develops	12–16	Migration and/or proliferation problems such as abnormal cortex or gyri
Neuronal migration Cortical sulci develop Glial proliferation Mostly postnatal, but some other myelination Formation of synapses	16–40	Hemorrhage or other destructive events

craniorachischisis occurs. The CNS is opened on the dorsal surfaces of the head and spinal area. Other CNS malformations cause anencephaly, spina bifida, and other outcomes.

Neural tube defects can be detected via elevated levels of alpha-fetoprotein (AFP), or through clinical imaging. AFP is a primary component of the fetal serum. When the neural tube is open, some AFP leaks into the amniotic fluid. It eventually reaches the maternal circulation, in which it can be detected. Neural tube defects often cause the death of the fetus. Fortunately, most neural tube defects can be prevented if the mother has enough *folic acid* in her diet at the time the neural tube should be closing—at the end of the first month of gestation. The rostral neuropore closes at day 25 and the caudal neuropore closes at day 27. Since the mother may not know she is pregnant at this time, all potentially childbearing women should take routine folic acid supplements. Folic acid is regularly added to fortified cereals and other grain products in the United States.

Since folic acid supplementation began, the birth prevalence of conditions such as spina bifida has decreased by 28%. Today, it is recommended that 400 μg of folic acid be supplemented every day. However, the recommended dosage for pregnant women is *1 mg* per day. Incidence of congenital malformations is 1.9 per 10,000 live Caucasian or Hispanic births, and 1.7 per 10,000 live African American births. Additional research has shown that neural tube defects are linked to an interaction between genetics, teratogens, and folic related metabolic disorders. Folic acid supplements are the safest, most effective way of preventing these defects. Couples who previously had a child with spina bifida (see below) have a recurrence rate of between 3% and 8%. Also, rates are higher in children with *trisomy* 13 or 18, as well as in *chromosome 13q deletion syndrome*. Teratogens related to neural tube defects include excessive vitamin A, alcohol, aminopterin, valproic acid, carbamazepine, clomiphene, glycol ether, herbicides, and lead.

There are two primary groups of neural tube defects: *occulta*, meaning "hidden," and *aperta*, meaning "visible." Most vertebral defects (75%) occur in the lumbosacral region, usually at the L5 to S1 levels. Loss of motor function varies, with different effects upon the spine and limbs.

Clinical considerations

There are a variety of malformations related to embryology. These include spina bifida of various forms, including spina bifida occulta, meningocele, and myelomeningocele; anencephaly; holoprosencephaly; and congenital hydrocephalus.

Spina bifida

In normal individuals, the spinal cord and cauda equina are encased in a sheath of bone and meninges that are protective. The three most common neural tube defects, usually in the lumbosacral area, include *spina bifida occulta, spina bifida with meningocele,* and spina bifida with *myelomeningocele* (see Fig. 2.4).

Spina bifida generally occurs as a failure of the posterior neuropore to close.

Spina bifida occulta is the mildest form of spina bifida and is defined as an incomplete fusion of the posterior arch of the vertebrae. There is no visible protrusion, but often, there is a depression or dimple of the skin, a small patch of dark hair, port-wine nevi,

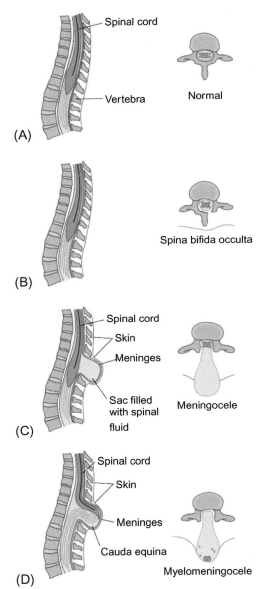

Fig. 2.4 Various degrees of spina bifida. (A) Normal anatomic structure. (B) Spina bifida occulta results in only a bony defect, with the spinal cord, meninges, and spinal fluid intact. (C) Meningocele involves the bifid vertebra, with only a cerebrospinal fluid (CSF)-filled sac protruding; the spinal cord or cauda equina (depending on the level of the lesion) remains intact. (D) Myelomeningocele is the most severe form because the spine is open and the protruding sac contains CSF, the meninges, and the spinal cord or cauda equina.

soft fatty deposits such as dermoid cysts or subcutaneous lipomas, or a combination of these. Usually, no neurologic dysfunction occurs. Sometimes, there will be disturbances of bowel or bladder control or foot weakness. All spinal cord malformations are referred to under the collective term *myelodysplasia*.

> **Focus on spina bifida complications**
> The higher that a malformation occurs along the back, the more nerve damage there will be, as well as the loss of muscle function and sensation. With Chiari II malformations, there is compression of the spinal cord, and related difficulties feeding, swallowing, and breathing. There are often symptoms such as choking, stiffness and weakness in the upper arms, hydrocephalus, and learning disabilities.

Spina bifida with meningocele

Spina bifida with meningocele is external protrusion of the meninges through a vertebral defect. In both meningocele and myelomeningocele, both described as spina bifida aperta, there is a sac-like cyst protruding outside of the spine. Meningocele rarely causes any neurologic deficits. The protruding sac contains spinal fluid, but no neural tissue. It may be covered with skin or meninges. Meningocele is the least common form of spina bifida. Minimally invasive spine surgery techniques are used to close meningoceles.

Spina bifida with myelomeningocele

This is defined as the external protrusion of both the meninges and the spinal cord through a vertebral defect. As mentioned earlier, in all forms of spina bifida there will be an increased level of AFP.

Clinical manifestations

In myelomeningocele, the alar and basal plates of both sides are visible as four unique bands on the exposed neural plate. On each side, the sulcus limitans, as well as the midline ventral groove between the basal plates, can be seen. The vertebrae do not form over the malformation, and the caudal neural tube walls are continuous with the skin of the back. A sac-like cavity exists on the back, which contains part of the spinal cord and meninges (see Fig. 2.5). Myelomeningocele causes permanent neurologic impairment, which varies based on the level of the malformation.

A Chiari type II malformation accompanies myelomeningocele. The cerebellum and caudal brainstem are excessive in length and pushed downward into the foramen magnum. The flow of **CSF** is often obstructed, resulting in an enlargement of the ventricles known as hydrocephalus. It is uncertain as to why these two deformities often occur together. Decreased pressure in the developing neural tube, caused by the tube being open,

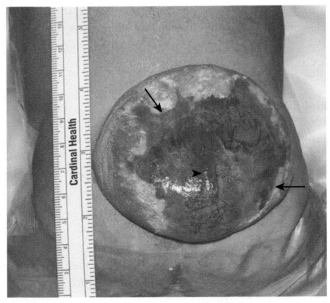

Fig. 2.5 Myelomeningocele in a newborn. The neural placode is visible at the surface *(arrowhead)* in this lumbosacral myelomeningocele. A placode is an area of thickening in the embryonic epithelial layer in which the spinal cord develops later. Abnormal epithelium lines the edges of the cerebrospinal fluid-filled cyst *(arrows)*. *(From Swaiman, K.F.; Ashwal, S.; Ferriero, D.M. Swaiman's Pediatric Neurology: Principles and Practice, 5th ed.; Saunders: Edinburgh, 2012.)*

results in abnormal cerebellum and rhombencephalon development. Today, new surgical techniques can be used to improve outcomes for many caudal neural tube closure defects.

Myelomeningocele may occur along with paralysis that is flaccid or spastic, bowel and/or bladder incontinence, hydrocephalus, cognitive limitations, and a variety of musculoskeletal deformities. These deformities may include hip dysplasia, scoliosis, clubfoot, hip dislocation, and hip or knee contractures. Nearly all patients have bladder or bowel problems since these functions are controlled at the S2–S4 spinal levels.

During the first 2 years of life, the patient often has varying degrees of truncal hypotonia and delayed automatic postural reactions. About 90% of affected children have related hydrocephalus, which usually occurs along with a type I or type II Arnold-Chiari malformation. Type I is usually asymptomatic and defined as the downward displacement of the cerebellar tonsils through the foramen magnum. Type II is usually symptomatic and defined as the downward displacement of the cerebellar vermis and *medulla* through the foramen magnum.

When the cranial nerves are involved, there may be choking, feeding problems, aspiration, pooling of secretions, and stridor. In older children, ataxia, spasticity, vertigo, pain, diplopia, or progressive weakness can occur.

Treatment

Treatment involves spinal surgery that addresses both the abnormal protrusion as well as any spinal curving that is present. The specific surgical technique is performed while the baby is still in the mother's womb. The uterus is opened, and the opening in the baby's back is surgically closed. Performing this surgery prior to birth provides significantly better results, since additional spinal cord damage may occur the longer the baby gestates.

Anencephaly

Anencephaly occurs when there is a failure of the anterior neuropore to close. In this condition, a large part of each cerebral hemisphere is absent, as well as parts of the skull (see Fig. 2.6). The walls of the neural tube can be continuous with the skin of the head. The neural tube's central cavity may be open to the outside of the body. This is a relatively common disorder. Anencephaly can be detected during early gestation via amniocentesis, which analyzes maternal serum AFP or ultrasonography. In this condition, the brain does not develop and therefore it is incompatible with life. Nearly all babies born with the condition will die soon after birth.

According to the **CDC (Centers for Disease Control and Prevention)**, about three of every 10,000 pregnancies have anencephaly annually. The causes are not proven

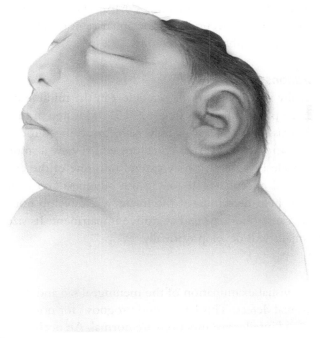

Fig. 2.6 Anencephaly.

but believed to be related to changes in genes or chromosomes, environmental factors, and even medications.

There is no cure or standard treatment for anencephaly.

Section review
1. Which two CNS malformations occur during week 4 of embryonic development?
2. What is craniorachischisis?
3. What is the role of folic acid during the embryonic period?
4. What are the clinical manifestations of myelomeningocele?
5. How can anencephaly be detected during gestation?

Encephalocele

Encephalocele is a herniation or protrusion of parts of the brain and meninges through a skull defect. This causes a sac-like structure to develop. Encephalocele occurs in about one of every 10,000 live births in the United States annually.

Pathophysiology
Encephalocele occurs in the early weeks of pregnancy. It is referred to as a *cranial meningocele* when the defect only contains meninges. Most encephaloceles contain meninges along with neural tissue and occur in the occipital region. The remainder occurs in the frontal, parietal, or nasopharyngeal regions.

Clinical manifestations
Encephalocele usually is visible at birth, as a defect in the midline of the skull, through which there is a large protruding mass (see Fig. 2.7). If located in the nasopharynx there will be no visible mass, but the infant may have breathing difficulties due to airway obstruction. Examination using a nasal speculum will review a smooth, round mass. If a frontal encephalocele is present, it may extend into one of the eye orbits, producing proptosis. An occipital encephalocele may cause blindness and cognitive impairment. Its size and location help determine if the infant will develop normally. An occipital encephalocele may cause blindness and cognitive impairment. Its size and location help determine if the infant will develop normally.

Treatment
Easily identified by visual examination of the meningeal sac and defect, surgery may be performed on a cranial defect. This has a good prognosis for normal recovery as long as the infant's motor and intellectual functions are normal. An occipital encephalocele may cause blindness and cognitive impairment. Its size and location help determine if the infant will develop normally.

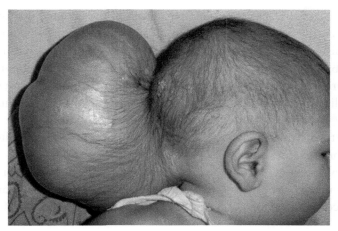

Fig. 2.7 Encephalocele.

Microcephaly

Microcephaly is a congenital condition caused by reduced proliferation or accelerated apoptosis. It is a rare defect concerning brain growth. The size of the cranium is significantly smaller than normal for the infant's age, gender, race, and gestation. The small skull size suggests the presence of microcephaly and is not treatable. *True microcephaly* or *primary microcephaly* can be caused by autosomal dominant, autosomal recessive, or X-linked genetic abnormalities. It may also be caused by various chromosomal abnormalities. Environmental causes of microcephaly include exposure to radiation or toxins, as well as intrauterine infections in the period of induction and migration of neural cells.

Secondary microcephaly occurs after birth. Head size is normal, but it then does not grow at normal rates. This is linked to many different causes, including infection, metabolic disorders, trauma, and maternal anorexia in the third trimester, and genetic conditions such as Rett syndrome. A microcephalic brain, in either type of microcephaly, may be as small as 25% of normal size. There may be diminished numbers and size of the cortical gyri. There is severe undergrowth of the frontal lobes while the cerebellum is often larger than normal. When caused by perinatal or postnatal disorders, there may be neuronal loss and gliosis in the cerebral cortex. Most children with microcephaly have developmental delays.

Holoprosencephaly

Even though the neural tube may close completely, the forebrain can still develop abnormally. It develops via many signaling molecules and genes with different expressions. Just one disrupted signaling molecule can cause extensive defects in the forebrain. A group of malformations called holoprosencephaly occurs due to partial or total failure of the prosencephalon to divide into the diencephalon and two telencephalic vesicles.

In this condition, there is an incomplete separation of the cerebral hemispheres, which results in one ventricle of the telencephalon. This process normally begins in the fourth week, and is finished during the second month. Holoprosencephaly is common, affecting one in 250 embryos, and usually fatal. Holoprosencephaly is associated with trisomy 13.

Due to interrelated processes involved with facial formation, there are usually serious facial abnormalities—sometimes as severe with a single eye in the midline of the face, with a crudely developed "nose" above it. Other factors during early development, sometimes due to environment, may cause less serious structural abnormalities, but major neurological abnormalities.

Later in development, disruptions may alter neuronal migration or proliferation. There may be related abnormal gyral patterns or ectopic gray matter areas. Even so, all basic CNS components may still be present.

Congenital hydrocephalus

Hydrocephalus is a condition in which there is accumulated CSF within the brain. This usually causes increased pressure inside the skull. Though it can be acquired later in life, *congenital hydrocephalus* is the result of birth defects such as neural tube defects or those related to aqueductal sentosis. Additional causes include brain tumors, meningitis, intraventricular hemorrhage, subarachnoid hemorrhage, and traumatic brain injury. The four types of hydrocephalus include communicating, noncommunicating, *ex vacuo*, and normal pressure hydrocephalus. Examination and medical imaging usually diagnose all forms.

Approximately one to three newborns per 1000 have congenital hydrocephalus. In infants, the CSF builds up in the CNS. The fontanelles of the head begin to bulge. Additional early symptoms include downward gazing of the eyes, irritability, separated skull sutures, seizures, sleepiness, and vomiting. Hydrocephalus can injure the brain and affect thinking and behaviors. About one in four patients will develop epilepsy.

Congenital hydrocephalus develops in utero during fetal development. The most common cause is aqueductal stenosis. This occurs when the passage between the third and fourth ventricles of the brain is blocked, or too narrow to allow enough CSF to drain. Fluid accumulates in the upper ventricles. Other causes of congenital hydrocephalus include arachnoid cysts, neural tube defects, Dandy-Walker syndrome, and Arnold-Chiari malformation. Since the cranial bones fuse by the end of the third year, hydrocephalus must occur before that time in order for head enlargement to occur. Causes are usually genetic, but can be acquired in the first few months of life, from the following:
- **Intraventricular hemorrhages**—in premature infants
- **Various infections**

- **Type II Arnold-Chiari malformation**—structural defect of the cerebellum
- **Aqueduct atresia and stenosis**—closure, aqueduct absence, or lack of flow
- **Dandy-Walker malformation**—enlarged fourth ventricle, absence of cerebellar vermis, cyst formation near the internal base of the skull

There are two different types of hydrocephalus; *communicating* (nonobstructive) and *noncommunicating* (obstructive). Basically, the communicating form occurs when there is no obstruction to the flow of CSF. Instead, there is an abnormality in the reabsorption of the CSF, such as when infectious particles obstruct the reabsorbing CSF granulation. Noncommunicating hydrocephalus occurs when there is an obstruction to the flow of the CSF, such as when caused by a tumor and other abnormalities.

In affected newborns and toddlers, head circumference enlarges quickly. Bulging, firm fontanelles may be present even when the patient is standing upright. The child will cry more often, have poor interest in feeding, be uninterested in stimuli, have retracted upper eyelids, and turn the eyes downward due to the increased pressure and paralyzed upward gaze. Movements are weak and vision may be reduced. Approximately 80%–90% of fetuses or newborns with spina bifida will develop hydrocephalus. A detailed discussion of various forms of hydrocephalus is included in Chapter 4.

Abnormalities may affect the autonomic nervous system, hearing, skin pigment, and cardiac function. Folic acid is extremely important for preventing neural tube defects, which may be hidden or visible. Malformations include various types of spina bifida, anencephaly, holoprosencephaly, and congenital hydrocephalus.

Section review
1. What are the causes of primary microcephaly?
2. What is holoprosencephaly?
3. What is an Arnold-Chiari malformation?

Clinical cases

Clinical case 1
1. Related to this case, are neonates with this type of spina bifida usually paraplegic?
2. Why can a normal function be restored if fetal surgery is performed early enough in gestation?
3. Can Arnold-Chiari malformations be reversed?

A 27-year old woman is referred for assessment of a 23-week gestation fetus, due to lumbosacral spina bifida. Serial ultrasound shows good leg movements and amniocentesis reveals a normal karyotype. An ultrafast fetal **magnetic resonance imaging** (MRI) reveals the lesion to extend from thoracic level 11 to sacral level 1,

Arnold-Chiari type II malformation, and borderline hydrocephalus. The mother wants to continue her pregnancy and agrees to surgery, which is successful in excising the lesion, and in closing the skin over the spinal cord of the fetus. The fetus is delivered via Cesarean section at 30 weeks gestation, with the Arnold-Chiari malformation and hydrocephalus no longer present. The baby had absent plantar flexion of one foot.

Answers:
1. Yes, neonates with thoracolumbar spina bifida are almost always paraplegic and have associated hydrocephalus that requires a lifelong shunt. The early fetal surgical repair greatly saved neurological function.
2. Normal function can be restored if performed early enough in gestation, since it arrests neural destruction. Also, the plasticity of the developing nervous system encourages healing and recovery of function.
3. Yes, in utero spina bifida repair can reverse complications of Arnold-Chiari malformations, including the need for shunts after birth.

Clinical case 2
1. For a baby with the conditions described, why is Cesarean section usually performed?
2. What are the clinical manifestations in this case?
3. How is myelomeningocele related to hydrocephalus?

At week 32 of gestation, a fetus was diagnosed via ultrasound with hydrocephalus and myelomeningocele. At week 38, the baby was delivered by Cesarean section and transferred to the pediatric intensive care unit. His weight was 7.5 pounds, with a normal temperature. His cranial fontanels were bulging and he had a high-pitched cry. Head circumference was 15.8 in, compared to his chest circumference of just 13.5 in.

Answers:
1. Cesarean section is usually performed when a baby has hydrocephalus because the size of the skull will likely not permit a normal delivery. Also, the integrity of the myelomeningocele is better protected from the stress of labor via a vaginal delivery.
2. The classic manifestations of hydrocephalus include bulging fontanels, a high-pitched cry, and a head circumference that is larger than chest circumference.
3. Myelomeningocele is a congenital neural tube defect due to incomplete closure of the spinal column. Since all components of the CNS develop early, hydrocephalus usually occurs in conjunction with neural tube defects.

Clinical case 3
1. What is the clinical description of anencephaly?
2. When and how can anencephaly be identified during gestation?
3. What are the possible factors that may cause anencephaly?

A woman's baby was prematurely delivered, dying soon because of anencephaly. The condition had been diagnosed in utero, so even though it was a sad situation, the woman was prepared because she already knew that the baby would never have survived with this condition. She talked with her doctors about the possible causes of anencephaly so that she would understand the future chances of having normal children with her husband.

Answers:
1. The clinical description of anencephaly is the absence of a major portion of the brain, skull, and scalp. Neural tissue is exposed and is not covered by the skull. The cerebral hemispheres are not developed. Anencephaly results from failure of the anterior neuropore to close.
2. Fetuses with anencephaly can be correctly identified when they are at 12–13 weeks of gestation. Ultrasound reveals this condition, as well as elevated maternal serum levels of AFP. It is usually associated with polyhydramnios.
3. Prevention of anencephaly includes adequate amounts of dietary folic acid and iron before pregnancy and during the first month, preventing chronic illnesses, and reducing drug or radiation exposure. Rarely, aneuploidy may be linked. Recurrence risk for anencephaly is 2%–5%.

Clinical case 4
1. What are the characterizations of holoprosencephaly?
2. What are other conditions often linked to holoprosencephaly?
3. Though of unknown cause, what risk factors are linked to holoprosencephaly?

A 15-month-old girl, with a history of severe hydrocephalus at birth, is brought to an eye specialist. She was also diagnosed with alobar holoprosencephaly at birth, with seizures. Referral to this eye specialist was because of her eyes being crossed. She has had several shunts placed to reverse her hydrocephalus. The family had no history of this condition. Over time, glasses and contact lenses were tried but failed to correct her eye condition. Strabismus surgery was indicated, and proved successful, but required several separate procedures.

Answers:
1. Holoprosencephaly is characterized by failure of the prosencephalon (embryonic forebrain) to develop, incomplete separation of cerebral hemispheres. This leads to single-lobed brain structures, along with severe skull and facial defects. Alobar holoprosencephaly occurs in two-thirds of patients and is the most serious form. Its facial anomalies include eye problems as seen in this case.
2. Associated comorbidities of holoprosencephaly include pituitary gland and hypothalamus dysfunction that results in body temperature dysregulation, seizures, and varying degrees of mental retardation. Other possible conditions may include dystonia and hypotonia.

3. Risk factors for holoprosencephaly include maternal diabetes, infections during gestation, and exposure to alcohol, lithium, Thorazine, hormones, anticonvulsants, and retinoic acid. There is also a genetic basis, in both autosomal dominant and recessive patterns. Chromosomal anomalies are associated with trisomy 13 being most common.

Clinical case 5

1. Besides the Zika virus (ZIKV), what are the other causes of microcephaly?
2. What are the treatments for microcephaly?
3. What is the prognosis for microcephaly?

The prevalence of the ZIKV infection was first reported in 2015 in South and Central America, and the Caribbean. It has been positively linked to increased incidence of microcephaly in babies born to mothers infected with ZIKV. An expectant mother in Brazil had a febrile illness and rash at the end of her first trimester of pregnancy. At 29 weeks of gestation, ultrasonography revealed that her baby had microcephaly and calcifications of the brain and placenta.

Answers:

1. Microcephaly is also caused by genetic abnormalities, maternal use of drugs or alcohol, cytomegalovirus, rubella, varicella, exposure to certain toxic chemicals, or untreated phenylketonuria. Microcephaly is also associated with Down syndrome, chromosomal syndromes, and neurometabolic syndromes.
2. There is no treatment for microcephaly that can restore a baby's head to a normal size or shape. Treatments focus on decreasing impact of deformities and neurological disabilities. Physical, speech, and occupational therapies are often used. Medications may be needed for seizures, hyperactivity, or neuromuscular symptoms. Genetic counseling helps families understand risks for microcephaly in future pregnancies.
3. Some children with microcephaly only have mild disabilities, while others are severe. Fortunately, some children have normal intelligence and continue to meet their age-appropriate developmental milestones.

Key terms

Adenohypophyseal placode
Alar plate
Alpha-fetoprotein
Anencephaly
Anterior (ventral) horn
Basal ganglia
Basal plate
Cerebellar tonsils
Cerebellar vermis
Cerebral aqueduct
Choroid plexus
Cranial placodes

Craniorachischisis
Diencephalon
Dorsal median septum
Dorsal median sulcus
Encephalocele
Fontanelles
Foramen cecum
Fourth ventricle
Holoprosencephaly
Hydrocephalus
Hypothalamic sulcus
Hypothalamus
Insula
Interventricular foramen
Intumescences
Lamina terminalis
Lens placode
Medulla oblongata
Medulla spinalis
Medullary pyramids
Mesencephalon
Metencephalon
Microcephaly
Myelencephalon
Neural crest cells

Neural fold
Neural groove
Neural plate
Neural tube
Neuropores
Neurulation
Notochord
Obex
Pontine flexure
Posterior (dorsal) horn
Primary vesicles
Proprioceptive
Rett syndrome
Rhombencephalon
Rhombic lips
Rostral
Secondary neurulation
Secondary vesicles
Spina bifida aperta
Sulcus limitans
Telencephalon
Thalamus
Trisomy 13
Truncal hypotonia
Ventral median fissure

Suggested readings

1. Ammar, A. *Hydrocephalus: What Do We Know? And What Do We Still Not Know?* Springer, 2017.
2. Benzel, E. C. *Spine Surgery: Techniques, Complication Avoidance, & Management*, 3rd Ed.; Saunders, 2012.
3. Bianchi, D. W.; Crombleholme, T. M.; D'Alton, M. E.; Malone, F. *Fetology: Diagnosis and Management of the Fetal Patient*, 2nd Ed.; McGraw-Hill Education/Medical, 2010.
4. Carlson, B. M. *Human Embryology and Developmental Biology*, 5th Ed.; Saunders, 2013.
5. https://www.cdc.gov/ncbddd/birthdefects/anencephaly.html.
6. https://www.chop.edu/pages/fetal-surgery-spina-bifida-case-study.
7. Cochard, L. R. *Netter's Atlas of Human Embryology*, Updated Edition; Saunders, 2012.
8. Cramer, G. D.; Darby, S. A. *Clinical Anatomy of the Spine, Spinal Cord, and ANS*, 3rd Ed.; Mosby, 2013.
9. Daroff, R. B.; Jankovic, J.; Mazziotta, J. C.; Pomeroy, S. L. *Bradley's Neurology in Clinical Practice*, 7th Ed.; Elsevier, 2015.
10. www.delmarlearning.com/companions/content/1401897118/casestudies/01_newborn.pdf.
11. https://embryology.med.unsw.edu.au/embryology/index.php/Neural_-_Medulla_Oblongata_Development.
12. https://embryology.med.unsw.edu.au/embryology/index.php/neural_system_development.
13. Fritsch, M. J.; Meier, U.; Kehler, U. *NPH – Normal Pressure Hydrocephalus: Pathophysiology – Diagnosis – Treatment*. TPS, 2014.
14. Gabbe, S. G.; Niebyl, J. R.; Simpson, J. L.; Landon, M. B.; Galan, H. L. *Obstetrics: Normal and Problem Pregnancies*, 7th Ed.; Elsevier, 2016.
15. Gilbert, S. F. *Barresi, M.J.F. Developmental Biology*, 11th Ed.; Sinauer Associates/Oxford University Press, 2016.

16. Hogge, W. A.; Wilkins, I.; Hill, L. M.; Cohlan, B. *Sanders' Structural Fetal Abnormalities*, 3rd Ed.; McGraw-Hill Education/Medical, 2016.
17. www.ijss-sn.com/uploads/2/0/1/5/20153321/ijss_oct_cr17.pdf.
18. Kaur, C.; Ling, E. *Glial Cells: Embryonic Development, Types/Functions and Role in Disease*. Nova Science Publishers Inc, 2013.
19. Klein, A. *Neural Tube Defects: Prevalence, Pathogenesis and Prevention – Neurodevelopmental Diseases, Laboratory and Clinical Research*. Nova Science Publishers Inc, 2013.
20. https://www.meduweb.com/threads/27922-holoprosencephaly-and-strabismus-case-with-photos.
21. Moore, K. L.; Persaud, T. V. N.; Torchia, M. G. *Before We Are Born: Essentials of Embryology and Birth Defects*, 9th Ed.; Saunders, 2015.
22. Moore, K. L.; Persaud, T. V. N.; Torchia, M. G. *The Developing Human: Clinically Oriented Embryology*, 10th Ed.; Saunders, 2015.
23. Norton, M. E. *Callen's Ultrasonography in Obstetrics and Gynecology*, 6th Ed.; Elsevier, 2016.
24. Ozek, M. M.; Cinalli, G.; Maixner, W. *Spina Bifida: Management and Outcome*. Springer, 2008.
25. Rubenstein, J.; Rakic, P. *Patterning and Cell Type Specification in the Developing CNS and PNS: Comprehensive Developmental Neuroscience*. Academic Press, 2013.
26. Sadler, T. W. *Langman's Medical Embryology*, 13th Ed.; LWW, 2014.
27. Schoenwolf, G. C.; Bleyl, S. B.; Brauer, P. R.; Francis-West, P. H. *Larsen's Human Embryology*, 5th Ed.; Churchill Livingstone, 2014.
28. https://www.sciencedirect.com/topics/veterinary-science-and-veterinary-medicine/spinalis.
29. www.upmc.com/services/neurosurgery/brain/conditions/brain-tumors/pages/meningocele.aspx.
30. https://library.med.utah.edu/webpath/labs/embrylab/embrylab.html.
31. Woodward, P. J.; Kennedy, A.; Sohaey, R. *Diagnostic Imaging: Obstetrics*, 3rd Ed.; Elsevier, 2016.

CHAPTER 3

Gross anatomy of the brain

The human brain is the most complex organ in the body. It coordinates everything that we see, hear, evaluate, and distinguish. Nearly 97% of the body's nervous tissue is contained within the adult human brain. The four brain regions include the cerebral hemispheres, diencephalon, brainstem, and cerebellum. There is a unique distribution of gray and white matter throughout the brain. The gray matter is made up of short, nonmyelinated neurons and neuron cell bodies. The white matter consists of mostly myelinated and some nonmyelinated axons. Think of nerves as electrical wires where the bare copper is covered by a plastic sheet to decrease waste of energy and to increase the electrical conductivity. Nerves are no different, and axons are similar to the bare copper that is covered by myelin sheath in order to increase the electrical conduction. As mentioned in previous chapters, the *cortex* is the gray color outer layer of the brain, which is involved in the higher mental and physical functions. The cortex is connected to the spinal cord through a network of tracts. These tracts send signals to or from the cortex in a fraction of a second, and therefore, these tracts all consist of myelinated axons that conduct electricity (action potential) much faster than the unmyelinated axons. Since myelin protein is white in color, areas of the brain and spinal cord that mostly consist of these myelinated axons are also white in color and called the *white matter*. The brain weighs approximately 1600 g (3.5 pounds) in men, and 1450 g (3.2 pounds). The difference in brain sizes is proportional to body size, not intelligence.

While the spinal cord has the same basic pattern, extra gray matter consists of cell bodies and nuclei located at the center. The white matter and tracts surround this H-shaped gray matter area. The cerebral hemispheres and cerebellum also have an outer layer of gray matter cortex. The most *rostral* region of the central nervous system (CNS) is the cerebral cortex, while the most *caudal* region is the brainstem.

Please note that there are other deep areas in the brain that are mostly consistent of nuclei and gray matter such as the thalamus, caudate, putamen, and globus pallidus in the basal ganglia area:

Ganglia, nuclei, and cell bodies = gray matter
Sensory and motor tracts (myelinated axons) = white matter

Brain divisions

The divisions of the brain include the cerebral hemispheres of the cerebrum itself, the cerebral cortex, its white matter, basal nuclei, diencephalon (thalamus, hypothalamus), brainstem, and cerebellum.

Cerebral hemispheres

The two large masses of the cerebrum, the **cerebral hemispheres**, are nearly identical and side by side. They are connected by a wide, flat, heavily myelinated axon bundle called the **corpus callosum**. They are separated by the *falx cerebri*, which is a layer of dura mater. Their surfaces are marked by **gyri**, which are the many convolutions and ridges that are separated by grooves. Basically, any groove that is shallow to slightly deep is called a **sulcus**, while each extremely deep groove is called a **fissure**. Similar in all human brains, these elevations and depressions are very complex. The *longitudinal fissure* separates the right and left cerebral hemispheres. A *transverse fissure* separates the cerebrum and cerebellum. Sulci divide each hemisphere into *lobes*.

Lobes

The lobes of the cerebral hemispheres are named based on the skull bones that they underlie, and include the following:
- **Frontal lobe**—forms the anterior part of each cerebral hemisphere. It is bordered posteriorly by a *central sulcus*, also called the *fissure of Rolando or simply the Rolandic fissure*. This passes outward from the longitudinal fissure at a right angle. The frontal lobe is bordered inferiorly by the *lateral sulcus*, also called the *Sylvian fissure or lateral fissure*. This emerges from the underside of the brain, along its sides.
- **Parietal lobe**—located posterior to the frontal lobe. It is separated from it via the central sulcus.
- **Temporal lobe**—located inferior to the frontal and parietal lobes. It is separated from them via the lateral sulcus.
- **Occipital lobe**—forms the posterior part of each cerebral hemisphere. It is separated from the cerebellum by an extension of the dura mater known as the *tentorium cerebelli*. There are no distinct boundaries between the occipital, parietal, and temporal lobes.
- **Insula**—also called the *island of Reil*, it lies deep inside the lateral sulcus. It is named due to its being covered by portions of the frontal, parietal, and temporal lobes. The insula is separated from the other lobes by a *circular sulcus*.

The various lobes of the brain are shown in Fig. 3.1.

Section review
1. What is the difference in the gray and white matter patterns between the brain and spinal cord?
2. How is the brain divided?
3. How are the two cerebral hemispheres connected?

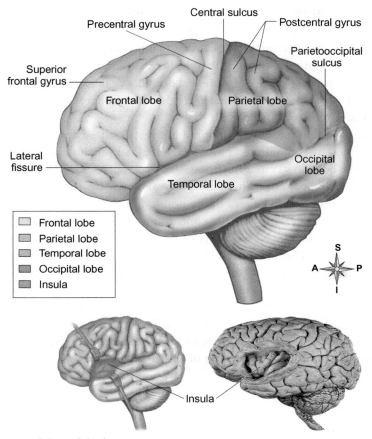

Fig. 3.1 The various lobes of the brain.

Functions of the cerebrum

The cerebrum functions concern higher brain activities. It interprets impulses from the sense organs and initiates voluntary muscular movements. The cerebrum also stores information as memories. It uses reasoning to retrieve this information. The cerebrum is responsible for intelligence as well as personality. The cerebrum is responsible for most of these activities since it contains the cerebral cortex.

Hemisphere dominance

Both of the cerebral hemispheres are involved in basic functions. These include the receiving and analyzing of sensory impulses, storing memories, and controlling the skeletal muscles on opposing sides of the body. One side is usually the *dominant hemisphere*.

For most individuals, the left hemisphere is the dominant hemisphere. Its functions include language, reading, writing, and speech. The left hemisphere is also dominant for complicated intellectual functions that require skills that are analytical, computational, and verbal. Less commonly, the right hemisphere is dominant in a percentage of left-handed people. In perhaps <1%, there is equal dominance between the two hemispheres. Note that over 90% of the population are left-hemisphere dominant regardless of whether they are right- or left-handed.

In each of us, the nondominant hemisphere is specialized for nonverbal functions, while still handling basic functions. These nonverbal tasks include motor skills requiring orientation of the body in its environment, visual experiences, and the understanding and interpreting of musical patterns. This hemisphere also provides thought processes involving emotions and intuition. The area of the nondominant hemisphere corresponding to the motor speech area does not actually control speech. Instead, it affects the emotional factors used when speaking. Nerve fibers of the corpus callosum are connected to the cerebral hemispheres. The fibers are able to transfer sensory information that reaches the nondominant hemisphere to the general area of the dominant hemisphere where the interpretation of information occurs. This is where information can be used for making decisions. The left hemisphere controls the right side of the body, and the right hemisphere controls the left side of the body. This is different in the cerebellum, where the left cerebellum controls balance in the *ipsilateral* (*same side*) of the body, and the right cerebellum controls balance in the right side of the body.

Cerebral cortex

The **cerebral cortex** is where the conscious mind is found and is considered the *executive suite* of our nervous systems. It allows for awareness, communication, memory, understanding, vision, language, and the voluntary movements. These are also called *higher cortical functions*. Its gray matter consists of neuron cell bodies, dendrites, and related glia and blood vessels. However, there are no white fiber tracts in the cerebral cortex. Billions of neurons make up the cortex, arranged in six layers. The cortex is only about 1/8 of an inch in thickness, yet makes up about 40% of the total mass of the brain, due to the many convolutions on its surface, which almost triple its total area.

The metabolic activities of the brain can be visualized by imaging known as *positron emission tomography (PET) scans*, and recently, with *functional magnetic resonance imaging (MRI) scans*. Certain motor and sensory functions are localized in specific *domains* of the cerebral cortex. Many higher mental functions, including language and memory, are different, spread over large areas of the cortex, and overlapping various domains.

The cerebral cortex has *motor areas*, *sensory areas*, and *association areas*. The motor and sensory areas are not the same as motor and sensory *neurons*. All of the neurons in the cerebral cortex are *interneurons*. Each hemisphere is mostly involved with sensory and motor functions of the opposite, or **contralateral** side of the body, as mentioned above. The two hemispheres are basically symmetrical in structure. However, their functions are not equal, and there is specialization (known as *lateralization*) of the cortical functions based on each side. No functional area of the cerebral cortex functions alone. Conscious behaviors involve the entire cortex in various ways. The cerebral cortex is discussed in detail in Chapter 6.

Cerebral white matter

The internal **cerebral white matter** is deep to the cortical gray matter. It is responsible for communicating between areas of the cerebrum, and between the cerebral cortex and the lower centers of the CNS. The white matter is mostly made up of myelinated fibers that are bundled together, to form large tracts. The fibers and tracts are classified by the directions in which they run. These classifications are as follows:

- **Association fibers**—connect various areas of the same hemisphere. Adjacent gyri are connected by short association fibers. Long association fibers are bundled into tracts. They connect the different cortical lobes. A good example would be the bundle of nerve fibers called the *arcuate fasciculus*, which connects the two language centers; Broca's area in the inferior frontal lobe and Wernicke's area in the superior temporal lobe.
- **Commissural fibers**—connect the corresponding gray matter areas of the cerebral hemispheres. These **commissures** help coordinate the functions of the two hemispheres. The largest of the commissures is the *corpus callosum*, lying superior to the lateral ventricles, deep inside the longitudinal fissure. The **anterior commissure** and the **posterior commissure** are less prominent.

Projection fibers—may enter the cerebral cortex from the lower brain or spinal cord enters, or may descend from the cerebral cortex to the lower areas. The cortex receives sensory information, and motor output leaves the cortex through these projection fibers. They link the cortex with the remainder of the nervous system, and also to the receptors and effectors of the body. The projection fibers run vertically, while the association and commissural fibers run horizontally, as in the example of the arcuate fasciculus.

- Projection fibers at the superior brain stem form a compact band called the **internal capsule**. This capsule passes between the thalamus and parts of the basal nuclei. Superior to this, the fibers have a fan-like radiation through the cerebral white matter, to the cerebral cortex. This is called a *radiating crown*, or the **corona radiata**, and is basically the area between the cortex and basal ganglia. It is important to note that the white matter tracts underneath the cortex descend on each side in a flower bouquet fashion, forming the internal capsule while passing through the basal ganglia area. These white matter tracts continue their course in the brainstem all the way to the spinal cord, where they run as the white matter tract until the spinal cord ends at the level of the first or second vertebral bone. It is also important to remember that these tracts are sending signals in both directions from the environment to the cortex as well as sending signals from the cortex to the peripheral nervous system.

Basal nuclei

The **basal nuclei**, also called the *basal ganglia*, are located deep between the corona radiata and brainstem. It is a group of subcortical nuclei, and the term basal nuclei is preferred. This is because a nucleus is a collection of CNS neuron cell bodies. Though "basal ganglia" was used more often in previous times, ganglia are actually peripheral nervous system (PNS) structures, so it is no longer preferred. The basal nuclei of each hemisphere include the **caudate nucleus**, **putamen**, and **globus pallidus**.

The caudate nucleus is shaped like a "C." It is arched superiorly over the diencephalon. Along with the putamen, it forms the **striatum**, named because fibers of the internal capsule pass through, creating a striped appearance. The caudate and putamen have different functions. However, histopathologically, they are very similar, and therefore appear similar on imaging like CT or MRI. The putamen and globus pallidus form a mass shaped like a lens, which is sometimes referred to as the *lentiform nucleus*. However, these nuclei functionally separate. The basal nuclei are related functionally with the *subthalamic nuclei* and the *substantia nigra* of the midbrain. The subthalamic nuclei are located in the lateral diencephalon "floor." The basal nuclei receive input from all of the cerebral cortex, from the other subcortical nuclei, and from each other. By relaying through the thalamus, the output nuclei of the basal nuclei, called the globus pallidus, and substantia nigra project to the premotor and prefrontal cortices. They influence the muscle movements controlled by the primary motor cortex. There is no direct access of the basal nuclei to the motor pathways.

The basal nuclei are relatively inaccessible and not fully understood. Their motor functions overlap with the cerebellar motor functions. They also play roles in emotion and cognition. The basal nuclei appear to filter out responses that are inappropriate or incorrect and pass the "best" responses on to the cerebral cortex. Related to motor activities, the basal nuclei are very important for beginning, ending, and monitoring the intensity of movements from the cortex. This is especially true for movements that are, in relation, stereotypes or slow, such as when the arms swing as we walk. The basal nuclei also inhibit unnecessary or antagonistic movements. To simplify, imagine the CNS as a sophisticated sound system where the thalamus serves as an amplifier that can increase or decrease the amplitude and frequency of signals. The basal ganglia nuclei serve as a tuning system to adjust the tone of different signals. Disorders of the basal nuclei include Parkinson's disease and Huntington's disease. The basal nuclei are discussed in detail in Chapter 7.

> **Focus on basal ganglia degeneration**
> The basal ganglia are in control of smooth movements, and when damaged, the ability to inhibit contradictory movements is affected. In patients with Parkinson's disease, there is difficulty in initiating movements, rigidity, and slowed movements. In Huntington's disease, overactivation leads to excessive, jerky, and writing involuntary movements.

Diencephalon

The **diencephalon** is made up of three paired structures: the thalamus, hypothalamus, and epithalamus. It forms the center portion of the forebrain and is surrounded by the cerebral hemispheres. The gray matter of the three paired structures enclose the third ventricle of the brain. The diencephalon only accounts for <2% of the brain's weight. There is also an inner chamber known as the *subthalamus* that is not visible from the brain exterior. The diencephalon and its substructures are discussed in detail in Chapter 8.

Brainstem

The **brainstem**, also spelled *brainstem*, is divided into three regions: the superior *midbrain*, the middle *pons*, and the inferior *medulla oblongata*. Each of these is about 1 in in length, and together, they make up only 2.5% of the total brain mass. The brainstem tissues are similar to those of the spinal cord. There is a deep gray matter surrounded by white matter fiber tracts. Differently, there is a nucleus of gray matter inside the white matter, which does not occur in the spinal cord. The brainstem is discussed in detail in Chapter 9.

Cerebellum

The **cerebellum** includes a midline **vermis**, seen when a human brain is *hemisected*, and a larger, lateral *hemisphere* on each side. Each hemisphere has thin, transverse, and parallel folds called **folia** that are separated by shallow *sulci*. The surface cortex of the cerebellum consists of gray matter, and there is a deeper layer of white matter. A sagittal section of the cerebellum shows the white matter branching into fern-like patterns called the **arborvitae**. Each hemisphere has four gray matter masses within the white matter. These are known as the **deep nuclei**. All input to the cerebellum travels to the cortex. All output comes from the deep nuclei. The cerebellum is discussed in detail in Chapter 17.

Section review
1. Which part of the brain is considered to be the executive suite of the CNS?
2. What are the three areas of the cerebral cortex?
3. What is the responsibility of the internal cerebral white matter?
4. What are the three parts of the basal nuclei?
5. What are the three parts of the diencephalon?

Clinical considerations

The brain, as well as the rest of the CNS, can be infected by bacteria, fungi, viruses, mycobacteria, and parasites. Infectious microorganisms gain entry via hematogenous spread, through the arterial blood, or by direct extension from another area of infection. Neurologic infections can cause disease by direct glial or neuronal infection, inflammation accompanied by edema, formation of mass lesions, interrupted cerebrospinal fluid (CSF) pathways, damage to vessels (**vasculopathy**), and the secretion of neurotoxins. Examples of such infections include encephalitis, brain abscess, and helminthic brain infections.

Encephalitis

Encephalitis is inflammation of the brain's parenchyma, which can be due to infection. *Acute disseminated encephalomyelitis* is inflammation of the brain as well as the spinal cord. Often, with the herpes simplex or West Nile viruses, the patient can develop either encephalitis or meningitis. Clinically, meningitis and encephalitis are distinguished based on the conscious state of the patient. The presence of altered mental status indicates the involvement of the brain parenchyma in encephalitis. Many different types of viruses may be linked. Viruses causing *primary encephalitis* directly invade the brain. They may be epidemic or sporadic. Causes of encephalitis include **arbovirus**, coxsackievirus, echovirus, poliovirus, herpes simplex, mumps, rabies, varicella-zoster, enteroviruses, Epstein-Barr virus, cytomegalovirus, adenovirus, rubella, measles, Murray Valley encephalitis (MVE) virus, Kunjin virus, West Nile virus, St. Louis encephalitis virus, Eastern equine encephalitis virus, Western equine encephalitis virus, and Japanese encephalitis virus.

Pathophysiology

It is difficult to determine the true incidence of encephalitis, due to a lack of standardized reporting methods. In acute encephalitis, there is cerebral edema and there may be even petechial hemorrhages. These occur throughout the cerebral hemispheres, brainstem, cerebellum, and sometimes, the spinal cord depending on the cause. A direct viral invasion of the brain usually damages neurons. A severe infection—for example, untreated herpes simplex virus (HSV) encephalitis, can cause brain hemorrhagic necrosis. Acute disseminated encephalomyelitis is signified by **demyelination.** This type of encephalomyelitis involves widespread inflammation of the white matter of the brain and spinal cord.

Clinical manifestations

The clinical manifestations of encephalitis include fever, **altered mental status**, and headache. These symptoms are often accompanied by focal neurologic deficits and seizures. The meningeal involvement symptoms are usually the same with the exception of altered mental status, which is by definition indicating the presence of encephalitis in addition to meningitis. Severe brain inflammation and a poor prognosis are suggested by status epilepticus—especially convulsive status epilepticus—or coma. There may be olfactory seizures, which manifest as an aura of bad odors, such as burnt meat or rotten eggs. These indicate involvement of the temporal lobe and suggest HSV encephalitis.

> **Focus on West Nile virus**
> Before 1994, outbreaks of West Nile virus were not common. They mostly occurred in the Mediterranean countries, Africa, and Eastern Europe. After 1994, West Nile virus occurred with more severity, primarily affecting the nervous system. By 2002, it affected four of every 100,000 people in the Midwestern United States. The virus causes encephalitis, meningitis, and poliomyelitis, with significant morbidity and mortality. As many as 20% of infected patients suffer permanent neurologic damage, with 2% of cases being fatal.

Diagnosis

In any patient with unexplained alterations of mental status, encephalitis should be suspected. Though various diagnostic tests may be suggested by clinical presentation and differential diagnoses, CT brain scan and CSF analysis, including polymerase chain reaction (PCR) for HSV, are usually performed. Sometimes other tests are also used to help identify the causative virus. However, some cases of encephalitis due to infections remain of unknown cause, despite extensive workup.

For HSV encephalitis, MRI is sensitive later on, and can show edema in the temporal areas. These are the areas usually infected by HSV. It may show basal ganglia and thalamic abnormalities in *eastern equine* and *West Nile* encephalitis. An MRI cannot exclude lesions mimicking viral encephalitis, such as brain abscess or sagittal sinus thrombosis. CT scan is helpful because of its rapid availability to exclude other abnormalities such as severe cerebral edema, hydrocephalus, or mass lesions, which can be contraindications indicating the need to perform a lumbar puncture to obtain CSF. If there is going to be a delay in obtaining CSF, the patient should be started on empiric antiviral and antibacterial treatment within the first 30 min of arrival to the emergency department.

Upon lumbar puncture, the CSF is checked for glucose, the number of white cells, proteins, and red blood cells. Different etiologies have different patterns of CSF profiles. Viral encephalitis causes the CSF to have lymphocytic pleocytosis with increased white blood cells in the hundreds (usually under 1000), mildly elevated proteins, normal glucose, and an absence of pathogens when using Gram stain and culture. This is similar to the CSF seen in aseptic meningitis. Hemorrhagic necrosis can bring many red blood cells and some neutrophils into the CSF, elevate protein, and slightly lower glucose. Today, PCR testing of the CSF for many viruses is more readily available.

The PCR tests for HSV in the CSF is very sensitive and specific. Results can be available within 4 hours, indicating treatment for HSV given its almost 100% specificity and very low false-negative and false-positive rates. Sometimes, for diagnosing acute infection, CSF viral immunoglobulin M (IgM) titers are often useful, especially for West Nile encephalitis. There are also universal bacterial and viral PCR panels that can check CSF for virtually all known pathogens. These tests may take several weeks to become available and are not used for acute treatment decision-making.

For patients who are worsening or not responding well to treatment with antimicrobials, sometimes suppressing the immune system may be indicated. This is also true for those with undiagnosed lesions. This is achieved with steroids in case of bacterial causes, and with immunoglobulin or specialized dialysis called **plasmapheresis** in treating refractory viral or bacterial encephalitis. The reason behind the effectiveness of steroids and other immune-suppressive agents is not entirely understood. It may be related to the fact that our immune system reaction against offending pathogens is responsible for most symptoms seen in encephalitis and is not a direct effect of the pathogens.

Treatment

Supportive therapy includes treatment of dehydration, fever, electrolyte disorders, and seizures. It is important to maintain euvolemia. When HSV encephalitis is suspected, acyclovir is started quickly and usually continued for 2 weeks. This drug is mostly nontoxic but can cause renal failure. Nephrotoxicity is able to be prevented in part by administering the drug slowly, with adequate fluids.

Brain abscess

A **brain abscess** is an intracerebral collection of pus (see Fig. 3.2). It may form when an area of cerebral inflammation becomes necrotic and is encapsulated by fibroblasts and glial cells. Intracranial pressure may increase due to edema around the abscess.

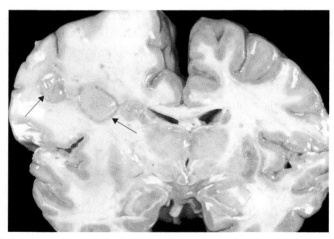

Fig. 3.2 Cerebral abscesses *(arrows)*.

Pathophysiology

Brain abscesses can be caused by a variety of factors. Direct extension of cranial infections may be causative, including mastoiditis, osteomyelitis, sinusitis, or subdural empyema. The cause may also be hematogenous spread, such as in bacterial endocarditis, congenital heart disease with right-to-left shunt, or intravenous drug abuse. Other causes include penetrating head wounds, including neurosurgical procedures, and unknown causes.

Clinical manifestations

Symptoms of brain abscess manifest from increased intracranial pressure and mass effect. They may include nausea, vomiting, lethargy, personality changes, seizures, focal neurologic deficits, and papilledema. Symptoms may develop over days to weeks. Prior to the infection being encapsulated, chills, fever, and leukocytosis may develop. These can be absent or subside over the disease progression.

Diagnosis

When symptoms suggest a brain abscess, contrast-enhanced CT or MRI is performed. The abscess will appear as an edematous mass with ring enhancement. It can be hard to distinguish from a tumor, and sometimes, from an infarction. CT-guided aspiration, culture, surgical excision, or a combination of these may be required in rare circumstances.

Treatment

Antibiotics used for brain abscess initially include those that cover anaerobic agents, as well as a wide range of Gram-negative and Gram-positive bacteria since most of these brain abscesses, are related to more than one pathogen. All patients receive antibiotics for 4–8 weeks intravenously. Both cefotaxime and ceftriaxone are effective against streptococci, Enterobacteriaceae, and most anaerobes. However, they are not effective against *Bacteroides fragilis*. This requires metronidazole. When *Staphylococcus aureus* is suspected, vancomycin is used until the patient's sensitivity to nafcillin can be determined. Response to antibiotics should be monitored by serial CT or MRI. Vancomycin is also used if the infection is related to a recent neurosurgery.

For increased intracranial pressure, the patient may benefit from brief use of high-dose corticosteroids such as dexamethasone. To prevent seizures, anticonvulsants may be recommended in individual cases.

Fungal meningoencephalitis

The majority of CNS fungal infections occur in patients who are immunocompromised. After widespread hematogenous fungal dissemination, the brain will usually be involved. The fungal species most commonly linked to these infections include *Candida albicans*, various Mucor species, *Aspergillus fumigatus*, and *Cryptococcus neoformans*. Also following immunosuppression, in endemic locations and after a primary pulmonary or cutaneous infection, the CNS may be infected by *Histoplasma capsulatum*, *Coccidioides immitis*, or *Blastomyces dermatitidis*. While the majority of fungi reach the brain via hematogenous dissemination, there may also be direct extension—especially in mucormycosis affecting a diabetic patient. A fungal infection of the CNS usually results in chronic meningitis, vasculitis, and parenchymal invasion. *Mucormycosis* and *Aspergillosis* directly invade blood vessel walls and are the most common fungal causes of vasculitis, though it sometimes occurs with candidiasis and other infections. Vascular thrombosis then develops, producing an infarction that often hemorrhages significantly and becomes septic due to ingrowth of the fungus.

Parenchymal infections are usually granulomas or abscesses. These may occur with most types of fungi and often occurs along with meningitis. *Candida* and *Cryptococcus* are the fungi that most often invade the brain. Candidiasis usually causes multiple micro-abscesses, with or without the formation of granuloma. *Cryptococcal meningitis* is a common opportunistic infection in AIDS patients. It may be fulminant and fatal in as few as 2 weeks, or it may be indolent and evolve slowly, over months or years. The CSF usually has a high concentration of protein but may only contain small amounts of cells. In the CSF, the mucoid-encapsulated yeasts can be visualized using certain stains or detected by using assays for cryptococcal antigens. CSF opening pressure on lumbar puncture is often high in these patients and due to persistent increased pressure in the brain. They often required serial lumbar punctures in order to decrease the intracranial pressure by removing CSFs daily.

Helminthic brain infections

Helminthic brain infections involve parasites that infect the CNS. These infections harm millions of people in developing countries. Americans who visit these countries may return home, not knowing immediately that they are infected. Worms can cause encephalitis, meningitis, cerebral masses, hydrocephalus, myelopathy, and stroke. About 20 different helminths can cause neurologic disorders. However, the pork tapeworm *Taenia solium* causes many more cases in the Western Hemisphere than any other worm, resulting in the disorder called **neurocysticercosis**. Food contaminated with the worm's eggs, if eaten, allows larvae to migrate to the brain, spinal cord, and CSF pathways as well as other tissues. Cysts form, usually <1 cm in diameter in the neural parenchyma. However, in the CSF spaces, these cysts can be >5 cm. Parenchymal brain cysts do not cause any symptoms until the worms die. This triggers gliosis, local inflammation, and edema, usually causing seizures, but also cognitive or focal neurologic deficits or personality changes. Neurocysticercosis is the most common cause of seizure outside the United States and is usually caused by eating undercooked pork.

Neurocysticercosis is suspected in travelers from developing countries who have eosinophilic meningitis and new-onset unexplained seizures. The disease is suggested by multiple calcified cystic lesions seen on MRI or CT. A contrast agent may help enhance the lesions. Diagnosis is based on serum and CSF serologic tests as well as the visualization of cysts and calcification in brain imaging. Drugs include albendazole (Albenza) is the drug of choice for helminths. Other drugs include praziquantel and dexamethasone to decrease edema and inflammation. Anthelminthic therapy has the potential to cause serious morbidity in patients with many cysts and may not be helpful for patients with just one cyst. Treatment is therefore highly individualized. Short- or long-term anticonvulsants may be needed. Surgical excision of cysts and ventricular shunting may be needed in some cases.

In the disease called **schistosomiasis**, necrotizing eosinophilic granulomas develop within the brain. This causes seizures, increased intracranial pressure, and diffuse, focal neurologic deficits. Focal deficits and sometimes, seizures, can be caused by large, solitary **echinococcal cysts**. Tapeworm larvae can cause **coenurosis**, which usually produces grape-like cysts able to obstruct the outflow of CSF in the fourth ventricle. A rare infection called **gnathostomiasis** causes necrotic tracts that are surrounded by inflammation along nerve roots, the spinal cord, or the brain. It can also cause subarachnoid hemorrhage, resulting in low-grade fever, photophobia, stiff neck, headache, and migratory neurologic deficits. These occasionally affect the sixth or seventh cranial nerves. Paralysis may also occur.

Protozoal brain diseases

Protozoal brain diseases include amebiasis, malaria, toxoplasmosis, and trypanosomiasis. *Cerebral amebiasis* is quickly fatal, a type of necrotizing encephalitis caused by *Naegleria* species. Infection with *Acanthamoeba* causes a chronic, granulomatous meningoencephalitis. It is sometimes hard to morphologically distinguish activated macrophages from the amebae. Methenamine silver or periodic acid-Schiff (PAS) stains help to see the microorganisms. However, correct identification is based on immunofluorescence studies, cultures, and molecular studies.

Cerebral malaria is a form of encephalitis that spreads quickly and is the complication of *Plasmodium falciparum* infection that causes the most deaths. It is usually a result of vascular dysfunction. In the brain, malaria is accompanied with reduced cerebral blood flow. This causes ataxia, seizures, and coma during the acute phase. Long-term cognitive deficits remain in up to 20% of affected children.

Toxoplasmosis is an opportunistic infection usually seen in patients with HIV who are immunosuppressed. The causative organism is *Toxoplasma gondii*. Symptoms are subacute and evolve over a 1- or 2-week period, with both focal and diffuse signs. Via CT scan and MRI, multiple ring-enhancing lesions may be seen, but this appearance is not specific for toxoplasmosis. CNS **lymphoma**, tuberculosis (less frequently), and fungal infections produce similar images. When the patient is not immunosuppressed, cerebral toxoplasmosis is most common when a primary maternal infection develops early in pregnancy. These infections often spread to the developing fetal brain, causing severe damage from multifocal necrotizing lesions that are able to calcify.

Toxoplasmosis of the CNS produces abscesses in the brain. These usually occur near the gray-white junction of the cerebral cortex and in the deep gray nuclei. Less commonly, they develop in the cerebellum and brainstem, and only rarely in the spinal cord. Acute lesions have central necrosis and petechial hemorrhaging surrounded by acute and chronic inflammation. They also have macrophage infiltration and vascular proliferation. Free **tachyzoites** and encysted **bradyzoites** are often seen at the peripheral areas of the necrotic foci. Routine **hematoxylin and eosin** (H&E) or **Giemsa** stains often reveal the microorganisms, but **immunohistochemical** methods are preferred. Nearby blood vessels may reveal significant intimal proliferation or vasculitis, with fibrinoid necrosis and thrombosis. Once treated, the lesions have large, clearly separated areas of coagulative necrosis that are surrounded by macrophages carrying many lipids. Near these lesions, there may also be cysts and free tachyzoites, which may be greatly reduced or even absent if therapy has been effective. Chronic lesions have small cystic spaces that contain small amounts of lipid- and hemosiderin-laden macrophages surrounded by gliotic brain tissue. In these older lesions, microorganisms are harder to detect.

Clinical cases

Clinical case 1
1. What are the complete signs and symptoms of St. Louis encephalitis?
2. Could this patient's job have had any relationship to his contracting of this disease?
3. What are the treatment options and survival rates?

A 27-year-old bird breeder in Florida was treated for flu-like symptoms, severe headache, nausea, and seizures. Tests revealed that he had contracted St. Louis encephalitis.

Answers:
1. St. Louis encephalitis causes a sudden onset of drowsiness, severe headache, nausea, abdominal pain, a rash, swollen glands, flu-like symptoms, and seizures.
2. This disease is caused by a virus transmitted from birds to the common Culex mosquito, and then to humans. This patient's job as a bird breeder definitely put him at a higher risk.
3. Treatment options are only supportive, since there is no vaccine against the infection or specific antiviral treatment for it. Serologic and other diagnostic tests help to determine the components of the disease. It is not usually fatal, but most deaths occur in infants, children under age 5, elderly people, and those with an impaired immune system.

Clinical case 2

1. After imaging studies, how is a definitive diagnosis made for a brain abscess?
2. What are some common treatments for a brain abscess?
3. How does the causative infection of a brain abscess actually enter the brain?

A 25-year-old woman collapsed at work and experienced a seizure with convulsions. Upon arriving at the emergency department, she was irritable, confused, had a bad headache, unable to speak without slurring her words, nauseous, and alternating between feeling feverish or cold. She underwent a general neurological evaluation including a CT scan with contrast. She was diagnosed with a brain abscess.

Answers:

1. Definitive diagnosis of a brain abscess involves drainage of the abscess. This allows the collection of the purulent fluid to be examined microscopically, and cultured, so that the causative organism can be identified.
2. If the infection is very early in its course, intravenous antibiotics may be all that is required. However, since systemically administered antibiotics usually do not enter the abscess cavity, some brain abscesses require surgical drainage. The infectious material is removed and washed out thoroughly. Any abnormal infectious tissue is also removed. IV antibiotics are usually continued after surgery. Additional treatment of the source of infection (endocarditis, mastoiditis, sinusitis, etc.) may be indicated.
3. The infection causing a brain abscess may enter by three primary routes: the bloodstream; via an infection within the nose, ear, or other parts of the skull; and because of direct trauma, such as from a head injury, or neurosurgery.

Clinical case 3

1. What areas of the body are most affected by the cysts that develop in neurocysticercosis?
2. How can the eggs of the pork tapeworm be spread?
3. What are the treatment options for this condition?

A college student went out with his friends on the weekend, and attended a backyard barbecue. While there, he ate a pork chop that was not sufficiently cooked, saying that he liked his meat very rare. Within 2 months, he began to experience headaches, confusion, vision changes, and eventually, seizures. He was taken by his family to an emergency department for evaluation of a prolonged seizure. He was diagnosed with the helminthic infection known as neurocysticercosis.

Answers:

1. The cysticerci, which are the cysts caused by neurocysticercosis, affect the nervous system. When the nervous system is not affected, the condition is called "cysticercosis," and it affects the skeletal muscles, eyes, and skin.
2. The pork tapeworm's eggs are usually spread via food, water, or surfaces contaminated with infected feces. An infected food handler may spread them when the hands are not

clean, and they can also be spread from fruits or vegetables that are fertilized or watered when infected human feces is present in the area.
3. Treatment of neurocysticercosis is varied. If the parasite is dead, treatment focuses on anticonvulsants for the seizures. Viable cysts are treated with anticysticercal drugs. If the parasite is viable or active and the patient has vasculitis, arachnoiditis, or encephalitis, then steroids or immunosuppressants are used first. For some forms, antiparasitic treatments such as albendazole are indicated. Surgical placement of a ventricular shunt is recommended for any hydrocephalus, followed by removal of cysts and their treatment. Antiepileptic medications may also be needed.

Clinical case 4

1. Before highly active antiretroviral therapy (HAART), was dementia common in HIV patients?
2. As HIV encephalitis progresses, what are the common signs and symptoms?
3. Since this patient also had HSV-2, are there any treatment options for this virus?

A 38-year-old HIV patient was admitted to the hospital because of generalized seizures and fever in the past 24 hours. He was showing signs of dementia and was unable to walk. Laboratory tests revealed increased CSF protein and cells with lymphocyte prevalence, but normal CSF glucose. Both the HIV and herpes simplex-2 viruses were present in his blood. MRI revealed lesions of the medial areas of both temporal lobes, without bleeding.

Answers:

1. Prior to HAART, dementia was a common source of morbidity and mortality in HIV patients. It was usually observed in the later stages of AIDS, in up to 50% of patients prior to their deaths.
2. HIV encephalitis is characterized by progressive alteration of mental status, confusion, slowness of thought processes and speech, poor concentration, memory loss, inability of self-care, meningeal-like symptoms, seizures, lethargy soon before death, and intracranial symptoms.
3. Usually, a 14- to 21-day course of intravenous acyclovir is used for HSV-2, which often results in consistent improvement of clinical findings and disappearance of the virus from the CSF. Other medications include abacavir, lamivudine, darunavir, and ritonavir.

Key terms

Anterior commissure	Bradyzoites
Arbor vitae	Brain abscess
Arbovirus	Brainstem
Basal nuclei	Caudate nucleus

Cerebellum	Gyri
Cerebral cortex	Helminthic brain infections
Cerebral hemispheres	Hematoxylin and eosin
Cerebral white matter	Immunohistochemical
Coenurosis	Internal capsule
Commissures	Neurocysticercosis
Contralateral	Perivenous demyelination
Corona radiata	Plasmapheresis
Corpus callosum	Posterior commissure
Deep nuclei	Prodrome
Diencephalon	Putamen
Echinococcal cysts	Schistosomiasis
Encephalitis	Striatum
Fissure	Sulcus
Folia	Tachyzoites
Giemsa	Vasculopathy
Globus pallidus	Vermis
Gnathostomiasis	

Suggested readings

1. Agyen-Mensah, K. *Imaging Appearances, Diagnosis & Treatment of Atypical Brain Abscesses: Imaging Appearances, A Guide to Making Prompt Diagnosis and Recommended Treatment of Atypical Brain Abscesses;* Lap Lambert Academic Publishing, 2017.
2. Anderson, M. W.; Fox, M. G. *Sectional Anatomy by MRI and CT*, 4th Ed.; Elsevier, 2016.
3. Berkowitz, A. *Lange Clinical Neurology and Neuroanatomy: A Localization-Based Approach.* McGraw-Hill Education/Medical, 2016.
4. Blumenfeld, H. *Neuroanatomy through Clinical Cases*, 2nd Ed.; Sinauer Associates/Oxford University Press, 2010.
5. Broussard, D. M. *The Cerebellum: Learning Movement, Language, and Social Skills.* Wiley-Blackwell, 2013.
6. Carter, R. *The Human Brain Book: An Illustrated Guide to its Structure, Function, and Disorders.* DK Publishing, 2014.
7. DeArmond, S. J.; Fusco, M. M. *Structure of the Human Brain: A Photographic Atlas*, 3rd Ed.; Oxford University Press, 1989.
8. Eroschenko, V. P. *Atlas of Histology with Functional Correlations*, 13th Ed.; LWW, 2017.
9. Filley, C. *The Behavioral Neurology of White Matter*, 2nd Ed.; Oxford University Press, 2012.
10. Shier, D. N.; Butler, J. L. *Hole's Human Anatomy & Physiology*, 14th Ed.; McGraw-Hill Education, 2015.
11. Hugdahl, K.; Westerhausen, R.; Sun, T.; Gannon, P. J.; et al. *The Two Halves of the Brain: Information Processing in the Cerebral Hemispheres.* The MIT Press, 2010.
12. Ito, M. *The Cerebellum: Brain for an Implicit Self.* FT Press Science, 2011.
13. Jirillo, E.; Magrone, T.; et al. *Immunity to Helminths and Novel Therapeutic Approaches (Immune Response to Parasitic Infections).* Bentham Science Publishers, 2018.
14. Lalonde, R. *The Brainstem and Behavior.* Nova Science Publishers Inc., 2017.
15. Marieb, E. N.; Hoehn, K. *Human Anatomy & Physiology*, 11th Ed.; Pearson, 2018.
16. McCance, K. L.; Huether, S. E. *Pathophysiology: The Biologic Basis for Disease in Adults and Children*, 8th Ed.; Mosby, 2018.
17. Nolte, J. *The Human Brain in Photographs and Diagrams*, 4th Ed.; Saunders, 2013.

18. Pandya, D.; Seltzer, B. *Cerebral Cortex: Architecture, Connections, and the Dual Origin Concept*. Oxford University Press, 2015.
19. Rolls, E. T. *Cerebral Cortex: Principles of Operation*. Oxford University Press, 2016.
20. Ruiz, A.; Fleming, D. *Encephalitis, Encephalomyelitis and Encephalopathies: Symptoms, Causes, and Potential Complications*. Nova Biomedical, 2013.
21. Saladin, K. S. *Anatomy & Physiology: The Unity of Form and Function*, 8th Ed.; McGraw-Hill Education, 2017.
22. Smythies, J. R.; Edelstein, L.; Ramachandran, V. S. *The Claustrum: Structural, Functional, and Clinical Neuroscience*. Academic Press, 2013.
23. Soper, H. V.; Comstock, T.; et al. *Understanding the Frontal Lobe of the Brain: Fractioning the Prefrontal Lobes and the Associated Executive Functions,* Vol. 11. Fielding University Press, 2017.
24. Spetzler, R. F.; Kalani, M. Y. S.; Nakaji, P.; Yagmurlu, K. *Color Atlas of Brainstem Surgery*. Thieme, 2017.
25. Vanderah, T.; Gould, D. J. *Nolte's The Human Brain: An Introduction to its Functional Anatomy*. Elsevier, 2015.

CHAPTER 4

Meninges and ventricles

The central nervous system (CNS) requires support in order to keep its shape as we move throughout the environment. The brain and spinal cord are encased in the skull and vertebral column, which offer a good amount of protection. The CNS is also suspended inside three membranous coverings, called **meninges**. This word comes from the Greek word *meninx*, which means "membrane." The meninges stabilize the shape as well as the position of the CNS when the head and body move. The first method is the mechanical suspension of the brain inside the meninges, which are also anchored to the skull. This means that the brain is constrained in its movements within the head.

The second method of stabilization is the layer of **cerebrospinal fluid (CSF)** inside the meninges. This fluid is formed within the brain ventricles. It fills them, emerging from openings in the fourth ventricle to fill the subarachnoid space. CSF helps regulate the composition of fluid that coats CNS neurons and glial cells. Basically, the brain and spinal cord float in the CSF, which serves as a cushion. The CSF allows certain chemical messengers to be thoroughly distributed within the nervous system. The CSF significantly decreases the likelihood that gravity and other variables can distort the brain. While the human brain, outside of the body, weighs about 1500 g, within the CSF, it weighs <50 g. There is usually between 120 and 500 cubic centimeters (cc) of CSF at any given time. The CSF assists the brain in maintaining its normal shape. When the brain is removed, it lacks this normal support and becomes greatly distorted in appearance. It can even tear apart because of gravitational forces. As mentioned before, the choroid plexus constantly produces CSF, at a rate of 20 cc per hour, with the capability of renewing the whole amount of CSF within 24 h.

Meninges

The nervous tissue of the brain is separated, in a form of *biochemical isolation*, from the general blood circulation by the *blood-brain barrier*. Layers comprising the **cranial meninges** are continuous with the *spinal meninges*. These layers include the *dura mater, arachnoid mater,* and *pia mater*. Except for a few areas, there is no space on either side of the cranial dura, under normal conditions. This is because one side is attached to the skull, while the other side is attached to the arachnoid mater. Even so, there are two **potential spaces** associated with the dura mater. These are the **epidural space** and the **subdural space**.

Dura mater and dural folds

Outer and inner fibrous layers make up the **dura mater**, meaning "tough matter." It is the strongest meninx. The dura layer surrounds the brain and consists of two layers of fibrous connective tissue: the meningeal layer and periosteal layer. Its outer layer is connected with the periosteum of the bones of the cranium. The deeper layer (the meningeal layer) forms the true external covering of the brain and continues caudally in the vertebral canal as the *spinal dura mater*. Therefore, there is no superficial epidural space above the dura mater. In the spinal cord, there is such a superficial epidural space. The outer *periosteal cranial dura* and inner *meningeal cranial dura* are fused together.

There are several areas where the meningeal cranial dura extends into the cranial cavity. Where this occurs, a sheet is formed that angles inward and then back out. The **dural folds** give the brain more support and stabilization. Inside the dural folds are large collecting veins called the **dural venous sinuses**. Veins in the brain open into the dural venous sinuses, delivering venous blood to the neck veins. These two layers form the following:

- **Falx cerebri**—a fold of the two layers of the dura, in the longitudinal cerebral fissure, which projects between the cerebral hemispheres and separates them into two halves. The inferior areas are attached anteriorly to the ethmoid bone's **crista galli**. The same areas are attached posteriorly to the *internal occipital crest*, which is a ridge along the occipital bone's inner surface. Two large dural venous sinuses lie inside this dural fold. They are known as the **superior sagittal sinus** and **inferior sagittal sinus**. The tentorium cerebelli is intersected by the posterior margin of the falx cerebri.
- **Tentorium cerebelli**—this dural fold protects the cerebellum as well as separating the inferior occipital lobe from the cerebellum. It is extended, at right angles to the falx cerebri, across the cranium. A paired dural venous sinus called the **transverse sinus** lies along the occipital bone. This sinus allows blood to drain from the posterior portion of the head.
- **Falx cerebelli**—inferior to the tentorium cerebelli, this dural fold divides the two cerebellar hemispheres, along the midsagittal line.

Overall, the dura mater may be separated, via a narrow *subdural space*, from the next membrane layer, which is called the *arachnoid mater*.

Section review
1. What are the three layers of the cranial meninges?
2. What is different when comparing the dura mater of the brain and the dura matter of the spinal cord?
3. What are the names of the three largest dural folds?

Arachnoid mater

The **arachnoid mater** is named because it resembles a spider web (spiders are clinically identified as *arachnoids*). It is a thin, avascular membrane with a cell layer that is attached to the dural border cell layer of the dura mater. There is no natural space in this layer. The **subarachnoid space** contains fibers and cells of the arachnoid trabeculae. It is located between the arachnoid mater and the pia mater. The subarachnoid space is filled with CSF. It also contains the largest blood vessels that serve the brain. Since the arachnoid mater is thin and elastic, these blood vessels do not have significant protection.

Over the gyri surfaces, the subarachnoid space is extremely narrow. It is also small where the arachnoid mater bridges over the small sulci. However, it is much larger where it bridges over larger surface irregularities, such as the space between the inferior cerebellum surface and posterior medulla surface, which is called the **cerebellomedullary cistern**. Areas like this, which have a large CSF volume, are known as **subarachnoid cisterns**. The cerebellomedullary cistern is the largest cranial cistern, and is also called the *cisterna magna*, which means "great cistern." Other large cisterns include the following:

- **Pontine cistern**—around the anterior pons surface and medulla; continuous posteriorly with the cerebellomedullary cistern.
- **Interpeduncular cistern**—between the cerebral peduncles; it contains the posterior part of the cerebral arterial *circle of Willis*.
- **Superior cistern**—above the midbrain; it is used as a radiological landmark, and is also called the *quadrigeminal cistern* and the *cistern of the great cerebral vein*; this cistern is continuous laterally with the curved and thin layer of subarachnoid space on both sides, which partially encircles the midbrain, eventually opening into the interpeduncular cistern; the superior cistern, along with these sheet-like extensions, is known as the ambient cistern.

The finger-like extension of the subarachnoid space, called the **transverse cerebral fissure**, is between the fornix and roof of the third ventricle. It continues anteriorly from the superior cistern and remains in this area when the cerebral hemispheres grew backward during development, over the diencephalon.

Knob-like projections of the arachnoid mater are called **arachnoid granulations**. These protrude superiorly through the dura mater, into the superior sagittal sinus. The arachnoid granulations absorb CSF into the venous blood of the sinus. The subarachnoid space is very clinically important, since the more proximal and larger arteries run through it. Any rupture of these arteries will cause blood to accumulate in the subarachnoid space, which is called *subarachnoid hemorrhage*. This type of hemorrhage is usually due to trauma. However, it can also happen in the setting of aneurysm rupture. An aneurysm is a balloon-like outpouching of the artery wall, which can slowly grow and sometimes rupture (see Chapter 5).

Pia mater

The **pia mater** is anchored by astrocyte processes. It is located very close to the brain surface and covers all external surfaces of the CNS. The term "pia mater" means "tender matter." It is composed of delicate connective tissue and has many tiny blood vessels. The pia mater is the only layer that clings tightly to the brain and follows all of its convolutions. Cerebral arteries and veins travel in the subarachnoid space, completely enveloped by pia mater. The pia mater is also found near the cerebral blood vessel branches where they penetrate the brain surface and eventually reach its internal structures.

The pia mater meets the layer of astrocyte "end feet" located at the CNS surface. The arachnoid trabeculae that span the subarachnoid space actually merge with the pia mater in a nearly invisible way, making it difficult to determine the actual end of the arachnoid matter and beginning of the pia mater. Therefore, this entire *leptomeningeal complex* is sometimes called the *piaarachnoid*. There are two pial extensions worth noting:
- **Denticulate ligament**: providing horizontal support to the spinal cord.
- **Filum terminale**: attached to the coccygeal bone, providing vertical support to the spinal cord.

Barrier function of the arachnoid mater

The CNS is somewhat isolated from the remainder of the body. Its environment is extremely regulated, in part by various diffusion barriers. These are located between the extracellular spaces within and surrounding the nervous system, and also in the extracellular spaces in other areas of the body. Between the CSF in the subarachnoid space and the extracellular fluids of the dura mater lies one of these barriers. For example, when a substance is injected into the middle meningeal artery, it will spread through the dura mater, but will not enter the subarachnoid space. The barrier is within the cellular layers of the arachnoid mater where it interfaces with the dura mater. Here, the cells are connected via many tight junctions, which occlude extracellular space.

Small vessels enter or leave the brain, each with a sleeve of **perivascular space** that is also called *Virchow-Robin space*. This is an extension of the subarachnoid space. It extends inward and is filled with connective tissue as well as extracellular fluid, to a level at which the vessel becomes a capillary. It is not fully understood as to the structure and function of this microscopic space. It may provide a functional communication pathway, between the extracellular space surrounding neurons and the subarachnoid space. Via an arachnoid villus, CSF passes from the subarachnoid space into a dural venous sinus.

Dural nerve supply

The **supratentorial** dura mater membrane is supplied by the ophthalmic division of the trigeminal nerve. The dominant nerve that supplies the majority of the supratentorial dura is the **tentorial nerve** branch of the ophthalmic nerve. However, similar to its arterial supply, innervation depends on the area of the dura. It supplies the falx cerebri, **calvarial dura**, and superior surface of the tentorium.

Innervation for the **infratentorial** dura mater is by the upper cervical nerves. This area has abundant sensory innervation and is extremely sensitive to pain. As a rule, the brain parenchyma does not contain any pain receptors. However, the dura layer is innervated mainly by branches of the trigeminal nerve, as mentioned above. This is partially the mechanism behind the migraine pain sensation, which is also attributed to the trigeminal nerve. Dura in the anterior cranial fossa is supplied by the anterior and posterior ethmoidal nerves, with additional small supplies from the maxillary nerve. In the middle cranial fossa, it is supplied by the meningeal branch of the maxillary nerve in its anterior section. It is supplied by the meningeal branch of the mandibular nerve or **nervus spinosus**, in its posterior section. In the posterior cranial fossa, it is supplied by the meningeal branches of the hypoglossal and vagus nerves. All of these are C1 and C2 fibers carried by the cranial nerves. Around the foramen magnum, the dura mater is supplied directly, by the C2 and C3 cervical nerves. Pathologies related to innervation of the meninges by the trigeminal nerve are *migraines*. There is pain in the distribution of the ophthalmic branch of the trigeminal nerve, and activation of the **trigeminovascular system**. This leads to the release of more cytokines and chemicals involved in the generation of pain.

Pain sensation in the dura mater

The brain, arachnoid mater, and pia mater are not sensitive to pain due to physical stimulation, but there is pain sensation in the dura mater. As a result, certain neurosurgical procedures can be performed without general anesthesia. The proximal portions of blood vessels at the base of the brain are also sensitive to pain. Except for the posterior fossa, most of the cranial dura receives sensory innervation from the trigeminal nerve. The dural nerves follow the meningeal arteries. They end near the arteries or near the dural sinuses. Areas of dura between branches of the meningeal arteries have only sparse innervation if they are innervated at all—except in the floor of the anterior cranial fossa. If these dural endings are deformed, it may cause certain types of headaches.

Different pain perceptions are based on whether stimulation occurs to the endings near the meningeal arteries or those near the dural sinuses. If the endings near the meningeal arteries are stimulated, pain is very accurately localized to the area being stimulated. For those near the dural sinuses, pain will be referred to areas of the peripheral distribution of the trigeminal nerve—such as the forehead, eye, or temple. The posterior fossa dura is mostly supplied by fibers of the vagus nerve and the second and third cervical nerves. Similar to supratentorial dural innervation, pain-sensitive endings in the posterior fossa are primarily located near dural arteries and venous sinuses. Here, deformation causes pain that is referred to the back of the neck, or behind the ear. The term *dermatome* refers to the area of the skin that is innervated by a spinal nerve. Interestingly, while all the spinal nerves have their own dermatome, there is no C1 dermatome. Despite all the efforts to localize headaches, the pain is often a nonlocalizing symptom. Any new development of headache in patients above the age of 50 (some guidelines even use age 35 as the cutoff) warrants obtaining a magnetic resonance imaging (MRI) of the brain, to rule out a brain lesion as the cause. Isolated headache without any other symptoms for more than 6 months is most likely related to a primary headache disorder such as migraine, tension, cluster headache, etc.

Dural arterial supply

The dural arterial supply is primarily to the outer layer of the dura mater, instead of the arachnoid or pia mater layers, since these do not require a large supply of blood. Several arteries supply the dura in the anterior, middle, and posterior cranial fossae. The **anterior cranial fossa** is supplied by meningeal branches of the anterior ethmoidal artery, posterior ethmoidal artery, and ophthalmic artery. This fossa is also supplied by the frontal branch of the middle meningeal artery. This artery enters the middle cranial fossa via the foramen spinosum.

The **anterior ethmoidal artery** is a branch of the ophthalmic artery. It supplies the anterior and middle ethmoidal sinuses, the frontal sinus, the natural nasal wall, and the nasal septum. The **posterior ethmoidal artery**, another branch of the ophthalmic artery, supplies the posterior ethmoidal sinuses, the dura mater, and the nasal cavity. The **ophthalmic artery** is a branch of the C6 segment of the *internal carotid artery*. The **middle meningeal artery** is a branch of the first section of the maxillary artery. It passes vertically, through the **auriculotemporal nerve roots**, entering the middle cranial fossa via the foramen spinosum. At this location, it branches into the superior tympanic branch and the ganglionic branch. It then divides into anterior and posterior divisions.

The **middle cranial fossa** is supplied by the frontal and parietal branches of the middle meningeal artery, the accessory meningeal artery, the ascending pharyngeal artery, and direct branches from the internal carotid artery. Most of the supratentorial dura mater is supplied by the middle meningeal artery, which is a branch of the maxillary artery. The maxillary artery actually mostly supplies the **calvarium** instead of the meninges. This artery is important because of its position in the extradural space, and its anterior divisions' proximity to the **pterion**, which makes it susceptible to damage from head injuries. The **accessory meningeal artery** branches off of the maxillary artery but may also branch from the middle meningeal artery. The **ascending pharyngeal artery** is the smallest branch of the external carotid artery. It is long, thin, and located deeply in the neck under other branches of the external carotid artery, and also under the **stylopharyngeus**. The internal carotid artery is a terminal branch of the *common carotid artery*.

The **posterior cranial fossa** is supplied by the vertebral, occipital, and ascending pharyngeal arteries. The paired *vertebral arteries* arise from respective **subclavian arteries**. They ascend in the neck to supply the posterior fossa and occipital lobes, along with the segmental vertebral and spinal column blood supply. The **occipital arteries** arise from the external carotid artery, opposite to the facial artery. They supply the back of the scalp, sternomastoid muscles, and deep back and neck muscles. Pathology related to dural arterial supply includes **extradural hemorrhage.** For a detailed explanation of the CNS vascular supply, see Chapter 5.

Section review
1. What are the different meningeal layers, potential spaces in between these layers, and their clinic relevance?
2. What do the terms supratentorial and infratentorial mean?
3. Which parts of the dura mater receive more or less arterial blood?

Dural venous sinuses

The dural venous sinuses are located between the endosteal and meningeal layers of the dura mater. They run in their own paths that are not parallel to arteries. The areas drained by the intracranial veins are different from the areas supplied by the major cerebral arteries. The dural venous sinuses form the primary drainage pathways of the brain, ultimately draining into the internal jugular veins through the sigmoid veins.

The main dural venous sinuses are either unpaired or paired. The unpaired sinuses include the following:

Superior sagittal sinus—this is the largest dural venous sinus. It runs in a sagittal plane, from the anterior falx cerebri to its point of termination at the occipital protuberance.

Inferior sagittal sinus—this sinus runs along the inferior edge of the falx cerebri, from front to back. It drains into the straight sinus and receives blood from the falx cerebri as well as small veins of the medial surface of the cerebral hemispheres.

Straight sinus—this sinus is located between the falx cerebri and tentorium cerebelli, with a triangular appearance. It receives the inferior sagittal sinus, the *vein of Galen* (anteriorly), and certain superior cerebellar veins over its length. The straight sinus runs posteroinferiorly toward the **confluence of sinuses**. Most (56%) of its drainage is via the confluence of sinuses, with less (21%) from the left transverse sinus and even less (13%) from the right transverse sinus. The straight sinus may be duplicated or hypoplastic in certain patients. If absent, a *persistent falcine sinus* usually exists, directly draining into the superior sagittal sinus. In general, there are several variations when it comes to dural venous sinuses, and some patients congenitally lack some parts on one side.

Occipital sinus—the smallest dural venous sinus, lying on the inner surface of the occipital bone. Vessels from the foramen magnum area, which may connect with the sigmoid sinus and internal vertebral plexus, join to pass through the attached falx cerebelli margin, draining posterosuperiorly at the confluence of sinuses.

Intercavernous sinus—actually consisting of anterior and posterior sinuses, they connect the left and right cavernous sinuses as well as the basilar venous plexus. They are located in the anterior and posterior borders of the **diaphragm sellae**.

The paired sinuses include the following:

Transverse sinus—this sinus drains the superior sagittal sinus, occipital sinus, and straight sinus. It empties into the sigmoid sinus, which reaches the *jugular bulb*. There are actually two transverse sinuses, which arise at the confluence of sinuses. There may be highly variable anatomy of the transverse sinuses. The most common variation is hypoplasia of the left sinus.

Sigmoid sinus—this sinus is a paired structure that is a continuation of the transverse sinus, which begins as the tentorium ends. Here, it receives the **superior petrosal sinus** and ends at the jugular bulb.

Superior petrosal sinus—this sinus drains the cavernous sinus posterolaterally to the transverse sinus. It runs along the superior aspect of the petrous temporal bone, and receives the cerebellar veins, inferior cerebral veins, and the labyrinthine vein that drains the inner ear structures.

Inferior petrosal sinus—this sinus is often a plexus of venous channels instead of a true sinus. It drains the cavernous sinus to the jugular foramen, or sometimes via a vein passing through the hypoglossal canal to the suboccipital venous plexus. It lies in a shallow grove between the petrous temporal bone and basilar occipital bone, connect across the midline by the basilar plexus. It receives blood from the medulla oblongata, pons, and inferior cerebellum, as well as the labyrinthine veins.

Cavernous sinus—this sinus actually consists of paired sinuses on either side of the pituitary fossa and body of the sphenoid bone. They lie between the endosteal and meningeal dura layers. They are divided by many fibrous septa into small "cave-like" structures. As mentioned in previous chapters, the most important structures include the internal carotid arteries; cranial nerves three, four, and six; as well as the trigeminal nerve's ophthalmic and maxillary branches.

Sphenoparietal sinus—this sinus is located along the posteroinferior ridge of the lesser wing of the sphenoid bone. It also consists of the parietal portion of the frontal ramus of the middle meningeal vein. This sinus drains into the cavernous sinus. It receives blood from the superficial middle cerebral vein, middle meningeal vein, and anterior temporal **diploic** vein. It varies in size based on the sizes of its supplying veins.

Basilar venous plexus—this plexus lies between the endosteal and visceral dura layers, on the inner surface of the **clivius**. It connects the following: inferior petrosal sinuses, cavernous sinuses, *intercavernous sinuses*, superior petrosal sinuses, internal vertebral venous plexus, and the marginal sinus that is around the foramen magnum margins.

Pathology that is related to the venous sinuses includes *dural venous sinus thrombosis*. This is usually related to pregnancy, birth control pill use, blood disorders, etc. For more detail on sinus thrombosis see Chapter 5. There is also another clinical classification of the dural sinuses in deep and superficial sinus thrombosis. Two of the main superficial veins are the vein of Labbé (inferior anastomotic vein) and vein of Trolard (superior anastomotic vein).

> **Section review**
> 1. Where are the dural venous sinuses located?
> 2. Which are the largest and smallest dural venous sinuses?
> 3. What is the "confluence of sinuses"?

Meninges of the spinal cord

The spinal cord is isolated from other body structures by the vertebral column, and its surrounding ligaments, tendons, and muscles. All of these structures help insulate the spinal cord from trauma and other stressors to the back. Additional protection supports the delicate neural tissues from dangerous contact with the bones of the vertebral canal. Therefore, a series of **spinal meninges** surround the spinal cord and provide shock absorption and stability. The spinal cord receives nutrients and oxygen from the branched blood vessels that lie within these layers.

Like the brain, the spinal meninges also have a *dura mater*, *arachnoid mater*, and *pia mater*. The spinal meninges are continuous with the cranial meninges at the *foramen magnum* of the skull.

The spinal dura mater

In the spine, the dura mater makes up the outer covering and contains dense collagen fibers aligned down the spinal cords' longitudinal axis. An *epidural space* lies between the dura mater and vertebral canal walls. This space contains **areolar tissue**, a protective layer of adipose tissue, and blood vessels. The spinal dura mater has various forms of connections to the surrounding bones. Its outer portion is fused with the periosteum of the occipital bone of the cranium, around the foramen magnum margins. At this point, the spinal dura mater becomes continuous with the cranial dura mater. The spinal dura mater lacks firm, significant connections to surrounding vertebrae. Longitudinal stability is provided by attachment sites at each end of the vertebral canal.

Also, at each intervertebral foramen are the extensions of the dura mater between adjacent vertebrae. These extensions fuse with connective tissue that surrounds the spinal nerves. On the posterior surface of the sacrum and inside its sacral canal, the spinal dura mater is tapered, from a sheath into a dense collagen fiber cord. This cord blends with parts of the filum terminale and forms the **coccygeal ligament**. This ligament travels along the sacral canal to finally blend with the periosteum of the coccyx.

The spinal arachnoid mater

When a person is healthy, there is no subdural space separating the dura mater from the inferior meningeal layers. The dura mater contacts the outer surface of the *arachnoid mater*, which is the middle layer of the spinal meninges. Simple squamous epithelium covers the inner dura mater and the outer arachnoid mater. This *arachnoid membrane* of epithelia is included in the arachnoid mater, which also includes a delicate collagen and elastic fiber network called the **arachnoid trabeculae.** These trabeculae extend between the arachnoid membrane and the outer pia mater. The *subarachnoid space* in between is filled with CSF.

In the spine, the arachnoid mater is extended inferiorly up to the filum terminale. The anterior and posterior roots of the **cauda equina** are within the fluid-filled subarachnoid space. When CSF is withdrawn from an adult patient, a needle must be inserted into the subarachnoid space, in the inferior lumbar region, to avoid spinal cord injury. This is known as a *spinal tap*, which is clinically described as a *lumbar puncture*.

The spinal pia mater

The inner meningeal layer is the *pia mater*, which is a mesh-like collection of elastic and collagen fibers. The subarachnoid space extends between the arachnoid epithelium and the pia mater, which is tightly bound to underlying neural tissues. The connective fibers of the pia mater interweave with the fibers spanning the subarachnoid space. This means that the arachnoid mater and pia mater are firmly bound together. Blood vessels that serve the spinal cord travel along the pia mater, inside the subarachnoid space.

As mentioned earlier in this chapter, paired **denticulate ligaments** run down the spinal cord, which are extensions of the pia mater. They originate on either side of the spinal cord and keep the cord from moving laterally. Longitudinal (superior-inferior) movement is prevented by dural connections at the foramen magnum and coccygeal ligament. The spinal meninges follow the posterior and anterior roots as they pass through the *intervertebral foramina*.

Section review
1. How do the spinal meninges protect the spinal cord?
2. What structures are involved when a spinal puncture is performed?
3. What structures prevent the spinal cord from moving laterally or longitudinally?

Ventricles

Chambers in the brain known as **ventricles** are formed during development. These chambers are formed by the expansion of the neural tube inside the cerebral hemisphere, diencephalon, metencephalon, and medulla oblongata.

Lateral ventricles

There is a large **lateral ventricle** inside each cerebral hemisphere. The two lateral ventricles are separated by a thin plate of tissue called the **septum pellucidum**. There is no direct connection between the two lateral ventricles. They each follow a long, C-shaped

course through the various lobes of their cerebral hemisphere. There are five divisions of each lateral ventricle, as follows:
- **Anterior (frontal) horn**—in the frontal lobe, anterior to the interventricular foramen.
- **Body**—in the frontal and parietal lobes; it extends posteriorly, to the area of the splenium of the corpus callosum.
- **Posterior (occipital) horn**—projects posteriorly into the occipital lobe.
- **Inferior (temporal) horn**—curves inferiorly and anteriorly, into the temporal lobe.
- **Antrum (trigone)**—the area near the splenium, where the body, posterior horn, and inferior horn meet.

The body, atrium, and inferior horn make up the original C-shaped development of each lateral ventricle. The anterior and posterior horns are extensions from this more basic shape. Other structures create the borders of the lateral ventricle over its course through the cerebral hemisphere. Many of these can be seen in coronal sections.

The caudate nucleus also has a C-shape. It has an enlarged *head*, which forms the lateral wall of the anterior horn. Its *body* is slightly smaller and makes up most of the lateral wall of the body of the ventricle. Its attenuated *tail* is in the roof of the inferior horn and proceeds posteriorly, with the caudate nucleus becoming smaller as the thalamus becomes larger, forming the floor of the body of the ventricle. The body of the corpus callosum makes up the roof of the anterior horn and body of the ventricle. The *genu* or anterior end of the frontal lobe of the corpus callosum is curved inferiorly, forming the anterior wall of the anterior horn. The septum pellucidum creates the medial wall of the body and anterior horn. It terminates near the splenium, marking the point where the bodies of the ventricles diverge from midline, beginning to curve around, into the inferior horns. The posterior horn is developed later, often varies in size, and may be crudely formed.

Third ventricle

The paired lateral ventricles connect to the **third ventricle** of the diencephalon via the **interventricular foramina (of Monro)**. The third ventricle is a narrow slit occupying most of the midline area of the diencephalon. In a hemisected brain, its entire outline is visible, looking somewhat like a poorly formed doughnut. Its hole corresponds to the **interthalamic adhesion**. This joins the thalami and usually crosses the ventricle. The third ventricle is an important clinical landmark. Once it is located, the structures on either side are parts of the thalamus. Anteriorly, the third ventricle ends at the **lamina terminalis**, which was formed from the rostral neuropore. Much of the medial surface of the thalamus and hypothalamus form the wall of the third ventricle. Part of the hypothalamus forms its floor. Its thin, membranous roof contains the choroid plexus.

The third ventricle narrows quickly at the posterior end of the mammillary bodies. It becomes the cerebral aqueduct of Sylvius, which is throughout the midbrain. In the anterior portion of each wall of the third ventricle, the interventricular foramen is important for radiological visualization, because it is visible via several different methods. The interventricular foramen is anatomically related to many deep structures. It is at the anterior end of the thalamus, for example. If one interventricular foramina is blocked, obstructive or noncommunicating *hydrocephalus* often occurs. The third ventricle's outline reveals four **recesses**, which are protrusions that correspond to the structures that have evaginated (turned inside out) from the diencephalon. They are as follows:

- **Optic recess**—anterior to the optic chiasm, at the base of the lamina terminalis.
- **Infundibular recess**—immediately posterior to the chiasm.
- **Pineal recess**—superiorly invades the stalk of the pineal gland.
- **Suprapineal recess**—just anterior to the stalk of the pineal gland.

Fourth ventricle

The third ventricle connects to the **fourth ventricle** of the pons and medulla, via the thin **cerebral aqueduct (of Sylvius)**, of the midbrain. The fourth ventricle is caudally known as the *central canal of the caudal medulla and spinal cord*. It is not usually wide, over much of its structure. At the widest point, the fourth ventricle is surrounded by the pons. This is another important clinical point in order to differentiate the pons from the midbrain and medulla. At the point where the ventricle becomes widest is the level of the pons.

The superior area of the fourth ventricle is between the posterior pons surface and the anterior cerebellum surface.

The cranial ventricles are lined with a type of neuroglia known as ependymal cells. The cells produce CSF, filling the ventricles, and continuously circulating within the CNS. CSF passes between the interior and exterior CNS areas, via three foramina found in the roof of the fourth ventricle.

Cerebrospinal fluid

CSF has many important functions as it totally surrounds and coats exposed CNS surfaces. These functions include the following:

- *Brain support*—the brain basically floats in CSF, suspending it inside the cranium. When supported by CSF, the human brain weights about 1.8 ounces (50 g). This is extremely different from its weight outside of the body, which is 3.09 pounds (1400 g).
- *Cushioning of neural structures*—the delicate brain and spinal cord are cushioned by CSF against physical trauma.
- *Transportation of chemical messengers, nutrients, and wastes*—the ependymal cell lining is freely permeable, except in locations where CSF is produced. There is constant chemical information exchange between the CSF and interstitial fluid surrounding the CNS neurons and neuroglia.

Formation and circulation of cerebrospinal fluid

CSF is primarily produced within the **choroid plexus** between each ventricle. The choroid plexus is present in all four brain ventricles. It contains strands of extremely convoluted, vascular membranes. The capillaries of the choroid plexus are surrounded by specialized ependymal cells that have microvilli and are interconnected via tight junctions. The ependymal layer is specialized as a cuboidal, secretory epithelium called the **choroid epithelium**. The entire ependyma-pia-capillary complex makes up the choroid plexus. A long, continuous band of the plexus, similar to the C-shaped course of each lateral ventricle, extends from close to the tip of the inferior horn. It continues through the body of the ventricle and reaches the interventricular foramen. However, there is no choroid plexus in the anterior or posterior horn. The flow of CSF is illustrated in Fig. 4.1.

The choroid plexus is enlarged in the area of the atrium, where it is called the **glomus**. The choroid plexus, with aging, becomes calcified. In each lateral ventricle, the choroid plexus grows through the interventricular foramen. It forms part of its posterior wall and also becomes one of two narrow strands of choroid plexus in the roof of the third ventricle. On an axial CT scan of the brain at the level of the third ventricle, there are usually three hyperdense (bright) structures. These are the pineal gland in the middle, and the two calcified lateral ventricles' choroid plexus at each side of the gland. There is no continuation of the plexus through the cerebral aqueduct, which still has an ependymal lining, and is totally surrounded by neural tissue.

In the fourth ventricle, the choroid plexus is formed from another invagination of the inferior medullary velum, into the caudal half of the ventricle. It has a "T" shape, with the vertical portion of the "T" made up of two adjacent, longitudinal strands of plexus. They often extend all the way to the median aperture, which is where they are directly exposed to the subarachnoid space. The transverse part of the "T" has one strand of plexus, extending into each lateral recess. Both ends reach a lateral aperture, at the point where a small piece of plexus usually protrudes through, exposed directly to the subarachnoid space.

Since one side of the pia mater faces the subarachnoid space, the choroid plexus is always adjacent to the subarachnoid space, on its pial side, and to the intraventricular space on its choroid epithelial side. This is easily seen in coronal sections of the brain. The **choroid fissure** is the location of the invagination of the choroid plexus into the lateral ventricle. This fissure is a C-shaped slit of the subarachnoid space. It accompanies the fornix of the inferior horn to the interventricular foramen. The space above the roof of the third ventricle, continuing laterally into the choroid fissure, is also subarachnoid space called the **transverse cerebral fissure**. This is a long extension of subarachnoid space in the middle of the cerebrum, because of growth of the cerebral hemispheres, posteriorly, over the diencephalon and brainstem. The transverse cerebral fissure continues posteriorly into the superior cistern.

Functionally, the choroid plexus is a three-layered membrane between the blood and CSF. Its first layer is the endothelial wall of each choroid capillary. The wall is fenestrated

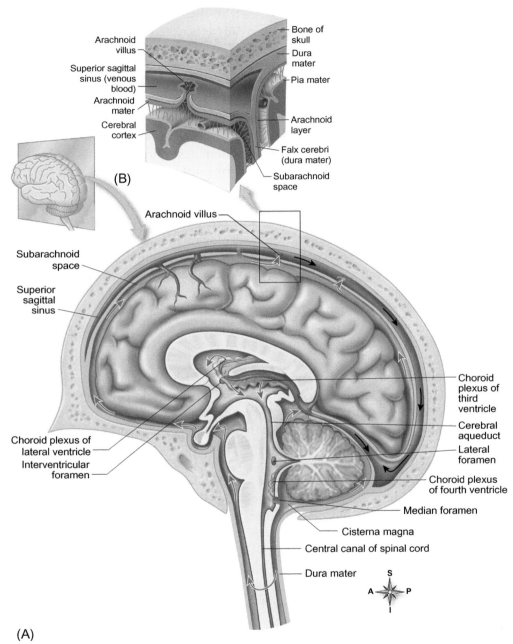

Fig. 4.1 The flow of cerebrospinal fluid (CSF).

and allows movement of substances out of the capillary to occur easily. Its second layer has scattered pial cells, collagen, and is comparatively fragmented. The third layer, or *choroid epithelium*, is derived from the same layer of cells that forms the ventricles' ependymal lining. Its cells appear to be specialized for secretion. They have a large amount of basal infoldings, many microvilli on the side that faces the CSF, and numerous mitochondria. Also, nearby cells are connected together via tight junctions, which occlude the extracellular space between them. They help limit the movement of substances across the choroid epithelium. Certain ions can diffuse across the tight junctions, but larger molecules such as peptides are blocked.

The surface of the choroid plexus is increased by the folding of individual cell membranes into the microvilli, and by the macroscopic folding of the choroid plexus into many villi and crevasses. Because of this extensive folding, the total surface area of the choroid plexus, not including the area provided by the microvilli, is more than $200\,cm^2$. This is about two-thirds of the overall total ventricular surface area.

As fluid leaks out of the choroid plexus capillaries, CSF is produced as a filtrate. This is followed by secretions of CSF by the ependymal cells into the ventricles. These cells also remove wastes from CSF and regulate its composition. Free exchange occurs between the brain's interstitial fluid and the CSF. Therefore, as the CNS has functional changes, there may be related changes in CSF composition. The reverse situation is also true. Differences in the composition of CSF as compared to blood plasma are significant. There are high amounts of soluble proteins in blood plasma, but not in CSF. The CSF also has different amounts of certain ions. It contains more sodium, chloride, and hydrogen ions than plasma, but less calcium and potassium ions. It also contains less lipids and glucose than plasma. This is why a lumbar puncture can reveal important information about CNS injury, disease, or infection.

The total volume of CSF in adults ranges between 140 and 270 mL. Approximately 600–700 mL are produced every day. When normal CSF circulation or resorption is interrupted, there may be many different complications. One example is *hydrocephalus*, commonly called *water on the brain*, in which an infant experiences faulty resorption of CSF. The skull of the infant may expand greatly because of the abnormally large volume of CSF. When hydrocephalus occurs in an adult because of poor resorption or blocked CSF circulation, the brain may be damaged and distorted.

> **Focus on indicators in CSF**
> Variations in the composition of CSF can indicate various pathologies. Increased protein content can suggest inflammatory processes such as meningitis, vasculitis, and even cerebral infarction. Increased white blood cells indicate infection. Numbers of cells between 800 and 1000 are more consistent with nonbacterial pathogens, while more than 1000 cells usually indicate a bacterial infection. Unusually low glucose levels in the CSF are seen in fungal infections, sarcoidosis, and tuberculosis. In a more specialized test, elevated levels of *tau protein* are seen in the CSF of Alzheimer's patients. This test can now detect

Alzheimer's disease even when the patient is asymptomatic. Unfortunately, CSF tau protein screening is still at the research stage and considered experimental. Therefore, it is usually not available outside of academic centers.

Focus on lumbar puncture

Since the meninges extend beyond the spinal cord, there is a good area for performing a lower **lumbar puncture** without damaging the spinal cord. This procedure allows CSF to be withdrawn from the subarachnoid space, in the lumbar region of the vertebral column. A needle is inserted just above or just below the fourth lumbar vertebra (see Fig. 4.2). The spinal cord ends about 1 in. above that level. The fourth lumbar

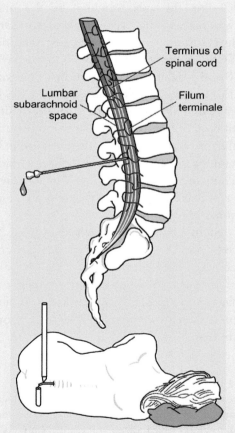

Fig. 4.2 Lumbar puncture. Cerebrospinal fluid is obtained by inserting a needle into the subarachnoid space in the lumbar region. *(From Mahon, C. R.; Manuselis, G. Textbook of Diagnostic Microbiology, 2nd ed.; Saunders: Philadelphia, 2000.)*

vertebra is located easily, since it lies on a line with the iliac crest. The patient is placed on one side, with the knees and chest drawn together, arching the back and separating the vertebrae in the fetal position. The needle is inserted once the patient's head is pointed toward the belly bottom, in a 45 degrees angle. When it enters the CSF, the thin nerve roots roll off the needle's tip, which allows CSF to be collected without any nerve tissue damage. Correct positioning is key to success when performing this procedure. CSF is then tested for the presence of bacteria, blood cells, or other abnormalities that can indicate injury or an infection such as meningitis (see Fig. 4.3). Sometimes, a *manometer* is attached to the needle, which is a sensory that determines CSF pressure in the subarachnoid space. Lumbar puncture is also used to inject diagnostic and therapeutic agents such as radiopaque dyes, as well as chemotherapy agents, into the subarachnoid space.

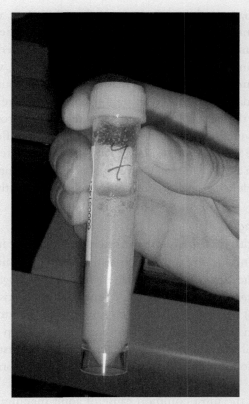

Fig. 4.3 Cerebrospinal fluid having a *pink color,* indicating the abnormal presence of blood.

> **Section review**
> 1. How many brain ventricles exist, and what are their passageways called?
> 2. Which cells produce CSF, and where does this occur?
> 3. What are the three basic functions of CSF?

Clinical considerations

There are a variety of conditions that affect the meninges and ventricles. These include various types of meningitis, hydrocephalus, hemorrhages, and meningioma.

Meningitis

Meningitis is inflammation of the meninges of the CNS. It is a very common CNS infection. When inflammation affects the meninges as well as the brain **parenchyma**, it is referred to as **meningoencephalitis**. Meningitis is classified as acute or chronic. Often, infectious meningitis begins with vague viral symptoms, which usually include fever, headache, and **nuchal rigidity**. This may develop over hours (especially in severe bacterial etiologies) to days (especially with viral etiologies). There is restriction and pain when passively flexing the neck. However, neck extension and rotation are less painful. When meningitis is severe, neck flexion may cause hip or knee flexion, which is called *Brudzinski's sign*. There may also be resistance to passive knee extension while the hip is flexed, which is known as *Kernig's sign*. Untreated bacterial meningitis will result in lethargy, confusion, focal deficits, seizures, and ultimately, death within hours

Acute bacterial meningitis

Acute bacterial meningitis is a **fulminant** and often fatal pyogenic infection that begins in the meninges. The most common bacteria that cause meningitis include *Group B streptococci* (during the first 2 months of life), *Neisseria meningitides* (meningococci), and *Streptococcus pneumoniae* (pneumococci). **Meningococci** exist in the nasopharynx of a small amount (5%) of people and may be spread via respiratory droplets and close contact. It is unknown why a small fraction of carriers develop meningitis, but most do not. Meningococcal meningitis usually occurs during the first year of life, but also in epidemics within closed populations such as college dormitories, boarding schools, or military barracks.

Pneumococci are the most common cause of adult meningitis, especially in alcoholics and along with conditions such as chronic otitis, **mastoiditis**, sinusitis, CSF leaks, pneumococcal pneumonia, recurrent meningitis, **asplenia**, or sickle cell disease. Because of routine vaccination, the incidence of pneumococcal meningitis is decreasing.

In infants, Gram-negative organisms—usually *Escherichia coli, Klebsiella* species, or *Enterobacter* species may cause meningitis. These organisms may also cause the disease in immunocompromised patients, post-CNS surgical patients, and those with CNS trauma, bacteremia, or hospitalization-acquired infections. In immunocompromised or colonized patients, *Pseudomonas* species sometimes cause meningitis. Today, *Haemophilus influenzae type b meningitis* is uncommon due to available vaccinations. It can still occur in immunocompromised patients and following head trauma in unvaccinated individuals. After penetrating head wounds or neurosurgical procedures, often as components of mixed infections, **Staphylococci** can also cause meningitis. These bacteria can also cause the disease after bacteremia, such as due to endocarditis, as well as following CNS surgeries.

Listeria is a type of bacteria that is more likely to cause meningitis in very young or very old patients. It may also be causative at any age when there is immunocompromised due to chronic renal failure, hepatic disorders, or the corticosteroid or cytotoxic therapies that follow organ transplantation. Table 4.1 summarizes the most common causative bacteria for acute meningitis, in different age groups.

Bacteria commonly affect the meninges via hematogenous spread, from areas of colonization in the nasopharynx or others, such as in pneumonia. Certain bacteria are more likely to colonize in the CSF. This is not fully understood, but binding pili and encapsulation appear to be related. Pili receptors and other bacterial surface component receptors in the choroid plexus make it easier for bacteria to penetrate the CSF. Bacteria can also enter the CSF via direct extension from localized infections, such as sinusitis or mastoiditis. They can also enter through exterior openings in CSF pathways that are normally closed. This may be due to **meningomyelocele**, penetrating injuries, neurosurgical procedures, or spinal dermal sinus.

Table 4.1 Most common causes of acute bacterial meningitis, by age.

Neonates (causative bacteria)	Adults 18–60 (causative bacteria)	Adults over 60 (causative bacteria)
Group B *Streptococcus*—50% of cases	*Streptococcus pneumoniae*—60%	*Streptococcus pneumoniae*—70%
Escherichia coli—25%	*Neisseria meningitides*—20%	*Listeria monocytogenes*—20%
Other Gram-negative rods—8%	*Haemophilus influenzae*—10%	*Neisseria meningitides*—3% to 4%
Listeria monocytogenes—6%	*Listeria monocytogenes*—6%	Group B *Streptococcus*—3% to 4%
Streptococcus pneumoniae—5%	Group B *Streptococcus*—4%	*Haemophilus influenzae*—3% to 4%
Group A *Streptococcus*—4%	–	–
Haemophilus influenzae—3%	–	–

Pathophysiology

The pathophysiology of meningitis involves neutrophils being drawn into the CSF space because of bacterial surface components, complement, and inflammatory cytokines. Neutrophils release metabolites that cause damage to cell membranes, including the vascular endothelium. Localized vasculitis, inflammation, and thrombophlebitis develop, which may cause infarction or focal ischemia. Brain edema occurs in the most severe cases. This inflammatory state disrupts the blood-brain barrier, increasing edema. CSF reabsorption is blocked by **purulent exudate**, resulting in hydrocephalus. As a result, intracranial pressure increases. Systemic complications include hyponatremia (because of altered amounts of antidiuretic hormone), disseminated intravascular coagulation, and septic shock. Sometimes, bilateral adrenal hemorrhagic infarction occurs.

According to the National Meningitis Association, about 600–1000 people contract meningococcal disease in the United States every year. Fortunately, this is the lower number of annual cases in history, partly due to the increased use of meningococcal vaccines. Of those who contract meningococcal disease, 10%–15% die. Of the survivors, about one of every five patients will have permanent disabilities such as brain damage, hearing loss, loss of kidney function, or limb amputations. Approximately 21% of all meningococcal disease cases occur in preteens, teens, and young adults ages 11–24. One in five US teens has not yet received their first dose of meningococcal vaccination and remains unprotected. Less than one-third of first-time vaccination recipients have received the recommended booster dose. Many teenagers have not received the meningococcal serogroup B vaccine since it was recommended by the Centers for Disease Control and Prevention in 2015. Acute bacterial meningitis may develop in children with cystic fibrosis or severe burns, with the causative organisms being *Staphylococcus aureus* or *Pseudomonas aeruginosa*.

Clinical manifestations

Acute meningitis is often preceded by a sore throat or respiratory illness. Adults often develop serious illness within 24 h, and children even more quickly. About 30% of patients have seizures. Less common symptoms include cranial nerve abnormalities, cranial nerve palsy, deafness, irritability, confusion, drowsiness, stupor, and coma. Dehydration is often seen, and shock can be caused by vascular collapse. Infection may spread to the joints, lungs, sinuses, and other areas. In meningococcal meningitis, a petechial or purpuric rash is common.

In children younger than 2 years, there may be no meningeal signs. In infants under 2 months of age, signs, and symptoms are often nonspecific, particularly in early disease. The common presenting symptoms include fever, hypothermia, poor feeding, irritability, lethargy, and vomiting. Symptoms that are possible, but usually occur later, include a high-pitched cry, bulging or tight fontanelles, or seizures. After several days, subdural effusions may develop. Typical signs include enlarging head size, persistent fever, and seizures.

The elderly can also have nonspecific symptoms. There may be no meningeal signs, or they may be absent. Arthritis may restrict motion of the neck, often in multiple directions. These should not be mistaken for **meningismus**.

Diagnosis

Diagnosis of acute meningitis is mainly clinical, with confirmation of the CSF analysis. Acute bacterial meningitis is suspected in children under 2 years of age who have lethargy, a high-pitched cry, progressive irritability, a bulging fontanelle, hypothermia, or meningeal signs. In children over 2 years of age, changes in consciousness, especially with fever or risk factors, must be assessed. As soon as acute bacterial meningitis, especially meningococcal, is suspected, it must be treated due to its rapid and potentially fatal progression. Blood is drawn and cultured. Gram staining and bacterial *DNA* **polymerase chain reaction (PCR)** is done if available. Lumbar puncture is required. However, extra attention needs to be made since these patients often have cerebral edema and increased intracranial pressure. Therefore, a lumbar puncture may cause herniation. However, immediate treatment with corticosteroids and antibiotics should occur as soon as possible. It should not be delayed waiting for imaging or lumbar puncture to be done.

Gram stain shows organisms in CSF in most (80%) of patients with acute bacterial meningitis. The white blood count is mostly neutrophilic—between 1000 and 10,000/μL. Glucose is usually <40 mg/dL due to impaired CNS glucose transport and consumption of glucose by bacteria and neutrophils. Protein is usually >100 mg/dL. In 90% of patients, cultures are positive but can be falsely negative if patients have been partially treated. In Gram-negative meningitis, the **limulus amebocyte lysate test** can detect **endotoxin**. CT scans may show normal or small ventricles, effacement of sulci, and contrast enhancement over any convexities. For subarachnoid inflammation, MRI with gadolinium is more sensitive. Scans must be checked for brain abscesses, mastoiditis, sinusitis, skull fractures, and congenital malformations. Blood cultures are positive in 50% of patients. Other tests include cell count with differential, electrolytes, coagulation tests, renal function tests, and PCR for bacterial pathogens, if available.

Alternate diagnoses include viral meningitis or encephalitis, subarachnoid hemorrhage, brain abscess, severe systemic infections, cerebellar tonsillar herniation, cerebral vasculitis, fungal meningitis or amebic meningoencephalitis, tuberculosis meningitis, and **Waterhouse-Friderichsen syndrome**. Tuberculosis meningitis is usually subacute or chronic, but sometimes acute. It may resemble acute bacterial or viral meningitis.

Treatment

Treatment for bacterial meningitis basically involves corticosteroids and antibiotics. When acute bacterial meningitis is suspected, these are given as soon as blood cultures are drawn. Dexamethasone is the corticosteroid of choice. Antibiotics such as ampicillin and ceftriaxone are used, with vancomycin as a broad bacterial coverage, initially administered intravenously. Acyclovir is always added to the regimen to cover for viral causes such as herpes simplex virus (HSV). Additional choices may be used, based on the age of the patient, the suspected pathogen, and patient allergies. These include cefotaxime, cefepime, rifampin, meropenem, metronidazole, and doxycycline. Supportive therapies include treatment of dehydration, fever, electrolyte disorders, shock, and seizures. For Waterhouse-Friderichsen syndrome, high doses of hydrocortisone are used. When brain herniation has occurred, medications include mannitol and hypertonic saline. Also, a barbiturate-induced coma may be considered.

Prevention of meningitis includes droplet precautions for bacterial causes during the acute treatment, various vaccinations, and chemoprophylaxis. Vaccinations include conjugated pneumococcal vaccine, routine vaccination against *H. influenzae type b*, and quadrivalent meningococcal vaccine. Chemoprophylaxis should be provided for anyone in close contact with the patient who has a definite diagnosis of meningococcal meningitis, using rifampin, ceftriaxone, ciprofloxacin, levofloxacin, or ofloxacin.

Viral meningitis

Viral meningitis is inflammation of the meninges with CSF lymphocytic **pleocytosis**. There is no apparent cause after routine CSF stains and cultures are utilized. Most of the time, a virus is implicated. Viral meningitis is also referred to as *aseptic meningitis*.

Pathophysiology

Most cases of viral meningitis are caused by **enteroviruses**, which include coxsackievirus, echovirus, and enteroviruses 68–71. The causative virus is transmitted via a fecal-oral route through various food products. It enters the gastrointestinal tract and spreads through the body via the bloodstream. The second most common causes of viral meningitis are arthropod-borne viruses, herpes simplex virus type 2 (HSV-2), and human immunodeficiency virus (HIV). Also, the mumps virus commonly causes the disease in places outside of the United States, but vaccination in this country has greatly reduced cases. **Mollaret's meningitis** is a syndrome that involves recurrent viral meningitis. This form is signified by large, atypical monocytes in the CSF. It may be related to HSV-2. Patients are usually unaware that they have been exposed. Viruses that are causative for encephalitis usually also cause low-grade viral meningitis.

Viral meningitis may occur in patients who are HIV-positive, at or near the time of **seroconversion**. The viral meningitis may represent the initial presentation of HIV, signified by headache, fever, and meningismus.

Clinical manifestations

Viral meningitis commonly follows a flu-like syndrome. It usually causes fever and headache, but there is no prominent **coryza**. Its signs are less obvious and develop more slowly than bacterial meningitis. The patient is not usually critically ill. Either systemic or nonspecific symptoms may exist, but there are no focal neurologic symptoms. For noninfectious forms, there is often no fever.

Diagnosis
Diagnosis of viral meningitis is through analysis of CSF, though there may be a CT scan or MRI performed prior to lumbar puncture. It must be determined whether the patient actually has acute bacterial meningitis and needs immediate administration of antibiotics. When symptoms are not severe, viral or other types of viral meningitis are considered. A head CT scan or MRI is done when a brain mass is suspected, based on **papilledema** or focal neurologic signs. Occasionally, idiopathic intracranial hypertension mimics viral meningitis. The distinction between acute bacterial and viral meningitis is based on CSF glucose, which is usually decreased in bacterial meningitis, along with protein being elevated. In viral meningitis, white blood cells in the CSF are mostly lymphocytes. If just a few neutrophils are present, fast consideration of early bacterial meningitis should be done, though neutrophils can be present in early viral meningitis. Bacterial meningitis forms that resemble viral meningitis include *Listeria meningitis*, tuberculosis meningitis, and also, partially treated bacterial meningitis. In viral meningitis, CSF pressure varies but is usually normal or slightly elevated. It is very high in bacterial meningitis.

The PCR test is the fastest way to identify the actual causative agent for viral meningitis. Other tests are used to diagnose the nonviral causes. There is no vaccine that protects against the enteroviruses, which most commonly cause viral meningitis. Good hygiene is essential, as well as avoiding close contact with affected patients.

Treatment
Treatment is supportive, using analgesics, antipyretics, and hydration. Antibiotics that are effective for bacterial meningitis are given prior to receiving results of cultures or repeated CSF tests, if early bacterial meningitis, partially treated bacterial meningitis, or *Listeria meningitis* cannot be excluded. Drug-induced chemical meningitis resolves once the causative drug is withdrawn. For HSV meningitis, intravenous acyclovir, for 14 days, is the treatment of choice. In the clinical setting, all patients are given IV acyclovir while waiting for CSF results.

> **Focus on other causes of meningitis**
> Lymphocytic meningitis can also be caused by bacteria such as rickettsiae—in **ehrlichiosis**, Rocky Mountain spotted fever, and typhus, as well as spirochetes—in leptospirosis, **Lyme disease**, and syphilis. Transient or chronic CSF abnormalities occur. The CSF may have characteristics of viral meningitis because an infection near the meninges can cause a sympathetic inflammatory response, without bacteria actually being present. Such infections include mastoiditis, brain abscess, infective endocarditis, and sinusitis.

> Noninfectious causes of meningitis are linked to leakage from intracranial cysts, neoplastic infiltration, **intrathecal drugs**, lead poisoning, and radiopaque agents. Uncommonly, inflammation can occur from some systemic medications. This is believed to be a hypersensitivity reaction. The drugs most often involved are ibuprofen and other NSAIDs, sulfa drugs and other antimicrobials, and the immune modulators. The immune modulators include cyclosporine, intravenous immune globulins, *ornithine ketoacid transaminase (OKT3) monoclonal antibodies*, and vaccines.

Chronic meningitis

The term *chronic meningitis* is used to describe meningeal inflammation that lasts for more than 1 month. It may be infectious or noninfectious and may be a form of viral meningitis.

Pathophysiology

Infectious causes include Lyme disease, fungal infections, AIDS, TB, ***Actinomyces*** infections, and syphilis. Noninfectious causes include sarcoidosis, **Behçet's syndrome**, vasculitis, and cancers such as leukemia, lymphomas, some carcinomas, melanomas, and gliomas. The most common causative gliomas include ependymoma, glioblastoma, and medulloblastoma. Chemical reactions to certain intrathecal injections can also be causative. Fungal meningitis has increased due to AIDS and the use of immunosuppressants. In AIDS, **Hodgkin lymphoma**, or lymphosarcoma, *Cryptococcus* species are usually causative. Also, the use of high-dose corticosteroids for a long time is implicated. Less common causes include *Candida*, *Coccidioides*, *Actinomyces*, *Aspergillus*, and *Histoplasma* species.

Clinical manifestations

Manifestations resemble acute meningitis but take weeks to develop, with fever sometimes being only slight. Common signs and symptoms include backache, headache, and cranial or spinal nerve root deficits. Hydrocephalus may develop, resulting in dementia. If intracranial pressure remains elevated, there may be days or weeks of decreased alertness, headache, or vomiting. Lack of treatment will result in death in just a few weeks or months, such as with TB or a tumor. Symptoms can continue for years, such as with Lyme disease.

Diagnosis

Diagnosis of chronic meningitis requires CT scan or MRI, and CSF analysis. Diagnosis is suspected if signs or symptoms develop over more than 2 weeks, with or without symptoms of cerebral dysfunction—especially if a potential cause, such as cancer or active TB exists. CT or MRI is performed to exclude mass lesions and to determine the safety of a lumbar puncture. CSF pressure is usually elevated but can be normal. Lymphocytes predominate in the CSF. Glucose is reduced but can be greatly decreased if there is a fungal infection or TB. CSF protein is high. Additional tests include fungal culture, acid-fast bacillus culture, and special staining methods. Fungi can be detected using wet mount microscopy or India ink preparations. CSF cryptococcal antigen is extremely specific and sensitive for the *Cryptococcus* species. Other tests are specific for syphilis, cancer, neurosarcoidosis, and other causes.

Treatment

The treatments of chronic meningitis are based on the cause. For meningitis resulting from Lyme disease, medications include doxycycline, cefuroxime, and amoxicillin. For fungal infections, amphotericin B, itraconazole, and fluconazole are used. For AIDS-related meningitis, amphotericin B, fluconazole, and rarely, surgery is used—the surgery involves placing a CSF shunt to relieve intracranial pressure. For tuberculous meningitis, rifampicin, isoniazid, pyrazinamide, and streptomycin are used. For *Actinomyces*-related meningitis, medications include tetracycline, clindamycin, and erythromycin. For syphilitic meningitis, intravenous antibiotics are administered, which may be followed by penicillin injections.

Section review
1. Which conditions often precede acute bacterial meningitis?
2. What are the infectious and noninfectious causes of viral meningitis?
3. What are the differences in the development of acute and chronic meningitis?

Chronic bacterial meningoencephalitis

Chronic bacterial meningoencephalitis is an infection of the meninges and brain. It may be caused by various species of *Mycobacterium tuberculosis*, *Treponema pallidum*, and *Borrelia*.

Tuberculosis

CNS tuberculosis, or *tuberculosis meningoencephalitis*, may be related to active disease in other areas of the body or can be isolated after seeding from silent lesions in the lungs or other areas. This form can involve the meninges or the brain. The meningoencephalitis is usually diffuse, with the subarachnoid space having a fibrinous or gelatinous exudate involving the base of the brain. This effaces the cisterns and encases the cranial nerves. White areas of inflammation may be scattered over the leptomeninges.

The involved areas contain mixed inflammatory infiltrates with lymphocytes, macrophages, and plasma cells. There may be well-formed granulomas with caseous necrosis and giant cells. There can be obliterative endarteritis and significant intimal thickening of the arteries that run through the subarachnoid space. Acid-fast staining often reveals the microorganisms present.

Infection may spread to the choroid plexus and ependymal surface via the CSF. Over long periods of time, a dense and fibrous adhesive arachnoiditis may develop, usually around the base of the brain, causing hydrocephalus. There may also be several rounded *Tuberculomas*, which are intraparenchymal masses linked to meningitis. Up to several centimeters in diameter, tuberculomas cause mass effects. The lesions usually have a center of caseous necrosis surrounded by granulomas. In inactive lesions, calcification may occur.

Tuberculosis meningoencephalitis causes headache, malaise, confusion, and vomiting. CSF often shows a pleocytosis consisting of mononuclear cells, or a mixture of these with neutrophils, an often highly elevated protein concentration, and a slightly reduce or even normal glucose. Serious complications include arachnoid fibrosis that causes hydrocephalus, and obliterative endarteritis, which causes arterial occlusion and brain infarction. If the spinal cord's subarachnoid space is involved, the nerve roots can be affected. It would cause symptoms similar to those of brain lesions, and must be distinguished from tumors of the CNS.

In AIDS patients, CNS tuberculosis is pathologically similar. However, there may be less host reaction than in patients with functioning immune systems. Those with HIV are also at risk for *Mycobacterium avium-intracellulare* infection, usually in disseminated infection settings. These lesions usually contain adjoining sheets of macrophages filled with microorganisms, and only a few granulomas, or none at all.

Neurosyphilis

Neurosyphilis is an outcome of the tertiary stage of syphilis. It only occurs in about 10% of patients with untreated infection. The primary patterns of CNS involvement are *meningovascular neurosyphilis*, *paretic neurosyphilis*, and *tabes dorsalis*. Effects can be incomplete or mixed, often involving tabes dorsalis along with paretic disease or *taboparesis*. HIV patients have increased risks for neurosyphilis, especially to *acute syphilitic meningitis* or *meningovascular disease*, due to impaired cell-mediated immunity. For the same reasons, this condition's progression and severity are both accelerated.

Meningovascular neurosyphilis involves the base of the brain, but it may affect the cerebral convexities and spinal leptomeninges. There may also be a related obliterative endarteritis called *Heubner arteritis*, along with a unique perivascular inflammatory reaction containing many lymphocytes and plasma cells. Plasma cell-rich mass lesions known as *cerebral gummas* may occur in the meninges, extending into the parenchyma.

Paretic neurosyphilis is usually insidious, with progressive cognitive impairment related to mood alterations such as delusions of grandeur, terminating in severe dementia (*general paresis of the insane*). Commonly, there is parenchymal damage of the cortex in the frontal lobe, but sometimes in other areas of the isocortex. There is a loss of neurons, gliosis, iron deposits, and proliferation of microglia (*rod cells*). Iron deposits can be shown via Prussian blue staining. These are believed to be the sequelae of small bleeding, due to microcirculation damage. The spirochetes can be seen in certain tissue sections.

Tabes dorsalis is due to damage to the sensory axons of the dorsal roots. It results in impaired joint position sense and ataxia, known as *locomotor ataxia*. There is also loss of pain sensation that leads to *Charcot joints*—damage to the skin and joints. Also, there can be other sensory disturbances, especially *lightning pains* and lack of deep tendon reflexes. Microscopic examination reveals loss of axons and myelin of the dorsal roots, with related pallor and atrophy in the dorsal spinal cord columns. Microorganisms cannot be seen in the spinal cord lesions.

Neuroborreliosis (lyme disease)

Lyme disease is caused by the spirochete called *Borrelia burgdorferi*. It is transmitted by various *Ixodes* tick species. When the nervous system is involved, this is called *neuroborreliosis*. There is much variation of the neurologic symptoms. These may include aseptic meningitis, facial nerve palsies and other polyneuropathies, and encephalopathy. Only rare cases have come to autopsy, which revealed focal proliferation of microglial cells in the brain and scattered extracellular microorganisms.

Hydrocephalus

Hydrocephalus is ventricular enlargement with excessive amounts of CSF. It is the most common cause of abnormal head enlargement in neonates. Intracranial pressure can greatly increase if hydrocephalus develops after the **fontanelles** have closed, but this does not increase head circumference.

Pathophysiology

Hydrocephalus is caused by either obstruction of CSF flow or impaired resorption of CSF. The rate of production of CSF is basically independent of blood pressure or intraventricular pressure. It continues to be produced even if its circulation is blocked or abnormal. When this happens, its pressure rises, and the ventricles expand, pushing against surrounding brain tissue. Therefore, hydrocephalus can be caused by excess CSF production, blockage of circulation, or deficient CSF reabsorption. Blockage of circulation is the overall most common cause.

A **papilloma** is a tumor of the choroid plexus and is sometimes causative. Sometimes, a papilloma directly causes a much larger production of CSF. An obstruction can occur anywhere in the CSF pathway. A tumor can occlude one or both interventricular foramina. The lateral ventricle that is involved becomes hydrocephalic, but the remainder of the ventricular system stays normal. Sometimes, pineal gland tumors push down on the midbrain. The aqueduct is squeezed shut, which causes hydrocephalus of both lateral ventricles and the third ventricle. Certain congenital abnormalities cause all three apertures of the fourth ventricle to be occluded or even fail to develop entirely. If this has occurred, there will be hydrocephalus of the total ventricular system.

Also, circulation may be obstructed in the subarachnoid space. In adults, this often occurs due to subarachnoid bleeding, clogging the arachnoid villi with red blood cells. Also, meningitis is sometimes followed by meningeal adhesions around the base of the brain. These block CSF flow through the tentorial notch, before it can reach the arachnoid villi of the superior sagittal sinus. Hydrocephalus of the entire ventricular system then develops. Though continuing defects in CSF reabsorption are not common, rarely, a congenital absence of the arachnoid villi is associated with hydrocephalus. *Congenital hydrocephalus* was discussed in Chapter 1. There may also be an obstruction of the superior sagittal sinus linked to the condition, probably because venous pressure rises enough to prevent CSF movement through the arachnoid villi.

Obstructive hydrocephalus, also called *noncommunicating hydrocephalus*, usually occurs in the *aqueduct of Sylvius*, but sometimes at the outlets of the fourth ventricle. The cerebral aqueduct, connecting the third and fourth ventricles, is often narrowed or has multiple channels that end without connecting to any other structures. One type of obstruction is due to *Dandy-Walker malformation*, a progressive cystic enlargement of the fourth ventricle. In this malformation, there is atresia of the *foramina of Luschka* or the *foramina of Magendie*. These normally allow emptying of the fourth ventricle into areas surrounding the brain. The defects cause ventricular flow of CSF into a "blind pouch." Another type of obstruction is due to *Chiari II type malformation*, which often accompanies spina bifida and **syringomyelia**. In this form, there is severe elongation of the cerebellar tonsils, causing them to protrude through the foramen magnum. The colliculi become beak-like in appearance, and the upper cervical spinal cord thickens. Additional obstructions may be due to brain tumors, cysts, arteriovenous malformations, trauma, blood clots, and infections.

Communicating hydrocephalus involves impaired resorption in the subarachnoid spaces. This usually occurs due to meningeal inflammation that is secondary to infection or to blood in the subarachnoid space. An example is a premature infant who has intraventricular hemorrhage.

According to the Honor Society of Nursing, about one in every 500 children born in the United States has hydrocephalus. About half of these cases involve congenital hydrocephalus—meaning that about one in every 1000 US births involves this form. This is equivalent to a one-tenth of 1% chance of a child being born with congenital hydrocephalus in the United States. This is about as common as Down's syndrome but more common than spina bifida or brain tumors.

Clinical manifestations

Signs and symptoms are based on increased intracranial pressure, or if this is not increased. In infants, higher pressure causes a high-pitched cry, irritability, lethargy, vomiting, bulging fontanelles, and strabismus. Older children who can speak may complain of decreased vision, headache, or both. A late sign of increased pressure is papilledema, but if this is initially absent, is still not a positive sign. Chronic hydrocephalus may cause a variety of outcomes. These include learning disorders such as difficulties with attention, information processing, and memory; precocious puberty (in females), and problems with abstracting, conceptualizing, generalizing, reasoning, and in the organization and planning of information in order to solve problems.

Diagnosis

Diagnosis of hydrocephalus includes prenatal ultrasonography, cranial ultrasonography for neonates, and CT scan or MRI in older infants and children. Diagnosis is suspected if a routine exam reveals increased head circumference, bulging fontanelles, or wide-spaced cranial sutures. Subdural hematoma, tumors, and **porencephalic cysts** must be considered. Macrocephaly may be due to **Alexander disease** or **Canavan disease** or can be benign, due to excessive CSF surrounding the brain. Cranial CT or ultrasonography monitors the progression of hydrocephalus once an anatomic diagnosis has been done. An electroencephalogram may be helpful if seizures occur.

Treatment

Treatment usually involves ventricular shunting but is based on etiology, severity, and progression of hydrocephalus. Ventricular taps or serial lumbar punctures can temporarily reduce CSF pressure in infants. Shunting is usually done for progressive hydrocephalus, which usually connects the right lateral ventricle to the peritoneal cavity. The first placement of a shunt in an infant or older child with closed fontanelles causes rapid fluid withdrawal, and possible subdural bleeding, as the brain is shrinking away from the skull. If the fontanelles are open, the skull can decrease its circumference and match the decrease in brain size. It is preferred to place a shunt before the fontanelles close.

Another treatment is the creation of an opening, via endoscope, between the third ventricle and the subarachnoid space. Ablation of the choroid plexus is often also performed. A ventricular shunt to the subgaleal space may be used for infants as a temporary procedure if a more permanent shunt is not needed. Shuts are rarely removed due to risks of bleeding and trauma. **Ventriculoperitoneal** shunts offer less complications than **ventriculoatrial** shunts. Complications can include infection and malfunction, which usually involves blockage at the ventricular end, or fracture of the tube. If intracranial pressure suddenly increases, regardless of the cause, this can be a medical emergency. Once a shunt is placed, head circumference and development are assessed, along with periodic imaging studies.

Meningioma

A **meningioma** is a benign tumor of the meninges that may compress nearby brain tissue. It is one of the most common intracranial tumors, making up almost 11% of all brain tumors. Meningioma is one of the few brain tumors that is more common in females than males. It usually occurs between ages 40 and 60, but can occur in childhood.

Pathophysiology

The tumor may develop in any dura, but usually over the convexities near the venous sinuses, along the base of the skull, in the posterior fossa. They rarely develop within the ventricles. It is possible for multiple meningiomas to develop. They compress, but do not invade brain parenchyma, and may invade and distort adjacent bones. Though histologic types vary, they all develop similarly, sometimes becoming malignant.

According to the American Society of Clinical Oncology and the Brain Science Foundation, meningiomas account for approximately 34% of all primary brain tumors. Malignant meningiomas account for between 2% and 3% of all meningiomas. The annual incidence is six cases per 100,000, with a peak incidence in the sixth and seventh decades. Meningiomas are more common in women than in men. The tumors are attached to the dura mater, and arise within the intracranial cavity, the spinal cavity, or rarely, one of the orbits (see Fig. 4.4).

Clinical manifestations

Signs and symptoms are based on which area of the brain is compressed by the tumor. In the elderly, midline tumors can cause dementia without any focal neurologic problems.

Diagnosis

The diagnosis of meningioma is by MRI, using a paramagnetic contrast agent. With CT scans or plain X-rays, bone abnormalities may be seen. These include brain atrophy,

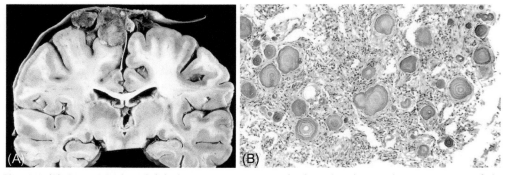

Fig. 4.4 (A) Parasagittal multilobular meningioma attached to the dura with compression of the underlying brain. (B) Meningioma with a whorled pattern of cell growth and psammoma bodies.

changes in the tuberculum sellae, and hyperostosis around the cerebral convexities. Meningioma is easily identified by homogenous contrast enhancement, which is often accompanied by a tail-like structure called the *dural tail*. Definite diagnosis of all brain lesions requires biopsy.

Treatment

For symptomatic or enlarging meningiomas, excision or radiation therapy is used. Serial neuroimaging is usually sufficient to monitor asymptomatic small meningiomas, especially in older adults. If possible, excision is preferred. If the meningioma is large, encroach on blood vessels (usually, the surrounding veins), or are very near the brainstem or other critical brain areas, surgery may cause more damage than that caused by the tumor. Surgery is then deferred. **Stereotactic radiosurgery** is performed for surgically inaccessible meningiomas, and electively for others. It is sometimes used when there is remaining tumor tissue following surgical excision, or when the patient is elderly. Radiation therapy may be useful if stereotactic radiosurgery is impossible, or if a meningioma recurs.

Section review
1. What are the two primary causes of hydrocephalus?
2. What are the primary causes of subarachnoid hemorrhage?
3. In which types of patients are meningioma most common?

Clinical cases

Clinical case 1
1. Based on the clinical case study below, what is the likely diagnosis?
2. What should the emergency department physician do to diagnose this patient?
3. If the patient had bacterial meningitis, what treatments should be given?

A 7-year-old boy awoke in the middle of the night. He complained to his parents about a bad headache, feeling nauseous, and having a sore neck. His mother took his temperature, which was 103.5°F. While preparing to go to the hospital, he vomited twice.

Answers:
1. The likely diagnosis is meningitis, but of what cause remains unknown.
2. Diagnosis of acute meningitis consists of CSF analysis, imaging tests, peripheral blood tests, serum sodium monitoring, and culturing of secretions and lesions.
3. As soon as acute bacterial meningitis, especially meningococcal, is suspected, it must be treated due to its rapid and potentially fatal progression. Blood is drawn and cultured. Gram staining and bacterial DNA PCR is done if available. Prompt lumbar puncture is

required. However, immediate treatment with corticosteroids and antibiotics should occur before these measures if possible. Treatment for meningitis basically involves corticosteroids and antibiotics. When acute bacterial meningitis is suspected, these are given as soon as possible.

Clinical case 2

1. What are the additional symptoms of Lyme disease?
2. How can physicians tell which of the three stages of Lyme disease are present?
3. What are the treatments for Lyme disease?

A 38-year-old male was assessed for pain and a burning sensation in his left foot, headaches, knee and joint pain, floaters across his visual field, neck stiffness, and periods of cognitive difficulties. He has been on treatments for rheumatoid arthritis. MRIs were negative for any demyelinating brain lesions. The patient regularly enjoys hiking through the woods near his house when he feels able to do so and has been bitten by several ticks. He is diagnosed with Lyme disease.

Answers:

1. Without treatment, Lyme disease can also cause rashes, drooping of one or both sides of the face, heart palpitations or an irregular heartbeat, and inflammation of the brain and spinal cord.
2. In early localized Lyme disease, there are fever, chills, headache, swollen lymph nodes, sore throat, and usually a rash with a "bull's-eye" appearance. In early disseminated Lyme disease, there is pain, weakness or numbness in the arms and legs, vision changes, heart palpitations, chest pain, rash, and facial paralysis. In late disseminated Lyme disease—weeks, months, or years after the tick bite, there is arthritis, severe fatigue, headache, vertigo, sleep disturbances, and mental confusion.
3. Antibiotics are usually for early-stage infections, usually with doxycycline, amoxicillin, or cefuroxime. For early disseminated infections, oral antibiotics are recommended, but IV antibiotics are used if there is meningitis or more severe heart problems. Both forms of antibiotics are used for late-stage Lyme disease. Patients with lingering arthritis received standard arthritis treatment. There is no treatment for *posttreatment Lyme disease syndrome*, and 10% of patients do not respond to antibiotics. As a general rule, any involvement of the CNS or peripheral nervous system (PNS) warrants a ceftriaxone treatment, opposed to the usual treatment with doxycycline.

Clinical case 3

1. What condition is signified by the symptoms listed in this case study?
2. What other conditions may be linked to this clinical case?
3. What are the appropriate tests and treatments?

An 8-month-old infant was brought to her pediatrician with delayed overall development and an enlarging head circumference. Her CSF pressure was elevated, and she had a bulging anterior fontanelle. Her mother complained that the infant was irritable, had a high-pitched cry, and had vomited several times.

Answers:
1. The likely diagnosis is hydrocephalus, though further diagnosis as being obstructive or nonobstructive hydrocephalus would require further testing.
2. Conditions that may be linked can be congenital or acquired and include Dandy-Walker malformation, Chiari II type malformation, spina bifida, syringomyelia, infection, and blood in the subarachnoid space.
3. Diagnosis of hydrocephalus includes prenatal ultrasonography, cranial ultrasonography for neonates, and CT scan or MRI in older infants and children. Treatment usually involves ventricular shunting but is based on etiology, severity, and progression of hydrocephalus. Ventricular taps or serial lumbar punctures can temporarily reduce CSF pressure in infants.

Clinical case 4
1. What are the most common causes of viral meningitis?
2. What are the diagnostic methods used for this condition?
3. What are the treatment options?

A 16-year-old high school student developed a severe headache, stiff neck, and nausea. His school principal called the boy's mother, and she took her son to the emergency room. After physical examination, the boy had a low-grade fever, said that the bright lights in the emergency room hurt his eyes, and seemed confused about where he was and what was happening. The emergency physician suspected meningitis and ordered a variety of tests to determine which type was involved. After these procedures, he diagnosed the boy with viral meningitis.

Answers:
1. In most cases this form of meningitis is caused by enteroviruses, which include coxsackievirus, echovirus, and enteroviruses 68–71.
2. Tests to determine the form of meningitis that exists include blood tests, CT scan of the brain, MRI of the brain or spinal cord, and most importantly, lumbar puncture.
3. The goal of treatment for viral meningitis is primarily supportive care, since viral infections do not respond to antibiotics, and most do not respond to specific antivirals. Analgesics, antipyretics, and hydration are supplied as needed. The exception is HSV and severe VZV infections, which are viral infections that require treatment with IV acyclovir.

Key terms

Alexander disease
Ambient cistern
Anterior cranial fossa
Arachnoid granulations
Arachnoid mater
Arachnoid trabeculae
Asplenia
Auriculotemporal nerve roots
Behçet's syndrome
Calvarial dura
Calvarium
Canavan disease
Cauda equina
Cerebellomedullary cistern
Cerebral aqueduct (of Sylvius)
Cerebrospinal fluid (CSF)
Choroid epithelium
Choroid fissure
Choroid plexus
Chronic bacterial meningoencephalitis
Clivius
Coccygeal ligament
Communicating hydrocephalus
Confluence of sinuses
Cranial meninges
Creutzfeldt-Jacob disease
Crista galli
Denticulate ligaments
Diaphragm sellae
Diploic
Dretrecogin alfa
Dura mater
Dural folds
Dural venous sinuses
Ehrlichiosis
Epidural space
Extradural hemorrhage
Filum terminale
Fourth ventricle
Fulminant
Glomus
Hydrocephalus
Hygroma
Infratentorial
Interpeduncular cistern
Interthalamic adhesion
Interventricular foramina (of Monro)
Lamina terminalis
Lateral ventricle
Meninges
Meningismus
Middle cranial fossa
Middle meningeal artery
Nervus spinosus
Nuchal rigidity
Obstructive hydrocephalus
Papilloma
Parenchyma
Perivascular space
Pia mater
Pleocytosis
Polymerase chain reaction (PCR)
Pontine cistern
Posterior cranial fossa
Potential spaces
Pterion
Recesses
Septum pellucidum
Seroconversion
Stylopharyngeus
Subarachnoid cisterns
Subarachnoid space
Subdural space
Superior cistern
Supernatant
Supratentorial
Syringomyelia
Tentorial nerve
Third ventricle
Transverse cerebral fissure
Transverse sinus
Trigeminovascular system
Waterhouse-Friderichsen syndrome
Xanthochromia

Suggested readings

1. Ammar, A. *Hydrocephalus: What Do We Know? And What Do We Still Not Know?* Springer, 2017.
2. Connolly, E. S.; McKhann, G. M. *Subdural Hematomas, An Issue of Neurosurgery Clinics of North America*, 28th ed.; Elsevier, 2017.
3. Deisenhammer, F.; Sellbjerg, F.; Teunissen, C. E.; Tumani, H. *Cerebrospinal Fluid in Clinical Neurology*. Springer, 2015.
4. DeMonte, F.; McDermott, M. W.; Al-Mefty, O. *Al-Mefty's Meningiomas*, 2nd ed.; Thieme, 2011.
5. Domino, F. J.; et al. *The 5-Minute Clinical Consult*, 18th ed.; Wolters Kluwer/Lippincott Williams & Wilkins, 2010.
6. https://www.earthslab.com/anatomy/dura-mater/.
7. Ganz, J. C. *Intracranial Epidural Bleeding—History, Management, and Pathophysiology*. Academic Press, 2017.
8. Hasbun, R. *Meningitis and Encephalitis—Management and Prevention Challenges*. Springer, 2018.
9. Irani, D. N. *Cerebrospinal Fluid in Clinical Practice*. Saunders, 2008.
10. https://www.merckmanuals.com/home/injuries-and-poisoning/head-injuries/intracranial-hematomas#v740117.
11. Meyers, S. P. *Differential Diagnosis in Neuroimaging: Brain and Meninges*. Thieme, 2016.
12. https://radiopaedia.org/articles/blood-supply-of-the-meninges.
13. https://radiopaedia.org/articles/dural-venous-sinuses.
14. https://radiopaedia.org/articles/innervation-of-the-meninges.
15. Rea, P. *Essential Clinical Anatomy of the Nervous System*. Academic Press, 2015.
16. Rinkel, G. J. E.; Greebe, P. *Subarachnoid Hemorrhage in Clinical Practice*. Springer, 2015.
17. https://www.sciencedirect.com/topics/neuroscience/meninges.
18. https://www.sciencedirect.com/topics/neuroscience/ventricular-system.
19. Watson, C.; Paxinos, G.; Kayalioglu, G. *The Spinal Cord: A Christopher and Dana Reeve Foundation Atlas*. Academic Press, 2008.

CHAPTER 5

Blood supply of the CNS

Human brain requires well-oxygenated blood supply on a continuous basis. If the blood supply to the brain stops, as in an ischemic stroke, two million neurons die every minute. The central nervous system (CNS) blood vessels, mostly in the gray matter, are densely arranged, forming a mesh-like structure. It is important to understand the blood supply to the brain, which creates further understanding about normal functions as well as the outcomes of cerebrovascular diseases.

The brain continuously demands oxygen and nutrients (mainly glucose), and has no energy reserves such as lipids or carbohydrates. Neurons cannot store oxygen. Because of this, the brain requires an extensive blood vessel network to meet its needs. Arterial blood comes to the brain via the *internal carotid arteries* and *vertebral arteries*. Most venous blood drains from the brain via the *internal jugular veins*, which drain the *venous dural sinuses*.

Arteries of the brain

Two pairs of arteries supply the brain and much of the spinal cord. These are the **internal carotid arteries** and the **vertebral arteries**. The internal carotid arteries, also known as anterior circulation, supply the anterior parts of the brain. They supply the majority of the telencephalon, and much of the diencephalon. The vertebral system, also known as posterior circulation, supplies the posterior parts of the brain. The vertebral arteries supply the brainstem, cerebellum, and parts of the diencephalon, spinal cord, and the occipital and temporal lobes of the brain.

Carotid arteries

On each side of the neck, an internal carotid artery ascends. It runs through the petrous bone, through the cavernous sinus, reaching the subarachnoid space at the base of the brain (see Fig. 5.1). After leaving the cavernous sinus, the internal carotid artery branches into the **ophthalmic artery**. This artery is the first intracranial branch of the internal carotid, and runs along the optic nerve to the orbit, supplying the eye, other structures in the orbit.

The internal carotid artery continues superiorly, along the optic chiasm. It splits into the **middle cerebral artery** and the **anterior cerebral artery**. Just prior to this split, it branches into the **anterior choroidal artery** and **posterior communicating artery**.

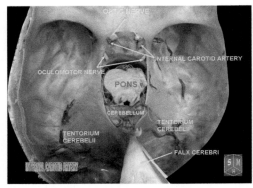 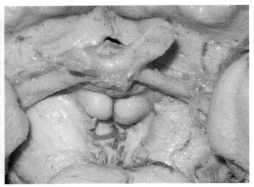

Fig. 5.1 Location of the internal carotid artery.

The anterior choroidal artery is along and thin. It is clinically important because it supplies many different structures.

The anterior choroidal artery supplies the optic tract and the choroid plexus of the inferior horn of the lateral ventricle. It also supplies deep brain structures, including parts of the internal capsule, and hippocampus. The posterior communicating artery runs posteriorly, but is inferior to the optic tract, toward the cerebral peduncle. It joins the **posterior cerebral artery (PCA)**, which is part of the vertebral artery system. Therefore, the posterior communicating artery connects the anterior circulation to the posterior circulation.

The anterior cerebral artery runs medially. It is superior to the optic nerve, entering the longitudinal fissure. The **anterior communicating artery** connects the two anterior cerebral arteries near the area where they enter the longitudinal fissure. The anterior cerebral arteries arch posteriorly and follow the corpus callosum. They supply the medial areas of the frontal and parietal lobes. Some smaller branches run onto the posterolateral surface of each hemisphere. Distal to the anterior communicating artery, the anterior cerebral artery becomes the **pericallosal artery**, which stays close to the corpus callosum. Near the corpus callosum's genu, the **callosomarginal artery** usually branches off from the pericallosal artery. It follows the cingulate sulcus.

Areas of the precentral and postcentral gyri extend superiorly, to the medial surface of the frontal and parietal lobes. Any occlusion of an anterior cerebral artery will cause restricted, contralateral somatosensory deficits (precentral gyrus) and motor deficits (postcentral gyrus). This will affect the leg more than the arm, due to the somatotropic structure known as **homunculus**.

The middle cerebral artery is large, and runs laterally into the lateral sulcus (Sylvian fissure). It divides into many branches supplying the insula, emerges from the lateral sulcus, and spreads, supplying the majority of the lateral surface of the cerebral hemisphere.

Most precentral and postcentral gyri are in this supply area. Therefore, occlusion of a middle cerebral artery will cause significant somatosensory and motor deficits. Also, when the left hemisphere is involved, language deficits nearly always occur. The left hemisphere is the dominant hemisphere in 90% of the population.

The middle cerebral artery branches to as many as 12 smaller branches, penetrating the brain near their origin. They supply the deep structures of the diencephalon and telencephalon. These are called the **lenticulostriate arteries**. Similar small branches are called the **perforating arteries**, also known as the *ganglionic arteries*. They arise from all arteries located around the base of the brain.

The perforated arteries are narrow and thin-walled vessels (see Fig. 5.2). This area can be involved in stroke. Because of the deep cerebral structures they supply, damage in these vessels causes neurological deficits that seem extensive in comparison to the vessel size. Since the somatosensory projection from the thalamus, to the postcentral gyrus, passes through the anterior parts of the internal capsule, damage to just a small area of the internal capsule (from rupture or occlusion of a perforating artery) results in deficits that are similar to those due to large areas of cortical damage.

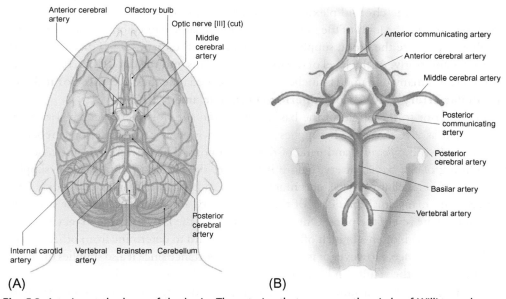

Fig. 5.2 Arteries at the base of the brain. The arteries that compose the circle of Willis are the two anterior cerebral arteries, joined to each other by the anterior communicating artery and two short segments of the internal carotids, off of which the posterior communicating arteries connect to the posterior cerebral arteries. ((A) from Moses, K. P., et al. Atlas of clinical gross anatomy, 2nd ed., Saunders: Philadelphia, 2013; (B) from Hagen-Ansert, S. Textbook of diagnostic sonography, 7th ed., St Louis: Mosby, 2012.)

Vertebral and basilar arteries

The vertebral arteries arise from the subclavian arteries, at the root of the neck. They branch to the vertebrae, cervical spinal cord, and certain deep neck structures, and join between the medulla and pons, to form the **basilar artery**. This artery continues along the ventral surface of the midline pons.

The two vertebral arteries run along the medulla oblongata. Prior to joining the basilar artery, each vertebral artery branches into the following:

- **Posterior spinal artery**—running caudally, along the posterolateral aspect of the spinal cord. It supplies the posterior third of the spinal cord.
- **Anterior spinal artery**—actually two arteries that join together, to form a single artery running caudally, along the anterior midline of the spinal cord. It supplies the anterior two-thirds of the spinal cord. Small in size, not enough blood is carried from the vertebral arteries to supply any more than the cervical areas of the spinal cord. Therefore, the artery must be refilled at various caudal points by the branches of the aorta.
- **Posterior inferior cerebellar artery**—abbreviated as *PICA*, it supplies much of the inferior cerebellar hemisphere. The PICA branches to other structures as it progresses toward the cerebellum. Curving around the brainstem, it supplies much of the lateral medulla, and the choroid plexus of the fourth ventricle.

By understanding the level of the brainstem where the large named branches emerge, it is possible to estimate the blood supply of each region of the brainstem. The basilar artery continues anteriorly. At the level of the midbrain, it splits into the two PCAs. Prior to this split, it creates many unnamed branches, but two named branches: the **anterior inferior cerebellar artery** (AICA) and **superior cerebellar artery (SCA).**

Just anterior to the origin of the basilar artery, the anterior inferior cerebellar artery, abbreviated as *AICA*, arises. It supplies more anterior areas of the inferior cerebellum surface, such as the *floccules*, and parts of the caudal pons. The SCA begins just caudal to the splitting of the basilar artery, before it becomes the left and right PCAs. It supplies the superior cerebellum surface, and much of the caudal midbrain and lateral pons. Collectively, the many small branches of the basilar artery are called the **pontine arteries**. They supply the rest of the pons. One of these arteries—the **labyrinthine artery** (or *internal auditory artery*) is often a branch off the AICA. It is hard to determine from other arteries just by its appearance. However, it is of functional importance since it also supplies the inner ear. If occluded, the result can be vertigo and ipsilateral (same side) deafness.

The PCA is curved around the midbrain. It passes through the superior cistern, with branches spreading to supply the medial and inferior surfaces of the occipital and temporal lobes. It also branches to the anterior midbrain and posterior areas of the diencephalon, including the thalamus. Exclusively the posterior circulation supplies the thalamus. It additionally branches to several **posterior choroidal arteries**. These supply the choroid plexus of the third ventricle as well as the body of the lateral ventricle. The anterior and posterior choroidal arteries anastomose near the glomus. Since the primary visual cortex is within the occipital lobe, any occlusion of a PCA, at its origin, causes bilateral visual field losses along with other deficits related to the midbrain and diencephalon.

Cerebral arterial circle

The **cerebral arterial circle** is shaped like a polygon, and is also known as the *circle of Willis*. Within this circle, the PCA is connected to the internal carotid artery via the posterior communicating artery. The anterior cerebral, internal carotid, and PCAs, of both sides, are interconnected here. The single anterior communicating artery is actually a short anastomosis, between the two anterior cerebral arteries. The arterial pressure in the internal carotid arteries is very similar to the pressure in the PCAs. Therefore, in normal circumstances, only small amounts of blood flow through the posterior communicating arteries.

If one vessel becomes occluded—within the cerebral arterial circle, or even near it—the communicating arteries may provide vital anastomotic blood flow, preventing neurological damage. This means it is possible, but not likely, for all of the brain to perfuse by a single artery out of the four major arteries normally supplying its blood flow. There are large variances in the sizes of the anterior and posterior communicating arteries. This means that establishing effective anastomotic flow after an arterial occlusion is also dependent on the extent of time that the occlusion has been present. Small communicating arteries are able to slowly enlarge, compensating for slowly developing occlusions. These newly formed arteries are also called *collaterals*. Only 30% of the population has a complete circle of Willis.

Asymmetries of the cerebral arterial circle often occur. Any one or several of the communicating arteries can be extremely small (*hypoplastic*) or even absent congenitally. At its origin, one anterior cerebral artery can be very small compared to the other. One PCA can continue to have its origin during embryonic development from the internal carotid artery, and be connected, via a posterior communicating artery, to the basilar artery (also known as fetal PCA). Such asymmetries lead to related asymmetries in how the brain is supplied with blood.

There are other collateral circulation routes, but the cerebral arterial circle is usually the most important. Sometimes there are anastomoses at the arteriolar and capillary levels. These occur between the terminal branches of the cerebral arteries. However, they are usually not able to maintain all of a major, adult cerebral artery should it become occluded. Sometimes they are able to maintain a large portion of the artery's blood flow. Also, competent arterial anastomoses can greatly enlarge, compensating for occlusions that develop slowly. Cases have been documented in which the PCA was supplied via the internal carotid artery on the same side by blood flow through the anterior choroidal artery. From there, additional supply was through one posterior choroidal artery, into the PCA.

There are also a small number of intracranial-extracranial anastomoses that are able to enlarge enough to be functionally competent. Most critical are anastomoses in the orbit between the ophthalmic artery and branches of the external carotid artery. Should an internal carotid artery become occluded, blood from the external carotid artery is able to flow backward (retrograde), reaching the internal carotid regions via the ophthalmic artery.

Meningeal arteries

The **meningeal arteries** mostly supply the outer layer of the dura mater, and not its inner layer. Recall that dura has two layers. Also, the arachnoid mater and pia mater do not require a large blood supply. Several arteries supply the dura mater in the anterior, middle, and posterior **cranial fossae**. These are divided as follows:

- **Anterior cranial fossa**—the meningeal branches of the *anterior ethmoidal artery, posterior ethmoidal artery,* and *ophthalmic artery*; also, the frontal branch of the *middle meningeal artery*. This artery enters the middle cranial fossa via the *foramen spinosum*.
- **Middle cranial fossa**—the frontal and parietal branches of the *middle meningeal artery;* the *accessory meningeal artery*; the *ascending pharyngeal artery*; and branches directly from the *internal carotid artery*. The supratentorial dura mater is mostly supplied by the middle meningeal artery, which is a branch of the *maxillary artery*. The middle meningeal artery actually mostly supplies the *calvarium* rather than the meninges. This artery is clinically important because of its location in the *extradural space,* and the position of its anterior portion to the *pterion,* which makes it susceptible to damage due to head injures, leading to *epidural hematoma*.
- **Posterior cranial fossa**—the *vertebral arteries, occipital arteries,* and *ascending pharyngeal arteries*.

Section review
1. Which paired artery supplies most of the tissues of the head, except for the brain and orbits?
2. Name the arterial anastomosis at the base of the cerebrum.
3. Name each of the vertebral artery branches.

Veins of the brain

The major venous drainage of the brain occurs through a system of cerebral veins. They empty into the venous dural sinuses and finally, into the **internal jugular veins**. Similar to cerebral arteries, these veins can be visualized by angiographic imaging. A collection of

emissary veins connects the extracranial veins and dural sinuses. They have only minor effects on the brain's normal pattern of circulation. However, they are clinically important as a path for infection to spread into the cranial cavity.

Basically, the cerebral veins are divided into **superficial veins** and **deep veins**. Each of these types of veins are organized into systems. The superficial veins usually lie on the surfaces of the cerebral hemispheres. Most of them empty into the superior sagittal sinus. Deep veins usually drain internal structures, eventually emptying into the straight sinus. The **basal vein (of Rosenthal)** drains various cortical areas. However, it is still categorized as a deep vein because it also drains certain deep structures, and eventually empties into the straight sinus. Cerebral veins lack valves. Unlike cerebral arteries, they are interconnected by many functioning anastomoses, both within one group of veins, and between superficial and deep vein groups.

Superficial veins

There are many variations in superficial veins. They consist of a superior group, which empties into the superior and inferior sagittal sinuses, as well as an inferior group, which empties into the transverse and cavernous sinuses. Between different human brains, only three veins are usually constant. These include the following:

- **Superficial middle cerebral vein**—running anteriorly and inferiorly, along the lateral sulcus. It drains the majority of the temporal lobe into the cavernous sinus, or into the proximal sphenoparietal sinus.
- **Superior anastomotic vein (of Trolard)**—usually running across the parietal lobe, connecting the superficial middle cerebral vein and superior sagittal sinus.
- **Inferior anastomotic vein (of Labbé)**—running posteriorly and inferiorly across the temporal lobe, connecting the superficial middle cerebral vein with the **transverse sinus**.

Deep veins

Compared to the superficial veins, the deep veins have a more constant appearance. Existing deep in the brain, in areas where arteries are small, they are used as clinically important radiological landmarks. The primary deep vein is the **internal cerebral vein**, formed at the interventricular foramen.

Formation of the internal cerebral vein is via the joining of two smaller veins. These are the **septal vein**, which runs posteriorly across the septum pellucidum, and the **thalamostriate vein**, also called the *terminal vein*. This vein runs in the groove between the thalamus and caudate nucleus, draining a large amount of blood from these structures. Close to the interventricular foramen, the thalamostriate vein is joined by the **choroidal vein**. This is a tortuous vessel, draining the choroid plexus of the body of the lateral ventricle.

Just after it forms, the internal cerebral vein has a sharp posterior bend. This bend is the **venous angle**, which is used in imaging evaluations to reveal the site of the interventricular foramen (see Fig. 5.3). The two internal cerebral veins run posteriorly through the transverse cerebral fissure. They join in the superior cistern, forming the unpaired **great**

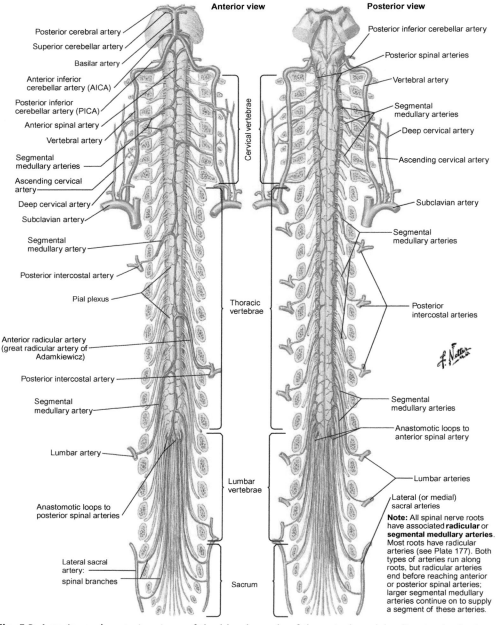

Fig. 5.3 Anterior and posterior views of the blood supply of the spinal cord, leading to the brain.

cerebral vein or the vein of *Galen*. This vein bends superiorly, joining the inferior sagittal sinus, forming the straight sinus. It has a short course, and is joined by the basal veins. On each side, the basal vein forms near the optic chiasm by the **deep middle cerebral vein**. This vein drains the insula. Several tributaries drain inferior areas of the basal ganglia as well as the orbital surface of the frontal lobe. The deep middle cerebral vein continues along the medial surface of the temporal lobe. It curves around the cerebral peduncle, entering the great vein.

Along with the superficial and deep veins, there is a separate and complicated collection of veins serving the cerebellum and brainstem. They drain into the great vein as well as the straight, transverse, and petrosal sinuses. There are nowhere near as many vascular problems related to the venous system as to the arterial system. Partially, this is true since occlusions or thrombosis occur less often in the venous system. However, hemorrhage is more common than ischemia in the venous system. This is also true regarding seizures caused by the occlusion of the veins, which is more common than arterial strokes. Also, veins have a great amount of functional anastomoses. Therefore, an occlusion that slowly develops in the anterior region of the superior sagittal sinus most likely would have no clinical manifestations. Even if an occlusion developed quickly, the only symptom might be a transient headache without any ischemic changes. The occlusion of veins in the venous system is called *venous sinus thrombosis*.

It should be noted that an occlusion in a more critical area, such as the posterior region of the superior sagittal sinus, is able to cause more serious problems. Clinical manifestations may include headache because of increased intracranial pressure, motor problems, and seizures.

Venous dural sinuses

The blood circulating in the brain is collected in the large, thin-walled **venous dural sinuses**. These are blood-filled spaces, between the dura mater layers. The dura mater has a periosteal layer, against the cranial bone, and a meningeal layer, against the brain. In some places, there is a space between these layers, which accommodates a sinus that collects blood. A vertical, sickle-shaped wall of dura called the *falx cerebri* is found between the two cerebral hemispheres. It contains two of the sinuses. Totally, there are about 13 venous dural sinuses. The most important sinuses are as follows:
- **Superior sagittal sinus**—within the superior margin of the falx cerebri; it overlies the brain's longitudinal fissure, beginning anteriorly near the crista galli and extending posteriorly to the level of the posterior occipital protuberance; it then bends, usually to the right, draining into one of the transverse sinuses. It is worth mentioning that the cerebrospinal fluid (CSF) is also reabsorbed in this area by the structures called *arachnoid granulations* (see Chapter 1).
- **Inferior sagittal sinus**—within the inferior margin of the falx cerebri, arching over the corpus callosum, and deep in the longitudinal fissure; it posteriorly joins the *great*

cerebral vein to form the *straight sinus*; this continues posteriorly where the superior sagittal and straight sinuses meet in a space known as the *confluence of the sinuses*.
- Right and left *transverse sinuses*—move from the confluence to encircle the inside of the occipital bone; they then lead to the ears, with their path indicated by grooves on the inner occipital bone surface; the right transvers sinus mostly receives blood from the superior sagittal sinus; the left transverse sinus mostly drains the straight sinus; each transverse sinus, laterally, take an S-shaped bend called the *sigmoid sinus*, and exits the cranium via the jugular foramen; blood then flows down the internal jugular vein emptying into the heart via the superior vena cava.
- **Cavernous sinuses**—blood-filled spaces on each side of the body of the sphenoid bone; they resemble *honeycombs*.

Cavernous sinus

The cavernous sinuses receive blood from the *superior ophthalmic vein* of the orbit, along with the *superficial middle cerebral vein* of the brain, and other sources. They drain through the transverse sinus, internal jugular vein, and facial vein. The cavernous sinuses are clinically important since infections can move from superficial sites, such as the face, into the cranial cavity via this pathway. If there is inflammation of a cavernous sinus, the structures that pass through it can be injured, including the internal carotid artery as well as cranial nerves III, IV, the ophthalmic branch of cranial nerve V, and the maxillary branches of cranial nerves V and VI. Remember that the mandibular branch of cranial nerve V does not pass through the cavernous sinus.

Section review
1. What are the three names of the superficial veins of the brain?
2. Which veins drain the dural venous sinus and where do these veins terminate?
3. Name the most important venous dural sinuses.
4. What arteries and nerves pass through the cavernous sinus?

Arteries of the spinal cord

The *vertebral arteries* are the primary source of blood to the spinal cord, as well as to the posterior part of the brain. These arteries merge anteriorly, forming a single artery called the *basilar artery of the circulus arteriosus (circle of Willis)*. The following arteries branch from the vertebral arteries, directly supply blood to the spinal cord:
- One *anterior spinal artery*—direct branch of the vertebral arteries.
- Two *posterior spinal arteries*—direct branches of the vertebral arteries.
- The *anterior radicular artery* and *posterior radicular artery*—these arteries originate from spinal branches of the vertebral arteries, and spinal branches of the ascending cervical arteries, deep cervical arteries, intercostal arteries, lumbar arteries, and sacral arteries.

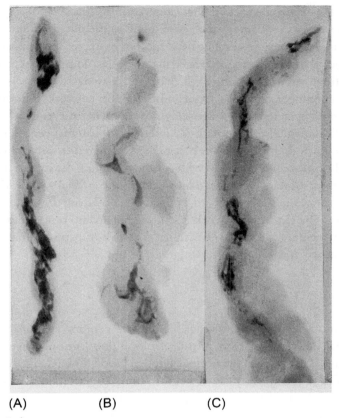

Fig. 5.4 Spinal arteries.

- The *artery of Adamkiewicz* is the largest radicular artery that is usually branched from the aorta at the level of the T11 vertebra.
- The *arterial vasocorona*, which is an anastomose between the spinal arteries.

The spinal arteries are shown in Fig. 5.4.

Venous drainage of the spinal cord

The spinal cord's venous system includes interior, small veins. They either drain into the posterior spinal vein or the anterior spinal vein, which drains into the **azygos system**. This is a vein that runs up the right side of the thoracic vertebral column. There are four basic components to the spinal cord's venous system. These are as follows:
- **Intradural veins**—extramedullary and intramedullary veins
- **Extradural veins**—epidural veins
- **Intraosseous veins**—within the marrow of the spinal bones
- **Paraspinal veins**—those adjacent to the spinal column

Classical names for these veins include the *anterior external, anterior internal, posterior internal,* and *posterior external venous plexuses*, which correspond to the intradural and extradural networks. The spinal veins are a capacitance network. For example, approximately 75% of the intracranial blood, at any given time, is located in these veins. The weakest part of this system involves the **radiculomedullary veins**, which bridge the two systems across the dura mater. They are fewer in number, and have no secondary support network. If thrombosis of these veins occurs, venous hypertension results—this is common in patients who have spinal dural fistulas. The radiculomedullary veins link the internal and external venous plexuses.

The spinal cord is drained by veins that extend to the cord surface. These intramedullary or radial veins are not normally visible. Once reaching the surface, the spinal veins are networked with more distribution than is seen in the arterial system. Extramedullary intradural veins run along the spinal cord, collecting blood from the intramedullary radial veins and others. The extrinsic network is made up of an anterior epidural plexus and a smaller posterior epidural plexus. A functional network of paravertebral veins and epidural plexi is known as the *Batson venous plexus*.

Clinical considerations

The most common causes of neurological deficits are cerebrovascular disease and accidents. In normal individuals, about 55 mL of blood flows through every 100 g of the CNS, per minute. This is called the normal cerebral blood flow (CBF). Any large reduction of this perfusion rate quickly causes neurons to malfunction, or even die. If the flow rate is reduced to about 20 mL per 100 g per minute, the neurons cease generation of electrical signals. Though neurons can survive in this situation for some time, timely restoration of the normal blood flow is required so that they can restore their functions. The area of the brain with reduced CBF to about 8–18 mL per 100 g per minute is called the *ischemic penumbra*, which is a salvageable tissue. Therefore, it is the main target of acute stroke treatment. If there is a reduction in flow of less than 8 mL per 100 g, per minute, irreversible death of the involved brain tissue will begin. In the brain, the most significant clinical considerations concerning blood flow include thrombosis, embolism, cerebral ischemia (ischemic stroke), transient ischemic attack (TIA), intracerebral hemorrhage (hemorrhagic stroke), and cerebral aneurysm ruptures. Cerebral angiography is a process used to assess the blood vessels and their function. Also, in the spinal cord, ischemia can be a significant clinical consideration. Ischemic stroke is caused by three main mechanisms: thrombosis, embolism, and reduction of blood flow.

Thrombosis

Thrombosis describes a condition in which a blood clot forms within a blood vessel. This condition, as well as *embolism*, can occlude an artery supplying the brain. The most common reason for thrombosis is *atherosclerosis*. When an atherosclerotic plaque ruptures, it can result in a blood clot. When arterial thrombosis occurs in a blood vessel of the brain, it can block the perfusion of blood and lead to a stroke. Think of thrombosis as a localized process, and emboli as a nonlocalized process.

Thrombi are predisposed by degeneration of the artery walls, also known as atheromas, especially if they are ulcerated. These can occur in any major cerebral artery. Atheromas are common in areas where blood flow is turbulent, especially at the carotid bifurcation. Partial or complete thrombotic occlusion usually occurs at the main trunk of the middle cerebral artery, and its branches. It is also common in the large arteries found at the base of the brain, in deep perforating arteries, as well as in small cortical branches. The basilar artery and the part of the internal carotid artery between the cavernous sinus and *supraclinoid process* often experience occlusions.

Thrombosis can also develop within the deep veins of the extremities or pelvis, or even within the venous system of the brain, as mentioned above. This condition usually includes inflammation of the vessel walls. Arterial thrombosis is more common than venous thrombosis. The incidence of arterial thrombosis is about 50–100 per 100,000 hospitalizations. It mostly occurs in the elderly, and is more common in males than females. The predominant age for **deep vein thrombosis (DVT)** is about 60 years. Its incidence is common in every 250,000 hospitalizations, and every 50,000 deaths are from complications of this condition. Its prevalence increases with aging, or due to long bone fractures, crushing injuries, and orthopedic and genitourinary surgery.

Pathophysiology

Thrombosis may affect veins or arteries. Risk factors for venous thrombosis include family or personal history of DVT, hormone therapy, use of birth control pills, pregnancy, vein injuries, lack of activity, inherited blood clotting disorders, a central venous catheter, older age, excessive alcohol use, dehydration, smoking, excessive weight, cancer, heart disease, lung disease, and Crohn's disease. Arterial thrombosis is caused by atherosclerosis, **microatheroma** (due to hypertension), fibromuscular dysplasia, and dissection, among other causes. Risk factors for atherosclerosis include diabetes, hypertension, cigarette smoking, high cholesterol, lack of activity, obesity, poor diet, family history, and older age. Less common causes of thrombosis include vascular inflammation that is secondary to processes such as: meningitis, blood vessel disorders, syphilis, antiphospholipid syndrome, hyperhomocysteinemia, polycythemia, thrombocytosis, sickle cell anemia, hemoglobinopathies, plasma cell disorders, and *moyamoya disease*.

Clinical manifestations

The clinical manifestations of deep venous thrombosis include pain in one leg—usually the calf or inner thigh, swelling in the leg or arm. For cerebral thrombosis, there is often severe headache as well as seizures.

Diagnosis

Diagnosis of thrombosis includes methods such as ultrasonography of blood flow, blood tests to determine clotting processes, venography using contrast media, and imaging procedures such as magnetic resonance imaging (MRI), CT scan, **magnetic resonance angiography (MRA),** and **magnetic resonance venography (MRV).** Prior to testing, history and physical examination help assess probability of DVT or sinus thrombosis. Ultrasonography, along with Doppler flow studies, is known as *duplex ultrasonography.* The need for other types of testing is based on pretest probability and the results of this ultrasonography, which directly visualizes the venous lining. Ultrasonography demonstrates abnormal vein compressibility. Doppler flow studies demonstrated impaired venous flow. The Doppler test is more than 90% sensitive and also more than 95% specific for femoral and popliteal vein thrombosis. However, it is less accurate for iliac or calf vein thrombosis.

D-Dimer is the smallest fibrinolysis-specific degradation product in the circulation. It is named because it contains two D fragments of the *fibrin* protein, joined by a cross-link. Elevated levels of D-Dimer suggest recent presence and lysis of thrombi. Assays will vary in specificity and sensitivity. Most, however, are sensitive while not being as specific. Only the most accurate tests are used. An example is the *enzyme-linked immunosorbent assay* or *ELISA*, which has about 95% sensitivity for DVT, and for superficial sinus thrombosis of the brain. This technology is used for diagnosis of several different diseases. When pretest DVT probability is low, the condition is safely excluded if the D-Dimer level is normal. Therefore, a negative D-Dimer test can identify patients with a low DVT probability, excluding the need for ultrasonography. However, positive results are nonspecific since levels can be elevated by liver disease, pregnancy, trauma, positive rheumatoid factor, inflammation, cancer, or recent surgery. Further testing is then required. The other major problem with D-Dimer is that the test results often take time to complete.

When pretest DVT is moderate to high, D-Dimer testing can be done along with duplex ultrasonography. Regardless of the D-Dimer level, a positive ultrasound result confirms diagnosis. If it does not reveal DVT, a normal D-Dimer level helps excludes DVT.

Venography has mostly been replaced by ultrasonography, which is noninvasive, has more availability, and is nearly as accurate for detecting DVT. Venography may still be used when ultrasonography results are normal, but pretest suspicion for DVT is high. Mostly because of contrast dye allergies, complications from venography occur in 2% of patients.

Additional tests include magnetic resonance (MR) venography, direct MRI of thrombi using T2-weighted gradient-echo sequencing, and a water-excitation radiofrequency pulse. The latter test may provide simultaneous views of thrombi in deep veins as well as subsegmental pulmonary arteries.

Treatment

Treatments of thrombosis vary and include anticoagulants, vessel catheterization, **stenting**, and other medications, based on the location and whether the thrombosis is in the venous or arterial system. Anticoagulants are commonly used for cerebral venous thrombosis. Complications of thrombosis are based on the location of the blockage. The most serious complications include stroke, heart attack, and pulmonary embolism (PE). Recall that thrombosis is a localized process, however, when a formed thrombosis travels to another site it will be called an embolism. Prevention of DVT includes physical activity, and exercising the legs during long trips. Smoking cessation, lowering weight, and management of related conditions such as diabetes, hypertension, and high cholesterol can achieve prevention of arterial thrombosis.

The most fearful complication of a DVT is when it travels (embolizes) to another artery or vein. On occasions, a DVT can embolize into the brain. This can happen when there is a connection between the right and left side of the heart, allowing the thrombus to travel from the venous system to the arterial system. This type of embolization involves *paradoxical emboli*. Therefore, all patients with DVT need to be anticoagulated acutely. Injectable unfractionated or low molecular weight heparin, followed by warfarin, will stabilize the clot and prevent it from further embolization.

The most common anticoagulants include low molecular weight heparins, unfractionated heparin (UFH), fondaparinux, and warfarin. A recently new generation of anticoagulation medications has been developed, which are called *non-vitamin K oral anticoagulants* (NOACs). The LMHWs include enoxaparin, dalteparin, and tinzaparin. These are the initial treatment of choice since they can be given to outpatients. They are as effective as UFH for reducing DVTs, thrombus extension, and risk of death due to PE. All of these agents catalyze the action of antithrombin, leading to inactivation of coagulation factor Xa and, not as significantly, factor IIa. Patients with renal insufficiency may receive a decreased dose of most of these medications. Except for warfarin, most of the newer generation anticoagulated medications have a predictable dose response.

An *inferior vena cava filter* or IVCF is an interesting device that can help us understand the practical clinical use of how blood clots may be prevented from traveling to the brain and lungs from the lower extremities. The filter is placed in the vein of the vascular system called the inferior vena cava (IVC). This is one of the major vessels that brings most of the blood from the lower half of the body back to the heart. Therefore, putting a mesh-like net device in this large vein, before it joins the heart, acts as filter of the blood from the heart for large particles like embolized materials from the lower half of the body.

Embolism

An **embolism** occurs when an **embolus**, a particle of foreign matter such as a part of a blood clot (thrombus), or an atherosclerotic plaque, travels through the bloodstream, and eventually blocks a vessel. Recall that in contrast to thrombosis, embolism is a nonlocalized process. A thrombosis can become an embolism when it travels from its localized origin.

Pathophysiology

Emboli can lodge anywhere in the cerebral arteries. They may be of cardiac thrombi origin, for example, due to atrial fibrillation, mitral stenosis, or other rheumatic heart diseases, after a myocardial infarction (MI), due to vegetations on heart valves from bacterial endocarditis or **marantic endocarditis**, and due to prosthetic heart valves. Additional causes include clots forming after open-heart surgery, and aortic arch atheromas that can embolize to the downstream vessels. Only rarely do emboli consist of fat from long bone fractures, or even air from decompression sickness. Recall that venous clots from the lower half of the body drain to the heart via a large vein called the IVC (see Thrombosis section). This clot can flow from the right to the left side of the heart via a patent foramen ovale and eventually travel to the arteries in the brain (*paradoxical emboli*). Emboli can spontaneously dislodge, or dislodge after invasive catheterization and other cardiovascular procedures. Thrombosis of the subclavian artery rarely results in embolic stroke in branches of the vertebral artery.

While a blood clot is referred to as a **thrombus**, a moving blood clot is called a *thromboembolus*. As an embolus moves through the blood vessels, it may become lodged in a smaller vessel, causing the blood to "back up" behind it. This causes the cells supplied by this vessel to receive insufficient oxygen, and die. There are several types of embolism, as follows:

- **Pulmonary embolism**—usually from a leg embolus lodging in one of the lung arteries. Though many emboli are broken down by normal processes, a serious PE may be fatal. The best way to prevent PE is to prevent DVTs from forming or starting to move in the blood vessels.
- **Brain embolism**—causing an ischemic stroke or TIA.
- **Retinal embolism**—blockage of tiny vessels at the back of the eye, usually resulting in sudden blindness of the affected eye.
- **Septic embolism**—when particles created by an infection block blood vessels. An example would be bacterial endocarditis in the heart, causing vegetation and emboli of infectious material into the brain.
- **Air embolism**—actually, bubbles in the bloodstream that can block arterial flow, such as when deep-sea divers rise to the surface too rapidly; this generates the air bubbles in the bloodstream.
- **Fat embolism**—particles of fat or bone marrow introduced into the bloodstream.

Clinical manifestations

The clinical manifestations of embolisms are somewhat similar, yet there are variances between the different types. PE causes shortness of breath, rapid breathing, wheezing, bloody sputum, cough, lightheadedness, dizziness, fainting, and sharp chest or back pain. Brain (cerebral) embolism causes pain, weakness, numbness, tachycardia, chest pain, vomiting or severe nausea, tingling, dimness or loss of vision often in only one eye, seizures, pale skin, altered mental state, and swallowing abnormalities. Septic embolism causes fever, chills, sharp chest or back pain, numbness, shortness of breath, sore throat, fatigue, persistent cough, spleen tenderness, lightheadedness, dizziness, fainting, and inflammation.

Air embolism generally causes difficulty breathing or respiratory failure, chest pain or heart failure, muscle or joint pain, stroke, mental status changes such as confusion or loss of consciousness, hypotension, and cyanosis of the skin. Fat embolism often causes rapid breathing, shortness of breath, mental confusion, lethargy, coma, fever, anemia, and pinpoint rash on the chest, head, and neck.

Diagnosis

Diagnosis of PE is a *ventilation perfusion (V/Q)* scan, CT spiral scan, and pulmonary angiography. Diagnosis of cerebral embolism is confirmed via CT scan, MRI. Workup for septic embolism includes complete blood count (CBC), blood cultures, angiogram, MRI, echocardiography, and chest X-ray. Diagnosis of air embolism is based on the progressive symptoms. For fat embolism, differential diagnoses include dyspnea, hypoxia, and abnormal chest X-rays, which can occur with thromboembolism or pneumonia. Continuous pulse oximetry may detect fat embolism before it becomes clinically apparent. Brain MRI and echocardiography may be helpful for diagnosis.

Treatment

Treatments for PE include anticoagulants such as heparin or warfarin, and thrombolytics such as streptokinase in acute treatment settings. There is no generally beneficial treatment for retinal embolism. However, using glaucoma medications to decrease internal eye pressure, and the inhalation of 5% carbon dioxide gas, followed by ocular massage, are used at some centers. Treatment of septic embolism includes antibiotics, and sometimes surgery to remove vegetation (infectious material that deposits on a heart valve). Treatment of air embolism is focused on stopping the source of the embolism, usually via use of a *hyperbaric chamber*. The patient may also be placed in the *Trendelenburg position*, keeping the pelvis higher than the head.

Section review
1. What are the risk factors for thrombosis?
2. What is the most common cause of thrombosis?
3. What methods are used to diagnose thrombosis?
4. Name the types of embolism?

Stroke

An abrupt incident of vascular insufficiency in the brain is referred to as **cerebral ischemia**, commonly known as a **stroke**, or *brain stroke*. This necrotic tissue is called an **infarct** (see Fig. 5.5). There are two broad categories of stroke: *ischemic* (thrombotic, embolic, or due to hypoperfusion) and *hemorrhagic*. It is important to understand that the brain requires 20% of the body's oxygen consumption, and receives about 15% of the resting cardiac output.

Strokes are a heterogeneous group of disorders. They involve sudden and focal interruption of CBF, which causes neurologic deficits. While hemorrhagic strokes make up only 13% of strokes, *ischemic strokes* make up the 87% majority. In Western countries, stroke is the third most common cause of death. In the United States, stroke is the most common cause of severe disability in the elderly. Stroke generally involves either the anterior circulation or the posterior circulation of the brain. The anterior circulation includes branches of the internal carotid artery, while the posterior circulation includes branches of the vertebral and basilar arteries.

Ischemia, as well as hypoxia and infarction, may result from impaired blood supply and oxygenation of CNS tissue. Ischemia of the brain can be categorized by the etiology of the stroke into three major causes: thrombosis, embolus, or strokes that are caused by decreased blood perfusion to the brain. Clinical manifestations of the ischemic or dead brain tissue are determined by the brain area affected, and mainly based on the role of the tissue that is now infarcted. The symptoms are sudden and may manifest as sudden

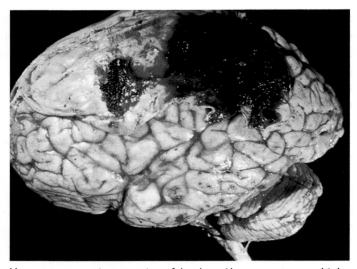

Fig. 5.5 Epidural hematoma covering a portion of the dura. Also present are multiple small contusions in the temporal lobe. *(Courtesy the late Dr. Raymond D. Adams, Massachusetts General Hospital, Boston, Mass.)*

one-sided weakness in the arms, loss of ability to talk, loss of vision, dizziness and vertigo, sudden problems balance or walking, and others. Strokes can also be caused by *hemorrhages* from various CNS blood vessels. Common etiologies include hypertension and vascular aneurysms or malformations.

The brain mostly requires oxidative metabolism in order for adenosine triphosphate (ATP) to be generated. It only has a slightly ability to use glycolysis or energy substrates, except for circulation-delivered glucose. In ischemia, blood supply stops. This ceases the flow of oxygen to the brain tissues and ATP becomes depleted. There is a loss of membrane potential required for neuronal electrical activity due to abnormal sodium levels within the cells. Recall from Chapter 1 that sodium is required for generation of the action potential. Sodium ions are balanced between the inside and outside of the neurons with the help of a channel that uses ATP to import sodium ions into the cells, to generate an action potential. Elevation of cytoplasmic calcium levels due to other dysfunction of membrane channels activates an enzymatic cascade, leading to cellular injury. Metabolic energy depletion related to ischemia can results in an inappropriate release of excitatory amino acid neurotransmitters, such as glutamate. This can increase cell damage via allowing excessive calcium ion influx through N-methyl-D-aspartate (NMDA)-type glutamate receptors. The "at-risk" area lies between the normal brain tissue and necrotic tissue. This area is called the *penumbra*. The penumbra is an area of interest for stroke researchers as salvaging this tissue can improve the overall outcome of the stroke significantly.

Focal cerebral ischemia

Focal cerebral ischemia occurs after reduced or stopped blood flow to a localized brain area, caused by arterial occlusion (due to embolus or thrombosis in most cases) or hypoperfusion, as mentioned before. Time is of the essence in the setting of an acute stroke, since an infarction process is now started in the area of the compromised vessel. The length of ischemia and ability for collateral flow to perfuse the area of the penumbra until the clot is removed correlates with the severity of the stroke. Other factors influencing the severity of the stroke are its size and location, the shape of the infarct, and the extent of local tissue damage. These brain tissues are at higher risk of an impending stroke, unless the area is perfused again by the opening of the flow through the occluded vessels. The primary (proximal) source of collateral flow is the circle of Willis, followed by the external carotid-ophthalmic pathway. This may be assisted by the distal branches of the cerebral arteries via cortical-leptomeningeal anastomoses. There is, however, not much collateral flow for deep penetrating vessels that supply the thalamus, basal ganglia, or deep white matter, except in long-term chronic stenotic processes that would allow production of diffuse collateral vessels along the entire CNS. Occlusive vascular disease able to cause a cerebral infarction may be caused by emboli from a distant source, in situ thrombosis, or **vasculitides**.

Global cerebral ischemia

Global cerebral ischemia is also called *diffuse ischemic/hypoxic encephalopathy*. It is due to a generalized reduction of cerebral perfusion, such as in severe hypotension, cardiac arrest, or prolonged shock. Clinical outcomes vary from transient postischemic confusional states to irreversible CNS damage. Neurons are the cells most able to be damaged, followed by astrocytes and oligodendrocytes. The most sensitive neurons are within the pyramidal cell layer of the hippocampus, cerebellar Purkinje cells, and pyramidal neurons of the cerebral cortex, especially the occipital area of the brain since the neurons in this area require higher levels of oxygen, due to their extensive activity. The occipital lobe is mainly responsible for vision. Some basal ganglia structures such as the globus pallidus are also very sensitive to the lack of oxygen and energy (ATP). In severe cases, there is widespread neuronal death, often causing the patient to be in a state with no high-level neurological functioning. Functions such as cognition, vision, language, and comprehension require more energy. Therefore, areas responsible for these higher cortical functions are more prone to ischemia. Patients often remain in a state in which they are not aware of their surroundings.

Previous neurologic opinion stated that in order for a person to be conscious he or she needed to be "awake and aware." Today, these patients are usually described as being in a "vegetative state" (or "minimally conscious state") if 3–12 months have passed, and they are awake, but not aware of their surroundings. However, this depends on the cause of the global hypoxemic ischemia. Preventive measures are often the most effective treatment for severe neurological diseases. Some patients will be clinically progressed to become *brain dead*, defined by irreversible diffuse cortical injury and/or severe brainstem damage. There is an absence of brain reflexes, respiratory drive, and cerebral perfusion. Diagnosis of brain death often requires examination of at least two physicians to validate that no brainstem reflex is preserved in the patient. Spinal reflexes may continue to be intact even after brain death is declared. Nonmedical professionals can interpret this as a sign that the brain is not dead, since the leg or arm can move due to spinal reflexes even in brain dead patients. Therefore, education of the patient's family is very important, to explain that these spinal reflexes are not indicative of brain activity. The reflexes are basically a closed electrical circuit between the spinal cord and the muscle, whereas the electricity is generated by motor neurons in the spinal cord. The brain is not involved in these types of spinal reflexes. In some patient, overtime spent on a respirator, the brain undergoes autolysis and becomes liquefied.

Pathophysiology

When there is inadequate blood flow in a single artery of the brain, this can often be compensated for via an efficient collateral system. This is especially possible between the carotid and vertebral arteries, via anastomoses at the circle of Willis. To a lesser extent, it is possible between the major arteries supplying the cerebral hemispheres. However, there can be normal variations in the circle of Willis, various collateral vessels, development of atherosclerosis, and other acquired arterial lesions. All of these variations may interfere with collateral flow. Sometimes these variations can increase the chance of brain ischemia. The formation of collateral vessels also happens after the chronic narrowing of the large vessels, due to atherosclerosis. Atherosclerosis causes large vessel disease, and not small vessel disease.

Risk factors for stroke include prior stroke, older age, hypertension, cigarette smoking, hypercholesterolemia, diabetes, sleep apnea, tendency to form clots (such as in certain cancers), and drugs such as cocaine, amphetamines, among others. For academic and practical purposes, stroke is categorized into four groups, based on etiology: large vessel disease (mainly due to atherosclerosis), cardioembolic, small vessel disease, and others such as some rare genetic disease such as **Fabry disease** that predisposes you to have stroke. Small vessel disease may be explained as small strokes in the basal ganglia, due to thrombosis of the small vessels. The various types of stroke are predisposed by certain risk factors, as mentioned above. Thrombotic stroke is predisposed by atherosclerosis. Embolic stroke can be predisposed by atrial fibrillation or a state of hypercoagulability due to blood disorders, cancers, infections, etc. Remember that the thrombosis process involves a local vessel occlusion, whereas the embolism process is nonlocal. A thrombosis within a large vessel such as the internal carotid artery (also known as the *donor vessel*) can dislodge and move further into the smaller arteries, causing an occlusion of the vessel. This is then called an embolus.

Clinical manifestations

Clinical manifestations of stroke are "sudden" as a general rule. It may include weakness of the face and/or limbs; aphasia (inability to speak and/or comprehend); loss of balance or gait, and numbness of the face or limbs. There may also be nonspecific symptoms such as dizziness, altered mental status, etc. The area of the brain that is involved influences the neurologic deficits that occur. Unilateral symptoms are often caused by anterior circulation stroke. With posterior circulation stroke there may be unilateral or bilateral deficits, since the left and right vertebral arteries join together to form the basilar artery, which perfuses most of the posterior circulation. The posterior circulation is more likely to affect consciousness, especially if the basilar artery is involved, which can lead to infarction of the arousal system in the pons. Some manifestations are more diagnostic for a specific type of stroke. For instance, a sudden and severe headache suggests a subarachnoid hemorrhage, which is one of the causes of a hemorrhagic stroke. If there is impaired consciousness or coma, often with headache, nausea, and vomiting, increased intracranial pressure is suggested. Basically, any lesions or masses within the cranium can cause an increase in the intracranial pressure. These can be tumors, inflamed tissue due to a previous stroke, or blood from a bleeding vessel (as in brain trauma). Increased intracranial pressure can occur

48–72 h after a large ischemic stroke, but earlier with many hemorrhagic strokes. In these cases, fatal brain herniation may occur.

Complications of stroke include confusion, incontinence, depression, pneumonia, and dysfunction of swallowing. This final complication may lead to aspiration, undernutrition, or dehydration. When the patient becomes immobile, there may be deconditioning, thromboembolic disease, urinary tract infections (UTIs), **sarcopenia**, contractures, and pressure ulcers.

Diagnosis

Evaluation of stroke focuses on determination of the type that occurred and if immediate treatment is needed. A stroke should be suspected if there are sudden neurologic deficits such as weakness, numbness, vertigo, inability to speak, loss of vision, facial droop, slurred speech, etc. Sudden severe headache is mostly related to hemorrhage strokes, or other types of sudden bleeding in the brain. When stroke is suspected, immediate neuroimaging studies are needed, to differentiate between hemorrhagic and ischemic strokes, as well as to detect signs of increased intracranial pressure. A CT brain scan, without contrast, is the imaging method of choice. *Multimodal CT arteriogram* of the head can show the cause of the stroke, such as a blockage of a large artery, or simply a large vessel occlusion (LVO). Also, CT scans are not sensitive enough to detect small posterior circulation strokes. MRI can detect intracranial blood, as well as small or acute ischemic stroke not detected by CT scan. However, CT scan is usually performed more quickly. When CT scan does not confirm a clinically suspected stroke, diffusion-weighted MRI can usually detect ischemic stroke.

Once a stroke is identified as ischemic or hemorrhagic, tests are performed to determine the cause or mechanism. Patients are evaluated for coexisting, acute general disorders such as dehydration, infection, hyperglycemia, hypoxia, and hypertension. They are questioned about depression, which often occurs after a stroke. Dysphagia, or inability to swallow, is evaluated in all strokes by simple bedside swallowing evaluation. Some patients lose their ability to swallow after strokes.

Treatment

Treatment of stroke is divided into two different categories: acute and chronic. Treatment of chronic stroke focuses on prevention of the same mechanism from causing another stroke. Stabilization may be needed before a complete evaluation can be performed. For a comatose or weakened patient, airway support may be required. The type of stroke determines which specific acute treatments are provided. Acute stroke treatment can be offered within 4.5 h of the initial symptoms. This acute treatment involves use of the clot-dissolving medication called *tissue plasminogen activator* or tPA.

Supportive care, correcting coexisting conditions, and preventing and treating complications are crucial during the acute phase as well as during convalescence. Coexisting abnormalities may include hypoxia, fever, hyperglycemia, dehydration, and sometimes, hypertension. Clinical outcomes are highly improved because of adequate care. While the patient is convalescing, there must be attempts to prevent deep venous thrombosis, aspiration, UTIs, undernutrition, and pressure ulcers. Passive exercises, especially of paralyzed limbs, and breathing exercises must be started early. This prevents atelectasis, contractures, and pneumonia.

The majority of patients need occupational and physical therapy in order to maximize recover of normal functions. Additional speech therapy and feeding restrictions may be required. Depression may require use of antidepressants, and many patients respond well to counseling. Rehabilitation is best when various disciplines are combined. The delay or prevention of future strokes occurs by understanding the mechanism of the stroke, and address the underlying mechanism. For example, in a stroke caused by thrombosis due to atherosclerosis, management of hypertension, diabetes, smoking, and hyperlipidemia can be helpful.

Ischemic stroke

Ischemic strokes are caused by sudden blockage of blood flow to an area of the CNS. They are commonly caused by a thrombus, embolus, or hypoperfusion. If an occlusion occurs within, or proximal to the cerebral arterial circle, there may be enough collateral circulation. This is particularly true when the involved artery was slowly occluded prior to a stroke occurring. Oppositely, anastomoses between arteries that are distal to the cerebral arterial circle are varied in nature. Collateral circulation is usually not adequate. Therefore, occlusion of any of these vessels usually results in an infarct in a predictable location.

Pathophysiology

An infarct's size is related to the size of the occluded vessel. They can be tiny lesions known as **lacunes**, due to occlusion of a small perforating artery. They can also be much larger, such as in infarcts affecting large regions of a cerebral hemisphere. It is possible for a very small lesion in the brainstem or internal capsule to have larger effects than damage to certain larger areas of the cerebellum or cerebral hemispheres. Infarct size and distribution are linked to the location of the occlusion along an artery's course. If a middle cerebral artery is occluded in the lateral sulcus, there will be a large infarct. If the same artery is occluded where it leaves the cerebral arterial circle, blood flow will be blocked into the lenticulostriate arteries on the same side, and deep structures will also be damaged.

Ischemic brain damage involves different processes and periods of time. If ischemia is profound, there is fast depletion of energy stores of neurons and glial cells. Processes that maintain membrane potentials fail, and the cells become depolarized. There is then excessive release of excitatory neurotransmitters, and additional depolarization. With multiple destructive events, the involved cells die quickly. At the same time, slower destruction, via inflammatory responses begin occurring over a period of days. Recall that an ischemic penumbra is the area at risk that can become irreversibly damaged if acute treatment is not done within the first few hours of the onset of symptoms.

Clinical manifestations

Clinical manifestations of ischemic stroke are based on the brain area affected. In embolic stroke, deficits usually become maximal during onset. Less commonly, deficits evolve slowly over 24–48 h. This is known as *evolving stroke* or *stroke in evolution*, and is typical in thrombotic stroke. The majority of evolving strokes involve unilateral neurologic dysfunction that often begins in one arm and then spreads *ipsilaterally*. Progression is interrupted by periods of stability, but occurs in steps. A stroke is termed *submaximal* when it is completed and there is residual function in the affected area. This suggests viable tissue that is likely to become damaged, or simply the penumbra.

Embolic strokes often occur at daytime, with headache preceding neurologic deficits. Thrombi usually occur during the morning hours, and are first noticed when the patient awakens. Lacunar infarcts may produce classic syndromes, such as pure sensory hemianesthesia, pure motor hemiparesis, dysarthria or *clumsy hand syndrome*, or ataxic hemiparesis. Lacunar infarcts are due to hypertension. Signs of cortical dysfunction, such as aphasia, are absent in lacunar strokes since these usually happen in the subcortical and deep structures. Therefore, the cortex is spared. If there are multiple lacunar infarcts overtime, this may result in vascular dementia. Worsening of symptoms during 48–72 h, especially progressive consciousness impairment, is usually due to cerebral edema instead of extension of the infarct area. Function usually improves in a few days unless the infarct is large or extensive. Additional improvement occurs gradually, for up to 1 year.

Diagnosis

Diagnosis of ischemic stroke is suggested by sudden neurologic deficits that refer to a certain arterial area. Ischemic stroke must be distinguished from other causes of similar focal deficits, also known as stroke mimics. These include hypoglycemia, postictal (Todd's) paralysis following a seizure, and rarely, migraine headache. A hemorrhagic stroke is more likely to cause headache than an ischemic stroke.

Neuroimaging and bedside glucose testing are mandatory even though diagnosis is clinical. CT scan is performed first, to exclude hemorrhage. Hemorrhage needs to be ruled out since treatment of ischemic stroke with tPA can worsen a hemorrhage and cause death. It takes up to 12 h for an acute stroke to show up on a CT scan. Therefore, acutely, the CT scan can be normal even in a large stroke. Early changes on CT scan can include effacement of sulci or the insular cortical ribbon, a dense middle cerebral artery sign, and loss of gray-white junctions between the cortex and white matter. After 12 h, infarcts are usually visible as **hypodensities**. Small, lacunar infarcts may only be visible via MRI. Diffusion-weighted MRI is very sensitive for early ischemia. It can be done immediately after the initial CT neuroimaging, and can show evidence of stroke within 30 min of the onset of the symptoms.

Since distinction between embolic, lacunar, and thrombotic stroke, based on history, examination, and neuroimaging is not always reliable, other tests are needed. These include carotid duplex ultrasonography, ECG, transesophageal echocardiography, and a variety of blood tests, including CBC, platelet count, fasting blood glucose, PT/PTT, homocysteine, lipid profile, ESR, and for at-risk patients, and antiphospholipid antibody. In order to detect concomitant MI, troponin I level is measured. Often, MRA or CT angiography is also performed to help identify the mechanism of the stroke, or to see whether there is a large clot within the anterior circulation that can be mechanically removed by a procedure called *thrombectomy*. Thrombectomy can be done up to 24 h from the onset of the stoke, in some cases.

Treatment

Treatments for stroke are focused on restoring blood flow to the area of ischemia as well as protecting neurons from additional destruction due to the ischemia. Drugs such as *tPA* break up clots and allow blood to flow back into the ischemic area. Unfortunately, when there is blood vessel damage, tPA can cause excessive brain bleeding, and more damage. It must be used very carefully, and therefore, a CT scan is always needed prior to administration of tPA to make sure there is no hemorrhage in the brain. The primary area of ischemia, or *core zone*, can be treated effectively for no more than a few hours after the occlusion occurs. However, the neurons in the *penumbra* can still be saved for days after occlusion (see above for more detail on penumbra). New research is proceeding to prevent penumbra from becoming an actual irreversible stroke.

For acute ischemic stroke, hospitalization is required and supportive measures are initiated during evaluation and stabilization. Perfusion of an ischemic brain area may require a high blood pressure since autoregulation is lost. Therefore, BP should not be decreased except in these situations:

- Blood pressure over 220 mmHg systolic or over 120 mmHg diastolic on two successive readings, more than 15 min apart
- Other present signs of end-organ damage: acute MI, aortic dissection, hypertensive encephalopathy, pulmonary edema, acute renal failure, and retinal hemorrhages
- Likely use of recombinant tPA

If indicated, nicardipine is given intravenously, or labetalol can be substituted.

If the patient is presumed to have thrombi or emboli, this may be treated with tPA, antiplatelet drugs, thrombolysis in situ, and/or anticoagulants. Most patients will not require thrombolytic therapy since they present after 4.5 h, and tPA cannot be given beyond this window. They should receive aspirin or another antiplatelet drug upon hospital admittance. Contraindications to antiplatelet drugs include aspirin- or nonsteroidal anti-inflammatory drug (NSAID)-induced asthma or urticaria, other hypersensitivities to aspirin, acute gastrointestinal (GI) bleeding, and G6PD deficiency, among others.

For acute ischemic stroke with no contradinciations to tPA, and of less than 4.5 h duration, *tPA* may be used, thought its use may cause symptomatic or even fatal brain hemorrhage in 3% of patients. When tPA is used with extreme care, tPA results in a higher chance of functional neurologic recovery in almost 33% of patients. Only physicians who are experienced in stroke management should use tPA for acute stroke patients. It is important to begin the timing from the moment that the patient was last observed to be normal, prior to the first symptom. This is often known as the *last known well*. Before tPA treatment, brain hemorrhage must be excluded by CT scan, as mentioned above. Systolic and diastolic BP must be less than 180 and 110, respectively, if a tPA is being considered. Antihypertensive drugs can be administered. Dosing of tPA is via rapid IV bolus injection of 10% of the dose, with the rest given by constant infusion over 1 h. Vital signs are closely monitored for the next 24 h, and BP must be maintained below 180/110. Bleeding complications must be managed closely. Anticoagulants and antiplatelet drugs must be avoided during the first 24-h period. The exclusion criteria for using tPA in stroke patients is listed in Table 5.1.

Thrombectomy is angiographically directed intra-arterial removal of a thrombus or clot. It can sometimes be used for major strokes, if symptoms began in the past 24 h. It is best used for strokes caused by large occlusions in the middle cerebral artery or LVO. Basilar artery clots may be intra-arterially lysed up to 24 h after stroke onset. This may sometimes be done later, based on clinical factors.

Table 5.1 Exclusion criteria for using tPA in stroke patients.

Criteria	Factors
CT scan	Intracranial hemorrhage present; multilobar infarct present (hypodensity in more than 1/3 of area supplied by middle cerebral artery)
Presentation	Subarachnoid hemorrhage suggested, even with negative CT scan
History	Any history of intracranial hemorrhage, stroke, or head trauma within past 3 months; arterial puncture at noncompressible site or lumbar puncture in past 7 days; major surgery or serious trauma in past 14 days; GI or urinary tract hemorrhage in past 21 days
BP	Systolic above 185 mmHg or diastolic above 110 mmHg after antihypertensive treatment
Platelets	Less than 100,000 per microliter
Heparin	Used within 48 h and elevated PTT
Oral anticoagulants	Currently being used, INR more than 1.7, or PT more than 15, recent use of NOAC agents
Plasma glucose	Less than 50 or more than 400 mg per deciliter (less than 2.78 or more than 22.2 milliosmoles per liter)
Other conditions	Bacterial endocarditis or suspected pericarditis

For stroke caused by cerebral venous thrombosis, anticoagulation is done using heparin or LMWH even in the presence of a hemorrhage. Warfarin can be used in some strokes, for example, in strokes caused by arterial fibrillation. Warfarin needs to then be used indefinitely in most patients, based on their risk of developing clots in the future. Before this, hemorrhage must be excluded via CT scan. Constant weight-based heparin infusion is used to increase PTT to 1.5–2 times baseline values, until the warfarin becomes therapeutic. In some situations, anticoagulation is needed immediately, such as in patients who need constant anticoagulated medication to prevent clots from being formed on a mechanical heart valve. Since warfarin predisposes the patient to bleeding and is continued following discharge from the hospital, it should only be prescribed to patients likely to comply with dosing and monitoring, who are not prone to falling.

Over the long term, supportive care continues as the patient is convalescing. Controlling hyperglycemia and fever, as well as other risk factors, can limit the amount of brain damage following the stroke. *Carotid endarterectomy* is used for patients with recent non-disabling, submaximal stroke, or TIA due to an ipsilateral (same side) carotid obstruction of 70%–99% of the arterial lumen, provided that the patient's life expectancy is 5 or more years. Basically this means that the mechanism of the stroke is due to an emboli from the atherosclerotic plaque, within the ipsilateral carotid artery. The plaque became dislodged and caused occlusion of a downstream vessel. Some studies suggest that symptomatic patients, including those with stokes and TIAs, may benefit from endarterectomy with antiplatelet therapy if the carotid obstruction is 60% or more, provided that patient's life expectancy is at least 5 years.

Secondary prevention for additional strokes utilizes oral antiplatelet drugs. These include aspirin, clopidogrel, or the combination drug aspirin/extended-release dipyridamole. For certain high-risk patients, aspirin is sometimes used along with warfarin. A recent method is the combination of clopidogrel and aspirin, for a short period of time after an acute stroke.

Transient ischemic attack

A **TIA** is similar to an ischemic stroke. However, TIA deficits usually only continue for a few minutes, but may less often continue for a few hours, and are followed by a total recovery from the symptoms. The cause of TIAs is tiny emboli originating from atherosclerotic plaques or a thrombi, among others. They partially occlude the arteries of the brain, but then resolve due to variety of reasons. One common example of a TIA syndrome is **transient monocular blindness**, also called *amaurosis fugax*. This is usually caused by emboli that have detached from a plaque in the internal carotid artery on the same side. They enter the ophthalmic artery, and cause a temporary loss of blood perfusion to the retina. Historically, TIAs were defined as neurological symptoms, which completely resolve within 24 h. With the advent of MRI, the definition of TIA is now changed to transient neurological symptoms with no evidence of stroke on diffusion-weighted imaging (DWI) MRI.

Pathophysiology
Most TIAs occur in middle-aged and elderly people, and are markedly increased risks for actual stroke, starting in the first 24 h. Less common causes of TIAs include impaired perfusion due to severe hypoxemia, profound anemia, carbon monoxide poisoning, or severe polycythemia especially in brain arteries that have had preexisting stenosis.

Clinical manifestations
Clinical manifestations of TIAs include transient monocular blindness that usually lasts for less than 5 min. This may occur when the ophthalmic artery is affected. The symptoms are sudden, last 2–30 min, then totally resolve. A patient can have several daily TIAs, or only two to three TIAs over several years.

Diagnosis
Diagnosis of TIAs involves resolution of stroke-like symptoms within 1 h, neuroimaging studies, and evaluation to identify the cause. These attacks are not suggested by isolated peripheral facial nerve palsy, and impaired consciousness or loss of consciousness. However, they must be distinguished from conditions that have similar symptoms (stroke mimics), including migraine aura, hypoglycemia, and postictal (Todd's) paralysis following a seizure. Neuroimaging is required since infarcts, small hemorrhages, and mass lesions cannot be clinically excluded. CT scans are usually the most immediately available, but often cannot identify infarcts within the first 12 h. MRI can detect infarction within 30 min of symptom onset. Diffusion-weighted MRI is the most accurate method to rule out infarcts in patients with presumed TIA, but this method is not always available.

Causes of TIA are determined by assessing cardiac sources of emboli, carotid stenosis, atrial fibrillation, hematologic abnormalities, and screening for stroke risk factors. Since risk of future ischemic stroke is high, immediate evaluation and/or treatment needs to take place quickly, usually with 24 h of the patient's hospitalization.

Treatment
Treatment of TIAs and strokes is focused on prevention of further attacks, via antiplatelet drugs and controlling the risk factors. Some patients may even require anticoagulation to prevent further episodes. Sometimes, invasive procedures such as endarterectomy or arterial angioplasty are required, in refractory cases. If there are present cardiac sources of emboli, such as atrial fibrillation, warfarin, or NOACs may be indicated. Prevention of stroke may also involve modification of stroke risk factors.

Section review
1. Explain the risk factors for stroke.
2. Describe the complications of stroke.
3. Differentiate between stroke and TIA.

Epidural hematomas

In the cranium, epidural hematomas occur when there is bleeding from an artery between the skull and the dura mater (see Fig. 5.6). Epidural hematomas are often caused by a head trauma resulting in the rupture of the middle meningeal arteries. There is usually a resultant severe headache, occurring immediately or within hours after the injury. This may disappear, but often returns with more severity in a few hours. This is also known as "lucid time." Increasing confusion, sleepiness, and a deep coma can follow very quickly. Remember that this is arterial blood. Therefore, the blood loss occurs quickly due to higher volume of blood flow in the arteries in

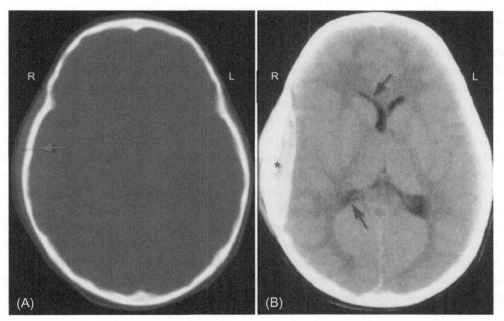

Fig. 5.6 Epidural hematoma in a 3-year-old girl who hit her head in an automobile accident. About 2 h after the accident, she complained of a severe headache and nausea and became lethargic. Computed tomography (CT) set to show bony details (A) revealed a fracture of the right temporal bone (arrow). CT set to reveal soft tissue details (B) revealed a lens-shaped epidural hematoma (*) and compression of the right lateral ventricle (arrows). After rapid neurosurgical treatment, she made a full recovery. L, left side; R, right side. *(Courtesy Dr. Raymond F. Carmody, University of Arizona College of Medicine.)*

comparison to bleedings from veins, such as in subdural hematoma (see below). Paralysis may develop on the side of the body opposite the hematoma, along with speech or language impairment and other symptoms based on site of the lesions. In the case of language dysfunction, this would be the left side of the cerebrum, which is the dominant hemisphere responsible for language.

> **Focus on early diagnosis of epidural hematomas**
> *Early diagnosis of epidural hematomas is essential.* This is usually based on CT scan results, with management of blood pressure and the patient's airway. Sometimes, open brain procedures are required. The CT scan of an epidural hematoma shows a classic lens shape of the blood along the dura and skull.

Subdural hematomas

Subdural hematomas are usually caused by bleeding veins, as opposed to arteries, such as in epidural hematoma. These "bridging veins" lie between the dura mater and arachnoid mater (see Fig. 5.7). Subdural hematomas may be acute, with symptoms developing within minutes to days after the bleeding.

Pathophysiology

Acute (or subacute) subdural hematomas may be caused by rapid bleeding following a head injury. Acute subdural hematomas are often caused by falls or motor vehicle crashes. Acute subdural hematomas are caused by bridging veins. Blood pushes neighboring brain tissues, and intracranial pressure increases if the bleeding continues. Since there is no additional space for the bleeding mass, the brain tissues are squeezed between the skull and the blood mass. Symptoms are mainly related to the amount of the bleed and the age of the patient. Interestingly, brain mass shrinks with aging. Therefore, patients over 65 have more space in their skulls, in case of acute subdural hematomas. This means that small to moderate subdural hematomas can be tolerated in older patients.

According to Medscape, acute subdural hematomas range in occurrence at between 5% and 25% of patients with severe head injuries, based on various studies. Annual incidence in the United States, of chronic subdural hematoma, is one to 5.3 cases per 100,000 population. Recent studies have shown a higher incidence of this disease, which is most likely due to an increase of in patients being diagnosed, because of better imaging techniques.

Clinical manifestations

In subacute subdural hematomas, symptoms develop over hours or days (48 h). They are most common over the lateral parts of the cerebral hemispheres. They sometimes occur bilaterally. The common neurologic findings are caused by pressure on the brain tissues, and include headache. In chronic subdural hematomas, symptoms develop over weeks to

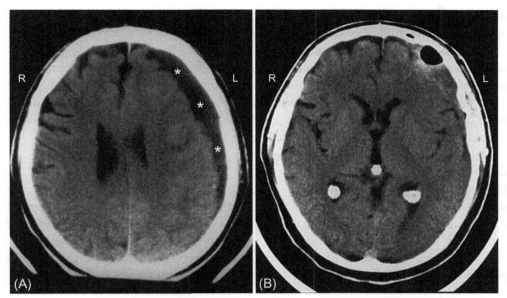

Fig. 5.7 (A) Crescent-shaped subdural hematoma (*) over the surface of the left cerebral hemisphere, compressing its subarachnoid spaces and lateral ventricle and shifting midline structures to the right. L, left side; R, right side. Although subdural hematomas are commonly caused by a blow to the head, they can also be caused by other mechanical disturbances (as in the shaken-baby syndrome). This is the case of a 64-year-old man whose headaches developed gradually after he began riding a roller coaster at an amusement park. The roller coaster, he reported, "swings people upside down as many as six times." He rode the roller coaster on 11 different occasions until his headaches became so severe that he was unable to continue. After surgical removal of the hematoma (B), subarachnoid spaces and ventricular symmetry returned and the patient recovered uneventfully. ((A) from Bo-Abbas, Y.; Bolton, C. F. N. Engl. J. Med. **1995**, 332, 1585. Copyright © 1995 Massachusetts Medical Society. All rights reserved. (B) Courtesy Dr. Y. Bo-Abbas, Victoria Hospital, London, Ontario, Canada.)

months. Chronic subdural hematomas are more common among alcoholics, older adults, and patients taking anticoagulant or antiplatelet medications. Alcoholics and older adults, who are more likely to fall or have bleeding issues, may ignore or forget head injuries that are not extremely severe. If so, they may develop small subdural hematomas that become chronic. By the time symptoms are noticed, a chronic subdural hematoma can be very large. Chronic hematomas are less likely than acute hematomas to cause a fast increase of intracranial pressure.

In older adults, the brain shrinks slightly. This stretches the bridging veins and makes them more likely to rupture if any type of injury occurs. Bleeding usually continues longer since the atrophied (or shrunken) brain tissue exerts less pressure on the bleeding vein and allows more blood loss. Any remaining blood is slowly reabsorbed. Sometimes, a fluid-filled space called a **hygroma** may develop, which can refill with blood and enlarge, since the bridging veins within it are even more prone to rupture.

Symptoms of a subdural hematoma include a persistent headache, varying drowsiness, confusion, memory changes, paralysis on the side of the body opposite the hematoma, and impairment of speech or language. Other symptoms depending on the area of the brain that is damaged. In infants, a subdural hematoma can cause the head to enlarge, since the skull is soft and pliable. As a result, pressure within the skull increases less in infants than it does in older children and adults.

Diagnosis
Diagnosis is based on CT or MRI. Chronic subdural hematomas are more difficult to diagnose because of the longer time period between the injury and development of symptoms. An older adult with gradual symptoms such as memory impairment and drowsiness may be mistakenly diagnosed with other causes of altered mental status. A CT scan can detect acute, subacute, and many chronic subdural hematomas. MRI is often very accurate for diagnosis of chronic subdural hematomas.

Treatment
For small hematomas, there is often no treatment needed except supportive care, since the blood will be absorbed on its own. For large hematomas, surgical drainage may be indicated. If the hematoma is large and is causing significant symptoms, surgical drainage is performed via a small hole in the skull. However, sometimes a larger opening is needed, such as when bleeding has occurred very recently, or if the blood is too thick to drain through a smaller hole. During surgery, a drain is usually inserted, then left in place for several days in case of hematoma recurrence. The patient is monitored closely for such recurrences. In infants, the hematoma is usually drained. Only about 50% of patients treated for a large acute subdural hematoma survive. Those with chronic subdural hematomas usually improve, or at least, their conditions do not become worse.

Spinal epidural or subdural hematomas
The spinal cord is also covered by meningeal layers. A *spinal* epidural or subdural hemorrhage, or more specifically, *hematoma*, is basically the accumulation of blood in the spinal subdural or spinal epidural space. This can mechanically compress the spinal cord. Most cases occur in the thoracic or lumbar region, but overall the condition is relatively rare.

Pathophysiology
Causes include anticoagulant or thrombolytic therapy, back trauma, or very rarely, following a therapeutic or diagnostic lumbar puncture (mostly in patients with bleeding **diatheses**).

Clinical manifestations

Signs and symptoms begin with back pain that is localized or radicular, and tenderness to percussion of the spine. A radicular low back pain is a shooting pain that radiates down the back of the leg below the level of the knees. This pain can be very severe. Spinal cord compression can develop. If the lumbar spinal roots are compressed, there may be resultant **cauda equina syndrome** and lower extremity paresis, progressing over minutes to hours.

Diagnosis

Diagnosis of epidural or subdural hemorrhage in the spine is via MRI and CT scan. Hematoma is suspected with acute, nontraumatic spinal cord compression, or sudden and unexplained lower extremity paresis. This is especially true if trauma or bleeding diathesis is present.

Treatment

Treatment of a spinal hematoma involves immediate surgical drainage in most cases. For patients taking anticoagulants, vitamin K1 (phytonadione) is administered. Fresh frozen plasma is given as well, to normalize the **international normalized ratio (INR)**. For patients with thrombocytopenia, platelets are administered.

Subarachnoid hemorrhage

Subarachnoid hemorrhage is sudden bleeding into the subarachnoid space, between the arachnoid and pia mater (see Fig. 5.8). Head trauma is the most common cause of subarachnoid hemorrhage. However, traumatic subarachnoid hemorrhage is usually

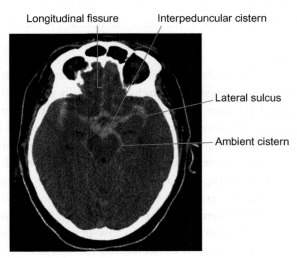

Fig. 5.8 Subarachnoid hemorrhage.

considered to be a separate disorder from the aneurysmal subarachnoid hemorrhage since they have different pathophysiology and complications.

Pathophysiology

Primary (spontaneous) cases are usually due to ruptured aneurysms. There are certain congenital connective tissues that are strongly associated with aneurysm formation. Bleeding can continue, causing catastrophic brain damage, or even can stop spontaneously. The most common ages for aneurysmal hemorrhage are between 40 and 65 years, though it can occur at any age. Other causes of subarachnoid hemorrhage include arteriovenous malformations (AVMs), bleeding disorders, cancer-related aneurysms, and septicemia.

When there is blood in the subarachnoid space, for unclear reasons, a chemical pathologic process develops within the brain arteries. This chemical reaction results in a disruption of physiologic blood flow, which is responsible for most of the complications of aneurysmal subarachnoid hemorrhage, within a few weeks. Focal brain ischemia may result from secondary vasospasm and disruption of the blood flow. About one in four patients develop signs of a TIA or an ischemic stroke at some point after the rupture (mostly in the first 4 weeks). Subarachnoid hemorrhage can cause disruption in the flow of the CSF as well. This disruption usually occurs at the reabsorption site of the CSF, and therefore, causes a communicating hydrocephalus. The other main complication of subarachnoid hemorrhage is seizure. Due to unknown reasons that are perhaps related to the release of chemicals from the ruptured aneurysm, the other arteries sometime suffer from spasm and narrowing of their lumens (vasospasm) that can lead to subsequent infarction if lasting more than 1 h. Vasospasm usually occurs between 72 h and 10 days from the day of the rupture. Secondary acute (communicating) hydrocephalus is another common complication, as mentioned above. If the aneurysm is not stabilized by intervention, rebleeding may occur, usually within 7 days. It is most dangerous on the first day.

Clinical manifestations

Signs and symptoms include a sudden onset of headache that is usually severe and peaks within a few seconds. The aneurysmal headache is classically described by patients as the "worst headache" they have ever had. If loss of consciousness occurs it is most often immediate, but can occur in a few hours. Within minutes or several hours, severe and irreversible neurologic deficits may occur from the mechanisms mentioned above. In the authors' own experience, patients who survive the first acute 48 h often experience a fluctuating clinical course during the first month. Seizures may occur in this period as well. The neck can be stiff, and sensitivity to light can also be seen. Within 1 day, chemical meningitis causes variant **meningismus** and vomiting. During the first 5–10 days, confusion, constant headache, and noninfectious low-grade fever are common. Recurrent or new symptoms may be caused by rebleeding.

Diagnosis

Diagnosis is usually with non-contrast CT scan in the active setting. If this is negative, a lumbar puncture is performed in patients who are highly suspected to have subarachnoid hemorrhage. CT scan within the first 6 h of symptom onset detects even a small subarachnoid hemorrhage (up to 99% sensitivity is reported). Testing must be done very quickly. Non-contrast CT is very sensitive, as is MRI, but CT scan is more commonly available. Lumbar puncture can help with diagnosis in patients with no signs of bleeding on their initial CT scan. Lumbar puncture is contraindicated if increased intracranial pressure is suspected, since the sudden decrease in CSF pressure can lessen the tamponade of a clot on the ruptured aneurysm, causing more bleeding. It can even cause shifting of the brain tissues toward the site of lumbar puncture (herniation).

In the CSF, diagnosis requires the presence of red blood cells, and the presence of **xanthochromia (see below)**. Traumatic lumbar puncture (trauma to the small vessels during the procedure) may be a cause for the presence of RBCs in the CSF. In about 6–12 h after a subarachnoid hemorrhage, RBCs start to lyse. Xanthochromia is referred to as a yellowish discoloration of the CSF due to the presence of blood byproducts such as bilirubin. Xanthochromia is diagnostic for subarachnoid hemorrhage. In the emergency setting, **MRA** or CT angiography should be done to identify the exact location of the aneurism. Conventional cerebral angiography should be performed in all patients with subarachnoid hemorrhages regardless of the location or severity since as many as 20% of patients (mostly women) have multiple aneurysms.

Treatment

Treatment for subarachnoid hemorrhage is better performed in a comprehensive stroke center. Studies show that hospitals that treated more than 40–60 acute aneurysmal subarachnoid hemorrhages per year have better outcomes. Therefore, it is now recommended that those hospitals with less than 40 of these cases should transfer patients in the first few hours to a referral center. Treatment requires a multidisciplinary approach in a dedicated neurological ICU known as an NICU. Hypertension should only be treated if the mean arterial pressure is higher than 140 mmHg. **Euvolemia** is the main concept of the treatment, to prevent disruption of the blood flow due to vasospasm. The patient must receive adequate bed rest. Restlessness and headache are treated as needed. This is one of the few areas in neurology in which a headache can be treated with narcotics. To prevent constipation and avoid straining, which can increase the pressure in the cranium, stool softeners are given. Anticoagulants and antiplatelet drugs are contraindicated. Nimodipine or labetalol can be administered to keep the systolic blood pressure below 140. Ventricular drainage should be considered if there are clinical signs of acute (communicating) hydrocephalus, due to defects in the CSF

reabsorption system. Aneurysms can be treated by three different procedures. These procedures are:
- Surgical clipping of the aneurysm via an open brain surgery, by a neurosurgeon.
- Coiling of the aneurysm via an intravascular angiographical procedure, by an interventional neurologist.
- Radiation treatment for some small aneurysms, by an interventional neuroradiologist. Radiation is targeted at the aneurysm location and begins a process that eventually will lead to the closure (by way of embolization) of small aneurysms within a few years of treatment. Detachable endovascular coils can be inserted to secure the aneurysm. Either the coiling or the clipping should be performed within 24 h of symptom onset to secure the aneurysm.

Intracerebral hemorrhage

Intracerebral hemorrhage is commonly considered a type of stroke, but its pathophysiology is not the same as seen in ischemic stroke. It is also called *intraparenchymal hemorrhage*, usually resulting from a rupture of the lenticulostriate artery, or other small perforating arteries. It is always confusing to differentiate brain hemorrhages that are considered hemorrhagic stroke from the types that are not. Basically, all subarachnoid, intraparenchymal, or lobar hemorrhages (usually due to amyloid angiopathy) are considered to be hemorrhagic strokes. Epidural and subdural hematomas are formed outside the brain tissue, and therefore, are not considered to be hemorrhagic strokes. One of the most common cause of *intraparenchymal* or intracerebral hemorrhage is hypertension. Another common cause of cerebral hemorrhage is bleeding that sometimes happens in the area of an ischemic stroke, also known as *hemorrhagic transformation*. When an ischemic stroke occurs, ischemic brain tissues are not the only structures damaged. Small vessels in the area of the stroke are also damaged, and die. These necrotic vessels can result in blood leaking into the area of infarct in up to half of patients (hemorrhagic transformation).

The most common cause of intracerebral hemorrhage is hypertension. This hypertensive bleeding usually occurs within the small arteries supplying the deep structures of the brain. These extremely thin-walled vessels are very prone to spontaneously rupture, especially in patients who experience a sudden increase in their blood pressure (see Fig. 5.9). Certain deep cerebral structures are supplied by these small arteries, also called lenticulostriate arteries. Hemorrhage can quickly progress and even be fatal in some cases. Less common causes of intracerebral hemorrhage include head trauma, tumors, blood clotting deficiencies, abnormalities of the blood vessels (vascular malformations), and illicit drug use (such as cocaine or methamphetamine). Intracerebral hemorrhage can happen at any age, with the average age being lower than for ischemic strokes. However, the risks continue to increase with age.

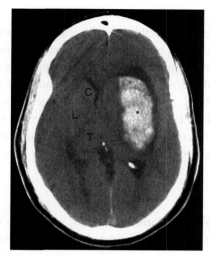

Fig. 5.9 Intracerebral hemorrhage.

Intracerebral hemorrhages are usually single, and at times can be large and catastrophic. Less common causes can be lobar intracerebral hemorrhages due to amyloid deposition in the cerebral arteries. This condition is known as *cerebral amyloid angiopathy*, and affects elderly people over age 65. Lobar hemorrhages in this condition may be multiple, and often recur. Amyloid angiopathy is a unique disease than can cause different types of bleeding such intracerebral hemorrhage, subarachnoid hemorrhage, and even micro-hemorrhages.

Pathophysiology

Intracerebral hemorrhage causes blood to accumulate as a mass and dissect through and compress nearby brain tissues. This can lead to neuronal dysfunction. Intracranial pressure is increased by large hematomas. Supratentorial hematomas and their accompanying edema cause pressure that may result in transtentorial brain herniation. It compresses the brainstem and often causes secondary hemorrhages within the midbrain and pons, as in the case of an **uncal herniation**. A hemorrhage that ruptures into the ventricular system is called an *intraventricular hemorrhage,* which is often associated with a poor prognosis. The blood in the ventricular system can disrupt the flow of the CSF, leading to acute hydrocephalus. Hematomas in the cerebellum may expand, blocking the fourth ventricle, causing a noncommunicative or obstructive hydrocephalus, or pressure on the brainstem structures, which can be very dangerous. Large hematomas may cause a shift of brain tissues to the other side of the brain, which is also called midline shift or herniation.

Clinical manifestations

Clinical manifestations of intracerebral hemorrhage include headache, weakness, confusion, difficulty swallowing, vision problems, loss of balance and coordination, language difficulties, and altered mental status, Hemorrhage puts pressure upon the brain and interferes with its oxygen supply, quickly causing brain and neuronal damage. Neurologic deficits usually occur suddenly and progressively. Large hemorrhages in the posterior fossa can be fatal within a few days in some patients. There may be very few neurologic deficits, since hemorrhage is actually less destructive to brain tissues than ischemic infarction. Clinically, hemorrhagic stroke patients are not easily distinguishable from ischemic stroke patients.

Diagnosis

Diagnosis is by neurological examination, CT scan, and MRI. This condition sometimes requires *immediate treatment*. There will be a sudden onset of headache, focal neurologic deficits, and impaired consciousness, especially in patients with significant risk factors.

Treatment

Treatment requires supportive measures and controlling general risk factors. Anticoagulants and antiplatelet drugs must be stopped acutely in all patients with acute brain bleeding. In patients who previously used anticoagulants, effects are attempted to be reversed by administering fresh frozen plasma, vitamin K, or platelet transfusions, and *protein complex concentrate* (PCC) depending on the type of anticoagulation that was used. Hypertension is treated only if systolic BP is higher than 160. Nicardipine or labetalol are commonly administered intravenously, in closely monitored doses to decrease systolic BP by no more than 15% in the first 8 h.

Sometimes, surgery is indicated to relieve pressure on the neighboring brain tissues and even to repair damage to blood vessels. Early evacuation of large lobar cerebral hematomas can save the lives of some patients. However, rebleeding occurs often, which may increase neurologic deficits. Evacuation procedures in the setting of acute cerebral bleeding have lost their popularity in the recent years, due to higher rates of neurological complications in patients who were operated on immediately. Long-term treatment is based on the location of the hemorrhage and the extent of the damage. Such treatment may include physical, speech, and occupational therapy. Prevention of intracerebral hemorrhage involves stopping smoking, treating heart disease and hypertension, controlling diabetes, adequate physical activity, and good diet. Often, permanent disability occurs, and those who recover fully may require months or years of therapy.

Cerebral aneurysm

Cerebral **aneurysms** are local dilations in arteries. They are most common at points where the arteries branch. When close to the brain, they usually occur in the anterior half of the cerebral arterial circle, or very close to it. They can occur in other locations as well. There are two ways in which an aneurysm causes neurological deficits. Aneurysms can become large. As they grow, they can push against brain structures, compressing them. They can also rupture, causing damage to the nearby brain tissues.

Pathophysiology

An aneurysm can also rupture, as mentioned above. This causes a **subarachnoid hemorrhage**. Based on the size of such a hemorrhage, and where it is located, the outcomes can vary. However, many aneurysms, especially when found before becoming too large, can be surgically repaired.

The incidence of brain aneurysm is less than 5% in the general population. Common contributing factors include arteriosclerosis, hypertension, dissection, and hereditary connective tissue disorders. These disorders include Ehlers-Danlos syndrome, **pseudoxanthoma elasticum**, and polycystic kidney disease. Less common causes include septic (infectious) emboli, which cause *mycotic aneurysms*. These usually occur distal of the first bifurcation of the arterial branches of the circle of Willis. Usually, brain aneurysms are less than 2.5 cm in diameter. They are either *saccular*, which is also described as *noncircumferential,* or non-saccular (fusiform).

> **Focus on vascular malformation**
> There are four main vascular malformations: **AVM**, telangiectasia, **cavernous angioma**, and venous anomaly. AVM is a congenital vascular malformation involving large anastomoses between arteries and veins. They usually have a circumscribed area that circumvents the capillary system. AVMs can become larger as the patient ages. Neurological problems develop by two different primary mechanisms. The first is when AVMs pull blood away from nearby normal brain tissues, causing stroke in the nearby tissues. The second mechanism is due to bleeding in this high flow network of arteries and veins that are interconnected.
>
> **Vein of Galen malformation** is the most common congenital vascular malformation. This condition is an example of a venous anomaly. It is basically a connection between the cerebral artery and cerebral vein. This connection (also known as a shunt) can cause such a high flow of blood between these two vessels that it often causes heart failure in neonates.

Clinical manifestations

Clinical manifestations, when aneurysms are symptomatic, include diplopia, and ocular palsies (especially with aneurysm of the posterior communicating artery, which can compress the oculomotor nerve). Aneurysms can cause subarachnoid hemorrhage due to bleeding into the subarachnoid space. They sometimes cause headaches before they rupture due to leakage of blood. This type of headache is called a *sentinel headache*. The actual aneurysm rupture causes a sudden, severe headache also known as a **thunderclap headache**.

Diagnosis

Detection of aneurysms may occur via neuroimaging—often while searching for other conditions. Diagnosis requires angiography, CT angiography, or MR angiography.

Treatment

Asymptomatic aneurysms less than 7 mm in size are less likely to rupture, and no immediate treatment is required in most cases. They are monitored with serial imaging studies in the following years (usually every 3 years). When the aneurysm is larger, located in the posterior circulation, or causing symptoms due to bleeding or compression of nearby neural structures, endovascular, or surgical therapies may be indicated.

Section review
1. What is cerebral amyloid angiopathy?
2. Explain AVM.
3. What are the clinical manifestations of aneurysms?

Focus on cerebral angiography

Most imaging procedures allow visualization of blood vessels. **Cerebral angiography** uses intravenous injection of iodinated dyes. This results in the blood appearing much more opaque, in X-rays, than the rest of the brain. Usually, a cerebral angiogram is produced following introduction of a catheter into the femoral artery. The catheter is treaded, using fluoroscopic visualization, up to the aorta, into the aortic arch. It is then moved into the artery being examined. Therefore, contrast material can be injected into either vertebral or internal carotid arteries.

When the dye is introduced, X-rays are taken very quickly, following the dye as it flows through the artery, into the capillaries, and finally, into the veins. Also, reconstructive photography and digital procedures can be used to remove images of bones, and to reveal blood vessels in controlled isolation.

Angiography was the first procedure developed that made images of both normal and diseased blood vessels. For many decades, it was used to assess brain changes that resulted in distortion of vessels. Today, it still produces images of great detail concerning cerebral vessels. However, *computed tomography angiography (CTA)* and *MRA* are less invasive, and also show the CNS in great detail. Therefore, they are now widely used for blood vessel imaging as first-line diagnostic tests in the majority of the patients.

Spinal cord ischemia

Spinal cord ischemia is a relatively uncommon form of spinal cord injury. Less than 2% of central neurovascular events affect the spinal cord. Less than 8% of acute myelopathies involve ischemia. The reason that spinal cord ischemia is uncommon is that the cord has many collateral sources of blood supply.

Pathophysiology

Several conditions can cause reduced spinal cord blood flow. These include hypotension, thoracoabdominal surgery, and mechanical spinal cord compression. Global hypotension can cause a large amount of injury to various body organs, and their dysfunction can mask the spinal cord injury. Some patients with hypoxic ischemic encephalopathy also have spinal cord ischemia, but this is often hidden by injury to the brain.

In thoracoabdominal surgeries, the risk of permanent paraplegia as a complication is relatively rare. However, a degree of permanent neurologic deficit can affect some patients. Sometimes, rheumatoid arthritis and degenerative bone changes of the neck bones can predispose an individual to ischemia during neck extension. Neck extension is required during intubation and airway protection. Therefore, it is extremely dangerous in patients with degenerative bone diseases of the neck. Hemorrhage into the spinal cord, or around it, can also cause compression of the arterial blood supply, resulting in ischemia.

Clinical manifestations

Spinal cord ischemia may involve weakness, usually following within minutes or even hours. It is an acute process in almost all patients. Urinary retention is common, but the weakness is the most often-seen symptom, and is usually greater than the sensory deficits, depending on the vascular territory of the ischemia. Weakness is bilateral. Recovery varies, based on the severity of the ischemia. Also, the absence of motor recovery after the first weeks usually indicates that the patient will not be fully independent in the future.

Diagnosis

Diagnosis is mainly based on history, physical examination, and clinical setting. Ischemia may be confirmed by MRI with diffusion-weighted sequences. The sensitivity of spinal MRI is less than brain MRI. Acute MRI of the spine is mostly available in academic settings.

Treatment

Management of blood pressure is very important to ensure adequate blood perfusion to the ischemic tissue. Placing a lumbar drain for CSF, to reduce pressure upon venous outflow, together with skilled neurointervention, prevents spinal cord ischemia in the majority of the cases. The mean arterial pressure needs to be monitored. In almost all patients without heart disease, a compensational physiologic process starts in the body, which increases the mean arterial pressure. This promotes perfusion. Fluids are administered if the blood pressure is low, and medications including vasopressors, such as phenylephrine are rarely used. Rehabilitative physical therapy should begin as soon as possible after other treatments are completed.

In spinal cord ischemia, if motor deficits are not severe, some patients regain the ability to walk. Repeat imaging is sometimes indicated if there is any worsening of the patient's condition.

Clinical cases

Clinical case 1
1. What is the likely diagnosis of this patient's condition?
2. Based on this patient's age and deficits, what type of medications should be given?
3. Based on this case, which area of her brain had lesions?

A 72-year-old woman with a history of uncontrolled hypertension developed difficulty walking in the morning. Her husband called 911, and the ambulance took her to a primary stroke center. She was lethargic but able to understand questions. However, she could not stand without being assisted, and was much weaker on the left side of her body. The patient was previously taking three medications: atenolol, clonidine, and aspirin. Her blood pressure was 160/100, her pulse was normal, and EKG showed no significant information. The attending physician ordered a CT scan, which revealed acute ischemic changes within the pons. All of the patient's lab tests were within normal limits.

Answers:
1. Probable ischemic stroke.
2. Thrombolytic therapy (tPA), via intravenous administration.
3. Because of the patient's left-sided symptoms, the right side had ischemic lesion. Lethargy probably was caused by the involvement of the arousal system in the pons.

Clinical case 2
1. What is the likely diagnosis of this patient's condition?
2. How did the neurologist differentiate between a TIA and a stroke?
3. What type of lifestyle modifications should be discussed with this patient?

A 58-year-old man was brought to the emergency department with slurred speech and a facial droop on the right side. The patient complained of numbness on the right side of the face as well as in the right arm. He was previously diagnosed with hypertension and hypercholesterolemia, and had previously used tobacco products for 25 years. He had a positive family history of heart disease, and did not exercise regularly. Blood pressure was 188/97, pulse rate 81, and EKG was normal. The patient did not complain of headache, nausea, vomiting, or chest pain, and vision was normal. A CT scan and MRI of the brain were normal. The patient was admitted so that neurological evaluation, MR angiography, fasting serum cholesterol, and BP monitoring could occur. Symptoms resolved within few minutes.

Answers:
1. The likely diagnosis is a TIA.
2. The fact that the patient's symptoms resolved soon after arrival indicate a TIA and not a stroke. TIA does not show an evidence of stroke on the DWI sequence of the MRI.
3. The patient must have regular exercise, and change his diet in order to lower his cholesterol, due to his previous hypercholesterolemia diagnosis. He must lose weight and eat healthier. However, controlling blood pressure is the most modifiable risk factor for stroke.

Clinical case 3
1. What type of radioimaging can be used to diagnose cerebral venous thrombosis?
2. What medications will be used to treat this patient?
3. How can excessive alcohol intake result in this condition?

A 40-year-old male patient is hospitalized following a weekend of binge drinking with his college friends. He has a severe headache and weakness of his left arm and left leg. Imaging studies revealed flow limitations in the superior sagittal sinuses. There is a hyperdensity in the left parietal sulcal space. There is also slight hypodensity in the right parasagittal frontal lobe, with subtle hyper dense areas. The hospital physician diagnosis this patient with cerebral venous thrombosis, as well as an early evolving hemorrhagic infarct.

Answers:
1. Radioimaging techniques for cerebral thrombosis include MRI, CTV, and MRV.
2. Anticoagulants are used for cerebral thrombosis.
3. Excessive alcohol intake leads to dehydration, which is a major risk factor for cerebral venous thrombosis.

Clinical case 4
1. Is atherosclerosis involved in spinal cord ischemia?
2. Is the anterior or posterior part of the spinal cord most often involved?
3. Which type of imaging study is the preferred for diagnosing this condition?

A 67-year-old man was taken to the emergency department with sudden, severe pain on the right side of his body, loss of bladder control, and bilateral upper limb weakness. Examination revealed an atypical pattern of sensory deficit. The patient's pain was assessed, and after imaging studies, he was diagnosed with acute spinal cord ischemia syndrome.

Answers:
1. Yes, in adults, atherosclerosis is usually the cause of spinal cord ischemia.
2. The anterior spinal cord is usually involved. However, the effects of anterior spinal cord ischemia are mostly bilateral because of the single, midline anterior spinal artery. In this patient's case, the posterior spinal cord was involved, since he had unilateral symptoms, due to the paired posterior spinal arteries.
3. MRI is the best method of diagnosing spinal cord ischemia. It is used for all patients with suspected spinal cord infarction, to confirm the diagnosis, but also to exclude a diagnosis of spinal cord compression or other types of impairment.

Clinical case 5
1. Related to the patient in this case study, how is an epidural hematoma differentiated from other types of conditions?
2. What are the appropriate treatments for this patient?
3. What complications can occur from his condition?

A 23-year-old college student was returning home on his bicycle. At a street corner, he was struck by a car. His head hit the curb and he was knocked unconscious. He was brought to the emergency room, where he awakened and complained of a severe headache. He had right-sided weakness, and an inability to communicate, which was getting worse as the emergency physician was examining him. Various tests were ordered *stat*, the student was diagnosed with an epidural hematoma in his left frontal lobe.

Answers:
1. A sudden severe headache is always a symptom that should make the health-care professional consider the possibility of acute bleeding in the brain. The most common examples are epidural, subdural, and subarachnoid hemorrhage, as well as hemorrhagic stroke. Whenever acute bleeding in the brain is a concern, an emergency CT scan of the brain, without contrast, should be obtained immediately. The bleeding is lens or convex shaped in epidural hematomas, but crescent shaped in subdural hematomas. The reason behind the shape of the bleeding in epidural hematoma is the location. This blood is confined between the dural layer and the skull bone in an epidural bleed. The dural layer is the most dense layer of the three meningeal layers. It covers the brain tightly, and follows the brain tissue grooves such as the sulci and fissures. Therefore, a bleed between the skull and this dense layer is confined tightly within, making the edges of the bleed sharp, causing a lens-shaped or convex appearance.
2. Acute surgical treatments of traumatic brain hemorrhages are sometimes considered on a case-by-case basis. Surgical drainage is usually by drilling into the skull, either using one or multiple small holes, which are also called burr holes.
3. Complications include increased intracranial pressure, permanent brain damage, coma, and even death.

Clinical case 6
1. What are the risk factors for a cerebral aneurysm to rupture?
2. If the patient in this case was not diagnosed and treated quickly, what would the likely complications be?
3. What are the diagnostic methods for cerebral aneurysm?

A 56-year-old woman experienced lightheadedness, and a sensation of something she later described as "popping" in the back of her neck. This was followed by a sudden, severe headache. The patient was restless due to the severity of the headache. She had left-sided weakness as well as severe slurred speech at the hospital. CT brain scan without contrast and CTA of the head and neck with contrast were obtained within 30 min of the patient's arrival to the emergency room, which revealed a ruptured right middle cerebral artery, bleeding into the subarachnoid area within the right frontoparietal lobes.

Answers:
1. Risk factors that increase the likelihood of a rupture of a cerebral aneurysm include uncontrolled hypertension, smoking, large size of the aneurysm, fast aneurysm growth, and family history of ruptured aneurysm.
2. Without prompt diagnosis and treatment, a ruptured cerebral aneurysm can cause three severe complications; hydrocephalus, vasospasm (which can lead to stroke), and seizures.

3. The diagnosis of cerebral aneurysms is made by CTA or MRA, and in some cases by cerebral angiography. Also, analysis of CSF should be done, to look for blood in the CSF (xanthochromia), when the initial CT brain scan is normal in the acute setting of sudden severe headache and focal neurological findings.

Key terms

Aneurysms
Anterior cerebral artery
Anterior choroidal artery
Anterior communicating artery
Anterior inferior cerebellar artery
Anterior perforated substances
Arteriovenous malformation (AVM)
Azygos system
Basal vein (of Rosenthal)
Basilar artery
Basilar venous plexus
Callosomarginal artery
Cavernous angioma
Cerebral angiography
Cerebral arterial circle
Cerebral ischemia
Cerulea dolens
Choroidal vein
Cranial fossae
Deep middle cerebral vein
Deep vein thrombosis (DVT)
Deep veins
Embolism
Embolus
Emissary veins
Epidural venous plexus
Fabry disease
Great cerebral vein
Hemianesthesia
Hypodensities
Infarct
Internal carotid arteries
Internal cerebral vein
Internal jugular veins
Intracerebral hemorrhage
Ischemic strokes
Labyrinthine artery
Lacunes

Lenticulostriate arteries
Magnetic resonance angiography (MRA)
Magnetic resonance venography (MRV)
Marantic endocarditis
Meningeal arteries
Microatheroma
Middle cerebral artery
Ophthalmic artery
Perforating arteries
Pericallosal artery
Phlegmasia alba
Pontine arteries
Posterior cerebral artery
Posterior choroidal arteries
Posterior communicating artery
Posterior perforated substances
Pseudoxanthoma elasticum
Radiculomedullary veins
Sarcopenia
Septal vein
Spinal cord ischemia
Stenting
Stroke
Subarachnoid hemorrhage
Superficial veins
Superior cerebellar artery
Thalamostriate vein
Thrombosis
Thrombus
Thunderclap headache
Transient ischemic attack (TIA)
Transient monocular blindness
Transverse sinuses
Uncal herniation
Vasculitides
Venous angle
Venous dural sinuses
Vertebral arteries

Suggested readings

1. Bianchi, D. W.; Crombleholme, T. M.; D'Alton, M. E.; Malone, F. *Fetology: Diagnosis and Management of the Fetal Patient*, 2nd ed.; McGraw-Hill Education/Medical, 2010.
2. Bradac, G. B.; Boccardi, E. *Applied Cerebral Angiography: Normal Anatomy and Vascular Pathology*, 3rd ed.; Springer, 2017.
3. Browning, W. *The Veins of the Brain And Its Envelopes*. Sagwan Press, 2018.
4. Campagnolo, D. I.; Kirshblum, S.; Nash, M. S.; Heary, R. F. *Spinal Cord Medicine*, 2nd ed.; LWW, 2011.
5. Connolly, E. S.; McKhann, G. M. *Subdural hematomas, An Issue of Neurosurgery Clinics of North America*, 28th ed.; Elsevier, 2017.
6. Cramer, G. D.; Darby, S. A. *Clinical Anatomy of the Spine, Spinal Cord, and ANS*, 3rd ed.; Mosby, 2013.
7. Dodds, J. A.; Anderson, A. P. *Carotid and Vertebral Artery Dissection: A Guide for Survivors and Their Loved Ones*. CreateSpace Independent Publishing Platform, 2017.
8. Dolenc, V. V.; Rogers, L. *Cavernous Sinus: Developments and Future Perspectives*. Springer, 2009.
9. Farooqui, A. A. *Ischemic and Traumatic Brain and Spinal Cord Injuries: Mechanisms and Potential Therapies*. Academic Press, 2018.
10. Goodman, C. C.; Marshall, C. *Pathology for the Physical Therapist Assistant*, 2nd ed.; Saunders, 2016.
11. https://www.healthline.com/health/lobar-intracerebral-hemorrhage.
12. https://www.hopkinsmedicine.org/healthlibrary/conditions/hematology_and_blood_disorders/thrombosis_85,p00105.
13. Icon Group International. *Intracranial Hemorrhage: Webster's Timeline History, 1885–2007*. ICON Group International, Inc., 2010
14. Joseph, R. S. *Brain Damage: Thrombi, Emboli, Hemorrhage, Aneurysms, Atherosclerosis, TIA, CVA, Disturbances of Cognition, Memory, Language, Visual Perception, Emotion, Physical Sensation*. University Press, 2011.
15. https://www.kenhub.com/en/library/anatomy/blood-supply-of-the-spinal-cord.
16. Kitchens, C. S.; Konkle, B. A.; Kessler, C. M. *Consultative Hemostasis and Thrombosis*, 4th ed.; Elsevier, 2018.
17. Marieb, E. N.; Hoehn, K. *Human Anatomy & Physiology*, 10th ed.; Pearson, 2015.
18. https://medbroadcast.com/condition/getcondition/embolism.
19. Mendelow, A. D.; Lo, E. H.; Sacco, R. L.; Wong, L. K. S.; Grotta, J. C.; et al. *Stroke E-Book: Pathophysiology, Diagnosis, and Management*, 6th ed.; Elsevier, 2015.
20. http://neuroangio.org/spinal-vascular-anatomy.
21. Park, M. S.; Taussky, P.; Albuquerque, F. C.; McDougall, C. G. *Flow Diversion of Cerebral Aneurysms*. Thieme, 2017.
22. Porter, R. S. *The Merck Manual*, 19th ed.; Merck, 2011.
23. https://radiopaedia.org/articles/blood-supply-of-the-meninges.
24. Ringer, A. J. *Intracranial Aneurysms*. Academic Press, 2018.
25. Rinkel, G. J. E.; Greebe, P. *Subarachnoid Hemorrhage in Clinical Practice*. Springer, 2015.
26. Saladin, K. S. *Anatomy & Physiology: The Unity of Form and Function*, 8th ed.; McGraw-Hill Education, 2017.
27. http://www.strokecenter.org/patients/about-stroke/intracerebral-hemorrhage/.
28. Thron, A. K. *Vascular Anatomy of the Spinal Cord: Radioanatomy as the Key to Diagnosis and Treatment*, 2nd ed.; Springer, 2016.
29. Ullah, F.; Raza, S. S.; et al. *The Stroke Manual: ABCs of the Cerebrovascular Accident*. CreateSpace Self Publishing/Amazon, 2017.
30. U.S. Department of Health and Human Services. *The Surgeon General's Call to Action to Prevent Deep Vein Thrombosis and Pulmonary Embolism*. CreateSpace Independent Publishing Platform, 2012.
31. Vanderah, T.; Gould, D. J. *Nolte's The Human Brain: An Introduction to its Functional Anatomy*, 7th ed.; Elsevier, 2015.

CHAPTER 6

Cerebral cortex

The cerebral cortex is a layer of neurons and synapses (gray matter) on the surface of the cerebral hemispheres. This is folded into gyri, and about two-thirds of the cortex's area is buried inside fissures. The cerebral cortex integrates higher mental functions, general movements, functions of the viscera, perceptions, and behavioral reactions. It has many different classifications. The cerebral cortex has 47 separate function areas, with differing cellular designs.

Stimulation of the precentral cortex or motor area, using electrodes, causes contractions of the voluntary muscles. If the motor speech area in the inferior frontal **gyrus** is destroyed, this causes motor aphasia or speech defects, even though the vocal organs are healthy and intact. If the brain cortex is stimulated as in case of a seizure, this stimulation will affect circulation, respiration, reactions of the pupils, and other visceral activities.

Cerebrum

The **cerebrum** is the largest and most obvious portion of the human brain. It forms from the embryonic structure called the *telencephalon*. The cerebrum is the center of voluntary motor control and complex mental processes.

Gross anatomy

The cerebrum is much larger than the other portions of the brain. It is divided into two *cerebral hemispheres* that are separated by the *longitudinal fissure*. A prominent tract of fibers called the **corpus callosum** also connects the hemispheres. Each hemisphere has conspicuous *gyri*, which appear as "wrinkles" that are separated by grooves known as *sulci*. The surface of the cerebrum folds into the gyri in a way that allows for a larger amount of cortex to fit inside the cranium and because of the gyri, the cerebrum has about 2500 cm^2 of surface area. Without the gyri, it would be one-third smaller and is significantly less able to process information. The human brain is unique compared to most other mammalian brains because of this extensive folding of the cerebrum.

Though the gyri of the human brain are usually very similar between individuals, there can be variances. Even in the same person, the gyri of the right cerebral hemisphere may be different than those of the left slightly. Each hemisphere is subdivided into five lobes with distinct anatomical and functional differences by some sulci that appear more prominently. The first four lobes are superficially visible. They are named for each cranial bone overlying them. There is a fifth lobe described by some, called the *insula*, which cannot be seen from the brain's surface. The cerebral lobes often work together in functions such as memory, speech, emotion, and vision. Certain lobes assist with only part of the overall cerebral function. The five cerebral lobes of the brain are as follows:

- **Frontal lobe**—lies just behind the frontal bone, superior to the eyes. It extends from the forehead caudally to a curved vertical groove known as the **central sulcus**. The frontal lobe is the center of abstract, conscious thought, and declarative or explicit memory. It is also where emotional and cognitive processes occur. These include mood, foresight, motivation, decision-making, planning, emotional control, the judgment of appropriate behaviors, speech production, and other voluntary motor controls. Frontal lobe lesions leave the patients disinhibited often with impulsive sexual conversations and behaviors.
- **Parietal lobe**—the upper part of the brain that underlies the parietal bone. It begins at the central sulcus and extends caudally to the **parieto-occipital sulcus**. The parietal lobe is visible on each hemisphere's medial surface. It handles visual processing (through the optic radiations that pass through the parietal lobe and temporal lobe, until they reach their final destination at the occipital lobe). The parietal lobe also controls spatial perception, which is often called "visuospatial perception." Large lesions such as a stroke in the right parietal lobe cause the patient to neglect the left side of his or her world. Atrophy and shrinking of the parietal lobe are the reason behind Alzheimer's patients' difficulties with navigation and getting lost, since the visuospatial tasks are disrupted due to atrophy of the parietal lobe.
- **Occipital lobe**—located at the back of the head, underlying the occipital bone, and caudal to the parieto-occipital sulcus. The occipital lobe is the main visual center.
- **Temporal lobe**—deep to the temporal bone, this lobe is lateral and horizontal. It is separated from the frontal and parietal lobes above it via the deep **lateral sulcus**. The temporal lobe is utilized in hearing, emotion, smell, language comprehension, learning, memory related to grammar and vocabulary, formation of new long-term memories (*memory consolidation*), and the storage of auditory, verbal, and visual memories. Temporal lobe lesions can cause memory problems, which are frequently seen in patients with Alzheimer's disease (AS) who have shrinking or atrophy of the temporal lobe. Recall that in AD, the temporal and parietal lobes are the main parts of the brain that are affected and atrophied.
- **Insula**—part of the temporal lobe. It is a small mass of cerebral cortex, deeply below the lateral sulcus. Retracting or removing some of the cerebrums above it can only see it. The insula is used to process sensations of pain and taste, in visceral sensation, emotional responses and empathy toward others, consciousness, and balancing heart rate and blood pressure when exercising, as well as other activities of cardiovascular homeostasis.

Section review
1. What is the basic structure of the cerebrum?
2. What are the five cerebral lobes called?
3. Where is the main visual center of the brain located?

Cerebral cortex

The **cerebral cortex** is the layer of the brain that covers the cerebral hemispheres. It makes up about 40% of the mass of the brain, containing 14–16 billion neurons. The two main types of neurons processed by the cerebral cortex are *stellate cells* and *pyramidal cells*. The **stellate cells** have rounded cell bodies, and short dendrites and axons that project in every direction. They are mostly involved with localized reception of sensory input and information processing.

The **pyramidal cells** are cone-shaped and tall in structure, with their apex pointing toward the brain surface. They have one thick dendrite with multiple branches, and smaller, knob-like *dendritic spines*. Their base has horizontally aligned dendrites and an axon passing into the white matter. Some pyramidal cells are the cerebrum's output neurons, the only ones with fibers leaving the cortex and connecting with other areas of the central nervous system (CNS). The pyramidal cell axons also have collaterals synapsing with deeper regions of the brain, or other cortical neurons.

Neocortex

The **neocortex** is a six-layered tissue that makes up most (about 90%) of the human cerebral cortex. The six layers of neurons have different thicknesses between the various parts of the cerebrum. The layers are also different in composition of the cells, connections of the synapses, neuron size, and where the axons end. All axons leaving the cortex and entering the white matter arise from layers III, V, and VI. Layer V is thickest in motor regions, while Layer IV is thickest in sensory regions.

Certain areas of the cerebral cortex have fewer layers. The type of cortex that appeared early in the evolution of vertebrates was called the **paleocortex**, which consisted of between one and five layers. In humans, it was limited to only part of the insula and some areas to the temporal lobe that processed the sense of smell. A three-layered **archicortex** then evolved in the hippocampus, which is the temporal lobe's memory-forming center. The neocortex was the last type of cortex to evolve.

Functional cerebral cortex areas

Some areas process motor information while others process sensory information. These two types of areas are separated by the central sulcus. There are also *association areas*,

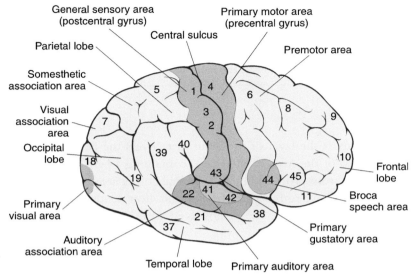

Fig. 6.1 The motor, sensory, and association areas of the cerebral cortex.

coordinating data going in and out of the motor and sensory areas. The motor, sensory, and association areas of the cerebral cortex are shown in Fig. 6.1.

Each cerebral hemisphere cortex sends and receives information from the opposite side of the body. An example is the left cerebral motor areas controlling the body's right-sided muscles. It is not known why this crossover of activities exists. One area of the cerebral cortex often has functions that overlap with other areas. Consciousness and other cortical functions are not the role of just one area. The functions of each lobe or area of the cerebral cortex are further summarized in Table 6.1.

Motor areas

The **primary motor cortex** is the surface of the precentral gyrus. Its neurons control voluntary movements, by regulating somatic motor neurons of the brainstem and spinal cord. These neurons have pyramidal cells. When a certain motor neuron is stimulated in the primary motor cortex, a contraction is generated in a specific skeletal muscle.

Sensory areas

At each sensory area of the cerebral cortex, information is reported in the pattern of neuron activity that is present. The **primary somatosensory cortex** is contained in the surface of the postcentral gyrus. Its neurons receive general somatic sensory information from pain, pressure, temperature, touch, and vibration receptors. We become aware of these sensations whenever nuclei of the thalamus relay information to the primary somatosensory cortex. Other areas of the cerebral cortex receive sensations of sight, smell, sound, and taste.

Table 6.1 The functions of the cerebral cortex.

Lobe or area	Functions
Frontal lobe	
Primary motor cortex	Voluntary skeletal muscle control
Parietal lobe	
Primary somatosensory cortex	Conscious perception of pain, pressure, taste, temperature, touch, and vibration; visuospatial tasks
Occipital lobe	
Visual cortex	Conscious perception of visual stimuli
Temporal lobe	
Auditory cortex, olfactory cortex	Conscious perception of hearing (auditory) smell (olfactory) stimuli as well as memory
All lobes	
Motor, sensory, association, and somatosensory areas	Sensory data integration and processing; motor activity processing and initiation

In the occipital lobe, the **visual cortex** receives visual information. In the temporal lobe, the **auditory cortex** receives information about hearing, and the **olfactory cortex** receives information about the smell. In the anterior insula and nearby frontal lobe areas, the **gustatory cortex** receives information from taste receptors of the pharynx and tongue. Therefore, any seizures coming from the insula may cause gustatory hallucinations or abnormal tastes in the mouth.

Association areas

Nearby **association areas** connect with the sensory and motor areas of the cerebral cortex. They interpret incoming data or coordinate motor responses. *Sensory association areas* monitor and interpret sensory information, and include the *somatosensory association cortex*, *visual association area*, and *premotor cortex*. In the primary somatosensory cortex, the **somatosensory association cortex** monitors these activities. It allows the recognizing of light touch, such as an insect landing on the skin.

The special senses of hearing, sight, and smell involve individual association areas. The **visual association area** is one example. It monitors and interprets activity in the visual cortex. It processes letters seen by the eyes into recognizable words. If the visual association area becomes damaged, a person could still see printed letters as symbols, but there would be no meaning obtained from them. Likewise, the **auditory association area** monitors activities in the auditory cortex, and this is where word recognition occurs. In general, these association areas are involved in a higher level of function and more complex tasks. For instance, if the primary visual cortex is responsible for recognizing and visualizing simple objects, the association cortex is associated with forming more complex pictures.

The **premotor cortex**, also known as the *somatic motor association area*, coordinates learned movements. The primary motor cortex does nothing by itself. Its neurons must be stimulated by other cerebral neurons. A voluntary movement occurs when the premotor cortex transmits instructions to the primary motor cortex. The proper pattern of stimulation is stored in the premotor cortex when actions are repeated. Therefore, the movement becomes easier and smoother since the *pattern* is triggered instead of the individual neurons being controlled. The frontal eye field of the premotor cortex controls learned eye movements. If this field is damaged, the individual can understand written letters and words but is unable to read since the eyes cannot follow the lines that are printed.

Integrative centers and higher mental functions

Information from a large number of association areas is received by *integrative centers*. These centers control complicated motor activities and perform analysis of information. From the sensory association areas, the *prefrontal cortex* integrates information and performs intellectual functions that include prediction of possible outcomes of various responses. The integrative centers are located in each cerebral hemisphere's lobes and cortical areas. They regularly handle mathematical computations, speech, understanding of spatial relationships, and writing. However, for these functions, the integrative centers are mostly restricted to either the right or left hemisphere. *Hemispheric lateralization* refers to the differing functions between the cerebral hemispheres. Related regions on the opposite hemisphere are active but have less well-defined functions. Even so, the cerebral hemispheres appear to be almost the same. For instance, a left-side weakness in the setting of stroke lateralizes the stroke lesion to the right side of the brain.

Specialized language areas

Language processing is very complicated and occurs in both cerebral hemispheres. However, it varies between individuals. Related to primarily the left hemisphere are *Wernicke's area* and *Broca's area*. The first, **Wernicke's area**, is close to the auditory cortex, and located in the superior temporal gyrus. It is associated with language comprehension, receiving information from the sensory association areas. It is important in a person's personality since it integrates sensory information and coordinates the access to auditory and visual memories.

Broca's area is also known as the *motor speech area*. It is near the motor cortex and utilized in speech production, located in the inferior frontal gyrus. This area regulates breathing patterns while speaking and vocalizations required for normal speech. It coordinates the activities of the muscles of respiration, the larynx, the pharynx, as well as those of the cheeks, lips, jaws, and tongue. If a person has damage to Broca's area, sounds can be made, but words cannot be formed. The *receptive speech area* is another name for the auditory association area. It utilizes feedback to adjust motor commands from the motor speech area. Many different speech-related problems can occur because of damage to a specific sensory area. Some patients have problems with speaking but understand the usage of correct words, while others can speak consistently yet use many incorrect words. Please note that these two (Broca's and Wernicke's areas) are the main language centers. However, in order to be able to read, speak, and write, other areas of the brain need to function in coordination.

Prefrontal cortex

In the frontal lobe, the **prefrontal cortex** coordinates information from all cortical association areas. It handles abstract functions such as predicting outcomes of actions or events. This cortex is not fully developed until a person is in his or her early 20s. If damaged, there will be difficulties in estimating relationships between specific events, such as understanding when something occurred in the past, or which events happened in which specific order.

There are complex connections between the prefrontal cortex and other brain areas. As the prefrontal cortex interprets events and predicts outcomes, there may be feelings of anxiety, frustration, or tension. However, if the connections between these areas are severed, these negative feelings disappear. This is why, in the mid-20th century, a procedure called a **prefrontal lobotomy** was used, especially upon violent or antisocial individuals. After a lobotomy, the patient would no longer be troubled by major psychological (hallucinated) problems or feel severe pain. Unfortunately, this procedure also caused inabilities to be tactful with others, be respectful of proper behaviors, or to even maintain their own sanitary needs. Today, lobotomies are no longer used to alter behavior, since the discovery of medications that target specific CNS areas and pathways. As mentioned before, patients with frontal lobe lesions are usually disinhibited and act impulsively.

Hemispheric lateralization

In the majority of people, the left cerebral hemisphere contains specialized language areas. The premotor cortex utilized for hand movements is larger on the left side for people who are right-handed than for people who are left-handed. The left hemisphere is also crucial for logical decision-making, mathematical calculations, and for other analytical functions.

The right cerebral hemisphere processes sensory information and helps the body relate to its environment. It allows the identification of familiar sights, smells, tastes, or touches. It is the dominant hemisphere for face recognition and for understanding three-dimensional images. The right hemisphere analyzes how emotions are used by others to whom we are speaking. When the right hemisphere is damaged, the individual may not be able to add emotional inflections to their speech, which is also called *prosody* and can happen with nondominant frontal lobe lesions. To assess language in a patient, there is testing for fluency, repeating ability, and comprehension, along with reading and writing skills.

Only about 9% of the human population is left-handed. Usually, the right hemisphere's primary motor cortex controls motor function for the dominant left hand. However, the centers for analytical speech and function are in the left hemisphere, exactly as they are in right-handed people. It is interesting to realize that a very high percentage of artists and musicians are left handed. Also, as a person begins to favor one hand over the other, the connection with the opposite side of the brain strengthens. As mentioned before, up to 90% of the population are left hemisphere dominant regardless of whether they are left or right handed.

Section review
1. What are the functions of the motor, sensory, and association areas of the brain?
2. What is the function of the primary somatosensory cortex?
3. What are the locations and functions of Wernicke's area and Broca's area?

Electroencephalogram

An **electroencephalogram** (EEG), is a printed recording of the electrical activity of the brain. The observed electrical patterns are called *brain waves*. The brains of people undergoing brain surgery, and changes that follow localized strokes or injuries, have been studied to better understand the brain's electrical activities. A *positron emission tomography (PET) scan* allows visualization of activities in specific brain regions, as does *functional magnetic resonance imaging (fMRI)*.

Brain activity is assessed using EEG by placing electrodes on the brain or outer skull surface. The billions of brain neurons and their activities generate a measurable electrical field that changes constantly when areas are stimulated or reduce their activities. A neural function is based on the electrical events of the plasma membrane of neurons.

In the two hemispheres, electrical activity is usually synchronized via a *pacemaker mechanism* involving the thalamus. This is the most acceptable mechanism. When there is a lack of this synchronization, there may be localized damage or a cerebral abnormality, such as a tumor or an injury to just one hemisphere. A **seizure** is a temporary cerebral disorder that involves abnormal movements, inappropriate behavior, unusual sensations, or a combination of these, as a result of electrical discharges. Seizure disorders, or *epilepsies*, are characterized by seizures that happen more than once. Based on the involved cortex area, the symptoms are different.

If the primary motor cortex is affected by a seizure, movements of the contralateral side will occur. If the auditory cortex is affected, the patient may hear odd sounds. Regardless of type, all seizures involve a significant change in electrical activities, which can be seen on an EEG. The changes may begin in one cortical area and then spread across the entire cortical surface. This is called focal seizure with secondary generalization.

Brain waves

Any number of patterns of rhythmic electric impulses are produced in different parts of the brain. There are four types of typical brain waves: *alpha*, *beta*, *theta*, and *delta*. These are similar and relatively stable in all normal individuals. Brain waves help in the diagnosis of certain neurologic disorders, including brain lesions and the likelihood of having a seizure.

Alpha waves

Alpha waves in healthy, awake adults occur while resting with the eyes closed. They disappear during sleep and vanish when there is concentration on a specific task. The rhythm of alpha waves may have a frequency between 8 and 13 Hz. Alpha waves are maximal over the occipital region. In most cases, a neurologist measures the alpha rhythm in the back of the head while the patient's eyes are closed.

Beta waves

Beta waves replace alpha waves during attention to tasks or stimuli and are of higher frequency. They are common while concentrating, or when a person is under stress or experiencing psychological tension. The rhythm of beta waves is usually between 13 and 40 Hz. Beta waves are present over the remaining frontocentral areas of the scalp besides the regions described under the alpha waves. If a patient opens the eyes or begins the mental activity, the alpha waves decrease or attenuate, to be replaced by beta waves all over the scalp. Beta waves from different sites that are out of phase are called *desynchronized*.

Theta waves

Theta waves in normal adults may appear transiently during sleep. However, theta waves are most common in children. If theta waves occur at other times, they may indicate cerebral dysfunction. The rhythm of theta waves is usually between 4 and 7.9 Hz. For example, in a coma caused by sedative medications or severe infection, the background EEG becomes "slow" and mostly consists of theta and delta waves. The neurologist sometimes refers to this as "slow background," which indicates a "nonspecific" dysfunction of the brain in that area.

Delta waves

Delta waves are low-frequency waves. In all ages of patients, they normally occur during deep sleep. They are also present in infant brains that are still developing. Additionally, they are present in adults who are awake, when areas of the brain have been damaged by inflammation, a tumor, or vascular blockage. The rhythm in delta waves are usually between 0.1 and 3.9 Hz.

Split-brain syndrome

Split-brain syndrome is also called *callosal disconnection syndrome*. It is a condition involving a cluster of neurological abnormalities caused by partial or complete severing or lesioning of the corpus callosum. The most common cause is a surgical procedure called a *corpus callosotomy*, but this is rarely performed today, only for severe cases of congenital refractory epilepsies. It is reserved as the last measure of treatment for extreme and uncontrollable epilepsies. By stopping propagation of seizure activity across the cerebral hemispheres, the patient's quality of life can be greatly improved. However, the procedure causes acute hemispheric disconnection symptoms lasting for days or weeks, and often, permanent chronic symptoms.

Additional and less-common causes of the split-brain syndrome include infectious lesions, ruptured arteries, stroke, tumors, and multiple sclerosis. Rarely, agenesis of the corpus callosum may cause split-brain syndrome. The syndrome causes the affected individual to be unable to learn to perform new tasks that require interdependent movement of each hand. Eye movements also remain coordinated. Unusual behaviors concerning speech and object recognition occur. These syndromes are also called *disconnection syndromes* since the white matter connection between two parts of the brain is disrupted.

White matter of the cerebrum

There is mostly white matter inside the cerebrum. The axons of the white matter are either classified as *association fibers*, *commissural fibers*, or *projection fibers* (see Fig. 6.2).

- **Association fibers**—in one cerebral hemisphere, these fibers interconnect areas of the cerebral cortex. The shorter **arcuate fibers** curve in an arc-like fashion, passing from one gyrus to another. Longer fibers form bundles called *fasciculi*. The **longitudinal fasciculi** connect the frontal lobe to the other lobes within the same cerebral hemisphere.
- **Commissural fibers**—between both hemispheres, these fibers interconnect to allow communication. Their bands, which link the hemispheres, include the corpus callosum and the **anterior commissure**. There are more than 200 million axons in the corpus callosum, which can carry approximately four billion impulses every second.
- **Projection fibers**—these fibers link the cerebral cortex to the diencephalon, brainstem, cerebellum, and spinal cord. They all pass through the diencephalon, where ascending axons linking the sensory areas pass near descending axons from the motor

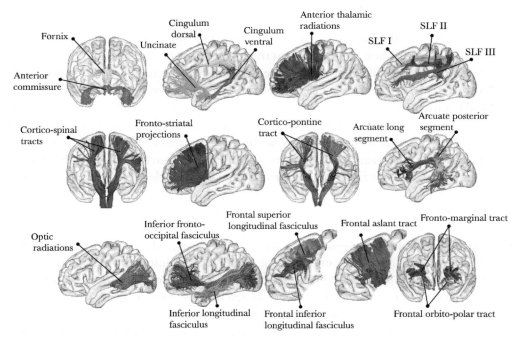

Fig. 6.2 Cerebral tracts.

areas. A dissected brain reveals the similarity of the ascending and descending fibers. The **internal capsule** is the complete collection of projection fibers.

Section review
1. Which type of brain wave has a higher frequency and is common when a person is under stress?
2. What is split-brain syndrome?
3. What are the names of the axons of the white matter?

Higher cortical functions

The higher cortical functions include language, vision, recognizing objects in space (visuospatial recognition), and awareness. The three characteristics of all higher-order functions are as follows:
- *The cerebral cortex must be involved*—complex interactions occur within the cortex and between it and other brain areas
- *Both conscious and unconscious information processing occurs*
- *Higher-order functions are adjusted over time*—they are not inborn (innate), fixed, or reflexive behaviors

Memory

To remember studied information, phone numbers, addresses, taste sensations, and many other types of information, we require **memories**. These are stored groups of information collected through life experience. The different types of memories are as follows:
- **Fact memories**—specific pieces of information such as colors or odors
- **Skill memories**—learned motor behaviors, such as how a key is used to open a door, or how to mix coffee with milk. Over time, skill memories become unconsciously controlled. The skill memories related to eating, an innate behavior, are stored in specific parts of the brainstem. Complex skill memories, such as playing the piano, require integration of motor patterns in the basal nuclei, cerebral cortex, and cerebellum.

There are two classes of memories: *short-term* and *long-term*, which are further explained as follows:
- **Short-term memories**—last for a short time, during which information can be immediately recalled. They only involve small pieces of information such as a name of someone or a street address. Repeating the information reinforces the original short-term memory. It can then be converted to long-term memory.
- **Long-term memories**—last for a longer time, even throughout the lifespan. Conversion between short-term and long-term memories is called **memory consolidation**. Long-term memory has two subtypes. *Secondary memories* fade over time, and possibly requiring great effort to recall. *Tertiary memories* remain with us throughout life, such as our own name and appearance.

> **Focus on memory**
> The ability of humans to retain memories begins just 20 weeks after conception. The brain's ability to store information is practically limitless. Sleep is significant to memory; it helps to retrieve and store long-term memories. The memory loss experienced by older people is not due to aging—it is because they participate much less in engaging activities than earlier in life. This lack of "brain exercise" results in loss of memory. When we think about our past experiences, stronger connections between active neurons are created.

Brain regions used in memory

Memory consolidation is associated with the *amygdaloid body* or *amygdala*, and the *hippocampus*. Both of these are parts of the limbic system. When the hippocampus is damaged, there will be an inability to convert short-term memories to new long-term memories. However, existing long-term memories remain intact. The tracts that lead from the amygdaloid body to the hypothalamus may link memories to certain emotions. Near the diencephalon, the **nucleus basalis** plays some role in storage and retrieval of memories. It is connected via tracts with the hippocampus, amygdaloid body, and all of the cerebral cortex. If damaged, there are changes in intellectual function, emotional states, and memory.

The cerebral cortex stores most long-term memories. Appropriate association areas handle conscious sensory and motor memories. Visual memories, for example, are stored in the visual association area. The premotor cortex stores memories of voluntary motor activity. The memories of words, voices, and faces require special areas of the occipital and temporal lobes. Specific memories usually require the activity of just a single neuron. In one area of the temporal lobe, an individual neuron responds to a specific word but ignores others. The neuron may also be activated by a correct combination of sensory stimuli related to a certain individual.

Many different areas of the brain collect information on a certain subject. The memory of an old friend exists in the visual association area (his/her face), the auditory association area (his/her voice), the speech center (his/her name), and the frontal lobes (his/her house, hobbies, likes, dislikes). Other areas store additional related information. When one area is damaged, the memory of the old friend is diminished in some way.

Cellular memory formation and storage

Neurons and synapses change as memories are consolidated. There are three primary processes involved in cellular memory formation and storage:

- *Increases in neurotransmitter release*—frequently active synapses store more neurotransmitter and release more with every stimulation. The more that is released, the more the effect on the postsynaptic neuron.
- *Synapse facilitation*—repeated activation of a neural circuit cause axon terminals to continuously release small amounts of neurotransmitter. It binds to postsynaptic membrane receptors. This causes a graded depolarization, bringing the membrane closer to the threshold. The resulting facilitation affects all neurons in the same circuit.
- *Forming of more synaptic connections*—repeated communications between a neuron and others cause axon tips to branch, forming more synapses on the postsynaptic neuron. Therefore, presynaptic neurons stimulation has an increased effect on the membrane potential of the postsynaptic neuron.

These three processes create anatomical changes, increasing communication along the neural circuit—the basis of memory storage. A **memory engram** is a single circuit corresponding to a single memory and is a functional definition. Memory engrams form from experience and repetition, which is very important in learning.

It takes at least 1 h for short-term memory to be converted into a memory engram. This is based on the original stimulus and its type, intensity, and frequency. It is easiest to convert strong, repeated, and very pleasant or unpleasant events to long-term memories. CNS stimulants such as caffeine or nicotine may enhance memory consolidation via facilitation.

The hippocampus is important in memory consolidation. It is believed that *N-methyl-D-aspartate (NMDA) receptors*, which are chemically gated calcium ion channels, are linked. Once activated by *glutamate*, a neurotransmitter, the gates open. Calcium ions enter the cell. The blocking of NMDA receptors in the hippocampus means that long-term memories cannot form. In general, the formation of memory is through the **Papez circuit**, which consists of the hippocampus, mammillary body, thalamus, and cingulate gyrus.

States of consciousness

Consciousness is described as a state of alertness and attentiveness. The individual is aware of external stimuli and events. Yet, a healthy person who is unconscious can still be wide-awake, nearly asleep, or very excited. The term *unconscious* refers to deep unresponsive, such as when under sedation, or simply a deep sleep.

In general, a patient in a coma is not alert or aware of surroundings. In clinical settings, an *altered level of consciousness* describes a patient that is not alert, not aware, or both. This is a very important concept.

Ongoing CNS activity is indicated by a person's wakefulness. A sleeping individual is unconscious, but may be awakened by noises or other normal stimuli. A healthy individual moves between an alert and conscious state and sleeps every day. Abnormal or depressed CNS function alters the state of wakefulness. A person in a deep *coma* is unconscious, and cannot become alert or aware by any degree of stimuli.

Apraxia

Apraxia is the inability to execute previously learned, purposeful motor activities, even though the willingness to execute them is present, and there is sufficient physical ability. This is often the result of brain damage, including infarction, trauma, or a tumor. It can also occur from degeneration of the parietal lobes and their connections, which retain the memory of learned movements, and from damage to other areas of the brain. These other areas may include the premotor cortex of the frontal lobe, anterior to the motor cortex, other frontal lobe portions, the corpus callosum, or from diffuse damage due to degenerative dementia.

A patient with apraxia cannot conceptualize or complete learned complex motor tasks, even though having the ability to make the individual movements needed. *Constructional apraxia* occurs when the patient cannot copy simple geometric shapes, even though they have normal sight, recognize the image, can hold and use a pen, and understand what they are being asked to do. Often, these patients do not recognize that they have a deficit.

The patient with apraxia is assessed by having him or her do certain learned tasks, or imitate another person doing the tasks. Tests may involve starting to walk and then stopping, combing the hair, saluting, striking a match and blowing it out, using a key to open a lock, taking a deep breath and holding it, or using scissors or a screwdriver. To exclude possible motor weakness and musculoskeletal abnormalities, strength and range of motion must be assessed. For less obvious apraxias, physical or occupational therapies may use neuropsychologic testing and assessment. The patient's caregivers should be asked about how the patient handles daily living tasks, especially to eat, prepare a meal, use a toothbrush, write, or to use various household tools.

Aphasia

Aphasia is a type of language dysfunction that can involve reduced comprehension or expression of words, as well as the nonverbal equivalents of words. It may also simply involve difficulty with language fluency. It occurs due to damage to any part of a basically triangular area in the brain, which includes the following:
- *Posterosuperior temporal lobe*—including Wernicke's area in the superior temporal gyrus
- Adjacent inferior parietal lobe
- *Posteroinferior part of the frontal lobe*—just anterior to the motor cortex, containing Broca's area in the inferior frontal gyrus
- *Subcortical connection between these regions*—usually in the left hemisphere regardless of whether the patient is right or left handed

The related damage may be from infarction, trauma, degeneration, or a tumor. The quality of rhythm and emphases that adds more meaning to our words is known as **prosody**. This is usually a functional part of both hemispheres, yet sometimes affected by dysfunction of the nondominant hemisphere by itself.

Aphasia is not a developmental language disorder, or caused by **dysarthria**, a dysfunction of the motor pathways and muscles involved in speech production. The two basic divisions of aphasia are *receptive* and *expressive*. Receptive aphasia is subdivided into categories called *sensory*, *fluent*, or *Wernicke's*. The patient is unable to comprehend words, or to recognize auditory, tactile, or visual symbols. This condition is due to a disorder of the posterosuperior temporal gyrus of the Wernicke's area in the language-dominant hemisphere. There is often another condition present, called **alexia**, which is the loss of the ability to read words.

Expressive aphasia is subdivided into categories called *motor*, *nonfluent*, or *Broca's*. There is an impaired ability to create words, yet no changes in comprehension or the ability to conceptualize. It is caused by a disorder of the dominant left frontal or frontoparietal area, including Broca's area. There is often a loss of the ability to write, called **agraphia**, and impairment of oral reading. Other, sometimes overlapping types of aphasia include *anomic*, *conduction*, *global*, *transcortical motor*, and *transcortical sensory*.

The patient with *Wernicke's aphasia* speaks normal words easily, often with meaningless word pieces including incorrect first letters of rhyming words (*phonemes*). The patient does not know the meaning or relationships of the words being used. What comes out is *word salad*—a group of incomprehensible words. Since the affected area of the brain is near the visual pathway, there is often a reduction in the right visual field. In *Broca's aphasia*, the patient comprehends and conceptualizes sufficiently, but has an inability to form words. Speech production and writing are usually both affected, often causing the patient frustration. There may also be an inability to name objects and impaired intonation and language rhythm.

Simply talking with the patient usually identifies gross aphasia. The condition should be differentiated from communication problems caused by impaired hearing or vision, impaired motor writing ability, or severe dysarthria. Wernicke's aphasia is sometimes mistaken for delirium, but it is a pure language disturbance without the same symptoms of delirium. The patient has normal consciousness, is attentive, and does not experience any type of hallucinations. Testing for aphasia involves normal and spontaneous speech patterns, naming objects, repeating complicated phrases, pointing to specific objects, answering simple questions, and demonstrating reading and writing skills.

> **Focus on aphasia**
> More people have aphasia than other common conditions such as cerebral palsy, multiple sclerosis, muscular dystrophy, or Parkinson's disease. There are more than two million Americans with aphasia. It affects everyone differently but does not affect intelligence. Most people improve over time, especially if speech therapy is provided. The condition can be improved even 10 or more years after onset. With time, the brain can create new networks and heal.

Dysprosodies

A **dysprosody** involves failed signaling or identification of various cues in normal speech signals. *Alexithymia* involves defective retrieval of emotion-related language, restricting the range of affective expression. Another explanation of dysprosody is that the "melodic" aspects of speech become abnormal. Broca's aphasia is designated by dysprosody, lack of fluency in conversation, decreased verbal output, the use of only short sentences, and *agrammatism*, which is the use of only nouns, verbs, and adjectives without the use of any additional "filler" words. Dysprosodies are usually attributed to neurological damage, including brain trauma or tumors, vascular damage, stroke, and severe head injuries.

Dysprosody is characterized by changes in the intensity of speech, the timing of speech segments, and in the ways, words are spoken, including their rhythm and pitch. The two types of dysprosody are *linguistic* and *emotional*. Linguistic dysprosody involves reduced ability to verbally convey the stress that is placed on certain words for emphasis, or to use normal patterns of intonation. Emotional dysprosody concerns the ability to express emotions through speech, as well as to understand emotions expressed verbally by others.

Testing for dysprosody usually begins by asking the patient to repeat a specific sentence, but each time, say it in a different way. This means the patient is required to say, for example, "I need to go the bank today," first in a happy tone, then an angry tone, then a sad tone, then a bored or indifferent tone. The patient with dysprosody will generally sound "monotone," unable to change the affective tone of voice as requested. Though changes in softness and loudness of speech still occur, the actual tonality does not vary.

Section review
1. What are skill memories?
2. Which part of the brain plays a role in storage and retrieval of memories?
3. What is constructional apraxia?

Clinical considerations

There are many conditions that affect the cerebral cortex. These include AD, delirium, dementia, seizure disorders, sleep disorders, headache, and brain infections.

Alzheimer's disease

AD was first described by physician Alois Alzheimer in 1907. Today, it is the most common form of dementia in the United States. The exact causes of Alzheimer's are not fully understood. The disease is not curable, and its progression cannot be prevented or delayed. The patient with AD will become totally dependent for all care, which heavily burdens spouses and family members. While it is ultimately fatal disease, the most common causes of death are related to its complications, and not the disease itself.

The strongest factor for developing dementia and AD is advancing age. However, there are some reversible causes of dementia or cognitive impairment that are sometimes called pseudodementia. These include deficiencies of vitamin B12, folate, or other vitamins; hypothyroidism and other thyroid disorders; depression; and normal pressure hydrocephalus, which is commonly known as *water on the brain*. These reversible conditions should always be ruled out in any patient with suspected dementia. At diagnosis, the average life expectancy of a patient with AD is approximately 8 years. However, this can vary anywhere from 3 to 20 years.

More than 80% of dementias in the elderly are linked to AD, which affects women more than men, partially due to their longer lifespans. It affects 50% of patients over the age of 85, yet only about 4% of individuals between 65 and 74 years. The prevalence of AD is 5.2 million in the United States. As the elderly population of industrialized nations increases, the prevalence of AD is also expected to triple by the year 2050.

Risk factors include aging, low education level, Down syndrome, and positive family history. Positive family history exists in 50% of cases. However, 90% of cases are sporadic.

Pathophysiology

AD is characterized by senile plaques, beta (β)-amyloid deposits, and neurofibrillary tangles in the cerebral cortex as well as the subcortical gray matter. It causes progressive cognitive deterioration. Pathological findings in Alzheimer's patients include diffuse cerebral atrophy in the hippocampus, amygdala, and subcortical nuclei. Imaging studies using a computed tomography (CT) or magnetic resonance imaging (MRI) do not reveal these findings. Only upon autopsy are neuritic senile plaques, pyramidal cell loss, neurofibrillary tangles revealed. These result in the characteristic decreased cholinergic innervation, with variable decreases in other neurotransmitters. There is also degeneration of the **locus ceruleus** and basal forebrain **nuclei of Mynert**, and possibly, **amyloid angiopathy**. Today, ongoing studies are focusing on agents that decrease production or reduce deposits of amyloid in the brain, as ways to hopefully prevent or treat AD. The overproduction of amyloid is linked to earlier onset AD, which occurs prior to the age of 65. This is believed to be genetically related, with alterations of chromosomes 1, 14, and 21.

Late-onset AD occurs after age 65 and is the most common form. Apolipoprotein E (Apo E) is related to the neurofibrillary tangles that become present. This protein's production is influenced by a gene located on chromosome 19, with three primary alleles: E2, E3, and E4. One copy of the gene is inherited from each parent. The E3 variant is the normal, beneficial variant of Apo E, with this protein believed to help nerve cells via its nutritional support. When one copy of E4 is inherited in a heterozygote (E3/E4) or two copies in a homozygote (E4/E4), there is a higher risk for AD. Those who are homozygous for E4 have a higher risk than those who only inherit one copy. While having one or two copies of the E4 allele is not diagnostic for AD, the risk is definitely increased. Inheritance of the E2 allele appears to be protective for AD, but this is a rarer occurrence.

Factors also related to increased risk for AD include low educational level, smaller brain size, head injuries, and lower levels of mental and physical activity in later life. Anything that causes recurrent head trauma, such as professional sports, is believed to increase the risks for developing AD. Vascular dementia carries underlying risks similar to other forms of vascular disease, including hyperlipidemia, hypertension, obesity, diabetes, and smoking.

Rare cases of vitamin deficiencies have also been mistaken for AD, but fortified foods have widely reduced such relationships. However, between 5% and 15% of people over age 65 years of age have vitamin B12 deficiency. Most of this vitamin is consumed from animal proteins, such as beef, chicken, pork, eggs, and milk. Because of reduced stomach acid production (atrophic gastritis), or because of chronic use of proton pump inhibitors and other acid-blocking medications, many older people cannot absorb vitamin B12 efficiently from food sources.

Approximately 10% of adults over age 65 have hypothyroidism, which may cause similar cognitive symptoms to those of early AD, such as pseudodementia since it is reversible. When the underlying thyroid problem is corrected, symptoms usually improve. Also, depression can cause *pseudodementia* since symptoms may be somewhat similar to those with actual dementia with one key difference; reversibility. Depression treatments usually improve cognitive symptoms.

The most common finding in Alzheimer's patients is the loss of cholinergic neurons, which are the nerve cells that use acetylcholine to communicate. Therefore, acetylcholine has less activity in the brain, which is linked to severe AD symptoms. Most proposed treatments for AD have focused on acetylcholine. Since they have only slightly been effective, it is now understood that loss of acetylcholine is due to underlying pathophysiology, and is not the actual cause of AD.

> **Focus on Alzheimer's disease**
> Alzheimer's disease is the sixth leading cause of death in the United States. More than 16 million Americans provide unpaid care for people with Alzheimer's or other dementias. Between 2000 and 2015, deaths from heart disease *decreased* 11%, while deaths from Alzheimer's *increased* 123%. It kills more people than breast cancer and prostate cancer combined. In 2018, Alzheimer's and other dementias cost the country $277 billion. By 2050, these costs could be as high as $1.1 trillion.

Clinical manifestations

As pathophysiologic changes begin to manifest in an Alzheimer's patient, cholinergic neurons and other neuronal pathways are lost. There is a decrease in brain size and weight. These progressive changes lead to reduced intellectual functioning that becomes severe, interfering with normal daily activities and relationships. Symptoms may be slight at first, with the patient forgetting more things than usual. Often the patient does not recognize this, but spouses or family members start to notice changes in memory or other behaviors. The patient can be easily disoriented, failing to remember names or locations. Cognitive testing may reveal deficiencies in short-term memory, dates, months, years, names of buildings, calculation abilities, visual-spatial relationship abilities, and judgments.

Progression of AD makes symptoms more easily recognizable. Paranoia may be demonstrated by the patient, such as in accusations of stealing or being threatened by others. Depressive symptoms or psychotic symptoms may include hallucinations or delusions. The patient may become verbally or physically aggressive, start to wander and become lost, or develop mannerisms of combativeness, repetitiveness, or uncooperativeness. Eventually, he or she is unable to handle bathing, dressing, feeding, or toileting tasks. More complicated tasks become completely foreign to the patient, such as taking medications correctly, managing money, using telephones, or functioning in public society. Regardless of what is done, most patients can no longer be well-cared for at home and must be moved to a skilled nursing facility with 24-h care. The symptoms correlate with the location of brain atrophy. In AD, the parietal and temporal lobes are primarily affected. Therefore, memory and visuospatial task abnormalities are seen first in this disease. Language is preserved early in the disease until later stages when the areas responsible for language also become affected.

Diagnosis

Basically, a diagnosis of exclusion is required for AD, once all other possible explanations have been considered. This is a clinical diagnosis. The disease is often diagnosed as being *probable* since no definitive tests exist except for biopsy of the brain—a major surgical procedure that is very uncommon. Diagnosis is assisted by full clinical evaluation and presentation history. Gradual symptoms showing progressive declines over months are consistent with AD or other forms of dementia. The sudden onset of symptoms, over days or weeks, is not consistent with a diagnosis of AD.

The pathophysiology of AD starts long before the patient becomes symptomatic. Therefore, by the time the patient manifests symptoms, the disease has been already present for years. Researchers are now focusing on identifying AD patients in their asymptomatic stages, in order to be able to stop the disease from progression at early and asymptomatic stages.

The *Mini-Mental State Examination* (*MMSE*) and other cognitive screening tests are usually done to diagnose cognitive deficits, but not to diagnose dementia or AD. Screening for depression helps rule out pseudodementia. Laboratory tests can help rule out possible causes of symptoms, and include thyroid function tests, and checking for levels of serum folate and vitamin B12. Imagining studies can help rule out stroke, tumors, infections, and other CNS disorders. The patient's medications should be evaluated to determine if they are contributing to symptoms. Medications that can cause deficits similar to those of AD include those with anticholinergic effects. These include antihistamines, diphenhydramine, amitriptyline, other tricyclic antidepressants, and oxybutynin. It has been shown that 10% of patients being evaluated for dementia were taking medications linked to cognitive impairment. The American Academy of Neurology (AAN) recommends checking for depression, B_{12} deficiency, and thyroid disorder in every patient presenting with cognitive dysfunction. Either a CT scan or an MRI scan of the brain is needed to rule out structural causes of dementia.

> **Focus on medications that may affect cognition**
> Many medications, especially if they have CNS adverse effects, can affect cognition. These include benztropine, carisoprodol, clorazepate, chlordiazepoxide, chlorpheniramine, chlorzoxazone, cimetidine, clonidine, cyclobenzaprine, desipramine, diazepam, dicyclomine, doxepin, flurazepam, guanadrel, guanethidine, hydroxyzine, imipramine, indomethacin, lorazepam, meperidine, methocarbamol, nortriptyline, oxazepam, phenobarbital, propoxyphene, temazepam, thioridazine, and tolterodine.

The term *mild cognitive impairment (MCI)* describes pre-dementia or pre-AD. There are no criteria available to classify patients with this condition. Many, but not all, will eventually develop dementia or AD at the rate of approximately 15% per year. As of today, there is no Food and Drug Administration (FDA)-approved medication to stop or slow the progression from MCI to AD.

Treatment

The focus of treatment for all types of dementias is to preserve cognition and physical functioning as much as possible. This can reduce or delay requirements for skilled nursing care, and minimize psychiatric and behavior symptoms that compromise care. When treating these patients, their safety is of utmost importance, and the reduction of stress to caregivers must also be emphasized.

Many nonmedication methods help improve care and quality of life for the patient and all caregivers. It is vital to educate everyone about the disease, its course, legal issues, and treatment options. The local Alzheimer's Association Chapter is an invaluable resource for education, information, and support. Often, nonmedication interventions are used before any medications are given. These interventions are described in Table 6.2.

The US FDA has approved two distinctly different classes of medications, aimed at treating the cognitive symptoms related to AD. These include the following: *acetylcholinesterase inhibitors (AChEIs)* and *NMDA receptor modulators*. The AChEIs were approved in the early 1990s and are based on the effects of acetylcholine. They allow more acetylcholine to remain between nerve cells, increasing their communication with other nerve cells. However, the AChEIs have only shown a slight effect against AD symptoms. Cognitive improvement is minimal, and patients, as well as caregivers, must understand what to expect from these agents. The first AChEI was *tacrine*, which is only rarely used today

Table 6.2 Nonmedication interventions for Alzheimer's disease.

Factor	Intervention
Activities	Should be appealing and pleasant (such as safe areas to walk)
Environment	Consistent, structured; optimal use of stimuli to reduce confusion or upset (radios, televisions, speakers)
Interaction	Be calm, firm, and supportive with the patient—especially when he or she is upset; when upset occurs, use redirection to calm the patient
Monitoring	Observe patients for new behaviors or symptoms, and alert health-care team
Orientation	Use of explanations, reminders, visual cues
Tasks	Use simplified choices and uncomplicated tasks
Additional information and support	Alzheimer's Association (http://www.alz.org)—Alzheimer's Research Form (http://www.alzforum.org)—National Family Caregivers Association (http://www.thefamilycaregiver.org)

due to its dosage quantity and potential liver damage, as well as other adverse effects. Nearly all AChEIs cause gastrointestinal effects. Additional examples of these agents include donepezil, galantamine, and rivastigmine. Donepezil is different from the others because it does not require dosage titration. These agents are generally used with extreme caution in patients also taking beta-blockers, calcium channel blockers, or digoxin.

The other class of medications for AD, the NMDA receptor modulators, actually only contain one current medication, which is called *memantine*. It works by blocking the effects of glutamate in the brain. It is believed that AD patients have an excess of glutamate, which causes large amounts of calcium to move into the nerve cells, leading to their death. Complete blockage of the effects of glutamate is not helpful since this can lead to psychosis. Memantine, therefore, is useful because it reduces the effects of glutamate, but does not totally block its binding. Its adverse effects usually involve dizziness, constipation, headache, and confusion. Because the two classes of AD medications work in different ways, they can be used in combination. It has been shown in various studies that combinations of these agents have given effects that were better than when the drugs were used alone. Again, patient and caregiver information and education are essential so that the effects of combination therapies are understood.

Additional symptoms of AD may include anxiety, depression, and behavioral symptoms. Treatments for such symptoms must be handled with care since they can cause additional adverse effects. Depression can worsen the appearance of cognitive symptoms. Newer types of antidepressants such as the selective serotonin reuptake inhibitors (SSRIs) and serotonin/norepinephrine reuptake inhibitors (SNRIs) are generally better for patients with AD and other dementias. The older *tricyclic antidepressants* are especially associated with worse cognitive impairment, due to strong anticholinergic adverse effects.

For dementia patients with anxiety, the SSRIs may be helpful since they not only reduce depression but anxiety as well. Buspirone is a non-benzodiazepine that may be helpful, but only in adequate doses, and is a drug that is often underdosed. It is safer for dementia patients since it causes less sedation than the drugs known as benzodiazepines. These drugs are either long-acting or short-acting and if used, must be carefully controlled due to their potential to endanger the safety of patients. Generally, they are used only when absolutely necessary. Remember that some dementia patients are extremely sensitive to sedative medications. This is very true in the case of Lewy body dementia. These patients are extremely sensitive to even tiny amounts of sedatives.

For dementia patients with behavior symptoms such as delusions, hallucinations, aggression, combativeness, poor self-care, or wandering, medications are commonly administered, but not highly effective. The FDA has never actually approved medications to treat such behavioral symptoms of AD or dementia. Poor self-care or wandering does not respond well to any medication. For delusions or hallucinations, antipsychotic medications are sometimes used, but this must only occur with extreme care, and only if benefits exceed risks. These agents may increase the risk of stroke or death in AD or dementia patients. Additional forms of treatment being used or studied for AD include gingko biloba (a common dietary supplement), vitamin E, medical food products (which only include Axona, Cerefolin N-acetyl cysteine (NAC), and Souvenaid—products that may alter brain chemistry), nonsteroidal antiinflammatory drugs (NSAIDs), lipid-lowering agents (especially the statin drugs), passive immunization with the monoclonal antibody *bapineuzumab*, and insulin.

Delirium and dementia

Delirium is also known as an *acute confusional state*. Along with **dementia**, it is one of the most common causes of cognitive impairment. Affective disorders such as depression can also disrupt cognition. Delirium and dementia are separate disorders that can easily be confused. In both, cognition is disordered. The major difference is that delirium mostly affects *attention*, while dementia mostly affects *memory*. The basic differences between delirium and dementia are listed in Table 6.3.

It is important to understand that delirium often develops in patients who have dementia. The two conditions must not be mistaken for each other—especially in the elderly. There is no definitive laboratory test to establish the cause of cognitive

Table 6.3 Basic differences between delirium and dementia.

Factors	Delirium	Dementia
Cause	Usually another condition, such as dehydration, infection, use of or withdrawal from certain drugs	Usually a chronic brain disorder, such as Alzheimer's disease, Lewy body dementia, vascular dementia
Onset	Sudden, with a definite start point	Slow, gradual, with an uncertain start point
Duration	Days to weeks; can be longer	Usually permanent
Effects	Usually reversible, almost always worse at night	Slowly progressive, often worse at night
Attention	Significantly impaired	Not impaired until condition is severe.
Orientation (time and place)	Varied	Impaired
Language use	Slowed, inappropriate, often incoherent	Sometimes difficult to "find the right words"
Memory	Varied	Lost, especially for recent events
Consciousness level	Variable impairment	Not impaired until condition is severe
Medical attention requirement	Immediate	Needed, but not as urgently

impairment. Therefore, it is essential to conduct a thorough patient history and physical examination, along with knowledge of baseline function of the patient.

Delirium is acute, transient, usually reversible and *fluctuating disturbance* of attention, cognition, and consciousness. While most common in the elderly, it may occur at any age. More than 10% of elderly hospitalized patients have delirium and between 15% and 50% experience the condition at some time during hospitalization. It is also common in nursing home residents. When occurring in younger patients, it is usually linked to drug use or life-threatening systemic disorders. Delirium is sometimes also called *toxic encephalopathy* or *metabolic encephalopathy*.

Dementia is chronic, global, and usually irreversible deterioration of cognition. It can occur at any age, but mostly affects the elderly. For people 85 or older, about 40% have dementia. For those aged 65–74, about 5% have the condition. More than 50% of all nursing home admissions involve dementia. More than five million Americans have the condition. Dementia is classified as Alzheimer's or non-Alzheimer's; cortical or subcortical; irreversible or potentially reversible; and common or rare.

Pathophysiology

Delirium may involve reversible impairment of cerebral oxidative metabolism, generation of cytokines, and multiple neurotransmitter abnormalities. Any type of stress causes **upregulating** of the sympathetic tone while **downregulating** the parasympathetic tone. This impairs cholinergic function, which contributes to the condition. The elderly have a higher vulnerability to reduced cholinergic transmission, increasing risks. No matter what the cause, there is impairment of the cerebral hemispheres, or the arousal mechanisms of the thalamus and brain stem reticular activating system. Neurologists use the words "delirium" and "toxic metabolic encephalopathy" interchangeably.

Primary diseases of the brain and other conditions can result in dementia. Patients can also develop *mixed dementia*, in which one or more types are involved. The pathophysiology of dementia may be related to normal-pressure hydrocephalus, subdural hematoma, hypothyroidism, vitamin B12 deficiency, delirium, and toxins such as lead. Cognition deteriorates slowly and can be resolved if treated adequately. Up to half of all patients who have MCI will develop dementia within 3 years. Drugs that worsen cognitive deficits include benzodiazepines, certain tricyclic antidepressants, antihistamines, antipsychotics, benztropine, and alcohol. For example, **Wernicke-Korsakoff syndrome** is a form of dementia caused by chronic, long-term alcoholism, which leads to vitamin B1 deficiency.

However, the leading cause of dementia is AD, which results from death of cerebral cortex cells. Abnormal proteins, believed to be *beta-amyloids*, form lesions in the cerebral cortex that eventually disrupt and destroy surrounding cells. *Vascular dementia* is caused by atherosclerosis in the brain, in which inadequate blood flow supplies insufficient oxygen. Areas of dead tissue then form, and vascular dementia develops. This is usually linked to previous cardiovascular conditions such as diabetes, heart disease, hypertension, and high cholesterol. *Parkinson's disease* results in the buildup of **Lewy bodies** in the brain, which eventually affect memory and cognition, causing dementia. *Lewy body dementia* usually affects people with no family history, and its causes are unclear. *Huntington's disease* also results in dementia and is caused by a genetic mutation resulting in death of nerve cells in the basal ganglia. Infectious dementias include *Creutzfeldt-Jakob disease* (CJD), AIDS-induced dementia, and **neurosyphilis**.

Clinical manifestations

The clinical manifestations of delirium are primarily difficult in focusing, maintaining, or shifting attention. There is a *fluctuating* consciousness level. The patient may be disoriented to time, and sometimes to place or person. They may hallucinate. There is confusion about daily events and routines, along with changes in personality and affect. The thought patterns become disorganized, and speech may by disordered, slurred, rapid, with **neologisms**, aphasic errors, or chaotic patterns.

Symptoms change over minutes to hours and may be reduced during the day but worse at night. Symptoms include fearfulness, paranoia, and inappropriate behaviors. The patient may be agitated, irritable, *hyperalert*, and hyperactive—then becoming lethargic, quiet, and withdrawn. The extremely elderly often become quiet and withdrawn, which is often mistaken for depression. Some patients alternate between the two states. There are common distortions of sleeping and eating patterns. Due to the cognitive disturbances, the patient has poor insight and impaired judgment. Other signs and symptoms are based on the cause.

The signs and symptoms of dementia develop slowly, usually beginning with the loss of short-term memory. Symptoms are divided into early, intermediate, and late classifications. Personality changes and altered behaviors can develop early or late in the progression of dementia. Based on the type of dementia, motor and other focal neurologic deficits occur at different stages. In vascular dementia, these occur early, while in AD, they develop late. In all stages, the chance for seizures is slightly increased. In about 10% patients with dementia, psychosis occurs, involving delusions, hallucinations, or paranoia. A higher percentage of patients may have temporary symptoms of psychosis. Table 6.4 explains the classifications of various dementias.

Early symptoms of dementia include short-term memory impairment, with the learning and retaining of new information becoming difficult. The patient will have difficulty finding the right word when speaking and experience mood swings and personality changes. It may become harder to handle daily living tasks such as remembering where objects were placed, handling banking activities, and even getting lost in a familiar location. There may be impairment of abstract thought, judgment, or insight. The patient often becomes agitated, hostile, and irritable because of these changes. Other manifestations include **agnosia**, apraxia, and aphasia. *Agnosia* is impairment of identifying objects

Table 6.4 Classifications of dementias.

Type of dementia	Characteristics
Alzheimer's disease	Beta-amyloid abnormalities
Alcohol-related dementia; dementia due to heavy metal exposure	Ingestion of drugs or toxins
Brain tumor dementia; chronic subdural hematoma dementia; normal-pressure hydrocephalus dementia	Structural brain disorders
Creutzfeldt-Jakob disease; variant Creutzfeldt-Jakob disease	Prions (protein-related infectious disease-causing agents—neither bacterial or fungal)
Depression-related dementia; hypothyroidism dementia; vitamin B12 deficiency dementia	Potentially reversible disorders
Frontotemporal dementias including Pick's disease; corticobasal ganglionic degeneration dementia, progressive supranuclear palsy dementia	Tau abnormalities (these are proteins inside neurons essential for normal CNS function)
Fungal dementia (caused by cryptococcosis); spirochetal dementia (due to syphilis or Lyme disease); HIV-associated dementia and postencephalitis syndromes	Infections
Huntington's disease dementia	Huntingtin abnormalities, which include caudate nucleus atrophy, small cell degeneration, and decreased levels of gamma-aminobutyric acid (GABA) and substance P
Lacunar disease dementia, Binswanger's disease dementia, multiinfarct dementia, strategic single-infarct dementia	Vascular
Lewy body dementia; Parkinson's disease-related dementia	Alpha-synuclein abnormalities (these are proteins mostly found at the tips of neurons, in the presynaptic terminals, which release critical neurotransmitters)

while sensory function is normal. *Apraxia* is impairment in performing learned motor activities while motor function is still intact. *Aphasia* is impairment of language use or comprehension. Family members of dementia patients often report odd behaviors along with emotional changes.

The intermediate stage of dementia involves a reduction in memory of remote events, but not a complete loss of this memory. The patient may need help bathing, dressing, eating, and toileting. Personality changes may worsen, with the patient more easily becoming anxious, irritable, inflexible, self-centered, angry, passive, unaffected, depression, indecisive, and lacking in spontaneity. Patients often withdraw from social situations, and behavior disorders may develop. They may wander, become quickly hostile or uncooperative, and even physically aggressive. In this stage, there is loss of social and environmental cues, and the patient usually does not know where they are, the hour, the day, the month, or even the year. Patients get lost easily, even within their own homes. They are at higher risk of falling or experiencing accidents. Psychosis may develop with hallucinations and delusions, and sleep patterns are often disrupted.

The late, severe stage of dementia makes the patient unable to walk, feed himself or herself, or perform other daily living activities. Incontinence develops, all memories are lost, and there may be an inability to swallow. The patient is at risk of pneumonia mostly because of aspiration, pressure ulcers, and undernutrition. It is usually necessary to place the patient into a long-term care facility. Eventually, the patient will no longer speak. When the patient appears ill, it is up to physicians to determine treatments, because there will be no febrile or leukocytic reactions to infections. In most cases, the patient will develop an infection, go into a coma, and die.

Diagnosis

Clinicians often overlook delirium, especially in the elderly. The condition should be considered for any elderly patient with impaired attention or memory. A formal mental status examination is required, with attention being assessed first. The patient should immediately repeat the names of three objects, repeat seven numerical digits forward and then five backward, name the days of the week forward and backward. When the patient does not register directions or other information, this inattention must be distinguished from poor short-term memory, which is when information is registered but quickly forgotten. Additional cognitive testing is useless when the patient cannot register information.

The next diagnostic steps involve the use of the *Diagnostic and Statistical Manual of Mental Disorders (DSM)* or *Confusion Assessment Method (CAM)*. For a positive diagnosis of delirium, there must be the following two components in existence: acute cognition change that fluctuates during the day, and inattention—the inability to focus or follow what is spoken. Also, one of the following two components must exist: disturbance of consciousness (involving less clarity) or an altered level of consciousness (being hyperalert, lethargic, stuporous, or comatose) or having disorganized thinking such as rambling statements, irrelevant conversation, or illogical idea flow.

Patient history involves interviews with family members, caregivers, and friends, to determine recent mental status changes—this is distinct from baseline dementia. History helps differentiate a mental disorder from delirium. Mental disorders almost never cause inattention or fluctuating consciousness, and their onset is almost always subacute. Behavior deterioration during evening hours, known as **sundowning**, is common in institutionalized dementia patients but may be difficult to differentiate. New symptomatic deterioration should be presumed to be delirium until it can be proven otherwise.

History should include use of alcohol and all types of drugs regardless of their legality. This should especially focus on drugs with CNS effects and on new drug additions, discontinuations, or dose changes of any kind.

Physical examination—especially in patients not cooperating completely—should include vital signs, hydration status, possible foci for infection, and examinations of the skin, head, neck, and neurology. Fever, meningismus, **Kernig signs**, or **Brudzinski signs** suggest CNS infection. Tremor or myoclonus suggest liver failure, uremia, drug intoxication, or electrolyte disorders such as hypomagnesemia or hypocalcemia. Ataxia or ophthalmoplegia suggest Wernicke-Korsakoff syndrome. Focal neurologic abnormalities such as a motor or sensory deficits, or cranial nerve palsies, as well as papilledema suggest structures CNS disorders. Head trauma is suggested by lesions, swellings, and related findings.

Testing includes CT scan, MRI, complete blood count (CBC), blood cultures, chest X-rays, urinalysis, and measurements of blood urea nitrogen (BUN), electrolytes, creatinine, plasma glucose, and any potential toxic levels of drugs. Additional testing includes pulse **oximetry**, liver function tests, albumin and serum calcium measurements, thyroid-stimulating hormone (TSH) measurements, vitamin B12 levels, erythrocyte sedimentation rate, antinuclear antibody testing, and also, tests for syphilis. If still unclear, CSF analysis can rule out encephalitis, meningitis, or subarachnoid hemorrhage. Serum ammonia levels can be measured, and checking for heavy metals should be done. Rarely, nonconvulsive status epilepticus can cause delirium. If this is suspected based on history, slight motor twitches, automatisms, or steady patterns of bewilderment and drowsiness, and EEG should be done.

A definitive diagnosis of dementia is difficult, and often requires postmortem pathologic examination of the patient's brain tissue. Dementia must be distinguished from delirium and other disorders. If possible, the cerebral areas affected are identified and it is determined if the condition may be reversible. The patient's attention is assessed first. If inattentive, it is probably delirium and not dementia, though advanced dementia severely impairs attention. Dementia is not the same as age-associated memory impairment, because people with this condition can learn new information if given enough time. With MCI, memory is affected while daily functions and other cognitive activities are not. When depression exists, treatment of the condition can relieve any associated dementia. Depressed people rarely forget important events, appointments, or other information. Neurological exams will show psychomotor slower but no other abnormalities. Depressed patients make only slight efforts to respond to testing, while dementia patients usually try hard but still respond incorrectly. If depression and dementia are coexistent, treatment of the depression will not fully restore cognitive abilities.

A short-term memory test is first conducted, such as registering three different objects and then recalling them after 5 min. Dementia patients will forget such simple information within 3–5 min. A different test assesses the naming of categorized objects, such as animals, furniture, or plants. The dementia patient will struggle to name just a few objects. A dementia diagnosis requires at least one of the following: aphasia, apraxia, agnosia, or an impaired ability to organize, plan, sequence, or think abstractly. This is known as *executive dysfunction*. Each deficit must greatly impair function and show a clear decline from a previous functional level. The deficits must not occur only during delirium.

Next, a formal mental status examination is performed. The presence of multiple deficits, especially in a patient with an average or higher level of education, suggests dementia. Patient history and physical examination then focuses on treatable disorders such as vitamin B12 deficiency, depression, hypothyroidism, and neurosyphilis. Laboratory tests include levels of B12 and TSH. Sometimes, CBC and liver function tests are performed, but usually with little positive results. Testing for HIV or syphilis may be indicated. Lumbar puncture can be done if a chronic infection or neurosyphilis is suspected.

For initial evaluation of dementia, or after a sudden change in cognition or mental status, CT or MRI should be performed. Neuroimaging can identify structural disorders and metabolic disorders such as **Hallervorden–Spatz disease** or Wilson's disease. An EEG may be helpful to evaluate attention lapses or bizarre behaviors. Single-photon emission CT or functional MRI can identify cerebral perfusion patterns, helping to diagnose AD due to frontotemporal dementia, and Lewy body dementia. Some patients require formal neuropsychologic testing to evaluate mood and mental functions. This helps to differentiate between age-associated memory impairment, MCI, amnesia, aphasia, apraxia, visuospatial problems, and dementia.

Treatment

Delirium can be resolved by treating causes, such as infection or dehydration, and by addressing pain or drug use. Good nutrition and hydration should be provided, and deficiencies of thiamin or vitamin B12 corrected. The environment should be kept stable, quiet, and with good lighting. Calendars, clocks, and family photographs will help to orient the patient. Hospital staff and family members can frequently reorient and reassure the patient. If hearing aids or eyeglasses are used by the patient, they must be checked for appropriate functioning. Treatment must be interdisciplinary, helping to enhance mobility, range of motion, reduce pain and discomfort, prevent skin breakdown, assist with incontinence, and minimize aspiration risks.

Patient agitation can harm anyone involved in treatment, including the patient. Drug regimens should be simplified. Intravenous lines, bladder catheters, and physical restraints should be avoided, as much as possible. Sometimes physical restraints are required, but these should only be applied by trained staff members and released every 2 h to prevent injury. They must be discontinued as soon as possible. Instead of restraints, hospital-appointed "sitters" may be provided to continually observe the patient. Family members should receive counseling about the effects of delirium and told that it is usually reversible but this may take weeks to months after the acute illness is resolved.

Medications for delirium include haloperidol that can reduce agitation and psychotic symptoms. Drugs must be administered with care because they have the potential to prolong or exacerbate delirium. Second-generation, atypical antipsychotics may be preferred due to fewer extrapyramidal adverse effects, but their long-term use in dementia patients can increase risks of stroke and death. Benzodiazepines have a faster onset of action than antipsychotics, but often worsen confusion and sedation in delirium patients. While these drugs are equally effective for agitation, the antipsychotics have fewer adverse effects. Benzodiazepines are preferred when delirium is caused by sedative withdrawal, and for those who cannot tolerate antipsychotics. Doses of these drugs must be reduced as quickly as possible.

For dementia, treatment begins with occupational and physical therapists evaluating the living place for safety, to prevent falls and accidents, manage behavior problems, and plan for the future progression of the disease. Sharp objects may need to be removed, appliances unplugged, and car keys taken away—all measures that help safe-proof the home or other living areas. In some states, physicians must report dementia patients to the Department of Motor Vehicles so that they can no longer drive a vehicle. For wandering patients, there are signal monitoring systems available, and a program called *Safe Return* into which they can be registered. Eventually, home health aides, housekeepers, and others may be needed for assistance. The patient may need to be moved into a living facility that does not have stairs, or an assisted-living or skilled nursing facility.

The living area should be designed to help preserve the patient's feelings of self-control and dignity. This is accomplished by frequent reinforcing the patient about orientation within the living space, having a bright and positive environment that the patient is familiar with, keeping new stimuli to a minimum, and offering regular activities of low-stress level. Large clocks and calendars are helpful, and medical staff members can wear large nametags, and reintroduce themselves on a regular basis. Any changes that must occur should be explained to the patient simply and clearly. They need time to adjust to changes and to familiarize themselves with new routines. If feeding or bathing is scheduled, it is often good to discuss what is going to happen with the patient so that they do not become resistant or violent. Frequent visits by caregivers and staff help patients to remain as social as possible. Radios, televisions, and night-lights are all good objects to keep the patient oriented and to focus attention. Private rooms that are dark and quiet are not conducive to these patients. Activities that relate to the patient's interests in life—prior to dementia—are suggested. They should not be too challenging but should be stimulating and enjoyable. Exercise should occur once per day, which improves cardiovascular tone and overall balance, as well as reducing restlessness, manage behaviors, and improve sleep. Music and occupational therapy help maintain fine motor control and provides nonverbal stimulation. Reminiscence therapy, socialization activities, and other group therapies help maintain interpersonal and conversational skills.

Regarding medications, sedatives, and anticholinergic drugs should be avoided since they worsen dementia. The elimination or limiting of CNS drugs often helps improve function. For Alzheimer's or Lewy body dementia, and possibly other forms, cholinesterase inhibitors may improve cognitive function. These drugs include donepezil, galantamine, and rivastigmine. They inhibit acetylcholinesterase, which increases ACh levels in the brain. Moderate to severe dementia may be slowed in its progression by using memantine, which can be used with cholinesterase inhibitors. Antipsychotics and other drugs have been used for behavior disorders. Dementia patients who show signs of depression should be given non-anticholinergic antidepressants—especially SSRIs.

Regarding care, since family members are often most responsible for these patients, training can be provided from nurses and social workers. This training must be continued, and there are also dementia support groups, websites, and educational materials available. Stress levels for caregivers are high, involving worry, exhaustion, frustration, anger, and resentment. Health-care workers should monitor for caregiver stress and, when needed, suggest support services such as nutritionists, social workers, nurses, and home health aids. Any dementia patient with an unusual injury may be experiencing elder abuse, and this must be investigated.

Near the end of a dementia patient's life, the overseeing of the patient's legal and financial matters must be handled by others, including family members, guardians, and lawyers. When the patient is in the early stages of dementia, his or her wishes about care should be discussed and documented. Financial and legal documents should be drawn up, such as durable power of attorney forms or paperwork related to the durable power of attorney for health care. Once signed, the patient's cognitive capacity must be evaluated, with the results recorded. It is important to make decisions about artificial feeding and treatment of acute disorders before these things are actually needed. For advanced dementia, palliative treatments may be more appropriate than aggressive interventions or hospitalizations.

Focus on frontotemporal dementia
The form of frontotemporal dementia known as "behavior variant" is characterized by prominent personality and behavioral changes. These often occur in people in their 50s and 60s, but may develop as early as in their 20s, or as late as in their 80s. Nerve cell loss is most prominent in areas controlling conduct, empathy, foresight, and judgment. This form progresses steadily and often quickly, often becoming full-blown within 2 years. There is also a "primary progressive aphasia" form, which affects language, speaking, writing, and comprehension.

Prion diseases

Prion diseases are also referred to as *transmissible spongiform encephalopathies*. They are progressive, untreatable, and fatal degenerative brain diseases. Examples include *CJD, variant CJD, Gerstmann-Sträussler-Scheinker disease, familial* fatal insomnia, and *kuru*. These diseases are usually sporadic, affecting about one of every one million people. There is misfolding of the normal cell-surface brain *prion protein* or *PrP*. Misfolded prions cause previously normal PrP to misfold. They are similar to beta-amyloid, and highly resistant to degradation. The cause leads to slow but significant intracellular accumulation and neuronal cell death. Changes include gliosis and histologic vascular or *spongiform* changes. This causes dementia and other neurologic deficits. Signs and symptoms develop in months to years following exposure.

Prion diseases can be spontaneous or hereditary. The PrP gene is contained in the short arm of chromosome 20. Prion diseases may be transmitted by infected tissue, such as via organ transplants. Rarely, the transmission is by blood transfusion. They also can be transmitted via food sources. Prion diseases should be considered in all dementia patients, especially if they are progressing quickly. Treatment is based on symptoms, though prions resist most disinfection methods. Medical professionals treating these patients must be protected from contact with contaminated tissues and instruments.

> **Focus on prion diseases**
> A prion consists of a misfolded protein and is an infectious agent that does not contain DNA or RNA. It is the smallest of all infectious agents—even smaller than a virus. It causes diseases that are among the most difficult to cure and is the infectious agent most resistant to sterilization.

Creutzfeldt-Jakob disease

CJD is a sporadic or familial prion disease, of which *bovine spongiform encephalopathy* or *mad cow disease* is one form. People over age 40, with a median age of about 60, are usually affected by CJD. This disease is rapidly progressive, usually being fatal within 6 months. Though occurring worldwide, it is most common in North African Jews. Only 5%–15% of cases are familial, with the autosomal dominant transmission. Age of onset is earlier in the familial form, with duration being longer. It can be transmitted after cadaveric corneal or dural transplants, use of stereotactic intracerebral electrodes or use of growth hormone prepared from human pituitary glands.

Variant CJD is most common in the United Kingdom, developing more in people under 30 years of age. A *mad cow disease* outbreak occurred in the early 1980s due to use of animal products infected with *scrapie*, a prion disease. Thousands of cows were infected, and people who ate their meat developed this form of CJD. In general, the term *sporadic* means that it is not acquired.

Pathophysiology
There are three major categories of CJD pathophysiology. In the sporadic form, the disease appears without any known risk factors. This is the most common type, accounting for about 85% of cases. In hereditary CJD, the patient will test positive for the related genetic mutation. In acquired CJD, the disease is transmitted via exposure to infected brain or nervous system tissue. Less than 1% of cases are acquired.

Clinical manifestations
Approximately 70% of patients with CJD have confusion and memory loss when initially evaluated, but these symptoms develop eventually in all patients. Early in the disease, 15%–20% of patient swill have ataxia and incoordination. Myoclonus is provoked by sensory stimuli such as noises, often developing in the middle of late stages. Other neurologic abnormalities can occur. These include hallucinations, neuropathy, seizures, and various movement disorders. Other common symptoms include diplopia, visual field defects, dimness or blurring of vision, and visual agnosia.

Diagnosis
CJD should be considered in the elderly who have quickly progressing dementia, especially with ataxia or myoclonus. However, other causes need to be ruled out first. Variant CJD is considered in younger people who have ingested processed beef in the United Kingdom. **Wilson's disease** must be excluded in these cases. Diagnosis is often difficult and takes time. MRI imaging may show cerebral atrophy. Diffusion-weighted MRI often shows abnormalities of the cortex and basal ganglia. While the CSF is often normal, a characteristic *14-3-3 protein* is often detected. Recently, a newer CSF study has been developed which is extremely specific for CJD and is now available in most academic centers. This test is called the Real-Time Quaking-induced Conversion assay, or simply, the "RT-QuIC."

EEG may show complex periodic sharp waves, which are diagnostic. Brain biopsy is the gold standard. However, it is rarely needed these days especially after the development of the RT-QuIC. Again, the diagnosis of this disease is also clinical.

Treatment
There is no effective treatment, so prevention of CJD is essential. Health-care professionals must use personal protective equipment when handling fluids and tissues, and avoid mucous membrane exposure. Contaminated skin can be disinfected by 5–10 min of exposure to 4% sodium hydroxide, then extensive washing with water. Contaminated materials and instruments should be steam autoclaved using sodium hydroxide or sodium hypochlorite. Standard sterilization methods are not effective. In the United States, repeated testing for mad cow disease has only revealed a few positive cases.

Gerstmann-Sträussler-Scheinker disease

Gerstmann-Sträussler-Scheinker disease is an autosomal dominant prion brain disease. It begins in middle age, worldwide, but is about 100 times less common than CJD. While developing between ages 40 and 60, the average life expectancy is about 5 years. There is cerebellar dysfunction, dysarthria, nystagmus, and unsteady gait when walking. Common symptoms include **gaze palsies**, deafness, dementia, hyporeflexia, extensor plantar responses, and Parkinsonism. Myoclonus is not as common as in CJS. This syndrome should be considered in patients with characteristic signs and symptoms, and family history, especially if they are 45 years of age or older. Genetic testing can confirm diagnosis.

Fatal insomnia

Fatal insomnia is usually a hereditary prion disorder. It causes difficulty in sleeping, motor dysfunction, and death. Extremely rare, fatal insomnia usually is caused by an autosomal dominant mutation related to chromosome 20, though a few sporadic cases have been seen. It usually begins between a person's late 30s and early 60s, with age 40 being median. Early symptoms include difficulty falling asleep and intermittent motor dysfunction such as myoclonus or spastic paresis. This may last for months. Eventually, there is progression to severe insomnia, myoclonus, sympathetic hyperactivity, and dementia. Sympathetic hyperactivity includes hypertension, tachycardia, hyperthermia, and sweating. Death occurs in approximately 13 months. Symptoms are mainly due to lesions in the thalamus (anterior and dorsal medial parts). Fatal insomnia should be considered when there is motor dysfunction, sleep disturbances, and family history. Genetic testing can confirm the condition.

Kuru

Kuru is a very rare, slowly progressive, fatal infection of the CNS. It is a transmissible spongiform encephalopathy that became an epidemic in New Guinea in the 1950s. It affected between 5% and 10% of the population. The incubation period can be 30 or more years, but death usually occurs within months of the onset of symptoms. Kuru was transmitted by ritual cannibalism of brain tissue during funeral rites. Initially, it causes unsteady gait, progressive trembling and shivering, and dysarthria. Ataxia progressively worsens, and the patient becomes unable to walk or stand. Muscle tremors and rigidity become pronounced. Incontinence and dysphagia develop, and the patient will become mute and unresponsive. It is fatal within 1 year of onset. Unlike other prior diseases, severe dementia does not occur. Today, this disease has nearly been eradicated due to the stopping of tribal cannibalism. The last known death from kuru occurred either in 2005 or 2009, which is under debate.

Section review
1. What are the pathological findings in AD?
2. How are delirium and dementia differentiated?
3. What is the definitive diagnosis of dementia?
4. What are the three categories of CJD?

Seizure disorders

A seizure is an abnormal, uncontrolled electrical discharge in the brain's cortical gray matter, which transiently interrupts normal function. Seizures usually cause abnormal sensations, altered awareness, involuntary focal movements, or generalized **convulsions**. Approximately 2% of all adults will have a seizure during their lifetimes, but two-thirds of these people will never have another seizure. **Epilepsy** is a chronic brain disorder that is also called *epileptic seizure disorder*. It mostly involves two or more recurrent, unprovoked seizures that are not related to reversible stressors. It is often idiopathic. *Nonepileptic seizures* are caused by temporary stressors or disorders. Fever can provoke seizures in children. They are most common in neonates and the elderly. *Psychogenic seizures* or *pseudoseizures* are older terms for nonepileptic seizures that are most often seen in patients with psychiatric disorders, but there are no abnormal electrical discharges in their brains. The causes of seizures are summarized in Table 6.5.

Pathophysiology

Generalized seizures involve electrical discharges that affect the cortex of both hemispheres (and are not focal to one hemisphere), usually causing loss of consciousness. They are most often due to metabolic disorders, but sometimes genetic disorders. Generalized seizures include infantile spasms, absence seizures, tonic-clonic seizures, atonic seizures, and myoclonic seizures. *Partial seizures* or focal seizures involve neuronal discharges in just one cerebral cortex, usually due to structural abnormalities. They may be simple, with no impairment of consciousness, or complex, with reduced by the incomplete loss of consciousness. Partial seizures can be followed by a generalized seizure, which is known as *secondary generalization*. It happens when a partial or focal seizure spreads to the other hemisphere, activating the entire cerebrum bilaterally. This can occur so quickly that the initial partial seizure is very brief, or not even clinically apparent.

Clinical manifestations

Seizures may be preceded by an **aura**, which can consist of autonomic, sensory, or psychic sensations. These may include paresthesia, sensing abnormal odors, a rising epigastric sensation, fear, or *déjà vu*—the feeling that something has happened before in exactly the same manner. Most seizures end on their own within 1–2 min. Generalized seizures are often followed by a *postictal* state, including deep sleep, confusion, headache, and muscle

Table 6.5 Causes of seizures.

Causes	Examples
Autoimmune disorders	Autoimmune encephalitis, such as NMDA encephalitis
Cerebral edema	Eclampsia; hypertensive encephalopathy
Cerebral hypoxia or ischemia	Carbon monoxide toxicity; cardiac arrhythmias; near suffocation; nonfatal drowning; stroke; vasculitis
CNS infections	AIDS; brain abscess; falciparum malaria; meningitis; neurocysticercosis; neurosyphilis; rabies; tetanus; toxoplasmosis; viral encephalitis
Congenital or developmental abnormalities	Cortical malformations; genetic disorders (such as fifth day fits, which are benign neonatal seizures in otherwise healthy infants; lipid storage diseases such as Tay-Sachs disease); neuronal migration disorders such as heterotopias; phenylketonuria
Drugs and toxins	(when given in toxic doses, these drugs cause seizures): camphor, cocaine, and other CNS stimulants; cyclosporins; lead; pentylenetetrazol; picrotoxin; strychnine; tacrolimus. (these drugs lower the seizure threshold): aminophylline, antidepressants (especially tricyclics); antihistamines with sedative properties; antimalarials; antipsychotics such as clozapine; **buspirone;** fluoroquinolones; theophylline
Expanding intracranial lesions	Hemorrhage, hydrocephalus, tumors
Head trauma	Birth injuries; blunt or penetrating injuries. (posttraumatic seizures occur in 25%–75% of patients with brain contusions, skull fractures, intracranial hemorrhage, extended coma, or focal neurologic deficits)
Hyperpyrexia	Drug toxicity such as with amphetamines or cocaine; fever; heatstroke
Metabolic disturbances	Common: hypocalcemia, hypoglycemia, hyponatremia. Less common: aminoacidurias, hepatic or uremic encephalopathy, hyperglycemia, hypernatremia, hypomagnesemia. In neonates: vitamin B6 deficiency
Pressure-related	Decompression illness, hyperbaric oxygen treatments
Withdrawal syndromes	Alcohol, anesthetics, barbiturates, benzodiazepines

soreness. Symptoms last from minutes to hours. Sometimes, this period will include **Todd's paralysis**, which is a transient neurologic deficit. It usually involves weakness of the limb contralateral to the seizure focus.

Most patients are neurologically normal between seizures. High doses, however, of drugs used for seizure disorders—especially anticonvulsants—can reduce alertness. If there is progressive mental deterioration, this is usually due to the neurologic disorder causing the seizures, and not the seizures themselves. Only rarely are seizures uninterrupted and of long duration. Table 6.6 classifies the various types of generalized and partial seizures, and Table 6.7 explains possible causes of symptomatic seizures.

Table 6.6 Classifications of generalized and partial seizures.

Generalized	Partial
Infantile spasms: Sudden flexion, adduction of arms and forward flexion of trunk. Seizures last a few seconds, recurring many times per day. Only occur in first 5 years of life, and replaced by other types of seizures. Developmental defects common.	*Simple partial seizures*: motor, sensory, psychomotor symptoms. **No altered consciousness.** Symptoms reflect affected brain area. *Jacksonian seizures/march*: focal motor symptoms begin in one hand, moving up the arm. Other focal seizures affect the face initially, spreading to an arm and sometimes a leg. Some partial motor seizures begin with one arm raising and the head turning toward that arm.
Typical absence seizures: 10- to 30-s loss of consciousness, eyelid fluttering, axial muscle tone may or may not be lost. Patients abruptly stop activity, then abruptly resume it. No postictal symptoms or awareness that the seizure occurred. These seizures are genetic, mostly occurring in children. Without treatment, likely to occur many times per day. Often occur while sitting quietly, precipitated sometimes by hyperventilation. Rarely occur during exercise. Neurologic, cognitive exam results usually normal.	*Epilepsia partialis continua*: A rare disorder causing focal motor seizures of usually an arm, hand, or one side of the face. Seizures recur every few seconds or minutes—for days to years at a time. These focal seizures can be very refractory to treatment. In children, usually caused by a focal cerebral cortical inflammatory process such as *Rasmussen encephalitis*, but possibly caused by a chronic viral infection or autoimmune condition. In adults, the cause is usually a structural lesion such as a stroke.
Atypical absence seizures: Usually occur as part of *Lennox-Gastaut syndrome*, a severe form of epilepsy beginning before 4 years of age. Last longer than typical absence seizures, with more pronounced jerking or automatic movements, and less complete loss of awareness. Often, there is a history of damage to the nervous system, developmental delay, abnormal neurologic exam results, and other seizure types. Atypical absence seizures usually continue into adulthood.	*Complex partial seizures*: Often preceded by an aura. Patient may stare during the seizure. Consciousness is impaired, but with some awareness of environment, such as noxious smells. There may be involuntary chewing or smacking of the lips, automatic but purposeless hand movements, saying unintelligible sounds that even the patient does not understand, resisting help from others, tonic or dystonic posturing of the extremity contralateral to the seizure focus, deviation of the head and eyes usually contralateral to the seizure focus, and pedaling or bicycling movements of the legs if the seizure occurred in the medial frontal or orbitofrontal regions. Motor symptoms resolve within 2 min, but confusion and disorientation may continue for another 2 min, and postictal amnesia is common.

Continued

Table 6.6 Classifications of generalized and partial seizures—cont'd

Generalized	Partial
	The patient may be very aggressive if restrained during the seizure, or while recovering consciousness if the seizure generalizes. Unprovoked aggression, however, is not common. Left temporal lobe seizures may cause verbal memory deficits. Right temporal lobe seizures may cause visual spatial memory deficits.
Atonic seizures: Most common in children as part of Lennox-Gastaut syndrome. Brief, complete loss of muscle tone and consciousness, causing falling or pitching to the ground—risking head trauma and other injuries.	
Tonic seizures: Most common in children while sleeping—usually from Lennox-Gastaut syndrome. Sustained contraction of axial muscles, beginning quickly or slowly, spreading to proximal limb muscles. Usually last 10–15 s. In longer seizures a few, rapid clonic jerks can occur as the tonic phase ends.	
Tonic-clonic seizures: Primarily or secondarily generalized. Primarily generalized seizures usually begin with an outcry, continuing with loss of consciousness and falling, then tonic contraction, then clonic—rapidly alternating contraction and relaxation—motion of muscles of extremities, trunk, and head. There may be tongue biting, frothing at the mouth, and urinary or fecal incontinence. Seizures usually last up to 2 min, with no aura. Secondarily generalized seizures start with a simple partial or complex partial seizure.	
Myoclonic seizures: Brief, fast jerks of a limb, several limbs, or the trunk. May be repetitive and lead to a tonic-clonic seizure. Jerks may be bilateral or unilateral. Consciousness is not lost unless it progresses into a generalized tonic-clonic seizure.	

Table 6.6 Classifications of generalized and partial seizures—cont'd

Generalized	Partial
Juvenile myoclonic epilepsy: Syndrome of myoclonic, tonic-clonic, and absence seizures, usually beginning in adolescence. A few bilateral, synchronous myoclonic jerks begin, followed in 90% of patients by generalized tonic-clonic seizures. Often occur upon awakening, especially after sleep deprivation or alcohol use. Absence seizures may occur in one third of patients.	
Febrile seizures: Due to a fever, in the absence of intracranial infection. Considered to be *provoked seizures*. Affect about 4% of children between 3 months to 5 years of age. Benign febrile seizures are brief, single, and generalized tonic-clonic in appearance. Complicated febrile seizures are focal, last more than 15 min, or recur two or more times within 24 h. About 2% of patients develop a subsequent seizure disorder. Incidence of seizure disorders and risk of recurrent febrile seizures are much greater in children with complicated febrile seizures, other neurologic abnormalities, onset before 1 year of age, or family history of seizure disorders.	
Status epilepticus: Involves tonic-clonic seizure activity lasting 5–10 min OR two or more seizures between which the patient dose not fully regain consciousness. Untreated generalized seizures lasting more than 60 min may cause permanent brain damage, and can be fatal. Heart rate and temperature are increased. Causes include rapid withdrawal of anticonvulsants and head trauma. Complex partial status epilepticus and absence status epilepticus often develop as extended episodes of mental status changes.	

Table 6.7 Symptomatic seizures and possible causes.

Description	Possible causes
Fever, stiff neck	Meningitis, meningoencephalitis, subarachnoid hemorrhage
Focal neurologic defects such as asymmetry of reflexes or muscle strength	Postictal paralysis, structural abnormality such as tumor or stroke
Generalized neuromuscular irritability such as hyperreflexia or tremulousness	Certain metabolic disorder such as hypocalcemia or hypomagnesemia, drug toxicity such as with sympathomimetics, or withdrawal syndromes from alcohol or sedatives
Loss of spontaneous venous pulsations	Increased intracranial pressure with less sensitivity; about 20% of patients with normal intracranial pressure lose spontaneous venous pulsations
Papilledema	Increased intracranial pressure
Skin lesions, such as axillary freckling or café-au-lait spots, hypomelanotic skin macules, or shagreen patches	Neurocutaneous disorders such as neurofibromatosis or tuberous sclerosis

Diagnosis

Diagnosis of seizure disorders is focused on determining if the event was truly a seizure or another cause of obtundation, a pseudoseizure, or syncope. New-onset seizure patients are evaluated in an emergency department and may be discharged after thorough evaluation. Patients with a known seizure disorder may be evaluated in a physician's office. Unusual sensations that suggest an aura and seizure must be evaluated as well as other manifestations. Conditions such as quickly decreasing brain circulation, causing syncopal convulsions, sometimes related to ventricular arrhythmia, must be ruled out. History should include information about the first seizure, subsequent seizures, duration, frequency, longest and shortest interval between them, sequential evolution, postictal state, and any precipitating factors.

Risk factors are assessed. These include CNS infections, prior head trauma, known neurologic disorders, drug use or withdrawal—prescribed or recreational, alcohol withdrawal, stopping anticonvulsant medications, family history of neurologic disorders and seizures. Rare triggers must also be assessed, which include flashing lights, repetitive sounds, video games, alcohol, and sleep deprivation.

Please note both alcohol and sleep deprivation are known triggers for seizures, and perhaps the reason behind high indices of first-time seizures during college years. Young patients such as these are often sleep-deprived while consuming alcohol.

The physical examination must assess tongue injury due to biting, incontinence, and prolonged confusion after a possible seizure. In fact, a lateral tongue bite is almost

95% sensitive in case of true epileptic seizure. A nonepileptic seizure can suggest a generalized tonic-clonic seizure because of generalized muscular activity and lack of response to verbal stimuli. The ways to distinguish pseudoseizures include the following:
- Intensity may wax and wane
- The patient often actively resists passive eye opening
- Postictal confusion is usually absent
- Pseudoseizures often last longer
- The typical tonic phase, followed by a clonic phase, usually does not occur
- Progression of muscular activity does not correspond to true seizure patterns, such as jerking that moves from one side to the other and back—nonphysiologic progression, or exaggerated pelvic thrusting
- Vital signs are usually normal, including temperature
- Back arching strongly suggests nonepileptic seizure
- Bilateral movements are also suggestive of nonepileptic seizures, except for rare cases of frontal lobe seizures, which can present with bilateral symptoms

When seizures are idiopathic, physical exam rarely indicates the cause. It can, however, provide clues when seizures are symptomatic.

Testing is performed, though normal results do not exclude a seizure disorder, so diagnosis may require video EEG monitoring to capture one of these nonepileptic episodes. This can prove that there are no EEG findings indicative of seizure activity while experiencing the nonepileptic seizure. For patients with a known seizure disorder but normal or unchanged test results, blood anticonvulsant levels are usually checked, along with signs and symptoms of trauma, infection, or a metabolic disorder. Neuroimaging is performed if seizures are newly occurring, or if exam results are abnormal for the first time. Head CT is usually performed immediately to exclude hemorrhage or a mass. This may be deferred or avoided in children with typical febrile seizures when neurologic status quickly normalizes.

If CT is negative, a follow-up MRI is recommended, providing better resolution for various structural abnormalities. It can detect cerebral venous thrombosis, cortical dysplasias, and other epileptiform lesions. High-resolution coronal T1 and T2 sequences are used for epilepsy protocols, which can detect sclerosis or hippocampal atrophy. MRI can detect common seizure causes, including malformations of cortical development in young children, as well as mesial temporal sclerosis, traumatic gliosis, and small tumors in adults.

EEG is vital for the diagnosis of epileptic seizures, especially for complex partial or absence status epilepticus. EEG can detect **epileptiform** abnormalities or discharges. These include sharp waves, spikes, slow-wave and spike complexes, and polyspike and slow-wave complexes. These abnormalities may be bilateral and generalized in

patients with generalized seizures. They may be localized in patients with partial seizures or simply focal seizures. Possible findings include
- Bilateral polyspike and wave abnormality in juvenile myoclonic epilepsy, at a rate of 4–6-Hz
- Abnormalities in temporal lobe foci that is interictal—between seizures—in complex partial seizures that begin in the temporal lobe
- Focal epileptiform discharges in secondarily generalized seizures
- Interictal symmetric bursts of 4–7-Hz epileptiform activity—in primarily generalized tonic-clonic seizures
- Slow spike and wave discharges, usually at less than 2.5 per second
- Three Hertz spikes and slow-wave discharges, at three per second, in typical absence seizures

It is important to understand that a normal EEG does not exclude a diagnosis of epileptic seizures. Diagnosis of epilepsy is clinical. If seizures are infrequent, EEG is less likely to detect abnormalities. An initial EEG can detect an epileptiform abnormality in just 30%–55% of patients who have a known seizure disorder. However, serial EEG can detect these abnormalities in 80%–90% of seizure disorder patients. Generally, serial EEG, using extended recording times, along with tests performed during sleep deprivation, greatly increase chances of detecting abnormalities in patients with seizures. One method is to use video and EEG monitoring for a few days, which is the most sensitive form of EEG available and helps differentiate epileptic from nonepileptic seizures.

Patients with nonepileptic seizures have seizure-like episodes that can be very hard to distinguish from an epileptic seizure. Video EEG can capture one of these nonepileptic episodes and show that there was no EEG correlation with the episode; therefore, diagnosing a nonepileptic seizure.

Laboratory tests are done to check for metabolic disorders. These include BUN, plasma glucose, creatinine, calcium, sodium, magnesium, phosphate, and liver function tests. If meningitis or CNS infection is suspected, a head CT is done. If results are normal, a lumbar puncture must be performed. Drug screening may be done to check for recreational drug use, but positive results do not indicate whether such use was causative for seizures, and results may be inaccurate.

In specialized epilepsy centers, advanced imaging tests can be done if seizures are refractory, and surgical resection is being considered. Functional MRI identifies functioning cortex and helps to guide surgical resection. When EEG and MRI cannot clearly identify the epileptic focus, *magnetoencephalography* (*MEG*), along with EEG, may help localize the lesion. MEG is also called *magnetic source imaging* (*MSI*). This may avoid the need for invasive intraoperative mapping procedures. During the postictal period, *single-photo emission CT* (*SPECT*) may detect increased perfusion in the seizure focus. This may help localize the area that will be surgically removed. Injection of contrast dye is required at the time of a seizure. Therefore, the patient is admitted for continuous

EEG and video monitoring when SPECT is performed during the peri-ictal period. Neuropsychologic tests can help identify functional deficits prior to and following surgery. They may help predict the prognosis and chances for rehabilitation.

Treatment

As a general rule, when the chance of having another seizure after a first seizure episode is higher than 60%, treatment with anticonvulsive medications is indicated. The usefulness of these medications after just one seizure is under controversy. Risks and benefits must be discussed with the patient. Risks for a subsequent seizure are low. Therefore, drugs may be withheld until the second seizure occurs—especially in children. Also, certain anticonvulsants cause behavioral and learning problem in children.

When a generalized tonic-clonic seizure occurs, injury is prevented by loosening clothing around the neck, and by placing a pillow under the patient's head. There should be no attempt to protect the tongue because of possible harm to the patient and rescuer. The patient should be rolled onto the left side to prevent aspiration. Family members and coworkers of seizure patients must be taught these procedures.

Patients with partial seizures must understand precautions to take should their seizures become generalized. They should avoid activities in which loss of consciousness could be dangerous, such as driving, climbing, swimming, bathing in a bathtub, or operating power tools. Once their seizures are totally controlled, ideally for 6 months or more, many of these activities can be resumed as long as appropriate safety methods are used. Exercise and social activities can be encouraged. Most states allow such patients to resume driving automobiles after being seizure-free for 6 months to 1 year. Rules for driving a vehicle after having a seizure vary significantly from state to state. However, most states allow the individual to drive again if there are no seizures 6 months to 1 year.

The patient must be advised to avoid illicit drugs such as cocaine, phencyclidine, alcohol, and amphetamines. Many illicit drugs trigger seizures. Some drugs lower the seizure threshold and should be avoided. These include tramadol (pain medication), meropenem (antibiotic), and bupropion (antidepressant), which are notorious for lowering seizure threshold. They should be avoided in patients with epilepsy. Family members must be taught not to overprotect the patient, be sympathetic, and to avoid making the patient feel inferior, self-conscious, or invalid. The only time a seizure patient should be institutionalized is if he or she is severely cognitively impaired, or if the seizures are so regular and violent, even with drug treatment, that much more care is required. Most centers observe a first-time seizure patient for few hours in the emergency room and complete the workup on an outpatient basis unless a reversible cause is identified while the patient is in the emergency department.

For prolonged seizures and status epilepticus, medications are needed to terminate the seizures, along with monitoring of respiration. If there is any indication of airway compromise, endotracheal intubation is necessary. Intravenous lorazepam is usually administered in varying doses. If seizures continue despite approximately 10 mg of intravenous lorazepam within the first 10 min, fosphenytoin, valproate or levetiracetam should be used as a secondary medication to abort the prolonged seizure. If intravenous access cannot be obtained, fosphenytoin can be given intramuscularly, or benzodiazepines can be given sublingually or rectally.

If the seizure persists, refractory status epilepticus is a possibility. Other anticonvulsants may be needed, such as propofol, midazolam, or pentobarbital. Intubation and general anesthesia are usually required if the seizure continues for more than 20 min. There are tertiary anti-seizure medications that include propofol, midazolam, or pentobarbital. These are used until the EEG manifestations of seizure activity have been suppressed. Later, the cause of status epilepticus must be identified and treated.

Use of antiseizure medications as preventive measures for post-traumatic seizures is controversial. Some experts prophylactically start medications if a head injury causes significant structural injuries. These may include brain lacerations, large contusions, or hematomas, or depressed skull fracture. Risks of seizures during the first week after injury are reduced by various medications, but they do not prevent permanent posttraumatic epilepsy from developing months or even years later. Unless seizures occur, these drugs should usually be stopped after 1–2 weeks.

The general principles of long-term drug treatment for epilepsy are as follows:
- With one anti-seizure medication, about 60% of patients are seizure-free.
- When seizures are hard to control from the beginning (in 30%–40% of patients), two or more drugs may eventually be needed.
- When seizures are intractable—refractory to an adequate trial of two or more drugs— patients should be sent to an epilepsy center to determine possibilities for surgery.

Doses are tailored to the patient's tolerance. The appropriate dose is the lowest dose that stops seizures with the fewest adverse effects, regardless of blood drug level.

The general goal of treatment is "no seizure, no side effect."

When toxicity develops before seizures are under control, doses are reduced and another drug is added at a low dose, gradually increased until effective. Since the two drugs can interact, close monitoring is required. Anticonvulsant drug levels are altered by many other drugs and vice versa. All potential drug-drug interactions must be carefully checked. When seizures are controlled, the drug is continued uninterrupted until the patient has had no seizures for at least 2 years. Then, stopping the drug *may* be an option in some patients. Relapses occur most often when the patient has had a seizure disorder since childhood, needs more than one drug to be without seizures, had previous seizures while taking an anticonvulsant, has partial or myoclonic seizures, has an underlying static encephalopathy, or has had an abnormal EEG in the last year. For relapses, 60% occur within 1 year and 80% occur within 2 years.

The newer anticonvulsants include clonazepam, felbamate, lamotrigine, levetiracetam, oxcarbazepine, pregabalin, tiagabine, topiramate, and zosinamide. For infantile spasms, atonic seizures, and myoclonic seizures, valproate is preferred, followed by clonazepam. Also, corticosteroids are often effective for infantile spasms. A ketogenic diet is one that is high in fats and induces ketosis. This type of diet may be helpful in children with refractory epilepsy and recently has been tried in adults with promising results as well. Drugs are not recommended for febrile seizures unless the child has a subsequent seizure when the fever has subsided. Drugs are also not recommended for seizure due to alcohol withdrawal. Treatment of the withdrawal syndrome, usually including a benzodiazepine, often prevents seizures.

Intractable seizures refractory to medical treatment occur in 10%–20% of patients, indicating that surgery may be an option. Resection of the epileptic focus usually improves seizure control greatly. After surgery, some patients do not need any further anticonvulsants after the first 2 years. Surgery is contemplated after extensive tests and monitoring, which is best performed in epilepsy centers.

Intermittent electrical stimulation of the left vagus nerve, using an implanted vagus nerve stimulator, may be successful for pediatric patients with intractable seizures who are not candidates for surgery.

Section review
1. What is the pathophysiology of partial or focal seizures?
2. What is Todd's paralysis?
3. What test is vital for the diagnosis of epileptic seizures?

Sleep disorders

Nearly 50% of Americans have some type of sleep disorder. This can cause emotional disturbances, poor motor skills, memory problems, decrease efficiency while working, and more likelihood of accidents, including traffic accidents. Sleep disorders can lead to cardiovascular disease in patients with sleep apnea. **Insomnia** is difficulty in falling asleep or staying asleep. The patient often complains of not feeling refreshed after sleep. **Excessive daytime sleepiness** (EDS) is the tendency to fall asleep during normal waking hours. Insomnia and EDS are symptoms of a variety of sleep-related disorders. *Parasomnias* are abnormal events related to sleep.

Pathophysiology
The two states of sleep each have physiological characteristics, as follows:
- *Nonrapid eye movement (NREM)*—this type of sleep makes up 75%–80% of total sleep time in adults. It has four stages, with increased depth of sleep. There are slow, rolling eye movements that characterize quiet wakefulness and early stage-one sleep. These disappear in the deeper sleep stages. Muscle activity also decreases. Stages three and four are referred to as *deep sleep* since the arousal threshold is high. People often perceive these stages as "high-quality" sleep.

- *Rapid eye movement (REM)*—this type of sleep follows every cycle of NREM sleep. There is low-voltage, fast activity, when measured on an EEG, and postural muscle **atonia**. Dramatic fluctuations of respiration rate and depth occur, and most dreams happen at this time.

There is a progression through the sleep stages, usually followed by a short interval of REM sleep. This occurs in cycles, five to six times every night (see Fig. 6.3). Each person needs varying amounts of sleep, ranging from 6 to 10 h per day. Infants sleep for a large part of each day. As we age, total sleep time and deep sleep usually decrease, and there are more sleep interruptions. Elderly people may actually not enter stages three or four of NREM. These changes may explain increasing EDS and fatigue with aging, but are of unclear clinical significance.

Insomnia is most often caused by adjustment sleep disorder and psychophysiologic problems, inadequate sleep hygiene, psychiatric disorders—especially mood, anxiety, and substance use disorders; and miscellaneous disorders, including chronic pain, cardiopulmonary disorders, and musculoskeletal conditions. EDS is most often caused by circadian rhythm disorders such as shift work sleep disorders or jet lag, insufficient sleep syndrome, obstructive sleep apnea (OSA) syndrome, narcolepsy, and miscellaneous medical, neurologic, and psychiatric conditions.

Diagnosis

Patient history should take into account the duration and age of the patient when symptoms began. Also, events such as new medical conditions, new medications, or life or work changes that coincided with onset must be considered. Symptoms during sleeping and waking hours should be recorded. Quality and quantity of sleep are assessed by

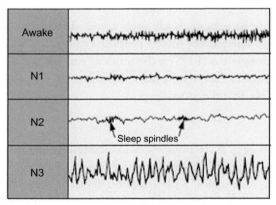

Fig. 6.3 Electroencephalogram (EEG) stages of wakefulness and NREM sleep. *Awake*, low-voltage fast activity; *stage N1*, falling asleep; *stage N2*, light sleep with sleep spindles; *stage N3*, slow delta waves. Rapid eye movement (REM) sleep looks similar to awake and stage N1 sleep. Sleep spindles are bursts of brain activity associated with onset of sleep.

determining bedtime, the latency of sleep—how long it takes to fall asleep after going to bed, how many times the patient awakens and when these occur, the last morning awakening time, the time when the patient gets out of bed, and if the patient naps, how often and for how long. Patients are advised to keep a "sleep log" for several weeks, which is better than just responding to questions about sleep patterns. Prior to bedtime, intake of food or alcohol, or any physical or mental activities that are performed must be evaluated. Also considered are drug intake or withdrawal, and the same for alcohol, caffeine, and nicotine. For EDS, the assessment focuses on likelihood of falling asleep during rest periods or when alertness is required. The **Epworth Sleepiness Scale** is used to assess EDS.

The patient's symptoms of snoring, interrupted breathing during sleep, symptoms of sleep apnea, anxiety, depression, mania, and hypomania—mental sleep disorders, leg restlessness, and restless legs syndrome (RLS) must be evaluated. Also, narcolepsy, cataplexy, and sleep paralysis must be considered. Spouses or other family members may help to identify the manifested symptoms. A past medical history should consider asthma, chronic obstructive pulmonary disease (COPD), heart failure, gastroesophageal reflux, hyperthyroidism, neurological movement, and degenerative disorders, as well as rheumatoid arthritis (RA) and other painful disorders. OSA may be linked to heart disorders, obesity, hypertension, stroke, smoking, snoring, and trauma to the nose.

Physical examination should check for obesity with fat distribution around the neck or stomach; excessively large neck circumference; mandibular hypoplasia and **retrognathia**; enlargement of the tongue, tonsils, soft palate, or uvula; nasal obstruction; decreased patency of the pharynx; increased obstruction of the soft palate and uvula by the tongue; and redundant pharyngeal mucosa. The chest is examined for **kyphoscoliosis** and expiratory wheezing. Right ventricular failure should be assessed, and a complete neurologic exam performed. Concerning signs and symptoms are:
- Breathing interruptions or awakening with gasping
- Falling asleep while driving or in other situations requiring alertness
- Frequent sleepwalking or other activities involving mobility while asleep
- History of violent behaviors while asleep
- Recent stroke
- Repeated episodes of falling asleep without warning
- Status cataplecticus—continuous attacks of catalepsy
- Unstable cardiac or pulmonary status

Treatment
Initially, specific underlying conditions are treated. Good sleep hygiene is taught, which is important for all causes and often solves all mild problems with sleep. Hypnotic medications may be used if controlled well. These include the benzodiazepines, non-benzodiazepines, and melatonin receptor agonists. Patients who experience daytime

sedation, incoordination, and other effects during the day must avoid activities needing alertness. Doses should be reduced, or the medication should be stopped or changed. Additional adverse effects include amnesia, hallucinations, and falling. Hypnotics must be used cautiously in those with pulmonary insufficiency and monitored in the elderly. Prolonged use is not advised since tolerance can develop. Abrupt discontinuation may cause rebound insomnia, anxiety, tremor, and seizures. The lowest effective dose should be used only for brief periods, then tapered off before stopping the medication.

Insomnia
A variety of sleep disorders may cause EDS. They may be intrinsic or extrinsic. Sleep is impaired by the use of caffeine or stimulants close to bedtime, exercise or excitement late in the evening, and irregular sleep-wake schedules. Sleeping late or taking naps to make up for lost sleep actually may harm normal nocturnal sleep even more. Insomniacs must have a regular awakening time and avoid naps, no matter how much nocturnal sleep they are getting. Psychophysiologic insomnia involves anxiety about yet another night of being unable to sleep properly.

Pathophysiology
Insomnia may be related to stressors, be psychophysiologic, physical, mental, due to sleep deprivation, and caused by drugs. Loss of a job, hospitalization, or other acute stressors are linked to insomnia. This type of insomnia is usually brief and resolves after the stressors are past. Physical sleep disorders include those that cause pain or discomfort, such as arthritis, cancer, and herniated spinal discs. They also include those that are worse when the patient moves, cause transient awakenings, and poor sleep quality. Nocturnal seizures are included. Insufficient sleep syndrome or *sleep deprivation* exists when the patient does not sleep enough at night, and cannot remain alert during waking hours. This is usually caused by job-related or social commitments that the patient must fulfill. It is likely to be the most common cause of EDS, which disappears when the patient gets more sleep, such as on weekends. Drugs that may lessen the ability to sleep include CNS stimulants, hypnotics, other sedatives, antimetabolite chemotherapy, anticonvulsants, oral contraceptives, methyldopa, propranolol, alcohol, and thyroid hormones. Popular hypnotics may cause apathy, irritability, and reduced alertness. Psychoactive drugs often cause abnormal movements while sleeping. Insomnia can also develop during withdrawal of CNS depressants, tricyclic antidepressants, monoamine oxidase inhibitors, or illicit drugs. When hypnotics or sedatives are withdrawn abruptly, there may be nervousness, tremors, and seizures.

Clinical manifestations
The clinical manifestations of insomnia including sleepiness during the day, general tiredness, irritability, and problems with concentration or memory. It may be idiopathic or

linked to a variety of stressors, health conditions, pain, medications, or other substances. Acute insomnia lasts from one night to a few weeks. Chronic insomnia is signified by lasting at least three nights per week, for 1 month or longer.

Diagnosis

Diagnosis of insomnia involves physical examination, medical history, and sleep history. Sleep diaries, kept for several weeks, may be extremely beneficial for diagnosis. The patient's spouse or family members may need to be interviewed about the patient's quality and quantity of sleep. Sometimes, there will be a referral to a sleep center for specialized testing.

Treatment

For insomnia linked to stressors, short-term treatment with hypnotics may be needed. Persistent anxiety may require specific treatment. Psychophysiologic insomnia requires use of hypnotics along with cognitive-behavioral strategies. The effects of these psychological therapies last longer and include sleep hygiene—restriction of the amount of time in bed, training about sleep habits and how to relax, controlling stimuli, and cognitive therapy. Hypnotics are used only when the patient needs quick relief, or when insomnia has caused fatigue and EDS. Treatment of physical sleep disorders is focused on the underlying condition, usually using bedtime analgesics.

For mental sleep disorders, antidepressants that provide more sedation may be used, at regular doses. However, they can cause EDS, weight gain, and other adverse effects. All antidepressants can be used with a hypnotic. When depression is present along with EDS, antidepressants with activating qualities are used.

> ### Focus on insomnia
> Rates of insomnia increase as we age. However, usually, it is attributable to some other medical condition. Also, people who are separated, divorced, or widowed report more frequent insomnia. Effects are potentially severe since sleep is just as important as diet and exercise. Regardless of the inability of "sleeping pills" to cure insomnia, their rates of use regularly increase, every year. Today, about one in four Americans take some type of medication to help them sleep. Only cognitive therapy has been shown to be curative for insomnia.

Sleep apnea

Sleep apnea involves periods of breathing cessation during sleep. There are two forms: central sleep apnea and OSA. Collectively, these disorders are referred to as *sleep-disordered breathing*.

Pathophysiology

Central sleep apnea involves repeated episodes of breathing cessation, or shallow breathing during sleep that last at least 10 s. It is caused by reduced respiratory effort, and usually manifests as insomnia, or disturbed, refreshing sleep.

OSA involves episodes of closure of the upper airway during sleep. This may be partial or complete. It leads to cessation of breathing for more than 10 s. The patient may awake, gasping for air. This disrupts sleep and results in unrefreshed sleep and EDS. OSA affects 2%–9% of adults but is often not diagnosed even in symptomatic patients. It is about four times more common in men than in women. It is also about seven times more common in obese patients, with a body mass index of 30 or higher. Severe cases increase the risk of death for middle-aged men.

Clinical manifestations

Though most people who snore do not have OSA, loud, disruptive snoring is reported by 85% of patients with OSA. Additional symptoms include choking, gasping, snorting, restless sleep, and difficulty staying asleep. Most patients are unaware of these symptoms but alerted to them by spouses, family members, and other people in the house. During the day, sleep apnea patients are often fatigued, **hypersomnolent**, and have impaired concentration. Sleep problems and degrees of daytime sleepiness are usually not closely related to the amounts of time the patient awakens during the night.

Diagnosis

Sleep apnea diagnosis is based on identified risk factors, symptoms, or both. There must be documentation of more than five episodes of hypopnea and apnea per hour. Regarding symptoms, there should be one or more of the following:
- Daytime sleepiness, fatigue, insomnia, unintentional sleep episodes, or unrefreshing sleep
- Awakening with breath-holding, choking, or gasping
- Reports by another person of breathing interruptions, loud snoring, or both

An extended sleep history is taken for the patient over the age of 65, those who report daytime symptoms, are overweight, or who have poorly controlled hypertension, heart failure, stroke, or diabetes. Physical examination of the nose, tonsils, and pharynx must be done. There should be assessment of tonsillar hypertrophy, or features of acromegaly and hypothyroidism.

Confirmation is via polysomnography, in which breathing is continuously measured, using **plethysmography**, nose and mouth airflow, oximetry of oxygen saturation, EEG of sleep architecture, chin electromyography to assess hypotonia, and *electrooculograms* to assess REMs. Polysomnography helps classify sleep stages and periods of sleep apnea. The patient is observed using video, with ECG monitoring to determine any occurring arrhythmias. There may also be an assessment of limb muscle activity,

RLS, periodic limb movement disorder, and certain sleep position effects. Additional testing may be needed, which includes upper airway imaging, thyroid-stimulating hormone tests, and more.

Treatment

Treatment of sleep apnea is focused on reducing episodes of hypoxia and fragmented sleep. A cure is defined as a resolution of symptoms to below 10 per hour. Treatment is directed initially at risk factors, then the sleep apnea itself. Specific treatments include *continuous positive airway pressure (CPAP)*, oral appliances, and airway surgery. The treatment of choice is CPAP for most patients. It improves upper airway patency via application of positive pressure to the collapsible upper airway segment. Effective pressures range between 3 and 15 cm of water. The pressure can be titrated during monitoring via repeated polysomnography. The CPAP device has been proven to reduce cognitive impairment and blood pressure. When it is withdrawn, symptoms recur over several days. Generally, the duration of CPAP therapy is indefinite. The most significant factor about CPAP is that many patients do not adhere to its use because of discomfort with the oxygen mask, dryness, and nasal irritation. CPAP can be better tolerated when warm, humidified air is used, and can be augmented with bi-level positive airway pressure for patients with obesity-hypoventilation syndrome.

Oral appliances may be used to alter the position of the mandible or to prevent it from retracting during sleep. Some of these devices pull the tongue forward. Both forms are becoming more common in the treatment of OSA. When the airway is obstructed by enlarged tonsils or nasal polyps, surgery may be an option. Another option is surgery for **macroglossia** or **micrognathia**. Surgery is the first-line treatment only if there is an anatomic obstruction; otherwise, it is a secondary treatment. The most common type of surgical procedure for sleep apnea is *uvulopalatopharyngoplasty (UPPP)*. This involves resection of the submucosal tissue, from the tonsillar pillars to the **arytoenoepiglottic folds**, including the adenoids, to enlarge the upper airway. This procedure may not be effective in morbidly obese patients, or in those with anatomic narrowing of the airway. Also, recognition of sleep apnea following UPPP is difficult since there will be a lack of snoring. These silent obstructions can be as severe as the episodes of apnea prior to surgery.

Other surgical procedures include hyoid advancement, midline **glossectomy**, and **mandibulomaxillary advancement**. As a last resort, *tracheostomy* may be performed, to bypass the site of obstruction. This effective but radical cure is indicated for severely affected patients, such as those with *cor pulmonale*. Modafinil is a medication that can be used for residual sleepiness in OSA patients that effectively use CPAP. Supplemental oxygen may improve blood oxygenation but can cause respiratory acidosis and a morning headache. Tricyclic antidepressants and theophylline may stimulate ventilatory drive, but either has limited effectiveness, a low therapeutic index, or both.

Narcolepsy

Narcolepsy is a condition of sudden and chronic sleepiness during the daytime, often with **cataplexy**, a sudden loss of muscle tone that is often triggered by an emotional stressor. It may also involve sleep paralysis and hallucinations.

Pathophysiology

Narcolepsy is of unknown cause. It is equally common in both sexes and affects between 0.2 and 1.6 of every 1000 people. It is strongly linked with specific human leukocyte antigen (HLA) haplotypes, which are groups of alleles inherited together from a single parent. Children of narcoleptics have a 40 times higher risk of having this condition. Since its occurrence in twins is only 25%, there is likely an environmental role in triggering the condition. Usually, the neuropeptide *hypocretin-1* is deficient in the CSF. Therefore, the cause may be HLA-related autoimmune destruction of the hypocretin-containing neurons located in the lateral hypothalamus. There is dysregulation of timing and controls of REM sleep, which then intrudes into wakefulness, and into the transition between wakefulness and sleep. Many symptoms result from the REM-characteristics of vivid dreaming and postural muscle paralysis. The diagnosis is mainly clinical and there is no need to check the hypocretin level in most cases.

Clinical manifestations

The primary symptoms of narcolepsy are EDS, cataplexy, hypnagogic and hypnopompic hallucinations, and sleep paralysis. About 10% of narcoleptics have all four types of symptoms. Nocturnal sleep is usually disturbed, and some patients have hypersomnia, which means extended sleep times. The condition usually begins in adolescents or in young adults, with no prior illness, though it may be precipitated by illness, stress, or sleep deprivation. Once the condition is established, it usually lasts throughout the patient's life and does not affect longevity.

An episode can occur anytime, from a few times to many times per day, lasting for minutes up to hours. The patient can only temporarily resist the need to sleep but is easily aroused when they do fall asleep. It usually manifests during conditions of monotonous activity, such as during meetings, when reading, or watching television. However, it can occur when driving, eating, speaking, or writing. *Sleep attacks* occur in some patients, with no warning. The patient often feels refreshed upon awakening, yet falls asleep again in just a few minutes. During the night, sleep may be interrupted by frightening, vivid dreams, and be unsatisfying. The condition causes low productivity, low motivation, poor concentration, depression, reduced quality of life, increased chance for physical injury, and often affects interpersonal relationships.

Cataplexy is brief muscular weakness or paralysis, without loss of consciousness. It is caused by sudden emotional reactions. The muscular weakness can be only in the limbs, or can cause the patient to fall. These attacks resemble muscle tone loss that happens during REM sleep. Cataplexy occurs in about 75% of narcolepsy patients. Triggers for cataplexy include sudden anger, happiness, fear, joy, or surprise. The patient may also experience a type of sleep paralysis, being temporarily unable to move as they are falling asleep, or immediately after awakening. Both of these occurrences can be extremely frightening to the patient. This sleep paralysis only occurs in about 25% of narcolepsy patients.

Hypnagogic hallucinations occur just as the patient is falling asleep, while hypnopompic hallucinations occur just after awakening. Hypnagogic hallucinations are much more common, occurring in about 33% of narcoleptic patients. In both types, the hallucinations are very vivid and may be auditory or visual. They resemble vivid dreams, which are normal during REM sleep.

Diagnosis

History of cataplexy strongly suggests narcolepsy in patients who have EDS. Nocturnal polysomnography is performed, followed by multiple sleep latency tests. For diagnosis, the following must be found:
- REM episodes during at least two of every five daytime nap opportunities
- Average time to fall asleep, or *sleep latency*, of less than or equal to 8 min, after a minimum of 6 h of nocturnal sleep
- No other abnormalities revealed by nocturnal polysomnography

The *maintenance of wakefulness test* helps to monitor the effectiveness of treatments but is not diagnostic. It measures the patient's ability to stay awake for a certain time period, using isolation from temperature, light, noise, hypnotic medications, alcohol, or smoking.

Patient history and physical examination usually reveal other causative disorders. Confirmation of the diagnosis is via brain imaging, and blood and urine tests. Causative disorders may include lesions of the hypothalamus or upper brainstem, increased intracranial pressure, and encephalitis. With or without hypersomnia, other disorders that can cause EDS include hyperglycemia, hypothyroidism, anemia, hypoglycemia, hypercapnia, uremia, hepatic failure, hypercalcemia, and seizure disorders. Acute systemic disorders such as influenza are linked to acute but brief EDS and hypersomnia. In teenage boys, the rare Kleine-Levin syndrome may cause episodic hyperphagia and hypersomnia, which may be an autoimmune response to an infection.

Treatment

Treatment of narcolepsy includes stimulant and *anticataplectic* drugs. The patient is advised to get adequate nocturnal sleep and to take brief naps of less than 30 min at the same time each day, usually in the afternoon. Modafinil is a long-acting wake-promoting drug that is beneficial for those with mild to moderate EDS. Dosing must be monitored closely because additional doses in the early afternoon can interfere with nocturnal sleep. Other medications include derivatives of amphetamines such as methylphenidate, methamphetamine, or dextroamphetamine. However, since abuse potential is high and there are many adverse effects, close monitoring is required. Sodium oxybate can be used, taken at bedtime. It also must be monitored since it has a high abuse and dependence potential. Additional medications include tricyclic antidepressants, monoamine oxidase inhibitors, SSRIs, and the anticataplectic called *clomipramine* may be used.

Idiopathic hypersomnia

Idiopathic hypersomnia is EDS with or without long sleep times. It is different from narcolepsy because it does not involve cataplexy, hypnagogic hallucinations, or sleep paralysis. It is believed to be caused by a dysfunction of the CNS. For idiopathic hypersomnia with long sleep times, history or sleep logs may reveal nocturnal sleep of more than 10 h. For idiopathic hypersomnia without long sleep times, the patient usually sleeps more than 6 h but less than 10 h. Polysomnography, for both forms, shows no other sleep abnormalities. Multiple sleep latency tests will reveal short sleep latencies of less than 8 min, with less than two REM periods. Treatments are similar to those for narcolepsy, except that anticataplectic drugs are not needed.

Parasomnias

Parasomnias are abnormal, undesirable behaviors occurring as the patient falls sleep, remains asleep, or is aroused from sleep. History and physical examination often confirm the diagnosis. Types of parasomnias include **somnambulism**, sleep terrors also called night terrors, nightmares, REM sleep behavior disorder, or sleep-related leg cramps.

In somnambulism, the patient walks, sits, or performs other complicated behaviors while still sleeping. The eyes are usually open, but the patient does not recognize people or objects. It is most common in late childhood and in adolescence. It occurs after and during arousal from NREM stages three or four. Prior sleep deprivation and poor sleep hygiene increase the chances of these episodes. Risks are higher for first-degree relatives of somnambulistic patients. Some patients continuously mumble incoherent words and may be injured by obstacles or staircases. The patient is not actually dreaming during the episode, and after awakening, usually does not remember it. Treatment is focused on protecting the patient from being injured, including alarms that go off when the patient leaves the bed, use of beds at lower heights, and removing obstacles from the bedroom. Bedtime benzodiazepines, especially clonazepam, are often helpful.

Sleep or night terrors may accompany sleepwalking. Patients display fear, screaming, flailing, and are difficult to awaken. Sleep terrors are more common in children, occurring after arousal from NREM stages three or four—therefore, they do not represent nightmares. In adults, they may be linked to alcoholism or mental problems. When daily activities are affected, bedtime administration of oral benzodiazepines such as clonazepam or diazepam may be helpful. Nightmares, which are most common in children, occur during REM sleep, often due to fever. In adults, alcohol often triggers nightmares. Treatment is focused on underlying mental distress.

Our body is designed in a way that we are not able to move excessively, while we are asleep. In some patients, this inhibitory system is disrupted causing a variety of movements and abnormal behaviors during sleep. In REM sleep behavior disorder, the patient may use bad language or make other verbal statements, and have violent body movements, such as punching, kicking, or waving the arms. The patient may be acting out dreams while not having the normal atonia present during REM sleep. This disorder is most common in the elderly, especially people with Alzheimer's or Parkinson's disease, *olivopontocerebellar* degeneration, vascular dementia, multiple system atrophy, or progressive supranuclear palsy. It may also occur in narcoleptics, and if using norepinephrine reuptake inhibitors. The cause is usually unknown. Diagnosis is suspected based on the reports of a bed partner or the patient and is usually confirmed by polysomnography. Excessive motor activity during REM may be detected. Audiovisual monitoring is used to document symptoms. Neurologic examination rules out neurodegenerative disorders, sometimes followed by CT or MRI. It is treated with bedtime clonazepam, often taken indefinitely. Bed partners may be advised to sleep elsewhere until symptoms resolve, and sharp objects should be removed from the bedroom.

Muscles of the calf or foot muscles often experience cramping during sleep, even in healthy middle-aged or elderly patients. Diagnosis is based on history and lack of disability or physical signs. Stretching the affected muscles for several minutes before sleep is often preventative. When a cramp occurs, stretching as soon as possible quickly relieves symptoms, and is preferred over medications. Though many drugs have been used, the effects are minimal while adverse effects are common. Avoiding sympathetic stimulants such as caffeine may be helpful.

Restless legs syndrome
RLS, involves abnormal motions of, or sensations in the lower extremities that can interfere with sleep. It is more common in middle or older age. More than 80% of patients also have *periodic limb movement disorder*, a similar condition that affects the upper extremities. The clinical term for RLS is *Willis-Ekbom Disease*.

Pathophysiology
While of unknown cause, it may involve abnormal dopamine neurotransmission in the CNS. It can occur as an isolated condition, or related to drug withdrawal, when using stimulants or certain antidepressants, in pregnancy, because of chronic renal or hepatic failure, from iron deficiency or anemia, and from other disorders. In primary RLS, it may be inherited. More than 33% of patients have a family history. Risk factors include a sedentary lifestyle, obesity, and smoking. It is now known that REM sleep behavior diseases are associated with developing Parkinson's dementia several years later.

Clinical manifestations
RLS is a sensorimotor disorder. There is an irresistible urge to move the legs. It is usually accompanied by paresthesia, and sometimes pain in the lower or upper extremities. These become more intense when the patient is laying down or inactive, and severity is usually worst at bedtime. The patient is advised to move the affected leg via walking, stretching, or kicking. The condition often makes it hard to fall asleep, causes continued nocturnal awakenings, or both.

Diagnosis
Diagnosis is based on the history provided by the patient or a bed partner. Once RLS is diagnosed, polysomnography may be performed to determine if the patient also has periodic limb movement disorder. However, polysomnography is not required for RLS diagnosis. The patient should be evaluated for contributing disorders, via blood tests, and hepatic or renal function tests.

Treatment
Dopaminergic drugs are the only medications specifically administered for RLS. Usage must be monitored for adverse effects, including reoccurrence of symptoms, worsening of symptoms, nausea, orthostatic hypotension, and insomnia. Pramipexole and ropinirole are effective and offer few serious adverse effects. Benzodiazepines help sleep continuity but do not reduce limb movements and must be used with caution. Gabapentin at bedtime can be helpful if there is accompanying pain. Opioids are only used as a last resort due to abuse potential and adverse effects and may require ferrous sulfate and vitamin C supplementation. Good sleep hygiene is required as an aid to treatment.

Section review
1. Which state of sleep makes up the majority of total sleep and how can it be explained?
2. What is the Epworth Sleepiness Scale, and how is it used to assess various sleep disorders?
3. What are the causes of insomnia?

Headache
Headache is a pain in any area of the head, including the face, scalp, and interior of the head. It is one of the most common reasons that patients seek medical treatment throughout the world.

Pathophysiology
Headache is caused by activation of pain-sensitive structures located in or around the brain, skull, face, sinuses, and teeth. It may be a primary disorder, or secondary to another

condition. Primary headache disorders include migraine, cluster headache, and tension-type headache. The most common causes of headache are tension-type headache and migraine. Secondary headache may be caused by extracranial disorders, intracranial disorders, systemic disorders, drugs, and toxins. Extracranial disorders include carotid or vertebral artery dissection, dental disorders, glaucoma, and sinusitis. Intracranial disorders include masses such as brain tumors, Chiari type I malformation, CSF leak, hemorrhage, idiopathic intracranial hypertension, infections, noninfectious meningitis, obstructive hydrocephalus, and vascular disorders. Systemic disorders include acute severe hypertension, bacteremia, fever, giant cell arteritis, hypercapnia, hypoxia, viral infections, and viremia. Drugs and toxins include analgesics, caffeine, carbon monoxide, hormones, nitrates, and proton pump inhibitors.

Clinical manifestations

The various types of headache often have unique identifiers. *Migraine headache* involves frequent unilateral or bilateral, pulsating, and long-lasting pain. This is sometimes accompanied by aura, nausea, photophobia, *sonophobia*, or *osmophobia*. It is worsened with activity. The patient prefers to lie down in the dark, and it is often resolved by sleeping. *Tension-type headache* is frequent or continuous, mild, bilateral, with squeezing occipital or frontal pain spreading through the entire head. It is usually worse at the end of the day. *Cluster headache* involves unilateral orbitotemporal attacks, at the same time of the day. It is deep and severe, lasting between 30 and 180 min. There is often lacrimation, flushing of the face, restlessness, or Horner's syndrome, which involves partial ptosis, miosis, and hemifacial anhidrosis (absence of sweating).

Diagnosis

Diagnosis of headache is initially based on ruling out a secondary cause. History of present illness and symptoms is assessed. Factors such as head position, sleep, time of day, light, sounds, odors, chewing, and physical activity are considered. If the headache has been recurrent, previous diagnoses are identified and compared to current symptoms. Age of first onset, episode frequency, temporal patterns, and treatment responses are noted. Review of systems takes into account vomiting, fever, red dye or other visual symptoms, visual field deficits, diplopia, blurred vision, lacrimation, facial flushing, rhinorrhea, pulsatile tinnitus, preceding aura, focal neurologic deficits, seizures, syncope at headache onset, and myalgias or vision changes in older adults.

History should identify exposure to drugs, caffeine and other substances, toxins, recent lumbar puncture, immunosuppressive disorders, IV drug use, hypertension, cancer, dementia, trauma, coagulopathy, and use of anticoagulants or alcohol. Family and social history is considered. Often, migraine headache has been undiagnosed in family members. Physical examination includes vital signs, general appearance, head and neck exam, and a full neurological exam. The scalp is checked for swelling or tenderness. The ipsilateral temporal artery and both temporomandibular joints are palpated. The eyes and periorbital areas are inspected for signs. Pupil size and light responses, visual fields, and extraocular movement are assessed. Fundi are checked for papilledema and spontaneous venous pulsations. Visual acuity is measured if there are vision symptoms or eye abnormalities. A slit lamp is used and intraocular pressure is measured if there is reddening of the conjunctiva. The nose is inspected for purulent discharge. The oropharynx is inspected for swelling. The teeth are checked for tenderness. The neck is flexed to assess stiffness, discomfort, or both, and the cervical spine is palpated for tenderness. Some serious disorders require immediate testing.

In patients with **thunderclap headache**, altered mental status, meningismus, papilledema, signs of sepsis such as rash or shock, acute focal neurologic deficits, and severe hypertension of more than 220/120, CT or MRI is indicated. When meningitis, subarachnoid hemorrhage, or encephalitis are being considered, lumbar puncture and CSF analysis are performed if not contraindicated by results of imaging. If acute narrow-angle glaucoma is suggested, tonometry is performed. MRI is suggested if the patient has focal neurological deficits of subacute or uncertain onset, is over 50 years of age, has lost weight, has cancer, has HIV or AIDS, has had changes in headache patterns, or has diplopia. Erythrocyte sedimentation rate is done if there are visual symptoms, tongue or jaw claudication, temporal artery signs, or other findings that may suggest giant cell arteritis. CT scan of the paranasal sinuses can rule out complicated sinusitis if there are systemic signs such as high fever, dehydration, prostration, or tachycardia. Findings that suggest sinusitis include frontal, positional headache, epistaxis, and purulent rhinorrhea. If the headache is progressive and findings suggest idiopathic intracranial hypertension or chronic meningitis, lumbar puncture, and CSF analysis are done.

> **Focus on diagnostic red flags for headache**
> The following findings are of importance when assessing headache:
> altered mental status, weakness, diplopia, papilledema, and focal neurologic deficits; cancer; immunosuppression; meningismus; age over 50 years; thunderclap headache; visual disturbances, jaw claudication, fever, weight loss, temporal artery tenderness, proximal myalgias; progressively worsening symptoms; red eye or halos around lights. In general, any focal neurological findings in addition to the headache, suggest that a full neurological workup be performed, including imaging.

Treatment

Treatment of headache is based on the cause. For migraines or cluster headaches, amitriptyline, beta-blockers, divalproex, lithium, topiramate, and verapamil may be preventative. Abortive treatment is with dihydroergotamine, valproate, and various triptans. Tension-type headache is treated with analgesics, and sometimes, behavioral and psychologic intervention. Low-pressure headaches are treated with hydration, analgesics, and sometimes, an epidural blood patch.

Clinical cases

Clinical case 1
1. How is the diagnosis of AD made?
2. What support can be offered to the patient's husband?
3. What are the risks for Alzheimer's when a close relative has had the disease?

A 68-year-old woman visits her physician. Her husband describes that she has been having trouble remembering people's names, the location of keys on her computer and that she has a 90-year-old uncle with dementia. Her only current conditions are uncomplicated shingles, temporal arteritis, and osteoporosis. She is tested for mental status and passes the test. However, upon revisiting the physician a year and a half later, her memory is worse. She has become more irritable and nervous. A neurological exam reveals buccolingual and limb apraxia. She is diagnosed with early AD.

Answers:
1. The diagnostic criteria for dementia are impairment of two or more of the following functions: memory, language, executive functioning (such as paying bills), visuospatial impairment PLUS decline from the previous level of functioning. In AD, memory and visuospatial functions are impaired early in the course of the disease manifesting as difficulty with remembering names and getting lost in familiar places.
2. Respite or day care can be arranged by the physician to help the patient's husband be able to care for his wife while still handling his own responsibilities. There are many community-based support groups and organizations for AD. He must be assisted in planning for his wife's institutionalization if the disease worsens, which it likely will do.
3. Individuals with a close family member who had Alzheimer's are more likely to develop it themselves. However, fewer than 1% of people with Alzheimer's inherited the disease. Genetics are a better way to predict the risk of developing this disease. The apolipoprotein E (APOE)-e4 gene is searched for in the patient, which is the gene most linked to Alzheimer's.

Clinical case 2
1. What type of delirium would this patient be diagnosed with?
2. What other type of behavioral changes may signify delirium?
3. What are the descriptions of hyperactive, hypoactive, and mixed delirium?

An 83-year-old woman fell at home and severely injured her hip, but did not break it. She was treated at the local hospital and released. She is given opioids to help reduce her pain as she heals, and a home nurse begins visiting her for rehabilitative physical therapy. Sometimes, the nurse has administered more medications on days when the pain was severe. After a few weeks, the son, who regularly checks on his mother, informs the family doctor that she is drowsy most of the day and very confused while awake. At night becomes restless and even agitated. The physician schedules a follow-up appointment and determines that the woman is dehydrated. He reduces her pain medications and assesses her mental capacity—she is experiencing delirium likely related to her medications. Over time, her delirium clears and her mobility improves.

Answers:
1. This patient has mixed delirium since she has some signs of both hypoactive and hyperactive delirium. Being drowsy or sleeping most of the day is not normal, and her chances of becoming restless and agitated at night are opposing signs to those of simply hypoactive delirium.
2. Delirium can also be signified by hallucinations, arguing, combativeness, making sounds for no apparent reason, being quiet and withdrawn, having slowed movements, and reversal of night-day cycles.
3. Hyperactive delirium is easily recognized and includes restlessness, agitation, rapid mood changes, hallucinations, and refusal to cooperated. Hypoactive delirium includes inactivity, reduced motor activity, sluggishness, abnormal drowsiness, or being "in a daze." Mixed delirium includes hyperactive as well as hypoactive signs and symptoms. The patient may quickly switch between them.

Clinical case 3
1. Could an EEG show abnormal electrical activity even if an MRI was normal?
2. Though not used often, if needed, what type of invasive testing may be performed for diagnosis?
3. How common is drug-resistant epilepsy?

A 29-year-old man was taken to a seizure clinic with drug-resistant epilepsy or refractory epilepsy. His seizures began at age 16, and he was on antiepileptic medications for 3 years, then stopped. Two years later, his seizures returned. He started taking different seizure medications again, but without relief. His sister described his symptoms as staring, decreased responsiveness, difficulty speaking, and automated hand movements. These occur about three times per day. About once per month, he would experience a generalized tonic-clonic convulsion.

Answers:
1. Yes, while an EEG might show an epileptiform focus, an MRI may not be able to show anything that appears abnormal. For example, focal cortical dysplasia is an important cause of drug-resistant epilepsy that may be subtle or invisible on MRI.
2. Most likely, subdural electrodes can be implanted to offer good spatial resolution of the epileptogenic zone in relation to the cerebral cortex, or stereotactic depth electrodes will allow very precise recording from areas below the cortical surface.
3. Drug-resistant epilepsy affects 25% of all epileptic patients. Early detection is vital in order to establish potential treatment options, and determine if the patient is a candidate for surgery.

Clinical case 4
1. What are the likely recommendations that this patient should follow in order to treat her insomnia?
2. Why is exercise recommended to help treat insomnia?
3. What is cognitive behavioral therapy for insomnia?

A 42-year-old woman is referred to a sleep center, complaining of chronic, severe insomnia affecting her daytime functioning. She first experienced the condition 8 years ago when she nearly lost her family business. However, the insomnia has remained, almost unchanged, ever since. She has been prescribed various sleep medications and antidepressants, all of which have been unsuccessful. All day, every day, she feels tired and stressed out, and drinks up to five strong cups of coffee daily to stay awake. She also tries to take short naps throughout the day. When it is time to go to bed, it can take up to several hours to fall asleep, and she usually wakes up three to four times per night.

Answers:
1. She should decrease her caffeine intake, adopt a rigid sleep hygiene schedule, try enrolling in a relaxation course, exercise regularly, and undergo cognitive therapy.
2. Exercise significantly improves sleep in chronic insomniacs. Postexercise drops in temperature may help promote sleepiness. Exercise also decreases arousal, anxiety, and depressive symptoms. It also helps regulate the circadian rhythms.
3. Cognitive-behavioral therapy for insomnia is a four-to-six-session treatment program that teaches skill sets to be used for insomnia, throughout life. Most participants report improved sleep satisfaction. It can be administered by a psychologist, psychiatrist, or other medical doctor with specialized training. Each patient's triggers for insomnia are identified and "defused" by developing healthy, effective behaviors. A list of individuals who offer this therapy can be found at www.absm.org/bsmspecialists.aspx.

Clinical case 5
1. What type of headache does this patient most likely have, and are there any concerning signs or symptoms?
2. What types of visual symptoms may be involved in an aura?
3. What are the common triggers of migraines?

A man in his 60s has been experiencing severe headaches that usually occur in the day, but occasionally at night. They are so bad that they have awakened him from sleep. His pain builds up over 1 or 2 h, and then lasts usually for about 8 h. He frequently experiences flashes of light affecting his vision prior to the headache onset. Movement or any physical activity worsens it, and he usually just lies down on his bed with all lights off and the window blinds closed, with the ceiling fan turned on. After an attack, he has sometimes felt so weak he could not go to work. He is always nauseous during an attack but has only vomited once or twice since the headaches began. His physician assesses the signs and symptoms, orders tests, and starts the patient on a treatment plan.

Answers:
1. This patient is most likely suffering from a migraine. His headache is bothersome, lasting more than 4 h, accompanied by nausea as well as photophobia (he prefers to be in a quiet, dark room). Therefore, it meets the criteria for a migraine headache. His age, as well as the episodes of headache that woke him up from sleep, are both concerning signs and symptoms. Other causes of secondary headaches should be ruled out prior to calling this headache a primary headache such as a migraine.
2. Visual symptoms related to auras may include field defects, central scotomas, tunnel vision, altitudinal visual defects, complete blindness, and often, a scintillating scotoma—this consists of an arc or band of absent vision with a shimmering or glittering irregular border. It may be accompanied by flashes of light or varied visual hallucinations.
3. Migraines are commonly triggered by any change in a patient's personal routine such as sleep time, using alcohol, hormonal changes, severe exercise, sleep changes, stress, and a variety of medications. Also, a large number of patients with migraine have a first-degree relative with a history of migraines.

Key terms

Agnosia	Cataplexy
Agraphia	Central sulcus
Alexia	Cerebrum
Alzheimer's disease	Convulsions
Amyloid angiopathy	Corpus callosum
Anterior commissure	Creutzfeldt-Jakob disease
Aphasia	Delirium
Apraxia	Delta waves
Archicortex	Dementia
Arcuate fibers	Downregulating
Arytoenoepiglottic folds	Dysarthria
Association areas	Dysprosody
Atonia	Electroencephalogram
Auditory association area	Epilepsy
Auditory cortex	Epileptiform
Aura	Epworth Sleepiness Scale
Beta waves	Excessive daytime sleepiness
Broca's area	Gaze palsies
Brudzinski signs	Glossectomy

Gustatory cortex
Hallervorden-Spatz disease
Hypersomnolent
Insomnia
Internal capsule
Kernig signs
Kyphoscoliosis
Lateral sulcus
Lewy bodies
Locus ceruleus
Longitudinal fasciculi
Macroglossia
Mandibulomaxillary advancement
Memories
Memory consolidation
Memory engram
Micrognathia
Narcolepsy
Neocortex
Neologisms
Neurosyphilis
Nuclei of Meynert
Nucleus basalis
Olfactory cortex
Operculum
Oximetry
Paleocortex
Papez circuit
Parasomnias
Parieto-occipital sulcus
Plethysmography
Prefrontal cortex
Prefrontal lobotomy
Premotor cortex
Primary motor cortex
Primary somatosensory cortex
Prosody
Pyramidal cells
Retrognathia
Seizure
Sleep apnea
Somatosensory association cortex
Somnambulism
Split-brain syndrome
Stellate cells
Sundowning
Theta waves
Thunderclap headache
Todd's paralysis
Upregulating
Visual association area
Visual cortex
Wernicke's area
Wernicke-Korsakoff syndrome
Wilson's disease

Suggested readings

1. Alpha Waves—https://www.sciencedirect.com/topics/neuroscience/alpha-wave.
2. Barker, R. A.; Cicchetti, F.; Robinson, E. S. J. *Neuroanatomy and Neuroscience at a Glance*, 5th ed.; Wiley-Blackwell, 2017.
3. Biller, J.; Gruener, G.; Brazis, P. *DeMyer's the Neurologic Examination: A Programmed Text*, 7th ed.; McGraw-Hill Education/Medical, 2016.
4. Blumenfeld, H. *Neuroanatomy through Clinical Cases*, 2nd ed.; Sinauer Associations/Oxford University Press, 2010.
5. Braak, H.; Del Tredici, K. *Neuroanatomy and Pathology of Sporadic Alzheimer's Disease*. Springer, 2015.
6. Chinthapalli, K.; Magdalinou, N.; Wood, N. *Challenging Concepts in Neurology: Cases with Expert Commentary*. Oxford University Press, 2016.
7. Crossman, A. R.; Neary, D. *Neuroanatomy: An Illustrated Colour Text*, 5th ed.; Churchill Livingstone, 2014.
8. Dehn, M. J.; Kaufman, A. S.; Kaufman, N. L. *Essentials of Working Memory Assessment and Intervention*. Wiley, 2015.
9. Dysprosody—https://www.sciencedirect.com/topics/neuroscience/dysprosody.
10. Filley, C. *The Behavioral Neurology of White Matter*, 2nd ed.; Oxford University Press, 2012.

11. FitzGerald, M. J. T.; Gruener, G.; Mtui, E. *Clinical Neuroanatomy and Neuroscience* E-*Book*, 6th ed.; Saunders, 2011.
12. Haines, D. E. *Neuroanatomy in Clinical Context: An Atlas of Structures, Sections, Systems, and Syndromes*, 9th ed.; LWW, 2014.
13. Henry, G. L.; Little, N.; Jagoda, A.; Pellegrino, T. R.; Quint, D. J. *Neurologic Emergencies*, 3rd ed.; McGraw-Hill Medical, 2010.
14. Jacobson, S.; Marcus, E. M.; Puglsey, S. *Neuroanatomy for the Neuroscientist*, 3rd ed.; Springer, 2018.
15. Lassone, M.; Jeeves, M. A. *Callosal Agenesis: A Natural Split Brain*. Springer, 2011.
16. Lechin, F.; van der Dijs, B.; Lechin, M. E. *Neurocircuitry and Neuroautonomic Disorders: Reviews and Therapeutic Strategies*. S. Karger, 2002.
17. Lewis, P. A.; Spillane, J. E. *The Molecular Pathology of Neurodegenerative Disease*. Academic Press, 2018.
18. Mai, J. K.; Majtanik, M.; Paxinos, G. *Atlas of the Human Brain*, 4th ed.; Academic Press, 2015.
19. Martin, J. H. *Neuroanatomy Text and Atlas*, 4th ed.; McGraw-Hill Education/Medical, 2012.
20. Oishi, K.; Faria, A. V.; van Zijl, P. C. M.; Mori, S. *MRI Atlas of Human White Matter*, 2nd ed.; Academic Press, 2010.
21. Petrides, M. *Atlas of the Morphology of the Human Cerebral Cortex on the Average MNI Brain*. Academic Press, 2018.
22. Simpkins, C. A.; Simpkins, A. M. *Neuroscience for Clinicians: Evidence, Models, and Practice*. Springer, 2012.
23. Split Brain Syndrome—https://www.britannica.com/science/split-brain-syndrome.
24. Thompson, M.; Thompson, L. *Functional Neuroanatomy*. Association for Applied Psychophysiology & Biofeedback, 2015.
25. Waxman, S. G. *Clinical Neuroanatomy*, 28th ed.; McGraw-Hill Education/Medical, 2016.

CHAPTER 7

Basal nuclei

Islands of gray matter lie deep within the white matter of each cerebral hemisphere. These are collectively called the *basal nuclei*, which were previously referred to as the *basal ganglia*. The basal nuclei include the caudate nucleus, lentiform nucleus, and amygdaloid body. Though the complete functions of the basal nuclei are not yet fully understood, they play vital roles in regulating voluntary motor functions, including those involved in posture, walking, and other repetitive or gross movements. The basal nuclei may also even be involved in learning and thinking.

The cerebral cortex consciously and voluntarily regulates complex movements and intellectual activities. Parts of the cerebrum, diencephalon, and brainstem process sensory information and send out motor commands that we are not aware of. Many of these subconscious activities are controlled by the basal nuclei, which are formed as the basal wall of the telencephalon thickens and forms masses of gray matter.

Structure of the basal nuclei

The **basal nuclei** consist of gray matter in each brain hemisphere, located deep to the floor of the lateral ventricle. The basal nuclei are embedded within the cerebrum's white matter. Around or between the basal nuclei are radiating projection fibers and commissural fibers. The basal nuclei have long been considered to be part of a more extensive functional group called the *basal ganglia*. The group included the basal nuclei as well as related motor nuclei located in the diencephalon and midbrain. Since the term *ganglia* is primarily restricted to the peripheral nervous system (PNS), the term *basal nuclei* is preferred over the term *basal ganglia*. There is a strong physical continuity above the orbital surface of the brain's frontal lobe.

The **caudate nucleus** consists of a large *head* and a thin, curved *tail* following the lateral ventricle's curve. It basically appears to be shaped like a "comma," with a "C"-shape, similar to the shape of the lateral ventricles. The head, located deep in the frontal lobe, is anterior to the **lentiform nucleus**, a lens-shaped nucleus that is made up of a lateral **putamen** and a medial **globus pallidus**. The head of the caudate nucleus actually merges with the *nucleus accumbens*, which then merges with the anterior putamen. The tail of the caudate nucleus is located in the temporal lobe. The caudate nucleus and putamen have similar histology and are sometimes referred to as the *corpus striatum* or simply *striatum*, which means "striated body." It has a striped appearance of its internal capsule, as the fibers pass through these nuclei. The lentiform (or *lenticular*) nucleus is lateral and partially anterior to the thalamus but separated from the thalamus and most of the head of the caudate nucleus by a thick fiber sheet called the **internal capsule**. This capsule contains most of the fibers that interconnect the cerebral cortex, thalamus, basal nuclei, and brainstem. The easiest way to locate the head of the caudate is to locate the frontal horn of the lateral ventricle. The structure adjacent to it is the head of the caudate.

The basal nuclei also consist of a *subthalamic nuclei* in the lateral floor of the diencephalon, as well as the **substantia nigra** of the midbrain. A thin layer of gray matter called the **claustrum** lies close to the putamen. A part of the limbic system called the **amygdaloid body** lies anterior to the tail of the caudate nucleus, as well as inferior to the lentiform nucleus. The amygdaloid body has an "almond" shape and is also called the *amygdala*. The formation of the basal nuclei begins during week five of gestation. The nuclei become prominent during weeks six to seven of gestation. The basal nuclei, including the caudate nucleus, putamen, and globus pallidus, have a total volume of approximately $8\,cm^3$. The basal nuclei receive most of their blood supply from tiny perforating branches of the middle cerebral artery (MCA) known as the lenticulostriate arteries. The thalamus receives its blood supply from the tiny perforating branches of the posterior cerebral artery (PCA), also called lenticulostriate arteries.

Functional considerations

The subconscious control of skeletal muscle tone and coordination of learned movements involve the basal nuclei. Normally, the nuclei do not initiate certain movements. However, when a movement has begun, the basal nuclei regulate general patterns and rhythms. During walking, the basal nuclei control arm and thigh movements occurring between the beginning of the walking process and the time we decide to stop walking. When any voluntary movement starts, muscle tone, especially in the appendicular muscles, is controlled by the basal nuclei in order to "set" the body position. When picking up an object, there is a conscious reaching and grasping utilizing the muscles of the forearm, wrist, and hand. The basal nuclei are operating subconsciously to position the shoulder and stabilize the arm during these maneuvers. However, the basal nuclei have no direct access to the motor pathways. They also play roles in emotion and cognition.

The basal nuclei regulate motor commands from the cerebral cortex by using a feedback loop. As information reaches the caudate nucleus and putamen from the cerebral cortex, lateral processing occurs in the basal nuclei and the nearby globus pallidus. The putamen is believed to be centrally involved in most motor functions of the basal nuclei. The majority of basal nuclei output moves from the globus pallidus and synapses in the thalamus. The thalamic nuclei project this information to the correct cerebral cortex areas. Therefore, the basal nuclei act by influencing descending pathways to affect movements. The electrical activity between the cerebral cortex, basal nuclei, and thalamus occurs in the forebrain. Since there are two halves of the forebrain, damage of the basal nuclei on one side interferes with movements on the opposite (contralateral) side. Historically, neuroscience students are asked to know the direct and indirect dopamine pathways. Recent studies showed that the functions of the dopamine pathways are very complex and cannot be explained simply as *direct* or *indirect*.

Basal nuclei activities are inhibited by the neurotransmitter *dopamine*, released by neurons of the substantia nigra, in the midbrain. When the substantia nigra is damaged, or there is less neuronal secretion of dopamine, the basal nuclei increase their activity. As a result, a gradual and generalized increase in muscle tone occurs, with symptoms that signify **Parkinsonism**. This is defined by having two of the three following symptoms: **bradykinesia**, tremor, and rigidity. Parkinson's disease (PD) is the most common cause of parkinsonism. In this disease, there is difficulty beginning voluntary movements. This is because opposing muscle groups are not able to relax. Therefore, they must be forced to move. Once a movement begins, every factor of the movement requires voluntarily control, which means the individual has to make intense efforts and concentrate on the movements. This difficulty with initiation of movement is also called "freezing."

There are widespread dopaminergic projections from the compact area of the substantia nigra to other parts of the basal nuclei—especially the striatum. There are also many inhibitory interconnections between various parts of the basal nuclei. Excitatory projects come from the subthalamic nucleus (STN) to the globus pallidus. There are also interconnections of the thalamic **intralaminar nuclei** with the globus pallidus and striatum. These connections form a normal *basal ganglia circuitry*. This circuitry consists of *direct* and *indirect pathways*. As mentioned above, this is far more complicated than just these two pathways. There is a nucleus in the brain that is constantly inhibiting movements called the *globus pallidus internal (GPi)*. It is a key exiting point for both pathways. The main difference between the direct and indirect pathways is that in the direct pathway, the GPi is being inhibited. In the indirect pathway, the GPi is being stimulated. Since the GPi is constantly inhibiting movements, inhibiting this nucleus in the direct pathway will *facilitate* movement. On the contrary, in the indirect pathway, the inhibitory function on movement is stimulated, resulting in movement inhibition. The thalamus acts as an amplifier and constantly stimulates the cortex. Both the *globus pallidus external (GPe)* and STN are specific to the indirect pathway. Dopamine is secreted from the substantia nigra pars compacta (SNC) and inhibits the indirect pathway (via the dopamine-2 or *D2* receptor) while exciting the direct pathway (via the D1 receptor). Let's summarize what we mentioned in four basic rules:

1. The SNC secretes dopamine; if it attaches to the D1 receptor, the direct pathway will be excited. If dopamine attaches to the D2 receptor, the indirect pathway is inhibited.
2. The GPi constantly inhibits the thalamus and therefore inhibits movement.
3. The STN and GPe are specific to the indirect pathway.
4. The D1 and D2 receptors are located within the striatum (caudate/putamen).

Section review
1. Why is the term *basal nuclei* preferred over the term *basal ganglia*?
2. Why does the corpus striatum appear to be striated?
3. How do the basal nuclei normally function in walking movements?
4. How can damage to the basal nuclei and decreases in dopamine cause impairments of movement?

Focus on dopamine
The *dopamine system* is a group of nerve cells in the midbrain and frontal brain areas. It includes the frontal lobe, responsible for thoughts and memories, the striatum—associated with movements, and the nucleus accumbens, which controls emotions and how the brain handles "rewards." Therefore, addiction to substances is caused by a flood of dopamine entering the brain's limbic system. Dopamine is essential for normal sleep patterns, sex drive, and overall health. When it is present in insufficient amounts, there will be uncoordinated movements throughout the body. Many foods contain the amino acids L-tyrosine and L-phenylalanine in quantities that boost dopamine levels since they eventually convert into dopamine. These food sources include fish, chicken, nuts, eggs, beans, and cheese.

Clinical considerations

Damage to the basal nuclei is associated with movement disorders such as those seen in PD. Many basal nuclei disorders cause abnormalities of movement. When the indirect pathway activity is reduced, hyperkinetic movement disorders develop. Involuntary movements of the **hyperkinetic disorders** are divided into *tremors*, and various states of **chorea**, **athetosis**, and **ballismus**. Tone disturbances may involve increases in flexors and extensors, such as in the rigidity of PD, or only in certain muscles. When only certain muscles are affected, the body may be twisted into abnormal postures that remain relatively fixed. This is known as *dystonia*.

Parkinson's disease

PD was named for Dr. James Parkinson, who wrote "An Essay on the Shaking Palsy" in 1817. In this paper, he described symptoms of tremor, increased muscle tone, and difficulty in starting to make voluntary movements. Over time, the condition was named *Parkinson's disease* because of his contribution. This disease along with related disorders signified by slowed or reduced movements, altered muscle tone, and involuntary movements are linked to basal nuclei damage. For a long time, these were called *extrapyramidal*

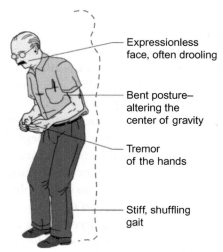

Fig. 7.1 Typical posture that results from Parkinson's disease.

disorders, distinguishing them from disorders of the pyramidal (corticospinal) system. This older term is no longer used since many involuntary movements that develop after damage to the basal nuclei are affected by the corticospinal tract.

PD is the most prominent of the **hypokinetic disorders**, and is idiopathic, slowly causing progressive degeneration of the central nervous system (CNS) (see Fig. 7.1). It is also the most common and familiar disease that involves the basal nuclei. It primarily involves muscle rigidity, slow and reduced movements, a *stooped* posture, and tremors.

Pathophysiology

The major causative factor of PD is the destruction of the **nigrostriatal** pathway. The SNC degenerate in PD, leading to reduction of thalamocortical excitation. As mentioned above, the SNC plays a modulatory role by inhibiting the indirect pathway and stimulating the direct pathway. Therefore, damage to the SNC leads to decreased dopamine secretion, which ultimately leads to decreased movement (bradykinesia). Please note that most of the movement disorder combines hypokinetic and hyperkinetic movements. For example, people with PD also have tremor, which is a hyperkinetic movement disorder due to the decreased activity of the indirect pathway.

Alpha-synuclein is a presynaptic glial and neuronal cell protein. It may form insoluble fibrils called **Lewy bodies**. Basically, Lewy bodies are aggregated, misfolded proteins (synuclein) in neuronal cells. Lewy bodies are the main pathologic feature of PD, accumulating in the substantia nigra of the midbrain. This synucleinopathy can also occur in other nervous system areas, including the basal nucleus of Meynert, dorsal motor nucleus of the vagus nerve, hypothalamus, neocortex, olfactory bulb, myenteric plexus of the GI tract, and sympathetic ganglia. Lewy bodies appear as part of a temporal sequence. PD may be a later development of a systemic synucleinopathy. This can also include **Lewy body dementia**. Often, the PD patient also has Alzheimer's disease, and these diseases, along with Lewy body dementia share several similar features. Every year, new research is helping us to better clarify the relationships between these diseases.

Mitochondrial dysfunction contributes to PD, based on mutations of the genes encoding the proteins *DU-1*, *PINK1*, and *parkin*. The DJ-1 protein acts as a transcriptional regulator. Oxidative stress causes it to move to the mitochondria, where it has cytoprotective effects. The PINK1 protein is a kinase degraded in the mitochondria. However, with mitochondrial dysfunction, it recruits parkin, an *E3 ubiquitin ligase*. Normally, PINK1 and parkin combine to clear dysfunctional mitochondria via **mitophagy**. Levels of mitochondrial complex I, part of the oxidative phosphorylation cascade, are reduced in the brains of patients with sporadic PD.

Mutations of the gene encoding *leucine-rich repeat kinase 2 (LRRK2)* are more usual causes of autosomal dominant PD, but also in some sporadic cases of the disease. LRRK2 is a cytoplasmic kinase. Pathogenic mutations that increase its activity suggest that functions gains may contribute to PD. These gains may involve **hyperphosphorylation** of normal targets or the emergency of unique targets.

In PD, the pigmented neurons of the locus ceruleus, substantia nigra, and other dopaminergic cells of the brainstem are lost. Dopamine is depleted in the caudate nucleus and putamen when substantia nigra neurons are lost. In some cases, there is a genetic predisposition. Various abnormal genes have been revealed. Inheritance may be autosomal dominant or autosomal recessive.

Clinical manifestations

PD, is usually of insidious onset, with **resting tremor** of one hand often the first symptom. Symptoms are unilateral at first. Strength remains almost normal, and most reflexes are not highly affected. The rigidity may be uniform throughout the range of movements. This is called *plastic* or *lead-pipe rigidity*. Rigidity may also be interrupted by brief relaxation periods, known as *cog-wheel rigidity*. Spasticity is defined as muscle tone that is selectively increased in the leg extensors or arm flexors. Rigidity may develop without tremor.

Slow movements are described as *bradykinesia*. Movements that are hard to initiate are called **akinesia**. When the amount of movements is reduced, it is referred to as **hypokinesia**. The face may lack expression, called **hypomimia**, and normal arm swings are decreased while walking. The mouth may hang open, and the patient may be unable to control drooling. Bradykinesia and hypokinesia are fundamental deficits that are not only caused by rigidity. Patients without extreme rigidity may still have significant trouble moving. Resting tremor in PD typically involves the hands in a "pill-rolling" movement. Tremor becomes reduced during voluntary movements and increases in times of emotional stress. The tremor is slow, coarse, absent during sleep, increased by fatigue and may also involve the wrists. The hands, then arms, then legs are affected eventually. While the tongue and jaw can be affected, the voice is not. As the disease progresses, the tremor sometimes becomes less prominent. Rigidity and hypokinesia may cause fatigue and muscular aches. Head tremor can be due to another disease called *essential tremor*, which is highly similar to PD.

The speech patterns become **hypophonic**. They are usually monotonous, with stuttering dysarthria. Due to the lack of finger muscle control, the patient's handwriting becomes smaller in size. Daily living activities are affected. There can also be episodes of *freezing*, wherein voluntary movements suddenly stop altogether. **Postural instability** develops, affecting walking. The patient may have problems beginning to walk, turning, and stopping. He or she starts to shuffle, taking only short steps. The arms are kept flexed to the *waist and do not normally swing. The steps can uncontrollably speed up, causing the patient to run in order to keep from falling. This is called **festination**. Loss of postural reflexes may cause falling* forward (propulsion) or falling backward (retropulsion) due to displacement of the center of gravity.

Sleep disorders often occur with PD, including insomnia (due to nocturia, or inability to turn over while in bed). Rapid eye movement (REM) sleep behavior disorder can be present years prior to the development of PD. There may be violent physical movements during REM sleep. Other neurologic symptoms often develop. This is because synucleinopathy develops in other parts of the CNS, PNS, and autonomic nervous system. These effects may include nearly universal sympathetic denervation of the heart, causing orthostatic hypotension. There may be dysmotility of the esophagus, causing dysphagia and greater chances of aspiration. Lower bowel dysmotility may result in constipation. This is also one of the earliest symptoms of PD, which can precede the disease by decades. Often, urinary hesitancy or urgency develop, as well as *anosmia*, which is an inability to smell. Other signs and symptoms include seborrheic dermatitis. It can also cause various dystonias, autonomic instability, personality changes, and depression.

Diagnosis

Diagnosis of PD is clinical. It is suspected in all patients who have the characteristic unilateral resting tremor, rigidity, or decreased movements. The tremor will disappear or attenuate during finger-to-nose coordination tests. In a neurologic examination, the patient cannot rapidly alternate movements or make rapid successive movements very well. Sensation is usually normal, as are reflexes, but they can be hard to perform due to increased tremor or rigidity. The patient may not be able to suppress eye closure when the frontal muscle is tapped between the eyes. This is called the **glabellar reflex**. If this is persistent, it is called **Myerson's sign**.

Slowed, decreased movements must be differentiated from the changes caused by corticospinal tract lesions. These cause weakness or paralysis (paresis) usually of the distal antigravity muscles; hyperreflexia; and Babinski's sign, which involves extensor plantar responses.

Corticospinal tract lesions cause spasticity, which increases deep tendon reflex responses and muscle tone. The increase in muscle tone occurs proportionally to the rate and degree of stretching placed upon a muscle until resistance suddenly stops (the clasp-knife phenomenon). Rigidity is different since resistance is not altered through the entire range of motion.

Diagnosis is confirmed by other characteristic signs. These include infrequent blinking, impaired postural reflexes, lack of facial expressions, and abnormal gait. A *pull test* is used to assess impairment of postural reflexes. If there is tremor without other signs, this suggests an early disease process or another diagnosis. As mentioned earlier, PD is different from **Parkinsonism**, which can be secondary to medications or damage to the basal ganglia from causes other than PD. Diagnosis is based on whether the use of levodopa results in a significant improvement. Neuroimaging is performed when needed. Dopamine transport (DaT) scan and single-photon emission computerized tomography (SPECT) imaging can be used to differentiate between non-PD and PD tremor when the diagnosis is not certain. There are some signs and symptoms that if present, make a diagnosis of PD unlikely. These include early dementia, headache, visual or sensory loss, paresthesias, apraxia, early gait or postural abnormalities, and early or severe autonomic dysfunction.

Treatment

There are many orally administered drugs for PD. There is no set treatment strategy, and not every patient responds the same way to medications. However, there are some general rules based on the experience of the author:

1. The treatment should be saved for the time that the patient is developing a disability from the symptoms. The available medications cannot change the course of the disease and should only be started when the symptoms are bothersome and disabling in the view of the patient.
2. There are no set guidelines on which medication to start. However, experts recommend initiating levodopa-carbidopa in patients over age 65.
3. In patients below age 65, the approach is based on severity, with levodopa-carbidopa still recommended for patients with severe symptoms.
4. If the patient is below age 65 and the symptoms are mild, then a dopamine agonist is an option, with levodopa-carbidopa used later.
5. If tremor is the main bothersome symptom, then anticholinergic medications are a good option.

Levodopa is the metabolic precursor of dopamine. It crosses the blood-brain barrier into the basal ganglia. Here, it is decarboxylated and forms dopamine. When the peripheral decarboxylase inhibitor *carbidopa* is administered with levodopa, it prevents the catabolism of levodopa. This lowers the dosage requirements of levodopa and reduces adverse effects. Levodopa has the most effective for relieving bradykinesia and rigidity, and often reduces tremor.

The adverse effects of levodopa are mainly dizziness, nausea, vomiting, lethargy, orthostatic hypotension, and dyskinesia. The term dyskinesia basically involves abnormal involuntary movements that can occur in two different settings: either as a side effect of levodopa-carbidopa when the level is peaked in the blood (known as "peak-dose dyskinesia") or in a setting called "biphasic dyskinesia."

This form of dyskinesia happens when the medications start to take effect or when the effects start to wean off, hence the term *biphasic*. Peak-dose dyskinesia is usually a choreaform movement that can later spread to other body parts. Biphasic dyskinesia usually presents as leg kicking. Sometimes, delirium or hallucinations occur, usually in the elderly and in patients with dementia. If dosage is sufficient to cause dyskinesias, these usually decrease with continued treatment. Sometimes, the lowest dose that lessens Parkinsonian symptoms also causes dyskinesias. For carbidopa/levodopa, dosages are increases every 4–7 days, as tolerated, until maximum effects are reached. Adverse effects can be lessened by increasing doses gradually, and by administering of this combination drug with meals or after them. It must be noted that a high-protein meal can impair the absorption of levodopa. When peripheral adverse effects are significant, the amount of carbidopa should be increased. Most patients with PD receive divided doses every 5–8 h. There is a dissolvable immediate-release form of carbidopa/levodopa that is taken without water and is useful when the patient has problems swallowing. Doses are identical to immediate-release carbidopa/levodopa.

Sometimes, levodopa is required to maintain motor function even if it induces delirium or hallucinations. Psychosis may be occasionally treated with oral quetiapine, which aggravates Parkinsonian symptoms significantly less (or not at all) compared to other antipsychotics such as olanzapine or risperidone. Haloperidol must be avoided. Quetiapine must be increased in small increments every 1–3 days, up to its maximum dose, as tolerated. Clozapine is of limited use since agranulocytosis develops in 1% of patients. A complete blood count must be done every week for 6 months, and every 2 weeks after that. Clozapine surprisingly has the least extrapyramidal side effects of all the antipsychotic medications.

After 2–5 years, the patient will have fluctuations in responses to levodopa. This is called the *on-off effect* or *biphasic dyskinesia*. It is not fully understood if dyskinesias and the on-off effect are from levodopa itself or the underlying disease. Over time, the improvement period that follows each dose becomes shorter. Drug-induced dyskinesias occur in cycles, from intense akinesia to uncontrolled hyperactivity. These common "swings" are controlled by levodopa doses that are as low as possible, and by using dosing intervals as short as every 1–2 h. Other methods include adjunctive dopamine-agonists, *catechol O-methyltransferase* (COMT) and/or MAO inhibitors, controlled-release carbidopa/levodopa, and amantadine.

Amantadine is an effective treatment for biphasic dyskinesia. It can also be used as an augmentation with levodopa. The most common side effects of amantadine are hallucinations, swelling of the feet, and formation of small clots in the skin capillaries, known as *livedo reticularis*.

Dopamine agonists work by directly activating dopamine receptors within the basal ganglia. Oral medications include bromocriptine, pramipexole, and ropinirole. These can be used a monotherapy but are usually not sufficient for any more than a few years. They are used for all stages of PD. When used early in treatment along with small levodopa doses, they may delay dyskinesias and the on-off effects. This may be because dopamine agonists stimulate dopamine receptors for a longer time than levodopa. This stimulation is primarily physiologic and can preserve receptors better. The dopamine agonists are very useful in later disease stages when responses to levodopa decrease or the on-off effects are significant.

Dopamine agonists have serious adverse effects that limit their use. However, reducing levodopa doses may minimize these effects. The dopamine agonists can cause, in 1%–2% of patients, hypersexuality, compulsive gambling, or overeating. This may result in a change in the drug used, or a reduction in dosage. Bromocriptine is not used widely due to its links to cardiac valvular fibrosis and pleural fibrosis. Apomorphine is an injectable dopamine agonist. It is used when "off" effects are severe as rescue therapy. It has a rapid onset of action, from 5 to 10 min, and a short duration of action, from 60 to 90 min. Apomorphine is given subcutaneously up to five times per day. Before the first dose, a small test dose is administered to check for orthostatic hypotension. Blood pressure is measured when the patient is standing and lying down prior to treatment, and then 20, 40, and 60 min afterward. The adverse effects are similar to the other dopamine agonists. Nausea is prevented by administering trimethobenzamide orally 3 days before apomorphine and then continuing it for the first 2 months of treatment. Recently, apomorphine was put back into the US market.

Selegiline is a selective monoamine oxidase type B (MAO-B) inhibitor. It inhibits one of the two primary enzymes that break down dopamine in the brain. This prolongs the action of each levodopa dose. In some patients who have slight on-off effects, selegiline will help prolong the effects of levodopa. Selegiline can reduce symptoms by decreasing oxidative metabolism of dopamine in the brain. An oral dose twice per day will not cause a hypertensive crisis, which can sometimes triggered by consuming tyramine in certain cheese and other foods during MAO inhibitor therapy. This adverse effect is often seen with nonselective MAO inhibitors because they block A and B isoenzymes. Selegiline, however, has nearly no adverse effects. However, it can potentiate levodopa-caused dyskinesias, nausea, and both mental and psychiatric adverse effects. Therefore, levodopa doses may need to be reduced. Selegiline is also available in a buccal-absorption form called *Zydis selegiline*. Rasagiline is an MAO-B inhibitor that is not metabolized to amphetamine. It is effective and useful in early and late PD. It may have purely symptomatic effects or neuroprotective effects. Newer studies have shown that rasagiline may change disease progression.

Anticholinergic drugs may be used as monotherapy in early disease with tremor as the main symptom, and later on to supplement levodopa. The common anticholinergic drugs include benztropine and trihexyphenidyl.

The COMT inhibitors include entacapone and tolcapone. They inhibit dopamine breakdown and are useful adjuncts to levodopa. For example, levodopa, carbidopa, and entacapone can be combined. Dosing may be as often as 8 times per day. Tolcapone is very potent and less often used due to rare reports of liver toxicity.

When a patient has levodopa-induced dyskinesias, serious motor fluctuations, or severe tremor, a procedure called deep brain stimulation of the STN or globus pallidus may be helpful, especially for treatment of tremor. If the patient only has tremor, there can be stimulation of the ventral is intermediate nucleus of the thalamus. However, since the majority of patients have other symptoms, stimulation of the STN is preferred since it relieves tremor and other symptoms, to a lesser degree.

It is the focus of treatment to maximize the patient's ability to perform activities. They should perform daily activities to the greatest possible extent. If they cannot do this, physical or occupational therapy may be needed. This can involve regular exercise to keep them physically fit. Adaptive strategies may include the installation of *grab bars* to reduce chances of falling. A combination of PD, antiparkinsonian drugs and inactivity can lead to constipation. Therefore, the patient should consume a high-fiber diet, exercise as often as possible, and drink sufficient amounts of fluids. Dietary supplements such as psyllium and stimulant laxatives such as bisacodyl may be helpful. Exercise is a very important and very effective treatment for PD. Therefore, the authors recommend everyone diagnosed with PD should start exercising on a daily basis for at least an hour.

Section review
1. What are the primary identifying characteristics of PD?
2. What type of cells are mostly lost in PD?
3. What is the differentiation between PD and secondary Parkinsonism based on?
4. What is the on-off effect, as seen in the treatment of PD?

Focus on PD
According to the Parkinson Association, approximately one million Americans live with PD. This is more than the combined number of people diagnosed with multiple sclerosis, muscular dystrophy, and amyotrophic lateral sclerosis (ALS) (Lou Gehrig's disease). About 60,000 Americans are diagnosed with PD every year, but this number does not reflect the amount of cases that are undetected. Worldwide, an estimated 7 to 10 million people are living with PD. Medication costs for an individual person with this disease average $2500 per year. Therapeutic surgery can cost up to $100,000 per patient. Combined direct and indirect costs of PD, including treatment, social security payments, and lost income due to inability to work, is estimated to be nearly $25 billion annually, in the United States alone.

Parkinsonism

Parkinsonism means rigidity, bradykinesia, and tremor. As mentioned previously, two of these symptoms must be present to qualify as Parkinsonism. Remember that this is different from PD because Parkinsonism is due to degeneration of the substantia nigra in the midbrain. Basically, Parkinsonism is defined as clinical symptoms and is not a disease by itself.

Pathophysiology
Many causes lead to Parkinsonism, including neurodegenerative conditions, drugs, metabolic diseases, toxins, and neurological conditions other than PD. There are two forms: *primary Parkinsonism* and *secondary Parkinsonism*. The causes of both forms of Parkinsonism are discussed in Table 7.1.

Clinical manifestations
The clinical manifestations of Parkinsonism including resting tremor, bradykinesia, and rigidity. As mentioned above by definition, parkinsonism is when two or all three of these signs are present.

Diagnosis
The diagnosis of Parkinsonism is clinical. There must be a thorough family, medication, and occupational history taken. There must be an evaluation of neurologic deficits that characterize other disorders, such as neurodegenerative disorders. Neuroimaging is performed when needed.

Treatment
If possible, the cause is corrected or treated. This sometimes results in the reduction or disappearance of symptoms. The medications used for PD are often not effective or only have slight benefits. Amantadine or an anticholinergic drug such as benztropine can treat Parkinsonism secondary to use of antipsychotics.

Huntington's chorea

Huntington's chorea, also called *Huntington's disease*, is a rare hereditary disorder involving neuronal degeneration. This degeneration is extremely severe in the striatum, primarily in the caudate nucleus, leading to atrophy of the caudate nucleus. It affects the neurons of the cerebral cortex, and other neurons in various areas of the brain, to a lesser degree. There are various involuntary, choreiform movements, with altered mood or cognitive functions. Movements become more pronounced slowly over time. There is then gradually worsening personality changes and dementia.

Table 7.1 Causes of primary and secondary Parkinsonism.

Primary Parkinsonism	Secondary Parkinsonism
Sporadic (idiopathic): most common form	**Neurodegenerative disorders related to alpha-synuclein pathology**: multiple system atrophies (glial and neuronal inclusions): motor neuron disease with Parkinson's disease features, nigrostriatal degeneration, olivopontocerebellar atrophy, Shy-Drager syndrome; dementia with Lewy bodies (cortical and brainstem neuronal inclusions)
Genetic: autosomal dominant; autosomal recessive	**Neurodegenerative disorders related to primary tau pathology ("tauopathies")**: corticobasal degeneration, frontotemporal dementia, progressive supranuclear palsy
Phenotype: influenced by gene-environment interactions	**Neurodegenerative disorders related to primary amyloid pathology ("amyloidopathies")**: Alzheimer's disease with parkinsonism
	Genetically-mediated disorders with some Parkinsonian features: Chediak-Higashi syndrome, Fragile X permutation with ataxia-tremor-Parkinsonism syndrome, Hallervorden-Spatz disease, Huntington's disease (Westphal variant), Prion disease, SCA-3 spinocerebellar ataxia, Wilson's disease, X-linked dystonia-Parkinsonism (DYT3)
	Miscellaneous acquired conditions: catatonia, cerebral palsy, normal pressure hydrocephalus, vascular Parkinsonism (atherosclerosis, amyloid angiopathy)
	Repeated head trauma (dementia pugilistica with Parkinsonism features)
	Infectious and postinfectious diseases: Creutzfeldt-Jakob disease, neurosyphilis, postencephalitic Parkinson's disease
	Metabolic conditions: hypoparathyroidism or pseudohypoparathyroidism with basal ganglia calcifications, non-Wilsonian hepatolenticular degeneration
	Multiple sclerosis
	Neoplastic disease
	Drugs: antiemetics (compazine, metoclopramide), alpha-methyldopa, dopamine-depleting agents (reserpine, tetrabenazine), fluoxetine, lithium carbonate, neuroleptics (typical antipsychotics), selected atypical antipsychotics, valproic acid
	Toxins: 1-methyl-1,2,4,6-tetrahydropyridine (MPTP), carbon disulfide, carbon monoxide, cyanide, hexane, manganese, methanol

Pathophysiology

Huntington's chorea is inherited in an autosomal dominant pattern. The defective gene is on the short arm of chromosome 4 (the *HTT* gene). Genetic tests are available today to determine individuals who are carriers and who will develop the disease. Therefore, it is possible to determine an individual's children are at risk. The disease usually begins, however, in middle age. There is atrophy of the caudate nucleus. There is also degeneration of the inhibitory medium spiny neurons located in the corpus striatum. Levels of two neurotransmitters decrease—these are gamma-aminobutyric acid (GABA) and glutamate.

The gene mutation that is causative results in abnormal repetition of the DNA sequence CAG, which codes for the amino acid glutamine. A large protein called **huntingtin** is produced. This has more **polyglutamine** residues, leading to disease from unknown causes. When more CAG repetitions are present, the disease beings earlier and has more severe effects. These repetitions can increase over successive generations, eventually leading to a worsened phenotype within a family.

Normally, there are 6–39 copies of the CAG repetitions in a normal *HTT* gene. When the number of repetitions increases beyond 39, the disease develops. The next generation will develop symptoms earlier when the number of copies is high. This is called *anticipation*. There is no sporadic form of this disease. Even though huntingtin exists in all body tissues, the deleterious effects of mutant huntingtin only occur in certain parts of the CNS. It appears that huntingtin can be taken up by neurons. Other possible pathways for the development of this disease may involve altered expression of the growth factor called *brain-derived neurotrophic factor (BDNF)* and the deleterious effects of protein aggregates.

Clinical manifestations

The signs and symptoms of Huntington's chorea develop insidiously, usually begin 35 and 50 years of age. However, it can develop before adulthood. Increased motor output often manifests as **choreoathetosis**. Before, or along with the movement abnormalities, there can be dementia or psychiatric disturbances. These psychiatric disturbances may include apathy, depression, anhedonia, irritability, antisocial behavior, and full-blown schizophreniform or bipolar disorder. Abnormal movements may include a lilting gait that resembles how a puppet moves, myoclonic jerks or irregular extremity movements, ataxia, facial grimacing, and motor acts not being able to be sustained, such as protruding the tongue. This inability is called *motor impersistence*.

Over time, the disease causes walking to become impossible, swallowing to be difficult, and results in severe dementia. The patient usually must be institutionalized. Death usually occurs within 13–15 years after the start of symptoms, most often from pneumonia or coronary artery disease.

Diagnosis

The diagnosis of Huntington's chorea is based on the common signs and symptoms, plus positive family history. However, up to 8% of patients have no clear family history despite abnormal genetic findings. Confirmation is by genetic testing. Neuroimaging will help exclude other disorders. When Huntington's disease is advanced, magnetic resonance imaging (MRI) or computed tomography (CT) coronal scans will reveal *boxcar ventricles*, in which their edges are squared-off because of caudate head atrophy.

Treatment

End-of-life care must be discussed with the patient and his or her caregivers early because of the disease's progressiveness. Treatment is only supportive. Antipsychotics such as chlorpromazine or haloperidol are used to partially suppress chorea and agitation. Doses are increased until they cannot be tolerated, or when adverse effects such as lethargy or Parkinsonism occur. Tetrabenazine may be an alternative medication. Doses may increase weekly from a single oral dose up to three divided doses until adverse effects become intolerable or chorea resolves. These may include akathisias, sedation, depression, and Parkinsonism.

Experimental therapies are focused on reducing glutamatergic neurotransmission via the *N*-methyl-D-aspartate receptor and promote mitochondrial energy production. Supplementation of GABA in the brain has been ineffective. If the patient has a first degree relative with Huntington's chorea, genetic testing and counseling can be offered.

Section review
1. How is Huntington's chorea inherited?
2. What are the clinical manifestations of Huntington's chorea?
3. What is the prognosis for Huntington's chorea, and which medications can be used for this disease?

Focus on Huntington's chorea
According to the National Institutes of Health, Huntington's chorea (also commonly referred to as Huntington's disease) affects about 3–7 per 100,000 people of European ancestry. In the United States alone, about one in every 10,000 people are affected. However, at least 150,000 others have a 50% risk of developing this disease. The general duration of Huntington's chorea ranges from 10 to 30 years. However, those with the adult-onset form generally survive about 15–25 years after onset.

Hemiballismus

Hemiballismus is a dramatic basal nuclei sign. It can occur in older individuals after a stroke, involving the STN, usually due to a lacunar stroke. Each STN is related, via the globus pallidus/**pars reticulata** (GPi-SNr) and ventral anterior nucleus/ventral lateral nucleus (VA/VL), to the ipsilateral motor cortex. This cortex handles movements of the contralateral side of the body, related to the dopamine pathways described above.

Pathophysiology

Hemiballismus is usually caused by a lesion in the contralateral STN. This is usually an infarct around the nucleus. This condition is very rare and is classified as a type of chorea. Additional causes of hemiballismus include traumatic brain injury, ALS, neoplasms, demyelinating plaques, and others.

Clinical manifestations

Hemiballismus is characterized by wild, uncontrolled movements of one arm or one leg. The limbs experience failing, ballistic, and undesired motion. Movements are usually continuous and violent. They may involve proximal or distal muscles and sometimes involve the facial muscles. The more a patient is stressed, the more the movements increase. Similar to all movement disorders, symptoms resolve during sleep.

Diagnosis

Diagnosis of hemiballismus is clinical, based on the patient's abnormal movements. These must be distinguished from other hyperkinetic movement disorders such as tremor, akathisia, and athetosis. All of these conditions involve less violent movements that use lower amplitude. A diffusion-weighted magnetic resonance imaging (DW-MRI) sequence may help in the setting of acute stroke to localize the lesion to the STN.

Treatment

Although hemiballismus is disabling, it is usually self-limiting and only lasts for 6–8 weeks. Treatment with various antipsychotic medications is usually effective. It is important to treat any underlying causes. Other medications include dopamine blockers, anticonvulsants, intrathecal baclofen, botulinum injections, and tetrabenazine. Surgery should only be used when the patient does not respond to medications. Lesioning of the globus pallidus or deep brain stimulation of this structure may be successful. Usually, lesioning is preferred because of the maintenance needed to continue stimulating the brain properly.

Essential tremor

Essential tremor is one of the most common of all movement disorders. It affects about 1% of the worldwide population. Incidence increases with age. There is no major difference in prevalence between males and females. Age of onset can be as early as childhood, with bimodal distribution, but onset peaks in the second and sixth decades of life. There is a slow progression of tremor intensity. The term *essential* means that the tremor is of unknown cause, which is the only symptom, and the critical aspect of the condition. Clinically, the term has been used in cases that have other features, including dystonia, or isolate head or voice tremors.

Pathophysiology

Essential tremor is often genetically linked, with usually an autosomal dominant pattern. Mutations may be causative but have not always been found. Several single-nucleotide polymorphisms are related to this condition. The only one, however, that has been well replicated is related to the gene that encodes *LINGO1*, a protein believed to inhibit cell differentiation during development, along with axonal regeneration and synaptic plasticity. Cerebellar dysfunction may be implicated. Magnetic resonance spectroscopy has shown decreased *N-acetylaspartate* in the cerebellum, indicating loss or dysfunction of neurons. Also, losses of cerebellar Purkinje cells have been seen. Increased LINGO1 levels and GABA dysfunction have been reported. The condition may also involve slight incoordination that is similar to symptoms seen in ataxia. As mentioned before, tremor-dominant PD and essential tremor can present very similarly.

The pathophysiology of essential tremor likely involves rhythmic activity in the cortico-ponto-cerebello-thalamo-cortical loop. However, the cause of this oscillation is not understood.

Clinical manifestations

Essential tremor is a syndrome characterized by isolated, often bilateral upper-limb action tremor. This is a highly frequent tremor that is absent or decreased during rest. It is not present during walking, which differs from the tremor of PD. The other characteristic features of PD tremor are the slow frequency and the fact that the tremor will restart once the arms are stretched out. It may occur with or without tremor in other locations, such as the head, larynx (voice tremor), or lower limbs, face, or jaw. There is an absence of other Parkinsonism symptoms such as rigidity and bradykinesia. However, there are sometimes additional, mild neurologic signs of mild ataxia, abnormal limb posturing, or impaired memory. When these accompany the tremor, the condition is described as *essential tremor plus*. There may be an impaired tandem gait. Essential tremor with additional tremor at rest is also classified as essential tremor plus.

Diagnosis

Complete patient history plus neurologic examination help to provide a phenotypic characterization of essential tremor. Considerations must include age of onset, family history, temporal evolution, and exposures to tremor-inducing drugs and toxins. These drugs include selective serotonin-reuptake inhibitors, valproate, lithium, and sympathomimetic agents, while the toxins include lead, manganese, and mercury. Neurologic examination must assess the distribution of the tremor and activation condition, such as whether it appears during rest, isometric extension of body parts against gravity, or goal-directed movements. There should be a visual estimation of the tremor frequency range (low, medium, or high), and assessment of signs suggesting systemic illness or other neurologic disease.

Rating scales are used to assess severity and effects of tremor upon activities of daily living, and health-related quality of life for the patient. The *Essential Tremor Rating Assessment Scale (TETRAS)* is most commonly used. According to the *International Parkinson and Movement Disorder Society*, if a cause is revealed for essential tremor the diagnosis will be "essential tremor due to" that cause. Other causative conditions must be considered. Differential diagnoses include enhanced physiologic tremor, isolated focal tremors of the head, voice, or palate; and orthostatic tremors. Tremor syndromes with prominent neurologic signs include dystonic tremors, intention tremor syndromes, tremors combined with Parkinsonism, Holmes tremor, and myorhythmia. Holmes tremor involves combined low-frequency rest, posture, and intention tremor, caused by lesions within the cerebellar outflow tract, which often leads to dystonia and ataxia as well.

If laboratory testing is also required, single-photon-emission CT that uses the tracer ^{123}I-*ioflupane* provides a measurement of the striatal presynaptic dopamine-transporter density. DaT scan is another way to distinguish between the tremor of PD and tremor of ET. Also, the patient can be followed over time, to determine if more definitive signs of Parkinsonism develop.

Treatment

Treatments for essential tremor are either pharmacologic, surgical, or other methods. Propranolol and primidone have shown the best results in treating essential tremor, by reducing the severity of upper-limb symptoms. Propranolol is a nonselective beta-blocker. Primidone is metabolized to phenylethylmalonamide and phenobarbital. It has been effective at similar levels to propranolol, also helping patients to be able to perform many daily tasks normally. Other medications that have shown some good results in treating essential tremor include alprazolam, topiramate, and some newer anti-epileptic medications.

Unilateral and bilateral deep-brain stimulation and unilateral thalamotomy target the thalamic nucleus. These procedures are used for medically intractable upper-limb tremor in essential tremor that is not responding to any treatments. This condition can be very disabling and hard to control in rare circumstances. Deep-brain stimulation has shown greater functional improvement than thalamotomy. It also causes fewer adverse events, such as dysarthria, gait disturbances, and sensory disturbances. After 5 years, about 50% of patients with essential tremor who underwent deep-brain stimulation reported a reduced effect, attributed to disease progression or the development of tolerance to the stimulation. Adverse events occur more often with bilateral than with unilateral deep-brain stimulation. They include reversible stimulation-induced ataxia, dysarthria, impaired balance, paresthesias, and tonic muscle contractions.

A newer focused ultrasound therapy was approved in 2016 by the Food and Drug Administration, based on a randomized trial. Unilateral thalamic thermoablation uses focused ultrasound with MRI guidance. It significantly reduced hand tremor and gave patients a better quality of life over 12 months. The most common adverse effects were intraprocedural discomfort, postoperative paresthesia or numbness, and gait disturbance. At 12 months after the procedure, paresthesia or numbness was present in 14% of patients, and gait disturbances were present in 9%.

Section review
1. What is the primary cause of hemiballismus?
2. What does the term "essential" describe in the condition known as "essential tremor"?
3. What are the diagnostic procedures used for essential tremor?

Focus on tremor
Some people with essential tremor may experience shaking of the legs and trunk, and they may experience a feeling of *internal tremor*. It is up to eight times more common than PD. The tremor may be more prominent when the affected individual is highly active or anxious. Due to embarrassment in public, many patients limit social interactions and even develop social phobias. Alcohol use helps with the tremor of ET. *Dystonic tremor* occurs in those affected by dystonia. *Cerebellar tremor* is usually a slow, high-amplitude tremor of the extremities and other parts of the body. *Enhanced physiologic tremor* is more noticeable and is usually due to a reaction to a drug, caffeine, or various medical conditions. *Orthostatic tremor* is rare, characterized by rapid muscle contractions in the legs occurring when standing.

Dystonias
Dystonia is defined as loss of CNS inhibitor function, with sustained, involuntary muscular contraction. There is sustained, involuntary twisting of the body, and repetitive movements or abnormal posture. Dystonia is a form of **hypertonia**, along with spasticity, rigidity, and **paratonia**. When muscular contractions are sustained for several seconds, they are described as **dystonic movements**. Examples include the **choreoathetoid movements** related to high levels of levodopa, known as peak-dose dyskinesia. If these contractions last for longer periods of time, they are referred to as **dystonic postures**, such as occur in torticollis. These postures can last for weeks, resulting in permanent, fixed contractures. Dystonia is related to basal nuclei abnormalities, but the actual pathophysiologic mechanisms are unclear. Dystonias can be primary or secondary and are also described as generalized, focal, or segmental.

Pathophysiology

Dystonia is caused by slow muscle contraction, and possibly by a failure in normal reciprocal inhibition of the muscles. Injury to the putamen or its outflow tracts is related to **hemidystonia**. Primary (idiopathic) dystonia is different from secondary dystonia, which is linked to degenerative or metabolic CNS disorders. These include Hallervorden-Spatz disease, Wilson's disease, multiple sclerosis, cerebral palsy, brain hypoxia, or stroke. Causative medications are usually the phenothiazines, butyrophenones, thioxanthenes, and antiemetics such as prochlorperazine (Compazine), which is frequently used in emergency rooms.

Clinical manifestations

Generalized dystonia is also called *dystonia musculorum deformans*. It is rare, progressive, and characterized by movements that cause sustained postures that may be extremely odd in appearance. Clinical manifestations usually begin in childhood. There is inversion and plantar fixation of the foot during walking. Just the trunk or one leg may be affected, but dystonia can affect the whole body. The most severe form may cause grotesque, twisted, and fixed postures, eventually causing the patient to be confined to a wheelchair. If symptoms begin in adulthood, usually only the face or arms are affected. Mental function is usually normal.

Focal dystonias only affect one body part, and most common being during a patient's 30s or 40s, with females affected more than males. Spasms may be at first only periodic, occurring at random or because of stress. The spasms are triggered by specific movements of the affected body part and are not present during sleep. They may progress over days, weeks, or many years. After this, the spasms may be triggered by movements of unaffected body parts and can continue even during rest. Eventually, distortion of the affected body part develops, which may be painful because of the position candy cause severe disability. The symptoms are based on the actual muscles involved.

Occupational dystonia involves focal dystonic spasms caused by performing skilled activities, and are described as *writer's cramp*, *typists' cramp*, and *yips* experienced by athletes, which are losses of fine motor skills used in athletic movements. **Spasmodic dystonia** involves the voice becoming strained, creaky, or hoarse because of abnormal involuntary contraction of the laryngeal muscles. **Torticollis** begins with a sensation of "pulling" that is followed by sustained torsion and deviation of the head and neck. It may be genetically linked, due to a medication side effect, or of unknown cause. In its early stages, torticollis can be overcome voluntarily by the patient. Certain motions, such as touching the face on the side contralateral to the deviation, can stop the spasms. This condition can also be caused by haloperidol and other dopamine-blocking drugs.

Segmental dystonias affect two or more continuous body parts. *Meige's disease* is one form, also described as *blepharospasm-oromandibular dystonia*. It consists of involuntary blinking, grimacing, and jaw grinding that usually begins in later middle age. This condition may mimic tardive dyskinesia's buccal-lingual-facial movements. Prevalence of dystonia is high in patients with Huntington's chorea.

Diagnosis

Diagnosis is clinical, based on the visible signs and symptoms, and may be difficult because of how the disorder is manifested. Misdiagnoses have included PD, carpal tunnel syndrome, essential tremor, temporomandibular joint dysfunction, Tourette's syndrome, conversion disorder, and other neuromuscular movement disorders.

Treatment

Treatments for dystonia are often ineffective. For generalized dystonia, high doses of trihexyphenidyl or benztropine are most common, often along with baclofen or another muscle relaxant, clonazepam or another benzodiazepine, or both. Severe generalized dystonia, or nondrug-responsive generalized dystonia may require deep brain stimulation of the *GPi*.

For focal or segmental dystonias, or for generalized dystonia that severely affects only certain body parts, purified botulinum toxin type A is the treatment of choice. This is injected into the affected muscles by experienced practitioners and weakens muscular contractions while not altering the abnormal neural stimulus. These injections are most effective for blepharospasm and torticollis, with varied dosages. Treatments must be given every 3–6 months. The treatment of medication-induced dystonia is usually diphenhydramine (an anticholinergic), in an IV form followed by few days of benztropine.

Section review
1. What are the classifications of dystonia?
2. What are the clinical manifestations of generalized dystonia?
3. What treatments may be used for the various forms of dystonia?

Focus on acquired dystonia
According to the National Institute of Neurological Disorders and Stroke, dystonias are either idiopathic, genetic or acquired. The acquired form, also called secondary dystonia, may be linked to hypoxia, neonatal brain hemorrhage, infections, drug reactions, heavy metal or carbon monoxide poisoning, trauma, and stroke. Acquired dystonia often plateaus and does not spread to other parts of the body. Recent studies have focused on mapping a gene for early-onset dystonia to chromosome 9. This DYT1 gene codes for a previously unknown protein called *torsin A*. Via prenatal testing, physicians can make a specific diagnosis for some cases of dystonia and investigate molecular and cellular mechanisms that lead to disease.

Clinical cases

Clinical case 1
1. What is the likelihood that the patient in this case study is depressed?
2. What do you think is happening in regard to his medications?
3. Are there any changes in this patient's treatment that may be helpful?

A 64-year-old man was diagnosed with PD. Over a few years, his gate when walking has become impaired, causing him to shuffle and become tired very easily. His wife feels that he is not interested in her or their family. She says that his face doesn't seem to respond when people talk to him. He has retired from work due to his health condition. He begins to find that his medication is wearing off at least 30 min before the next dose is due to be taken.

Answers:
1. Depression, as well as anxiety, are more common in patients with PD than in those without. At least 50% of those diagnosed with this disease will experience some form of depression.
2. Medications such as levodopa begin to wear off more and more quickly over time, which is known as the "on-off effect." Once this begins in a Parkinson's patient, the patient feels better as a new dose starts to take effect, but worse before the next dose is due. Over time, the duration of the "on" state becomes shorter and the wearing "off" happens sooner.
3. Many patients respond to a change in medications, to controlled-release forms of levodopa and other medications. Another option is to shorten the interval between doses by about 30–60 min. This is especially effective for advanced PD. Dopamine agonists, COMT inhibitors, and MAO-B inhibitors may be added to levodopa to help regulate its effects for better results. Please remember that the examination of PD patients should be done once during the "on" time and once while coming off of the medication until the effect is completely gone. This helps to assess the effectiveness of the symptoms at the medication's peak plasma level. There must be monitoring for the development of any dyskinesia while the drug level in the plasma is rising, at peak plasma level, and while the effect of the medication is wearing off.

Clinical case 2
1. What do radiographic studies show in relation to this disease?
2. What is the prognosis for the patient in this case study?
3. Are there any medications specifically approved to treat this disease?

A 34-year-old woman presents with a history of abnormal, involuntary, and stereotyped movements. She has had memory loss and emotional instability. Her husband reports that she has become progressively rigid in her ability to move and that she has said things to him that made him feel like she was another person entirely, sometimes making no sense, and sometimes really scaring him with her statements. After testing to rule out various diagnoses, this patient is diagnosed with Huntington's chorea.

Answers:
1. Radiographic studies show atrophy of the heads of caudate, with enlargement of the frontal horns, making them "box-like" in appearance, as well as generalized cortical atrophy.
2. Unfortunately, Huntington's chorea that is of adult onset usually leads to the patient's death within 15 years.
3. Today, only tetrabenazine has been specifically approved by the FDA for the treatment of Huntington's chorea. Many other drugs approved for depression, psychosis, PD, and Alzheimer's disease have been used, but this is called "off-label prescribing," and only has limited benefits.

Clinical case 3
1. What are the treatment options for this patient's condition?
2. What is the prognosis for this condition?
3. What are the less common causes of this condition?

A 70-year-old man presents to his physician following a minor stroke. He has a wild flailing, involuntary movement of his upper left arm and less violent movement of his left leg. Neurologic examination reveals hemiballismus but no other neurological findings. A brain CT shows a hyperdense lesion at the level of the right cerebral peduncle, which is interpreted as a hemorrhagic stroke. The lesion is approximately 1 cm in diameter. There is also cerebral atrophy.

Answers:
1. Treatments include dopamine blockers, anticonvulsants, intrathecal baclofen therapy, botulinum injections, tetrabenazine, antipsychotics, and neurosurgery.
2. Most patients with hemiballismus go into spontaneous remission. With today's treatments, most patients respond very well, and the symptoms are mostly controlled.
3. Other causes of hemiballismus include traumatic brain injury, non-ketotic hyperglycemia, ALS, neoplasms, vascular malformations, tuberculomas, demyelinating plaques, and complications of HIV infection.

Clinical case 4
1. What is the likely diagnosis of the patient in this case study?
2. How does this patient's condition differ from other similar diseases?
3. What neurologic examination techniques are used for this condition?

A 64-year-old woman visits her physician. She has a tremor in both of her hands. She explains that it began, only slightly, about 10 years previously. It started slowly, was symmetric, and has gradually worsened. Her fine-motor movements are now affected, which has changed her handwriting. She is having severe trouble eating, some trouble getting dressed, and has noticed that the tremor becomes worse when she feels stressed. The physician asks if any close relative had a similar tremor, and she tells him that her mother had the same condition. Recently, she was treated for depression and is currently taking bupropion and fluoxetine. Neurologic examination reveals a postural and action tremor of both hands, with a medium frequency range. Otherwise, there are no significant signs or symptoms. She is also reporting more alcohol consumption in the past few months as she thinks alcohol helps with her symptoms.

Answer:
1. The likely diagnosis is essential tremor. The condition usually worsens over time and can be very severe. It is sometimes confused with PD.
2. Essential tremor mainly involves the hands, head, and voice. PD tremors usually begin in the hands and can affect the legs, chin, and other parts of the body. Essential tremor usually occurs when using the hands, while tremors from PD are most prominent when the hands are not being used. Essential tremor can develop along with ataxia and other neurological signs and symptoms, however.
3. Neurologic examination for essential tremor involves checking tendon reflexes, muscle strength and tone, ability to feel certain sensations, posture, coordination, and gait.

Clinical case 5
1. What are the possible causes of this patient's condition?
2. What techniques may help to relieve some symptoms of this condition?
3. Can physical rehabilitation be helpful for this condition?

A 51-year-old woman began to experience changes in her ability to walk or run. It affected her left foot initially, but then the right foot as well, followed by her knee and hip joints. Eventually, the patient required the use of a wheelchair since she could no longer stand for any extended time period. She complained of poor circulation and the feeling of pins and needles in her feet. The patient was diagnosed with multifocal dystonia.

Answers:
1. Dystonia may be hereditary, or due to physical trauma, infection, poisoning, or reactions to neuroleptics and other drugs.
2. Reducing movements that trigger or worsen dystonic symptoms give some relief. Also, the patient should reduce stress, get adequate sleep, have moderate exercise, and practice relaxation techniques.
3. Physical rehabilitation may be beneficial for dystonia patients. It can help manage changes in balance, mobility, and overall function. Potential treatment

interventions include splinting, therapeutic exercise, manual stretching, soft tissue and joint mobilization, postural training and bracing, neuromuscular electrical stimulation, constraint-induced movement therapy, activity and environmental modification, and gait training.

Key terms

Accelerometry	Hypokinetic disorders
Alpha-synuclein	Hypomimia
Amygdaloid body	Hypophonic
Akinesia	Internal capsule
Athetosis	Intralaminar nuclei
Ballismus	Juvenile Parkinsonism
Basal nuclei	Lentiform nucleus
Bradykinesia	Lewy bodies
Caudate nucleus	Lewy body dementia
Chorea	Mitophagy
Choreoathetoid movements	Myerson's sign
Choreoathetosis	Nigrostriatal
Claustrum	Paratonia
Dystonic movements	Parkinsonian
Dystonic postures	Parkinsonism
Festination	Parkinson's disease
Focal dystonias	Pars reticulata
Glabellar reflex	Polyglutamine
Globus pallidus	Postural instability
Hemiballismus	Putamen
Hemidystonia	Resting tremor
Huntingtin	Segmental dystonias
Huntington's chorea	Spasmodic dystonia
Hyperkinetic disorders	Substance P
Hyperphosphorylation	Substantia nigra
Hypertonia	Torticollis
Hypokinesia	Ventral intermediate nucleus

Suggested readings

1. Ahlskog, J. E. *The New Parkinson's Disease Treatment Book: Partnering With Your Doctor to Get the Most From Your Medications*, 2nd ed.; Oxford University Press, 2015.
2. Bates, G.; Tabrizi, S.; Jones, L. Huntington's Disease. In *Oxford Monographs on Medical Genetics;* 4th ed.; Oxford University Press, 2014.
3. Brin, M. F.; Comella, C. L.; Jankovic, J. *Dystonia: Etiology, Clinical Features, and Treatment.* LWW, 2004.
4. Colosimo, C.; Riley, D. E.; Wenning, G. K. *Handbook of Atypical Parkinsonism.* Cambridge University Press, 2011.

5. Dressler, D.; Altenmuller, E.; Krauss, J. K. *Treatment of Dystonia*. Cambridge University Press, 2018.
6. Groenewegen, H. J.; Voorn, P.; Berendse, H. W.; Mulder, A. B.; Cools, A. R. The Basal Ganglia IX. In *Advances in Behavior Biology;* Springer, 2009.
7. Hayden, M. R.; Bruyn, G. *Huntington's Chorea*. Springer, 2011.
8. Hewitt, J.; Gabata, M. *Huntington's Disease: Causes, Tests, and Treatments*. CreateSpace Independent Publishing Platform, 2011.
9. Jones, S. *Dopamine-Glutamate Interactions in the Basal Ganglia*. CRC Press, 2011.
10. Kanovsky, P.; Bhatia, K. P.; Rosales, R. L. *Dystonia and Dystonic Syndromes*. Springer, 2015.
11. Mai, J. K.; Paxinos, G. *The Human Nervous System*, 3rd ed.; Academic Press, 2011.
12. Mancall, E. L.; Brock, D. G. *Gray's Clinical Neuroanatomy: The Anatomic Basis for Clinical Neuroscience*. Saunders, 2011.
13. Meara, J.; Koller, W. C. *Parkinson's Disease and Parkinsonism in the Elderly*. Cambridge University Press, 2000.
14. Morel, A. *Stereotactic Atlas of the Human Thalamus and Basal Ganglia*. CRC Press, 2007.
15. Nolte, J. *Essentials of the Human Brain*. Mosby, 2009.
16. Plumb, M.; Bain, P. *Essential Tremor: The Facts*. Oxford University Press, 2006.
17. Politis, M. Imaging in Movement Disorders: Imaging in Atypical Familial Movement Disorders. In *International Review of Neurobiology;* Vol. 142; Academic Press, 2018.
18. Rana, A. Q.; Hedera, P. *Differential Diagnosis of Movement Disorders in Clinical Practice*. Springer, 2013.
19. Rana, A. Q.; Chou, K. L. *Essential Tremor in Clinical Practice*. Springer, 2015.
20. Singer, H. S.; Mink, J.; Gilbert, D. L.; Jankovic, J. *Movement Disorders in Childhood*. Saunders, 2010.
21. Smythies, J. R.; Edelstein, L.; Ramachandran, V. S. *The Claustrum: Structural, Functional, and Clinical Neuroscience*. Academic Press, 2014.
22. Soghomonian, J. J. *The Basal Ganglia: Novel Perspectives on Motor and Cognitive Functions*. Springer, 2016.
23. Steiner, H.; Tseng, K. Y. *Handbook of Basal Ganglia Structure and Function*, 2nd ed.; Academic Press, 2016.
24. Verstreken, P. *Parkinson's Disease: Molecular Mechanisms Underlying Pathology*. Academic Press, 2017.
25. Weiner, W. J.; Shulman, L. M.; Lang, A. E. *Parkinson's Disease: A Complete Guide for Patients and Families*, 3rd ed.; Johns Hopkins University Press, 2013.
26. Zid, D.; Russell, J.; OhioHealth. *Delay the Disease—Exercise and Parkinson's Disease*, 2nd ed.; OhioHealth, 2017.

CHAPTER 8

Diencephalon: Thalamus and hypothalamus

The **diencephalon** is part of the *prosencephalon*, also known as the *forebrain*. It develops from the primary cerebral vesicle. It then differentiates, forming the rostral *telencephalon* and the caudal diencephalon. Out of the side of the telencephalon, the cerebral hemisphere forms, containing the lateral ventricles. The lateral ventricles and third ventricle are able to communicate through the foramen of Monro, also called the *interventricular foramen*. The diencephalon, therefore, is largely related to the structures developing lateral to the third ventricle. The lateral diencephalon walls superiorly form the *epithalamus*, and centrally, the *thalamus*. Inferiorly, they form the *subthalamus* and *hypothalamus*. As an anatomical marker, the structures on either side of the third ventricle make up the thalamus.

Most sensory, motor, and limbic pathways stop somewhere in the diencephalon (thalamus). Most limbic and motor pathways also involve telencephalic structures. Almost all connections between the cerebral cortex and subcortical structures, especially the diencephalon, course through the internal capsule. The only part of the diencephalon visible on an intact human brain is the inferior surface of the hypothalamus, including the mammillary bodies and infundibulum.

Epithalamus

The **epithalamus** is the most distal part of the diencephalon. It forms the roof of the third ventricle. The **pineal gland** extends from its posterior border and is externally visible. The caudal border of the epithalamus is formed by the *posterior commissure*. The epithalamus is also made up of the anterior and posterior *paraventricular nuclei*, medial and lateral *habenular nuclei*, and the *stria medullaris thalami*.

Pineal gland

The pineal gland is also called the *epiphysis cerebri* or *pineal body*. It has a shape like a pinecone. The pineal gland is small, red-gray in color, and is located within a depression between the superior colliculi. It lies inferiorly to the splenium of the corpus callosum. It is separated from this structure by the tela choroidea of the third ventricle and cerebral veins. The pineal gland is enveloped within the lower layer of the tela, which is reflected in the tectum. The pineal gland is only about eight millimeters in length. Its base is directed anteriorly and attached via a peduncle. This divides into the superior and inferior laminae, which are separated by the third ventricle's *pineal recess*, containing the posterior and habenular commissures. Unusual commissural fibers sometimes invade the gland but do not end near the parenchymal cells.

Septa enter the pineal gland from surrounding pia mater, dividing the gland into lobules while carrying blood vessels and very thin, unmyelinated sympathetic axons. There is a rich blood supply to the pineal gland. The pineal arteries branch from the medial posterior choroidal arteries, which also branch, from the posterior cerebral artery. In the pineal gland, arterial branches supply fenestrated capillaries that have endothelial cells on a tenuous, often incomplete *basal lamina*. These capillaries drain into many pineal veins. The veins open into the internal cerebral veins, the great cerebral vein, or both. Postganglionic adrenergic sympathetic axons develop from neurons in the superior cervical ganglion. These enter the dorsolateral (DL) part of the gland, from the area of the tentorium cerebelli as the **nervus conarii** (single or paired). The nerve is deep to the endothelium of the straight sinus' wall. It is related to the parenchymal cells and blood vessels in the pineal gland.

There are clusters and cords of *pinealocytes* in the gland, related to astrocyte-resembling neuroglia, the primary cellular component of the pineal gland's stalk. The extremely modified pinealocytes are neurons containing many synaptic ribbons. They are distributed, randomly, between nearby cells, coupled by gap junctions. From each cell body, two or more processes extend, ending in rounded expansions near capillaries, or rarely, on the ependymal cells of the pineal recess. The expansions surround mitochondria, rough endoplasmic reticulum, and densely cored vesicles storing the hormone *melatonin*.

The pineal gland excretes this hormone, which is an antioxidant and sleep-inducing substance, in response to variations in circadian rhythm sensed by the hypothalamus. The pineal gland and hypothalamic nuclei help to regulate the sleep-wake cycle. The precursor to melatonin is *serotonin*, which is synthesized from tryptophan by the pinealocytes, then secreted into the fenestrated capillaries. The pineal gland has major regulatory actions. This endocrine gland regulates activities of the adenohypophysis, endocrine pancreas, neurohypophysis, parathyroids, suprarenal cortex and medulla, and the gonads. Its effects are primarily inhibitory. From the pinealocytes, **indolamine** and polypeptide hormones may reduce the synthesis and release of hormones from the pars anterior. This may be by direct action upon its secretory cells, or by indirect inhibition of production of hypothalamic releasing factors. Pineal secretions can reach target cells through the cerebrospinal fluid or bloodstream. Some indolamines, such as melatonin and enzymes needed for their biosynthesis (such as N-acetyltransferase and 5-HT), reveal circadian rhythms in concentration. Levels rise in darkness and fall during the daytime. This is when secretion may be inhibited by sympathetic activity. The intrinsic rhythmicity of an endogenous circadian oscillator, within the suprachiasmatic hypothalamus nucleus, may control cyclical pineal actions.

Once a person reaches his or her 20s, calcareous deposits accumulate within the pineal extracellular matrix. Here, they are deposited concentrically as **corpora arenacea**, also known as "brain sand." Calcifications of the pineal gland and choroid plexus are the most common areas of intracranial calcifications, as seen on CT scans of the skull. The pineal gland is a good indicator of a midline on CT, and can be used in the assessment of midline shifts in the setting of increased intracranial pressure.

> **Focus on melatonin**
> The body makes less melatonin as we age, so melatonin supplementation may help the elderly to sleep better. Naturally, melatonin levels begin to rise in the blood at approximately 9:00 p.m., causing alertness to be reduced. Melatonin levels remain in the blood for about 12 h, until the morning light influences them to reduce to daytime levels, which are barely detectable.

Habenular nuclei

The **habenular nuclei** are located posteriorly near the **dorsomedial (DM)** aspect of the thalamus. They are immediately deep to the third ventricle's ependyma. The stria medullaris thalamus is located above and laterally. The primary habenular outflow goes to the interpeduncular nucleus, mediodorsal thalamic nucleus, mesencephalic tectum, and reticular formation. Most of this is from the fasciculus retroflexus to the interpeduncular nucleus, with the latter creating relays to the midbrain reticular formation. From here, the **tectotegmentospinal tracts** and dorsal longitudinal fasciculi are connected with autonomic preganglionic neurons that control gastric and intestinal secretions and motility, salivation, and the motor nuclei for deglutition and mastication. The stria medullaris crosses over the superomedial thalamic aspect. It is slightly medial to the habenular trigone, sending extensive amounts of fibers into the ipsilateral habenular. Additional fibers cross in the anterior pineal lamina. These decussate as the habenular commissure, reaching the contralateral habenular. They interconnect the hippocampal cortices and amygdaloid complexes, accompanied by crossed **tectohabenular fibers**.

The physiological functions of the habenular nuclei are poorly understood. They may be related to the sleep cycle. However small the habenular is in size, it is still the main integration area of olfactory, somatic, and visceral afferent pathways. When lesions are present in this area, regulation of visceral and neuroendocrine functions is affected. If the habenular is ablated, there are significant alterations in metabolism, and in thermal and endocrine regulation.

Section review
1. From which part of the diencephalon does the pineal gland extend?
2. What is the precursor to the hormone melatonin, and what are the effects of melatonin?
3. What is the location of the habenular nuclei?

Thalamus

The **thalamus** is the final relay point for ascending sensory information that is projected to the primary sensory cortex. It filters information, acts as an amplifier, and passes on desired sensory information. The thalamus also regulates the basal nuclei and cerebral cortex activities, relaying information between these areas. Recall that the thalamus ventral anterior/ventral lateral (VA/VL) nucleus is the last part of the dopamine pathway. The thalamus has bilateral, egg-shaped nuclei that are located on either side of the third ventricle (see Fig. 8.1). The thalamus is about 4 cm in length. The thalamus is deep within the brain and makes up most (80%) of the diencephalon.

Thalamus also forms part of the floor of the lateral ventricle. The superior thalamic surface's lateral border is signified by the *stria terminalis* and the overlying *thalamostriate vein*. These structures separate the thalamus from the caudate nucleus' body. There is

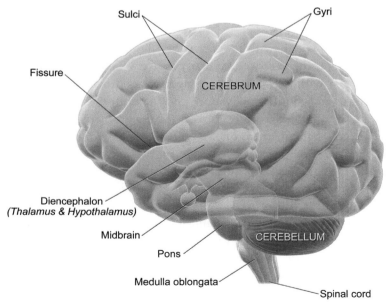

Fig. 8.1 Brain structures.

a lateral, thin sheet of white matter called the *external medullary lamina*. This separates the primary body of the thalamus from the *reticular nucleus*. More lateral, the thick posterior limb of the internal capsule lies between the *lentiform complex* and the thalamus.

The medial thalamic surface is the superior (dorsal) area of the lateral wall of the third ventricle. In most individuals, the thalamic nuclei are connected by an intermediate mass called the **interthalamic adhesion**. This adhesion is present in most younger brains but absent in most older brains. The boundary with the hypothalamus is signified by a small *hypothalamic sulcus*. This curves away from the upper part of the cerebral aqueduct to the interventricular foramen. The thalamus is continuous with the tegmentum of the midbrain, as well as the subthalamus and hypothalamus.

Inside, the major part of the thalamus is divided into anterior, medial, and lateral nuclear portions. The division is made by a vertical, "Y"-shaped sheet of white matter called the *internal medullary lamina*. Recall that the thalamus mostly consists of gray matter. Also, there are intralaminar nuclei inside the internal medullary lamina. Midline nuclei meet the lateral walls of the third ventricle. The reticular nucleus makes up a lateral covering of the main nuclear mass. This covering is shaped like a shell. An external medullary lamina made up of nerve fibers separates the lateral covering and main nuclear mass.

The thalamic nuclei are primarily named based on their locations. Each nucleus has functional specializations, and project fibers into, while receiving fibers from specific regions of the cerebral cortex. The *stria medullaris* is the line of attachment of the thalamus to the roof of the third ventricle. The thalamus receives afferent impulses from all of the body. The impulses synapse with one or more of its nuclei.

The thalamus sorts and filters information. Impulses related to similar functions are relayed in groups via the internal capsule. They reach appropriate areas of the sensory cortex as well as specific cortical association errors. Arriving afferent impulses are crudely determined to be pleasant or unpleasant. The localization of specific stimuli and their discrimination occur in the cerebral cortex. Almost all other inputs rising to the cerebral cortex move through the thalamic nuclei. They include inputs assisting regulation of emotion and visceral functions from the hypothalamus, via the anterior nuclei (AN). Also included are instructions aiding in the direction of activity of motor cortices from the cerebellum and basal nuclei. These are via the VL nuclei (cerebellum) and ventral AN (basal nuclei). Also, inputs for memory or sensory integration are sent to certain association cortices, via the pulvinar, lateral dorsal, and lateral posterior (LP) nuclei. The thalamus is vital for mediating sensation, arousal of the cortex, motor activities, learning, and memory. The understanding of the thalamus has helped in the treatment of many disorders, including epilepsy, various types of pain, Parkinson's disease, and psychiatric disorders. The functions of different thalamic nuclei are a very high-yield topic that must be memorized:

- **Dorsomedial (DM)** nucleus—involved in motivation and emotions
- **Dorsolateral (DL), lateral posterior (LP)** nuclei—involved in emotions as well
- **Anterior nuclei (AN)** —involved in memory formation

- **Ventroanterior (VA), Ventrolateral (VL)** nuclei—involved in motor movements
- **Pulvinar** nucleus—involved in visuospatial and spatial attention
- **Reticular** nucleus—involved in arousal as well as generation of sleep spindles on EEG; this is the only thalamic nuclei that *do not project* to the cortex, but *do project* to the dorsal thalamic nucleus
- **Ventro posterolateral (VPL), Ventro posteromedial (VPM)** nuclei—involved in sensory functions of the body and face, respectively
- **Medial geniculate nucleus (MGN)** and **Lateral geniculate nucleus (LGN)**—the auditory and visual relay nuclei, respectively.

Subthalamus

The **subthalamus** is a complicated area, with nuclear groups and fiber tracts present. The primary nuclear groups include the subthalamic nucleus, **fields of Forel**, zona incerta, and **pregeniculate nucleus**. Rostral extensions of the red nucleus and substantia nigra are also located in the subthalamus. The primary subthalamic tracts include the following:
- *Upper portions of the medial, spinal, and trigeminal lemnisci, along with the* **solitariothalamic tract**—these all approach their terminations in the thalamic nuclei
- **Dentatothalamic tract**—from the contralateral superior cerebellar peduncle, along with ipsilateral **rubrothalamic fibers**
- *Fasciculus retroflexus*
- **Fasciculus lenticularis**
- **Fasciculus subthalamicus**
- *Ansa lenticularis*
- *Fascicles from the* **prerubral field**—*also called the H field of Forel*
- *Continuation of the fasciculus lenticularis*—in the H_2 field of Forel
- **Fasciculus thalamicus**—also called the H_1 field of Forel

Subthalamic nucleus

The **subthalamic nucleus** is closely related to the basal ganglia. Its neurons are the only *glutamatergic* neurons within the basal ganglia network. The subthalamic nucleus is lens shaped, making up the largest portion of the subthalamus, known as the *ventral thalamus*. The nucleus is immediately inferior to the zona incerta and rostral to the substantia nigra. It receives projections from the **pallidosubthalamic fibers** of the lateral **pallidal division**, the **corticosubthalamic fibers** of the cerebral cortex, the **nigrosubthalamic fibers** of the nigral complex, and the **parabrachial pontine reticular formation**.

The subthalamic nucleus projects to the **subthalamopallidal fibers** of both pallidal divisions and to the **subthalamonigral fibers** of the substantia nigra. These connections make up an essential portion of the indirect pathway that underlies basal nuclear function. Subthalamic neurons utilize the excitatory neurotransmitter called *glutamate*. Predominantly, the subthalamic cells are inactive. This is because of constant inhibition from cells of the external pallidal segment. When this inhibition is removed, such as in Parkinson's disease, the subthalamic neurons have an increased level of activity, causing characteristic motor deficits. This is partially mediated by a large cortico-subthalamic projection. Treatment of Parkinson's disease, therefore, may involve delivering electrical impulses to the subthalamic nucleus via surgically inserted electrodes. This deep-brain stimulation interferes with the subthalamic nucleus functions, reducing its output.

Zona incerta

The **zona incerta** is a collection of small cells between the ventral area of the external medullary lamina in the thalamus and the cerebral peduncle. It is dorsolaterally linked to the reticular nucleus. The zona incerta receives fibers from the pregeniculate nucleus, sensorimotor cortex, deep cerebellar nuclei, spinal cord, and trigeminal nuclear complex. It then projects to the **pretectal region** and spinal cord, with unknown functions. The subthalamic fasciculus is connected to the globus pallidus via the subthalamic nucleus. It has a large two-way structure of fibers through the internal capsule, mixing with it at right angles.

Section review
1. What are the functions of the thalamus?
2. What structure is the largest portion of the subthalamus?
3. From where does the zona incerta receive fibers?

Hypothalamus

The **hypothalamus** is located below the thalamus, capping the brainstem. It forms the inferolateral walls of the third ventricle. As it merges into the midbrain inferiorly, the hypothalamus is extended from the optic chiasma, where the optic nerves cross over, to the posterior margin of the **mammillary bodies**. Each of these bodies contains three to four *mammillary nuclei* that primarily relay signals from the limbic system to the thalamus. The mammillary bodies are paired round nuclei. They bulge ventrally from the hypothalamus and act as information relay areas in the olfactory pathways. The hypothalamus is separated from the thalamus by a shallow *hypothalamic sulcus* in the third ventricle's wall. The inferior surface of the hypothalamus is one of only a few areas of the diencephalon visible on the brain surface. The **infundibulum** is a stalk of hypothalamic tissue between the mammillary bodies and optic chiasma. It connects the **pituitary gland** to the base of the hypothalamus. The hypothalamus is only $4\,cm^2$ of neural tissue, making up only 0.3% of the entire brain.

There are many functionally important nuclei in the hypothalamus, much like the thalamus. Though small, the hypothalamus is the primary visceral control center and extremely important for homeostasis of the body. It influences nearly all body tissues. The homeostatic activities of the hypothalamus are listed in Table 8.1.

Hypothalamic lesions are linked with many unusual endocrine disorders as well as with metabolic, motor, emotional, and visceral disturbances. Normally, the hypothalamus controls the endocrine system through **magnocellular** neurosecretory extensions to the posterior pituitary. It acts through parvocellular neurosecretory projections to the median eminence as well, and through the autonomic nervous system. It controls the

Table 8.1 Homeostatic activities of the hypothalamus.

Control of autonomic nervous system (ANS)	By controlling activity in brainstem and spinal cord	Blood pressure, rate and force of heartbeat, motility of digestive tract, eye pupil size, many other visceral activities
Initiation of physical responses to emotions	Located in the limbic system, contains nuclei for perceiving pleasure, fear, anger, and those involved in sex drive, other biological rhythms	Initiates most physical expressions of emotion such as high blood pressure, pallor, sweating, pounding heart, and dry mouth
Regulation of body temperature	Acts as body's thermostat, with neurons monitoring blood temperature and receiving input from thermoreceptors	Initiates cooling via sweating, or heat-generation via shivering, to maintain relatively constant temperature
Regulation of food intake	Responds to changing blood levels of glucose, amino acids, or hormones such as cholecystokinin, ghrelin	Regulates hunger and satiety
Regulation of water balance and thirst	As body fluids become too concentrated, osmoreceptors are activated; they excite nuclei that trigger release of antidiuretic hormone (ADH) from posterior pituitary	ADH causes kidneys to retain water; neurons in the thirst center are stimulated, increasing thirst and fluid intake
Regulation of sleep-wake cycles	Along with other brain regions, helps regulate sleep	Suprachiasmatic nucleus sets timing of sleep cycle in response to daylight and darkness, via visual pathways
Control of endocrine functions	Releasing and inhibiting hormones control secretion of anterior pituitary gland hormones	Supraoptic and paraventricular nuclei produce ADH and oxytocin
Memory	Via mammillary nuclei in the signal pathway from the hippocampus to the thalamus	This is important in memory; lesions to the mammillary nuclei cause memory deficits

endocrine output of the anterior pituitary and peripheral endocrine organs. Vasopressin and oxytocin are the posterior pituitary hormones mostly involved in the control of osmotic homeostasis and reproductive functions, respectively. The hypothalamus influences the secretions from the thyroid gland, suprarenal cortex, gonads, mammary glands, and also processes of growth and metabolic homeostasis.

The hypothalamus affects the parasympathetic and sympathetic divisions of the ANS. Generally, parasympathetic effects mostly occur when the anterior hypothalamus is stimulated. The sympathetic effects are more related to the posterior hypothalamus. When the anterior hypothalamus and paraventricular nucleus are simulated, there may be decreased BP and heart rate.

Damage to the anterior hypothalamus can cause an uncontrolled increase in body temperature. Projections to the ventromedial hypothalamus help regulate the intake of food. When damaged, these structures can cause uncontrollable eating and obesity. When the posterior hypothalamus is stimulated, it causes sympathetic arousal. There is **piloerection**, vasoconstriction, increased metabolic heat production, and shivering. The control of shivering lies in the DM posterior hypothalamus. This does not mean that there are special parasympathetic and sympathetic centers, however. Many different parts of the hypothalamus, when stimulated, greatly alter the cardiac output, heart rate, peripheral resistance, vasomotor tone, differential blood flow, respiration depth and frequency, alimentary tract motility and secretion, erection, and ejaculation.

> **Focus on pituitary gland function**
> The pituitary gland significantly regulates body metabolism. It affects appetite, and therefore, weight gain and weight loss. It also regulates mood, behavior, and many mental and emotional health issues, along with thirst, urinary output, and regulation of the thyroid and other endocrine glands.

Hypothalamic-releasing hormones

Small peptides known as **hypothalamic-releasing hormones** and *hypothalamic-inhibiting hormones* (such as dopamine) are secreted from the cells of the arcuate nucleus, which inhibits secretions of prolactin. These peptides move down the axons of these cells. They are released into the first capillary bed and enter the bloodstream. The median eminence is an example of the **circumventricular organs**. It has fenestrated capillaries. From this point, the releasing and inhibiting hormone dopamine moves down the hypophyseal portal vessels to the adenohypophysis, which is where its actions occur. In general, the releasing hormones promote the release of certain hormones, with the inhibiting hormones having the opposite effect. All axons carrying these factors make up the **tuberoinfundibular** or *tuberohypophyseal* tract.

The *parvocellular neurosecretory system* consists of small cells secreting these releasing and inhibiting factors. This system is different from the *magnocellular neurosecretory system*, which has larger neurons located in the supraoptic and paraventricular nuclei, secreting oxytocin and vasopressin. Identical, small hypothalamic peptides are also present in neurons of many different central nervous system (CNS) areas, and in other body cells. Therefore, the brain may use one chemical as a locally acting neurotransmitter upon a postsynaptic neuron, and also as a hormone that acts over a larger distance. Both actions, however, are targeted upon the same physiological outcome.

Neurohypophysis

The infundibulum and posterior lobe, combined, make up the **neurohypophysis**. Its medial surface extends anteriorly to the lamina terminalis. Its superior surface extends to the hypothalamic sulcus, and its posterior surface extends to the diencephalon's caudal edge. Like the brainstem, the anterior and posterior boundaries are sometimes revealed by using transverse planes. Functionally similar neural tissue that is just in front of the anterior plane is both structurally and functionally continuous with the hypothalamus. This region, known as the *preoptic area*, is considered to be part of the anterior hypothalamus.

The majority of cells of the supraoptic nucleus, along with many cells of the paraventricular nucleus, secrete hormones traveling down axons of the supraoptic and paraventricular nuclei, to be released in the neurohypophysis. There is a separate control system for each lobe of the pituitary gland. One is a neural projection to the neurohypophysis and the other is a vascular link with the adenohypophysis.

Section review
1. What are the major functions of the hypothalamus?
2. What is the result of damage to the anterior hypothalamus?
3. What hormones are released from the hypothalamus?

Blood supply to the diencephalon

Branches of the posterior communicating, posterior cerebral, and basilar arteries chiefly supply the thalamus. Please note that the blood supply to the thalamus is entirely from the posterior circulation. The medial branch of the posterior choroidal artery supplies the habenular region, posterior commissure, pineal gland, and medial areas of the thalamus such as the pulvinar. Hypophyseal arteries from the internal carotid artery supply the pituitary gland. The lamina terminalis is supplied by the anterior cerebral and anterior communicating arteries. Branches of the internal carotid and posterior cerebral arteries supply the choroid plexuses of the third and lateral ventricles.

Internal capsule

The **internal capsule** is a compact fiber bundle in the cleft between the lenticular nucleus, the thalamus, and the head of the caudate nucleus. Nearly all information in and out of the cerebral cortex passes through the internal capsule. The **corticopontine fibers** descend from the cortex to the corona radiata, and from there, through the internal capsule, through the cerebral peduncle, and reach the pontine nuclei. The *corticobulbar fibers* reach the motor nuclei of the cranial nerves and other brainstem areas. The *corticospinal fibers* reach the spinal cord motor neurons and interneurons. Additional fibers extend from the cerebral cortex through the internal capsule to other subcortical areas, including the putamen, caudate nucleus, and other parts of the basal nuclei. These fibers spread out as the *corona radiata*, above the internal capsule, mixing with various fiber bundles that connect different cortical areas in the **centrum semiovale** of each brain hemisphere.

The internal capsule is a continuous fiber sheet forming the medial edge of the lenticular nucleus. It contuses around, posteriorly and inferiorly, partially containing this nucleus. Inferiorly, many of its fibers run down into the cerebral peduncle. Superiorly, they spread out into the corona radiata through the centrum semiovale and reach their cortical origins and destinations. The overall system of fibers is somewhat shaped like a trumpet, with a hole cut into its "bell end". The thin part of the "trumpet" corresponds to the cerebral peduncle. There are five regions of the internal capsule, as follows:

- **Anterior limb**—between the lenticular nucleus and head of the caudate nucleus; it contains fibers connecting the anterior nucleus and cingulate gyrus, and most connecting the DM nucleus and prefrontal cortex; there are also the *frontopontine* fibers that project from the frontal lobe to the ipsilateral pontine nuclei
- **Posterior limb**—between the lenticular nucleus and thalamus; contains fibers connecting the VA nucleus and VL nucleus with motor regions of the cortex; also contains corticospinal and corticobulbar fibers, along with somatosensory fibers from the ventral posterolateral nucleus and ventral posteromedial nucleus to the postcentral gyrus; for most of their length, the fibers are located in the posterior third of the posterior limb, near the somatosensory projections; corticobulbar fibers to the cranial nerve motor nuclei are anterior to the corticospinal fibers, near the genu, but mostly back in the posterior limb
- **Genu**—at the junction of the anterior and posterior limbs, near the interventricular foramen and venous angle; the demarcation of the anterior and posterior limbs is distinct at the genu, which is a transitional zone; it contains frontopontine fibers and other fibers that connect the DM nucleus with the prefrontal cortex
- **Retrolenticular part**—posterior to the lenticular nucleus; contains most of the fibers that connect the thalamus with the posterior areas of the cerebral hemisphere, including fibers in both directions, between the parietal and occipital association areas as well as the pulvinar-LP complex; these include part of the *optic radiation*, a large group of visual system fibers linking the LGN to the calcarine sulcus; the area of the

optic radiation in the retrolenticular part of the internal capsule terminates in the superior part of the calcarine sulcus, conveying information from inferior areas of the visual fields; there are also more corticopontine fibers, primarily from the parietal lobe
- **Sublenticular part**—inferior to the lenticular nucleus; this part contains the rest of the optic radiation, meaning the fibers that end in the inferior part of the calcarine sulcus that carries data about the superior visual fields; there are connections between the temporal association areas and the pulvinar; also present is the *auditory radiation*, with fibers passing laterally from the MGN, under the lenticular nucleus, that turns superiorly and ends in the transverse temporal gyri

The transition between the retrolenticular and sublenticular parts is very gradual, usually with no clear division. Due to the curved nature of the internal capsule, all of its parts are not visible in any particular section.

Functions of the thalamic nuclei and their major connections

In the thalamus, the **thalamic nuclei** include *relay nuclei* that receive distinct input bundles. They project to discrete functional areas of the cerebral cortex. The role of the relay nuclei is to deliver information from certain functional systems to appropriate areas of the cerebral cortex. *Association nuclei* are mostly interconnected with the association cortex, but also with some subcortical structures. Additionally, there are nuclei with diffuse cortical projections and even one type that has no projections to the cortex. The *intralaminar* and *midline nuclei* appear to function as part of the basal nuclei and limb system but are not fully understood. Specific inputs come from many areas. They project to the cerebral cortex, but even more significantly to the basal nuclei and limbic system. Anatomical loops of many motor systems involve pathways between the cerebellum and cerebral cortex, and also between the basal nuclei and cerebral cortex. These usually involve the thalamic nuclei, which can be distinguished from each other by location in the thalamus, and by patterns of their inputs and outputs.

Projection neurons make up more than 75% of the neurons in the majority of thalamic nuclei. However, proportions of projection neurons and interneurons are different in various nuclei. Sources of regulatory inputs are similar between various thalamic nuclei. Most are from the cerebral cortex, primarily the cortical area projected toward by a thalamic nucleus. Some of these come from the thalamic reticular nucleus, with the rest including many cholinergic, noradrenergic, dopaminergic, and serotonergic endings from the brainstem's reticular formation. The anterior thalamic nucleus projects to the cingulate gyrus. This completes a large loop through the diencephalon and telencephalon. This loop (also called the *Papez circuit*) allows interactions between the neocortex, limbic structures, and hypothalamus. The hypothalamic nuclei functions are as follows:
- **Paraventricular nucleus**—gives rise to the **preganglionic sympathetic nerve**
- **Paraventricular/supraoptic lesion**—causes diabetes insipidus—involved in secretions of antidiuretic hormone (ADH), oxytocin, corticotropin-releasing hormone (CRH)

- **Preoptic (median/lateral) nucleus**—involved in GnRH secretion (median preoptic nucleus); the **lateral preoptic nucleus**—secretes GABA, and is involved in sleep onset
- **Arcuate nucleus**—secretes dopamine
- **Mammillary nucleus lesion**—causes *Wernicke encephalopathy*
- **Anterior nucleus**—involved in parasympathetic thermal regulation; a lesion of this nucleus causes hyperthermia
- **Posterior nucleus**—also involved in thermal regulation; a lesion in this nucleus causes hypothermia
- **Posterolateral**—secretes *orexin* (involved in *narcolepsy*), and has cholinergic and monoaminergic projections to the *lateral preoptic nucleus* and brainstem
- **Dorsomedial**—involved in behavioral actions; a lesion may lead to violent behavior
- **Ventromedial**—the satiety center; therefore, a lesion will lead to obesity
- **Lateral hypothalamic nucleus**—the hunger center; therefore, a lesion will lead to a loss of appetite and weight loss

Section review
1. What are the five regions of the internal capsule?
2. What is the role of the thalamic relay nuclei?
3. Which part of the hypothalamic nuclei causes Wernicke encephalopathy?

Functional considerations of the hypothalamus

The hypothalamus is the primary autonomic control center of the brain. It is involved in many functions, including regulation of temperature, visceral responses, and some functions of the limbic system. Neurosecretory neurons of the hypothalamus produce oxytocin and vasopressin. The hypothalamus gives rise to many fibers concerned with autonomic regulation. Many are involved in sympathetic control. These are located through the brainstem, near the spinothalamic tract, reaching the intermediolateral cell column of the spinal cord in a primarily uncrossed pattern. Dopamine in the hypothalamus participates in the control of prolactin secretion, and in various functions of the hypothalamus.

The suprachiasmatic nucleus in the hypothalamus is the primary control for most circadian rhythms. Direct input from the retinas to this nucleus provides information for regulation of these rhythms into the 24-h cycle. Therefore, the hypothalamus has significant effects regarding sleep patterns. Melatonin from the pineal gland increases as evening approaches, acting upon melatonin receptors on the **suprachiasmatic nucleus** of the hypothalamus. This results in an overall decrease in awakeness. Neurons in the anterior hypothalamus (preoptic area) send inhibitory projections to most or all parts of the wakefulness-promoting network. The hypothalamus also plays an important role in controlling REM sleep.

Another function of the hypothalamus concerns responses to temperatures. When the environment is cold, there are autonomic responses coordinated by the hypothalamus, which include cutaneous vasoconstriction and shivering. More complex functions include regulation of emotions, anger, and even sexual behavior. There are three primary categories of hypothalamic interconnections involved, which include:

- Interconnections with components of the limbic system
- Outputs influencing the pituitary gland
- Interconnections with motor and sensory, nuclei of the visceral and somatic systems in the brainstem and spinal cord

The hypothalamus also contains neurons that directly respond to physical stimuli. These cells may be sensitive to the temperature of the hypothalamus itself. Activity of other cells is sensitive to blood osmolality, glucose concentration, and certain hormones in the blood that pass through the hypothalamus.

The hypothalamus has projections to the hippocampus, septal nuclei, amygdala, brainstem, and spinal cord via the same fiber bundles carrying afferents to the hypothalamus. The hypothalamus controls both lobes of the pituitary gland, with a separate control system for each lobe. Hypothalamic connections with the pituitary, limbic system, and brainstem allow for control of many drive-related and emotional states. The hypothalamus is therefore implicated in drinking and feeding behavior, gut motility, and other actions that are mostly organized outside of the hypothalamus. Its functions are similar to a collection of upper motor neurons. Parts of most behaviors resulting from stimulation of a certain hypothalamic location can be brought about by stimulation of appropriate brainstem sites. The prominent parts of this hypothalamus-brainstem network include the solitary tract nucleus, the parabrachial nuclei, the periaqueductal gray, and the ventrolateral reticular formation of the rostral medulla. Each of these has a unique primary function, yet they are all interconnected.

The solitary tract nucleus is the main visceral sensory nucleus of the brainstem. It projects into visceral reflex arcs and to more rostral CNS levels. The rostral solitary nucleus is involved in taste sensations, via CN VII, IX, and X. The caudal solitary nucleus is involved in baroreceptor reflexes via CN IX and X. The parabrachial nuclei have a more generalized role in transmitting data related to "well-being." They send input to forebrain structures involved in emotions and drives—including the hypothalamus, amygdala, and thalamus. The periaqueductal gray is the origin of a descending pain-control pathway and also regulates complicated responses, mostly to threatening stimuli. It causes piloerection, tachycardia, aggressive behaviors, and sometimes, analgesia. The ventrolateral reticular formation regulates cardiovascular and respiratory functions, micturition, swallowing, defecation, and sexual function.

Regulation of the autonomic nervous system

The hypothalamus has integrative centers for autonomic activity. The neurons are similar to upper motor neurons of the somatic nervous system. The hypothalamus is the control center for many functions of the peripheral nervous system. Its connections with endocrine and nervous structures allow it to play an important role in maintaining homeostasis. The neural signals to the ANS come from the lateral hypothalamus, projecting to the lateral medulla.

The most important hypothalamic nucleus of the central autonomic network is the paraventricular nucleus. This structure has two classes of neurons but three functional categories. The classes include the magnocellular neurons, which are larger, and the parvocellular neurons, which are smaller. The magnocellular neurons contain vasopressin and oxytocin. They project their axons into the posterior pituitary gland, and the hormones are directly released into the bloodstream. The parvocellular neurons have a neuroendocrine-related functional subset, which projects to the median eminence, secreting releasing hormones into the hypophyseal portal bloodstream. This controls anterior pituitary hormone secretion. There is also a group of parvocellular neurons making up the third functional group, which is involved in central autonomic control. Unlike any other brain area, the paraventricular nucleus has direct influence over sympathetic and parasympathetic outflow. The parasympathetic input is from the solitary tract nucleus, as explained above. This makes the paraventricular nucleus the only site, in a closed efferent-afferent reflex loop, which is part of both the sympathetic and parasympathetic nervous systems.

Clinical considerations

Clinical considerations that involve the structures of the diencephalon are quite varied. They include hypothalamic syndromes, diabetes insipidus, pituitary insufficiency, hypothermia, hyperthermia, Cushing disease, and Cushing syndrome.

Hypothalamic syndromes

There are two unique types of hypothalamic syndromes. In the first type, many hypothalamic functions, or all of them, are disordered. There are often combinations of diseases in contiguous structures, referred to as *global hypothalamic syndromes*. The second type involves a selective loss of hypothalamic-hypophyseal function. This is due to a discrete lesion of the hypothalamus. It often causes a deficiency or overproduction of just one hormone, known as a *partial hypothalamic syndrome*.

Pathophysiology
Many lesions are able to invade or destroy most or all of the hypothalamus. These include sarcoid diseases, other granulomatous diseases, germ-cell tumors, other tumors, idiopathic inflammatory diseases, and some forms of bacterial encephalitis. In about 5% of sarcoidosis cases, the hypothalamus is implicated. This may be the primary disease manifestation but usually is combined with hilar lymphadenopathy and facial palsy. An magnetic resonance imaging (MRI) may reveal the lesion. Tumors of the **hypothalamopituitary axis** include lymphoma, metastatic carcinoma, **craniopharyngioma**,

and many germ-cell tumors. The germ-cell tumors include **terminomas**, embryonal carcinoma, choriocarcinoma, and teratomas. These develop in childhood, usually invade the posterior hypothalamus, and are sometimes accompanied by increased serum alpha-fetoprotein or increased beta subunits of chorionic gonadotropin. A hamartomas of the hypothalamus causes a unique syndrome of **gelastic epilepsy**. Of the inflammatory conditions, *infundibulitis* or *infundibuloneurohypophysitis* is a cryptogenic inflammation of the pituitary stalk and neurohypophysis. These areas become thicker, with infiltrates of T cells and other lymphocytes, and of plasma cells. This condition is believed to be autoimmune in nature. *Histiocytosis X* is a group of diseases including **Letterer–Siwe disease**, **Hand–Schüller–Christian disease**, and eosinophilic granuloma. Multiple organs are affected, including the hypothalamus, nearby structures, and the leptomeninges. This often causes cells, mostly histiocytes, to be present in the cerebrospinal fluid (CSF). In children, these conditions develop slowly. An obscure inflammatory condition known as **Erdheim–Chester disease** can also affect this region, sometimes along with proptosis. However, this is mostly a bone disease.

Clinical manifestations

The clinical manifestations of hypothalamic syndromes are unique for each type. Pituitary insufficiency and diabetes insipidus are discussed below. Tertiary hypothyroidism causes fatigue, cold intolerance, weight gain, depression, memory and cognitive problems, body aches, stiffness, inflammation, hoarseness, slowed speech, hearing problems, constipation, menstrual cycle changes, skin changes, hair and nail changes, bradycardia, shortness of breath during exercise, weakness, hypertension, and hypercholesterolemia. Related developmental disorders are signified by stunted growth and sexual development, precocious puberty, rapid weight gain, low thyroid hormone levels, and low sex hormone levels.

Diagnosis

Diagnosis of hypothalamic syndromes involves patient history, physical examination, and a large variety of tests. Blood and urine tests will assess levels of cortisol, estrogen, growth hormone, pituitary hormones, prolactin, testosterone, thyroid hormones, sodium, and the osmolality of the tested body fluids. Other possible tests include hormone injections followed by timed blood samples, brain MRI or computed tomography (CT), and if a tumor is present, a visual field eye examination.

Treatment

Treatment is based on the cause of the syndrome. For tumors, surgery or radiation may be required. For hormonal deficiencies, the missing hormones will need to be replaced. Many causes of these syndromes are treatable, and usually, hormone replacement is very successful. Without treatment, however, complications can be severe and life threatening.

Diabetes insipidus

Diabetes insipidus usually occurs when the pituitary gland's posterior lobe stops released enough ADH. There is impaired water conservation by the kidneys. Excessive amounts of water are lost via the urine. Consistent thirst occurs, yet the body does not retain fluids that are consumed.

Pathophysiology
This condition is associated with destructive lesions of the hypothalamus, resulting in a lack of vasopressin. The patient then develops polyuria, reduced blood volume, and polydipsia. In cases of acquired diabetes insipidus, the most established causes include brain tumors, head injury, infiltrative granulomatous diseases, and less commonly, intracranial surgical trauma. In young patients, causes are most often granulomatous infiltration of the base of the brain by sarcoid, eosinophilic granuloma, Hand-Schüller-Christian disease, or Letterer-Siwe disease. Of primary tumors, the most common are glioma, craniopharyngioma, hamartomas, choristoma, large chromophobe adenomas, and pinealoma. Metastatic tumors from the lung or breast or leukemic and lymphomatous infiltration may also be linked. Mild global hypothalamic dysfunction following brain irradiation for glioma may cause diabetes insipidus. Severe cases of hypothalamic destruction occur in brain death, with diabetes insipidus being a common component. Pituitary tumors are seldom related unless they become very large, invading the infundibulum and pituitary stalk.

Clinical manifestations
The clinical manifestations of diabetes insipidus include polydipsia, polyuria with the urine being very diluted, reduced blood volume, dehydration, unexplained weakness, lethargy, muscle pain, irritability, nocturia, bed-wetting, sleeping problems, fever, vomiting, diarrhea, delayed growth, and weight loss.

Diagnosis
The diagnosis of diabetes insipidus involves a water deprivation test to measure changes in body weight, urine output, urine concentration, and blood. The physician may also measure levels of ADH or administer synthetic ADH during this test. Other diagnostic methods include urinalysis and skull MRI. If an inherited form of this condition is suspected, the family history of polyuria should be considered. Genetic screening may be suggested.

Treatment
Mild cases of diabetes insipidus may not require treatment as long as fluid and electrolyte intake is balanced with urinary losses. However, severe cases can cause 10 L of fluid to be lost every day. This is extreme since normal urinary output is only 1–2 L per day. Without

treatment, dehydration and electrolyte imbalances will be fatal. This may be treated effectively with a synthetic form of ADH known as desmopressin.

Section review
1. Which part of the diencephalon is the primary controller of the autonomic nervous system?
2. What is the main visceral sensory nucleus of the brainstem?
3. What are clinical manifestations of hypothalamic syndromes?
4. How is diabetes insipidus diagnosed?

Pituitary insufficiency

Pituitary insufficiency is less-than-normal production of pituitary hormones. It is also known as *hypopituitarism* and involves the subnormal function of the pituitary gland in which one or more pituitary hormones are deficient. It may be temporary or permanent. *Panhypopituitarism* refers to complete loss of all pituitary function and is also referred to as the *pituitary failure*.

Pathophysiology

The genetic basis of familial hypopituitarism is linked to pituitary transcription factor in the anterior pituitary gland, from early fetal development and throughout life. Various mutations of the genes for the transcription factor result in an insufficient expression of these factors in the anterior pituitary gland. This causes combined pituitary hormone deficiency, which involves growth hormone, prolactin, and thyrotropin.

Clinical manifestations

In hypopituitarism, signs and symptoms include acromegaly; visual field defects and double vision; headache; pituitary apoplexy; **lymphocytic hypophysitis**; skin, hair, and nail changes; oligomenorrhea or amenorrhea; infertility, reduced libido, and loss of sexual function; osteoporosis; delayed puberty; hair loss; decreased muscle mass; anemia; central obesity and other weight changes; impaired attention and memory; growth retardation and short stature; tiredness and fatigue; failure to thrive; hypoglycemia; hyponatremia; collapse and shock; vomiting; hyperpigmentation of the skin; cold intolerance; constipation; slowed cognition; bradycardia; hypotension; cretinism; and problems with breastfeeding.

Diagnosis

Diagnosis of hypopituitarism requires an assessment of the previous head injury or radiation therapy. Blood testing is often diagnostically sufficient, but a *simulation test* may be required to determine if there is a hormone deficiency. There may need to be an evaluation of the levels of cortisol, growth hormone, thyroid hormones, thyroid-stimulating hormone, luteinizing hormone, follicle-stimulating hormone, testosterone, ADH, and sodium. For men, a semen analysis may be performed if there are concerns about fertility.

Treatment
Treatment focuses on the underlying cause. Hormonal replacement is usually the first method. If successful, it may be required for the life of the patient. Hormone replacement medications include corticosteroids, levothyroxine, sex hormones, growth hormone, LH, and FSH. If there is a tumor, surgery, and radiation therapy are indicated.

Hypothermia and hyperthermia

Hypothermia is body temperature that is below normal. The heat-gain center of the brain functions to prevent this condition from developing. When the preoptic area temperature drops excessively, the heat-loss center is inhibited, and the heat-gain center is activated. Heat is conserved by the sympathetic vasomotor center decreasing blood flow to the dermis. This reduces losses by convection, conduction, and radiation. The skin then cools.

Hyperthermia is an elevated body temperature because of abnormal thermoregulation. It occurs when the body produces or absorbs more heat than it is able to dissipate. Extreme temperature elevation is a medical emergency that requires immediate treatment in order to prevent disability or death.

Pathophysiology
Hypothermia may be primary, with the cold injury being the major pathology, or secondary because of another condition. It is classified as either mild, moderate, or severe. Many pathophysiological changes are caused by reduced enzyme actions.

The pathophysiology of hyperthermia begins in the preoptic region of the anterior hypothalamus. In response to an infection, some white blood cells release pyrogens directly affecting the anterior hypothalamus. Body temperature then rises as the rate of electrical discharge of warm-sensitive neurons increases progressively.

Clinical manifestations
As blood flow is restricted, the skin may appear pale or slightly bluish in color, in lighter-skinned people. There is no damage to epithelial cells since they can stand extended periods at temperatures as low as 77°F or as high as 120°F. Blood returning from the limbs is moved into a deep vein network. In warm conditions, blood flows through a superficial vein network. When it is cold, blood is moved to a vein network lying deep to an insulating subcutaneous fat layer. The venous network wraps around deep arteries. Heat is moved from the warm blood flowing out to the limbs, to cooler blood returning from

peripheral areas. This traps heat near the body core, restricting heat loss. This exchange between oppositely-moving fluids is called *countercurrent exchange*. In *shivering thermogenesis*, there is a gradual increase in muscle tone. This increases the energy consumption of skeletal muscle tissue in the body. The more energy used, the more heat produced. When the heat-gain center is highly active, muscle tone increases and stretch receptor stimulation produces short contractions of antagonizing muscles, resulting in shivering. This elevates oxygen and energy consumption even more. Shivering is able to raise the temperature by increasing heat generation up to 400%. *Non-shivering thermogenesis* is a slower heat-gain process, involving hormones released that increase all tissues' metabolic activities.

Early stages of hyperthermia are described as *heat exhaustion, heat prostration,* and *heat stress*. Symptoms include extreme sweating, rapid breathing, and a fast, weak pulse. If this progresses to *heatstroke*, the skin becomes hot and dry as the blood vessels dilate to try to increase heat loss. The inability to cool the body through perspiration may cause the skin to feel dry. Additional manifestations include dehydration, nausea, vomiting, headache, hypotension, fainting, dizziness, confusion, hostility, an intoxication-like state, tachycardia, tachypnea, pale or bluish skin color, seizures, organ failure, unconsciousness, and eventually, death.

Diagnosis
Severe hypothermia causes decreased cardiac output and respiratory rate. If the core temperature falls below 82°F, the cardiac arrest often occurs. Body temperature continues to decline, with skin and sometimes mucous membranes becoming pale or blue, and cold. The patient appears dead, but since metabolic activities have decreased through all systems, it is possible for the patient to be saved, even following several hours.

Hyperthermia is usually diagnosed via the body temperature and history supporting this condition instead of a fever. The lack of fever-related symptoms suggests hyperthermia. Medications that can raise body temperature are evaluated by patient history.

Treatment
For the hypothermic patient, emergency services should be contacted immediately. Cardiopulmonary support and gradual external and internal rewarming is required. Warm baths and blankets are used. A warm saline solution may be introduced into the peritoneal cavity. The key factor here is to restore body warmth *slowly*. When CPR is needed, it should be done while the warming procedure is occurring. If possible, the patient should be given warm fluids. The patient must be constantly monitored.

For hyperthermia, the underlying cause is removed. If caused by a medication, this must be stopped immediately. Other medications may be required to counteract the effects of the causative medication. Antipyretics are not effective for hyperthermia, only for fever. Mechanical cooling measures include removal of clothing, increased water consumption, and resting in a cool area. Active cooling methods include sponging of the head, neck, and trunk with cool water; removal of any heat-inducing factors; use of fans or air condition units; immersing the patient in iced, cool, or even tepid water. For severe cases, the patient must be treated in a medical facility, which can utilize intravenous hydration, gastric lavage with iced saline, and hemodialysis if needed.

Cushing disease and Cushing syndrome

Cushing disease is caused by elevated levels of cortisol, due to abnormal release of hormones from an anterior pituitary tumor, and is most common in adults between the ages of 20 and 50 years. However, children can also be affected. It is estimated to occur in 10–15 million people globally, and for unknown reasons, affects females more often than males. **Cushing syndrome** occurs when the body is exposed to high levels of cortisol for a long period of time and is also known as *hypercortisolism*.

Pathophysiology

Somatic mutations may be linked to Cushing disease. They are acquired during a person's lifetime and present only in certain cells. The involved genes often play a role in regulating hormone activities. Cortisol produced by the adrenal glands is triggered by the release of adrenocorticotropic hormone (ACTH) from the pituitary gland. Cushing disease occurs when an adenoma forms in the pituitary gland, causing excessive release of ACTH (via CRH), and then, elevated cortisol production. Cushing disease usually occurs alone, but rarely, as a symptom of genetic syndromes with pituitary adenomas as a feature, including *multiple endocrine neoplasia type 1 (MEN1)* or *familial isolated pituitary adenoma (FIPA)*.

Cushing disease is a subset of Cushing syndrome, which occurs due to chronically increased cortisol levels from adenomas in other locations than the pituitary gland. These include adrenal gland adenomas. Also, certain prescription drugs increase cortisol production and lead to Cushing syndrome.

Clinical manifestations

The first sign of Cushing disease is weight gain around the trunk and in the face. There may be stretch marks on the thighs and abdomen, and bruising occurs easily. Abnormal fat deposits may cause a *hump* on the upper back. Other symptoms include muscle weakness, severe fatigue, and progressive thinning of the bones, making them prone to osteoporosis. The immune system becomes weak, increasing the chance of infections. Mood disorders may include anxiety, depression, and irritability. Concentration and memory may be affected, and there is an increased chance for hypertension and diabetes mellitus. Women may have irregular menstruation and excessive hair growth (hirsuitism). Men may experience erectile dysfunction. Children usually have slowed growth patterns.

Cushing syndrome includes all of the signs and symptoms of Cushing disease, as well as progressive obesity, especially around the midsection and upper back, a moon-shape of the face, thinning of the skin, slow healing, acne, decreased libido and fertility, and loss of emotional control.

Diagnosis
Diagnosis of Cushing disease is by patient history, physical examination, and tests to measure levels of ACTH. *Inferior petrosal sinus sampling* may be performed to distinguish an ACTH-producing tumor on the pituitary from one in another area of the body. Additional testing includes dexamethasone suppression and CRH stimulation tests. An MRI may be used, with contrast, to reveal a causative tumor.

Diagnosis of Cushing syndrome is difficult based simply on signs and symptoms. Laboratory tests help in diagnosis, and also to determine if the syndrome developed as a long-term outcome of Cushing disease. Commonly, cortisol is measured from the saliva or urine, or the dexamethasone suppression test is used. Often, true Cushing syndrome is revealed after tests that exclude other conditions affecting cortisol production are performed.

Treatment
Treatment of Cushing disease is based on the cause. If a tumor is present, surgery may be required. Other treatments include radiation, chemotherapy, and hormone-inhibiting drugs. For Cushing syndrome, if the use of corticosteroids is related, these must be stopped by tapering or reduced. As in Cushing disease, surgery is performed to remove the causative tumor, and sometimes the entire affected gland. Radiation therapy may be needed. Medications include ketoconazole, mitotane, metyrapone, mifepristone, and pasireotide. After surgery, the patient may need cortisol replacement medications.

Section review
1. What is the pathophysiology of pituitary insufficiency?
2. What is heatstroke?
3. How is Cushing disease differentiated from Cushing syndrome?

Clinical cases

Clinical case 1
1. What other tests may be performed to rule out various differential diagnoses?
2. Which medication is commonly prescribed for central diabetes insipidus, usually with good clinical results?
3. What are the four types of diabetes insipidus and their causes?

A 45-year-old woman presented complaining of polydipsia, polyuria, nocturia, and weight loss over the previous several months. Testing revealed that she did not have diabetes mellitus. A water deprivation test was suggestive for diabetes insipidus. An MRI revealed infundibular hypophysitis and no hyperintense signal in the neurohypophysis. This patient did not have an autoimmune disease, infection, or any infiltrative disease. The diagnosis was central diabetes insipidus.

Answers:
1. Other tests may include chest x-ray, abdominal ultrasound, mammography, breast ultrasound, and thoracoabdominal CT scan.
2. An oral form of desmopressin is usually prescribed for diabetes insipidus. This medication is also used to treat bed-wetting, hemophilia A, von Willebrand's disease, and high blood urea levels.
3. Central DI is due to a lack of vasopressin, often from damage to the hypothalamus or pituitary gland, or from genetics. Nephrogenic DI occurs when the kidneys do not respond correctly to vasopressin. Dipsogenic DI is due to abnormal thirst mechanisms in the hypothalamus. Gestational DI only occurs during pregnancy.

Clinical case 2
1. What types of medications are likely going to be needed for this patient because of his hypopituitarism with CDI?
2. How common is central diabetes insipidus with post-traumatic hypopituitarism?
3. What is the overall prognosis for hypopituitarism?

A 9-year-old boy was injured in a car accident. A CT scan revealed a subdural hematoma, pneumocephalus, and several facial bone fractures. He received surgery for his trauma and was hospitalized in stable condition. His urine volume suddenly increased, 6 h following surgery and he had an extremely low urine-specific gravity. This form of diabetes insipidus was successfully treated. A year later, he developed a form of hypopituitarism, and water hydration tests also revealed central diabetes insipidus.

Answers:
1. This boy will need levothyroxine sodium hydrate and vasopressin.
2. Central diabetes insipidus occurs in one of every three-to-five patients with post-traumatic hypopituitarism.
3. Hypopituitarism is related to an increased risk of cardiovascular disease, and also an increased risk of death that is 50%–87% higher than in the normal population. Nearly all patients have a growth hormone deficiency. Hypopituitarism is usually permanent and requires lifelong medications.

Clinical case 3
1. If this patient's core body temperature was between 30°C and 34°C, what would the classification of her hypothermia be?
2. What are the other signs and symptoms of hypothermia?
3. What are other types of treatment for hypothermia?

A 70-year-old woman was discovered by her family on the floor of her bedroom, and taken to the emergency department. She was assessed has having hypothermia, and was very drowsy when her family tried to communicate with her. Active rewarming was started with warmed intravenous fluids. She was admitted to the intensive care unit.

Answers:
1. Her classification would have been "moderate hypothermia." The other classifications include "mild" (core body temperature of 34°C or higher), "severe" (less than 30°C), and "profound" (less than 20°C).
2. Other signs and symptoms of hypothermia include lethargy, confusion, pupil dilation, depressed respirations, heart rhythm changes, diuresis, volume depletion, hyperglycemia, increased plasma viscosity, coagulopathy, renal impairment, and electrolyte imbalances.
3. Other treatments include peritoneal lavage, extracorporeal rewarming, and esophageal warming tubes.

Clinical case 4
1. Based on this case study, what is the likely diagnosis?
2. What are the other signs and symptoms of this condition?
3. Would physical therapy be beneficial to this patient?

A 33-year-old woman with asthma and bursitis, obesity, hypertension, and type II diabetes mellitus was examined with complaints of muscle weakness and lower back pain. The muscle weakness has been worsening of the last month. She complains of becoming fatigued with only very little exertion and has had regular headaches at night. She has been taking inhaled corticosteroids for the past 7 years. She also reports irregular menstrual cycles in the past 2 years, with unexplained abdominal weight gain.

Answers:
1. The likely diagnosis is Cushing syndrome.
2. Other signs and symptoms include reddening of the cheeks, a moon-shaped face, fat pads on the upper back, thin skin, bruisability and ecchymoses, hypertension, red striations, thinning of the arms and legs, and poor wound healing.
3. Yes, Cushing syndrome patients have great benefit from physical therapy. They benefit from muscular strengthening and endurance programs to a high degree.

Key terms

Centrum semiovale	Letterer-Siwe disease
Circumventricular organs	Lymphocytic hypophysitis
Corpora arenacea	Magnocellular
Corticopontine fibers	Mammillary bodies
Corticosubthalamic fibers	Nervus conarii
Craniopharyngioma	Neurohypophysis
Cushing disease	Nigrosubthalamic fibers
Cushing syndrome	Pallidal division
Diabetes insipidus	Pallidosubthalamic fibers
Diencephalon	Parabrachial pontine reticular formation
Dorsomedial	Piloerection
Epithalamus	Pineal gland
Erdheim-Chester disease	Pituitary gland
Fasciculus lenticularis	Preganglionic sympathetic nerve
Fasciculus subthalamicus	Pregeniculate nucleus
Fasciculus thalamicus	Prerubral field
Fields of Forel	Pretectal region
Gelastic epilepsy	Solitariothalamic tract
Habenular nuclei	Subthalamic nucleus
Hand-Schüller-Christian disease	Subthalamonigral fibers
Hyperthermia	Subthalamopallidal fibers
Hypothalamic-releasing hormones	Subthalamus
Hypothalamopituitary axis	Suprachiasmatic nucleus
Hypothalamus	Tectohabenular fibers
Hypothermia	Tectotegmentospinal tracts
Indolamine	Terminomas
Infundibulum	Thalamic nuclei
Internal capsule	Thalamus
Interthalamic adhesion	Tuberoinfundibular
Lateral preoptic nucleus	Zona incerta

Suggested readings

1. Abla, O.; Janka, G. *Histiocytic Disorders*. Springer, 2017.
2. Ammari, R. *Role of the Subthalamic Nucleus in the Basal Ganglia Network*. Lap Lambert Academic Publishing, 2012.
3. Beck-Peccoz, P. *Syndromes of Hormone Resistance on the Hypothalamic-Pituitary-Thyroid Axis (Endocrine Updates Book 22)*. Springer, 2004.
4. Cardinali, D. P. *Autonomic Nervous System: Basic and Clinical Aspects*. Springer, 2017.
5. Colombo, J.; Arora, R. *Clinical Autonomic Dysfunction: Measurement, Indications, Therapies, and Outcomes*. Springer, 2014.

6. Dharani, K. *The Biology of Thought: A Neuronal Mechanism in the Generation of Thought—A New Molecular Model*. Academic Press, 2014.
7. Dudas, B. *The Human Hypothalamus: Anatomy, Functions and Disorders (Neuroscience Research Progress)*. Nova Science Publishers, Inc, 2013.
8. Galella, L. *Vasopressin: Mechanisms of Action, Physiology and Side Effects (Neuroscience Research Progress)*. Nova Science Publishing Inc, 2013.
9. Gattass, R.; Soares, J. G. M.; Lima, B. *The Pulvinar Thalamic Nucleus of Non-Human Primates and Functional Subdivisions (Advances in Anatomy, Embryology and Cell Biology)*. Springer, 2018.
10. Geer, E. B. *The Hypothalamic-Pituitary-Adrenal Axis in Health and Disease: Cushing's Syndrome and Beyond*. Springer, 2017.
11. Hewitt, J.; Gabata, M. *Cushing's Syndrome: Causes, Tests, and Treatments*. CreateSpace Independent Publishing Platform, 2011.
12. Holtzman, R. *Surgery of the Diencephalon (Contemporary Perspectives in Neurosurgery)*. Springer, 2012.
13. Laws, E. R.; Pace, L. *Cushing's Disease: An Often Misdiagnosed and Not So Rare Disorder*. Academic Press, 2016.
14. Leng, G. *The Heart of the Brain: The Hypothalamus and Its Hormones*. The MIT Press, 2018.
15. Melmed, S. *The Pituitary*, 4th ed.; Academic Press, 2016.
16. Reichlin, S. *The Neurohypophysis: Physiological and Clinical Aspects*. Springer, 2012.
17. Robertson, D.; Biaggioni, I.; Brunstock, G.; Low, P. A.; Paton, J. F. R. *Primer on the Autonomic Nervous System*, 3rd ed.; Academic Press, 2011.
18. Rojo, G. G. *Hyperthermia and Hypothermia in Medicine*. Ediciones Bohodon, 2016.
19. Sherman, S. M.; Guillery, R. W. *Functional Connections of the Cortical Areas: A New View from the Thalamus*. The MIT Press, 2013.
20. Song, J. L. *Thalamus: Anatomy, Functions and Disorders (Neuroscience Research Progress)*. Nova Biomedical, 2011.

CHAPTER 9

Brainstem

The brainstem is positioned between the cerebrum and spinal cord and consists of the midbrain, pons, and medulla. A pathway for fiber tracts runs between the higher and lower neural areas. Also, brainstem nuclei are associated with 10 of the cranial nerve pairs and they are greatly involved with innervation of the head. The cranial nerves and their functions are discussed in detail in Chapter 10. The brainstem is perhaps the most strategic structure of the brain. In neuroanatomy, identifying a lesion is the most important aspect of diagnosis (see Chapter 22). By the end of this chapter, you will be able to identify any neurological lesions within the brainstem by just following a simple method of localization. The whole point of learning neuroanatomy is to gain the ability to identify the exact location of the lesion based on the history and exam (signs and symptoms). Please remember that cranial nerves II, III, and IV originate from the midbrain. Cranial nerves V, VI, VII, and VIII originate from the pons. The rest of the cranial nerves (IX, X, XI, XII) originate from the medulla. Another important fact is that CN III, IV, VI, XII exit the brainstem at the middle or *medial* portion. There are different pathways and elongated nucleuses located within the brainstem. They are categorized into two groups: *lateral* and *medial*.

The rule of 4 of the brainstem

In 2005, Dr. Gates from Australia published an article titled "The rule of 4 of the brainstem: a simplified method for understanding brainstem anatomy and brainstem vascular syndromes for the non-neurologist." This is perhaps the best method to understand the anatomy of the brainstem and other various brainstem syndromes. These four rules are:
1. There are four structures in the "midline" beginning with the letter **M**.
2. There are four structures to the side (lateral) beginning with the letter **S**.
3. There are four cranial nerves in the medulla, four in the pons, and four above the pons (two in the midbrain) as explained above.
4. The four motor nuclei that are in the midline are those that divide equally into 12, *except for numbers I and II*. These include CN III, IV, VI, and XII. CN V, VII, IX, and XI are in the lateral brainstem or on the **S**ide.

The structures that start with the letter **M** are:
1. The **M**otor pathway (or corticospinal tract): contralateral weakness of the arm and leg.
2. The **M**edial lemniscus: contralateral loss of vibration and proprioception in the arm and leg.
3. The **M**edial longitudinal fasciculus (MLF): ipsilateral internuclear ophthalmoplegia (failure of adduction of the ipsilateral eye toward the nose and nystagmus in the opposite eye as it looks laterally).
4. The **M**otor nucleus and nerve: ipsilateral loss of the cranial nerve that is affected (III, IV, VI, or XII).

The structures that start with the letter **S** are:
1. The **S**pinocerebellar pathways: ipsilateral ataxia of the arm and leg.
2. The **S**pinothalamic pathway: contralateral alteration of pain and temperature affecting the arm, leg and rarely, the trunk.
3. The **S**ensory nucleus of CN V: ipsilateral alteration of pain and temperature on the face in the distribution of the fifth cranial nerve (this nucleus is a long vertical structure that extends in the lateral aspect of the pons down into the medulla).
4. The **S**ympathetic pathway: ipsilateral Horner's syndrome, which involves partial ptosis and a small pupil (miosis).

These pathways pass through the entire length of the brainstem and can be likened to 'meridians of longitude' whereas the various cranial nerves can be regarded as 'parallels of latitude'. If you establish where the meridians of longitude and parallels of latitude intersect, then you have established the site of the lesion.

Midbrain

The **midbrain**, at the top of the brainstem, has two bulges called **cerebral peduncles** on its ventral aspect. The brainstem is located below the *diencephalon*. The diencephalon was discussed in detail in Chapter 8. The cerebral peduncles form vertical pillar-like structures to appear to hold up the cerebrum. The term *cerebral peduncles* means *little feet of the cerebrum*. Each peduncle has a *crus cerebri* that contains a large corticospinal or *pyramidal* motor tract that descends to the spinal cord. Fiber tracts called the *superior cerebellar peduncles* connect the midbrain to the cerebellum dorsally. Some descending fibers carry voluntary motor commands from the cerebral hemispheres.

A hollow *cerebral aqueduct* runs through the midbrain, and through the entire brainstem. It connects the third and fourth ventricles. It defines the cerebral peduncles ventrally from the roof of the midbrain, which is known as the **tectum** (which means "roof" in Latin) and is located posterior to the cerebral aqueduct. Inside the tectum are two pairs of sensory nuclei described collectively as the **corpora quadrigemina**. Separately, they are known as the *superior colliculi* and *inferior colliculi*. The **superior colliculi** act as visual reflex centers, coordinating eye movements when visually following moving objects, even without conscious visualization. The **inferior colliculi** act as auditory relays from the hearing receptors to the sensory cortex. They also help react reflexively to sounds, such as in the *startle reflex*. One simple way to memorize this is to note that our eyes are located superior to our ears. Therefore, the superior colliculi are for eye movement, and the inferior colliculi are for the auditory pathway.

The periaqueductal gray matter surrounds the aqueduct and plays a part in pain suppression. This gray matter includes nuclei controlling the nuclei of the *oculomotor nerve* and *trochlear nerve* in the midbrain. The fibers controlling micturition also pass through the periaqueductal gray matter that surrounds the aqueduct.

A band-like **substantia nigra** is deep to the cerebral peduncles. It has a dark color due to a large amount of melanin pigment. The substantia nigra is linked functionally to the basal nuclei and is the largest midbrain nucleus. It inhibits the activity of the cerebral basal nuclei, which are involved in the subconscious control of learned movements and muscle tone. Degeneration of the dopamine-releasing neurons causes Parkinsonian features, due to the decrease in dopamine. The oval **red nucleus** is deep to the substantia nigra. Its color is due to a rich blood supply and iron pigment within its neurons. The red nuclei act as relays in certain descending motor pathways related to *limb flexion*. They are embedded in the *reticular formation*, nuclei throughout the brainstem's core. Both the red nucleus and the substantia nigra are contained within an area anterior to the cerebral aqueduct that is known as the **tegmentum**.

While the reticular formation extends through the brainstem, it is primarily within the midbrain, which also contains the *reticular activating system (RAS)*. This is a specialized system that, when stimulated, causes alertness and awakening. Damage to the RAS causes unconsciousness, resulting in not being awake or aware. Along the length of the brainstem, there are midline **raphe nuclei** and two lateral types of nuclei: the **medial group of nuclei** and the **small group of nuclei**. The medial group is also referred to as the *large cell group*, while the small group is also referred to as the *small cell group*. The unique aspect of the reticular neurons is that their axonal connections are so long. Individual reticular neurons reach the hypothalamus, thalamus, cerebral cortex, cerebellum, and spinal cord. Therefore, they are well-suited for regulating brain arousal.

The RAS is inhibited by hypothalamic sleep centers and those of other areas. It is depressed by sleep-inducing drugs, tranquilizers, and alcohol. An irreversible coma can occur if this system is severely injured, such as due to trauma. Some RAS nuclei are also involved in sleeping patterns. There is also a motor arm to the reticular formation. Coma is a state of not being awake or aware of surroundings. Some nuclei project to motor neurons within the spinal cord via the reticulospinal tracts. They help to control skeletal muscles' coarse limb movements. Other nuclei are autonomic centers regulating visceral motor functions. These include the vasomotor, cardiac, and respiratory centers of the medulla. Remember that the oculomotor nerve (CN III) exits the midbrain ventrally in the middle. Cranial nerves II and IV emerge from the lateral midbrain. Cranial nerve IV is the only cranial nerve that exits the brainstem dorsally.

Section review
1. What is the difference between the functions of the superior colliculi and the inferior colliculi?
2. What is the explanation of the color of the substantia nigra?
3. What is the RAS?

Pons

The **pons** is the bulging area of the brainstem that is separated from the cerebellum dorsally by the fourth ventricle. It consists mostly of conduction tracts in two directions, linking the cerebellum with the midbrain, diencephalon, cerebrum, and spinal cord. The deep projection fibers of the pons run longitudinally, between the higher brain centers and the spinal cord. Its more superficial ventral fibers are transverse as well as dorsal. They form the *middle cerebellar peduncles*, connecting the pons bilaterally with the two sides of the cerebellum, dorsally. The fibers come from many *pontine nuclei*, which relay information between the cerebellum and the motor cortex. The *trigeminal*, *abducens*, and *facial* nerve pairs issue from the pontine nuclei. Other nuclei are part of the reticular formation, while some aid the medulla oblongata in normal breathing rhythm control. The four groups of components within the pons are as follows:

- **Sensory and motor nuclei of the cranial nerves** V, VI, VII, and VIII—the jaw muscles, anterior facial surface, the lateral rectus muscle of the eye, and vestibular and cochlear nuclei of the internal ear are innervated by these cranial nerves
- **Nuclei that aid in controlling respiration**—the *apneustic center* and the *pneumotaxic center* are the pontine centers that process information from the respiratory rhythmicity centers of the medulla oblongata; the apneustic center is in the middle and lower pons, while the pneumotaxic center is in the rostral pons
- **Nuclei and tracts that handle information to or from the cerebellum**—this follows the linkage between the pons and the cerebellum, brainstem, cerebrum, and spinal cord
- **Ascending, descending, and transverse pontine fibers**—other portions of the central nervous system (CNS) are interconnected via longitudinal tracts; the middle cerebellar peduncles of the cerebellum are connected to the **transverse pontine fibers**; these cross the anterior pons surface, and are axons linking the pontine nuclei with the cerebellum of the opposing side

Table 9.1 describes the primary components and functions of the pons.

Table 9.1 Primary components and functions of the pons.

Region or nucleus	Functions
Gray matter	
Apneustic and pneumotaxic centers	Adjust activities of respiratory rhythmicity centers within medulla oblongata
Reticular formation	Automatically processes incoming sensations, outgoing motor commands
Nuclei of cranial nerves V, VI, VII, and VIII (partially)	Relay sensory information; issue somatic motor commands
Other nuclei or relay centers	Relay sensory and motor information to cerebellum
White matter	
Descending tracts	Carry motor commands from higher centers to cranial or spinal nerve nuclei
Ascending tracts	Carry sensory information from nuclei of brainstem to the thalamus
Transverse pontine fibers	Interconnect the cerebellar hemispheres

Medulla oblongata

The **medulla oblongata**, also simply called the *medulla*, is the most inferior part of the brainstem. It is a cone-shaped area that blends with the spinal cord near the foramen magnum of the skull. When viewed in a sectional view, the inferior part of the medulla oblongata looks very much like that of the spinal cord, with the more complex organization of the gray and white matter. As the central canal of the cord continues up into the medulla it becomes wider, forming the fourth ventricle's cavity. The medulla and pons both help form the ventral wall of the fourth ventricle. At this point, the canal looks almost identical to that of the spinal cord.

There are three groups of nuclei in the medulla. The first group includes nuclei and processing centers controlling visceral functions. The coordination of complex autonomic reflexes occurs in the medulla. The second group contains sensory and motor nuclei of the CNS. The third group contains relay stations, which conduct communications between the brain and spinal cord, via ascending or descending tracts through the medulla. These groups are further explained in the following section.

Reflex centers

The *reflex centers* handle autonomic and reflex activity. Nuclei are embedded within the tightly mingled mass of gray and white matter known as the **reticular formation**, as mentioned earlier. This extends through the medulla's central core, into the hypothalamus. The reticular formation is not only responsible for arousal. It also aids in autonomic functions, including blood pressure, breathing, and thermoregulation. It also assists with endocrine functions, body posture, alertness, sleep, and skeletomuscular reflex activity. The two major groups of reflex centers within the medulla oblongata are cardiovascular and respiratory, as follows:

- **Cardiovascular centers** (baroreceptor reflex)—regulate heart rate, cardiac contraction strength, and blood flow through the peripheral tissues. Functionally, these centers are subdivided into *cardiac centers* and *vasomotor centers*. However, it is difficult to identify their actual anatomical boundaries. In this reflex, blood flow is regulated by monitoring blood pressure and heart rate through the receptors that are located within the carotid arteries and the aortic arch. Cranial nerve IX carries the information from the carotid arteries while CN X (the vagus nerve) carries the information from the aortic arch. These signals then synapse in the medulla's sympathetic-parasympathetic center, resulting in an increase or decrease in the heart rate and blood pressure, in order to maintain normal blood flow.
- **Respiratory rhythmicity centers**—regulate the basic pace of respiratory movements. They are controlled by input from the respiratory centers located in the pons. Please note that both the medulla and midbrain are also involved in the respiration process. Injury to the midbrain may result in hyperventilation. Injury to the bilateral pons may cause cessation of breath during inspiration, while injury to the medulla can cause disorganized or *ataxic breathing patterns*.

Sensory and motor nuclei of the cranial nerves

Inside the medulla, there are sensory and motor nuclei related to cranial nerves VIII, IX, X, XI, and XII. These nerves send motor commands to the muscles of the pharynx, neck, and back. They also send motor commands to the visceral organs in the thoracic and peritoneal cavities. Cranial nerve VIII carries sensory information, from receptors located in the internal ear to the vestibular and cochlear nuclei. These nuclei extend from the pons, into the medulla. The vestibular nuclei are involved with balance, while the cochlear nuclei are involved with hearing.

Relay stations

Somatic sensory information is relayed to the thalamus by the **gracile nucleus** and the **cuneate nucleus**. The tracts that exit these brainstem nuclei cross to the opposite side of the brain prior to reaching their destinations. This crossing over is described as a *decussation*. The site where it occurs is called the **decussation of pyramids**. There are also some important paired nuclei inside the medulla. The **solitary nuclei** are also called the *nucleus solitaries* (Fig. 9.1). These are the visceral sensory nuclei. They receive information from the cranial and spinal nerves. The information is processed and then forwarded to other autonomic centers, located in the medulla and in other areas. We have two solitarius nucleuses; caudal and rostral. The rostral nucleus solitarius is the center for taste sensation (via CN VII, IX, and X). The caudal nucleus solitarius is responsible for the baroreceptor reflex.

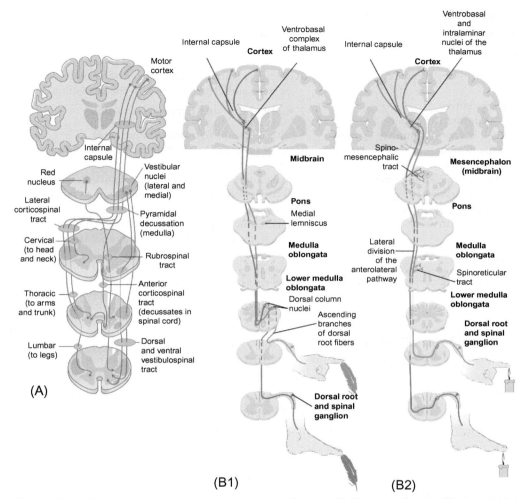

Fig. 9.1 Examples of somatic motor and sensory pathways. (A) Motor pathways. The pyramidal pathway through the lateral corticospinal tract and the extrapyramidal pathways through the rubrospinal, reticulospinal, and vestibulospinal tracts. (B) Sensory pathways. (B1) The dorsal column-medial lemniscal pathway for transmitting critical types of tactile signals: touch/proprioception. Note the later corticospinal tract decussation is in the lower medulla. The corticobulbar tract is not shown. (B2) Anterior and lateral divisions of the anterolateral sensory pathway: pain/temperature. Note the decussation is in the spinal cord. *(A: From Compston, A. et al. McAlpines's Multiple Sclerosis, 4th ed.; Churchill Livingstone: London, 2006; B: From Hall, J. E. Guyton and Hall Textbook of Medical Physiology, 13th ed.; Saunders: Philadelphia, PA, 2016.)*

Table 9.2 Primary components and functions of the medulla oblongata.

Region or nucleus	Functions
Gray matter	
Inferior olivary complex	Relays data from the red nucleus, other midbrain centers, and cerebral cortex to the cerebellum
Reflex centers: Cardiovascular centers	Regulate heart rate, force of contraction
Respiratory rhythmicity centers	Set basic pace of respiratory movements
Gracile nucleus Cuneate nucleus	Both of these nuclei relay somatic information to the thalamus
Additional nuclei or centers: Cranial nerves VIII (partially), IX, X, XI (partially), and XII	Sensory, motor nuclei of the five cranial nerves; nuclei relay ascending data from spinal cord to higher centers
Reticular formation	Contains nuclei and centers regulating vital autonomic functions; extends into pons and midbrain
White matter	
Ascending and descending tracts in the funiculi	These tracts link the brain and spinal cord

The **inferior olivary complex** is made up of three nuclei collectively forming the *inferior olivary nucleus*. These nuclei relay information to the cerebellar cortex. This information is about somatic motor commands when they are issued by the motor centers at higher levels. Most of the olivary nuclei create the **olives**. These are prominent bulges, shaped like actual olives, located along the ventrolateral surface of the medulla.

Table 9.2 summarizes the primary components of the medulla oblongata, along with their functions.

Section review
1. What are the four groups of components within the pons?
2. What are the three groups of nuclei within the medulla oblongata?
3. What senses are the vestibular and cochlear nuclei involved in?
4. What term describes a "crossing over of tracts"?

Focus on the brainstem
Damage to the brainstem can be fatal, since the patient may be unable to breathe, swallow, or perform other basic motor functions. As mentioned earlier in this chapter, most of the unconscious physiologic body functions such as breathing, heartbeat, and digestion are

> **Focus on the brainstem—cont'd**
>
> dependent on a normal brainstem. The brainstem structures continually function, at all hours of the day, without any need for input from the remainder of the brain. Therefore, people who experience brain damage can still have nearly normal spontaneous body functions (normal breathing, normal heart rate, and normal blood pressure) as long as the medulla and other parts of the brainstem are not harmed. The patient may be considered brain dead, from which there is no possibility of recovery when the brainstem is irreversibly damaged. Opiates and alcohol can cause dysfunction in different parts of the brainstem. In cases of overdose, it is possible to die if the brainstem cannot function normally due to the effects of drugs or alcohol. Opiates and alcohol both suppress the respiratory center in the medulla, which can result in breathing cessation and death.

Corticobulbar tracts

The **corticobulbar tracts** are one of the three descending tracts of the **corticospinal pathway**, which is also called the *pyramidal system*. The corticobulbar tracts, along with the *lateral corticospinal tracts* and *anterior corticospinal tracts* enter the white matter of the internal capsule. Then, they descend into the brainstem, emerging on either side of the midbrain as the *cerebral peduncles*.

The axons of the corticobulbar tracts synapse on lower motor neurons, of the motor nuclei of cranial nerves III, IV, V, VI, VII, IX, XI, and XII. They provide conscious control over the skeletal muscles that move the eyes, face, and jaw, as well as certain muscles of the neck and pharynx. These tracts also provide innervation to the motor centers of the medial and lateral pathways.

Corticospinal tracts

In the corticospinal tracts, axons synapse on lower motor neurons within the spinal cord's anterior horns. Along their descent, these tracts can be visualized along the anterior surface of the medulla oblongata. They appear as a pair of thick bands called **pyramids**. Along the pyramids' length, about 85% of the axons decussate and enter the descending **lateral corticospinal tracts** on the opposing side of the spinal cord. The remaining 15% continue without crossing, along the spinal cord as the **anterior corticospinal tracts**. At the targeted segment of the spine, one axon in the anterior corticospinal tract decussates within the anterior white commissure. It synapses on the lower motor neurons in the anterior horns.

Functional considerations

The brainstem links the cerebrum, spinal cord, and cerebellum. Nerve tracts traveling through the brainstem relay signals from the cerebellum to parts of the cerebral cortex that are involved in motor control. This allows coordination of fine motor movements needed for many different activities. Functionally, the brainstem controls several important body functions. These include alertness, arousal, breathing, blood pressure regulation, digestion, heart rate, other autonomic functions, and relay of information between peripheral nerves and spinal cord to the upper parts of the brain. Therefore, when the brainstem is injured, there can be difficulties with mobility and coordination of movements. Walking, writing, eating, and many other activities may become very difficult. A brainstem stroke can destroy tissue required for the direction of respiration, heart rhythm, and swallowing. It can also affect hearing, speech, limb movement, and normal sensations throughout the body.

Section review
1. What are the three tracts of the corticospinal pathway?
2. In the corticospinal tracts, where do the axons synapse?
3. Which body functions are regulated by the brainstem?

Clinical considerations

There are a variety of clinical considerations concerning the brainstem. The most common include Arnold-Chiari phenomenon, pontine hemorrhage, infarctions of the pons, trauma to the midbrain, blockage of the cerebral aqueduct, Weber syndrome, and Benedikt syndrome.

Arnold-Chiari phenomenon

Arnold-Chiari phenomenon is also called *Arnold-Chiari malformation* and is usually only seen in children born with *spina bifida (type II)*, the incomplete development of the spinal cord and often, its protective covering. Type I is the most common, and mostly **asymptomatic;** while type II is often **symptomatic** and associated with obstructive hydrocephalus (due to compression of the fourth ventricle), meningomyelocele (spinal cord and meninges herniated through a vertebral defect), and syringomyelia. In type II Arnold-Chiari malformation (see Fig. 9.2), the cerebral tonsils extend into the foramen magnum for more than 5 mm. Other conditions sometimes associated with these malformations include tethered cord syndrome and spinal curvatures. Most children born with type II malformations have obstructive hydrocephalus, due to the compression of the fourth ventricle. There are a total of four types of this malformation, with types III and IV not being compatible with life.

Pathophysiology
Arnold-Chiari phenomenon usually occurs along with **myelomeningocele.** There can be many complications, including interference with the cerebrospinal flow, resulting in

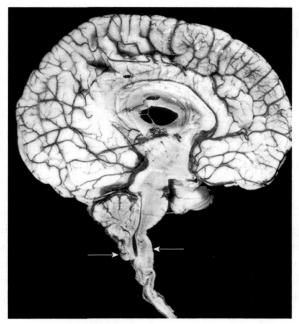

Fig. 9.2 Arnold-Chiari malformation. Midsagittal section showing small posterior fossa contents, downward displacement of the cerebellar vermis, and deformity of the medulla (*arrows* indicate the approximate level of the foramen magnum).

hydrocephalus. There is a genetic link to the development of this condition, though ongoing research on the pathophysiology continues.

Clinical manifestations

In the Arnold-Chiari phenomenon, there are several abnormalities of the base of the skull and cranium, as well as displacement of the brainstem and cerebellum through the foramen magnum. This *type II* malformation differs from *type I* malformation in which the posterior fossa does not grow enough to accommodate the cerebellum. This forces the cerebellum through the foramen magnum. The portions extending outside the skull are called the **cerebellar tonsils**.

When the cerebellar tonsils are pushed through the foramen magnum, there may be compression of the lower brainstem and cranial nerve nuclei. This includes cranial nerve XII, the hypoglossal nerve, which innervates the tongue. Therefore, compression affects swallowing and speech. If the trigeminal nerve is compressed, there may be facial pain or numbness. When the medulla is compressed, problems with breathing can occur. When the pons is compressed, there can be problems with sleep, and sleep apnea may result. The affected individual may show changes in breathing, and weakness of the limbs. As mentioned in the beginning of this chapter, the brainstem is filled with strategically important structures that are in very close proximity to each other, within a confined space.

Therefore, most lesions in the brainstem result in more than just one sign or symptom, and in most cases present as a *syndrome* of signs and symptoms.

Diagnosis

Most babies born with Arnold-Chiari phenomenon are easily diagnosed due to the appearance of the malformation. Symptoms caused by this phenomenon are usually severe. A detailed history is required about the start of symptoms and their severity. Physical examination is followed by imaging studies, including magnetic resonance imaging (MRI) of the brain and spinal cord, to visualize the displaced brain structures. Once confirmed, repeated MRIs may be required to assess the status of the condition.

Treatment

Since the Arnold-Chiari phenomenon patient often has coexisting conditions, there may be a variety of surgeries required (only in symptomatic patients), which affect recovery time. They are often performed shortly after birth and are quite varied based on the severity of the malformation. Surgery is usually directed at relieving pressure within the brain and spinal cord. The most common type of surgery is *posterior fossa decompression*, in which a small portion of the bone at the back of the skull is removed to make room for the brain to expand, relieving the pressure. There may also be a need to remove part of the spinal column to relieve spinal cord pressure. There are many risks to this form of surgery, including risks of infection, fluid accumulation in the brain, leakage of cerebrospinal fluid (CSF), and delayed wound healing. Nearly all symptoms are usually relieved by surgery, but if there is spinal nerve injury, the surgery may be unable to reverse previous damage.

Section review
1. In Arnold-Chiari phenomenon, which structures extend into the foramen magnum?
2. What is the difference between type I and type II Chiari malformations?
3. What is the most common type of surgery for Arnold-Chiari phenomenon?

Focus on Arnold-Chiari phenomenon

Individuals with type II malformations (Arnold-Chiari phenomenon) have symptoms that are usually more severe than those with type I, and usually appear during childhood. Life-threatening complications can occur during infancy or early childhood, and treatment requires surgery. The term "Arnold-Chiari" refers to the two pioneering researchers who identified the condition. Ongoing research is aimed at genetic factors that increase the risks of the malformation, brain signaling mechanisms that regulate the formation of the brain structures, and more focused surgical methods to provide the optimal level of correction of the malformation.

Pontine hemorrhage

A **pontine hemorrhage** is a form of intracranial hemorrhage, occurring within the pons (see Fig. 9.3). It is most often caused by chronic, poorly controlled hypertension, and has a very poor prognosis. Pontine hemorrhage due to hypertension is the third most common type of hypertensive bleeding, after the basal ganglia (putamen and caudate) and thalamus. Another common location for hypertensive bleeding is the cerebellum.

Pathophysiology

Like the penetrating arteries into the basal ganglia, the penetrating arteries from the basilar artery that extend into the pons may experience **lipohyalinosis** because of poorly controlled hypertension. The vessel walls become more likely to rupture. The vessels usually involved are the larger paramedian perforating arteries. Also, pontine hemorrhage can be secondary to underlying lesions. These may include the following:
- **Vascular malformations**—which may be due to cavernous malformation or simply **cavernoma**.
- **Tumors**—which may be neuroepithelial (primary) brain tumors, or metastases
- **Downward herniation**—resulting in **Duret hemorrhage**
- **Lesions from supratentorial surgery**—causing remote hemorrhage

Clinical manifestations

Pontine hemorrhage usually causes sudden, severe neurological deficits. Based on the speed of enlargement of the hematoma and its exact location, the clinical manifestations may include decreased level of consciousness, long tract signs such as **tetraparesis**, cranial nerve palsies, seizures, and **Cheyne-Stokes respirations**, along with pinpoint pupils.

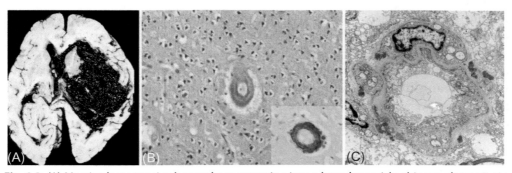

Fig. 9.3 (A) Massive hypertensive hemorrhage rupturing into a lateral ventricle; this can also occur in the pons. (B) Amyloid deposition in a cortical arteriole in cerebral amyloid angiopathy; *inset*, immunohistochemical staining for amyloid-beta shows the deposited material in the vessel wall. (C) Electron micrograph shows granular osmophilic material in a case of CADASIL (cerebral autosomal dominant arteriopathy with subcortical infarcts and leukoencephalopathy).

Diagnosis

A computed tomography (CT) scan of the brain is the first, and often the only imaging study needed for diagnosis of pontine hemorrhage. For an acute intraparenchymal hemorrhage, abnormalities are usually located centrally in the pons. The hematoma most often extends in a **rostrocaudal** direction, along the traversing long tracts, instead of laterally into the middle cerebellar peduncles. Usually, it does not extend past the **pontomedullary junction** inferiorly and the inferior midbrain superiorly. Often these hematomas rupture into the fourth ventricle. For patients with small-volume hemorrhage who may have an underlying lesion, an MRI may be helpful to identify vascular malformations. In general, intraventricular hemorrhage is an indication of poor prognosis.

Differential diagnosis is basically between this type of hemorrhage and those due to underlying lesions. Since the patient usually presents suddenly with severe impairment, the correct diagnosis is not difficult. It is worth considering unruptured or asymptomatic vascular malformations, as well as hemorrhagic metastasis. Vascular malformations may include the following:

- **Cavernous malformations**—these may have small hemorrhages that cause repeated symptoms
- **Arteriovenous malformation**—there is usually little to no mass effect, and serpentine irregular density is **isodense** to intravascular blood in another area
- **Developmental venous anomaly**—this is linear with no mass effect, and density is isodense to intravascular blood in another area

Hemorrhagic metastases may be linked to melanoma, renal cell carcinoma, thyroid carcinoma, and other cancers.

Treatment

Pontine hemorrhages are usually very dangerous and most large hemorrhages are fatal. Open surgical evacuation of the clot is not usually performed, though stereotactic clot aspiration has been performed successfully. For smaller hemorrhages, the patient's life can be saved by proper management and treatment of hydrocephalus using **extraventricular drains** though there are often significant neurological deficits caused by these procedures. Overall, mortality is between 30% and 90%. The overall volume of the hemorrhage and initial **Glasgow Coma Scale** score are related to the outcome.

Section review
1. What are the secondary causes of pontine hemorrhage?
2. What is the usual diagnostic method for pontine hemorrhage?
3. What is the overall prognosis for pontine hemorrhage?

Infarctions of the pons

Infarctions of the pons are usually caused by thrombosis or embolism of the basilar artery or its branches. When the paramedian section of the pons is involved, there may be damage to the corticospinal tracts, pontine nuclei, and the fibers that pass through the middle cerebellar peduncle to the cerebellum. The trigeminal nerve will be affected if the infarct is laterally situated, as will the **medial lemniscus** and middle cerebellar peduncle. It is also possible for the corticospinal fibers to the lower limbs to be affected. For different infarction syndromes please see the following section.

Trauma to the midbrain

Injury to midbrain with trauma can occur in many ways. For example, the sudden lateral head movement can cause the cerebral peduncle to forcefully strike the sharp, rigid free edge of the tentorium cerebelli. Sudden head movements that are due to trauma cause various brain regions to move at varying velocities, such as when the forebrain moves at a different velocity to the cerebellum. This type of movement can cause the midbrain to be twisted, torn, bent, or stretched.

When the oculomotor nucleus is involved, there will be resulting ipsilateral paralysis of the levator palpebrae superiors; the inferior oblique muscle; and the superior, medial, and inferior rectus muscles. The parasympathetic fibers of the oculomotor nerve are located at the periphery of the nerve. Therefore, any pressure over this nerve causes a dilated pupil that will be insensitive to light, and will not constrict upon accommodation. The eye with CN III palsy is angled down and out, and the pupil is dilated.

If the trochlear nucleus is involved, it will cause contralateral paralysis of the eyeball's superior oblique muscle. Therefore, when one or both of these nuclei are involved, or the **corticonuclear** fibers converging on them are involved, ocular movements will be impaired.

> **Focus on vertebrobasilar stroke**
>
> The vertebrobasilar arterial system perfuses blood into the medulla, cerebellum, pons, midbrain, thalamus, and occipital cortex. Occlusion of this system's larger vessels usually leads to major disability or even death. Vertebrobasilar stroke carries a high mortality and morbidity rate. Most survivors have multisystem dysfunction. This may include quadriplegia, hemiplegia, ataxia, dysphagia, dysarthria, gaze abnormalities, and cranial neuropathies, depending on the location of the stroke. Preceding stroke, there may be *vertebrobasilar insufficiency*. This basically means a reversible decrease in the blood flow of the vertebrobasilar arterial system. Atherosclerosis can narrow the large vessels of the posterior circulation, similar to all other large vessels. Risk factors for this condition include smoking, hypertension, diabetes mellitus, obesity, age above 50 years, family history, and hyperlipidemia. People with atherosclerosis or peripheral artery

Continued

> **Focus on vertebrobasilar stroke—cont'd**
>
> disease have an increased risk of developing vertebrobasilar insufficiency. Common symptoms include vision loss, double vision, dizziness or vertigo, numbness or tingling in the hands or feet, nausea or vomiting, slurred speech (dysarthria), changes in mental status (via the RAS), sudden and severe weakness, loss of balance and coordination, difficulty swallowing, and others.

Blockage of the cerebral aqueduct

The blockage of the cerebral aqueduct, or **aqueductal stenosis**, is a common cause of *obstructive hydrocephalus*. The normal cerebrospinal fluid flow requires the aqueduct to be open. When blocked, this is called *stenosis*, leading to symptoms of hydrocephalus.

Pathophysiology

The pathophysiology of aqueductal stenosis is diverse. It may be congenital, due to an infection or hemorrhage, idiopathically acquired, or due to a tumor. Some patients are born with a congenitally narrow or totally obstructed cerebral aqueduct. When the blockage is total, there is usually pediatric hydrocephalus. If the blockage is incomplete, there may be no initial symptoms or symptoms that may develop later in life. The obstruction may be a general narrowing or can involve small webs or rings of tissue across the aqueduct. Infections of the CSF or hemorrhage into the brain ventricles due to other causes can sometimes lead to scarring. This may result in aqueductal stenosis. Idiopathic cases may develop in adulthood, with new onset or gradual onset of hydrocephalus. Often, aqueductal stenosis is of unknown cause.

The blockage of the cerebral aqueduct may be from a brain tumor in the midbrain. This usually occurs from a tumor of the pineal region, including *pineoblastoma*, *pineocytoma*, or *pineal germinoma*. Since the pineal region is just behind the midbrain at the level of the cerebral aqueduct, the expansion of these tumors can externally compress and obstruct the aqueduct. Similarly, tumors of the midbrain itself, such as a *glioma*, can also compress the aqueduct.

Clinical manifestations

The clinical manifestations of aqueductal stenosis are varied, based on the cause and type of obstruction. Generally, symptoms resemble those of hydrocephalus, including headache, nausea, vomiting, and reduced consciousness, leading to coma and death if severe and untreated. The patient may experience visual symptoms such as **dysconjugate gaze**, or reduced visual acuity if hydrocephalus is chronic. There may be other neurological symptoms if a tumor is present, since other parts of the nervous system may be affected.

Diagnosis

The patient presenting with hydrocephalus-like symptoms must be given a complete neurological assessment, including imaging studies such as CT or MRI. A CT scan can show hydrocephalus but is less accurate in revealing the cause of the condition in some cases. Aqueductal stenosis usually causes tri-ventricular hydrocephalus, which is enlargement of the lateral and third ventricles, while the fourth ventricle appears normal. This pattern is usually diagnostic since it is linked to some obstruction at the level of the aqueduct. An MRI can clearly reveal a tumor as well as ventricle enlargement. The new MRI sequences, such as *constructive interference in steady state (CISS)* and *fast imaging employing steady-state acquisition (FIESTA)* are extremely sensitive in showing subtle abnormalities within the base of the skull and to assess structures around the cerebral aqueduct.

Treatment

Treatment of aqueductal stenosis is extremely varied between patients. Generally, treatments used for obstructive hydrocephalus are appropriate. This usually means diverting the flow of CSF with a shunt procedure or **endoscopic third ventriculostomy (EVD)**. However, for a patient with an obstructive brain tumor, other surgical or medical treatments may be needed. Sometimes, surgical removal of the mass will relieve the pressure and reverse the hydrocephalus. For others, the hydrocephalus may persist after treating the tumor, and either shunting or ventriculostomy may also be required. Treatment plans cannot be recommended without careful evaluation of the individual factors.

Section review
1. How can trauma to the midbrain occur?
2. What are the causes of aqueductal stenosis?
3. Treatments for which other conditions are usually appropriate for aqueductal stenosis?

Focus on hydrocephalus via the aqueductal stenosis

The cerebral aqueduct (of Sylvius) is a narrow channel connecting the third and fourth brain ventricles. Normally, cerebrospinal fluid flows through the aqueduct with no obstruction, but when a blockage occurs, stenosis can lead to symptoms of hydrocephalus. According to the Hydrocephalus Association, the condition affects about one million Americans in every stage of life. Of nearly 40,000 hydrocephalus operations performed every year, only 30% are the patient's first surgery to treat the condition. Medical costs for hydrocephalus are over $2 billion per year. One recent study estimates that 700,000 older Americans are living with *normal pressure hydrocephalus*, an often undiagnosed and untreated condition. Hydrocephalus also goes undiagnosed and untreated in younger adults, resulting in substantial workforce loss and health-care costs.

Weber syndrome

Weber syndrome is a midbrain stroke syndrome that primarily affects adults. It involves the cerebral peduncle and ipsilateral fascicles of the oculomotor nerve (cranial nerve III). This syndrome was first described by Sir Hermann Weber, and English physician originally born in Germany, in the year 1863. Weber syndrome is also referred to as *superior alternating hemiplegia*. It is important to note that Weber syndrome is not the same as *Sturge-Weber syndrome*, a rare congenital neurological and skin disorder that is present from birth.

Pathophysiology
Weber syndrome is usually caused by an ischemic stroke in the midbrain, which commonly involves branches of the posterior cerebral artery. Weber syndrome, like Benedikt syndrome, results from an insult to the oculomotor nuclear complex. Additional causes of Weber syndrome include brain tumors, traumatic injuries, or infections.

Clinical manifestations
The patient with Weber syndrome usually presents with ipsilateral CN III palsy, and contralateral hemiplegia or hemiparesis. Clinical manifestations differ based on which midbrain structures are affected. Weber syndrome mainly involves medial parts of the midbrain. Therefore, structures labeled with "M" will be affected by this syndrome. Damage to the corticospinal **m**otor fibers causes weakness on the other side of the body (contralateral arm or leg weakness). The oculomotor nerve fibers are also midline. Therefore, there is an ipsilateral oculomotor nerve palsy, with a drooping eyelid, a fixed and wide pupil that is pointed down and out, causing diplopia (double vision). Symptoms usually continue to get worse following the stroke, due to swelling in and around the brainstem. Other midline structures in the midbrain are not injured in Weber syndrome.

Diagnosis
Diagnosis is based on a complete neurological examination and complete patient history. MRI or CT can confirm the diagnosis.

Treatment
Over the first few weeks after a stroke-causing Weber syndrome, swelling in the brain diminishes, and there is some improvement of symptoms. Physical therapy is very helpful in healing the brain injury and optimizes the patient's ability to move the affected muscles. Ideally, the patient should be admitted to a multidisciplinary stroke unit for treatment. For persistent neurological symptoms, brain imaging is required immediately. Neurosurgery is indicated in the case of intracranial hemorrhage. However, if the hemorrhage is excluded, thrombolytic therapy can be initiated if the patient arrives within the first 4 h and half after the onset of the symptoms. Supportive care includes assessment of

swallowing and treatment of infections if occur. For paralysis, passive full range-of-motion exercises should be started in the first few days. Referral to physiotherapists or rehabilitation facilities is usually needed.

Benedikt syndrome

Benedikt syndrome is also known as *paramedian midbrain syndrome*. It is a midbrain stroke syndrome involving the fascicles of the oculomotor nerve (CN III) and the red nucleus. This syndrome was first described by the Hungarian-Austrian neurologist Moritz Benedikt in 1889. It is a rare condition affecting less than 1% of all people.

Pathophysiology
Benedikt syndrome is usually caused by an ischemic stroke that most often involves branches of the posterior cerebral artery. A range of neurological symptoms may affect the midbrain, cerebellum, and other structures. The corticospinal tracts, **brachium conjunctivum**, and superior cerebellar peduncle decussation may be affected by Benedikt syndrome. Causative lesions may include infarction, hemorrhage, or tumor in the tegmentum and cerebellum. The median zone is usually impaired. There may be occlusion of the posterior cerebral artery or the paramedian penetrating branches of the basilar artery.

Clinical manifestations
The patient with Benedikt syndrome usually presents with ipsilateral CN III palsy, crossed **hemiataxia**, and crossed choreoathetosis. Benedikt syndrome primarily affects either left or right eye movement. The pupils dilate and do not respond to any stimuli. The tendons are hyperactive. Ptosis and diplopia develop. Often, the patient cannot move an arm and shoulder while walking.

Diagnosis
Via radiographic imaging alone, it is difficult to distinguish Benedikt syndrome from Weber syndrome, unless clear involvement of the red nucleus is identified. If the diagnosis reveals a hemorrhage in the midbrain, thrombolytic therapy will most likely be needed.

Treatment
Treatment is based on the etiology of the stroke. Some of the symptoms of Benedikt syndrome may be relieved by deep brain stimulation. This is especially true for the related tremors. If the patient has had an acute ischemic stroke, tissue plasmin is activated with alteplase. Surgery can be performed if medications are ineffective, and nutritional support is required. Adequate exercise is helpful because of its effects in increasing blood circulation.

Section review
1. What is the most common cause of Weber syndrome?
2. What lesions may cause Benedikt syndrome?
3. How are the eyes affected by Benedikt syndrome?

The rule of four of the brainstem in summary
Remember that lesions in the brainstem often cause "crossed signs and symptoms," which basically means that the patient develops some symptoms in one side of the body and different symptoms on the other side. This type of presentation only happens when the lesion is located within the brainstem. The reason behind it is very simple. Some fibers in the midbrain don't cross, while other fibers cross to the other side.

Clinical cases

Clinical case 1
1. What is the likely diagnosis for the patient in this case study?
2. Is it possible for this diagnosis to be performed before a child is born?
3. What is the likely outcome for a patient of this age, with this diagnosis?

A 9-month-old girl had a variety of symptoms that worried her parents, which included headache, dizziness, nystagmus, difficulty swallowing, and impaired coordination. When she was born, she had myelomeningocele and hydrocephalus. An MRI revealed abnormality of her medulla, cerebellum, and fourth ventricle.

Answers:
1. The likely diagnosis is an Arnold-Chiari malformation. The hydrocephalus is the obstructive type.
2. Yes, the diagnosis of an Arnold-Chiari malformation can be made prenatally by using ultrasound. This is the only type also known as "type II" or "classic" Chiari malformation.
3. A better prognosis is seen with this condition with early recognition of symptoms and treatment. Today, better outcomes are seen because of the recognition of shunt malfunction prior to the development of ventricular dilatation.

Clinical case 2
1. Where is the lesion located and what is the name of this syndrome?
2. Which artery or arteries perfuse the lateral part of the medulla?
3. Explain the signs and symptoms based on the "rule of 4" method?

A 38-year-old man presented to the emergency room with 8 h' history of unsteadiness. Further history and exam revealed severe balance problems and ataxia on the right side of the body. His right eyelid was droopy. His voice was very hoarse, and he also had vertigo. There was a loss of pain and temperature on the left side of his body.

Answers:
1. This presentation is consistent with a stroke in the right lateral medulla, which is also known as Wallenberg's syndrome. In order to localize a lesion using the rule of 4, first identify whether the lesion is medial or lateral, then use the cranial nerves (except for the trigeminal nerve) to identify where the lesion is along the brainstem (midbrain, pons, or medulla). For example, in this case, having a hoarse voice localizes the lesion to the medulla.
2. This syndrome is usually associated with a posterior inferior cerebellar artery or vertebral artery occlusion.
3. Signs and symptoms can be easily remembered based on the rule of 4; since the lateral (side) medulla is involved, you would expect the spinothalamic pathway, sympathetic pathway, sensory nucleus of the trigeminal nerve, and spinocerebellar pathway to be injured. Therefore, the patient will have left side loss of body pain and temperature, right side Horner's syndrome, impaired sensation on the right side of the face, and right-sided ataxia respectively. Other symptoms are vertigo and nystagmus due to the involvement of CN VIII, right side paralysis of the palate and vocal cords, decreased gag reflex and dysphagia. These are all related to the corresponding CN nucleus, which is located on the lateral side of the medulla.

Clinical case 3
1. What is the likely treatment for this patient's condition?
2. How common are pineal cysts in relation to this condition?
3. What are the other possible causes of cerebral aqueduct blockage?

A 47-year-old woman visited her doctor complaining of a severe occipital headache throughout the last month. Imaging studies revealed a lobulated pineal cyst, which was obstructing the upper margin of the cerebral aqueduct. There were moderate hydrocephalus and mild **transependymal** signal abnormalities.

Answers:
1. The likely treatment is endoscopic resection of the mass.
2. Pineal cysts are very common. They are usually asymptomatic but can cause obstructive hydrocephalus, usually due to compression of the tectum. In this patient's case, the cyst has prolapsed into the cerebral aqueduct.
3. Additional causes include a structurally narrow aqueduct, forking of the aqueduct (splitting into multiple, separate channels), the formation of a septum of glial cells across the aqueduct, gliosis, and other medical conditions such as Brickers-Adams-Edwards syndrome and bacterial meningitis.

Clinical case 4

1. Based on this patient's symptoms, where did the stroke occur?
2. In this condition, which artery is usually the location of the stroke?
3. What are the risk factors for this syndrome?

A 54-year-old man with diabetes, hypertension, and a history of smoking had been diagnosed with Weber syndrome after experiencing a midbrain stroke that affected his right eye, and also his left arm and leg. He had blurred vision in the right eye, and the eyelid appeared to be drooping over the eye. His walking movements were abnormal, with extreme difficulty controlling his left arm and leg.

Answers:

1. The stroke occurred on the right side of the midbrain because this affects the same-side eye but the opposite-side limbs.
2. The posterior cerebral artery or one of its branches is usually the location of the strokes seen in Weber syndrome.
3. Risk factors for Weber syndrome include a history of smoking, diabetes, hypertension, hypercholesterolemia, and a family history of stroke.

Clinical case 5

1. What is the other name for this patient's condition?
2. Besides stroke, what are the other causes of Benedikt syndrome?
3. Are there any treatments that may relieve some of the symptoms?

A 63-year-old woman was hospitalized after symptoms of a stroke. Imaging revealed a lesion within the tegmentum of the midbrain. The patient had ipsilateral third nerve palsy and contralateral **hemitremor**, double vision, and ataxia. She had previously been diagnosed with hypercholesterolemia, atherosclerosis, and had a positive family history for stroke. She was diagnosed with Benedikt syndrome.

Answers:

1. Benedikt syndrome is also called "paramedian midbrain syndrome." It is a rare type of posterior circulation stroke, with neurological symptoms affecting the midbrain, cerebellum, and other structures.
2. Benedikt syndrome may also be caused by infarction, hemorrhage, tumors, or tuberculosis in the tegmentum of the midbrain and cerebellum. The median zone is specifically impaired, often resulting from occlusion of the posterior cerebral artery or the paramedian penetrating branches of the basilar artery.
3. Deep brain stimulation may provide relief of the tremors seen in Benedikt syndrome, but this is still experimental.

Clinical case 6

1. Based on this patient's symptoms, where did the stroke occur?
2. In this condition, which artery is usually the location of the stroke?
3. What are the risk factors for this syndrome?

An 87-year-old male presents to the emergency department with sudden onset of weakness on the right side, to the point that he cannot move his right arm or his right leg. On exam, there is a loss of vibration and proprioception on the right side as well. When the patient is asked to stick his tongue out, it deviates to the left.

Answers:

1. This presentation is consistent with a stroke in the left *medial* medulla. In order to localize a lesion using the rule of 4, first identify whether the lesion is medial or lateral, then use the cranial nerves (except for the trigeminal nerve) to identify where the lesion is along the brainstem. In this case, tongue deviation localizes the lesion to the medulla. The tongue is innervated by CN XII; therefore, this can also localize the lesion to the medial area. Another localizing point is the contralateral loss of position and vibration, which also points to a medial lesion due to MLF involvement.
2. This syndrome is usually associated with the anterior inferior cerebellar artery.
3. Signs and symptoms can be easily remembered based on the rule of 4; since the medial medulla is affected, you would expect all the pathways that start with "M" to be involved.

Key terms

Anterior corticospinal tracts
Aqueductal stenosis
Arnold-Chiari phenomenon
Benedikt syndrome
Brachium conjunctivum
Cavernoma
Cerebellar tonsils
Cerebral peduncles
Cheyne-Stokes respirations
Corpora quadrigemina
Corticobulbar tracts
Corticonuclear
Corticospinal pathway
Cuneate nucleus
Decussation of pyramids
Duret hemorrhage
Dysconjugate gaze
Endoscopic third ventriculostomy
Extraventricular drains
Glasgow Coma Scale
Gracile nucleus
Hemiataxia
Hemitremor
Inferior colliculi
Inferior olivary complex
Isodense
Lateral corticospinal tracts
Lipohyalinosis
Medial group of nuclei
Medial lemniscus
Medulla oblongata
Midbrain
Myelomeningocele

Olives	Solitary nuclei
Pons	Substantia nigra
Pontine hemorrhage	Superior colliculi
Pontomedullary junction	Tectum
Pyramids	Tegmentum
Raphe nuclei	Tetraparesis
Red nucleus	Transependymal
Reticular formation	Transverse pontine fibers
Rostrocaudal	Weber syndrome
Small group of nuclei	

Suggested readings

1. Aqueductal Stenosis. http://www.nervous-system-diseases.com/aqueductal-stenosis.html.
2. Ammar, A. *Hydrocephalus: What Do We Know? And What Do We Still Not Know?* Springer, 2017.
3. Arnold-Chiari Malformation. https://neckandback.com/conditions/chiari-malformation-type-arnold-chiari-syndrome/.
4. Arnold-Chiari Phenomenon. https://www.webmd.com/brain/chiari-malformation-symptoms-types-treatment#2.
5. Benedikt Syndrome. www.explainmedicine.com/article/neurology/benedikt-syndrome.
6. Benedikt Syndrome. https://infogalactic.com/info/Benedikt_syndrome.
7. Benedikt Syndrome. https://radiopaedia.org/articles/benedikt-syndrome.
8. Binder, D. K.; Sonne, D. C.; Fischbein, N. J. *Cranial Nerves: Anatomy, Pathology, Imaging*. Thieme, 2010.
9. Carmichale, S. W.; Stoddard, S. L. *Revival: The Adrenal Medulla 1986–1988*. CRC Press, 2019.
10. Cromarty, M.; Siddiqui, S. *Stroke—It Couldn't Happen to Me: One Woman's Story of Surviving a Brain-Stem Stroke*. CRC Press, 2018.
11. Czervionke, L. F.; Fenton, D. S. *Imaging Painful Spinal Disorders*. Saunders, 2011.
12. Di Giovanni, G.; Di Matteo, V.; Esposito, E. *Birth, Life and Death of Dopaminergic Neurons in the Substantia Nigra*. SpringerWien, 2009.
13. Hewitt, J.; Gabata, M. *Chiari Malformation: Causes, Tests, and Treatments*. CreateSpace Independent Publishing Platform, 2011.
14. Howard, J.; Singh, A. *Neurology Image-Based Clinical Review*. Demos Medical, 2016.
15. Icon Group International. *Midbrain: Webster's Timeline History, 1870–2007*. Icon Group International, Inc, 2010.
16. Icon Health Publications. *Arnold-Chiari Malformation—A Medical Dictionary, Bibliography, and Annotated Research Guide to Internet References*. Icon Health Publications, 2004.
17. Joseph, R. *Brainstem and Cerebellum: Medulla, Pons, Midbrain, Reticular Formation, Arousal, Vision, Hearing, Norepinephrine, Serotonin, Dopamine, Sleeping, Dreaming, REM, Cranial Nerves, Motor Control*. University Press, 2011.
18. Kobayashi, T.; Lunsford, L. D. *Pineal Region Tumors: Diagnosis and Treatment Options*. S. Karger, 2009.
19. Lal Gautam, P.; Paul, G.; Yadav, S. *Comparison of FOUR Score & Glasgow Coma Scale—Predicting Outcome in Neurosurgical Patients*. LAP Lambert Academic Publishing, 2016.
20. Mallucci, C.; Sgouros, S. *Cerebrospinal Fluid Disorders*. CRC Press, 2009.
21. Mohr, J. P.; Wolf, P. A.; Grotta, J. C.; Moskowitz, M. A.; Mayberg, M. R.; von Kummer, R. *Stroke: Pathophysiology, Diagnosis, and Management*, 5th ed.; Saunders, 2011.
22. Naidich, T. P.; Castillo, M.; Cha, S.; Smirniotopoulos, J. G. *Imaging of the Brain: Expert Radiology Series*. Saunders, 2012.
23. Pineal Cyst Obstructing Aqueduct. https://radiopaedia.org/cases/pineal-cyst-obstructing-aqueduct.
24. Pontine Hemorrhage. https://radiopaedia.org/articles/pontine-haemorrhage.
25. Pontine Hemorrhage. https://www.wjgnet.com/2218-6212/full/v3/i3/83.htm.

26. Rigamonti, D. *Adult Hydrocephalus*. Cambridge University Press, 2014.
27. Spetzler, R. F.; Kalani, M. Y. S.; Nakaji, P.; Yagmurlu, K. *Color Atlas of Brainstem Surgery*. Thieme, 2017.
28. Steiger, H. J.; Etminan, N.; Hanggi, D. *Microsurgical Brain Aneurysms: Illustrated Concepts and Cases*. Springer, 2015.
29. Stein, J.; Silver, J. K.; Pegg Frates, E. *Life After Stroke: The Guide to Recovering Your Health and Preventing Another Stroke*. Johns Hopkins University Press, 2006.
30. Tubbs, R. S.; Shoja, M. M.; Oakes, W. J. *Changes in the Cerebellum, Pons, and Medulla Oblongata due to Congenital Hydrocephalus of the Cerebrum*. Rhazes, LLC, 2016.
31. Tubbs, R. S.; Oakes, W. J. *The Chiari Malformations*. Springer, 2013.
32. Urban, P. P.; Caplan, L. R. *Brainstem Disorders*. Springer, 2011.
33. Valvavanis, A.; Schubiger, O.; Naidich, T. P. *Clinical Imaging of the Cerebello-Pontine Angle*. Springer, 2011.
34. Weber Syndrome. https://radiopaedia.org/articles/weber-syndrome.
35. Weber Syndrome. https://www.verywellhealth.com/webers-syndrome-3146475.
36. Wilson-Pauwels, L.; Steward, P. A.; Akesson, E. J.; Spacey, S. D. *Cranial Nerves: Function and Dysfunction*, 3rd ed.; PMPH USA, 2010.

CHAPTER 10

Cranial nerves

The **cranial nerves** are components of the peripheral nervous system (PNS). They are directly connected to the brain, in 12 pairs. The cranial nerves are named in relation to their function or distribution. Each nerve is numbered by its position along the brain's longitudinal axis, starting at the cerebrum. Each cranial nerve is abbreviated as "CN" and followed by its appropriate Roman numeral. For example, "CN I" refers to cranial nerve I, which is the olfactory nerve. When the full name of each cranial nerve is used, then just its Roman numeral is required, such as *olfactory nerve (I)*.

Every cranial nerve is attached to the brain adjacent to its associated nuclei, whether sensory, motor, or both. For the sensory nuclei, postsynaptic neurons relay information to other nuclei, or to centers in the cerebral cortex or cerebellar cortex, for processing. Motor nuclei information is similarly processed, as they receive input from higher brain centers, or from various nuclei found along the brainstem.

The cranial nerves are classified as follows:
- **Mostly sensory**—carrying somatic sensory information (pressure, touch, temperature, vibration, pain) or special sensory, such as sight, hearing, balance, and smell
- **Motor**—dominated by axons of somatic motor neurons
- **Mixed (sensory and motor)**—a mixture of sensory and motor fibers

Cranial nerves have not only these primary functions but also secondary functions.

Cranial nerves III, VII, IX, and X also contribute to the parasympathetic autonomic system. The 12 pairs of cranial nerves connect to the under portion of the brain, primarily on the brainstem (see Fig. 10.1). They pass through small holes called *foramina* in the cranial cavity within the skull. This allows them to extend between the brain and their peripheral connections. It is important to note that the sensory fibers of the cranial nerves are termed *proprioceptive*.

The classifications of the 12 pairs of cranial nerves are listed in Table 10.1.

Olfactory nerves (I)

The **olfactory nerves (I)** are special sensory nerves for the sense of smell. They originate in the receptors of the olfactory epithelium and pass through the olfactory foramina in the cribriform plate of the ethmoid bone, ending at the **olfactory bulbs**.

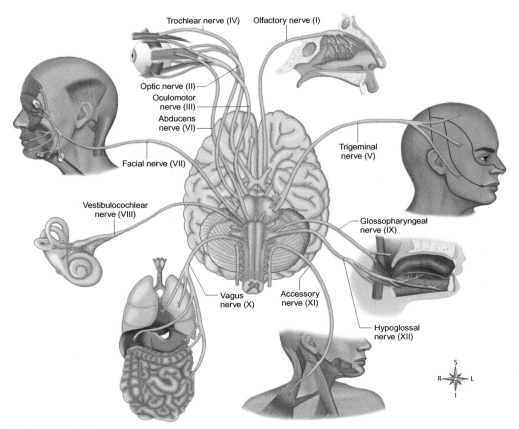

Fig. 10.1 Cranial nerves.

Specialized neurons called *olfactory receptors* are located in the epithelia that cover the roof of the nasal cavity, superior nasal conchae, and superior nasal septum (see Fig. 10.2). Axons combine forming 20 or more "bundles" that penetrate the cribriform plate. The bundles are parts of the olfactory nerves, which enter the olfactory bulbs after only a short distance. These bulbs are masses of neurons on both sides of the crista galli. *Olfactory afferents* synapse inside the olfactory bulbs. Postsynaptic neuron axons continue to the cerebrum along thin **olfactory tracts**. The olfactory nerves are the only cranial nerves that are directly attached to the cerebrum without stopping at the thalamus. The other cranial nerves synapse at the thalamus. The olfactory system will be discussed in detail in Chapter 15.

Table 10.1 Classifications of the cranial nerves.

Name and number	Function
Olfactory (I)	Sensory (special sensory: smell)
Optic (II)	Sensory (special sensory: vision)
Oculomotor (III)	Motor (eye movements)
Trochlear (IV)	Motor (eye movements)
Trigeminal (V)	Mixed (sensory and motor, to face)
Abducens (VI)	Motor (eye movements)
Facial (VII)	Mixed (sensory and motor, to face)
Vestibulocochlear (VIII)	Sensory (special sensory: balance and equilibrium via vestibular nerve, and hearing via cochlear nerve)
Glossopharyngeal (IX)	Mixed (sensory and motor, to head and neck)
Vagus (X)	Mixed (sensory and motor, widely distributed in thorax, abdomen)
Accessory (XI)	Motor (to muscles of neck and upper back)
Hypoglossal (XII)	Motor (tongue movements)

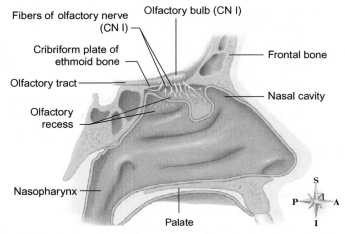

Fig. 10.2 Olfaction. Midsagittal section of the nasal area shows the locations of major olfactory sensory structures. *(From Patton, K.T.; Thibodeau, G.A.; Douglas, M.M. Essentials of Anatomy & Physiology, Mosby: St Louis, 2012.)*

Optic nerves (II)

The primary function of the **optic nerves (II)** is the special sense of vision. They originate in the retinas of the eyes, passing through the *superior optic fissure*, and end in the diencephalon, via the **optic chiasm**. Visual information is carried from special sensory ganglia within the eyes. There are about one million sensory nerve fibers in the optic nerves, which eventually converge at the ventral, anterior area of the diencephalon. At the optic chiasm, the fibers from each retina's medial half cross over to the opposite side of the brain.

The crossed-over axons then continue to the lateral geniculate body of the thalamus, where they are termed the **optic tracts**. Once they synapse in the lateral geniculate body, their projection fibers bring information to the visual cortex, in the occipital lobes. The visual pathway is discussed in detail in Chapter 14.

Oculomotor nerves (III)

The primary function of the **oculomotor nerves (III)** is motor, controlling eye movements. They originate in the midbrain, passing through the superior orbital fissures of the sphenoid bone. These nerves end in two different areas. The somatic motor components end in the superior, inferior, and medial rectus muscles; the inferior oblique; and the levator palpebrae superioris. The visceral motor components end in the intrinsic eye muscles.

The motor nuclei controlling cranial nerves III and IV are within the midbrain. Each oculomotor nerve innervates four of the six extrinsic muscles to move the eye, as well as the levator palpebrae superioris, which raises the upper eyelid. On either side of the brain, CN III emerges from the ventral midbrain surface to penetrate the posterior orbit wall, at the superior orbital fissures. When the oculomotor nerves are damaged, there is often pain above the eye, dropping of the eyelids, and double vision. This is because the movements of each eye cannot be properly coordinated.

The neurons of the **ciliary ganglion** receive preganglionic autonomic fibers of the oculomotor nerves. The neurons control the intrinsic eye muscles, which change pupil diameter, to adjust the amount of light that enters the eye. These muscles also change the lens' shape, focusing images on the retina.

> **Focus on third cranial nerve disorders**
> Third cranial (oculomotor) nerve disorders can impair ocular motility, pupillary function, or both. Signs and symptoms include ptosis, **diplopia**, and paresis of eye adduction, as well as of upward and downward gaze. When the pupil is affected, it will be dilated, with impaired light reflexes. When the pupil is affected or the patient is increasingly unresponsive, a CT scan must be performed as soon as possible.

Trochlear nerves (IV)

The primary function of the **trochlear nerves (IV)** is also motor, controlling eye movements. These nerves originate in the midbrain, passing through the superior orbital fissures of the sphenoid bone, to reach the superior oblique muscles. The trochlear nerves are the smallest of the cranial nerves. The *trochlea* is shaped like a *pulley* and is a sling of ligamentous tissue. Each superior oblique muscle passes through a trochlea, to its insertion on the eye surface. If CN IV is damaged, or its nucleus is damaged, there will be difficult when trying to look down or to the side.

> **Focus on cranial nerve disorders**
> Cranial nerve disorders involve dysfunction of smell, sight, chewing, facial sensation or expression, tasting, hearing, balance, swallowing, speaking, shoulder elevation, and tongue movements.

Trigeminal nerves (V)

The trigeminal nerves (V), are mixed nerves, with both sensory and motor activities. The trigeminal nerves (see Fig. 10.3) are divided into the *ophthalmic, maxillary,* and *mandibular* divisions, as follows:

- **Ophthalmic nerve (V_1)**—sensory, from the orbital structures, cornea, nasal cavity, forehead skin, supper eyelid, eyebrow, and part of the nose; passing through the superior orbital fissure
- **Maxillary nerve (V_2)**—sensory, from the lower eyelid, upper lip, gums, teeth, cheek, nose, palate, and part of the pharynx; passing through the foramen rotundum
- **Mandibular nerve (V_3)**—mixed; sensory from the lower gums, teeth, lips, palate, and part of the tongue; motor to the motor nuclei of the pons; passing through the foramen ovale. This is the only division of the trigeminal nerve that is not within the cavernous sinus

The ophthalmic, maxillary, and mandibular nerves reach the sensory nuclei of the pons. The mandibular nerve also innervates the muscles of mastication (chewing). The sensory (posterior) and motor (anterior) roots of the trigeminal nerves originate on the lateral pons surface. The sensory branch is larger, with the **semilunar ganglion** containing cell bodies of sensory neurons. The term *trigeminal* signifies the three subdivisions of these nerves. Nerve divisions are related to the *ciliary, sphenopalatine, submandibular,* and *otic* ganglia. These autonomic (parasympathetic) ganglia innervate facial structures. The trigeminal nerves do not contain any visceral motor fibers. The trigeminal nerves are discussed in greater detail in Chapter 11. The trigeminal nerves innervate the meninges and therefore are involved in the pathogenesis of many primary headache syndromes also known as **trigeminal autonomic cephalgia** (*TAC*).

> **Focus on trigeminal neuralgia**
> Tic douloureux or trigeminal neuralgia is severe, paroxysmal, facial pain caused by a disorder of the fifth cranial nerve. It can be idiopathic or affect adults with multiple sclerosis. It can also occur after trauma to the face. The pain is described as similar to an electrical shock, lasting for seconds, and often triggered by movement. There are several treatment options available including invasive and noninvasive surgical procedures. Carbamazepine is the most effective medication to treat this disorder. However, other antiepileptic medications are commonly used.

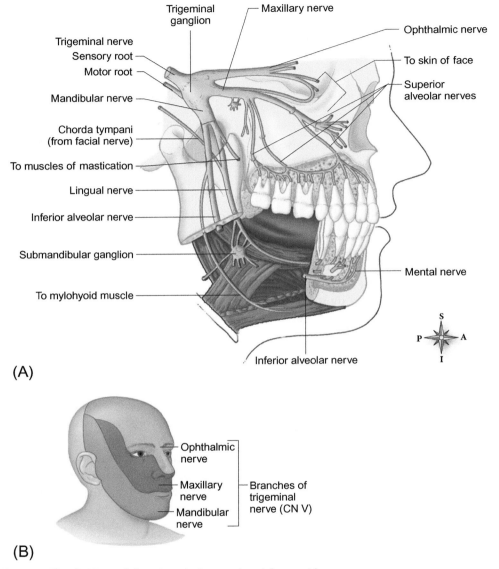

Fig. 10.3 The divisions of the trigeminal nerve (cranial nerve V).

Abducens nerves (VI)

The primary function of the **abducens nerves (VI)** is also motor, controlling eye movements. They originate in the pons, passing through the superior orbital fissures of the sphenoid bone, to reach the lateral rectus muscles. These muscles are the sixth pair of extrinsic eye muscles. When they contract, this causes the eyes to look to the side.

Basically, the abducens nerves cause *abduction* of the eyes. Each of them emerges from the inferior brainstem surface, at the border of the pons and medulla oblongata.

Movement of the eyes via the abducens nerves

The external eye muscles are of two types: extrinsic and intrinsic. The *extrinsic eye muscles* are skeletal muscles. They attach to the bones of the eye orbits and to the outside portion of the eyeballs. These voluntary muscles can move the eyeballs in any direction. There are four straight muscles and two oblique muscles attached to each eye. Their names describe their positions: *superior, inferior, medial,* and *lateral rectus muscles*; and the *superior* and *inferior oblique muscles* (Fig. 10.4). The extrinsic eye muscles or simply "*extraocular muscles*" have the following functions:

1. Medial rectus (via CN III): Adduction
2. Lateral rectus (via CN VI): Abduction
3. Superior rectus (via CN III): Elevation—Intorsion: inward rotation of the eye toward the nose
4. Inferior rectus (via CN III): Depression—Extorsion: outward rotation of the eye towards the ear
5. Superior oblique (via CN IV): Depression—Intorsion
6. Inferior oblique (via CN III): Elevation—Extorsion

Please note that CN III, or the oculomotor nerve, innervates all the extraocular muscles except for the superior oblique (CN IV) and lateral rectus (CN VI).

The intrinsic muscles of the eyes are smooth and involuntary. They are found inside the eyes and include the irises and ciliary muscles. The eyes are some of the only organs that have *both* voluntary and involuntary muscles. The functions of these muscles will be discussed in detail later in this chapter.

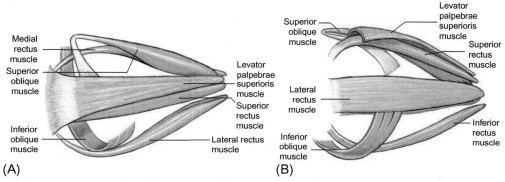

Fig. 10.4 Extrinsic muscles of the right eye. (A) Superior view. (B) Inferior view. *(From Dutton, J.J. Atlas of Clinical and Surgical Orbital Anatomy, 2nd ed.; Saunders: Philadelphia, 2011.)*

Facial nerves (VII)

The primary function of the **facial nerves (VII)** is mixed, sensory and motor, to and from the face. They originate in different areas. The sensory components originate on the anterior two-thirds of the tongue, while the motor components originate in the motor nuclei of the pons. The facial nerves pass through the internal acoustic meatus, to canals that lead to the stylomastoid foramina. The sensory components reach the sensory nuclei of the pons, while the somatic motor components reach the muscles of facial expression. The visceral motor components reach the lacrimal (tear) glands and nasal mucous glands, via the pterygopalatine ganglion. They also reach the submandibular and sublingual glands via the submandibular ganglion.

The cell bodies of the sensory neurons are within the **geniculate ganglia**, while the motor nuclei are in the pons. On both sides, sensory and motor roots emerge from the pons, entering the internal acoustic meatus within the temporal bone. Each facial nerve passes through the facial canal, reaching the face via the stylomastoid foramen. They then split, forming the *temporal, zygomatic, buccal, marginal mandibular,* and *cervical* branches.

Sensory neurons monitor proprioceptors of the facial muscles. They provide deep pressure sensations throughout the face and receive taste information from the anterior two-thirds of the tongue. The superficial scalp and face muscles, as well as the deep muscles close to the ears, are controlled by somatic motor fibers.

Preganglionic autonomic fibers are carried by the facial nerves to the pterygopalatine and submandibular ganglia. Postganglionic fibers from the **pterygopalatine ganglia** innervate the lacrimal and small glands of the nasal cavity and pharynx. The submandibular and sublingual glands are innervated by the **submandibular ganglia**. As a rule, facial glands are all innervated by the facial nerve, except for the parotid gland, which is innervated by CN IX. As a high-yield topic for neurology exams, just remember that the facial nerve innervates the posterior belly of the digastric muscle, while the trigeminal nerve innervates the anterior belly of the digastric muscle. The trigeminal nerve and facial nerve are both involved in the blink reflex, while the sensation over the cornea is sensed by the trigeminal nerve's afferent limb. The facial nerve serves as the efferent limb, causing the eye to blink. *Bell's palsy* is related to paralysis of the facial nerve and is discussed later in this chapter.

Vestibulocochlear nerves (VIII)

The primary function of the **vestibulocochlear nerves (VIII)** is a special sensory, but of two types. The vestibular nerve handles balance and equilibrium, while the cochlear nerve is responsible for hearing. The vestibulocochlear nerves originate in the monitoring receptors of the internal ear—the vestibule and cochlea. They pass through the internal acoustic meatus of the temporal bones, ending in the vestibular and cochlear nuclei of the pons and medulla oblongata.

The CN VIII are also known as the *acoustic*, *auditory*, and *stato-acoustic* nerves, but "vestibulocochlear" is the preferred term since it indicates the two major nerve branches involved. Each vestibulocochlear nerve is found posterior to the origin of the facial nerves. They link the boundary between the pons and medulla oblongata, reaching sensory receptors of the internal ear, via the internal acoustic meatus, which also allows passage of the facial nerves.

The **vestibular nerve** begins at the receptors of the *vestibule*, which is the area of the internal ear concerned with balance. Sensory neurons are in adjacent sensory ganglia. Axons target the **vestibular nuclei** of the pons and medulla. They convey information about orientation and movements of the head. The **cochlear nerve** monitors receptors of the *cochlea*, which is the area of the internal ear that provides the sense of hearing. Cell bodies of sensory neurons are within a peripheral (*spiral*) ganglion. Axons synapse in the **cochlear nuclei** of the pons and medulla. Axons that leave the vestibular and cochlear nuclei relay sensory information to other centers or cause reflexive motor responses. The auditory and vestibular system will be discussed in detail in Chapters 12 and 13.

Glossopharyngeal nerves (IX)

The primary function of the **glossopharyngeal nerves (IX)** is mixed, sensory and motor, to and from the head and neck. Recall that CN I, III, VII, and IX also have parasympathetic fibers. The sensory components originate at the posterior one-third of the tongue, parts of the pharynx and palate, and the carotid neck arteries. The motor components originate in the motor nuclei of the medulla oblongata. The glossopharyngeal nerves pass through the jugular foramina, between the occipital and temporal bones. The sensory components end in the sensory nuclei of the medulla. The somatic motor components end in the pharyngeal muscles that are used in swallowing (see Fig. 10.5). The visceral motor components end in the parotid gland, via the otic ganglion.

The medulla oblongata has both sensory and motor nuclei of cranial nerves IX, X, XI, and XIII, as well as the vestibular nucleus of CN VIII. That is why vertigo is not used to localize the lesion to the pons, as the nucleus is in the medulla. The same is true for the trigeminal nucleus and tracts, which are elongated inferiorly all the way to the C2 vertebra level. Therefore, sensory loss of the face is not used to localize the lesion to the pons. The glossopharyngeal nerves innervate the tongue and pharynx. They penetrate the cranium in the jugular foramen along with cranial nerves X and XI. Recall that CN IX innervates the parotid gland. Sensory fibers are the most abundant type. Sensory neurons on either side are in the **superior ganglion** (also known as the *jugular ganglion*) and the

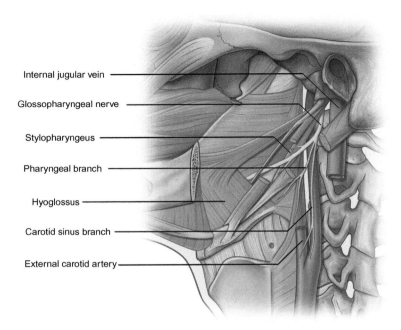

Fig. 10.5 The glossopharyngeal nerve in the anterior triangle of the neck.

inferior ganglion (also known as the *petrosal ganglion*). Sensory fibers carry information from the pharyngeal lining and soft palate to a nucleus of the medulla. Taste sensations are provided from the posterior third of the tongue. There are also special receptors monitoring blood pressure and dissolved gas concentrations of the carotid arteries, which are major blood vessels of the neck. Somatic motor fibers control the pharyngeal muscles of swallowing. Visceral motor fibers synapse in the **otic ganglion** and postganglionic fibers innervate the parotid gland of each cheek.

> ### Focus on glossopharyngeal nerve functions
> CN IX is among the cranial nerves that carry parasympathetic fibers. These fibers innervate the parotid gland and result in secretions from the gland, due to their parasympathetic activity. The glossopharyngeal nerve is also involved in autonomic regulation, by participating in the *baroreflex* to maintain blood pressure. Baroreceptors in the carotid sinus detect changes in blood interarterial pressure. Axons of these stretch receptors travel with the glossopharyngeal nerve to the medulla, in order to constantly monitor the heart rate and blood pressure. Baroreceptors are also present on the aortic arch, and their axons travel with the vagus nerve to the medulla. Another important reflex involving CN IX and CN X is the gag reflex, in which CN IX

serves as the afferent limb and CN X serves as the efferent limb of the reflex. The uvula is therefore innervated by the vagus nerve, and it deviates toward the opposite side of the lesion if damaged.

Vagus nerves (X)

The primary function of the **vagus nerves (X)** are mixed, both sensory and motor, to and from many areas of the thorax and abdomen (see Fig. 10.6). The sensory components originate from the pharynx, auricle and external acoustic meatus of the external ear,

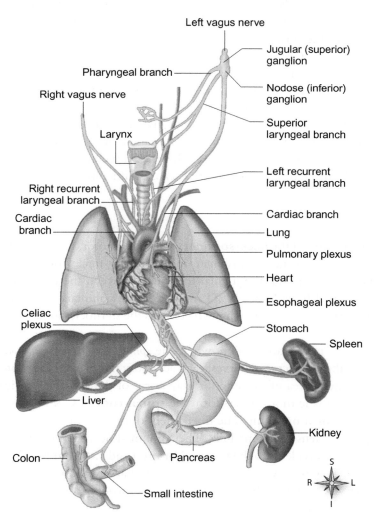

Fig. 10.6 Vagus nerve (cranial nerve [CN] X). The vagus nerve is a mixed cranial nerve with many widely distributed branches—hence the name vagus, which is the Latin word for "wanderer."

diaphragm, and visceral organs of the thoracic and abdominopelvic cavities. The motor components originate in the motor nuclei of the medulla. The vagus nerve pass through the jugular foramina. The sensory components reach the sensory nuclei and autonomic centers of the medulla. The visceral motor components reach the palate muscles, pharynx, and within thoracic and abdominal cavities—digestive, respiratory, and cardiovascular systems.

The vagus nerve begins just posterior to the area where the glossopharyngeal nerve attaches and has many small *rootlets*. The vagus nerve has extensive branches and radiations. Sensory neurons are in the superior ganglion (or *jugular ganglion*) and the inferior ganglion (or *nodose ganglion*). The vagus nerves provide somatic sensory information about the external acoustic meatus and diaphragm. They also provide special sensory information from the pharyngeal taste receptors. Recall that the remainder of tongue taste sensations are by CN V, VII, and IX). The majority of vagal afferents carry visceral sensory information from the esophagus, respiratory tract, and abdominal viscera, including distal portions of the large intestine. Autonomic control of visceral function must receive this visceral sensory information.

Regarding the motor components, each vagus nerve carries preganglionic autonomic (parasympathetic) fibers affecting the heart. They also control smooth muscles and glands in areas monitored by the sensory fibers. These areas include the stomach, gallbladder, and intestine. Damage to CN IX or CN X causes difficulty in swallowing.

Accessory nerves (XI)

The primary function of the **accessory nerves (XI)** is motor, to the muscles of the neck and upper back. These nerves originate in the motor nuclei of the spinal cord and medulla. They pass through the jugular foramina, between the occipital and temporal bones. The internal branch of CN XI innervates the voluntary muscles of the palate, pharynx, and larynx. The external branch controls the sternocleidomastoid and trapezius muscles.

The accessory nerves are also known as the *spinal accessory* or *spino-accessory* nerves. Different from the other cranial nerves, each accessory nerve has some motor fibers originating in the lateral area of the anterior gray horns of the first five cervical spinal cord segments. The somatic motor fibers form the **spinal root** of CN XI, entering the cranium via the foramen magnum. The fibers then join motor fibers of the **cranial root**, originating at a nucleus of the medulla. The combined nerve leaves the cranium via the jugular foramen. It divides into the **internal branch** and **external branch**. The internal branch joins the vagus nerve, innervating the voluntary swallowing muscles of the soft palate and pharynx, as well as the intrinsic muscles controlling the vocal cords. The external branch controls the sternocleidomastoid and trapezius muscles. Motor fibers originate in the lateral gray area of the anterior horns of cervical spinal nerves C_1 to C_5.

Hypoglossal nerves (XII)

Hypoglossal nerves (XII) is only motor, controlling tongue movements. These nerves originate in the motor nuclei of the medulla, passing through the hypoglossal canals of the occipital bone, to reach the tongue muscles. Each hypoglossal nerve exits the cranium and curves, reaching the skeletal tongue muscles. It provides voluntary motor control of tongue movements. Normal function of CN XII is verified by having a patient stick out the tongue. If there is damage to these nerves, the tongue will move toward the side of the lesion.

Section review
1. Name the cranial nerves that are mostly involved in each of the following:
   a. Moving the eyeball
      b. Sticking out the tongue
      c. Regulating the heart rate and digestive activity
2. List the cranial nerves that are exclusively sensory and state their functions.
3. Which cranial nerve is the only one that extends beyond the head and neck areas?
4. Which cranial nerve carries sensory signals from the largest area of the face?

Clinical considerations

Cranial nerve injuries may occur due to compression from increased intracranial pressure, or direct compression or destruction of a tumor, stroke, hemorrhage, vascular malformation, and others. Acceleration-deceleration injuries, shearing injuries, skull fractures, intracranial hemorrhage, aneurysms, lesions, brainstem herniation, meningitis, and multiple sclerosis may all cause damage to the cranial nerves. Conditions caused by cranial nerve injuries include anosmia, hyposmia, circumferential blindness, various types of hemianopia, strabismus, Bell's palsy, hemifacial spasm, and glossopharyngeal neuralgia.

Anosmia

Anosmia is the total lack of the sense of smell. **Hyposmia** is a partial loss of smell. With anosmia, the patient usually has a normal perception of salty, sweet, sour, and bitter substances, but cannot discriminate between flavors since this greatly depends on normal olfaction. There are common complaints of losing the sense of taste, known as **ageusia**, and of lack of enjoyment of food. Anosmia is often not recognized when it is unilateral.

Causes of anosmia include intranasal swelling or obstruction. Odors are unable to reach the olfactory area. It also occurs if the olfactory neuroepithelium, nerve filia, bulbs, tracts, or central connections are destroyed. In young adults, head trauma is a major cause. Previous upper respiratory infections, mostly influenza, as well as allergic diseases, are the main causes of anosmia or hyposmia. Certain medications may be related as well.

Additional causes include previous head or neck trauma or radiation, recent surgery to the nose or sinuses, or tumors of the nose or brain.

Important related clinical manifestations include nasal congestion, rhinorrhea, or both. Rhinorrhea must be asses such as whether secretions are bloody, mucoid, purulent, or watery. Past medical history must include the history of cranial trauma or surgery, sinus conditions, medications, and exposure to fumes or chemicals. The nasal passages must be inspected for discharge, inflammation, polyps, and swelling. The patient should breathe through one nostril while the other is held shut so that obstructions can be identified. A thorough neurologic examination, especially of mental status and cranial nerves, is performed.

If anosmia or hyposmia occur suddenly after head trauma or exposure to toxins, these causes are likely responsible. It is important to take into account all possible disorders that can be causative. For example, neurologic symptoms that wax and wane, and affect multiple areas, may suggest multiple sclerosis or another neurodegenerative disease. When anosmia progresses slowly in the elderly, and there are no other symptoms or findings, normal aging is probably the cause.

When no cause is obvious, patients should have a CT scan of the head, including the sinuses, with contrast. This will rule out tumors or fractures of the floor of the anterior cranial fossa. MRI can evaluate intracranial disease and may be better for those with no nasal or sinus pathology seen in the CT scan.

Even after successful treatment for sinusitis, the sense of smell does not always return. There are no actual treatments for anosmia or hyposmia. Adding concentrated flavoring agents to food can help patients enjoy eating somewhat more. Smoke alarms in patients' homes are even more important when these conditions are present. They should also avoid eating stored foods since they cannot detect when these may have gone bad. Natural gas should not be used for heating or cooking since patients may not be able to detect gas leaks. It should be noted that some loss of olfaction with aging is normal, usually by age 60, and becoming more noticeable following age 70.

Fourth cranial nerve palsy

The fourth cranial nerve is the *trochlear* nerve. Fourth cranial nerve palsy is often of unknown origin, with only a few causes being identified. This type of palsy impairs the superior oblique muscle. It causes paresis of vertical gaze, mostly in adduction.

Pathophysiology
The few identified causes of fourth cranial nerve palsy include closed head injury, which is the most common cause, and infarctions due to small-vessel disease, such as seen in diabetes mellitus. Closed head injuries may cause unilateral or bilateral palsies. Rarer causes of this type of palsy include aneurysms, myasthenia gravis, or tumors such as tentorial meningioma, and **pinealoma**.

Clinical manifestations

Since the superior oblique muscle becomes paretic, the eyes do not look down. The patient will see double images, with one above and slightly to the side of the other. This makes movements such as walking down a flight of stairs difficult, because it requires looking down and inward. The double images can be compensated for by tilting the head to the opposite side of the lesion.

Diagnosis

This condition is diagnosed clinically.

Treatment

The only treatment for fourth cranial nerve palsy is repetition of oculomotor exercises or wearing *prism glasses*, to adjust the images. These idiopathic cranial nerve palsies usually resolve on their own within a few months with excellent prognosis.

Sixth cranial nerve palsy

The sixth cranial nerve is the *abducens* nerve. Palsy of this nerve affects the lateral rectus muscle and impairs eye abduction on the same side. When the patient looks straight ahead, the eye may be slightly adducted.

Pathophysiology

Sixth cranial nerve palsy may be caused by diabetic small-vessel disease, and sometimes due to a condition called *mononeuritis multiplex* or *multiple mononeuropathy in the setting of infiltrative disease such as lymphoma*. Palsy may also be caused by compression of the abducens nerve due to lesions in the cavernous sinus, such as nasopharyngeal tumors—or tumors of the orbit or base of the skull. Other causes include head trauma, increased intracranial pressure, or both. Cranial nerve VI is the longest cranial nerve, and therefore, most susceptible to increased intracranial pressure. One of the signs of increased intracranial pressure is abducens nerve palsy. This sign should not be used for lateralization and only should be interpreted as a "concern" for increased intracranial pressure.

Clinical manifestations

Sixth cranial nerve palsy is signified by binocular horizontal diplopia when the patient looks to the side of the paretic eye. Since the tonic action of the medial rectus muscle is unopposed, the eye will be slightly adducted when looking straight ahead. It will abduct sluggishly and incomplete. Even when abduction is maximal, the lateral sclera will be exposed. If paralysis is complete, the eye will not be able to move past midline. Palsy may be caused by nerve compression due to a thrombus because of a stroke or head trauma, or a lesion in the cavernous sinus.

Diagnosis

While most sixth cranial nerve palsies are obvious, their causes are not. When retinal venous pulsations are seen by an ophthalmologist, there is probably no increased intracranial pressure. Since it is usually available, CT scan is often performed, but MRI is actually preferred because it gives greater resolution of the orbits, cavernous sinus, cranial nerves, and posterior fossa. When imaging is normal, but increased intracranial pressure is suspected, a lumbar puncture is performed to assess for idiopathic intracranial hypertension previously known as pseudotumor cerebri. This condition is more common in obese females and can also cause headache as well as vision abnormality due to the increased intracranial pressure.

Treatment

For most patients, sixth cranial nerve palsy is resolved by treating the underlying disorder. When the palsy is idiopathic, it usually resolves on its own within 2 months. If the palsy is due to increased intracranial pressure, then the underlying cause needs to be addressed. This can be due to a tumor, hemorrhage, idiopathic intracranial hypertension, severe meningitis, encephalitis, and others.

Strabismus

Strabismus is a misalignment of the eyes, causing deviation from a normal, parallel gaze (Fig. 10.7). Most cases of strabismus in children are caused by muscle imbalance or refractive errors. However, it can rarely be caused by retinoblastoma or other serious ocular defects, and neurological diseases. If untreated, a significant number of children with severe strabismus have some amount of vision loss due to **amblyopia**. Some types of strabismus are severe, based on the direction of the deviation, its constancy or

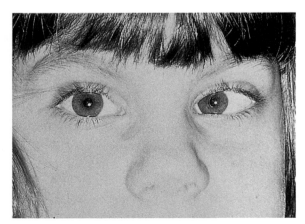

Fig. 10.7 Strabismus.

intermittent occurrence, and coexisting conditions. There are four prefixes used to explain the various types of strabismus:
- *Eso-* meaning toward the nose
- *Exo-* meaning toward the temples
- *Hyper-* meaning upward
- *Hypo-* meaning downward

The term *tropia* refers to manifest deviations that are detectable with both eyes open, with binocular vision. Tropia can affect one or both eyes and may be constant or intermittent. Latent deviation is detectable only when one eye is covered, and vision is monocular. This is called *phoria*. Deviations in phoria are latent since the brain uses the extraocular muscles to correct minor misalignments. Deviations that are the same in all directions are called *comitant*. Deviations that vary based on the direction of the gaze are called *incomitant*.

Pathophysiology

Strabismus is mostly congenital, developing during infancy. It may also be, in rare cases, acquired; developing after 6 months of life. Risk factors for infantile strabismus include family history, Down syndrome, Crouzon syndrome, prenatal drug or alcohol exposure, low birth weight or prematurity, congenital eye defects, and cerebral palsy. Acquired strabismus may develop acutely or gradually. It may be caused by retinoblastoma or other tumors, head trauma, cerebral palsy, spina bifida, or palsies of the III, IV, or V cranial nerves, encephalitis, meningitis, or acquired eye defects. Causes are varied based on the type of deviation that is present.

Esotropia is usually infantile, of an idiopathic nature, though it is believed to be caused by an anomaly of fusion. Accommodative esotropia is common, developing between 2 and 4 years of age. It is linked to *hyperopia*. Sensory esotropia develops when severe visual loss interferes with the brain's ability to maintain ocular alignment. Uncommonly, esotropia can be paralytic, due to a VI cranial nerve palsy. Esotropia can also be caused by Duane's syndrome, which is the congenital absence of the abducens nucleus, with abnormal innervation of the lateral rectus extraocular muscle by cranial nerve III. Additionally, esotropia may be due to Mobius' syndrome, involving anomalies of multiple cranial nerves.

Exotropia is usually intermittent and idiopathic. Rarely, it is constant and paralytic, as in cranial nerve III palsy. **Hypertropia** may be paralytic, due to cranial nerve IV palsy, congenital or following head trauma. Less often, it occurs due to cranial nerve III palsy. **Hypotropia** may be restrictive, due to mechanically restricted movement of the eye globe, instead of a neurological condition affecting eye movement. An example is restrictive hypotropia that results from a fracture of the floor or walls of the eye orbit. Less often, restrictive hypotropia is due to Graves' ophthalmopathy, cranial nerve III palsy, and Brown syndrome. This syndrome involves congenital or acquired tightness and restriction of the superior oblique muscle tendon.

Clinical manifestations

The **phorias** rarely cause symptoms, but the **tropias** sometimes cause symptoms. For example, *torticollis* may develop as a compensation for the brain being unable to fuse images because of the eyes being misaligned, an attempt to reduce diplopia. Some children who have tropias have normal, equal vision. Amblyopia, however, often develops with tropias. It is caused by cortical suppression of images in the deviating eye to avoid confusion and diplopia.

Diagnosis

Diagnosis of strabismus occurs during physical and neurologic examinations as part of well-child checkups. A good identifier is if the child uses one eye to fixate on an object. Physical examination should include tests of visual acuity, pupil reactivity, and extent of extraocular movements. Neurological examination should focus on the cranial nerves. Diagnostic tests include corneal light reflex, alternate cover, and cover-uncover tests. The corneal light reflex test is used to detect large deviations. The child looks at a light and the reflection, or *reflex*, from the pupil is observed. If normal, the reflex is symmetric—the same location on each pupil. For an exotropic eye, the reflex is nasal to the pupillary center. For an esotropic eye, it is temporal to the pupillary center. Vision screening machines are now being used to identify at-risk children.

In the alternate cover test, the child is asked to look at an object. One eye is then covered as the other eye is observed for movement. If alignment is normal, no movement should be detected. When strabismus is present, the unoccluded eye shifts to establish fixation on the object once the other eye, already fixed on the object, is occluded. The test is then performed on the other eye. In a similar cover-uncover test, the child is asked to look at an object while each eye is covered and uncovered alternately. Latent strabismus causes the affected eye to shift position when it is uncovered.

Tropia can also be measured by using prisms positions so that the deviating eye does not need to move in order to fixate. The prism's power needed to prevent deviation quantifies the tropia. It gives a measurement of the magnitude of misalignment of the visual axes. Ophthalmologists use the *prism diopter* as the unit of measurement. One prism diopter is a deviation of the visual axes of one centimeter at one meter.

Strabismus must be distinguished from *pseudostrabismus*—the appearance of esotropia when visual acuity of a child is good in both eyes, but when a wide nasal bridge or broad **epicanthal folds** obscure a large portion of the sclera nasally as the child looks laterally. In pseudostrabismus, the light reflex and cover tests will be normal.

Treatment

It is not clear when it is the right time to perform surgery. Most experts believe that the correction surgery should be done early, while amblyopia is not developed, perhaps before age 7. Others believe that the age of the patient regarding surgery does not change the outcome. All children must have formal vision screening when they are in preschool. Treatment is based on vision equalization, then aligning the eyes. The child may need an eye patch, or the normal eye to be instilled with eye drops, causing more use of the amblyopic eye. As vision improves, there is a better prognosis for developing binocular vision, and for stability if surgery is needed. Patching, without other treatments, does not resolve strabismus in most cases. If the amount of refractive errors is large enough to interfere with fusion, especially with accommodative esotropia, eyeglasses or contact lenses may be used. Specific eye exercises can aid in correcting intermittent exotropia with convergence insufficiency.

When nonsurgical methods fail, surgical repair may be performed, to loosen or tighten certain eye structures—usually the rectus muscles. These surgeries are usually performed on an outpatient basis, with a more than 80% success rate. Complications usually involve over- or under-correction of the condition, and recurrent strabismus later in life. Rare complications include excessive bleeding, infection, and vision loss.

Bell's palsy

Bell's palsy is due to a lower motor neuron lesion of the facial nerve. The cause is unknown or idiopathic, but it can be also due to infections such as the human immunodeficiency virus (HIV), varicella-zoster virus (VZV), and herpes simplex virus (HSV), Lyme disease, and sarcoidosis. Diabetes is a risk factor for idiopathic facial nerve palsy.

In peripheral facial droop due to Bell's palsy, both upper and lower sides of the face are paralyzed, including the ability to close the eye on the same side (Fig. 10.8). Clinical manifestations of Bell's palsy also include pain behind the ear, which may precede the facial paralysis in some patients. The paresis is often maximal within 48–72 h. The patient may have a numb or heavy feeling in the face (paresthesia). The side that is affected is flat and expressionless. The patient cannot partially or completely wrinkle the forehead, blink the eye, or grimace. If severe, the palpebral fissure will widen causing the eye to not be able to close. This can irritate the conjunctiva and dry the cornea. Though sensory examination is usually normal, there may be pain when the external auditory canal or an area behind the ear, over the mastoid, are touched. When the nerve lesion is proximal to the geniculate ganglion, there may be impairment of salivation, taste, and lacrimation. **Hyperacusis** may also be present if the lesion is proximal to the nerve of the stapedius muscle. The diagnosis of Bell's palsy is clinical. It is distinguished from hemispheric stroke or upper motor neuron lesions because these result in weakness most of the lower face. Patients who have central lesions can usually wrinkle their brow and close their eyes tightly.

There is no need for MRI or lumbar puncture in patients with typical isolated Bell's Palsy. Lyme disease is only tested if the patient has been in a geographical area where ticks are common, and when the presentation is suggesting Lyme disease.

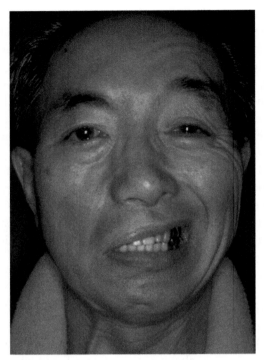

Fig. 10.8 Patient with the left facial paralysis. *(From Takushima, A.; Harii, K.; Hiortaka, A.; et al. Fifteen-year survey of one-stage latissimus dorsi muscle transfer for treatment of longstanding facial paralysis. J. Plast. Reconstr. Aesthet. Surg.* ***2013,*** *66 (1), 29–36.)*

Hemifacial spasm

Hemifacial spasm involves unilateral contractions of the facial muscles and is painless and synchronous. It is caused by dysfunction of the facial nerves (VII) or their motor nuclei. Hemifacial spasm is also known as *tic convulsif*, and may first appear as small, slightly noticeable tics around the eyelids, cheeks, or mouth. These contractions can later expand to other parts of the face. They are most common in women over age 40. Additional manifestations include changes in hearing, tinnitus, ear pain that is often behind the ear, and spasms that move down the entire face.

Diagnosis is clinical and takes into consideration **blepharospasm**, focal seizures, and various types of *tics*. Treatment is similar to that of trigeminal neuralgia, including muscle relaxants. However, botulinum toxin—either type A or type B—is also able to be used effectively for this condition. Additional treatments include plenty of rest, reducing caffeine intake, and dietary supplements of vitamin D and magnesium. Also, chamomile and blueberries have been effective for some patients.

If surgery is chosen, this usually involves *microvascular decompression*, in which a small opening is made in the skull, behind the ear. A small piece of artificial padding material is placed between the nerve and the blood vessels that are pushing on it. This procedure usually eliminates or at least greatly decreases hemifacial spasm.

Glossopharyngeal neuralgia

Glossopharyngeal neuralgia may also be caused similarly to the causes of trigeminal neuralgia and hemifacial spasm—from nerve compression by a pulsating blood vessel. A rare cause is a tumor in the neck or cerebellopointine angle. In many cases, there is no identifiable cause. Glossopharyngeal neuralgia is rare but usually affects men above the age of 40.

In this condition, paroxysmal attacks of unilateral severe pain, lasting only briefly, may occur spontaneously, or after talking, chewing, sneezing, or swallowing. The pain can last from a few seconds to a few minutes. It usually begins in the tonsillar region, or at the base of the tongue. The pain then radiates to the ipsilateral ear. Sometimes, increased vagus nerve activity causes sinus arrest accompanied by syncope. These episodes can be extremely infrequent.

Diagnosis is clinical and distinguished from trigeminal neuralgia by the pain's location. Swallowing or touching the tonsils with an applicator usually precipitates the pain. Applying lidocaine to the throat briefly eliminates spontaneous or evoked pain. MRI can exclude cerebellopointine angle, tonsillar, and pharyngeal tumors, as well as metastatic lesions in the anterior cervical triangle. Local nerve blocks performed by an ear-nose-throat specialist help distinguish between pain caused by tumors, **carotidynia**, and superior laryngeal neuralgia.

Treatments, like those for trigeminal neuralgia, involve oral anticonvulsants, but if these are not effective, topical cocaine can be applied to the pharynx for temporary relief. Surgery can be performed to decompress the nerve from a pulsating blood vessel. When pain is only in the pharynx, surgery can be performed just to the extracranial area of the nerve. When pain is widespread, surgery must be done on the intracranial area of the nerve.

Clinical cases

Clinical case 1
1. Why did the physician suspect multiple sclerosis?
2. What is the significance of horizontal diplopia in this case?
3. What is the likely outcome for this patient's internuclear ophthalmoplegia?

A 43-year-old man came to his doctor's office with weakness in his right leg and double vision for the past few days. A neurologic examination revealed horizontal diplopia during lateral gaze to the left. The patient had an adduction deficit in the right eye, and nystagmus in the left eye, when gazing to the left. He had an adduction deficit in the left eye, and nystagmus in the right eye, when gazing to the right. The physician surprised the patient when he told him that he was referring him to see a neurologist. When the patient asked why the physician told him he needed to be evaluated for multiple sclerosis.

Answers:
1. Internuclear ophthalmoplegia is caused by injury or dysfunction in the medial longitudinal fasciculus, which allows for conjugate eye movement. In younger patients with bilateral symptoms, multiple sclerosis is often the cause since the MLF is highly myelinated.
2. Horizontal diplopia, or seeing two images of an object side by side, is an important sign of neurological impairment being involved. In multiple sclerosis, it is caused by scarring in the brainstem and is a common symptom.
3. Usually, the outlook for MS-related internuclear ophthalmoplegia is good. Once treated, most patients have a complete recovery of eye function. Remember, in intranuclear ophthalmoplegia, the abnormality cannot be overcome by the oculocephalic reflexes. This condition is often bilateral, with the inability to adduct either eye.

Clinical case 2
1. What are the risk factors for infantile forms of strabismus?
2. Why did this patient probably develop the left-eye amblyopia as a result of her infantile esotropia?
3. If nonsurgical methods fail for this patient, what are the surgical options?

A 6-month-old girl is brought to an eye specialist by her parents, to assess her crossed eyes, which have been in this condition since birth. The girl appears to have normal vision regardless of the strabismus. She was a full-term infant without perinatal complications and no other medical problems. Upon examination, her left eye is clearly crossed inward (esotropic). With the left eye covered, her right eye could fix on objects and easily follow movements. When the cover is alternated from one eye to the other, there is always an outward shift of the opposite eye. She is referred to a pediatric ophthalmologist. At this examination, she is diagnosed with infantile esotropia and left-eye amblyopia.

Answers:
1. The risk factors for infantile forms of strabismus include family history, Down syndrome, Crouzon syndrome, prenatal drug or alcohol exposure, low birth weight or prematurity, congenital eye defects, and cerebral palsy.
2. This patient's left-eye amblyopia probably developed because of cortical suppression of images, in an attempt to avoid confusion and diplopia.
3. When nonsurgical methods fail, surgery can be performed to loose or tighten certain eye structures—usually the rectus muscles. These (mostly) outpatient surgeries have more than an 80% success rate.

Clinical case 3
1. Will this patient's condition likely resolve on its own?
2. If treatments are required, what would they include?
3. What is this patient's prognosis?

A 40-year-old man awoke one morning and discovered that something severe had changed in his facial appearance. The left side of his face seemed unable to move—he could not smile normally or even close his left eye all the way. Afraid that he had experienced a stroke overnight, he had his wife call their physician, who agreed to see him right away. The patient had a history of diabetes and had recently had influenza. After electromyography, blood tests, and a CT scan, there were no signs of a stroke or any other structural causes of pressure on his facial nerve. He was diagnosed with Bell's palsy.

Answers:
1. Bell's palsy is a facial paralysis caused by damage or trauma to the facial nerves. It is not related to stroke, though stroke is the most common cause of facial paralysis. The condition often subsides within 2 weeks, without treatment.
2. When treatments for Bell's palsy are indicated, they include corticosteroids, acyclovir, analgesics; keeping the affected eye moist, and protected from injury by eye patches, especially during the night; physical therapy, facial massage, and acupuncture. Rarely, decompression surgery may be performed.
3. Usually, the prognosis for Bell's palsy is excellent. Though recovery times vary, this usually occurs over 2 weeks to 6 months, though symptoms can last longer or never totally disappear. Rarely, Bell's palsy recurs, either on the same side of the opposite side of the face.

Clinical case 4
1. What is the difference in location of the pain that signifies glossopharyngeal neuralgia, in comparison to trigeminal neuralgia?
2. Which type of surgery has been successful in treating glossopharyngeal neuralgia?
3. What is the prognosis after this type of surgery?

A 43-year-old woman went to the emergency department with severe spasms of pain in her throat and neck. Upon examination, she was diagnosed with glossopharyngeal neuralgia, admitted to the hospital, and treated with a variety of medications and analgesics. She was afraid to eat, drink, or even talk because all of these activities seem to provoke the shooting pain. After a significant time in the hospital with no real relief, her family's research resulted in finding a qualified neurosurgeon who had successfully treated this condition. Surgery was performed, and the patient's health returned to normal.

Answers:

1. Glossopharyngeal neuralgia involves pain at the base of the tongue, which can radiate to the ear and neck. It is often triggered by eating, drinking, sneezing, speaking, laughing, or coughing. Trigeminal neuralgia involves pain in one side of the face, triggered by any touching of that side of the face.
2. Microvascular decompression is the type of surgery that has been successful in treating glossopharyngeal neuralgia. A small hole is made in the base of the skull, through which an operative microscope and microinstruments are inserted. The base of the glossopharyngeal nerve is located, and the microinstruments are used to lift the blood vessel loop away from the nerve, alleviating the pressure. The vessel is then wrapped in a form of plastic sheath to permanently maintain decompression.
3. Long-term success rates for microvascular decompression surgery are high—between 70% and 80% of patients experience a full recovery.

Key terms

Abducens nerves (VI)
Accessory nerves (XI)
Ageusia
Amblyopia
Anosmia
Bell's palsy
Bitemporal hemianopia
Blepharospasm
Carotidynia
Cerebellopontine angle
Ciliary ganglion
Circumferential blindness
Cochlear nerve
Cochlear nuclei
Contralateral homonymous hemianopia
Cranial nerves
Cranial root
Crouzon syndrome
Diplopia
Epicanthal folds
Esotropia
Exotropia
External branch
Eye stroke

Facial nerves (VII)
Geniculate ganglia
Glomus jugulare
Glossopharyngeal nerves (IX)
Glossopharyngeal neuralgia
Hemifacial spasm
Hemisensory
Hyperacusis
Hypertropia
Hypoglossal nerves (XII)
Hyposmia
Hypotropia
Inferior ganglion
Internal branch
Ipsilateral
Nasal hemianopia
Oculomotor nerves (III)
Olfactory bulbs
Olfactory nerves (I)
Olfactory tracts
Optic chiasm
Optic nerves (II)
Optic neuritis
Optic tracts
Otic ganglion

Phorias
Pterygopalatine ganglia
Retinitis pigmentosa
Reversible cerebral vasoconstriction syndrome
Semilunar ganglion
Spinal root
Strabismus
Sturge-Weber syndrome
Subclavian steal syndrome
Submandibular ganglia
Superior ganglion
Suprasellar
Trigeminal autonomic cephalgia
Trigeminal nerves (V)
Trochlear nerves (IV)
Tropias
Tubular vision
Vagus nerves (X)
Vestibular nerve
Vestibular nuclei
Vestibulocochlear nerves (VIII)

Suggested readings

1. https://now.aapmr.org/cranial-nerve-visual-and-hearing-dysfunction-in-disorders-of-the-cns/.
2. Binder, D. K.; Sonne, D. C.; Fischbein, N. J. *Cranial Nerves: Anatomy, Pathology, Imaging*. Thieme, 2010.
3. Chan, J. W. *Optic Nerve Disorders: Diagnosis and Management*. Springer, 2007.
4. https://my.clevelandclinic.org/health/diseases/15766-homonymous-hemianopsia.
5. Ellenbogen, R. G.; Abdulrauf, S. I.; Sekhar, L. N. *Principles of Neurological Surgery*, 3rd Ed.; Saunders, 2012.
6. https://www.emedicinehealth.com/script/main/art.asp?articlekey=24516.
7. Guntinas-Lichius, O.; Schaitkin, B. *Facial Nerve Disorders and Diseases: Diagnosis and Management*. TPS, 2015.
8. https://www.healthline.com/health/hemifacial-spasm.
9. https://www.healthtap.com/topics/homonymous-hemianopia-treatment.
10. Jones, H. R.; Srinivasan, J.; Allam, G. J.; Baker, R. A.; Inc Lahey Clinic. *Netter's Neurology*, 2nd Ed.; Saunders, 2011.
11. Kumbhare, D.; Robinson, L.; Buschbacher, R. *Buschbacher's Manual of Nerve Conduction Studies*, 3rd Ed.; Demos Medical, 2015.
12. Kushner, B. J. *Strabismus: Practical Pearls You Won't Find in Textbooks*. Springer, 2017.
13. Lalwani, A. *Current Diagnosis & Treatment – Otolaryngology – Head and Neck Surgery*, 3rd Ed.; McGraw-Hill Education/Medical, 2011.
14. Lambert, S. R.; Lyons, C. J. *Taylor and Holt's Pediatric Ophthalmology and Strabismus*, 5th Ed.; Elsevier, 2016.
15. Li, S.; Zhong, J.; Sekula, R. F. *Microvascular Decompression Surgery*. Springer, 2016.
16. Lifferth, A. *The Optic Nerve Evaluation in Glaucoma: An Interactive Workbook*. LWW, 2017.
17. Miloro, M. *Trigeminal Nerve Injuries*. Springer, 2013.
18. https://www.modasta.com/health-a-z/tubular-vision-tunnel-vision/.
19. Rea, P. *Essential Clinical Anatomy of the Nervous System*. Academic Press, 2015.
20. Ryugo, D. K.; Fay, R. R.; Popper, A. N. *Auditory and Vestibular Efferents (Handbook of Auditory Research)*. Springer, 2010.
21. https://www.sciencedirect.com/topics/neuroscience/bitemporal-hemianopsia.
22. https://www.sciencedirect.com/topics/neuroscience/homonymous-hemianopsia.
23. Sindou, M.; Keravel, Y.; Moller, A. R. *Hemifacial Spasm: A Multidisciplinary Approach*. Springer, 1997.
24. Skorkovska, K. *Homonymous Visual Field Defects*. Springer, 2017.
25. Slattery, W. H.; Azizzadeh, B. *The Facial Nerve*. Thieme, 2014.
26. Soodan, K. *Nerve Physiology and Trigeminal Nerve*. Lap Lambert Academic Publishing, 2016.
27. Steed, M. B. *Peripheral Trigeminal Nerve Injury, Repair, and Regeneration – Atlas of the Oral and Maxillofacial Surgery Clinics of North America*. Saunders, 2011.
28. Stennert, E.; Michel, O.; Kreutzberg, G. W.; Jungehulsing, M. *The Facial Nerve: An Update on Clinical and Basic Neuroscience Research (Reprint of 1994 Ed.)*. Springer, 2014.

29. Syed, S. *Physiotherapy Management for Bell's Palsy*. Lap Lambert Academic Publishing, 2014.
30. Tubbs, R. S. *Glossopharyngeal Nerve Stimulation: Potential Application in Patients with Epilepsy*. VDM Verlag, 2009.
31. Wilson-Pauwels, L.; Stewart, P. A. *Cranial Nerves: Function and Dysfunction*, 3rd Ed.; PMPH USA, 2010.
32. Yoshioka, N.; Rhoton, A. L. *Atlas of the Facial Nerve and Related Structures*. Thieme, 2015.
33. Zufall, F.; Munger, S. D. *Chemosensory Transduction: The Detection of Odors, Tastes, and Other Chemostimuli*. Academic Press, 2016.

CHAPTER 11

Trigeminal and facial nerves

The cranial nerves may be affected by many unique diseases. Certain cranial nerves and their disorders were discussed in Chapter 10. Some will be discussed later—especially the disorders of the vestibulocochlear nerves, in Chapters 12 and 13. The optic nerves were discussed in Chapter 14, while the olfactory and gustatory nerves were discussed in Chapter 15. This chapter focuses on the trigeminal (CN V) and facial (CN VII) nerves.

There are functional and anatomical relationships between cranial nerve V (the trigeminal nerve) and cranial nerve VII (the facial nerve), in both their sensory and motor divisions. Facial sensations are innervated by the trigeminal nerves, along with the muscles of mastication. However, the muscles of facial expression are mostly innervated by the facial nerves, along with the sensation of taste. There are many interactions between the sensory parts of these nerves.

Trigeminal nerve

The fifth cranial nerve is the **trigeminal nerve**, which is a mixed nerve: sensory and motor. It carries sensory impulses from the major parts of the face and head, as well as from the mucous membranes of the nose, mouth, and paranasal sinuses. Its sensory impulses additional come from the cornea and conjunctiva. The trigeminal nerve provides the sensory innervation of the dura in the anterior and middle cranial fossae. In the sensory part of the nerve, the cell bodies are within the **Gasserian ganglion**. In humans, this is the largest sensory ganglion. It lies in the **inferomedial** area of the middle cranial fossa, within a recess called **Meckel's cave**. The central axons of the ganglion cells become the sensory root of the trigeminal nerve.

The peripheral branches of the Gasserian ganglion form the trigeminal nerve's three sensory divisions. The first, or ophthalmic division, passes through the cavernous sinus as well as the **superior orbital fissure**. The second, or maxillary division, also passes through the cavernous sinus. It leaves the middle fossa through the **foramen rotundum**. The third, or mandibular division, does not travel through the cavernous sinus. Instead, it emerges from Meckel's cave inferiorly and passes through the **foramen ovale**.

Trigeminal sensory nuclei

The fibers of the sensory root of the trigeminal nerve run dorsomedially as they enter the pons, toward the **principal sensory nucleus**. About half of the fibers are divided into ascending and descending branches, while the other half ascend or descend without dividing. The descending fibers from the trigeminal nerve's spinal tract. This ends in the medially adjacent spinal nucleus of the trigeminal nerve. Around the small neurons in the principal sensory nucleus and some ascending trigeminal fibers, many of which are extremely myelinated, that synapse. The principal sensory nucleus lies lateral to the motor nucleus. It is medial to the middle cerebellar peduncle, and inferiorly continuous with the trigeminal nerve's spinal nucleus. The principal nucleus is believed to mostly regulate tactile stimuli.

Additional ascending fibers enter the **mesencephalic nucleus**. This cellular column contains the only primary sensory unipolar neurons that have somata in the central nervous system (CNS) instead of in a sensory ganglion. Nerve fibers ascending to the mesencephalic nucleus may have collaterals reaching the motor nucleus of the trigeminal nerve as well as the cerebellum. Most fibers arising in the trigeminal sensory nuclei cross over the midline. They ascend in the trigeminal lemniscus and terminate in the contralateral ventral posteromedial (VPM) thalamic nucleus. From here, third-order neurons project to the postcentral gyrus.

> **Focus on the trigeminal nerve**
> The trigeminal nerve is the principal sensory nerve of the head. It supplies most of the face but does not extend to the angle of the jaw, and only slightly onto the auricle. It is also the somatic afferent innervation of the nose, mouth, and eyes. It provides sensory innervation to parts of the dura matter. Its peripheral terminals release a variety of cytokines believed to be involved in the generation of headaches.

Trigeminal motor nucleus

The motor portion of the trigeminal nerve supplies muscles of mastication and tensor tympani of the inner ear. It originates in the trigeminal motor nucleus, in the mid pons. Its exiting fibers pass below the Gasserian ganglion. They are then incorporated into the mandibular nerve. The masseter and pterygoid muscles are used for chewing. They are involved in many brainstem reflexes—primarily the *jaw jerk*, which is discussed below.

The trigeminal nerve's motor nucleus is of ovoid shape. It lies within the upper pontine tegmentum, below the lateral area of the fourth ventricle's floor. It is medial to the principal sensory nucleus but separated from it via fibers of the trigeminal nerve. The motor nucleus forms the rostral portion of the pharyngeal efferent column, utilized for special visceral functions. The motor nucleus has large, multipolar neurons mixed with smaller, multipolar cells. The neurons have discrete subnuclei in their organization, with axons that innervate individual muscles. The motor nucleus receives fibers from both corticonuclear tracts. The fibers exit the tracts at the nuclear level, or higher within the pons as **aberrant corticospinal fibers**. They may end at motor neurons or interneurons. The motor nucleus collects afferents from the trigeminal nerve's sensory nuclei. These may include some from the mesencephalic nucleus, forming monosynaptic reflex arcs use for proprioceptive control of the masticatory muscles. The motor nucleus also receives afferents from the reticular formation, red nucleus, tectum, medial longitudinal fasciculus, and it is believed, from the locus ceruleus. Together, these are pathways through which salivary secretion (via CN VII) and mastication are coordinated.

Tensor tympani and the stapedius reflex

Loud sounds cause reflex contractions of the tensor tympani and stapedius. This attenuates movements of the tympanic membrane and ossicles of the middle ear. Afferent impulses are carried via the cochlear nerve, reaching the cochlear nuclei of the brainstem. Efferent fibers to the tensor tympani arise within the trigeminal nerve's motor nucleus. They travel in the mandibular division of the nerve. Efferent fibers to the stapedius begin in the facial nucleus. They travel via the facial nerve.

> **Focus on the tensor tympani muscle**
> Closing the eyes very tightly results in a loud, rumbling sound in the ears. This is caused by contraction of the tensor tympani muscle, which normally functions to dampen certain sounds, such as those created by chewing. Closing the eyes tightly contracts the tensor tympani and tightens the eardrum via a tendon that shifts the stirrup back from the oval window to the inner ear.

Trigeminocerebellar fibers

The **trigeminocerebellar fibers** lie within the inferior cerebellar peduncles. They transmit proprioceptive information from the face to the cerebellum. This information comes from the facial muscle spindles (proprioceptors). The primary cell bodies are in the mesencephalic nucleus of the trigeminal nerve. The fibers transmit information to secondary afferent cell bodies within the **oralis** and **interpolaris** areas of the **spinal trigeminal nucleus** and principal nucleus. Axons from the spinal nucleus, and a small number of axons from the principal nucleus form the trigeminocerebellar tract, ascending to the cerebellum. Spinal trigeminal nucleus is stretched down to the upper cervical levels.

Trigeminal reflexes

The trigeminal reflexes include the corneal (blink) reflex, and the jaw jerk (masseter) reflex. The **corneal reflex**, or *blink reflex*, is the involuntary blinking of the eyelids caused by something touching the cornea of the eye. It can also result from any peripheral stimulus. This utilizes the orbicularis oculi muscles, which are the facial nerve efferents. The reflex is rapid, occurring in just 0.1 s. Its purpose is to protect the eyes from foreign bodies and bright lights. When bright lights are involved, it is also described as the *optical reflex*, which occurs more slowly, mediated by the visual cortex in the occipital lobe of the brain. The optical reflex is absent in infants less than 9 months of age. The corneal reflex also occurs when sounds louder than 40–60 dB are made. The reflex is mediated by the nasociliary branch of the ophthalmic branch of the trigeminal nerve that senses corneal stimulation via afferent fibers. It is also mediated by the temporal and zygomatic branches of the facial nerve that initiate the motor response via efferent fibers. Additional mediation is from the nucleus in the pons. The use of contact lenses can reduce or stop testing of this reflex. Damage to the ophthalmic branch of the trigeminal nerve results in absent corneal reflex when the affected eye is stimulated. Stimulation of one cornea usually has a consensual response, with both eyelids closing as a result.

The masticatory nucleus is the relay for the **jaw jerk reflex**, which is the only important supraspinal monosynaptic reflex. This reflex is brought about by lightly tapping the relaxed and open jaw, in a downward direction. It is also described as the *masseter reflex* or *mandibular reflex*, and is used to test the status of the trigeminal nerve as well as to help distinguish an upper cervical cord compression from lesions located above the foramen magnum. The mandible is tapped just below the lips, at the chin, while the mouth is held slightly open. The response is an upward jerking of the mandible by the masseter muscles—a very slight or sometimes absent movement. In an individual with upper motor neuron lesions, this reflex may be extremely pronounced. This reflex is classified as a **dynamic stretch reflex**. Sensory neurons of the trigeminal mesencephalic nucleus send axons to the trigeminal motor nucleus, which innervates the masseter. While not part of a standard neurological examination, testing this reflex is done when there are other signs of damage to the trigeminal nerve. A normal jaw jerk reflex points diagnosis toward **cervical spondylotic myelopathy**, and away from multiple sclerosis or amyotrophic lateral sclerosis. The jaw jerk reflex is enhanced in spastic bulbar palsy.

> **Focus on the jaw jerk reflex**
> Fast stretching of muscles that close the jaw activates muscle spindle afferents. These travel via the trigeminal nerve's mandibular division to the brainstem. The muscles that close the jaw include the masseter, medial pterygoid, and temporalis. Cell bodies of the primary afferent neurons are within the mesencephalic trigeminal nucleus. Collaterals project, monosynaptically, to the trigeminal nerve's motor nucleus within the pons. Motor axons from the nucleus travel along the mandibular nerve, innervating muscles acting upon the temporomandibular joint, to close the jaw.

Facial nerve

The facial nerve, or seventh cranial nerve, is primarily a motor nerve. It supplies all muscles involved in facial expression on its side of the face. There is a small sensory portion called the *nervus intermedius of Wrisberg*. This carries taste sensations from the anterior

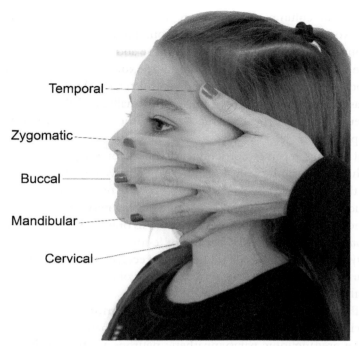

Fig. 11.1 The five major facial nerve branches. *(Courtesy of Morvarid Moini, DMD, MPH.)*

two-thirds of the tongue. It also variable carries cutaneous sensations from the anterior wall of the external auditory canal. Taste fibers initially travel sideways through the lingual nerve, which is a branch of the trigeminal mandibular portion. They later join the **chorda tympani**, carrying taste sensations through the facial nerve to the **tractus solitarius nucleus**. Secretory motor fibers begin in the superior salivatory nucleus. They innervate the lacrimal gland through the greater superficial petrosal nerve, as well as the sublingual and submaxillary glands via the chorda tympani. Facial nerve innervates all glands in the face except for the parotid gland that is being innervated by CN IX. The five major branches of the facial nerve are the temporal, zygomatic, buccal, mandibular, and cervical branches. These are shown in Fig. 11.1.

Facial motor nuclei

The facial nuclei are *motor*, lying within the caudal pontine reticular formation. They are posterior to the dorsal trapezoid nucleus and ventromedial to the spinal tract as well as the trigeminal nerve nuclei. Facial neurons are grouped to form columns innervating individual muscles or corresponding to facial nerve branches (refer to Chapter 10, Fig. 10.1). The neurons that innervate the muscles of the scalp and upper face are dorsal, while those that supply the lower facial musculature are ventral. Efferent fibers of large motor neurons of the facial nuclei form the facial nerve's motor root. The nucleus is part of the pharyngeal efferent column, yet lies more deeply within the pons that would be expected. Its axons have an unusual organization. They initially incline dorsomedially, toward the fourth ventricle. They are caudal to the abducens nucleus, ascending medial to it, near the medial longitudinal fasciculus. The axon's next curve around the rostral pole of the abducens nucleus. They descend ventrolaterally, through the reticular formation. After passing between their own nucleus, medially, and the spinal nucleus of the trigeminal nerve, laterally, they emerge between the olive and **restiform body**, at the cerebellopontine angle.

The facial nucleus receives corticonuclear fibers for intentional control. Neurons innervating the muscles in the scalp and upper face receive bilateral corticonuclear fibers. The neurons that supply the lower facial muscles receive mostly contralateral innervation. The upper and lower motor neuron lesions of the facial nerve can be clinically differentiated since the upper motor lesions result in paralysis only in the contralateral lower face, known as *supranuclear facial palsy*. Lower motor lesions result in total, ipsilateral paralysis, or *Bell's palsy*. The facial nucleus additionally receives ipsilateral afferents from the nucleus solitaries, as well as from the sensory nucleus of the trigeminal nerve. This establishes vital reflex connections.

Certain efferent facial nerve fibers begin in **visceromotor neurons** in the superior salivatory nucleus. This is located within the reticular formation, dorsolateral to the facial nucleus. These preganglionic parasympathetic neurons are part of the general visceral efferent column. Fibers from them reach the sensory root of the facial nerve, which is part of the intermediate nerve. The fibers continue, via the chorda tympani, to the submandibular ganglion. They also continue via the greater petrosal nerve and the pterygoid canal's nerve, reaching the pterygopalatine ganglion.

> **Focus on facial motor lesions**
> Contralateral hemiparesis suggests a pontine lesion near the facial motor nucleus, while hemiparesis ipsilateral to the facial paralysis suggests a cortical or subcortical lesion.

Clinical considerations

Clinical considerations related to the trigeminal and facial nerves include conditions known as *trigeminal neuralgia (tic douloureux), trigeminal neuropathies and neuritis, hemifacial spasm, Bell's palsy*, other facial palsies, *facial hemiatrophy*, and additional facial nerve disorders.

Trigeminal neuralgia

Trigeminal neuralgia is also referred to as *tic douloureux*. It is severe and paroxysmal facial pain caused by a disorder of the fifth (trigeminal) cranial nerve. The pain is described as *lancinating*, which is also described as a sensation of cutting, piercing, or stabbing. The condition is most prevalent in elderly adults. Trigeminal neuralgia may be caused by an intracranial artery or less commonly, a venous loop compressing the trigeminal nerve at its root entry zone into the brainstem. The involved vessels include the *anterior inferior cerebellar artery* and the **ectatic basilar artery**. Less common causes include compression from tumors, or sometimes, multiple sclerosis plaques at the root entry zone. These are usually revealed by accompanying sensory and other deficits. Disorders that can cause similar symptoms, such as multiple sclerosis, are occasionally believed to be trigeminal neuralgia, so identifying the actual cause of the pain is essential.

Pathophysiology
The mechanism of action of trigeminal neuralgia is unclear.

Clinical manifestations
The pain occurs along the distribution of one or more trigeminal nerve sensory divisions (Fig. 11.2). Usually, it involves the maxillary division. The paroxysmal pain lasts for seconds or as long as 2 min. Attacks can rapidly recur. The pain is excruciating and may incapacitate the patient. Pain may be precipitated by stimulating a facial trigger point,

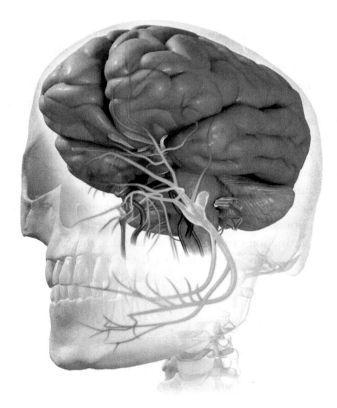

Fig. 11.2 The pain zones of trigeminal neuralgia.

involving activities such as brushing the teeth, chewing, or smiling. The patient may be unable to sleep on the affected side of the face.

Diagnosis
Diagnosis of trigeminal neuralgia occurs by differentiating from other disorders that cause facial pain. These include the following:
- *Chronic paroxysmal hemicranias*—involves more prolonged attacks lasting 5–8 min, with a dramatic response to a medication called indomethacin (both therapeutic and diagnostic)
- *Postherpetic pain*—which is of constant duration without paroxysms; there is a common antecedent rash, scarring, and an effect upon the ophthalmic division

With trigeminal neuralgia, the neurologic examination will be normal. Neurologic deficits—especially loss of facial sensation—suggest that the pain is being caused by something else. These causes may include a tumor, stroke, multiple sclerosis plaques, vascular malformations, and other lesions compressing the trigeminal nerve or disrupting its brainstem pathways.

Treatment
Treatment of trigeminal neuralgia is usually with anticonvulsants, including carbamazepine (the initial drug of choice), oxcarbazepine, gabapentin, phenytoin, baclofen, and amitriptyline. After 2 weeks of using carbamazepine, the hepatic enzymes and complete blood count should be assessed, then every 3–6 months. When carbamazepine is ineffective or has adverse effects, one of the other anticonvulsants is then tried. A peripheral nerve block may also provide temporary relief.

When pain is severe regardless of what is used, **neuroablative** treatments may be considered. These may only have a temporary effect. Improvement can be followed by recurring pain that is actually worse than previously. In *posterior fossa craniectomy*, a small pad is placed, separating the pulsating vascular loop from the trigeminal root. In radiosurgery, a gamma knife is used to cut the proximal trigeminal nerve. Via a percutaneous **stereotaxically** positioned needle, electrolytic or chemical lesions, or balloon compression of the trigeminal ganglion can be made. Sometimes, the trigeminal nerve fibers are cut between the ganglion and brainstem. As a last resort, the trigeminal nerve itself is sometimes destroyed.

> **Focus on trigeminal neuralgia**
> Trigeminal neuralgia is a form of neuropathic pain, which is associated with nerve injuries or nerve lesions. The typical or *classic* form, known as *type 1* or *TN1*, causes extreme, sporadic and sudden burning or shock-like facial pain. The *atypical* form, called *type 2* or *TN2*, has constant aching, burning, and stabbing pain that is slightly less intense.

> **Focus on trigeminal neuralgia—cont'd**
>
> However, both forms can occur in an individual, and sometimes at the same time. Rarely, both sides of the face may be affected at different times, or even more rarely, at the same time.

Trigeminal neuropathies

Facial and cranial injuries, as well as fractures, are probably the most common conditions that damage the trigeminal nerve branches. Trauma most often affects the superficial nerve branches, which include the **supratrochlear**, supraorbital, and infraorbital branches. Sensory loss develops from the time of injury. Partial regeneration may be accompanied by constant pain. **Herpes zoster** is the disease that most affects the trigeminal nerves. Persistent pain following a herpetic infection of the trigeminal nerve is serious and does not respond well to any treatments. A middle ear infection and osteomyelitis of the petrous bone apex may spread, reaching the ganglion and root, potentially involving the sixth cranial (abducens) nerve, as part of **Gradenigo syndrome**. Though HIV infection is not clearly involved in infection of the trigeminal nerve, reactivation of latent herpes zoster occurs with AIDS.

Compression or invasion of the trigeminal root may occur from intracranial meningiomas, trigeminal schwannomas, vestibular schwannomas, **cholesteatomas**, **chordomas**, and tortuous basilar artery branches. The nerve may also be infiltrated by sinus tumors and metastatic disease, resulting in pain and gradual progression of sensory loss. In cases of multiple sclerosis, demyelination at the trigeminal root's point of entry into the pons is another well-known cause. Trigeminal neuropathies can be bilateral in the setting of multiple sclerosis.

The trigeminal nerve's ophthalmic division may be implicated in the wall of the cavernous sinus, along with the third, fourth, and sixth cranial nerves. Recall that the third branch of trigeminal nerve V3 does not pass through the cavernous sinus. Many processes can be involved, including thrombosis of the cavernous sinus. Branches of the trigeminal nerve, at their foramina of entry or exit, may be affected by sphenoid bone tumors. These include metastatic carcinoma, myeloma, **lymphoepithelioma** of the nasopharynx, and squamous cell carcinoma. There is also a rarer **perineural infiltration** of superficial trigeminal nerve branches by squamous cell skin cancers of the face. The nerve's mandibular division can be compressed by the roots of an impacted wisdom tooth (third molar). Numbness of the chin and lower lip, due to infiltration of the mental nerve, may be the first sign of metastatic carcinoma of the breast, prostate, or multiple myeloma. This is known as the **numb-chin sign**.

Slowly evolving unilateral or bilateral trigeminal neuropathy also occurs, with sensory impairment confined to the area of the trigeminal nerve. There may be pain, paresthesias, or taste disturbances. These symptoms can occur as part of a widespread sensory neuropathy or **ganglionopathy**, which is a paraneoplastic effect of cancers, or with Sjögren's disease. There is also a common link between *isolated trigeminal neuropathy* and immune-mediated diseases of connective tissue. Studies have revealed diseases such as scleroderma or mixed connective tissue disease, sometimes with organ- or nonorgan-specific serum autoantibodies. Scleroderma is diagnosed by several specific antibody tests. Symptoms may involve the opposite side of the face, years after initial symptoms. There have also been cases of trigeminal neuropathy with scleroderma, Sjögren's disease, and lupus erythematosus. Some Sjögren's disease patients who had trigeminal neuropathy, also had related antibodies or inflammation of the minor salivary glands, a long time before characteristic manifestations occur. It is believed that an inflammatory lesion of the trigeminal ganglion or sensory root is implicated. Drugs such as stilbamidine and trichloroethylene cause sensory loss, burning, tingling, and itching only in the trigeminal sensory areas.

An isolated trigeminal neuropathy has been previously described as *Spillane's trigeminal neuritis*. In this condition, there is related paranasal sinusitis. Today, it is believed that connective tissue disease was actually interrelated. A less common type of idiopathic trigeminal sensory neuropathy has an acute onset, and usually completely or partially resolves, similar to Bell's palsy. Another form involves acute trigeminal symptoms of unknown origin. Facial numbness may occur as part of an upper cervical disc syndrome, including same-sided body numbness. The cervical spinal trigeminal nucleus or tract may be compressed in these cases. Facial numbness also happens with syringomyelia and other diverse conditions that affect the trigeminal nerve's spinal nucleus. There are other signs of brainstem or upper cervical cord disease.

A clinically rare condition is called *trigeminal motor neuropathy*, which is idiopathic and unilateral. There is an aching pain in the cheek and unilateral weakness of mastication. Electromyography shows denervation in the ipsilateral masseter and temporalis muscles, but outcomes are good. For most cases of trigeminal neuropathy, except for those due to tumors, herpes zoster, and demyelination, results of gadolinium-enhanced magnetic resonance imaging (MRI) will be normal, along with the cerebrospinal fluid. Recording of blink reflexes assesses the function of the trigeminal nerve. Some facilities have developed an evoked potential test just for the trigeminal nerve.

Hemifacial spasm

Hemifacial spasm involves unilateral, mostly painless, synchronized facial muscle contractions. It is caused by dysfunction of the facial nerve, its motor nucleus, or both. Hemifacial spasm can occur due to the compression of the nerve.

Diagnosis is clinical, focal seizures, blepharospasm, and tics should be excluded prior to diagnosis. Treatment with Botulinum toxin type A or B can be effectively used to treat hemifacial spasm. Hemifacial spasm was discussed in greater detail in Chapter 10.

Bell's palsy

Bell's palsy is subacute and idiopathic, unilateral palsy of the facial nerve. There is hemifacial paresis of the upper and lower parts of the face. While of unknown cause, the pathophysiology is believed to be swelling of the facial nerve because of an immune or viral disorder. Pain behind the ear often precedes the facial paresis. This condition is more common in diabetic patients. There are no specific diagnostic tests, so the condition is diagnosed by excluding other potential disorders. Treatments include corticosteroids, antivirals, eye lubrication, and use of eye patches. Bell's palsy was discussed in greater detail in Chapter 10.

Additional causes of facial palsy

The facial nerve may also be affected by **Lyme disease**, with no direct spirochetal infection of the nerve. Lyme-infected patients have shown almost simultaneous facial palsy and mild distal sensory polyneuropathy. Facial palsy may also be caused by HIV infection. The palsy of both Lyme and HIV is linked to a pleocytosis of the spinal fluid. Therefore, serologic and protective cerebrospinal fluid (CSF) examination can be helpful if either process is suspected. In rare cases, chickenpox in children may result, within 1 or 2 weeks, in facial paralysis. Another important cause of facial palsy is *Sarcoidosis*.

Ramsay-Hunt syndrome is caused by herpes zoster of the geniculate ganglion. It involves a facial palsy related to a vesicular eruption in the external auditory canal, other cranial integument areas, and the oropharyngeal mucous membrane. The infection may be identical, at first, to Bell's palsy, until the vesicles appear, though they sometimes do not appear. The eighth cranial nerve is also affected in some cases, resulting in deafness, nausea, and vertigo. The virus can be isolated before vesicles emerge by collecting exudate from the pinna skin upon a Schirmer strip, which is normally used to quantitate tearing. Polymerase chain reaction (PCR) techniques are then applied. Treatment involves acyclovir, famciclovir, or valacyclovir.

Facial palsy may be produced by tumors of the parotid gland, or those invading the temporal bone. These may be carotid body tumors, cholesteatoma, dermoid tumors, or **granulomatosis** (such as tuberculosis and sarcoidosis.

Facial hemiatrophy

Facial hemiatrophy is also known as *Parry-Romberg syndrome*. It is most common in females, causing the disappearance of fat within the dermal and subcutaneous tissues on one or both sides of the face. It therefore resembles **facial paresis**. The condition usually starts in adolescence or early adulthood, slowly progressing. When advanced, the affected side is gaunt as the skin becomes thin, wrinkled, and darkened. The hair may turn white and then fall out. The sebaceous glands become atrophic, but the muscles and bones are unusually not affected. The condition is actually a type of *lipodystrophy*.

A neural or growth factor is probably related due to localization within a **myotome**, of a known cause. In some cases, there is variegated coloration of the iris and **congenital oculosympathetic paralysis**. In rare cases, certain conjoined CNS abnormalities are linked to the ipsilateral hemisphere. These are primarily migraine, seizures, trigeminal neuralgia, and ventricular. The relationship between the conjoined CNS abnormalities is not fully understood. Facial asymmetry in adults frequently coexists with congenital or early-onset superior oblique palsy, and compensatory torticollis or head tile.

Problems with recovery from facial nerve palsy

When peripheral facial paralysis has been present for a long time, with some return of motor function, a contracture with diffuse **myokymic** activity may develop. There is narrowing of the palpebral fissure and deepening of the nasolabial fold. Facial muscle spasms may develop and continue indefinitely, resulting from any facial movements. Over time, the tip of the nose and corner of the mouth may pull over toward the affected side. This is a unique and acquired form of hemifacial spasm.

Aberrant or anomalous regeneration of the facial nerve fibers, after Bell's palsy or other injury, may cause other unique disorders that are limited types of **synkineses**. The most common type is the *jaw-winking phenomenon*, also known as the **Wartenberg sign** or *inverse Marcus-Gunn sign*. Jaw movements that are primarily lateral, engaging the pterygoid muscle, cause involuntary closing of the eyelid that is ipsilateral to the movement. If regenerating fibers that were originally connected with the orbicularis oculi later become connected with the orbicularis oris, eyelid closure may cause the corner of the mouth to retract. If visceromotor fibers that originally innervated the salivary glands later begin to innervate the lacrimal gland, there is anomalous tearing known as **crocodile tears**, which occurs when the patient salivates. A similar action explains gustatory sweating of the upper lip and cheek.

Additional facial nerve disorders

A fine, rippling activity of all muscles on one side of the face is known as **facial myokymia**. It is most often related to multiple sclerosis or a brainstem glioma. However, it can be related to neuromuscular junction disorders such as **neuromyotonia**. The condition has also developed following facial nerve disease such as Guillain-Barré variant (Fischer syndrome), which can be bilateral. Myokymia is usually related to the recovery stage instead of the early phase of this syndrome. The movements and their lack of rhythm seen in myokymia distinguish them from coarse and intermittent spasms as well as contracture, clonus, **tardive dyskinesia**, and tics. The electromyography pattern shows spontaneous **asynchronous discharge** of nearby motor units. These are either single, doubled, or tripled at rates between 30 and 70 cycles per second. It is believed that demyelination of the **intrapontine** area of the facial nerve, and maybe, supranuclear disinhibition of the facial nucleus, is causative.

The only sign of a focal cortical seizure may be a clonic or tonic contraction of one side of the face. Blepharospasm, which is involuntary recurrent spasms of both eyelids, may occur with nearly any type of dystonia and different from a seizure. There may be various degrees of spasm of the other facial muscles. Relaxants and tranquilizers help very little with this condition. However, botulinum toxin injections into the orbicularis oculi muscles can give temporary or lasting relief. Some patients have been helped by baclofen, clonazepam, and tetrabenazine, usually in increased dosages. In earlier times when these treatments failed, the periorbital muscles were destroyed by injecting doxorubicin, or by surgical **myectomy**. Since botulinum toxin came into use, these measures are no longer used, since they are rather extreme and cannot be reversed. Blepharospasm sometimes resolves on its own. Rhythmic, unilateral myoclonus, similar to palatal myoclonus, may only affect the facial, lingual, or laryngeal muscles. Hypocalcemic tetany causes hypersensitivity of the facial nerve. Spasm of the facial muscles is brought about by tapping in front of the ear, known as the **Chvostek sign**, but this phenomenon occurs in many normal patients.

Clinical cases

Clinical case 1
1. Is this probably a case of trigeminal neuralgia?
2. Could neurosurgical intervention be done for this patient?
3. What type of neurosurgical intervention is likely?

An 86-year-old female described 15 years of sharp, stabbing pain that radiated down the distribution of the second and third divisions of her right trigeminal nerve. She had two trigger points. One was on her right cheek, and the second was an intraoral trigger. Her symptoms were often triggered by eating, and she began to lose weight secondary to the pain. She had not contralateral pain, dysesthetic pain, or any burning pain sensations. She did not recently have any dental work and had no history of dental caries. Initial control of her symptoms had been with carbamazepine, but they had recently worsened even with increased dosages, and there was development of medication-related adverse effects.

Answers:
1. Yes, especially if there are no neurological problems, and brain MRI shows no masses or gross abnormalities.
2. Yes, neurosurgical intervention may be pursued since this patient was refractory to medical therapy and had worsening symptoms.
3. Noninvasive gamma knife radiosurgery is likely because of the patient's surgical risk profile.

Clinical case 2

1. If an adenoid cystic carcinoma of the parotid gland was present, what type of surgical procedure could be performed?
2. What are the other competing diagnoses for this patient?
3. Why can misdiagnosis of Bell's palsy cause further complications?

A 36-year-old man was referred for physical therapy evaluation and management of right facial muscle weakness. Idiopathic Bell's palsy had been previously diagnosed. The paralysis had lasted for 14 months, and there was also severe facial pain, regional facial numbness, Hyperacusis in the right hear, and excess tearing from the right eye. Brain MRI did not show a space-occupying lesion compressing the facial nerve on the right side. Two teeth were extracted on the right side of the mouth, which were believed to be the source of the facial pain, but the pain continued. The duration of the paralysis, pain, and numbness were inconsistent with Bell's palsy. The patient was referred to an ear-nose-throat specialist. An MRI of the base of the skull was ordered, revealing a diffuse, infiltrating lesion affecting the deep lobe of the parotid gland and lower division of the trigeminal nerve.

Answers:

1. For this carcinoma, there would likely be a total parotidectomy, partial right mandible resection, and facial nerve reconstruction.
2. Other competing diagnoses include parotid gland, facial nerve, or cranial base tumors, which all produce unilateral facial paralysis. Several tumors can be malignant, emphasizing the need for early recognition.
3. Since Bell's palsy is a diagnosis of exclusion, additional assessment, and testing that can reveal an alternative cause is usually delayed for three to 6 months, which is the expected time for signs of recovery from Bell's palsy.

Clinical case 3

1. What are the potential outcomes of trigeminal herpes zoster?
2. What is the description of the type of lesions caused by trigeminal herpes zoster?
3. What is usually the initial treatment for this condition?

A 57-year-old woman presented with severe herpes zoster infection that involved the maxillary and ophthalmic branches of her trigeminal nerves. She had severe facial skin eruptions, swelling, and pain. The patient's general practitioner presumed the condition to be facial cellulitis and had prescribed oral flucloxacillin followed by oral co-amoxiclav. These did not work, and the patient was taken to the emergency department because of intense worsening of her condition. Due to the initial delay in beginning the needed antivirals, the outcome could have been poor, but the patient recovered and had only mild scarring at 2 months postinfection.

Answers:
1. Potential outcomes include severe long-term neurological sequelae, including encephalitis, vision loss, and postherpetic neuralgia.
2. Trigeminal herpes zoster causes crusted, weeping vesicles over the course of the trigeminal nerve. The skin is usually reddened, warm, and does resemble the effects of cellulitis.
3. The initial treatment is usually intravenous acyclovir, but doses must take into account any autoimmune conditions as well as the possibility of bacterial superinfections that can be present.

Clinical case 4
1. What are the differential diagnosis for this patient?
2. Can Lyme neuroborreliosis involve the facial nerve?
3. What is the common course of treatments for this condition?

A 38-year-old man with hypertension and type 2 diabetes presented to the emergency department, with left-sided facial drooping and numbness that was present when he awakened in the morning. One month before, the patient had a flu-like syndrome with fever, myalgia, arthralgia, and fatigue as well as an ovoid-shaped rash on his thigh. He developed a diffuse headache and intermitted blurred vision 1 week after the initial symptoms. He tested negatively for Lyme disease and was treated for cellulitis with cephalexin for 5 days. The headache continued but the other symptoms resolved after 10 days. Eventually, an enzyme immunoassay was performed in the hospital, and the patient's Lyme antibodies showed very high levels of IgM, confirmed by a positive Western blot test.

Answers:
1. The differential diagnoses include idiopathic bell's palsy, multiple sclerosis, vascular insult, diabetic neuropathy, dural venous sinus thrombosis, and even intracranial masses.
2. Yes, Lyme neuroborreliosis can present as aseptic meningitis, recurrent meningoencephalitis, and cranial or spinal neuropathies, with the facial nerve being the most commonly involved. Sometimes, it affects other cranial nerves—primarily CN III, VII, IX, and X.
3. Treatment of Lyme disease is based on the clinical manifestations. It may include doxycycline, amoxicillin, cefuroxime, intravenous antibiotics, oral antibiotics, implantation of a temporary pacemaker, medications to treat meningitis if also present, and sometimes, corticosteroids.

Key terms

Aberrant corticospinal fibers	Lymphoepithelioma
Asynchronous discharge	Meckel's cave
Cervical spondylotic myelopathy	Mesencephalic nucleus
Cholesteatomas	Myectomy
Chorda tympani	Myokymic
Chordomas	Myotome
Chvostek sign	Neuroablative
Congenital oculosympathetic paralysis	Neuromyotonia
Corneal reflex	Numb-chin sign
Crocodile tears	Oralis
Dynamic stretch reflex	Perineural infiltration
Ectatic basilar artery	Principal sensory nucleus
Facial hemiatrophy	Ramsay-Hunt syndrome
Facial myokymia	Restiform body
Foramen ovale	Sjaastad syndrome
Foramen rotundum	Spinal trigeminal nucleus
Ganglionopathy	Stereotaxically
Gasserian ganglion	Superior orbital fissure
Gradenigo syndrome	Supratrochlear
Granulomatosis	Synkineses
Hemifacial spasm	Tardive dyskinesia
Herpes zoster	Tractus solitarius nucleus
Inferomedial	Trigeminal nerve
Interpolaris	Trigeminal neuralgia
Intrapontine	Trigeminocerebellar fibers
Jaw jerk reflex	Visceromotor neurons
Lyme disease	Wartenberg sign

Suggested readings

1. Alter, K. E.; Wilson, N. A. *Botulinium Neurotoxin Injection Manual.* Demos Medical, 2014.
2. Attlee, T.; et al. *Face to Face with the Face: Working with the Face and the Cranial Nerves through Cranio-Sacral Integration.* Singing Dragon, 2016.
3. Beider, S. *Getting out from the Funhouse Tunnel: How I Overcame Superior Oblique Myokymia.* Amazon Digital Services LLC, 2014.
4. Chowdhury, T.; Schaller, B. J. *Trigeminocardiac Reflex.* Academic Press, 2015.
5. Walker, H. K. Cranial Nerve V: The Trigeminal Nerve. In *Clinical Methods: The History, Physical, and Laboratory Examinations,* 3rd Ed.; Butterworths: Boston, 1990.
6. https://www.ncbi.nlm.nih.gov/books/NBK384/.
7. Esslen, E.; Fisch, U. *The Acute Facial Palsies: Investigations on the Localization and Pathogenesis of Meato-Labyrinthine Facial Palsies (Schriftenreihe Neurologie – Neurology Series).* Springer, 2011.
8. Ferreira, J. N. A. R.; Fricton, J.; Rhodus, N. *Orofacial Disorders: Current Therapies in Orofacial Pain and Oral Medicine.* Springer, 2017.
9. Golding-Kushner, K. J.; Shprintzen, R. J. *Velo-Cardio-Facial Syndrome Volume 2: Treatment of Communication Disorders.* Plural Publishing Inc., 2011

10. Guntinas-Lichius, O.; Schaitkin, B. M. *Facial Nerve Disorders and Diseases: Diagnosis and Management*. Thieme, 2015.
11. Icon Group International. *Hemiatrophy: Webster's Timeline History, 1885–2007*. Icon Group International Inc., 2010.
12. Kumbhare, D.; Robinson, L.; Buschbacher, R. *Buschbacher's Manual of Nerve Conduction Studies*, 3rd Ed.; Demos Medical, 2015.
13. Lustig, L. R.; Niparko, J.; Minor, L. B.; Zee, D. S. *Clinical Neurotology: Diagnosing and Managing Disorders of Hearing, Balance and the Facial Nerve*. Informa Healthcare, 2002.
14. Medifocus.com Inc. *Medifocus Guidebook on: Trigeminal Neuralgia – A Comprehensive Guide to Symptoms, Treatment, Research, and Support*. CreateSpace Independent Publishing Platform, 2018.
15. Miloro, M. *Trigeminal Nerve Injuries*. Springer, 2013.
16. Miodownik, C.; Lerner, V. *Tardive Dyskinesia: Current Approach*. Nova Science Publications Inc., 2018.
17. Okeson, J. P. *Bell's Oral and Facial Pain*, 7th Ed.; Quintessence Publishing Company, 2014.
18. Sindou, M.; Keravel, Y.; Moller, A. R. *Hemifacial Spasm: A Multidisciplinary Approach*. Springer, 2012.
19. Skouras, E.; Pavlov, S.; Bendella, H.; Angelov, D. N. *Stimulation of Trigeminal Afferents Improves Motor Recovery After Facial Nerve Injury – Functional, Electrophysiological and Morphological Proofs (Advances in Anatomy, Embryology and Cell Biology)*. Springer, 2013.
20. Slattery, W. H.; Azizzadeh, B. *The Facial Nerve*. Thieme, 2014.
21. Smouha, E. E.; Bojrab, D. I. *Cholesteatoma*. Thieme, 2011.
22. Strasheim, C.; Harris, S. *New Paradigms in Lyme Disease Treatment: 10 Top Doctors Reveal Healing Strategies That Work*. BioMed Publishing Group, 2016.
23. Usunoff, K. G.; Marani, E.; Schoen, J. H. R. *The Trigeminal System in Man. Advances in Anatomy Embryology and Cell Biology*, Vol. 136. Springer, 2012.
24. Yoshioka, N.; Rhoton, A. L. *Atlas of the Facial Nerve and Related Structures*. Thieme, 2015.
25. Zarins, U.; Kondrats, S. *Anatomy of Facial Expression*. Exonicus Inc., 2017

CHAPTER 12

Auditory system

The human ear has various sensory functions. It not only is used for hearing, but also for balance or *equilibrium*. For hearing and balance, the stimulation needed involves activation of **hair cells**, which are mechanoreceptors. Upon the hair cells, sound waves and fluid movement act to generate receptor potentials. After this, nerve impulses are generated that are soon perceived in the brain as either sound or balance. The eighth cranial nerve consists of the *cochlear nerve*, used for hearing and *vestibular nerve*, used for balance. There are three anatomical parts of the ear: the *external* ear, the *middle* ear, and the *inner* ear.

Anatomical structures

The anatomical structures of the human ear include the external, middle, and inner components (Fig. 12.1). Note that these structures are not illustrated by their actual sizes, with the middle and inner structures enlarged so that they are more easily seen.

External ear

There are two divisions of the external ear. These include the visible appendage known as the *auricle* or *pinna*, and the ear canal, which is also called the **external acoustic meatus**. The meatus is about 3 cm in length. At its beginning, it slants upward slightly, then curving downward through the temporal bone, in a direction that is both inward and forward. The ear canal ends at the **tympanic membrane**, which is commonly called the *eardrum*, which is stretched across the inner part of the ear canal and separates it from the middle ear. The external ear canal and outer tympanic membrane surface may be examined with an **otoscope**, a device that uses a light that can be pointed into the ear. Persistent perforation of the tympanic membrane, due to infection or trauma, may require a surgical procedure known as **myringoplasty**, which uses connective tissue to support healing of the perforation.

When changes occur in the ear canal and tympanic membrane, clinicians are able to determine a variety of pathologies. One example is a middle ear infection, known as **otitis media**. It causes the eardrum to be reddened and inflamed, with an outward bulging into the ear canal. This bulging is caused by pus and other fluids accumulating in the middle ear. Otitis media is usually caused by an ascending infection from the

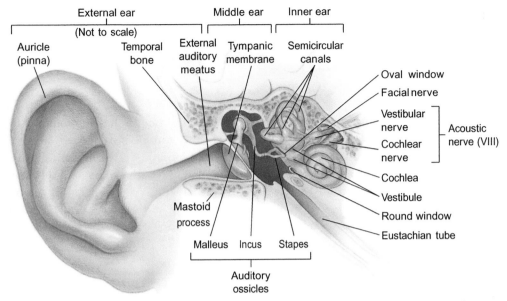

Fig. 12.1 The ear. External, middle, and inner ear structures. (Anatomic structures are not drawn to scale. Middle and inner ears enlarged for better visualization here.) *(From Patton, K. T.; Thibodeau, G. A. Anatomy & Physiology, 8 ed.; Mosby: St Louis, 2013.)*

nasopharynx, through the pharyngotympanic tube, to the middle ear cleft. This may also affect the **mastoid aditus** and antrum.

Cerumen (earwax), secreted by modified sweat glands within the auditory canal, can cause pain and temporary deafness. Common injuries to the external ear include blunt trauma, inflammation known as *cauliflower ear* (which may become permanent), and skin cancer due to its exposure to the sun. The *lobule* or lobe of the auricle may become infected because it is a common site for body piercings. It heals more slowly since it has a small blood supply. The *tragus* of the auricle is just in front of the ear opening in adults. The method of examining the external ear is shown in Fig. 12.2.

Middle ear

The **tympanic cavity** is also known as the *middle ear*. It is very small, lined with epithelial cells, and resides within the temporal bone. The tympanic cavity contains the three **auditory ossicles**, known as the *malleus, incus,* and *stapes*. The names of these tiny bones describe their shapes: malleus = hammer, incus = anvil, and stapes = stirrup. The "handle" portion of the malleus is attached to the inner tympanic membrane surface, while its "head" attaches to the incus. The incus then attaches to the stapes. There are tiny ligaments and muscles connected to the ossicles, helping to stabilize them and

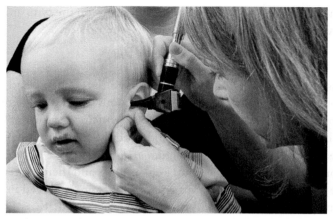

Fig. 12.2 Examining the external ear.

via reflex actions, dampen vibrations when sounds are severely loud. Several openings exist into the middle ear cavity. One is from the external acoustic meatus, and is covered with the tympanic membrane. There are also the **oval window**, into which the stapes is fitted, and the **round window**, which is membrane covered. Additionally, there is one opening into the auditory tube, also called the *eustachian tube*. This is partly made up of bone, cartilage, and fibrous tissue. The auditory tube is lined with mucosa, extending down, forward, and inward from the middle ear to the nasopharynx, also called the **pharyngotympanic tube**. This is the part of the throat located behind the nose.

The middle ear is posteriorly continuous with many mastoid air spaces within the temporal bone. These openings create routes through which infections can progress. Especially in children, head colds can lead to mastoid or middle ear infections, via the nasopharynx-auditory tube and the middle ear-mastoid pathway. **Mastoiditis** is a potentially dangerous and even life-threatening condition. It develops because of a bacterial infection that spreads from the tympanic cavity through the aditus to the mastoid antrum and related mastoid air cells. Sometimes, it may spread through the tegmen tympani to the dura mater of the middle cranial fossae. This can cause a temporal lobe abscess or meningitis. The condition can also spread into the posterior cranial fossa and cerebellum.

The auditory tube helps equalize pressures against the inner and outer surfaces of the tympanic membrane. This prevents pressure differences from rupturing the membrane and causing significant pain. Equalization of pressure occurs through swallowing or yawning, which causes air to spread quickly through the open tube. Then, atmospheric pressure presses against the inner surface of the tympanic membrane. Since atmospheric pressure is exerted against its outer surface on a continual basis, the pressures are equal. Another way to equalize pressures is to use a variation of the **Valsalva maneuver**, in which the mouth is closed, the nose is pinched, and the patient slightly exhales.

This pushes air into the auditory tube to equalize pressure on the tympanic membrane. It works if swallowing or yawning does not equalize the pressures.

Inner ear

The inner ear is also called the **labyrinth**. It has a complex shape, and consists of two primary parts—the bony labyrinth, and the membranous labyrinth. There are three parts of the bony labyrinth: the vestibule, cochlea, and semicircular canals (Fig. 12.3). The membranous labyrinth contains the **utricle** and **saccule** in its vestibule; the *cochlear duct* inside its cochlea, and the membranous *semicircular ducts* inside the bony canals. The vestibule and its contents, along with the semicircular canals and their contents are involved in balance. The cochlea and membranous cochlear duct are involved in hearing. Inside the membranous labyrinth, the **endolymph** is a clear fluid that has large amounts of potassium. The electrical potential of the endolymphatic compartment is about 80 mV more positive than the **perilymphatic compartment**. This is known as the **endolymphatic potential**.

Endolymph appears to circulate in the labyrinth and enter the endolymphatic sac. Here, it is moved into the adjacent vascular plexus through the specialized epithelium of the sac. In other labyrinthine regions, pinocytotic removal of fluid also occurs. Endolymph has more potassium and chloride, but less sodium and calcium concentrations,

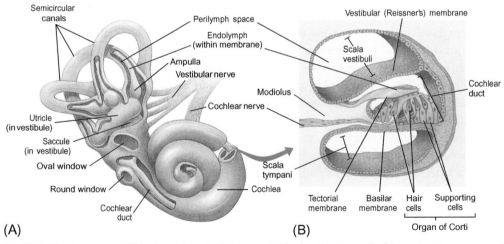

Fig. 12.3 The inner ear. (A) The bony labyrinth *(orange)* is the hard outer wall of the entire inner ear and includes semicircular canals, vestibule, and cochlea. Within the bony labyrinth is the membranous labyrinth *(purple)*, which is surrounded by perilymph and filled with endolymph. Each ampulla in the vestibule contains a crista ampullaris that detects changes in head position and sends sensory impulses through the vestibular nerve to the brain. (B) The inset shows a section of the membranous cochlea. Hair cells in the organ of Corti detect sound and send the information through the cochlear nerve. The vestibular and cochlear nerves join to form the eighth cranial nerve. *(From Patton, K. T.; Thibodeau, G. A.* Anatomy & Physiology, *8 ed.; Mosby: St Louis, 2013.)*

than perilymph. The high potassium levels are vital for function of mechanosensory hair cells. Concentrations are maintained by the lateral walls, which contain the **spiral ligament** and **stria vascularis**. These tissues collectively promote recirculation of potassium from perilymph, back to endolymph. This occurs via uptake through potassium channels and gap junction communication. The gap junctions form from **connexins**, which are significant in that their mutations are significant causes of hearing loss. Vestibular regions may not have any endolymphatic potential. This is because their lateral wall structure is much simpler than that of the cochlea. The difference in potassium concentration between endolymph and perilymph is clinically relevant. Endolymph is resorbed into the cerebrospinal fluid (CSF) from the endolymphatic sac, providing a drainage site for endolymph from all of the membranous labyrinth.

The differences in ionic potential and composition, regulated by the homeostatic labyrinth wall tissues, are required for the maximal sensitivity of mechanosensory hair cells. These cells convert vibrations starting in the inner fluids from head or sound movements, into electrical signals. The signals are transmitted through the vestibulocochlear nerve to the vestibular and cochlear nuclei, respectively, of the brainstem. The membranous labyrinth is surrounded by **perilymph**, which is similar to CSF. This fluid fills the space between the membranous labyrinth, its contents, and its surrounding bony walls.

Perilymph is most like CSF in the scala tympani, with greater concentrations of potassium, glucose, amino acids, and proteins that it contains in the scala vestibuli, from which the perilymph may be derived from plasma. The perilymph in the scala tympani contains some CSF, from subarachnoid spaces via the cochlear canaliculus. Perilymph homeostasis is mostly locally regulated.

A further explanation of the bony labyrinth is that it contains the vestibule, semicircular canals, and cochlea. The vestibule contains the saccule and utricle, also called the *saccules* and *utriculus*. All of these cavities are lined by periosteum, and contain the membranous labyrinth. The bone of the bony labyrinth is denser, and much harder than other parts of the petrous bone.

Section review
1. At which structure does the ear canal end?
2. What are the names of the three auditory ossicles?
3. What are the three parts of the bony labyrinth?

Hearing process

The actual organ of hearing is the **organ of Corti** or *spiral organ*. It is located upon the basal membrane, through the total length of the cochlear duct, which is discussed later. The organ of Corti has supporting cells and vital hair cells, projecting to the *endolymph*, which is produced by the marginal cells of the stria vascularis as well as the dark cells of the vestibule. The supporting cells and hair cells are covered by an adherent gelatinous membrane, known as the **tectorial membrane**. Sensory neuron dendrites, with cell bodies within the spiral ganglion of the modiolus, begin around the bases of the inner row of ciliated hair cells of the organ of Corti. Sound vibrations cause the cilia to bend. This causes the membrane potential to be altered, and transduces sound into a neural signal. The neurons' axons extend, forming the cochlear nerve, which is a branch of the eighth cranial nerve. They carry impulses, which produce the sensation of hearing. Outer rows of hair cells react to vibrations by shortening or lengthening, to amplify the vibrations and make the ear more sensitive. If extremely loud sounds occur, reflexive neural signals to the outer hair cell result in them reducing their amplification activities, protecting the delicate inner hair cells.

Organ of Corti

The *organ of Corti* is made up of many epithelial structures lying on the zona arcuata of the basilar membrane. Toward the center are two cellular rows: the internal and external **pillar cells**. Their bases are widened, resting on the basal membrane, yet their cell bodies, which appear as rods, are highly separated. The rows lean toward each other, contacting again at the heads of the pillar cells. They enclosing the **tunnel of Corti** between them and the basal membrane. The tunnel of Corti has a triangle-shaped cross section. A single row of inner hair cells lies internal to the inner pillar cells. Three or four rows of outer hair cells are external to the outer pillar cells. Supporting cells called *outer phalangeal cells* or **Deiters' cells** cup the bases of the outer hair cells—except for a gap area, in which cochlear axons synapse with these. The **reticular lamina** is formed by the apical ends of hair cells and the apical processes of the supporting cells. This lamina is blanketed by the tectorial membrane, which is gel like, and projects from the spiral limbus. The reticular lamina stops ions from penetrating, maintaining the electrochemical gradient between the surrounding apices, basolateral membranes of sensory hair cells, and fluids. The tectorial membrane is separated by a thin gap from the reticular lamina—except where the outer hair cells' apical stereocilia make contact.

The outer hair cells are also surrounded by the intercommunicating *spaces of Nuel*. Along with the tunnel of Corti, these spaces are filled with perilymph that diffuses through the basilar membrane's matrix. This fluid is sometimes referred to as **cortilymph**. Minor alterations in the perilymphatic composition may occur in it, since it is exposed to functions of specialized excitable cells and synaptic endings. Every pillar cells as a base or *crus*, a lengthened rod or *scapus*, and an upper end or *caput*, which means "head." Each caput and crus is intact. However, the scapi are separated via the tunnel of Corti. Many microtubules that are 30 nm in diameter are arranged in linked parallel bundles of 2000 or more in the scapus. These originate in the crus, and diverge above the scapus, terminating in the head region. The nucleus rests in an expansion upon the basal lamina.

The inner pillar cells number nearly 6000. Their bodies create an angle of about 60 degrees with the basilar membrane, with their bases resting on the basilar membrane near the tympanic edge of the internal spiral sulcus. Their heads look like the proximal end of the ulna. They have deep concavities for the out pillar cell heads, which they hand over, forming the top part of the tunnel of Corti. The outer pillar cells number almost 4000. They are more oblique and longer than the inner pillar cells, with an angle of about 40 degrees with the basilar membrane. Their heads are fitted into the concavities on the heads of the inner pillar cells. They project externally as slim processes between the first row of outer hair cells, which contact the processes of the Deiters' cells. Between the bases of the inner and outer pillar cells, the distances between the bases increase from the cochlear base to its apex. The angles they form with the basilar membrane are reduced.

The cochlea's sensory transducers are its hair cells. They collectively detect sound wave frequency and amplitude entering the cochlea. All of these hair cells have a consistent organization. They are lengthened cells, with modified apical stereocilia or microvilli, containing parallel groups of actin filaments. They are similar to the vestibular hair cells. Groups of synaptic contacts with cochlear nerve fibers are located at their bases, which are founded. Along the inner edge of the inner pillar cells and spiral tunnel is a single row of inner hair cells. The outer hair cells are arranged in three to five rows, mixed with supporting cells. These groups have unique roles in sound reception, with their structural differences reflecting this. There are 12,000 outer hair cells and only 3500 inner hair cells. These two sets lean toward each other apically, at nearly the same angles as the nearby inner and outer pillar cells. Their arrangement is precise, and closely related to the cochlea's sensory abilities.

The inner hair cells are slightly curved, with a pear shape. The thinner end is pointed toward the surface of the organ of Corti. The wider, basal end is located with a larger distance above the basilar membrane's inner end. Inner border cells and inner phalangeal cells surround the inner hair cells. These are externally attached to the heads of the inner pillar cells. Each inner hair cell's flat apical surface is elliptical when seen from above. Their long axes are directed toward the row of hair cells. The width of the apex is more than that of the inner pillar cells. Each inner hair cell is related to more than a single inner pillar cell. Each apex has 50–60 stereocilia in several groups of increasing height, with the tallest ones on the strial side. Of the shorter rows, their tips are diagonally connected to the sides of the nearby, taller stereocilia via thin filaments called *tip links*. Each stereocilium is also connected to every one of its neighboring structures by different lateral links. The stereociliary row's height is varied along the length of the cochlea. It is shortest at its base and tallest at its apex. The stereociliary bases are inserted into a transverse lamina of dense fibrillary material called the *cuticular plate*. This lies just beneath the apical surface of every inner hair cell. The plate has a small aperture containing a basal body. During development, a kinocilium with microtubules is anchored at this point. This condition is also present in the vestibular hair cells. Each inner hair cell, at its base, forms 10 or more synaptic contacts that have afferent endings. Each of these is marked with a presynaptic structure that is similar to the retina's ribbon synapses. Sometimes, an efferent synapse makes direct contact with a hair cell base. However, these are usually presynaptic to the afferent endings' terminal expansions instead of to the hair cells.

The outer hair cells are nearly two times as tall as the inner hair cells, and are in long, cylindrical cells. The outer row is longest in any one cochlear region. The hair cells of the cochlear apex are taller than those in the base, and are surrounded by the Deiters' cell processes (apical or phalangeal). However, on the internal side of the inner row, they are surrounded by the heads of the outer pillar cells. There may be up to 100 stereocilia per cell, in three rows of graded heights with the tallest on the outer side. The rows are arranged as a V or a W shape, based on the cochlear region, with the angles pointing externally. The stereocilia also have height gradations based on their cochlear region, with those of the cochlear base being shortest. Similar to the inner hair cells, the stereocilia have tip links and other filamentous connections with neighboring stereocilia. They are inserted, at their thin bases, into a cuticular plate. The tallest stereocilia are embedded into slight impressions on the underside of the tectorial membrane. The round nucleus is located near the base of the cell. Below it are a few synapses with a ribbon-like appearance. These are related to the afferent endings of the cochlear nerve. The afferent endings are of less number, and are smaller than the groups of efferent boutons contacting the base of the cell. At the afferent synapses of both inner and outer hair cells, the neurotransmitter is glutamate. In the efferent endings, it is acetylcholine. Other neurotransmitters or neuromodulators have also been revealed.

The cochlear hair cells are extremely and quickly sensitive to sound vibrations causing submicron deflections of the stereociliary hair bundles. The outer hair cells detect these vibrations and also generate force, increasing auditory sensitivity and frequency discrimination. Between the rows of outer hair cells are the Deiters' cells. Their enlarged bases are on the basilar membrane. Their apical ends partly surround the bases of the outer hair cells. Each of them has a phalangeal process extended diagonally upwards, between the hair cells, to the reticular membrane. Here, it forms a plate-like expansion filling the gaps between the hair cell apices.

External to the Deiters' cells are five to six rows of columnar supporting or external limiting cells, including *Hensen's cells* and *Claudius' cells*. Apices of hair and supporting cells that form the reticular lamina are joined by tight junction, desmosomes, and **adherens junctions**. The reticular lamina forms an extremely impermeable barrier to ions passage, except through the mechanotransducer channels of the stereociliary membranes. Additionally, it forms solid support between the hair cell apices. It links them mechanically to the underlying basilar membrane's movements. This causes lateral shearing movements between the overlying tectorial membrane and the stereocilia. If there is hair cell loss due to trauma, such as ototoxic drugs or excessive noise, the supporting cells quickly expand. They fill the gap and disturb the regular pattern of the reticular lamina via phalangeal scarring, but restore its function.

Tectorial membrane

The tectorial membrane is above the sulcus spiralis internus and organ of Corti. This stiff, gelatinous plate contains collagen types II, V, and IX. These collagens are interspersed with **tectorins**, a form of glycoproteins, contributing about 50% of the total proteins. When transversely sectioned, the tectorial membrane is easily identified by its almost flat underside and its convex upper surface. On the modiolar side, it is thinner, which is where it is attached to the vestibular labium of the spiral limbus. Its outer area forms a thickened ridge that overhangs the edge of the reticular lamina. Its lower surface is smoother, except where the outer hair cells' stereocilia are embedded. This leaves V- or W-shaped indentations. There is an S-shaped ridge is called **Hensen's stripe** that projects toward the inner hair cells' stereocilia. It is believed that the spiral limbus' interdental cells actually secrete the membrane.

Vestibule

The **vestibule** of the bony labyrinth is its egg-shaped, central cavity. It is about 5 mm from front to back and vertically, yet only 3 mm across. Posterior to the cochlea, the vestibule is also anterior to the semicircular canals, and medially flanks the middle ear. Its lateral wall contains the oval window or **fenestra vestibuli**. The two membranous labyrinth sacs (utricle and saccule) are suspended in the vestibular perilymph and connected by a small duct. Perilymph resembles CSF in its ionic composition, especially inside the scala tympani. It has high concentrations of potassium, amino acids, and proteins.

The saccule is slightly elongated. It is a globular sac in the spherical recess near the opening of the scala vestibuli, and is perforated by tiny holes called the *macula cribrosa media*. These transmit fine vestibular nerve branches to the saccule. The saccule is continuous with the membranous labyrinth, extended anteriorly into the cochlea as the cochlear duct. Behind the recess containing the saccule is the oblique **vestibular crest**. Its anterior end forms the **vestibular pyramid**. The crest is divided below, enclosing the small depression known as the **cochlear recess**. This is penetrated by the vestibulocochlear fascicles on their way to the vestibular end of the cochlear duct. The saccule is about 2.6 mm in length and 1.2 mm at its widest point. Unlike the utricular macula, the saccular macula is in a vertical plane on the wall of the saccule. It has an elliptical shape, with a small anterosuperior bulge. It is covered by an otolithic membrane, with a striola similar to that of the utricle. This is about 0.13 mm in width, extending along its long axis as an S-shaped strip, with functionally and anatomically polarized sensory cells.

The area above the striola is the **pars interna** and the area below is the **pars externa**. The saccule and utricle have similar operations. Due to its vertical orientation, however, the saccule is highly sensitive to linear head acceleration in the vertical plane, making it a major sensor of gravity, while the head is upright. It is also highly sensitive to movement along its anteroposterior axis.

The utricle is larger than the saccule, and continuous with the semicircular ducts that extend into the semicircular canals posteriorly. The utricle is irregular, dilated, and oblong. Its sac-like structure contacts the elliptical recess and the region inferior to this. The elliptical recess containing the utricle lies posterosuperior to the vestibular crest, in the roof and medial wall of the vestibule. The macula of the utricle is called the *utriculus*. It is a specialized neurosensory epithelium that lines the membranous wall. This is the largest vestibular sensory area of tall. It is heart shaped or triangular when viewed from its surface, lying horizontally. Its long axis is orientated anteroposteriorly, while its sharp angle points posteriorly. Except at the anterior edge, the utricle is flat. It is 2.8 mm in length and 2.2 mm in width. In adults, a bulge is usually present on its anterolateral border, with an indentation at the anteromedial border. Its epithelial surface has an **otolithic membrane**, and a gelatinous composition in which many small **otoliths** are embedded. The otolithic membrane has a curved ridge that corresponds to the narrow crescent of sensory epithelium called the striola, which is only 0.13 mm in width. Sensory hair cell density in this piece of epithelium is 20% less than in the remainder of the macula. The striola is laterally convex, running from the medial area of the anterior margin, in a posterior direction. It points to, but does not reach the posterior pole.

The *pars interna* is the area of the macula that is medial to the striola. It is slightly larger than the *pars externa*, which is lateral to it. Here, sensory cells are anatomically and functionally polarized compared to the midline of the striola. In each utricle, the macula is basically horizontal when the head is in the normal position. In any horizontal plane, linear head acceleration causes the otolithic membrane to "lag" behind movement so the membranous labyrinth, due to inertia produced by its mass. The membrane, therefore, maximally stimulates a group of hair cells via deflection of their bundles toward the striola. It inhibits others by deflecting the bundles away from it. As a result, every horizontal head movement will create a specific pattern of firing in the afferent nerve fibers of the utricle.

There are a number of holes called the *macula cribrosa superior* that perforate the pyramid and adjoining part of the elliptical recess. The holes in the pyramid transmit nerves to the utricle, while those in the adjoining part of the **elliptical recess** transmit nerves to the ampullae of the anterior and lateral semicircular canals. The area of the pyramid and elliptical recess corresponds to the superior vestibular area within the **internal acoustic meatus**. Below the elliptical recess, the vestibular aqueduct open, and reaches the posterior surface of the petrous bone. It contains one or more tiny veins as well as the endolymphatic duct, which is part of the membranous labyrinth. In the posterior vestibule, there are five openings of the semicircular canals. The anterior wall of the vestibule has an elliptical opening leading into the scala vestibuli of the cochlea. Both the utricle and saccule contain equilibrium receptor regions known as *maculae*. These respond to gravitational force, and report changes of head position to the brain.

Endolymphatic duct and sac

The endolymphatic duct becomes distally dilated to form the endolymphatic sac. It runs in the osseous vestibular aqueduct, and is of varying size. It can extend through an opening on the posterior petrous bone surface, and end between the two dural layers on the petrous temporal bone's posterior surface, close to the sigmoid sinus. Through all of the endolymphatic duct, the surface cells resemble the cells lining the unspecialized areas of the membranous labyrinth. They consist of low cuboidal or squamous epithelium. Where the duct dilates and forms the endolymphatic sac, the epithelia lining and subepithelial connective tissue become more complex. A distal sac and an intermediate or *rugose segment* are visible. In the intermediate segment, the epithelium has light and dark cylindrical cells. The light cells are regular in form. They have many long surface microvilli, with endocytic invaginations between them. In their apical region, they have vesicles that are large and clear. The dark cells are wedge shaped. They have narrow bases, dense and fibrillary cytoplasm, and much less apical microvilli.

The endolymphatic sac is important in maintaining vestibular function. Endolymph created in other parts of the labyrinth is absorbed in this region. This is believed to occur primarily by the light cells. If the sac is damaged, or its connection to the rest of the labyrinth it blocked, endolymph will accumulate. This causes *hydrops*, affecting cochlear as well as vestibular function. The epithelium is also permeable to macrophages and other leukocytes. These can remove cellular debris from the endolymph. The epithelium is also permeable to various immune system cells contributing antibodies to this fluid.

Section review
1. Where is the organ of Corti located?
2. What is the vestibule, and how is it shaped?
3. What is the endolymphatic sac important for?

Semicircular canals

The **semicircular canals** are posterior and lateral to the vestibule. Cavities of the bony semicircular canals are projected from the posterior aspect of the vestibule. Each of them is arranged in one of three planes. Therefore, there is an *anterior, posterior,* and a *lateral* semicircular canal in each internal ear. The anterior and posterior canals are at right angles to each other within the vertical plane. The lateral canal is horizontal. The anterior semicircular canal is 15–20 mm in length. The posterior semicircular canal is 18–22 mm in length. The lateral canal is only 12–15 mm in length. The canals are compressed from side to side. Each canal creates about two-thirds of a circle.

The **ampulla** of each canal is about twice the diameter of the canal itself. Each ampulla lies within the ampullae of the bony canal. This contains a region of equilibrium receptors, known as a **crista ampullaris**. The receptors respond to angular, rotational head movements. The semicircular canals are about one-quarter of the diameter of their osseous (bony) canals. Before entering the utricle, the medial ends of the anterior and posterior canals are fused, forming one common duct, called the **crus commune**. Each ampulla's membranous wall contains a transverse elevation called the **septum transversum**, upon the central area of which is a saddle-shaped sensory ridge called the **ampullary crest**. It contains hair cells and supporting cells. This is widely concave on the free edge, for most of its length, with a concave recess called the **planum semilunatum** at either end, between the ridge and duct wall. Across the ridge are crests of the lateral and anterior semicircular canals. They have rounded corners, while the posterior crest is more angled. The cupula, a vertical plate of gelatinous extracellular material, is attached on the crest's free edge (Fig. 12.4). This projects a long way into the ampulla's lumen. It is easily deflected by movements of endolymph from head rotations within the duct, which is how stimuli are delivered to the sensory hair cells. Therefore, the three semicircular canals can detect angular accelerations during head tilting or turning, in all three-dimensional (3D) planes of space.

A **semicircular duct** is threaded through each semicircular canal, and communicates anteriorly with the utricle. Each duct is much smaller than its containing canal, also elliptical, and believed to be linked to sensory control of body movements. This contains a region of equilibrium receptors, known as a *crista ampullaris*. The receptors respond to angular, rotational head movements.

Cochlea and cochlear duct

The term *cochlea* means "snail," and this is how the outer part of the bony labyrinth appears. It is 5 mm in length from its base to its apex, and 9 mm across the base. The cupula or *apex* of each cochlea points to the anterosuperior is of the medial wall of the tympanic cavity. The base of each cochlea points to the bottom of the internal acoustic meatus. It is perforated by many apertures for the cochlear nerve. When cut into sections, the cochlea appears as a tube wound in a spiral shape around a cone-shaped bone core called the *modiolus*.

Within the modiolus is the spiral ganglion, consisting of cell bodies of the first sensory neurons of the auditory relay. The modiolus has a broad base, near the lateral part of the internal acoustic meatus. Here, it corresponds to the spiral tract, or *tractus spiralis*

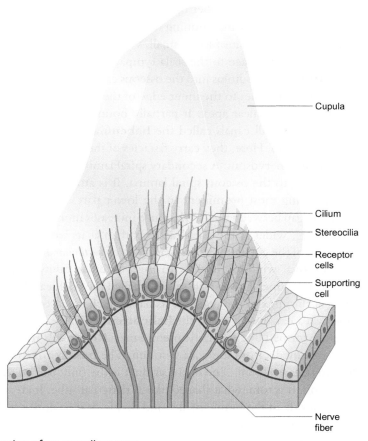

Fig. 12.4 A section of an ampullary crest.

foraminosus. Several openings are present for the fascicles of the cochlear nerve. The openings for the first 1.5 turns are through the small holes of the spiral tract. The openings of the apical turn are through the hole forming the center of the tract. From the spiral tract, canals continue through the modiolus, opening spirally into the base of the osseous spiral lamina. The small canals then enlarge, fusing to form **Rosenthal's canal**. This spiral canal in the modiolus courses similarly to the osseous spiral lamina. It contains the spiral ganglion. The primary tract continues through the center of the modiolus to the cochlear apex.

The osseous cochlear canal spirals for nearly three turns, around the modiolus. It is about 35 mm in length. At the first turn, the canal is bulged toward the tympanic cavity and underlies the promontory. The canal is 3 mm in diameter at the base of the cochlea. However, it becomes continually reduced in diameter as it spirals apically toward the cupula, where it ends. While the round and oval windows are the two main openings at its base, there is also a third and smaller opening, for the cochlear aqueduct or *canaliculus*. This is a tiny funnel-shaped canal running to the inferior surface of the petrous temporal bone. The canaliculus transmits a small vein to the inferior petrosal sinus, connecting the subarachnoid space to the scala tympani.

A ledge projects from the modiolus into the osseous canal, and is known as the osseous or *primary spiral lamina*. It attaches to the inner edge of the basilar membrane, ending in a hook-like *hamulus* at the cochlear apex. It partially boundaries the helicotrema. From Rosenthal's canal, many small canals called the **habenular perforata** radiate through the osseous lamina to its rim. Here, they carry fascicles of the cochlear nerve to the organ of Corti, via the foramen nervosum. A secondary spiral lamina is projected inward, from the outer cochlear wall, to the osseous spiral lamina. It is attached to the basilar membrane's outer edge, being most prominent in the lower part of the first turn. Toward the cochlear apex, the gap between the two laminae increases more and more. The basilar membrane is, therefore, wider at the cochlear apex than at its base.

The upper section of the cochlea is the *scala vestibuli* or *vestibular duct*. At the base of the scala vestibuli is the oval window, leading to the vestibular cavity, yet sealed by the "footplate" of the stapes bone. The middle section, called the cochlear duct or *scala media*, is "blind," ending at the apex of the cochlea. Its flanking channels open to each other at the modiolar apex via a narrow slit called the **helicotrema**. The scala media's upper and lower boundaries are formed by two elastic membranes. **Reissner's membrane** is thin, and separates the scala media from the scala vestibuli. The **basilar membrane** separates the scala media and scala tympani. The cochlear duct is the only part of the internal ear involved with hearing. It is somewhat shaped like a triangular tube. It creates a shelf-like structure across the inside of the bony cochlea and divides it into upper and lower areas. The organ of Corti sits on the inner surface of the basilar membrane. The lower section is the *scala tympani* or **tympanic duct**. The scala tympani is separated from the tympanic cavity by a secondary tympanic membrane located at the round window, or *fenestrae cochleae*. The scala vestibuli and scala tympani are filled with *perilymph*, while the cochlear duct is filled with *endolymph*.

Cochlear nerve

The cochlear nerve connects the cochlear nuclei and related brainstem nuclei to the organ of Corti. It lies inferior to the facial nerve, through the entire internal acoustic meatus. It is closely associated with the vestibular nerve's superior and inferior divisions. These are located in the posterior compartment of the canal. The cochlear nerve leaves the internal acoustic meatus in a common fascicle. The cochlear nerve contains 30,000–40,000 nerve fibers, with unimodal diameter distribution ranging from 1 to 11 μm (peaking at 4–5 μm). The nerve contains both functionally afferent and efferent somatic fibers, along with adrenergic postganglionic sympathetic fibers from the cervical sympathetic system.

The afferent fibers are myelinated axons. They have bipolar cell bodies in the spiral ganglion of the modiolus. Most ganglion cells (90%–95%) are large type I cells. The remainder is smaller type II cells. The type I cells have prominent spherical nuclei, many ribosomes and mitochondria. It is not fully understood if these are surrounded by myelin sheaths in humans. The type II cells are always unmyelinated, with lobulated nuclei. Their cytoplasm has neurofilaments, but less mitochondria and ribosomes than the type I cells. Basilar fibers are the peripheral processes of the type II cells, and are afferent to the outer hair cells. They have a spiraling course, turned toward the cochlear apex near the inner hair cell bases. They are as long as about five pillar cells, then turn radially to cross the floor of the tunnel of Corti, often diagonally. They then form part of the outer spiral bundle. The afferent fibers of the outer spiral group bundles course to the basal part of the cochlea. They continually branch along the way, supplying several outer hair cells. The outer spiral bundles also have efferent fibers.

In the cochlear nerve, the efferent fibers are derived from the **olivocochlear system**. In the modiolus, efferent fibers form the intraganglionic spiral bundle. This may be one or several groups of fibers at the periphery of the spiral ganglion. The two main groups of olivocochlear efferents are *lateral* and *medial*. The lateral efferents are from small neurons in or close to the lateral superior olivary nucleus. They primarily but not totally arise ipsilaterally, organized into inner spiral fibers through the inner spiral bundle. They terminate on afferent axons supplying the inner hair cells. The medial efferents originate from large neurons, near the medial superior olivary nucleus, with most arising contralaterally. They cross the tunnel of Corti, synapsing with the outer hair cells mostly by direct contact with their bases, with some synapsing with the afferent terminals. These medial efferents are myelinated. The efferent innervation of outer hair cells is decreased along the organ of Corti from the cochlear base to the apex, as well as from the inner row to the third row. These efferents use acetylcholine, gamma-aminobutyric acid (GABA), or both as their neurotransmitter. They can also contain other neurotransmitters and neuromodulators.

The medial efferents' activities inhibit cochlear responses to sound. With increasing sound levels, their activity strength increases slowly. They may regulate the cochlea's micromechanics by changing mechanical responses of the outer hair cells, which alters their participation in frequency sensitivity and selectivity. The lateral efferents that are related to the inner hair cells also have responses to sound. They mostly make contacts with the inner radial afferent fibers instead of with the inner hair cell base. These efferents are believed to modify transmission via the afferents. The radial fibers may be excited by the cholinergic fibers. Those containing GABA may inhibit them, those this is not fully understood. The autonomic nerve endings appear to be totally sympathetic. In the cochlea, there are two adrenergic systems. A perivascular plexus is derived from the stellate ganglion, and a blood vessel-independent system develops from the superior cervical ganglion. These systems both travel with afferent and efferent cochlear fibers, and appear to be only present in areas away from the organ of Corti. The sympathetic nervous system may cause cochlear effects that are primary and secondary, via remote alteration of the metabolism of various cells, as well as by affecting blood vessels and nerve fibers that it contacts.

> **Focus on cochlear damage**
> Damage to the cochlea usually causes permanent hearing loss that is sensorineural in nature. Potentially damaging factors include loud or extended noise exposure, powerful antibiotics, meningitis, Meniere's disease, acoustic tumors, and the natural aging process. Noises reaching 115 dB, with exposure as short as 15 min, can cause cochlear damage. For severe cochlear damage, a cochlear implants may be required in order to allow the patient to hear.

> **Section review**
> 1. What are the three planes of the semicircular canals?
> 2. What does the cochlea look like, and what is its modiolus?
> 3. What structures does the cochlear nerve connect?

Basilar membrane

The basilar membrane runs from the tympanic lip of the osseous spiral lamina to the spiral ligament's basilar crest, and has two zones. The thin **zona arcuate** runs from the spiral limbus to the outer pillar cells' bases, supporting the organ of Corti. It is made up of compacted bundles of collagenous filaments only 8–10 nm in diameter, primarily radial in orientation. The outer and thicker **zona pectinata** begins under the bases of the outer pillar cells. It is attached to the crista basilaris. In the zona pectinata, the basilar membrane is trilaminar. Its upper and lower layers fuse at its attachment point to the crista basilaris. The basilar membrane is 35 mm in length. Its width is 0.21 mm basally, to 0.36 mm apically. There is related narrowing of the osseous spiral lamina, and decreased thickness of the basal crest. The basilar membrane's lower or tympanic surface is covered by a layer of vascular connective tissue, and lengthened perilymphatic cells. The spiral vessel or *vas spirale* is larger than the others, and is just below the tunnel of Corti.

Bony labyrinth microstructure

Fibroblast-like perilymphatic cells and extracellular matrix fibers line the wall of the bony labyrinth. Cell morphology is different in various areas. Where the perilymphatic space is narrow, such as the cochlear aqueduct, the cells appear stellate or reticular. They have cytoplasmic extensions, in a sheet-like pattern, crossing the extracellular space. In areas where the space is greater, such as the scala vestibuli and scala tympani, as well as large parts of the vestibule, the perilymphatic cells upon the external surface of the membranous labyrinth and periosteum are flattened, resembling a squamous epithelium. On parts of the basilar membrane's perilymphatic surface, the cells have a cuboidal appearance.

In the bony surfaces lining the perilymphatic space, **canaliculi perforantes**, are present in a wide distribution. These micropores are between 0.2 and 23 µm in diameter. They are most numerous in the floor of the scala tympani, and in the peripheral and modiolar parts of the osseous spiral lamina. They are least numerous in the scala vestibuli's osseous wall. These canaliculi may provide a large fluid communication channel between the scala tympani and spiral canal, which could be used for drug-based cochlear therapies, or to deliver stem cells and others to cure deafness, or to implant electrodes.

Internal acoustic meatus

Also known as the *internal acoustic canal* or *internal auditory canal*, this meatus is separated from the internal ear, at its lateral fundus, via a vertical plate that is unequally divided by a falciform (transverse) crest. The facial, nervus intermedius, cochlear, superior, and inferior vestibular nerves pass through openings in the vertical plate of the internal acoustic meatus, above and below the transverse crest. Superior to the crest, canals are present in which the facial and superior vestibular nerves enter. The facial nerve lies anterior to the superior vestibular nerve. Here, it is separated at the lateral meatus end by **Bill's bar**, a vertical bone ridge. The nervus intermedius is between the superior vestibular nerve and facial motor root. It may adhere to the super vestibular nerve as well. In the superior vestibular area, there are openings for nerves to the anterior and lateral semicircular ducts, and to the utricle. Below the crest, an anterior cochlear region contains small holes in a spiral, known as the *tractus spiralis foraminosus*, which surrounds the central cochlear canal. Behind this area, the inferior vestibular region has openings for the saccular nerves. Most posteroinferiorly, a hole called the **foramen singular** allows the nerve to enter the posterior semicircular duct. Vascular loops in the internal acoustic meatus, from the anterior inferior cerebellar artery, may cause pulsatile tinnitus.

Vascular supply

The labyrinthine artery is the primary source of blood for the inner ear. The semicircular canals are also supplied by the stylomastoid branch of the occipital artery or posterior auricular artery. The labyrinthine artery divides off of the basilar artery, and sometimes, from the anterior inferior cerebellar artery. At the bottom of the internal acoustic meatus, it divides into the cochlear and vestibular branches. The cochlear branch further subdivides into 12–14 tiny branches through the modiolus canals. These are distributed as a capillary plexus to cochlear structures such as the basilar membrane, spiral lamina, and stria vascularis. The utricle, saccule, and semicircular ducts are supplied by vestibular arterial branches.

Veins that drain the vestibule and semicircular canals accompany the related arteries. They connect toward the utricle, forming the vein of the vestibular aqueduct. This empties into the sigmoid or inferior petrosal sinus. The inferior cochlear vein, of the cochlear aqueduct, usually drains into the inferior petrosal sinus or superior bulb of the internal jugular vein. It is created by a union of the common modiolar and vestibulocochlear vines, providing nearly all venous outflow from the cochlea. Near the basal cochlear turn, the common modiolar vein is formed by a joining of the anterior and posterior spiral veins. The vestibulocochlear vein is formed by a joining of the anterior and posterior vestibular veins, as well as the vein of the round window. When it is present, a labyrinthine vein drains the apical and middle cochlear coils into one of the following: the posterior area of the superior petrosal sinus, or the transverse sinus or inferior petrosal sinus.

Innervation

The tympanic cavity consists of the tympanic plexus and facial nerve. Branches from these structures supply other structures in the tympanic cavity, as well as facial structures. Nerves of the tympanic plexus ramify on the surface of the promontory. This is on the medial wall of the tympanic cavity. The nerves come from the tympanic branch of the glossopharyngeal nerve and the **caroticotympanic nerves**. From the cerebellopontine angle, the vestibulocochlear nerve emerges, continuing trough the posterior cranial fossa, close to the facial nerve, labyrinthine vessels, and nervus intermedius. Along with these, the vestibulocochlear nerve enters the petrous temporal bone through the **porus acusticus** of the internal acoustic meatus. The nerve then divides into an anterior trunk (the cochlear nerve) and a posterior trunk (the vestibular nerve). These trunks contain centrally directed axons of bipolar neurons, along with a lesser amount of efferent fibers arising from brainstem neurons. These terminate on the cochlear and vestibular sensory cells.

There are two histologically distinct parts of the vestibulocochlear nerve. The central glial zone is near the brainstem, and there is also a peripheral (nonglial) zone. In the central glial zones, axons are supported by central neuroglia. In the nonglial zone, they are surrounded by Schwann cells. The nonglial zone may extend into the cerebellopontine angle, medial to the internal acoustic meatus. During development, several weeks may be required between distal Schwann cell myelination and proximal glial myelination. This gap may coincide with the time of the organ of Corti's final maturation processes.

Vestibular nerve

The vestibular nerve contains cell bodies of bipolar neurons. They lie in the vestibular ganglion, within the trunk of the nerve, in the lateral end of the internal acoustic meatuses. Their peripheral processes innervate the ampullary crests of the semicircular canals, and the maculae of the utricle and saccule. In the vestibular nerve, the axons travel to the central nervous system (CNS). The vestibular nerve enters the brainstem at the cerebellopontine angle. It ends in the vestibular nuclear complex. The neurons of this complex are projected to motor nuclei in the brainstem and upper spinal cord, as well as to the cerebellum and thalamus. Thalamic efferent projects connect to a cortical vestibular area believed to be near the intraparietal sulcus in the second area of the primary somatosensory cortex.

Vestibular ganglion

In the vestibular ganglion, the neuron cell bodies are of many different sizes, with circumferences from 45 to 160 µm. Their sizes have no effect upon their distribution. The cell bodies have a large amount of granular endoplasmic reticulum, which forms Nissl bodies in certain areas, and prominent Golgi complexes. They are often arranged in pairs, covered by a thin layer of satellite cells. The cell bodies are so close together that just a thin endoneurium layer separates the nearby covering of satellite cells. The ganglion cells may, therefore, affect each other directly by an electrotonic spread known as **ephaptic transmission**. In the vestibular ganglion, there are two distinct sympathetic components. These include a perivascular adrenergic system, derived from the stellate ganglion; and the blood vessel-independent system from the superior cervical ganglion.

Intratemporal vestibular nerve

The vestibular ganglion cells' peripheral processes aggregate into definable nerves with specific distributions. The primary nerve divides with and at the ganglion, into superior and inferior divisions connected by an isthmus. The large superior division passes though small holes in the superior vestibular area, at the fundus of the internal acoustic meatus. It supplies the ampullary crests of the lateral and anterior semicircular canals, through the lateral and anterior ampullary nerves, respectively. A secondary branch supplies the utricle's macula. The larger part of the utricular macula is actually innervated by the utricular nerve, a separate branch of the superior division. Part of the saccule is supplied by a different branch of the superior division. The inferior division passes through small holes in the inferior vestibular area. It supplies the rest of the saccule and the posterior ampullary crest through saccular and singular branches, respectively. The singular branches pass through the foramen singular. Sometimes, the posterior crest is innervated by a small supplementary or accessory branch. This is probably a remnant of the crista neglecta, which is another area of sensory epithelium that is possible, but rarely found in humans.

In the inferior part of the vestibule nerve, afferent and efferent cochlear fibers are also present. They leave at the *anastomosis of Oort*, joining the main cochlear nerve. The *vestibulofacial anastomosis* is more centrally located, between the facial and vestibular nerves.

It is where fibers beginning in the intermediate nerve pass, from the vestibular nerve to the main facial nerve trunk. The vestibular nerve has about 20,000 fibers. Of these, 12,000 are in the superior division, and 8000 are in the inferior division. Fiber diameter distribution is bimodal, peaking at 4 and 6.5 μm. Small fibers mainly reach the type II hair cells. Larger fibers usually supply type I hair cells. Aside form the afferents, there are also efferent and autonomic fibers. Efferent fibers only synapse with the afferent calyceal terminals, around type I cells, and mostly with afferent boutons on type II cells. A few of them directly contact the cell bodies of the type II cells. The autonomic fibers terminate beneath the sensory epithelia, and do not contact the vestibular sensory cells.

> **Focus on vestibular disorders**
> The most commonly diagnosed vestibular disorders include benign paroxysmal positional vertigo, labyrinthitis or vestibular neuritis, Meniere's disease, and secondary endolymphatic hydrops. These disorders also include superior semicircular canal dehiscence, acoustic neuroma, perilymph fistula, ototoxicity, enlarged vestibular aqueduct, migraine-associated vertigo, and *mal de debarquement* (a sensation of continued movement, such as when on a boat, occurring after returning to land). Other related problems include complications from aging, autoimmune disorders, and allergies.

Section review
1. What is the primary source of blood for the inner ear?
2. What are the divisions of the vestibulocochlear nerve?
3. What is the definition of *ephaptic transmission*?

Perceiving sound

Sound is created by vibrations in the air, fluids, and even solid materials. As we speak, vibrations of our vocal cords create sound waves as they produce vibrations in the air that passes over them. There are a variety of terms used for the perception of sound, as follows:
- **Pitch**—the number of sound waves occurring during a specific unit of time (frequency) determines the pitch, or *tone*.
- **Hertz** (Hz)—a measurement of pitch, in waves per second; high frequencies are often expressed in thousands of waves per second, or *kilohertz (kHz)*; human hearing usually ranges between 20 Hz and 20 kHz, with most sensitivity being between 1 and 4 kHz.
- **Volume**—the perceived loudness of a sound wave, determined by its *amplitude* (height).

- **Decibels** (dB)—logarithmic units used to measure the volume of sound waves; many humans can hear pitches around 3 kHz as low as 0 dB, but pitches above and below 3 kHz require volumes ranging up to 60 dB or higher; sounds that are louder than 85 dB, when sustained, can cause irreversible hearing damage.

The ability to hear sound waves relies on pitch, volume, and other acoustic factors.

The effects of sound waves upon cochlear structures are shown in Fig. 12.5.

High-frequency sound waves cause the narrow part of the basilar membrane, which is near the oval window, to vibrate. Lower frequencies cause the membrane to vibrate near the apex of the cochlea. In this area, it is much thicker and wider. You can see that sound waves of different frequencies vibrate and cause displacement (bulging) of the basilar membrane at different areas. This explains how certain groups of hair cells respond to different sound frequencies. When a specific part of the basilar membrane bulges upward, cilia on hair cells attached to that area are stimulated. Then, the sound of a certain pitch can be perceived.

Different degrees of loudness of the same sound are perceived and determined by the movement (amplitude) of the basilar membrane at any certain point. The greater the upward bulge, the more that the cilia on the attached hair cells are simulated or bent, increasing the perceived loudness. A moving wave of perilymph is caused by the upward displacement of the basilar membrane. This is quickly dampened as it pushes through the cochlea. The sense of hearing is due to stimulation of the auditory area of the cerebral cortex. Sound waves must first be projected through air, bone, and fluids to stimulate nerve endings, then create impulse conduction over nerve fibers.

Pathway of sound waves

In the air, sound waves enter the external auditory canal, aided by the pinna. Reaching the inner end of the canal, they strike the tympanic membrane, and it vibrates. This moves the malleus, since its handle is attached to the membrane. The head of the malleus is attached to the incus, with the incus attached to the stapes. Therefore, vibration of the malleus moves the incus, and moves the stapes against the oval window into which it fits almost exactly. Then, fluid conduction of sound waves beings. The movement of the spates against the oval window results in pressure being created, inward into the perilymph, in the scala vestibuli of the cochlea. A rippling effect begins in the perilymph, transmitted via the vestibular membrane, which is the roof of the cochlear duct, to the endolymph inside, and then to the organ of Corti. The basal membrane also receives impulses, and transmits the ripples through the perilymph of the scala tympani, eventually reaching the round window.

Neural pathway of hearing

Neuron dendrites with cell bodies in the spiral ganglion, with axons forming the cochlear nerve, terminate around the bases of hair cells (of the organ of Corti), as well as the

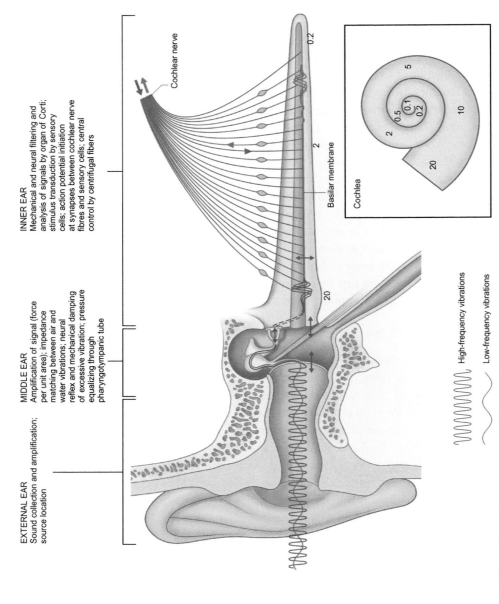

Fig. 12.5 Effects of sound waves upon cochlear structures.

tectorial membrane stuck to their upper surfaces. Hair cell movement against the adherent tectorial membrane stimulates the dendrites. Impulse conduction is initiated by the cochlear nerve, to the brainstem. Prior to reaching the auditory area of the temporal lobe, the impulses pass through *relay stations* in the nuclei of the medulla oblongata, pons, midbrain, and thalamus.

Hearing tests

Hearing loss can affect everyone, and often progresses very slowly, so that it is hard to notice its occurrence until significant symptoms develop. Hearing tests help determine not only hearing loss, but also its mildness or severity. They also help define the type of hearing loss, which is either conductive, sensorineural, or mixed. A thorough patient history is taken first, to determine the patient's background in regards to environmental noise exposure and family history of hearing loss. Also, medical conditions such as allergies, ear infections, impacted earwax, and head colds affect hearing ability. Trauma to the ears or head is also a consideration.

Hearing tests are painless and noninvasive. The patient is usually tested in a quiet, sound-treated booth that keeps out any extraneous noises. Headphones or soft earbuds are worn, connected to an *audiometer*, which supplies test tones at different frequencies and volumes. The patient is asked to raise on hand or push a button when each test tone is heard. The test measures the softest sounds that can be heard at each test frequency, and is known as *pure tone audiometry*. Another test called *speech audiometry* utilizes recorded or live speech instead of pure tones. It evaluates the softest speech sounds that can be heard and understood by the patient. If necessary, **tympanometry** may be performed, which determines if any hearing loss can be helped by hearing aids, or if another medical treatment is better to treat the hearing loss. The results of tympanometry are represented on a graph called a **tympanogram**. Another test is for acoustic reflexes, in which a soft plug that creates pressure changes and generates sounds is placed into each ear. It determines how well the eardrums move, and measures the reflexive responses of the middle ear muscles.

All newborn babies are required to receive hearing tests prior to leaving the hospital. These include *otoacoustic emission screening* and *automated auditory brainstem response*. The first test records tiny sounds made by the inner ear in response to clicks or chirps delivered through a small, flexible plug placed into the infant's ear. The second test records activity of the auditory nerve in response to the same type of clicks or chirps, and requires small electrodes to be taped to the baby's scalp. For older children, pediatric audiologists test patients 6 months of age and older using *visual reinforcement audiometry*, which measures reactions to sounds by turning the head toward them. Once a child is of school age, auditory testing is similar to the type provided for adults.

Hearing loss is measured in decibels, in the following categories:
- **Normal hearing**—0 to 25 dB hearing loss
- **Mild hearing loss**—26 to 40 dB
- **Moderate hearing loss**—41 to 70 dB
- **Severe hearing loss**—71 to 90 dB
- **Profound hearing loss**—greater than 91 dB hearing loss

> **Focus on hearing loss**
>
> Hearing impairment is a very common disabling sensory defect. Conductive hearing loss may result from trauma to the external or middle ears. It can also occur from blockage of the external acoustic meatus, disruption of the tympanic membrane, and infections associated with perforation of the tympanic membrane. The most common cause of a conductive hearing loss is middle ear effusion. Sensorineural hearing loss is the most prevalent type of hearing impairment. Usually, it is from loss or damage of the sensory hair cells or their innervation, but can also occur from lateral wall degeneration and loss of endolymphatic potential. At least 60% of hearing loss may be genetically linked.

Equilibrium

Within the vestibule and semicircular canals lie the sense organs used for the sense of *balance* or **equilibrium**. In the utricle and saccule are the sense organs functioning for **static equilibrium**. This function is required to sense head position in relation to gravity, or to sense the body's acceleration or deceleration, such as in a car or other vehicle. Inside the semicircular ducts are the sense organs associated with **dynamic equilibrium**. This is required to maintain balance when the head or body is suddenly moved or rotated.

Vestibule and semicircular canals

The vestibule is the central portion of the bony labyrinth, which opens into the oval and round windows from the middle ear, along with the three semicircular canals of the internal ear. The utricle and saccule are the membranous structures inside the vestibule, with walls made of simple cuboidal epithelium, and filled with endolymph. The three semicircular canals, inside the temporal bone, are at approximately right angles to each other. Inside these canals, separated from them by perilymph, are the membranous semicircular ducts. They each contain endolymph and join with the utricle inside the bony vestibule. Near the junction with the utricle, each of these canals enlarges, forming an ampulla.

The left and right semicircular canals of each pair of the left and right anterior canals respond oppositely to head movements affecting them. Some vestibular neurons receive bilateral input from the vestibular receptors. Therefore, they can "compare" between discharge rates of left and right canal afferents, which increases their sensitivity. Angular acceleration and deceleration of the head causes a counterflow of endolymph within the semicircular canals. This deflects the cupula of each crista and also bends the **stereociliary bundles** and **kinociliary bundles**. When a steady head movement velocity is reached, the endolymph very quickly reaches the same velocity as the surrounding structures. This is due to friction with the canal walls. Then, the cupula and receptor cells return to their resting state.

Maculae
The weight of otoconial crystals in the maculae causes a gravitational pull upon the otoconial membrane. It also pulls on the stereociliary bundles of sensory cells inserted into its base. Therefore, they can detect static orientation of the head related to gravity, and shifts in position based on the extent of deflection of the stereocilia. The two maculae are at right angles to each other. Their cells are functionally orientated in the opposite directions, across the striolar boundaries. Movement causes hair cell depolarization on one side of the striola, with hyperpolarization on the other side. The curved shape of the striola means that small hair cell groups on the macular epithelium respond to specific directions of head tilt or linear acceleration. Since the otoconia have collective inertia and momentum, any linear acceleration and deceleration, along the anteroposterior axis, can be detected. This is via overshooting or lagging of the otoconial membrane in relation to the epithelial surface. The saccular macula is able to signal such velocity changes.

Visual reflexes in equilibrium
The control of visual reflexes requires the vestibular system. These reflexes allow fixation of gaze upon an object no matter how the head is moving. They require coordination of eye, neck, and upper trunk movements. Constantly adjusting the visual axes is done mostly through the medial longitudinal fasciculus. This connects the vestibular nuclear complex with neurons of the oculomotor, trochlear, and abducens nuclei, and also with the upper spinal motor neurons and vestibulospinal tracts.

When there is abnormal vestibular input or central connection activity, the visual reflexes experience different effects, such as nystagmus. This can be brought about by the caloric test of vestibular function. In this, there is syringing of the external acoustic meatus with water, above or below body temperature. This appears to directly stimulate the cristae of the lateral semicircular canal. In Meniere's disease, spontaneous high activity in the afferent fibers of the vestibular nerve is seen. A range of disturbances occur, including dizziness and nausea, with the nausea reflecting the vagal reflex pathway's vestibular input.

Static equilibrium
Movements of the **macula**, a small strip of epithelium, provide information related to head position or acceleration. The macula is within the utricle and the saccule. This

sensory epithelium contains receptor hair cells and supporting cells, covered by a gelatinous matrix. Movement of the hair cells generates action potentials. This occurs as head changes position in relation to gravitational force. The way in that the maculae function is based on the weight of the **otoconial crystals**, which create a gravitational pulling effect upon the otoconial membrane and also the stereociliary bundles of sensory cells in its base. In the matrix of the macula, the otoliths are made up of protein and calcium carbonate. The sensory cells can detect static orientation of the head in relation to gravity. They additionally detect shifts in position based on the amount of deflection of the stereocilia. The two maculae are positioned at right angles. Their cells are functionally orientated in the opposite directions across their boundaries. With movement, there is depolarization of the hair cells on one side of the **striola**, with hyperpolarization of cells on the other side. Due to the curved shape of the striola, small groupings of hair cells on the macular epithelium respond to specific directions of head tilting or linear acceleration. Since the otoconial have collective inertia and momentum, acceleration and deceleration occurring linearly along the anteroposterior axis may be detected by changes in the otoconial membrane related to the surface epithelium. The saccular macula can then signal about these velocity changes.

Stimulation of the macula also can cause **righting reflexes**. These muscular responses restore the body parts to normal positions after being displaced. Righting reflexes are also activated by impulses from the eyes and proprioceptors. Disturbances of equilibrium, nausea, vomiting, and other symptoms can be caused by interruption of vestibular, visual, and proprioceptive impulses that cause righting reflexes to occur.

Dynamic equilibrium

Dynamic equilibrium is based on normal function of the crista ampullaris within the ampulla of each semicircular duct. This structure is also known as the *ampullary crest*. It is unique in its makeup, being a type of sensory epithelium that is quite similar to the maculae. Each crista appears as a ridge, marked with many hair cells that have processes embedded in the gelatinous **cupula**, which appears like a flap of tissue.

The cupula does not utilize otoliths, and does not respond to gravitation forces. Instead, it functions along with the flow of endolymph within the semicircular ducts, which are also at right angles with each other. Therefore, movements in every direction can be detected. Movement of the cupula causes the hairs within it to produce a receptor, then an action potential. This moves through the vestibular area of the eighth CN to the medulla oblongata, then to other brain and spinal cord areas to be interpreted and utilized. If the body is spun around, the semicircular ducts move similarly. However, inertia prevents the endolymph from moving at the same speed. The cupula, therefore, moves oppositely to the direction of head movement, until after the initial movement ceases. Dynamic equilibrium can detect changes in the direction, as well as the rate of a movement.

Vestibular system microstructure

In relation to gravity and changes in head movement, via the mechanosensitive hair cells, the maculae and crests detect head orientation. The hair cells are synaptically contacted with afferent and efferent endings of the vestibular nerve, upon their basolateral aspect. All of the epithelium lies on a thick, fibrous connective tissue that contains myelinated vestibular nerve fibers and blood vessels. The axons, as they perforate the sensory epithelium's basal lamina, lose their myelin sheaths.

There are two types of sensory hair cells within the vestibular system. Type I vestibular sensory cells are 25 µm in length. They have a free surface of just 6–7 µm in diameter. The basal care of the cell fails to reach the basal lamina of the epithelium. Every cell is usually shaped like a bottle, having a narrow neck, and a wider, rounded basal area that contains the nucleus. Its apical surface has 30–50 stereocilia, which are large and regular modified microvilli about 0.25 µm wide. Just one kinocilium is present, similar in arrangement to the microtubules of actual cilia. The kinocilium is long compared to the stereocilia, up to 40 µm. These are regularly arranged behind the kinocilium by descending height order, with the longest next to the kinocilium. Basally, the kinocilium emergency from a common basal body, with a centriole located just beneath it.

Every cell, near the inner surface of their basal two-thirds, has many synaptic ribbons related to synaptic vesicles. The postsynaptic surface of an afferent nerve ending closes the larger part of the sensory cell body as a cup-like *calyx*. Efferent nerve fibers synapse with the external calyx surface, and not directly with the sensory cell. The kinocilium give structural polarity to the bundle, which is related to functional polarity. Fine extracellular filaments *cross-link* the stereocilia and kinocilium. A specific one called the *tip link* connects each row of shorter stereocilia with adjacent stereocilia in the next taller row. This tip link occurs in all types of hair cells. It is believed to be required for transduction. Mutations in the proteins making up the tip link are significant in *Usher syndrome*, which involves visual and auditory abnormalities. When the bundle is deflected to the kinocilium, there is hair cell depolarization, increasing neurotransmitter release from its base. When deflected away from the kinocilium, hair cell hyperpolarization reduces neurotransmitter release.

Type II sensory cells have greater size variations, with some being up to 45 µm in length. They may span across all of the sensory epithelium, while others are smaller than the type I cells. Primarily, type II cells are cylindrical, yet resemble type I cells in their components. Their kinocilia and stereocilia are, however, shorter and of less variable length. The most significant difference is in their efferent nerve terminals. Type II cells receive several efferent nerve boutons that contain mixtures of small, clear, and sense-cored vesicles around their bases. Their afferent endings are not calyces, but instead, small expanded areas.

Polarization lets the hair cells have certain orientations optimizing their sensory functions. They are arranged symmetrically in the macula, on either side of the striola. In the utricle, kinocilia are on the side of the sensory cell closest to the striola, with the excitatory direction toward midline. In the saccule, functional and structural polarity is opposite—away from it. In ampullary crests, cells have their rows of stereocilia at right angles to the semicircular duct's long axis. In the lateral crest, kinocilia are on the side near the utricle. In the anterior and posterior crests, they are away from this. These differences are functionally important. Any acceleration of the head will maximally depolarize one hair cells group, and maximally inhibit another set.

Type I and II sensory cells lie within a supporting cell matrix, reaching from the epithelium base to its surface, forming surrounding structures around the sensory cells. Of irregular form, they are recognized by their nuclei position, usually lying below the level of sensory cell nuclei, just above the basal lamina. Supporting cells' apices are attached via tight junctions to nearby supporting cells, and to hair cells, producing the reticular lamina. This composite layer forms a relatively impermeable plate in relation to ions, except for the hair cells' mechanosensitive transduction channels. The otolithic membrane is made up of extracellular material, in two strata. The external layer has otoliths (otoconia) in heterogeneous distribution. These are attached to the more basal, gelatinous layer. Here, the stereocilia and kinocilia of the sensory cells are inserted. The gelatinous material mostly consists of glycosaminoglycans related to fibrous proteins.

Section review
1. What are the differences between static and dynamic equilibrium?
2. How do the otoconial crystals function in relation to changes in head position?
3. What are the two types of sensory hair cells in the vestibular system?

Clinical considerations

There are various clinical disorders of the auditory system that affect all three compartments of the ears. Our focus is primarily on the inner ear, such as otosclerosis, tinnitus, deafness, Meniere's disease, vertigo, vestibular neuronitis, drug-induced ototoxicity, and benign tumors of the inner ear. The clinical considerations concerning the auditory system are discussed in detail in Chapter 13.

Key terms

Adherens junctions
Ampulla
Ampullary crest
Auditory ossicles
Auricle
Basilar membrane

Bill's bar
Canaliculi perforantes
Caroticotympanic nerves
Cerumen
Cochlear recess
Connexins

Cortilymph
Crista ampullaris
Crus commune
Cupula
Deiters' cells
Dynamic equilibrium
Elliptical recess
Endolymph
Endolymphatic potential
Ephaptic transmission
Equilibrium
External acoustic meatus
Fenestra vestibuli
Foramen singular
Habenular perforata
Hair cells
Helicotrema
Hensen's stripe
Internal acoustic meatus
Kinociliary bundles
Labyrinth
Macula
Mastoid aditus
Mastoiditis
Myringoplasty
Olivocochlear system
Organ of Corti
Otitis media
Otoconial crystals
Otolithic membrane
Otoliths
Otoscope
Oval window
Pars externa
Pars interna
Perilymph
Perilymphatic compartment
Pharyngotympanic tube
Pillar cells
Planum semilunatum
Porus acusticus
Reissner's membrane
Reticular lamina
Righting reflexes
Rosenthal's canal
Round window
Saccule
Semicircular canals
Semicircular duct
Septum transversum
Spiral ligament
Static equilibrium
Stereociliary bundles
Stria vascularis
Striola
Tectorial membrane
Tectorins
Tunnel of Corti
Tympanic cavity
Tympanic duct
Tympanic membrane
Tympanogram
Tympanometry
Utricle
Valsalva maneuver
Vestibular crest
Vestibular pyramid
Vestibule
Zona arcuate
Zona pectinata

Suggested readings

1. Baldwin, C. L. *Auditory Cognition and Human Performance: Research and Applications*. CRC Press, 2012.
2. Carlson, M. L. *Tumors of the Ear and Lateral Skull Base: Part 2 (Otolaryngologic Clinics of North America, Clinics Review Articles)*. Elsevier, 2015.
3. Celesia, G. G.; Hickok, G. *The Human Auditory System, Volume 129: Fundamental and Clinical Disorders (Handbook of Clinical Neurology)*. Elsevier, 2015.
4. Dallos, P.; Fay, R. R. *The Cochlea (Springer Handbook of Auditory Research)*. Springer, 1996.
5. Eggermont, J. J. *Noise and the Brain: Experience Dependent Developmental and Adult Plasticity*. Academic Press, 2013.
6. Eshraghi, A. A.; Telischi, F. F. *Otosclerosis and Stapes Surgery, An Issue of Otolaryngology Clinics of North America*. Elsevier, 2018.

7. Fahy, F.; Thompson, D. *Fundamentals of Sound and Vibration*, 2nd ed.; CRC Press, 2015.
8. Ferrand, C. T. *Speech Science: An Integrated Approach to Theory and Clinical Practice*, 4th ed.; Pearson, 2017.
9. Gordon-Salant, S.; Frisina, R. D.; Fay, R. R.; Popper, A. N. *The Aging Auditory System (Springer Handbook of Auditory Research)*. Springer, 2010.
10. Harris, J. P.; Nguyen, Q. T. *Meniere's Disease, An Issue of Otolaryngologic Clinical Surgery*. Saunders, 2010.
11. Herdman, S. J.; Clendaniel, R. *Vestibular Rehabilitation (Contemporary Perspectives in Rehabilitation)*, 4th ed.; F.A. Davis Company, 2014.
12. Hussain, S. M. *Logan Turner's Diseases of the Nose, Throat and Ear; Head and Neck Surgery*. CRC Press, 2015.
13. Manley, G. A.; Gummer, A. W.; Popper, A. N.; Fay, R. R. *Understanding the Cochlea (Springer Handbook of Auditory Research)*. Springer, 2017.
14. Mansour, S.; Magnan, J.; Haidar, H.; Nicolas, K.; Louryan, S. *Comprehensive and Clinical Anatomy of the Middle Ear*. Springer, 2013.
15. Martin, F. N.; Clark, J. G. *Introduction to Audiology (Pearson Communication Sciences and Disorders Series)*, 12th ed.; Pearson, 2014.
16. Miller, J. *Hearing Loss and Equilibrium: Learn to Walk Again Like a Young Person*. Amazon Digital Services LLC, 2018.
17. Mills, S. E.; Stelow, E. B.; Hunt, J. L. *Tumors of the Upper Aerodigestive Tract and Ear (AFIP Atlas of Tumor Pathology, Series 14)*. American Registry of Pathology, 2014.
18. Motasaddi Zarandy, M.; Rutka, J. *Diseases of the Inner Ear: A Clinical, Radiologic, and Pathologic Atlas*. Springer, 2010.
19. Musiek, F. E.; Baran, J. A. *The Auditory System: Anatomy, Physiology, and Clinical Correlates*. Plural Publishing, 2016.
20. Ozkaya, N.; Leger, D.; Goldsheyder, D.; Nordin, M. *Fundamentals of Biomechanics: Equilibrium, Motion, and Deformation*, 4th ed.; Springer, 2017.
21. Ricketts, T. A.; Bentler, R.; Mueller, H. G. *Essentials of Modern Hearing Aids: Selection, Fitting, Verification*. Plural Publishing, 2017.
22. Silman, S.; Emmer, M. B. *Instrumentation in Audiology and Hearing Science: Theory and Practice*. Plural Publishing; PAP/CDR Edition, 2011.
23. Sokolowski, B. *Auditory and Vestibular Research: Methods and Protocols*, 2nd ed.; Springer Protocols, 2016.
24. Tos, M. *Manual of Middle Ear Surgery: Volume 1: Approaches, Myringoplasty, Ossiculoplasty and Tympanoplasty*. Thieme Medical Publishers, 1993.
25. Van Opstal, J. *The Auditory System and Human Sound-Localization Behavior*. Academic Press, 2016.

CHAPTER 13

Auditory system lesions and disorders

It is important to distinguish peripheral disorders related to the inner ear, from central disorders related to the brainstem or brain. Both *acute vestibular neuronitis* and *stroke* in the medulla cause vertigo, difficulty with balance, and ataxia. In order to distinguish the central causes of auditory and vestibular symptoms from peripheral causes, this chapter explains relevant clinical illnesses that are due to peripheral auditory and vestibular system disorders in detail, as they are very common in real practice.

Tinnitus

Tinnitus is defined as an abnormal noise in the ears. It occurs in 10%–15% of the population, in varying ages. There are two forms: *subjective* and *objective*. **Subjective tinnitus** is perceived sound without an acoustic stimulus, which is only heard by the patient. The majority of cases of tinnitus are subjective. **Objective tinnitus** is less common, resulting from noise generated by structures close to the ear. The condition can be loud enough to actually be heard by the examiner.

Pathophysiology

Subjective tinnitus is believed to be caused by abnormal neuronal activity in the auditory cortex. There may also be a loss of suppression of intrinsic cortical activity, and sometimes, creation of new neural connections. Conductive hearing loss, due to cerumen impaction, Eustachian tube dysfunction, or otitis media, may be linked to subjective tinnitus, via alteration of sound input to the central auditory system. Objective tinnitus involves actual noise generated by physiologic activities near the middle ear. This may be related to a turbulent flow of blood in abnormal vessels, such as caused by an arteriovenous malformation.

Clinical manifestations

Tinnitus can be described as buzzing, ringing, hissing, roaring, or whistling. It may be variable and complex. Objective tinnitus usually is in synchronization with the heartbeat, but may be intermittent. The condition is most noticeable in quiet environments and when there is a lack of other stimuli. Therefore, it is often reported as being worse at bedtime. Tinnitus may also be continuous, which is annoying, and can be distressing. Some patients experience depression because of it, and the condition is usually worsened by stress.

Diagnosis

Diagnosis of tinnitus is based on history, systems review, physical examination, and testing. Any factors that worsen or relieve it, such as head position changes and swallowing, should be noted. Important related symptoms include hearing loss, ear pain or discharge, and vertigo. The first step is to rule out hearing loss. History must assess noise exposure, sudden pressure changes, ear or central nervous system (CNS) infections or trauma, head radiation therapy, and recent major weight loss. Medications must be evaluated, especially salicylates, aminoglycosides, or loop diuretics.

The ear canal is inspected for cerumen, discharge, and foreign bodies. The tympanic membrane is inspected for signs of acute or chronic infection, and tumors. A bedside hearing test is performed. The cranial nerves—especially vestibular function—are tested, as well as peripheral strength, sensation, and reflexes. A stethoscope is used to listen for vascular noises over the carotid arteries, jugular veins, and dover as well as adjacent to the ear.

Red flags include bruit, especially over the ear or skull; unilateral tinnitus; pulsatile tinnitus, and accompanying neurologic symptoms or signs other than hearing loss. Considerations include acoustic neuroma, which is ruled out via gadolinium-enhanced magnetic resonance imaging (MRI). If there is a visible vascular tumor in the middle ear, computed tomography (CT), gadolinium-enhanced MRI, and referral to a subspecialist is used to confirm diagnosis. There may be a need to investigate the carotid, vertebral, and intracranial vessels. This usually beings with magnetic resonance angiography (MRA), sometimes followed by an arteriogram.

Treatment

Tinnitus may be reduced by treating the underlying disorder. Using a hearing air for corrective haring loss relieves tinnitus in about half of all patients. Stress and mental factors such as depression must be treated, and the patient can eliminate use of caffeine and other stimulants. Background sound-emitting devices may mask the tinnitus and aid in sleeping. Electrical stimulation of the inner ear, such as with a cochlear implant, is used for profoundly deaf patients. Tinnitus is a common complaint in the elderly since one of every four people over age 65 has significant hearing impairment.

Hearing loss

Almost 10% of people in the United States have some amount of hearing loss. Between one of every 800–1000 neonates are born with severe to profound hearing loss, and twice to three times as many are born with lesser hearing loss. In childhood, another two to three of every 1000 children develop moderate to severe hearing loss. Teenagers are at risk for excessive exposure to head trauma, noise, or both. Older adults usually have a progressive decrease in hearing called **presbycusis**, likely related to noise exposure and the aging process.

In early childhood, hearing deficits can result in impairments over the lifespan that will affect expressive and receptive language skills. Severity is based on the age at which hearing loss began, duration, sound frequencies affected, and the degree of impairment. Other considerations are intellectual disabilities, coexisting visual impairments, primary language deficits, and inadequate linguistic environments. Children with other sensory, cognitive, or linguistic deficiencies are most severely affected.

Pathophysiology

Hearing loss is classified as *conductive, sensorineural,* or *mixed* (involving a combination of both). **Conductive hearing loss** is secondary to lesions in the external auditory canal, eardrum, or middle ear. The lesions stop sound from being normally conducted to the inner ear (see Fig. 13.1). **Sensorineural hearing loss** is due to lesions of the inner ear, a sensory condition, or of the auditory nerve, a neural condition (Table 13.1). These are very different. *Sensory hearing loss* may be reversible, and is rarely life threatening. However, a *neural hearing loss* is rarely recoverable. It may be caused by a brain tumor; often this is a *cerebellopontine angle tumor*. **Mixed hearing loss** may be caused by severe head injuries, with or without fractures of the skull or temporal bone, from chronic infections, or by genetic disorders. This form can occur when a transient conductive hearing loss, often from otitis media, is superimposed upon a sensorineural hearing loss.

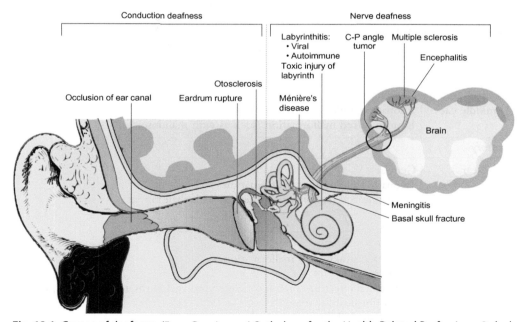

Fig. 13.1 Causes of deafness. *(From Damjanov, I. Pathology for the Health-Related Professions, 3rd ed.; Saunders: Philadelphia, 2006.)*

Table 13.1 Differences between neural and sensory hearing losses.

Test	Neural hearing loss	Sensory hearing loss
Acoustic reflex decay	Present	Absent or mild
Discrimination with increasing intensity	Deteriorates	Improves
Otoacoustic emissions	Present	Absent
Recruitment	Absent	Present
Speech discrimination	Severe decrement	Moderate decrement
Waveforms in auditory brainstem responses	Absent, or with abnormally long latencies	Well formed, with normal latencies

Hearing loss may be congenital or acquired, progressive or sudden, temporary or permanent, unilateral or bilateral, and mild or profound. It can also be caused by various medications. The most common causes of hearing loss, overall, are cerumen accumulation, noise exposure, aging, and especially in children and young adults, infections. Table 13.2 summarizes congenital causes of hearing loss. Many cases of congenital hearing loss involves mixed losses, a combination of conductive and sensory, with or without neural.

Clinical manifestations

Hearing loss often presents along with cerumen accumulation, foreign bodies obstructing the auditory canal, tinnitus, otitis media, and various autoimmune disorders.

Table 13.2 Congenital causes of hearing loss.

Area affected	Etiology (from most to least common)
Conductive	
External and middle ear	Genetic Idiopathic (unknown) malformation Drug-induced malformation (such as with thalidomide)
Sensory	
Inner ear	Genetic Idiopathic (unknown) malformation Drug-induced malformation (such as with thalidomide)
Neural	
CNS	Anoxia Idiopathic (unknown) malformation Genetic Congenital infection (such as rubella, cytomegalovirus, toxoplasmosis, syphilis) Neurofibromatosis (type 2) Rh incompatibility

Diagnosis

Diagnosis of hearing loss involves detection and quantification of the amount of loss and the possible causes. Screening should begin at birth so that the patient can develop optical language development. Specialists are required to properly evaluate hearing loss. Without screening early in life, severe bilateral losses may not be recognized until the age of 2 years. Mild to moderate, or severe unilateral losses may not be recognized until the child reaches school age. History must assess how long hearing loss has been present, its development, if unilateral or bilateral, if sounds are distorted and in what manner, and if speech discrimination is difficult. Head injury, barotrauma, new medications, ear pain, tinnitus, ear discharge, disorientation in darkness, vertigo, headache, facial weakness or asymmetry, ear fullness, and abnormal taste sensations must all be evaluated.

Medical history should consider CNS or repeated ear infections, noise exposure, head trauma, rheumatic disorders, and family history. Physical examination inspects the external ear. The eardrum is examined, and in the neurologic examination, attention is paid to the second through seventh cranial nerves, vestibular function, and cerebellar function. **Weber's test** and the **Rinne test** utilize a tuning fork, to differentiate between conductive and sensorineural hearing loss. Weber's test has the tuning fork placed on the midline of the head, with the patient indicating in which ear the tone is louder. The Rinne test involves hearing by bone and air conduction, with the two being compared. Red flags include unilateral sensorineural hearing loss, and abnormalities of the cranial nerves besides hearing loss.

Usually, cerumen, injury, infectious sequelae, medications, and significant noise exposure are easily apparent after examination. Patients with focal neurologic abnormalities are of highest concern. Children with delay in speech of language development of difficulties in school must be evaluated for hearing loss. Considerations also include aphasia, autism, and intellectual disability. Delayed motor development can signal vestibular deficit, often related with sensorineural hearing loss. Testing includes audiologic tests, and sometimes, MRI or CT.

Pure-tone audiometry quantifies hearing loss for various frequencies. An *audiometer* delivers pure tones of specific frequencies, at different intensities (see Fig. 13.2). It determines the patient's hearing threshold, which means how loud a sound must be in order to be perceived for each frequency. The hearing in each ear is tested from 125 or 250 Hz to 8000 Hz by air conduction, which uses earphones, and up to 4 kHz by bone conduction, which uses an oscillator in contact with the mastoid process or forehead. The test results are plotted on graphs called *audiograms*. These show the difference between the hearing threshold and normal hearing, at each frequency. The difference is measured in decibels. The normal threshold is 0 dB hearing level. Hearing loss is present if the patient's threshold is less than 25 dB hearing level. When hearing loss is severe enough to require loud

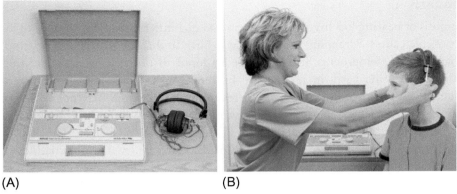

Fig. 13.2 (A) An audiometer. (B) Placing the headphones.

test tones, intense tones presented to one ear may be heard in the other ear. In these cases, a *masking sound*, usually *narrow band noise*, is presented to the ear not being tested to isolate it from the unwanted tone.

Tympanometry measures the impedance of the middle ear to acoustic energy. It is often used to screen children for middle ear effusions. A probe that contains a sound source, microphone, and air pressure regulator is placed tightly, with an airtight seal, into the ear canal. The probe microphone records reflected sounds from the tympanic membrane, as pressure in the canal is varied. Normally, the maximal compliance of the middle ear occurs when the pressure in the ear canal is equal to that of the atmospheric pressure. Abnormal compliance patterns reveal specific anatomic disruptions. If there is Eustachian tube obstruction and middle ear effusion, maximal compliance occurs with a negative pressure in the ear canal. When the ossicular chain is disrupted, such as in necrosis or dislocation of the long process of the incus bone, the middle ear is highly compliant. When the ossicular chain is fixed, such as in stapedial ankylosis in otosclerosis, compliance may be reduced or normal. The acoustic reflex is contraction of the stapedius muscle in response to loud sounds. This changes the compliance of the eardrum, and protects the middle ear from acoustic trauma. The reflex is tested by creating a tone, and measuring what intensity provokes a change in middle ear impedance, as revealed by movement of the eardrum. An absent reflex may indicate middle ear disease, or a tumor of the auditory nerve.

Speech audiometry includes the speech reception threshold and the word recognition score. The speech reception threshold is a measurement of the intensity at which speech is recognized. To determine this, the patient is presented with a list of words at specific sound intensities. The words usually have two equally accented syllables, known as *spondees*. Example words include *baseball*, *railroad*, and *staircase*. The examiner notes the intensity at which the patient repeats 50% of the words correctly. The speech reception threshold approximates the average hearing level at the speech frequencies, such as 500, 1000, and 2000 Hz. Word recognition score testing assesses ability to discriminate between various speech sounds, called *phonemes*. It is determined by presenting 50 phonetically balanced one-syllable words at an intensity of 35–40 dB above the patient's speech reception threshold. The list of words contains phonemes in the same relative frequency found in conversational English. The score is the percentage of words that the patient can correctly repeat, and shows the ability to understand speech under optimal listening conditions. Normal scores range between 90% and 100%. The score is normal with conductive hearing loss, but at a higher intensity level, yet can be reduced at all intensity levels with sensorineural hearing loss. Discrimination is worse in neural than in sensory hearing loss.

Advanced testing includes gadolinium-enhanced MRI of the head. This can detect lesions of the cerebellopontine angle, which may be required for patients with an abnormal neurologic examination, or those whose audiologic tests show poor word recognition, asymmetric sensorineural hearing loss, or a combination, when etiology is unclear. CT is performed if bone tumors or bony erosion is suspected. MRA is done if vascular abnormalities, such as arteriovenous malformations, are suspected. The auditory brainstem response is another useful test and utilizes surface electrodes, monitoring brain wave response to acoustic stimulation in patients who cannot otherwise respond.

Electrocochleography is used to assess dizziness. It measures cochlear and auditory nerve activity, with an electrode placed on or through the eardrum. Its advantage is that it can be used in patients who are awake. It is a useful tool in intraoperative monitoring. *Otoacoustic emissions testing* measures sounds created by the outer hair cells of the cochlea. It screens neonates and infants for hearing loss, and monitors patients taking ototoxic drugs. Some patients, such as children with reading or learning problems, and elderly patients who appear to hear but not comprehend, should have a central auditory evaluation. This measures discrimination of distorted or degraded speech, discrimination while there is a competing message in the opposite ear, the ability to fuse partial or incomplete messages delivered to each ear into a meaningful and complete message, and the ability to localize sound in space when acoustic stimuli are delivered to both ears at the same time.

Treatment

Treatments are based on the causes. Ototoxic drugs must be stopped, or their doses lowered as applicable to the patient's needs. Attention to peak and trough drug levels help minimize risks of ototoxicity. Middle ear effusion fluid can be draining via myringotomy, and prevented by inserting a tympanostomy tube. Benign growths and malignant tumors can be removed. Corticosteroids may help for hearing loss caused by autoimmune disorders. Damage to the eardrum or ossicles, or otosclerosis, may require reconstructive surgery. Brain tumors that cause hearing loss may sometimes be removed, preserving hearing. Many hearing loss causes have no cure. Treatment involves compensating for the loss with hearing aids. For severe to profound loss, a cochlear implant may be needed. Various coping mechanisms can also help.

Hearing aids can significantly improve communication abilities. Physicians should encourage use of hearing aids and help patient overcome any social stigma about their use—they are as helpful for hearing as eyeglasses are for seeing. Hearing aids utilize a microphone, amplifier, speaker, earpiece, and volume control. The location of these components varies widely between different models. An audiologist should be involved in hearing aid selection and fitting. Ideally, they are adjusted to the individual's particular pattern of hearing loss. Those with mostly high-frequency hearing loss usually need a hearing aid that selectively amplifies just the high frequencies. Some hearing aids have vents within the ear mold, allowing passage of high-frequency sound waves. Some use digital sound processors with multiple frequency channels. Therefore, amplification more closely matches hearing loss in relation to the patient's audiogram. Some people with hearing aids have difficulty using telephones. Often, the hearing aid will cause squealing when the ear is placed next to the phone handle. Some hearing aids, therefore, have a phone coil with a switch that runs the microphone off, linking the phone coil electromagnetically to the speaker magnet in the phone.

For moderate to severe hearing loss, a postauricular, ear-level hearing aid can be used. This fits behind the pinna and is coupled to the ear mold with flexible tubing. In-ear hearing aids are contained completely within the ear mold. They fit less conspicuously into the concha and ear canal, and are used for mild to moderate hearing loss. Some people with mild hearing loss of just high frequencies are comfortably fitted with postauricular aids, and completely open ear canals. Canal hearing aids are contained entirely within the ear canal. They are often preferred by people who would otherwise refuse to use a hearing aid, but are often difficult for elderly people to manipulate. The *contralateral routing of signals (CROS) hearing aid* is sometimes used for severe unilateral hearing loss. A hearing aid microphone is placed into the nonfunctioning ear. Sound is routed to the functioning ear via a wire or radio transmitter. This allows the patient to hear sounds from the nonfunctioning side, and allows for limited capacity in localizing sound. If the better ear also has some hearing loss, the sound from both sides can be amplified with the hearing aid. For profound hearing loss, the body aid type is worn in a shirt pocket or a body harness, and connected by a wire to the earpiece, coupled to the ear canal by a plastic ear mold.

A bone conduction aid can be used when an ear mold or tube cannot be used, such as in atresia of the ear canal, or persistent otorrhea. An oscillator is held against the head, usually over the mastoid, with a spring band. Sound is conducted through the skull to the cochlea. Bone conduction hearing aids use more power, cause more distortion, and are less comfortable than air conduction hearing aids. Some bone conduction aids are surgically implanted in the mastoid process, avoiding the discomfort and prominence of the spring band.

Cochlear implants are helpful for profoundly deaf patients, and those with some hearing but who cannot understand speech easily. They provide electrical signals directly into the auditory nerve, through multiple electrodes implanted in the cochlea (see Fig. 13.3). An external microphone and processor convert sound waves to electrical impulses. These are transmitted through the skin electromagnetically, from an external induction coil, to an internal coil implanted in the skull, above and behind the ear. The internal coil connects to electrodes inserted into the scala tympani. Cochlear implants aid in speech reading by providing information about the intonation of words and speech rhythms. Most adults with cochlear implants are able to discriminate words without visual clues, allowing them to talk on the telephone. Cochlear implants enable deaf people to hear and distinguish environmental sounds such as warning signals. They help the deaf to modulate their voices, making their speech more intelligible.

Brainstem implants are used for patients who have had both acoustic nerves destroyed, such as by bilateral temporal bone fractures or neurofibromatosis. This helps restore partial hearing. They utilize electrodes connected to sound-detecting and sound-processing devices, similar to cochlear implants.

Other supportive treatments include using devices that alert the patient to conditions that they may not be able to hear, such as someone ringing the doorbell, a smoke alarm

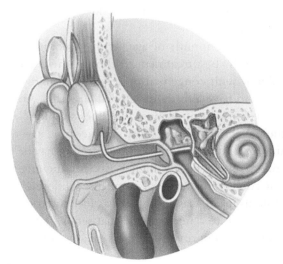

Fig. 13.3 Cochlear implant.

going off, and others. Special sound systems can transmit FM radio or infrared signals to help people hear in public gathering places, and other areas where there is competing noise. Telephone communication devices are also readily available. Learning lip reading or sign language is often very helpful. Most people get useful speech information from lip reading even when they have no formal training. Normal-hearing individuals can understand speech in a noisy place when they can see the speaker and visualize his or her lips. Therefore, health-care personnel should be aware of this, and position themselves appropriately when speaking to the hearing impaired. Lip reading can be learned in aural rehabilitation sessions, in which age-matched peers meet regularly for instruction and supervised practice, to optimize their communication skills.

Methods to help in hearing and communication include visiting public places when they are less busy, asking for booth seating to block out some amount of sound, facing people directly when speaking, and identifying yourself as hearing impaired when beginning a telephone conversation. Assistive listening systems can be used by speakers for their audiences. These use inductive loop, infrared, or FM technology to send sound through the microphone to the hearing aids of audience members. For profound hearing loss, communication is often through American Sign Language (ASL), and other versions, though this is the most common in the United States. Other forms include *Signed English, Signing Exact English,* and *Cued Speech.*

For children, there is an additional need for support of language development, with appropriate therapy. Since children must hear language to spontaneously learn it, most deaf children develop language only via special training. Ideally, this should start as soon as the hearing loss is identified, except for a deaf child with deaf parents who are fluent in sign language. Deaf infants must be provided with a form of language input. A visually based sign language can provide a basis for later development of oral language when a cochlear implant is not available.

For infants as young as 1 month of age, with profound, bilateral hearing loss, a cochlear implant may be the best treatment. Although these allow auditory communication for many children with congenital or acquired deafness, they are usually more effective for children who have already developed language skills. Children with postmeningitic deafness develop an ossified inner ear. They should receive a cochlear implant early to maximize their effectiveness. For those with acoustic nerves destroyed by tumors, implantation of brainstem auditory-stimulating electrodes may be helpful. Children with cochlear implants may have a slightly higher risk for meningitis than children with cochlear implants, or adults with cochlear implants. Children with unilateral deafness should use special systems in the classroom such as FM auditory trainers. These systems allow the teacher to speak into a microphone that sends signals to a hearing aid in the child's nonaffected ear. They improve the child's greatly impaired ability to hear speech against noisy backgrounds.

> **Focus on hearing loss**
>
> According to the Center for Hearing and Communication, hearing aids may offer dramatic improvement for most people with hearing loss. However, people with hearing loss wait an average of 7 years before seeking help. Only 16% of physicians routinely screen for hearing loss. Noise-induced hearing loss, while preventable, is permanent.

Section review
1. What are the various causes of tinnitus?
2. What is the difference between conductive hearing loss and sensorineural hearing loss?
3. What is the difference between audiometry and tympanometry?

Sudden deafness

Sudden deafness involves severe sensorineural hearing loss developing in less than 72 h. Initial hearing loss is usually unilateral, unless caused by medications. It may range in severity from mild to profound. Sudden deafness may be linked to tinnitus, dizziness, and vertigo. Most cases are idiopathic, but some occur because of an obvious cause. These may include: as blunt head trauma, large ambient pressure changes, ototoxic drugs, infections, acoustic neuroma, multiple sclerosis, Meniere's disease, small cerebellar strokes, syphilis reactivation in HIV-infected patients, **Cogan's syndrome**, systemic vasculitis, **Waldenström's macroglobulinemia**, sickle cell disease, and certain forms of leukemia.

Sudden deafness is evaluated by considering history of present illness, review of systems, past medical history, and physical examination. Red flags include abnormalities of the cranial nerves other than hearing loss. Traumatic, ototoxic, and certain infectious causes are usually apparent clinically. An audiogram is usually performed. Unless diagnosis is clearly drug toxicity or an acute infection, then gadolinium-enhanced MRI is performed. A perilymphatic fistula can be confirmed by tympanometry and **electronystagmography**. A CT will show bony characteristics in the inner ear. For some patients, serologic tests for possible HIV or syphilis, complete blood count and coagulation profiles for hematologic conditions, and erythrocyte sedimentation rate plus antinuclear antibodies for vasculitis will be required. Treatment is focused on the causative disorder when it is known. Fistulas are evaluated and repaired surgically. For idiopathic deafness, a short course of glucocorticoids and antiviral drugs effective against herpes simplex are given. Glucocorticoids may be given orally, or via transtympanic injection.

Vertigo

True **vertigo** involves an illusion or hallucination of motion in any plane. Diagnosis is not difficult if the patient states that environmental objects seem to be spinning or moving rhythmically in one direction, or that he or she senses the head or body are spinning. Sometimes, the description may be described as a back-and-forth or up-and-down body movement—usually of the head. The patient may state that the floor or walls are changing position. He or she may be unsteady when walking, or pull to one side. The rhythmic jerking movements of **oscillopsia** are additional effects of vestibular disorders, especially in patients with multiple sclerosis.

Pathophysiology

The vestibular system is the primary neurologic system involved in balance. It includes the vestibular apparatus of the inner ear, the vestibulocochlear cranial nerve, and the vestibular nuclei in the brainstem and cerebellum. Disorders of the inner ear and eighth cranial nerve are considered to be peripheral disorders. Those of the vestibular nuclei, and their pathways within the brainstem and cerebellum, are central disorders. Rotary motion causes flow of endolymph within the semicircular canal that is oriented in the plane of motion. Based on the direction of flow, the endolymph movement will either stimulate or inhibit neuronal output from the hair cells that line the canal. Similar hair cells in the saccule and utricle are embedded in a matrix of calcium carbonate crystals (otoliths). Deflection of the otoliths by gravity will stimulate or inhibit neuronal output from the attached hair cells.

Clinical manifestations

Dizziness may be described by the patient as a false sense of motion or spinning, light headedness, feeling faint, unsteadiness, loss of balance, a feeling of floating, wooziness, or heavy handedness. Vertigo is described as spinning, tilting, swaying, imbalance, and being pulled to one direction. Additional symptoms that may accompany vertigo include nausea, vomiting, nystagmus, headache, sweating, tinnitus, and hearing loss.

Diagnosis

The key to correct and prompt diagnosis of vertigo is to distinguish between the peripheral causes from central causes. Note that all cases of vertigo, regardless of the cause, are worsened by movement. This feature should not be used in distinguishing between central and peripheral causes.

Treatment

Treatment of vertigo is based on the cause. This may include stopping, reducing, or replacing any causative drugs. Medications include diazepam, meclizine, and other antihistamine/anticholinergics. For related nausea, prochlorperazine is used. The **Epley maneuver** may be used to reposition the otoliths. The patient's head is rotated by a clinician in an attempt to mobilize the calcium crystals within the inner ear.

Meniere's disease

Meniere's disease is also known as *endolymphatic hydrops*. It affects the inner ear, causing vertigo, tinnitus, and fluctuating sensorineural hearing loss. Pressure and volume changes of the labyrinthine endolymph affect functions.

Pathophysiology

The mechanisms of endolymphatic fluid buildup are unknown. Risk factors include family history, allergies, preexisting autoimmune disorders, ear or head trauma, and rarely, syphilis. The peak incidence of this condition is between ages 20 and 50.

Clinical manifestations

Meniere's disease causes sudden vertigo, usually with nausea and vomiting, which lasts for up to 24 h. There may be diaphoresis, diarrhea, and **gait unsteadiness**. Tinnitus can be constant or intermittent, with a buzzing or roaring sound not related to position or motion. Hearing impairment, usually of the lower frequencies, may follow. Prior to each attack, the patient senses fullness or pressure in the affected ear. In half of patients, just one ear is affected. In early stages, symptoms remit between episodes. Symptom-free interludes can last for over 1 year. With disease progression, hearing impairment continues and worsens, with the tinnitus becoming constant.

Diagnosis

Diagnosis is based on excluding other possible conditions with similar symptoms. These include acoustic neuroma and other cerebellopontine angle tumors, brainstem stroke, and viral labyrinthitis or neuritis. An audiogram and MRI with gadolinium enhancement of the CNS, with attention to the internal auditory canals, should be performed. An audiogram usually shows a low-frequency sensorineural hearing loss in the affected ear. During an acute attack, examination will reveal **nystagmus** and falls to the affected side. Between attacks, the *Fukada stepping test*, which involves marching in place while the eyes are closed, can be performed. If the patient has Meniere's disease, he or she often turns away from the affected ear, revealing a unilateral labyrinthine lesion. Sensorineural hearing loss may also be indicated using the Rinne test or Weber's test. As a rule when there is episodic vertigo, the next question should be whether the patient has hearing loss, since in Meniere's disease, hearing loss is usually present. However, in benign paroxysmal positional vertigo (BPPV), which is another cause of episodic vertigo, there is no hearing loss.

Treatment

Meniere's disease is usually self-limiting. Treatment of acute attacks is focused on symptom relief. Medications include prochlorperazine or promethazine, diphenhydramine, meclizine, cyclizine, diazepam, or a corticosteroid burst of prednisone. Other measures include avoiding alcohol and caffeine, a low-salt diet, and diuretics. If medical management is unsuccessful, intratympanic gentamicin may be used. Serial audiometry is recommended for follow-up. Surgery is only done for frequent and severely debilitating episodes that are otherwise unresponsive. This may involve endolymphatic sac decompression, vestibular neurectomy, or labyrinthectomy.

> **Focus on geriatrics**
> With aging, balance and equilibrium become less steady, involving vision, the inner ear, proprioception, and blood pressure. Older individuals are more likely to have cardiac or cerebrovascular disorders contributing to dizziness. They also take more drugs that may cause dizziness, including those for angina, hypertension, heart failure, anxiety, seizures, and drugs such as antibiotics, antihistamines, and sleep aids. Therefore, the elderly are at a higher risk for falling and experiencing fractures. Their fear of moving and falling often greatly decreases their ability to perform daily activities. However, physical therapy and exercises to strengthen muscles may be extremely helpful in maintaining independent ambulation for as long as possible.

Section review
1. What are the possible causes of sudden deafness?
2. What are the clinical manifestations of vertigo?
3. What other conditions may mimic Meniere's disease?

Benign paroxysmal positional vertigo

BPPV is also known as *benign postural vertigo* and *positional vertigo*. There are short episodes of vertigo triggered by certain head positions. These last for less than 60 s, and may accompanied by nausea and vomiting.

Pathophysiology

BPPV is the primary cause of relapsing otogenic vertigo. BBPV is usually idiopathic, however, development of this condition is also linked to otoconial crystal displacement, degeneration of utricular otolithic membranes, labyrinthine concussion, otitis media, ear surgery, recent viral infection, head trauma, lengthy anesthesia or bed rest, previous vestibular disorders, and occlusion of the anterior vestibular artery.

Clinical manifestations

When the patient moves his or her head, vertigo is triggered. Nausea and vomiting may occur, with no hearing loss. As mentioned above, episodic vertigo with no hearing loss is most likely BPPV.

Diagnosis

Diagnosis of this form of vertigo is based on nystagmus, using the **Dix-Hallpike maneuver**, and lack of other abnormalities upon neurologic exam. In this condition, the positional nystagmus begins slowly. It is susceptible to fatigue, and unidirectional. In contrast, in central causes of vertigo, the nystagmus is always present with no latency and is always bidirectional. In general, vertigo is one of the symptoms that should be taken seriously, especially if constant in duration. An MRI is usually ordered for patients with constant vertigo.

Treatment

This type of vertigo usually subsides on its own within several weeks or months, but can reoccur. The Epley maneuver is the treatment of choice.

Ramsay Hunt syndrome

Herpes zoster oticus is also known as Ramsay Hunt syndrome. It is an infection of the eighth cranial nerve ganglia and the geniculate ganglion of the facial nerve. Risk factors for herpes zoster infection include chemotherapy, immunodeficiency secondary to cancer, HIV infection, and radiation therapy.

Pathophysiology

Herpes zoster oticus is caused by the same virus that causes chickenpox. The virus remains dormant in the body for years. When it reactivates, it may affect the facial nerve, resulting in this condition.

Clinical manifestations

Signs and symptoms include severe ear pain, facial paralysis that may be transient or permanent, hearing loss that may be permanent or resolve partially or completely, and vertigo that lasts for days to weeks. Vesicles develop on the pinna, and in the external auditory canal, along the sensory branch of the facial nerve's distribution. Less often, headache, confusion, stiff neck, and involvement of other cranial nerves occurs.

Diagnosis

Diagnosis is usually clinical. If the viral cause is in question, vesicular scrapings are collected for viral cultures or direct immunofluorescence, followed by an MRI with contrast.

Treatment

Though of unknown value, antiviral drugs, corticosteroids, and surgical decompression have all been used. Medications include prednisone, acyclovir, and valacyclovir.

> **Focus on Ramsay Hunt syndrome (herpes zoster oticus)**
> Ramsay Hunt syndrome is a common complication of shingles, which is caused by the varicella-zoster virus. It occurs when the virus spreads to the facial nerves. In some cases, hearing loss may be permanent. Facial paralysis may be temporary or permanent, and vertigo may last for days to weeks.

Acute vestibular neuronitis

Acute vestibular neuronitis is an inflammation of the vestibular nerve. It is characterized by an acute onset of vertigo, nausea, and vomiting within hours. This may last for several days, and is not related to any auditory or other neurologic manifestations. Most patients gradually improve over 1–2 weeks, those recurrences do happen. Many patients report having a respiratory tract illness 1–2 weeks before onset of symptoms, which suggests a viral cause. Acute vestibular neuronitis can also occur in people who have herpes zoster oticus. Sometimes, attacks of acute vesibulopathy recur over months or years. There is no determination of whether an affected patient will experience recurrences.

Drug-induced ototoxicity

There are many different ototoxic drugs. The dose and duration of a drug, its infusion rate, the lifetime dose, concurrent use of other potentially ototoxic drugs, present renal failure, and genetic susceptibility are all factors. Ototoxic drugs should never be used for topical application to the ear if the tympanic membrane, since they might diffuse into the inner ear. Streptomycin usually causes more damage to the vestibular area than to the auditory area of the inner ear. Vertigo and the maintenance of balance are usually temporary, but severe loss of vestibular sensitivity can continue and even become permanent. When vestibular sensitivity is lost, the patient has difficulty in walking—especially in dark areas—and *oscillopsia*, which is a sensation of the environment "bouncing" with each step. Between 4% and 15% of patients receiving more than 1 g per day, for more than 1 week, develop hearing loss that is measurable. This usually occurs after a latent period of 7–10 days, slowly worsening with continued treatment. Deafness may follow, which is complete and permanent.

The greatest cochleotoxic effect of all the antibiotics is with neomycin. Large oral or colonic irrigation-administered doses can be absorbed in sufficient quantities to affect hearing—especially with present mucosal lesions. This drug should not be used for wound irrigation, intrapleural irrigation, or intraperitoneal irrigation. This is because extremely large amounts of the drug may be retained and absorbed, resulting in deafness. Other drugs with similar potential include kanamycin and amikacin. Viomycin has cochlear and vestibular toxicity properties. Gentamicin and tobramycin have the same. Vancomycin may cause hearing loss—especially when there is renal insufficiency.

Additional ototoxic drugs include chemotherapeutic drugs, especially cisplatin and carboplatin, which contain platinum. They cause tinnitus and hearing loss, which may be profound and permanent, beginning immediately after the first dose, or delayed for several months after treatment is completed. Sensorineural hearing loss occurs bilaterally. It progresses in decrements, and is permanent. Intravenous ethacrynic acid and furosemide have caused profound and permanent hearing loss in patients with renal failure who were previously receiving aminoglycoside antibiotics. High doses of salicylates cause temporary hearing loss and tinnitus. Quinine and similar synthetic substitute drugs may also cause temporary hearing loss.

Acoustic neuroma

An **acoustic neuroma** is a Schwann cell-derived tumor of the eighth cranial nerve (see Fig. 13.4). It is also known as *acoustic neurinoma, vestibular schwannoma*, and eighth nerve tumor. It nearly always arises from the vestibular division of the eighth cranial nerve, making up approximately 7% of all intracranial tumors.

Pathophysiology

As the tumor grows, it projects from the internal auditory meatus into the cerebellopontine angle. It then compresses the cerebellum and brainstem. The fifth and seventh cranial nerves are later affected. Bilateral acoustic neuromas are common in *neurofibromatosis type 2*.

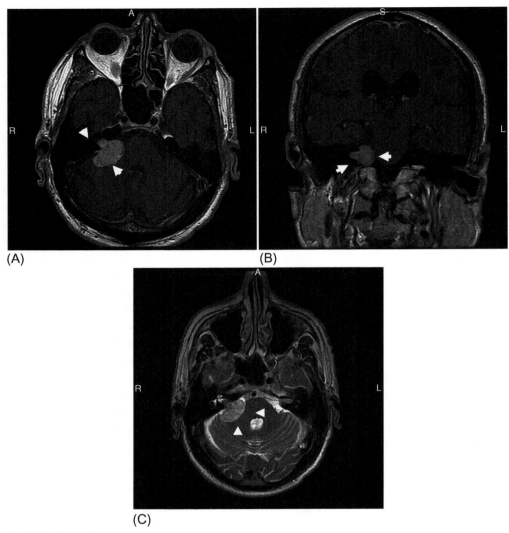

Fig. 13.4 Acoustic neuroma.

Clinical manifestations

The primary symptom is slowly progressive unilateral sensorineural hearing loss. Its onset may, however, be abrupt, with fluctuating degrees of impairment. Other early symptoms include dizziness, disequilibrium, unilateral tinnitus, headache, otalgia, a sensation of fullness or pressure in the ear, numbness, or weakness of the facial nerve, and trigeminal neuralgia.

Diagnosis

The first diagnostic test is an audiogram, which usually reveals asymmetric sensorineural hearing loss, and more impairment of speech discrimination than expected for the amount of hearing loss. These findings indicate the need for imaging, which is preferred to be gadolinium-enhanced MRI.

Treatment

For small tumors, microsurgery may be performed to preserve the facial nerve. When a middle cranial fossa or retrosigmoid approach is used, this may preserve remaining hearing. If no useful hearing remains, a translabyrinthine route may be used.

Labyrinthitis

Labyrinthitis can be autoimmune, viral, or bacterial. **Purulent labyrinthitis** is a bacterial infection of the inner ear (see Fig. 13.5) that often results in deafness and loss of vestibular function. It is also known as *suppurative labyrinthitis*. The condition usually occurs when bacteria spread to the inner ear along with severe acute otitis media, an enlarging cholesteatoma, or following meningitis. Symptoms include severe vertigo and nystagmus, nausea and vomiting, varying amounts of hearing loss, and tinnitus. Pain and fever are also common.

This condition is suspected if nystagmus, vertigo, sensorineural hearing loss, or a combination of these occurs during an episode of acute otitis media. A CT scan of the temporal bone can identify erosion of the otic capsule bone, or additional complications of acute otitis media such as coalescent mastoiditis. An MRI may be performed if there are symptoms of brain abscess or meningitis. These include altered mental

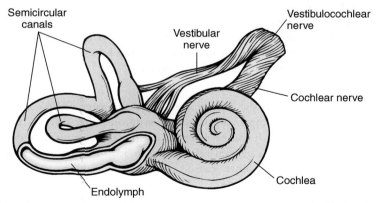

Fig. 13.5 Labyrinth or inner ear. Labyrinthitis is caused by a disturbance in the fluid in the semicircular canals.

status, higher fever, and meningismus. In these cases, blood cultures and a lumbar puncture are also performed.

Treatment utilizes intravenous antibiotics such as ceftriaxone, based on culture and sensitivity test results. To drain the middle ear, a myringotomy, and sometimes, tympanostomy tube placement, is performed. A mastoidectomy may also be required.

> **Focus on labyrinthitis**
> Certain people may suffer from labyrinthitis for months, known as *chronic labyrinthitis*. For this form, vestibular rehabilitation therapy (VRT) may be needed for treatment. This utilizes exercises that help "retrain" the brain and nervous system to compensate for abnormal signals coming from the vestibular system. It helps coordinate hand and eye movements, stimulate sensations of dizziness that can then be accommodated to, improves balance and walking ability, and improves strength and fitness.

Section review
1. How is BPPV diagnosed?
2. What is Ramsay Hunt syndrome?
3. What drugs may cause ototoxicity?
4. What are the symptoms of acoustic neuroma?

Clinical cases

Clinical case 1
1. Can migraine cause vertigo?
2. What are the complications of severe, untreated vertigo?

A 42-year-old woman who had frequent migraines reported recurrent episodes of rotational vertigo, about once per month, over the past 2 years. Each episode lasted for more than 4h, often with a right ear fluctuating hearing loss. The patient also reported the presence of mild confusion during these episodes.

Answers:
1. Yes, vestibular migraine is also known as migraine-associated vertigo. Some migraine patients have some accompanying vestibular syndrome that involves disruption in balance at various times. It can be prior to, during, after, or totally independent of the migraine events. As a general rule, the diagnosis of migraine is a diagnosis of exclusion and therefore, other causes of vertigo should be ruled out before diagnosing anyone with migraine.
2. Severe vertigo that is not treated is disabling. It may result in irritability, loss of self-esteem, depression, and injuries due to falling.

Clinical case 2
1. What is the diagnosis of this patient's condition?
2. What are the possible causes of this condition?
3. What are the various treatments for this condition?

A 54-year-old man woke up with sudden hearing loss in his left ear. The man went to his physician, who referred him to an ear specialist. A pure-tone hearing test was conducted, revealing significant left ear hearing loss, yet the eardrum was normal and there was no impacted wax. Tympanometry revealed normal middle ear function. Evaluation of the outer hair cells of the cochlea revealed normal function as well.

Answers:
1. Sudden sensorineural hearing loss is the most likely diagnosis. It is described as an abrupt loss of 30 dB, or greater, over at least three audiometric frequencies, occurring within 72 h or less.
2. Causes of this condition include autoimmune diseases such as AIDS, Lupus and Meniere's disease; vascular conditions such as cardiopulmonary bypass or sickle cell anemia; neurologic and neoplastic lesions such as auditory tumors, auditory surgeries, migraines, and multiple sclerosis; trauma, such as from concussion, ototoxicity, perilymph fistula, and temporal bone fracture; and infections such as herpes, mumps, measles, syphilis, and Lyme disease.
3. If autoimmune in nature, prednisone and other steroids are indicated. Other treatments include diuretics, a low-sodium diet, and restricted nicotine, caffeine, and alcohol consumption.

Clinical case 3
1. What other testing is performed for Meniere's disease?
2. What are basic measures, aside from medications that the patient may take?
3. What are potential long-term complications of Meniere's disease?

A 49-year-old woman complained of dizziness, with five attacks of vertigo over the past 4 months. Each episode lasted for 1–2 h, during which she was either nauseous, or actually vomited. She also complained of left ear fullness, hearing loss, and tinnitus that worsened before the onset of vertigo. She has no previous history or ear surgery or family history or ear disease, no history of noise exposure or head trauma, and has never used any ototoxic medications. She takes medication for hypertension and also has allergic rhinitis. Physical examination revealed Weber lateralizing to the right, and positive Rinne bilaterally. There are no effusions and the fistula test is negative. The rest of her head, neck, and neurological examinations are normal. An ear-nose-throat specialist confirms a diagnosis of Meniere's disease.

Answers:
1. Meniere's disease is also tested for by using audiometry, videonystagmography, vestibular-evoked myogenic potentials, and electrocochleography.

2. The patient should remember to sit or lie down when feeling dizzy, and to rest during and after attacks. Salt intake must be limited, and measures to manage stress undertaken.
3. Complications overtime include permanent hearing loss, pain, disruption in daily living, anxiety, depression, fatigue, emotional stress, falling, increased chances for accidents, and problems driving motor vehicles.

Clinical case 4
1. Is gentamicin linked to temporary or permanent hearing loss?
2. Since gentamicin is an aminoglycoside, what are other risk factors for ototoxicity caused by these agents?
3. Why are elderly patients such as this man especially at risk for ototoxicity from these agents?

A 76-year-old man was hospitalized with pyrexia, confusion, and rigor associated with positive blood cultures for *Enterococcus* species. His medical history included myocardial infarction, pulmonary embolism, and noninsulin-dependent diabetes mellitus. Treatments proved successful for his overall health, but part of his treatments involved the drug *gentamicin*. Soon, he could not understand what the hospital staff was saying to him, yet he had no obvious hearing problems upon admission. Audiometry revealed a mixed conduction and sensorineural deafness with notable high tone loss. His hearing continued to worsen. Eventually, the gentamicin was replaced with netilmicin, and his hearing returned to normal within several days.

Answers:
1. Yes, gentamicin ototoxicity, leading to hearing loss or deranged vestibular function, has been widely reported, and is usually reversible. Prevalence of clinically evident ototoxicity with multiple daily regimens of gentamicin has been estimated at 11% of patients.
2. Other risk factors include large cumulative doses, repeated courses of treatment, bacteremia, fever, hypovolemia, liver dysfunction, noise exposure, preexisting hearing disorders, and other ototoxic drugs—especially loop diuretics.
3. The elderly are especially at risk for ototoxicity from gentamicin and other aminoglycosides because their creatinine clearance may be greatly reduced without increase of serum creatinine.

Clinical case 5
1. For this case study, what is the likely treatment?
2. Is this type of tumor common?
3. What are the potential complications of surgery to remove this type of tumor?

A 30-year-old man reported sudden tinnitus, ear fullness, and hearing loss in his right ear. Overtime, he began to experience vertigo, headache, and progressive loss of balance. Audiometry revealed moderate sensorineural hearing loss in his right ear. The patient continued to be monitored. Two years later, he developed right facial paralysis. An MRI of his skull was performed, revealing a vestibular schwannoma, also known as an acoustic neuroma, on the right side, within the cerebellopontine angle. The tumor measured 4.9 cm in its greatest diameter.

Answers:
1. For acoustic neuroma, the likely treatment is surgical removal, followed by radiation therapy, specifically stereotactic radiosurgery.
2. Acoustic neuroma is the most common tumor of the cerebellopontine angle, followed by meningioma.
3. Surgical complications may include leakage of cerebrospinal fluid (CSF) through the wound, hearing loss, facial weakness or numbness, tinnitus, balance problems, persistent headache, meningitis, stroke, and brain hemorrhage.

Key terms

Acoustic neuroma	Nystagmus
Acute vestibular neuronitis	Objective tinnitus
Cogan's syndrome	Presbycusis
Conductive hearing loss	Purulent labyrinthitis
Dix-Hallpike maneuver	Rinne test
Dizziness	Sensorineural hearing loss
Electrocochleography	Subjective tinnitus
Electronystagmography	Tinnitus
Epley maneuver	Vertigo
Gait unsteadiness	Waldenström's macroglobulinemia
Meniere's disease	Weber's test
Mixed hearing loss	

Suggested readings

1. Ackley, R. S.; Decker, T. N.; Limb, C. J. *An Essential Guide to Hearing and Balance Disorders.* Psychology Press, 2018.
2. Anne, S.; Lieu, J. E. C.; Kenna, M. A. *Pediatric Sensorineural Hearing Loss: Clinical Diagnosis and Management.* Plural Publishing, 2017.
3. Babu, S.; Schutt, C. A.; Bojrab, D. I. *Diagnosis and Treatment of Vestibular Disorders.* Springer, 2019.
4. Berenson, D. *I Have WHAT??? Multiple Myeloma? Waldenström's Macroglobulinemia? Amyloidosis? MGUS? Written by Those Who Know!!!,* 2nd ed.; IMBCR, 2017.
5. Campbell, K. C. M. *Pharmacology and Ototoxicity for Audiologists.* Delmar Cengage Learning, 2006.
6. Campillo, R. V. *Tinnitology: Objective Acouphenometry: Picks Up the Sound Signal of Tinnitus.* Editorial Academica Espanola, 2017.
7. Conn, P. M. *Conn's Translational Neuroscience.* Academic Press, 2016.
8. Deruiter, M.; Ramachandran, V. *Basic Audiometry Learning Manual,* 2nd ed.; Plural Publishing, 2016.

9. Dillon, H. *Hearing Aids*, 2nd ed.; Thieme, 2012.
10. Eggermont, J. J. *Hearing Loss: Causes, Prevention, and Treatment*. Academic Press, 2017.
11. Goldberg, J. M.; Wilson, V. J.; et al. *The Vestibular System: A Sixth Sense*. Oxford University Press, 2012.
12. Hayat, M. A. *Tumors of the Central Nervous System, Volume 7—Meningiomas and Schwannomas*. Springer, 2012.
13. Icon Group International. *Tympanometry: Webster's Timeline History, 1969-2007*. Icon Group International, 2010.
14. Jacob, E. *Medifocus Guidebook on: Acoustic Neuroma*. Medifocus.com, 2013.
15. Lyon, R. F. *Human and Machine Hearing: Extracting Meaning From Sound*. Cambridge University Press, 2017.
16. McCaslin, D. L. *Electronystagmography/Videonystagmography (Core Concepts in Audiology)*. Plural Publishing, 2012.
17. Medifocus.com Inc. *Medifocus Guidebook on: Meniere's Disease*. CreateSpace Independent Publishing Platform, 2018.
18. Musiek, F. E.; Baran, J. A. *The Auditory System: Anatomy, Physiology, and Clinical Correlates*. Plural Publishing, 2016.
19. Musiek, F. E.; Baran, J. A.; Shinn, J. B.; Jones, R. O. *Disorders of the Auditory System*. Plural Publishing, 2011.
20. Nagaratnam, N.; Nagatnam, K.; Cheuk, G. *Diseases in the Elderly: Age-Related Changes and Pathophysiology*. Springer, 2016.
21. Oeding, K. A. M.; Listenberger, J.; Smith, S. *The Audiogram Workbook*. Thieme, 2016.
22. Silman, S.; Emmer, M. B. *Instrumentation in Audiology and Hearing Science—Theory and Practice*. Plural Publishing, 2011.
23. Toriello, H. V.; Reardon, W.; Gorlin, R. J. *Hereditary Hearing Loss and Its Syndromes (Oxford Medical Genetics)*, 2nd ed.; Oxford University Press, 2004.
24. Tos, M. *Surgical Solutions for Conductive Hearing Loss*. TIS, 2000.
25. Wolfe, J.; Schafer, E. *Programming Cochlear Implants (Core Clinical Concepts in Audiology)*, 2nd ed.; Plural Publishing, 2014.

CHAPTER 14

Visual system

The sense of sight and our visual system is important for almost everything required for normal living. The eyes allow response to all types of stimuli, as they convert stored photochemical energy into nerve impulses. The visual brain cortex can then interpret these. Our visual system functions by allowing light to enter through the lenses of the eyes, with focusing and inversion of the image onto the retina. Finally, the visual cortex creates the images that we perceive. Recall that visual function is one of the higher cortical functions of the human brain (see Chapter 6). Other predominant structures of the eyes include the cornea, sclera, iris, pupil, choroid, and retina.

Anatomy of the eyeball

Most of the eyeball lies within the eye socket, also known as the bony *orbit*. Just a small anterior surface is exposed. The eyeball is composed of three layers of tissues: the fibrous layer containing the sclera and cornea, the vascular layer containing the choroid, ciliary body, and iris, and the inner layer containing the retina, optic nerve, and retinal blood vessels.

Fibrous layer

The outermost layer of the eyeball is called the **fibrous layer**. It is made up of the white-colored *sclera* and the transparent *cornea*. The **sclera** is commonly referred to as the *white of the eye*. It covers the majority of the ocular surface. The sclera is opaque, made up of dense fibrous connective tissue, containing dense collagen fibers and thin layers of elastin.

The **cornea** is a transparent structure that is structurally continuous with the sclera and lies over the colored part of the eye, which is called the *iris* (see Fig. 14.1). It is the anterior portion of the fibrous layer. In between the cornea and sclera is a border called the **corneoscleral junction** or *corneal limbus*.

When the cornea is damaged, blindness can occur even if the photoreceptors and other functional parts of the eye are still normal. There is an extremely limited ability of the cornea to repair itself. Therefore, injury to the cornea requires immediate treatment, or significant vision loss will occur.

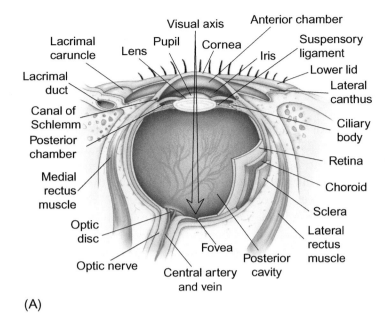

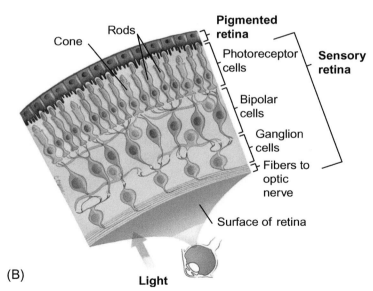

Fig. 14.1 Structure of the eyeball and cell layers of the retina. (A) Horizontal section through the left eyeball. The eye is viewed from above. (B) Pigmented and sensory layers of the retina. *(From Patton, K. T.; Thibodeau, G. A.* Anatomy & Physiology, *8th ed.; Mosby: St. Louis, 2013.)*

Vascular layer

The middle coat of the eyeball is called the **vascular layer** or *uvea*. It is pigmented and has three components: the *iris*, *ciliary body*, and *choroid*. The majority of the vascular layer is made up of the choroid. This layer has many blood vessels, lymphatic vessels, and the eye's smooth muscles, which are known as the *intrinsic muscles*. There are four functions of the vascular layer as follows:
- Providing a path for blood vessels and lymphatic to supply eye tissues
- Regulating how much light enters the eye
- Secreting and reabsorbing the aqueous humor circulating in the eye chambers
- Controlling the shape of the lens, which is essential in focusing

The **choroid** is a vascular layer separating the fibrous layer and the inner layer, posterior to the ora serrata. The sclera covers the choroid, which is attached to the outer layer of the retina. In the choroid, a large capillary network brings nutrients and oxygen to the retina. There are also many melanocytes in the choroid, especially close to the sclera. The eyes are perfused through the ophthalmic artery, which is a branch of the internal carotid artery, inside the cranium.

Inner layer

The **inner layer** of the eye is also known as the **retina**. It is a light-sensitive layer. The retina has a thin-pigmented *layer* and a thick *neural layer*. Inside the pigmented layer are pigment cells that are used by the *photoreceptors*. These are mostly located in the neural layer. While the two retinal layers are usually extremely close together, they do not have tight interconnections. The pigmented layer is continuous over the ciliary body and iris. The neural layer only extends anteriorly to the ora serrata. The neural layer is cup-shaped. It creates the posterior and lateral boundaries of the posterior cavity.

The retina originates from the neural tube and has a *blood-retina barrier system* that separates the retina from other structures. The *intraretinal capillaries* have bands of tight junctions that form the blood-retina barrier. The *ciliary epithelium* prevents diffusion of blood into the aqueous and vitreous humors. Substances in the sclera and choroid cannot reach the retina because *retinal pigment epithelial cells* also have tight junctions. These form a layer that is analogous to the blood-brain barrier. Movement between the choroidal capillaries and photoreceptors is controlled by movement across the pigmented epithelium.

There are five major neuronal cell types in the retina. Their somata are neatly arranged in three layers, with synapses in two additional layers, found between the layers of the cell bodies. Each synaptic layer has three different cell types; one type of cell carries visual information in, while another type carries information out. The third type of cell serves as an interconnector.

Beginning peripherally, photoreceptor cells that are stimulated by the light project to the first synaptic layer. Here, they terminate on the **bipolar cells** and **horizontal cells**.

The bipolar cells project to the next synaptic layer, while the horizontal cells spread laterally, interconnecting receptors, bipolar cells, and other horizontal cells. In the second synaptic layer, the bipolar cells terminate on **ganglion cells** and **amacrine cells**. Axons of ganglion cells join to leave the eye as the optic nerve. Amacrine cell processes spread laterally, interconnecting bipolar cells, ganglion cells, and other amacrine cells.

Retinal layers
The entire retina consists of 10 layers. These begin with the pigmented layer, and five layers are the layers of cell bodies and synapses described previously. In the names of the various layers, the term *nuclear* refers to cell bodies. The term *plexiform* refers to synaptic zones. The terms *inner* and *outer* refer to how many synapses are involved that separate a structure from the brain. Photoreceptors, for example, are considered *outer* in comparison to bipolar cells. The 10 layers of the retina are as follows: the *pigmented layer* (or *retinal pigment epithelium*), the *neural layer* (consisting of *rods* and *cones*), the *outer limiting membrane*, the *outer nuclear layer*, the *outer plexiform layer*, the *inner nuclear layer*, the *inner plexiform layer*, the *ganglion cell layer*, the *nerve fiber layer*, and *the inner limiting membrane*.

Pigmented layer
In the retina, the pigmented layer absorbs light passing through the neural layer. This prevents the light from reflecting back through the neural layer, which would produce visual effects described as *echoes*. The pigment cells interact biochemically with the photoreceptors of the retina. This epithelium is a single layer of polygonal cells. One side of each cell joins the choroid, as blood is supplied via choroid capillaries to the first two retinal layers. The other side of each cell forms many processes partially surrounding the outer parts of the receptor cells and filling the space that exists in embryonic development inside the optic cup wall. Pigmented epithelial cells are metabolically linked with the receptors. They also help absorb light that pass through the retina.

Neural layer
There are several layers of cells in the neural layer of the retina (see Fig. 14.2). The outer part of the retina contains the **photoreceptors**, which detect light. There are two main types of photoreceptors. **Rods** are extremely sensitive to light, and allow for vision in dim light, with a poor ability to see different colors. **Cones** provide color vision, and when there is bright light, they provide sharper and clearer vision than the rods. The outer segments of the rods and cones are continuously regenerated. The oldest parts of their tips are sloughed off and phagocytized by nearby pigmented epithelial cells. There are also inner segments, a cell body, and a synaptic terminal. It is important to understand that the terms "rod" and "cone" really only refer to the outer segment plus the inner segment of a photoreceptor cell. However, the terms are often used to refer to the entire receptors.

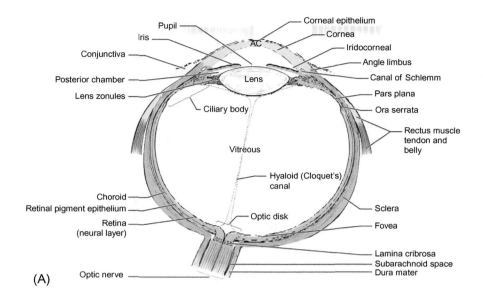
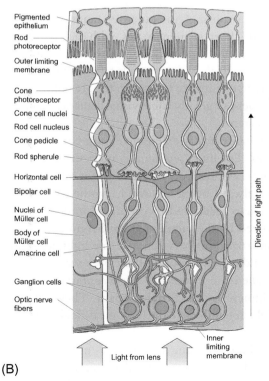

Fig. 14.2 Structure of the eyeball and cell layers of the retina. (A) The right eye is viewed from above (horizontal section). (B) The various layers and cells of the retina. *AC*, Anterior chamber. *((A) From Forrester, J. V., et al.* The Eye: Basic Sciences in Practice, *4th ed.; Elsevier: St. Louis, 2016. (B) From Gartner, L. P.* Textbook of Histology, *4th ed., Elsevier: Philadelphia, 2017.)*

The outer segment of each rod is long and cylindrical, while the outer segment of each cone is tapered and much shorter. Both of the outer segments have hundreds of flattened sacs, or *discs*. In the cones, most of these discs are continuous with the extracellular space. In the rods, nearly all discs are separated from the external membrane, and are totally intracellular.

The major protein of the outer segments of rods and cones is the visual pigment. In the rods, this is called **rhodopsin**. Since there is no universal name for the visual pigments of the cones, they are known as **cone pigments**. Rhodopsin and cone pigments are types of G protein-coupled receptors. They control many postsynaptic effects and certain other sensory transduction. In the rods and cones, the ligand of the receptor protein **opsin** is a derivative of vitamin A known as **11-cis retinal**. It allows photopigments to absorb available light. Differences between the rod opsins and those in each of the three cone types result in wavelengths that can be absorbed. In *phototransduction*, light isomerizes 11-cis retinal to *all-trans retinal*, which then dissociates from opsin. The isomerization of retinal causes opsin to activate molecules of **transducin**, a G-protein, which activates the enzyme *phosphodiesterase* to hydrolyze **cyclic guanosine monophosphate** (cGMP). A great amplification then occurs along this process, with more and more molecules of each substance being activated. Recall that deficiency of vitamin A can cause nighttime vision difficulties.

Photos of light moving through the outer segments of rods and cones pass through as many as several thousand sheets of membranes, with each being full of molecules of visual pigment. Therefore, the outer segments are where visual transduction occurs. Absorbed photos cause a *receptor potential*, which spreads to the remainder of the cell. In receptor cells, the photosensitive portions are located in the area of the neural retina farthest from incoming light. This does not greatly affect visual acuity or sensitivity since the retina is so thin and almost transparent.

Each of the outer segments has extensive cilium connected to their inner segments, via narrow ciliary stalks. The inner segments contain many mitochondria and other types of organelles. The mitochondria supply the energy for transduction and synthesis of visual pigments, which are always being renewed, synthesized within the inner segment, moved through the ciliary stalk, and incorporated into the disc membranes. Old discs are quickly phagocytized by the pigmented epithelium.

The eyes need time to adjust between extreme light conditions and poor light conditions, as the cones and rods function in their own ways. There is a third type of photoreceptor. This is called the *intrinsically photosensitive retinal ganglion cell* (*ipRGC*). In ipRGC, the photopigment is called **melanopsin**. These retinal ganglion cells respond to varying levels of light and influence the biological clock of the body—the *24-h circadian rhythm*.

Across the retina, there is an uneven distribution of the rods and cones. About 125 million rods create a wide band around the edges of the retina. Moving toward its center, the density of rods becomes less. Approximately six million cones are located in the area where visual images arrive after passing through the cornea and lens. This region is called the **macula lutea** or *macula*, in which there are absolutely no rods. The macula lutea is a yellowish area near the retina's center. Inside the macula, the largest number of cones is found in the **fovea centralis**, also called the *fovea*. This is the area of sharpest color vision, and is a small depression in the macula.

While looking at an object directly, its image is "placed" on this portion of the retina. If an imaginary line were drawn from the object's center, through center of the lens, to the fovea centralis, it would create the eye's **visual axis**. As long as there is sufficient light, a very good image can be seen, but since the cones cannot function in very dim light, the same image is not visible. Shifting the gaze to one side will help to see a dim object because it moves the image to the periphery instead of the fovea centralis, affecting the more sensitive rods.

Outer limiting membrane
This layer appears as a distinct line when microscopically viewed. It is actually a row of intercellular junctions. Glial cells that are elongated and specialized, known as **Muller cells**, span nearly all of the retina. They end distally at the bases of the inner rod and cone segments. Nearby Muller processes and inner segments are joined via junctional complexes. These form the outer limiting membrane.

Outer nuclear layer
This layer merely consists of the cell bodies of the rods and cones. The spherical rod granules are much more numerous than the stem-like cone granules. They are also located at different levels throughout this layer. The rod granule nuclei have a cross-striped appearance. Prolonged from each cell extremity is a fine process. The outer process is continuous with a single rod of the layer of rods and cones. The inner process ends in the outer plexiform layer in an enlarged extremity and is embedded in the area where the outer processes of the rod bipolar cells are divided. Along its course, there are many varicosities.

The cone granules are close to the membrana limitans externa, through which they are continuous with the cones of the layer of rods and cones. There are no cross-striations. The cone granules contain a piriform nucleus that almost completely fills them. From the inner extremity of each granule, a thick process passes into the outer plexiform layer. There, it expands into a pyramidal enlargement (footplate), from which there are many fine fibrils that contact the outer processes of the cone bipolar cells.

Outer plexiform layer
This thin synaptic zone is where receptors terminate on the horizontal and bipolar cells. Also, the processes of the horizontal cells spread laterally. The rods and cones synapse on distinct groups of bipolar cells and on different areas of certain horizontal cells.

Inner nuclear layer

This layer contains retinal interneuron cell bodies and the cell bodies of the Muller cells. The horizontal cell nuclei are near this layer's distal edge, while the bipolar cell nuclei are in the middle. The amacrine cell nuclei are near the proximal edge. Bipolar cells move visual information through this layer, containing on to the second synaptic zone.

Inner plexiform layer

This thicker synaptic zone is where bipolar cells terminate on the amacrine and ganglion cells. The amacrine cell processes spread laterally. Interconnection patterns are complex, and the amacrine cells are a good example. There are more than 30 different chemical and anatomical characteristics, each with unique functions, of these cells alone.

Ganglion cell layer

This layer contains the ganglion cell bodies, with dendrites branching in the inner plexiform layer, and axons exiting the eye as the optic nerve. This very thin layer is based on the fact that there are only about one million ganglion cells, far less than the number of cones and rods. A good amount of convergence is used for retinal processing, but this does not occur uniformly across the retina. Visual information travels in a variety of parallel streams. Axons of several distinct ganglion cell types share the same optic nerve as it continues toward the brain.

Nerve fiber layer and optic disc

This layer has axon and ganglion cells converging like the spokes of a wheel toward the **optic disc**, also called the *optic papilla*. This is a circular region that is medial to the fovea centralis and is the point of origin of the optic nerves (II). From here, the axons turn and penetrate the wall of the eye, proceeding to the thalamus. Our eyes are basically the only place where the brain is not covered by bones. Therefore, the eyes are a "window" to look at the brain with an ophthalmoscope. This is very true in the setting of papilledema, which is explained in detail later in this chapter. The retina is supplied with blood by the *central retinal artery (CRA)* and drained by the *central retinal vein (CRV)*. These vessels pass through the center of the optic nerve, emerging on the optic disc's surface. Occlusion of these arteries or veins can cause partial vision loss in one eye (see CRAO and CRVO in "Clinical considerations" section). There are no photoreceptors or other structures in the optic disc that are found elsewhere in the retina. Light striking the optic disc is not perceived. Therefore, it is called the eye's **blind spot**. However, we do not normally see a blank spot in our field of vision. This is because the involuntary movement of the eye keeps the visual image moving. The brain can then replace the missing visual information.

Inner limiting membrane

This layer is a thin basal lamina, between the vitreous and proximal ends of the Muller cells.

Focus on retinal scanning

A retinal scan is a biometric system that identifies a person by using the unique patterns of his or her retina. This system can even differentiate between identical twins. The retinal scanner uses the reflection of light that is absorbed by the retinal vein. It is more accurate than a fingerprint scanner.

Section review
1. What are the three layers of tissues in the eyeball?
2. What are the 10 layers of the retina?
3. What are the functions of the rods and cones?

Chambers of the eye

Both the anterior and posterior chambers of the eye contain *aqueous humor*. However, the much larger posterior chamber is mostly filled with a gelatinous substance known as the *vitreous body*. The **aqueous humor** circulates within the anterior cavity, passing into it from the posterior chamber via the pupil (see Fig. 14.3). The aqueous humor also diffuses freely through the posterior cavity, crossing the surface of the retina. It is similar to cerebrospinal fluid in its composition. Aqueous humor is clear, watery, and may leak out when the eye is injured.

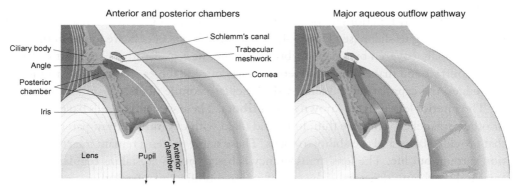

Fig. 14.3 Formation of aqueous humor.

Aqueous humor

Aqueous humor is formed via active secretion by the epithelial cells of the ciliary processes. The epithelial cells regulate its composition. The circulation of the aqueous humor allows nutrients and wastes to be transported and also creates a cushioning effect. The pressure created by the aqueous humor helps maintain the shape of the eye. The pressure stabilizes the retina's position and presses the neural layer against the pigmented layer.

The **intraocular pressure** of the eye is easily measured within the anterior chamber as the fluid pushes against the inner cornea surface. Measurement is usually done via *applanation tonometry*. Tension is measured and normally ranges between 12 and 21 mm Hg. The aqueous humor is secreted into the posterior chamber at a rate of between 1 and 2 mL/min. It also leaves the anterior chamber at the same speed. The connective tissue network near the base of the iris filters the aqueous humor, which then enters the **scleral venous sinus** or *canal of Schlemm*. This passage extends all the way around the eye, at the corneoscleral junction level. Via collecting channels, the aqueous humor is then brought to the scleral veins. In only a few hours of forming, the aqueous humor is removed and recycled. This is normally at the same speed it is generated in the ciliary processes.

Vitreous body

The gelatinous **vitreous body** is contained within the eye's posterior cavity. The **vitreous humor** is its fluid portion. The eye's shape is partially stabilized by the vitreous body and prevents distorting of the eye as the extrinsic eye muscles change its position. The vitreous body has specialized cells that produce collagen fibers and *proteoglycans*. These make up its gelatinous mass. The vitreous body forms during embryonic development. Unlike the aqueous humor, it is not replaced.

Lens

The **lens** of the eye is transparent and curves outward, a condition that is described as *biconvex*. The lens is a flexible disc, posterior to the cornea. It is held in place by the ciliary zonule, and mainly functions to focus visual images upon photoreceptors. It accomplishes this by changing its shape (see Fig. 14.4). Concentric layers of cells are found in the lens, and it is totally covered by a dense, fibrous capsule. Its capsular fibers are mostly elastic. They contract and make the lens spherical unless outside forces are applied. Around the lens edges, the capsular fibers blend with the ciliary zonule fibers.

The interior lens cells are known as **lens fibers**. These specialized cells have no nuclei or other organelles. They are thin, long, and filled with **crystallins**, transparent proteins that provide clarity and the ability to focus. These stable proteins remain intact and function throughout life. The lens' transparency is based on the crystallins regulating structural and biochemical factors. If the crystallins are modified over time, they can result in a loss of lens transparency, known as a **cataract**. A cataract may be caused by an injury, a drug reaction, or ultraviolet radiation, such as from exposure in the angiography suite,

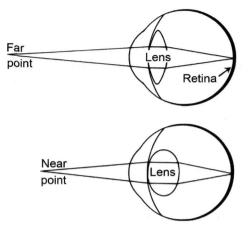

Fig. 14.4 Accommodation of the lens.

causing early cataracts in interventional cardiologists and neurologists, despite using protective glasses. The most common form of cataracts, however, are *senile cataracts*, which are only caused by aging.

With age, the lens becomes yellow-colored and loses transparency. With this clouding, more intense light is required for reading and clarity starts to fade. Once the lens is totally opaque, though the photoreceptors are still normal, the individual is *blind* in that eye. For a cataract, surgery can be performed to remove the lens, either intact or after it has been broken up by using high-frequency sound waves. The lens is then replaced with an artificial structure, and using glasses or contact lenses regulates the patient's vision.

Visual acuity

For the ability to see clearly, light rays must be *refracted*, which means bent. The lens of the eye must experience *accommodation*, and the pupil must be able to *constrict*. Also, the eyes must be in *convergence* with each other. For refraction, the cornea, aqueous humor, lens, and vitreous body must all be functioning. Light rays are refracted at the anterior cornea, the anterior lens, and the posterior lens. **Visual acuity** is the sharpness or clearness of visual perception. It is affected by the ability to focus, the state of the retina, and normal function of the visual pathway along with the processing centers of the brain.

For visual acuity, the standard vision rating is called "20/20." This means a person can see an object clearly at 20 ft away that should be able to be seen clearly at this distance. A person with 20/15 has better-than-average vision since he or she can clearly see objects at 15 ft that should be clearly seen at 20 ft. A person with 20/30 vision has a deficiency of sight since he or she cannot clearly see an object at 20 ft that should be clearly seen at 30 ft.

An individual is termed *legally blind* when visual acuity falls below 20/200, even when glasses or contact lenses are used. There are more than 1.3 million legally blind people in the United States, with more than 50% of them being age 65 or older. *Blindness* is defined as a total absence of vision, due to damage to the eyes or optic pathways. It is commonly caused by diabetes mellitus, glaucoma, cataracts, retinal detachment, scarring of the cornea, accidental injury, and hereditary factors.

Some people have an abnormal blind spot, called a **scotoma**, appearing in the field of vision in an area beside the optic disc. This can be a permanent condition that may be caused by optic nerve compression, photoreceptor damage, or central visual pathway damage. Temporarily, visual scotoma frequently happens during a migraine attack. Small spots that may drift across the visual field are called *floaters*, usually, a temporary condition caused by blood cells or cell debris within the vitreous body. They are most easily seen when staring at a white wall or piece of paper.

Pupillary light reflexes and pathway

Accommodation requires an increase in the curvature of the lens, pupil constriction, and convergence of the eyes. Accommodation, therefore, involves increased lens curvature in order to achieve greater refraction. As the ciliary muscle contracts, the choroid is pulled closer to the lens. This loses the suspensory ligaments, and the lens bulges. This is the process used for *near vision*. The opposite, *far vision*, requires relaxation of the ciliary muscle and flattening of the lens. Near vision produces eyestrain since the ciliary muscle must remain contracted for a long time. With aging, farsightedness often develops as the lens loses elasticity. This condition is called "old eye" or *presbyopia*.

As the outer radial fibers of the iris relax, the pupil is constricted, preventing divergent light rays from entering the eye via the peripheral areas of the cornea and lens. Peripheral rays such as these would not be refracted enough to be focused on the retina, and result in blurred images. As the pupil constricts for near vision, it is called the *near reflex*. It occurs at exactly the same time as accommodation of the lens.

Section review
1. What are the functions of the aqueous humor?
2. What are the descriptions of the terms accommodation and convergence?
3. How is a person determined to be "legally blind"?

Visual pathways

The visual pathways start at the photoreceptors, ending at the *visual cortex* within the cerebrum of the brain. Unlike other sensory pathways, in the visual pathways, messages are required to cross two different synapses. These are the photoreceptor to a bipolar cell, and a bipolar cell to a ganglion cell. After this, the messages can move to the brain. While the extra synapse prolongs the synaptic delay, it allows for processing and integration of visual stimuli prior to leaving the retina.

In the *striate cortex* (a part of the visual cortex), incoming information is broken up into components such as orientation, color, depth, and motion. This information is then distributed to various specialized areas for additional processing. It is believed to speed up visual perception and allows a large variety of information to be analyzed quickly to determine whether a significant bit of information is present. The visual pathway is explained in more detail in "Central processing" section.

Retinal processing

Each photoreceptor in the retina monitors a certain receptive field. A large amount of convergence must occur at the beginning of the visual pathway due to the presence of millions of bipolar and ganglion cells. Convergence is different between rods and cones. Cones lack convergence ability. Each ganglion cell monitors a certain part of the visual field, known as its *receptive field*. Up to 1000 rods may pass information via the bipolar cells to just one ganglion cell. The large ganglion cells monitoring the rods are called **M cells**. They provide general information about objects, motions, and shadows in dim light. Due to a large amount of convergence taking place, M cell activation indicates that light arrives in only a general area, not a specific one.

The activity of ganglion cells is varied, based on their activity patterns in the receptive field, which is usually circular. They usually respond differently to stimuli in the center of the field than to stimuli arriving on the edges. **On-center neurons** are ganglion cells excited by light in the center of the receptive field. **Off-center neurons** are those excited by light at the edges of the field. Both types provide information about which area of the field is illuminated. This retinal processing allows for better detection of the edges of objects.

In the cones, due to lack of convergence, the processing is different. Within the fovea centralis there is a one-to-one ratio of cones to ganglion cells. The **P cells** are the ganglion cells that monitor the cones. They are smaller and more abundant than M cells. The P cells are active in bright light, providing information about color, fine details, and edges of objects. Activation of a P cell means that light has arrived in a specific location, allowing the cones to provide more precise information about images than rods. Images produced by cones are of much higher resolution than those produced by rods.

> **Focus on color vision**
> The human eye can see 7,000,000 colors. Certain colors can irritate the eyes, while others are soothing. Pure, bright lemon yellow is the most fatiguing color. More light is reflected by bright colors, resulting in excessive stimulation of the eyes. The most soothing color for the eye is a mixture of green and blue.

Central processing

Axons from all ganglion cells converge at the optic disc. They penetrate the wall of the eye, proceeding to the diencephalon as the optic nerves (II) exiting the orbit superior to the optic fissure. From each eye, one of the optic nerves reaches the diencephalon and thalamus after partially crossing over at the **optic chiasm**, also known as the *optic chiasma* (see Fig. 14.5).

From here, about 50% of the fibers continue to the lateral geniculate body of the same side of the brain. The other 50% cross over, reaching the same structure on the other side. Visual information then travels from each lateral geniculate body to the **visual cortex**, located in the occipital lobe of the cerebral hemisphere on that specific side. The optic chiasm is a very important anatomical location since any lesions proximal to it cause a mono-ocular vision abnormality (affecting only one eye). Any lesions beyond the optic chiasm result in bilateral binocular vision abnormalities. The **optic tract** refers to the visual fibers of the optic nerves, once they cross and pass beyond the optic chiasm until they synapse at the lateral geniculate bodies.

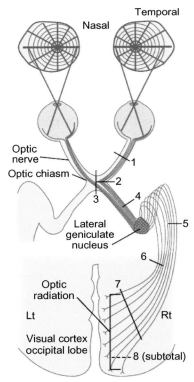

Fig. 14.5 Visual fields and neural pathways of the eyes.

Optic radiation is a term that describes the section or bundle of projection fibers that link the lateral geniculate bodies in the thalamus with the visual cortex in the occipital lobe. Collaterals from the fibers synapse in the lateral geniculate. They then continue to subconscious processing centers located in the diencephalon and brainstem.

Visual field

Visual images are perceived because of integrated information arriving at the visual cortices. Each eye receives visual images that is slightly different from the other eye. This is partially because the foveae of the eyes are 2–3 in. apart, and also because the nose and eye socket block the view of each opposite side. The ability to judge depth or distance by interpreting the three-dimensional relationships between objects is called **depth perception**. The brain compensates by comparing relative object positions within the two received images.

When looking straight ahead, there is overlapping of the left and right visual images. The **field of vision** is the combination of these two areas. The fovea centralis of each eye lies in the center of the area of overlap. Visual information from the left half of this combined field of vision reaches the right occipital lobe's visual cortex and vice versa.

Visual processing in the brainstem

Many visual centers in the brainstem receive information from either the lateral geniculate bodies, or via collaterals from the optic tracts. Collaterals bypassing the lateral geniculate bodies synapse in either the superior colliculi or hypothalamus. In the midbrain, the superior colliculi initiate motor commands that unconsciously control movements of the eye, head, and neck—in response to visual stimuli, also known as *reflexive saccade*. Pupillary reflexes and reflexes controlling eye movement are influenced by collaterals that carry information to the superior colliculi. The function of other brainstem nuclei is affected by visual inputs to the suprachiasmatic nucleus of the hypothalamus. This nucleus, and the *pineal gland* within the epithalamus, use visual information to help establish the body's **circadian rhythm**. This is a daily pattern of arousal linked to the day-night cycle. It affects endocrine function, metabolic rate, digestion, blood pressure, the sleep-wake cycle, and various behavioral and physiological processes.

Visual pigments

The photoreceptors use a light-absorbing molecule called **retinal**, which combines with proteins known as *opsins* to translate light into electrical impulses. Four types of visual pigments are then formed. Retinal absorbs different wavelengths of the visible light spectrum based on the type of opsin to which it is bound. Cone opsins are different from rod opsins, and also different from each other. The names of cones are based on the colors—actually wavelengths—of light that each type of cone best absorbs. Red cones have wavelengths close to 560 nm; green cones are close to 530 nm; and blue cones are close to 420 nm.

The ability to see other colors besides red, green, and blue come from the overlapping of cone absorption as well as the ability to see intermediate hues such as purple, orange, and yellow. This occurs from differential activation of more than one cone type at the same time. Since yellow light stimulates red as well as green cone receptors, it can be seen. When the red cones are stimulated more than the green cones, the color we perceive is orange. When all types of cones are equally stimulated, we see the color white.

Retinal is a chemical relative of vitamin A and is actually manufactured from this vitamin. In the pigmented retinal layer, cells absorb vitamin A from the blood and supply it to the rods and cones. Retinal assumes different three-dimensional forms called *isomers*. When it binds to opsin, retinal has a shape called *11-cis-retinal*. When it absorbs a photon of light, the retinal twists and snaps to become **all-trans-retinal**. This change also causes opsin to change shape and become activated. Capturing of light by visual pigments is the only stage that requires light. This is a simple photochemical occurrence that begins an entire chain of chemical and electrical reactions in rods and cones. This will cause electrical impulses to be transmitted along the optic nerve.

Section review
1. What are the functions of the M cells and P cells?
2. What is the pathway of the optic nerve?
3. What is all-trans-retinal?

Clinical considerations

This part focuses on clinical considerations of a variety of visual system disorders. In the authors' view, the best way to learn the neuroanatomy of the visual pathway is to understand the system through clinical syndromes and cases. These include cranial nerve disorders including palsies, total blindness, hemianopia, homonymous hemianopsia, papilledema, retinoblastoma, glaucoma, cataract, retinal detachment, color blindness, and nyctalopia.

Total blindness

Total blindness is the inability to see anything, even in lighted conditions. One eye can become totally blind without affecting the other eye (see Fig. 14.6). Any person who loses sight completely must ask another person to help them find emergency treatment, and not wait for the vision to return. If treatment is immediate, there may be a chance of restoring vision.

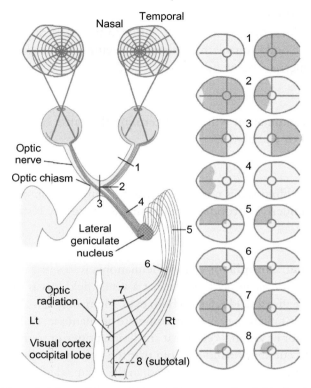

Fig. 14.6 Visual fields and defects that accompany damage to the right visual pathways. (1) Optic nerve: blindness. (2) Lateral optic chiasm: grossly incongruous, incomplete (contralateral) homonymous hemianopsia. (3) Central optic chiasm: bitemporal hemianopsia. (4) Optic tract: incongruous, incomplete homonymous hemianopsia. (5) Temporal loop of the optic radiation: congruous partial or complete (contralateral) homonymous superior quadrantanopia. (6) Parietal *(superior)* projection of the optic radiation: congruous partial or complete homonymous inferior quadrantanopia. (7) Complete parietooccipital interruption of the optic radiation: complete congruous homonymous hemianopsia with psychophysical shift of the foveal point, often sparing central vision and resulting in "macular sparing." (8) Incomplete (subtotal) damage to the visual cortex: congruous homonymous scotomas, usually encroaching at least acutely on central vision. *(From Goldman, C., Ed.* Goldman's Cecil Medicine, *24th ed., Vol. 2, Saunders: Philadelphia, 2012.)*

Pathophysiology

Glaucoma, macular degeneration, and cataracts having a lazy eye, optic neuritis, retinitis pigmentosa, tumors, diabetes mellitus, stroke, birth defects, eye injury, and complications from eye surgery, may cause blindness. Other people at risk for blindness are those who work with sharp objects or toxic chemicals.

Clinical manifestations

Though total blindness means "no vision," prior to this occurring, there may be partial vision loss that is cloudy, unable to see shapes, seeing only shadows, seeing poorly at night, or even experiencing *tunnel vision*.

Diagnosis

A complete eye examination will help determine the cause of blindness, regardless of whether it is total or partial.

Treatment

In most cases, total blindness is not treatable, but if caused by an injury or disease, early treatment of the underlying condition may partially or totally restore sight.

Homonymous hemianopia

Homonymous hemianopia, or simply, hemianopia is loss of half of a visual field (see Fig. 14.6). The term *homonymous* means that a visual field loss is similar for both eyes. Please note that the visual field refers to both eyes. Hemianopia is also called *hemianopsia*. Damage to the optic radiation may cause complete hemianopia; deficits more often affect a quadrant or a sector of vision. Recall that any lesion posterior to the optic chiasm causes *bilateral* visual field cut abnormalities. Note that the eyes are in the frontal area, and the visual cortex is located in the back of the head, in the occipital lobe. Therefore, any lesion in this long pathway can cause a vision abnormality. The optic radiations pass through the temporal and parietal lobes, depending on the visual field that they are carrying.

Pathophysiology

A lesion in one optic tract, interrupting all fibers carrying information from the contralateral visual fields, may cause homonymous hemianopia. When an injury to the left optic tract occurs, the patient will have visual difficulties (visual field cuts) in the right eye's inner field, but in the left eye's outer field. Similarly, if there is massive damage to the visual cortex of one occipital lobe, such as the occlusion of a posterior cerebral artery (PCA) close to where it originates from the basilar artery, contralateral homonymous hemianopsia would probably develop. The most common causes are stroke, brain tumor, and trauma. Recall that optic radiations and tracts cross at the optic chiasm level. Therefore, any lesions posterior to the optic chiasm are not only bilateral but also contralateral to the side of the lesion, as mentioned in the case of PCA occlusion stroke.

Clinical manifestations

Hemianopia often causes the patient to bump into objects, making movement difficult and increasing the likelihood of tripping or stumbling. It is easy to be startled by people or moving objects, and other symptoms include reading difficulties, clumsiness, dizziness, fear, anxiety, panic attacks, and avoiding outside events without explanation.

Diagnosis

Hemianopia is usually diagnosed following a series of vision tests. The depth of vision is tested for each eye individually. Blank areas of vision are identified to determine that hemianopia is present. An MRI, with or without contrast, may be ordered to determine the cause of the condition. The cause can be stroke (sudden in onset) in the visual cortex, or simply a brain tumor (with gradual onset of symptoms) in the occipital lobe, among many other differentials. As a rule, if the hemianopia is isolated and no other signs or symptoms are present, then the lesion is most likely located within the occipital lobe.

Treatment

Treatment is based on the underlying cause. If the cause is an acute stroke, then thrombolytic medications like tissue plasminogen activator (tPA) or thrombectomy procedures can remove the clot—if the patient presents early enough. Some of the visual abnormalities due to a compression by a tumor may be resolved after removal of the tumor.

Unilateral vision loss and abnormalities

Most of the reasons behind unilateral visual loss are related to the eye itself. There are, however, a few other etiologies that can cause a unilateral vision abnormality. A unilateral vision abnormality means the lesion responsible for the vision abnormality is anterior to the optic chiasm. For example, lesions of the optic nerves are linked to a variety of conditions. When there is compression of the optic nerve, affecting the noncrossing fibers on that side, the result is **unilateral visual loss** of the **ipsilateral** eye since the nerve is a noncrossing fiber.

Bitemporal hemianopia

Damage in the central area of the optic chiasm that affects the crossing fibers causes a heteronymous type of hemianopia, such as **bitemporal hemianopia**. This usually results from midline pressure, due to a pituitary gland lesion that is close to the chiasm. Recall that the optic nerves become the optic tract when they cross at the level of the optic chiasm. Remember that the optic nerve consists of two groups of fibers: nasal visual field fibers and temporal visual field fibers. Interestingly, fibers that carry the temporal visual field are the only fibers that cross at the optic chiasm. Therefore, any lesions at this level cause temporal visual field loss in both eyes (see Fig. 14.6). This condition is also known as bitemporal *hemianopsia*. It is the most common visual defect resulting from direct impingement of the optic chiasm by an expanding mass. Defects can be incomplete or subtle and detected only by a small difference in visual sensitivity—such as to color—across the vertical meridian of each visual field. Growth of many tumors is very slow, allowing crossing fibers to adapt. Therefore, a certain degree of upward pressure is needed before conduction slows or is interrupted. This means that visual loss takes some time to be noticed. Contrast-enhanced imaging may be helpful to identify locations of the anterior optic pathways in a patient with **suprasellar** tumors. This offers the ability to predict how the patient's vision will be, following decompression.

Localizing the lesion within the visual pathway

Any lesion within the eye or the optic nerve portion would cause ipsilateral and unilateral vision abnormalities of the same eye. Any lesion in the optic tract portion of the visual pathway will lead to bilateral, confraternal visual field loss in both eyes. Beyond lateral geniculate nucleus (LGN), the optic radiations divide into superior and inferior visual fibers. Since the parietal lobe is located superiorly to the temporal lobe, therefore, it is not a surprise that the *superior* fibers pass through the *parietal lobe* and the *inferior fibers*, or **Meyer's loop**, pass through the temporal lobe. Another important fact to localize is to remember that the superior optic radiations carry signals from the *inferior* visual field, while the Meyer's loop of the inferior fibers carry signals from the *superior* visual field. Below, the location of a lesion is listed, followed by its expected visual abnormalities.

    Left superior occipital lobe: Right homonymous inferior quadrantanopsia

    Right temporal lobe stroke involving Meyer's loop: Left homonymous superior quadrantanopsia (commonly known as a "pie in the sky" defect)

    Right LGN: Left homonymous hemianopia (see Fig. 14.6)

To summarize, the light passes through the lenses and pupils to the retina. From the retina, it moves to the optic nerves, through the optic chiasm, and continues along the optic nerves to the LGN, then to the optic radiations and the visual cortex.

Hereditary optic neuropathies

Hereditary optic neuropathies are caused by genetic abnormalities. They cause vision loss, and sometimes, cardiac or neurologic problems. These neuropathies usually appear during childhood or adolescence, with bilateral and symmetric central vision loss. Optic nerve damage is usually permanent and may be progressive. Once optic atrophy is found, there has always been a substantial injury to the optic nerve. Understand that these conditions involve both eyes, and therefore, there is bilateral vision loss.

Pathophysiology

There are two primary pathophysiologic forms of hereditary optic neuropathy. The first is called *dominant optic atrophy* and is inherited as an autosomal dominant condition. It is thought to be the most common form, occurring in one of every 10,000–50,000 people. It is also believed to be **optic abiotrophy**, which is premature degeneration of the optic nerve that causes progressive vision loss. This form occurs in the first decade of life.

The second primary form is called *Leber's hereditary optic neuropathy*. It involves a mitochondrial DNA abnormality, affecting cellular respiration. This is the primary manifestation, even though mitochondrial DNA in many areas of the body is affected. Between 80% and 90% of cases occur in males. It is inherited with a maternal inheritance pattern. Therefore, all offspring of a mother with this abnormality will inherit the condition. Only females can pass on the abnormality since the zygote only receives mitochondria from the mother.

Clinical manifestations

Dominant optic atrophy usually does not cause any related neurologic abnormalities. Nystagmus and hearing loss have been sometimes reported. There is slowly progressive, bilateral vision loss that is usually mild until worsening late in life. Either the entire optic disk or just the temporal area is pale, with no visible vessels. A blue-yellow color vision deficit usually occurs.

In the Leber form, vision loss usually starts between ages 15 and 35 years, though it can occur anywhere between 1 and 80 years. There is central painless vision loss, usually followed within weeks to months by loss in the other eye. Some patients have simultaneous vision loss. The majority of patients lose vision to worse than 20/200 acuity (which is legally blind by definition). Upon ophthalmoscopic examination, there may be **telangiectatic microangiopathy**, swelling of the nerve fiber layer surrounding the optic disk, and a lack of leakage seen in fluorescein angiography. Over time, optic atrophy manifests. Some patients have cardiac conduction defects. Others have slight neurologic abnormalities. These may include postural tremor, dystonia, loss of ankle reflexes, a multiple-sclerosis-like illness, or spasticity.

Diagnosis

Dominant optic atrophy is confirmed via molecular genetic testing. Diagnosis of Leber's hereditary optic atrophy is mostly clinical. An ECG should be performed to reveal occult cardiac conduction defects.

Treatment

There is no effective treatment for hereditary optic neuropathies. Some patients benefit from low-vision aids such as large-print devices, magnifiers, and talking watches. Genetic counseling should be performed. For Leber's hereditary optic neuropathy, many medications have been used without success. Quinone analogs such as idebenone and ubiquinone may be helpful early in the disease course. Avoiding alcohol and other oxidative stressors of mitochondrial energy production may be beneficial, but are not proven. Basically, the patient should avoid smoking and excessive intake of alcohol. If there are cardiac or neurologic abnormalities, the patient must see a specialist.

> **Section review**
> 1. What are the causes of homonymous hemianopia?
> 2. When does bitemporal hemianopia occur?
> 3. What is the function of Meyer's loop?

Anterior ischemic optic neuropathy

Anterior ischemic optic neuropathy (AION) is infarction of the optic disc. There are two types: nonarteritic and arteritic (also known as *giant cell arteritis*). The nonarteritic form is more common, usually affecting people age 50 years and older. Vision loss in the nonarteritic form is not permanent, and not as severe as in the arteritic form, which usually affects people age 65 years and older. Another form of optic neuropathy is simply called optic neuritis, which can be due to a variety of other causes, which will be discussed later.

Pathophysiology

The majority of ischemic optic neuropathy cases are unilateral, with bilateral, sequential cases occurring in some patients depending upon the cause (arteritic vs. nonarteritic). The pathophysiology of the nonarteritic form is not fully understood. However, atherosclerotic narrowing of the posterior ciliary arteries may increase risks for nonarteritic optic nerve infarction, especially after a hypotensive episode. The arteritic form is related to giant cell arteritis. This form is exclusively found in patients over age 50 with abnormal inflammatory markers.

Nerve edema is caused by acute ischemia, which worsens the ischemia itself. A risk factor for the nonarteritic form is a small optic cup to optic disk ratio. There is no common concurrent medical condition for the nonarteritic form. However, diabetes and hypertension are present in some cases and are believed to be risk factors. Vision loss when the patient awakens leads to suspicion of nocturnal hypotension as a likely cause of the nonarteritic form. The nonarteritic form usually resolves on its own within 12 weeks.

Clinical manifestations

For both the arteritic and nonarteritic forms of ischemic optic neuropathy, vision loss usually occurs quickly, over minutes, or even gradually, over days. The patient may notice the vision loss upon awakening. Additional symptoms present with the arteritic form include general malaise, a temple headache, muscle ache or pain, jaw claudication (pain with chewing), pain when combing the hair, and tenderness over the temporal artery. These symptoms may not develop until after vision is lost. Visual acuity is lessened, and an afferent pupillary defect appears due to the damage to the optic nerve on that side. The optic disk will be swollen with surrounding hemorrhages. Visual field examination often reveals defects in the inferior and central visual fields. The arteritic form can spread to the other eye.

Diagnosis
Diagnosis of ischemic optic neuropathy is mostly based on clinical evaluation, often with ancillary testing. It is important to exclude the arteritic form since the other eye is at risk if treatment is not begun quickly. The erythrocyte sedimentation rate (ESR) is almost always elevated to over 50 in the arteritic form. The C-reactive protein is also helpful for diagnosis. If temporal arteritis is suspected, a temporal artery biopsy should be performed. For isolated cases with progressive vision loss, CT scan or MRI may be performed to rule out any compressive lesions such as glioma, which can rarely be a cause of painless optic neuropathy.

Treatment
There is no effective treatment for ischemic optic neuropathy, but to protect the other eye from the condition, the arteritic variety is treated with oral corticosteroids such as prednisone. Treatment should not be delayed while biopsy results are being awaited. The abnormality of the temporal artery will last for weeks after steroid treatment. However, treatment with steroids would not interfere with diagnosis, treatment of the nonarteritic variety, using aspirin or corticosteroids, has not been successful. Risk factors must be controlled. For both forms, low-vision aids may be helpful. Most lost vision will never be recovered. As many as 40% of patients will spontaneously recover some amount of useful vision in the nonarteritic form.

Optic neuritis
Optic neuritis is an inflammation of the optic nerve. It is most common in adults between the ages of 20 and 40 years. It can be from multiple sclerosis or another inflammatory disease, or simply idiopathic. Optic neuritis is painful, where *nonarteritic anterior ischemic optic neuropathy (NA-AION)* is painless.

Pathophysiology
Optic neuritis can be the first manifestation of multiple sclerosis (as a clinically isolated syndrome). Other causes include neuromyelitis optica (NMO), sarcoidosis, Lupus, infections such as syphilis or Lyme disease, or even after vaccination for influenza. The cause may not be identified in more than half of the cases.

Clinical manifestations
The primary symptom of optic neuritis is vision loss that is often maximal in 1–2 days. It may vary, from a small central or paracentral scotoma to total blindness. There is usually eye pain (over 90% of the time, this condition is very painful), which is worsened by eye

movements. If there is swelling of the optic disk, the condition is called *papillitis*. Without this swelling, it is called *retrobulbar neuritis*. Usually, there will be reduced visual acuity, visual field deficits, and disturbed color vision (red color desaturation) that is often out of proportion to the amount of visual acuity loss. If the contralateral eye is not affected, or is involved to a lesser degree, there is usually a detectable afferent pupillary defect testing of color vision is helpful. In about 66% of patients, the inflammation is completely retrobulbar. This causes no visible alteration of the optic fundus. In the remainder of cases, there will be disk hyperemia, edema within or around the disk, vessel engorgement, or a combination of these factors.

Diagnosis

Diagnosis of optic neuritis is based on the characteristic pain and vision loss. Neuroimaging should be via gadolinium-enhanced MRI, which may show enlargement and enhancing of the optic nerve. An MRI may help to diagnose multiple sclerosis or those at high risk of developing MS. Fluid attenuating inversion recovery (FLAIR) sequences during MRI may show common demyelinating lesions, in a periventricular location, if the neuritis is related to demyelination.

Treatment

Corticosteroids should be used in the acute phase. The steroid should be given in an intravenous (IV) form for 3 days followed by 11 days of oral steroid tapering. Steroids do not change the course of the disease or the outcome. However, steroids decrease recovery time. Fortunately, most episodes resolve on their own, with a return of vision in 2–3 months. Most patients with a common history of optic neuritis and no other connective tissue or systemic disease will recover vision. If the MRI shows any MS-like lesions, then the chance of developing MS is increased to the point that now we treat these patients with MS disease-modifying medications.

Papilledema

Papilledema is edema of the optic disk (see Fig. 14.7). It is caused by increased intracranial pressure. This is related to the subarachnoid space being continuous with the optic nerve sheath.

Pathophysiology

Causes of papilledema include brain tumors, intracranial hemorrhage, hydrocephalus, cerebellar or cerebral abscess, Lyme disease, intracranial hypertension, brain infections, and reactions linked to kidney transplant medications. As cerebrospinal fluid pressure increases, the pressure is transmitted to the optic nerve. The optic nerve sheath compresses the nerve, reducing axoplasmic transport. Then, axoplasmic substances collect at the **lamina cribosa** in the sclera, causing swelling of the optic disc. The *physiologic*

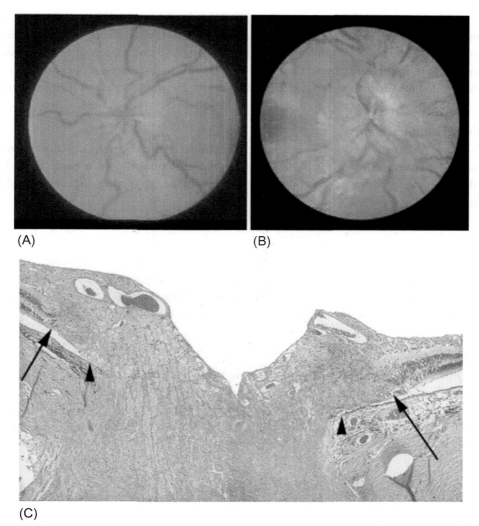

Fig. 14.7 The optic nerve in anterior ischemic optic neuropathy (AION) and papilledema. (A) In the acute phases of AION the optic nerve may be swollen, but it is relatively pale because of decreased perfusion. (B) In papilledema secondary to increased intracranial pressure, the optic nerve is typically swollen and hyperemic. (C) Normally, the termination of Bruch membrane *(arrowhead)* is aligned with the beginning of the neurosensory retina, as indicated by the presence of stratified nuclei *(arrow)*, but in papilledema the optic nerve is swollen, and the retina is displaced laterally. This is the histologic explanation for the blurred margins of the optic nerve head seen clinically in this condition. *((A,B) Courtesy Dr. Sohan S. Hayreh, Department of Ophthalmology and Visual Science, University of Iowa, Iowa City, IA. (C) From the teaching collection of the Armed Forces Institute of Pathology.)*

cup is a bright area in the center of the optic disc that becomes obliterated due to this swelling. The optic disc later is raised above the level of the remaining retina. The margins become blurred and nondistinct. Hemorrhaging and collections of white exudate, due to nerve infarcts, surround the margins of the optic disc, swelling becomes severe. The swelled nerves compress the retinal veins, leading to venous stasis and engorgement.

Clinical manifestations

The common manifestations of papilledema are similar of those seen in any cause of intracranial hypertension, which include headache, nausea, vomiting, short-term vision loss, reduced scope of vision, and vision changes such as blurring, graying, flickering, or double vision.

Diagnosis

Papilledema is usually easy to diagnose, utilizing ophthalmoscopic examination of the retina. Red spots in the retina indicate bleeding. A vision test may determine the extent of disease. MRI and CT may be used to exclude brain tumors. A lumbar puncture is performed to measure the CSF pressure if there is no concern for herniation.

Treatment

The cure for papilledema is based on treating the cause. Brain tumors may require radiation, or surgery. There may be the need for repeated lumbar punctures to remove excess CSF in the cranium. Medications like acetazolamide help to reduce spinal fluid accumulation, lowering its production, and reducing CNS pressure in the setting of idiopathic intracranial hypertension.

Toxic amblyopia

Toxic amblyopia is a loss of visual acuity. It is believed to result from a toxic reaction in the orbital portion, or *papllomacular bundle*, of the optic nerve. Toxic amblyopia is also called *nutritional amblyopia*.

Pathophysiology

Toxic amblyopia is usually symmetric and bilateral. Undernutrition may be the cause if the patient is an alcoholic. Actual tobacco-related amblyopia is rare. The optic nerve can be damaged by many agents, which include lead, chloramphenicol, methanol, digoxin, and ethambutol. Likely risk factors include deficiencies of antioxidants and protein. Other nutritional disorders may be related, including *Strachan's syndrome*, which involves **orogenital dermatitis** and polyneuropathy.

Clinical manifestations

Toxic amblyopia usually causes vision blurring and dimness after days to weeks. There is often a small central or pericentral scotoma that slowly enlarges. It usually involves the fixation and the blind spot (centrocecal scotoma), progressively reducing vision. Total blindness may occur from methanol ingestion. Other nutritional causes usually do not cause significant vision loss. Though retinal abnormalities are not usually present, temporal disk pallor can develop late in the disease course.

Diagnosis

Diagnosis is based on a history of undernutrition, or toxic or chemical exposures, along with typical bilateral scotomata upon visual field tests. Laboratory tests for toxins such as lead and methanol are performed. Once the optic nerve has atrophied, there will usually be no recovering of lost vision.

Treatment

Vision may improve if the cause is treated or removed quickly. Exposure to toxins should stop immediately. For lead poisoning, chelation therapy is performed. For methanol poisoning, dialysis, ethanol, fomepizole, or a combination of these is used. If oral or parenteral B vitamins are administered prior to severe vision loss developing, they can reverse the condition if undernutrition is the suspected cause. Low-vision aids may be helpful. The role of antioxidants is not fully understood and they have no proven efficacy.

Section review
1. How fast does AION develop?
2. What optic condition can be the first manifestation of multiple sclerosis?
3. What is the pathophysiology of papilledema?

Glaucoma

Glaucoma is the primary cause of visual impairment and blindness worldwide (see Fig. 14.8). It is signified by intraocular pressure, of the aqueous humor, greater than the normal range of 12–20 mm Hg. *Open-angle glaucoma* is the most common form and involves obstruction of outflow of aqueous humor at the trabecular meshwork or Schlemm canal. *Normal* or *low-tension glaucoma* is a type of open-angle glaucoma, with damage to the optic nerve that does not show symptoms. There is gradual vision loss when the intraocular pressure is still normal. In *narrow-angle* or *angle-closure glaucoma*, the iris is displaced forward, toward the cornea, with narrowing of the iridocorneal angle (see Fig. 14.9). This obstructs the outflow of aqueous humor from the anterior chamber. In *acute angle-closure glaucoma*, there is acute closure of the iridocorneal angle and a sudden

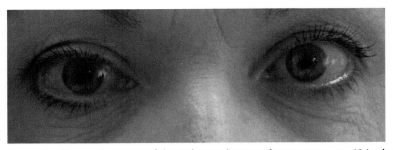

Fig. 14.8 Acute angle closure glaucoma of the right eye (intraocular pressure was 42 in the right eye). Note the mid-sized pupil on the left that was not reactive to light and conjunctivitis.

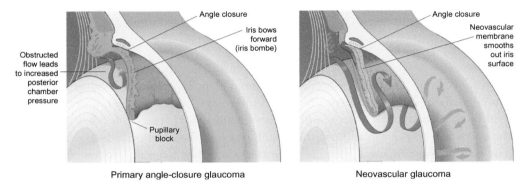

Fig. 14.9 *Left*: Primary angle-closure glaucoma. In anatomically predisposed eyes, transient apposition of the iris at the pupillary margin to the lens blocks the passage of aqueous humor from the posterior chamber to the anterior chamber. Pressure builds in the posterior chamber, bowing the iris forward (iris bombe) and occluding the trabecular meshwork. *Right*: A neovascular membrane has grown over the surface of the iris, smoothing the iris folds and crypts. Myofibroblasts within the neovascular membrane cause the membrane to contract and to become apposed to the trabecular meshwork (peripheral anterior synechiae). Outflow of aqueous humor is blocked, and the intraocular pressure becomes elevated.

rise in intraocular pressure, causing pain, redness, and visual abnormalities. In *chronic angle-closure glaucoma*, the anterior chamber angle is progressively and permanently closed. In *secondary glaucoma*, obstruction—either open- or closed-angle—is caused by uveitis, hemorrhage, lens rupture, or tumors. In *congenital glaucoma*, there is a trabecular meshwork malformation and excess extracellular matrix in the outer meshwork.

Pathophysiology

For glaucoma, family history is a risk factor since this is often inherited condition. A risk factor for open–angle glaucoma is **myopia**. Chronic increased intraocular pressure causes the death of retinal ganglions as well as optic nerve degeneration. This causes loss of peripheral vision, and then central vision impairment and blindness. When the pressure is extremely high, blindness can develop within hours to days. Loss of visual acuity is due to pressure upon the optic nerve. Lack of nutrients, ischemia, cytotoxic components, and deficient immunity may cause the death of affected neurons.

Clinical manifestations
There are often no early signs or symptoms of open-angle glaucoma. The patient may not notice vision changes initially because peripheral vision is often lost first, and clear vision remains until late in the disease process. Once vision loss is noticed, the disease is usually extremely advanced. This vision loss is not reversible with any treatment. Regular eye examinations are essential to prevent the development of glaucoma because there are treatments that help protect the vision if the disease is identified early enough. Symptoms of acute angle-closure glaucoma include hazy or blurred vision, rainbow-colored circles seen around bright lights, severe eye and head pain, nausea or vomiting occurring with the pain, and sudden monocular vision loss. In general, painful eye reddening and non-reactive pupils, with sudden vision loss, is most likely due to close-angle glaucoma.

Diagnosis
Early detection of glaucoma and prompt treatment can prevent optic neuropathy as well as visual impairment. For diagnosis, the pupils are dilated and a vision test is performed. The optic nerve is visualized, which will have a distinct appearance if glaucoma is present. Eye pressure will be performed with a tonometry test. A visual field test will determine if there is peripheral vision loss.

Treatment
Medicated eye drops are often used to reduce secretion or increase the absorption of aqueous humor. Surgery may be performed to open the trabeculae spaces and to reduce intraocular pressure. This can involve laser surgeries known as **trabeculoplasty**, to open the drainage area; **iridotomy**, to puncture the iris and allow better fluid flow; or *cyclophotocoagulation*, to cause the ciliary body to produce less fluid. A microsurgical technique called *trabeculectomy* creates a new channel to drain fluid and reduce eye pressure. Neuroprotective treatments are being studied.

> **Focus on glaucoma**
> Glaucoma is a leading cause of blindness in African Americans and Hispanics in the United States. Open-angle glaucoma is three to four times more common in African Americans than in non-Hispanic Caucasians. It is 15 times more likely to cause blindness in African Americans than in any other group.

Cataract

A *cataract* is an opaque or cloudy area in the ocular lens. It causes visual loss when it is located on the visual axis. Since the lenses of the eyes enlarge with aging, cataracts occur more often in older people. Cataracts can form in the center of the lens, at its edges, or in back of the lens.

Pathophysiology

Cataracts develop because of changes in metabolism and movement of nutrients in the lens. The most common form of cataracts is *degenerative*, but they may also occur congenitally, or because of infection, radiation, drugs, trauma, or diabetes mellitus.

Clinical manifestations

Cataracts cause reduced visual acuity, blurring of vision, glare from bright lights, and decreased perception of colors. The vision may also be clouded or dim, and it is often difficult for the patient to see at night. Brighter light will be required for reading and other activities. The patient may also see halos around lights, or experiencing yellowing of colors. Often, one eye will have a double vision while the other will not. There is usually the need for frequent changes in eyeglasses or contact lenses as vision worsens.

Diagnosis

Diagnosis of a cataract requires a visual acuity test, a **slit-lamp examination**, and a retinal exam. A slit-lamp uses an intense line of light to illuminate the cornea, iris, lens, and space between the iris and cornea. The retinal exam involves dilating eye drops so that the retina can be seen by the physician.

Treatment

Cataracts are treated by removing the entire lens and replacing it with an intraocular artificial lens. Sometimes, prescription glasses are able to help clear the vision, but in most cases, surgery is indicated.

Age-related macular degeneration

Age-related **macular degeneration** occurs because of damage to the macula. The two forms of this disease are referred to as "dry" or "wet" macular degeneration. In *dry macular degeneration*, the most common form, there is no **neoangiogenesis** present. The dry form always precedes the wet form, in which neoangiogenesis is present. In people aged 75 years or older, approximately 8% have some form of macular degeneration. This percentage increases in older groups of people.

Pathophysiology

In atrophic or dry macular degeneration, there are diffuse or discrete deposits in the *Bruch membrane* or **drusen**, along with widespread atrophy of the pigmented epithelium of the retina. This form makes up about 90% of all cases. Wet macular degeneration, also called *neovascular macular degeneration*, involves *choroidal neovascularization*. This is the presence of **angiogenic** vessels that are believed to originate from the **choriocapillaris**. They penetrate the Bruch membrane from beneath the retinal pigmented epithelium. This vascularization may also penetrate the epithelium and end up directly under the neurosensory retina. The membrane vessels may leak, with exuded blood becoming macular scars. Sometimes, these vessels hemorrhage. There is then localized bleeding that is sometimes mistaken for a tumor. The bleeding can also result in a large vitreous hemorrhage.

Also, neovascularization can occur in younger people, related to pathologic myopia or *Fuchs spot*, if the Bruch membrane is traumatized or otherwise disrupted, or because of an immune response to *systemic histoplasmosis*, known as *presumed ocular histoplasmosis syndrome*. The Bruch membrane contains the basement membrane of the retinal pigmented epithelium. A disturbance of the Bruch membrane or choriocapillaris affects the health of the photoreceptors and causes vision loss. Overall, 71% of macular degeneration cases are linked to genetics, primarily the **complement factor H gene**. The related causative genes appear to decrease in function, and macular degeneration may be due to excessive complement activity. Cigarette smoking and intense exposure to light also increase incidence in genetically predisposed individuals.

Clinical manifestations

In dry macular degeneration, there is early blurred vision, a need for brighter light to read, and difficulty in recognizing other people until they move very close to the affected person. When more advanced, this form causes a blurred spot or *scotoma* in the center of the visual field. Symptoms are usually in both eyes.

Wet macular degeneration symptoms usually develop quickly and get worse in a very short time. They include visual distortions, such as seeing straight lines as if they are curved, and reduced central vision in one or both eyes. There may be decreased intensity or brightness of colors, a general "hazy" appearance to the overall vision, and a well-defined blurred or blind spot in the field of vision. Usually, this form affects one eye at a time. There may also be retinal edema, a gray discoloration of the **subretinal** space, exudates in or around the macula, and detachment of the retinal pigmented epithelium. Macular degeneration rarely causes total blindness since it does not affect peripheral vision.

Diagnosis

Both forms of macular degeneration are diagnosed by using a **funduscope** to examine the eyes. Visual changes are often detected by using an **Amsler grid**. When wet macular degeneration is suspected, **fluorescein angiography** is performed. This will demonstrate and characterize the subretinal choroidal neovascular membranes and point out areas of atrophy. **Optical coherence tomography** helps to identify intraretinal and subretinal fluid. It can also assess responses to treatments.

Treatment

For dry macular degeneration, there is no effective treatment, but studies are ongoing about replacing diseases retinal epithelia with stem cells. In this form, daily supplements of zinc oxide, copper, vitamin C, vitamin E, and beta-carotene or vitamin A are suggested. For patients who smoke or have smoked in the past 7 years, beta-carotene and vitamin A must not be used since they increase risks of lung cancer. Patients must reduce cardiovascular risk factors by eating foods high in omega-3 fatty acids and dark green leafy vegetables.

For wet macular degeneration, the most common treatment is the injection of *anti-vascular endothelial growth factors* or *anti-VEGFs* into the vitreous body. These drugs include ranibizumab, bevacizumab, and aflibercept, which are able to restore reading vision in as many as one-third of patients. Daily supplements, as used for the other form of the disease, are also helpful. Thermal laser photocoagulation of neovascularization outside the fovea can prevent severe vision loss from occurring. In some patients, photodynamic laser treatment is helpful. Corticosteroids such as triamcinolone may be injected intraocularly with an anti-VEGF drug. Rarely used treatments include subretinal surgery, transpupillary thermotherapy, and macular translocation surgery.

> **Focus on age-related macular degeneration**
> Aging is a major risk factor for macular degeneration. Additional risk factors include smoking, which more than doubles the risk for AMD, the African American and Hispanic/Latino ethnicities, family history, lack of exercise, hypertension, and high cholesterol.

Retinal detachment

Retinal detachment occurs when fluid separates the photoreceptors from the pigmented epithelium of the retina. This fluid may be exudate, hemorrhage, or the vitreous humor. Retinal detachment often causes visual impairment and blindness. Risk factors for this condition include holes in the retina or **vitreoretinal traction**. When retinal detachment occurs, the outer parts of the retina are deprived of oxygen and nutrients, since the distance used in diffusion is increased. The pigmented epithelium and photoreceptors cannot communicate normally. The most common form of this disease is known as **rhegmatogenous retinal detachment**, in which retinal breaks occur because of vitreoretinal traction (see Fig. 14.10). Risk factors for this form include

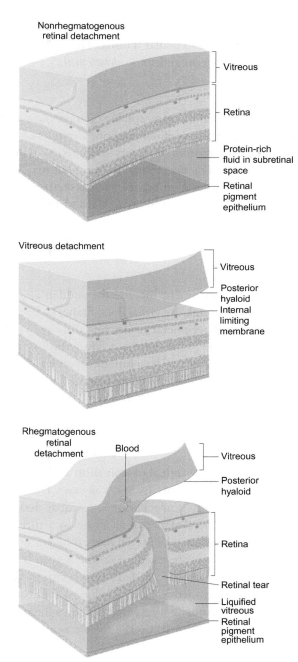

Fig. 14.10 Retinal detachment is defined as the separation of the neurosensory retina from the RPE. *Top*: In nonrhegmatogenous retinal detachment the subretinal space is filled with protein-rich exudate. *Middle*: Posterior vitreous detachment involves the separation of the posterior hyaloid from the internal limiting membrane of the retina and is a normal occurrence in the aging eye. *Bottom*: If during a posterior vitreous detachment the posterior hyaloid does not separate cleanly from the internal limiting membrane of the retina, the vitreous humor will exert traction on the retina, which will be torn at this point.

myopia, previous cataract surgery, and ocular trauma. The other forms of retinal detachment include *traction retinal detachment* and *serous detachment.*

Pathophysiology
Retinal detachment may be caused by intracapsular cataract extraction, lattice degeneration, severe myopia, trauma, or vitreoretinal traction. Contraction of the fibrous membranes may cause tractional separation of the layers of the retina, which is often a result of worsening diabetic retinopathy. Traction retinal detachment may be caused by vitreoretinal traction from preretinal fibrous membranes, such as seen in proliferative diabetic or sickle cell retinopathy. Serous detachment is due to transudation of fluid into the subretinal space. It is caused by choroidal hemangiomas, primary or metastatic choroidal cancers, and severe uveitis—especially in *Vogt-Koyanagi-Harada syndrome.*

Clinical manifestations
The warning signs of retinal detachment include the sudden appearance of floaters—small specks that seem to drift through the visual field, flashes of light in one or both eyes, blurred vision, gradually reduced peripheral vision, or a shadow over the visual field. It is most common in people over age 50, if there is a family history of the disease, or the patient is extremely nearsighted. Retinal detachment is painless. If the macula becomes involved, central vision becomes very poor. There may be a simultaneous vitreous hemorrhage. In the early stages, traction and exudative (serous) retinal detachments may cause blurriness but no other symptoms.

Diagnosis
Diagnosis of retinal detachment is by retinal examination and ultrasound imaging. Funduscopy can differentiate the various forms in almost all cases. Direct funduscopy, using a handheld ophthalmoscope, can miss some peripheral retinal detachments. Therefore, indirect ophthalmoscopy with scleral depression or use of a three-mirror lens should be performed. **B-scan ultrasound** may be used if there is bleeding in the eye, obscuring a clear view of the retina. Both eyes are usually examined even if only one has signs of retinal detachment. Multiple visits may be needed to assess vitreous separation beginning in the unaffected eye. Vitreous hemorrhage may be due to a retinal tear. If this or a cataract, corneal opacification, or traumatic injury that obscures the retina is present, retinal detachment must be suspected, confirmed by B-scan ultrasonography.

Treatment
If there is a retinal tear or hole that has not progressed to detachment, photocoagulation or **cryopexy** may be used. In photocoagulation, a laser is used to burn an area around the retinal tear, which usually connects the retina to the underlying tissue. In cryopexy, the

eye is numbed, and then a freezing probe is applied to the outer surface of the eye, directly over the tear. The freezing causes a scar, which helps secure the retina to the eye wall.

Once retinal detachment has occurred, there are several treatment options. In *pneumatic retinopexy*, a bubble of air or gas is injected into the vitreous cavity, which pushes the damaged retinal areas against the wall of the eye. Cryopexy is then used to repair the retinal break. In *scleral buckling*, a piece of silicone is sutured to the sclera over the affected area. This indents the eye wall, relieving some of the force created by the vitreous. It is also possible to create a scleral buckle that encircles the entire eye, which does not block vision and can remain in place permanently. In *vitrectomy*, the vitreous humor is drained, along with any tissue pulling on the retina, and replaced temporarily with air, gas, or silicone oil. This will help flatten the retina, and eventually, the air or gas will be absorbed and replaced by natural vitreous humor. If silicone oil is used, it is surgically removed in several months. Sometimes, vitrectomy is combined with scleral buckling.

Though they are usually localized, retinal detachments caused by retinal tears can expand, involving the entire retina if not treated quickly. Therefore, an ophthalmologist must examine any patient with a suspected or established retinal detachment urgently. Retinal tears without detachment may be sealed via laser photocoagulations or transconjunctival cryopexy. Almost all of these detachments can be surgically reattached. For nonrhegmatogenous detachments that are transudative, systemic corticosteroids or corticosteroid-sparing drugs, or a surgically implanted slow-release corticosteroid may be successful. Systemic corticosteroid-sparing drugs include azathioprine, methotrexate, and antitumor necrosis factor drugs. Primary and metastatic choroidal cancers must be treated. Choroidal hemangiomas may be treated with localized photocoagulation.

Section review
1. What are the differences between open-angle and angle-closure glaucoma?
2. What are the differences between dry and wet macular degeneration?
3. What are the risk factors for retinal detachment?

Central retinal artery occlusion

Central retinal artery occlusion (CRAO) is blockage of the CRA, usually by an embolism. Sudden, painless, unilateral blindness is the identifying symptom.

Pathophysiology
The causative emboli may come from atherosclerotic plaques within the ipsilateral carotid, or the aortic arch. They dislodge, causing a blockage in the CRA. Another possible cause is an embolism from the heart. Occlusion can also affect any branch of the retinal artery and is called *branch retinal artery occlusion (BRAO)*. Weeks to months after the occlusion, neovascularization of the retina or iris, with secondary, neovascular glaucoma can develop. Retinal neovascularization may cause vitreous hemorrhage.

Clinical manifestations

Retinal artery occlusion usually causes unilateral, sudden, blindness or visual field defects, without pain. For acute cases, funduscopy will reveal a pale, opaque fundus with a red fovea, which is known as a *cherry-red spot*. The arteries are usually attenuated and may appear bloodless. Sometimes, the embolic obstruction is visible. When a major arterial branch is occluded instead of the entire artery, the vision loss and fundus abnormalities will be limited to only that area of the retina that is supplied by that branch. In contrast to painless CRAO, BRAO, and NA-AION, giant cell arteritis is painful. It will often cause a headache, a temporal artery that is tender and palpable, fatigue, jaw claudication, or a combination of these symptoms.

Diagnosis

Acute, painless vision loss is indicative for diagnosis, usually confirmed by funduscopy. When fluorescein angiography is performed, it shows the obstruction of the vessel very clearly. After diagnosis, a routine acute stroke workup is recommended, including brain MRI without contrast, MRA or CTA of the head and neck, echoencephalography, lipid panels, and the hemoglobin A1C (HBA1C) level in the blood.

Treatment

There is no treatment for this condition. Ocular massage may be performed in the hope of dislodging the clot. Hyperbaric oxygen, as well as intraarterial thrombolytic medications, are all-experimental at this point and not proven to be beneficial in randomized trials. Intraarterial treatment is beneficial when thrombolytics are infused directly into the ophthalmic artery, in order to dissolve the clot. However, these treatments rarely restore the vision.

If the patient presents within 4.5 h of the symptom and has no contraindications to receive thrombolytics, the authors recommend treatment with tPA. While vision with a branch artery occlusion often remains good to fair, vision loss is often significant from a central artery occlusion regardless of treatment. Vision loss will be permanent once retinal infarction occurs, sometimes in less than 2 h, but nearly always by 24 h.

Central retinal vein occlusion

Central retinal vein occlusion (CRVO) is blockage of the CRV by a thrombus. There is painless, often sudden vision loss.

Pathophysiology

Causes of CRVO are linked to hypertension, age, glaucoma, diabetes, and increased blood viscosity. Occlusion can also be idiopathic. This type of occlusion is not common in younger people. It can also affect just one branch of the retinal vein. In weeks to months after the occlusion, neovascularization of the retina or iris (**rubeosis iridis**) with secondary, neovascular glaucoma can occur. Retinal neovascularization may cause vitreous hemorrhage.

Clinical manifestations
Usually, CRVO causes painless vision loss suddenly, but it may occur gradually, over days to weeks. Via funduscopy, there will be visible hemorrhages through the retina, retinal vein engorgement and tortuousness. Sometimes, there is significant retinal edema. If an obstruction is only in one branch of the vein, the changes will be limited to one quadrant.

Diagnosis
Diagnosis is suspected with painless visual loss, especially in at-risk patients, and confirmed by funduscopy. The patient with a central occlusion must be evaluated for hypertension and glaucoma, and tested for diabetes. Younger patients are tested for increases in blood viscosity, via a complete blood count and other coagulable factors as needed. Visual acuity upon presentation is a good indicator of what the patient's final vision will become. If it is at least 20/40, acuity will usually remain good, sometimes almost normal. If acuity is worse than 20/200, it will remain at that level, or become worse, in 80% of patients.

Treatment
There is no recommended treatment for CRVO. If neovascularization develops, panretinal photocoagulation is recommended, since this may decrease vitreous hemorrhaging and prevent neovascular glaucoma. Intravitreal injection of corticosteroids and anti-VEGF drugs are still being investigated. Most patients have some visual deficit. In less severe cases, there may be spontaneous improvement to the near-normal vision of varying amounts of time.

Diabetic retinopathy
Diabetic retinopathy is a term that encompasses intraretinal hemorrhage, microaneurysms, exudates, macular edema and ischemia, neovascularization, vitreous hemorrhage, and even traction retinal detachment in severe cases.

Pathophysiology
Diabetic retinopathy is a primary cause of blindness. The amount of retinopathy is greatly related to a patient's duration of diabetes, blood glucose levels, and blood pressure levels. Retinopathy can be worsened in pregnant women since this impairs blood glucose control. The two primary forms of retinopathy are called *nonproliferative* and *proliferative*.

Nonproliferative, or *background* retinopathy begins first. It causes increased capillary permeability, hemorrhages, microaneurysms, exudates, macular ischemia, and macular edema. The term *macular edema* describes retinal thickening due to fluid leakage from capillaries. Macular edema may be clinically significant with either nonproliferative or proliferative retinopathy. It is the most common cause of vision loss related to diabetic retinopathy.

Proliferative retinopathy follows nonproliferative retinopathy and is more severe. It may result in vitreous hemorrhage and traction retinal detachment. Proliferative retinopathy involves neovascularization on the inner, vitreous surface of the retina. It can extend into the vitreous cavity, causing vitreous hemorrhage. Neovascularization often appears with preretinal fibrous tissue. Along with the vitreous humor, this tissue can contract, causing traction retinal detachment. It is possible for neovascularization to occur in the eye's anterior segment, on the iris. Neovascular membrane growth in the angle of the eye at the peripheral iris margin can occur and lead to neovascular glaucoma. With proliferative retinopathy, vision loss may be severe.

Clinical manifestations

The visual symptoms of nonproliferative retinopathy accompany macular edema or ischemia. The patient may be unaware of vision loss. This form begins with capillary microaneurysms, dot and blot retinal hemorrhages, hard exudates, and soft exudates that are also known as cotton-wool spots. The hard exudates are yellow, discrete, and usually deeper than retinal vessels, suggesting retinal edema. Cotton-wool spots are microinfarctions leading to retinal opacification. They have fuzzy edges, with white, obscure underlying vessels. The later stages of nonproliferative retinopathy include macular edema, which is seen on slit-lamp biomicroscopy as elevations and blurring of the retinal layers. There may also be abnormal venous dilation and intraretinal microvascularity.

Symptoms of proliferative retinopathy may include black spots or flashing lights in the visual field, blurring, and sudden, severe, painless loss of vision. These symptoms may partially be due to vitreous hemorrhage or traction retinal detachment. Proliferative retinopathy causes newly developed capillaries to appear on the optic nerve or retinal surface. Funduscopy may reveal macular edema or retinal hemorrhage.

Diagnosis

Funduscopy is used for the diagnosis of retinal detachment. Fluorescein angiography can determine the amount of damage, aid in developing treatment plans, and help to monitor treatment results. Optical coherence tomography can assess the extent of macular edema and responses to treatment. All patients with diabetes must have an annual dilated ophthalmologic examination. Pregnant women with diabetes must be examined every trimester. Vision symptoms indicate referral to an ophthalmologist.

Treatment

It is vital to control the patient's blood glucose and blood pressure (BP). Intensive blood glucose control slows retinopathy progression. If macular edema is clinically significant, it is treated with focal laser procedures. In more severe cases, intravitreal injection of triamcinolone and anti-VEGF drugs may be helpful. For recalcitrant diabetic macular edema, vitrectomy may be helpful. For some patients with severe nonproliferative retinopathy, panretinal laser photocoagulation can be performed. Most patients, however, must be closely followed until proliferative retinopathy develops.

Panretinal laser photocoagulation is used for proliferative diabetic retinopathy if there are high risks of vitreous hemorrhage, significant preretinal neovascularization, or combined anterior segment neovascularization and neovascular glaucoma. This procedure can greatly reduce risks for severe vision loss. Vitrectomy may preserve or restore lost vision if the patient has either vitreous hemorrhage persisting for 3 months, extensive preretinal membrane formation, or traction retinal detachment.

Hypertensive retinopathy

Hypertensive retinopathy is vascular damage of the retina due to hypertension. Symptoms occur late in the disease course, and funduscopy is used to reveal the extent of the damage.

Pathophysiology

When the BP becomes acutely elevated, it usually causes reversible vasoconstriction of the retinal blood vessels. Hypertensive crisis can cause papilledema. When hypertension is more prolonged or severe, there are exudative vascular changes due to endothelial damage and necrosis. Thickening of arteriole walls and other changes usually require years of elevated BP to develop. Smoking intensifies the effects of hypertension upon the retina. Hypertension is also a primary risk factor for diabetic retinopathy and occlusion of the retinal artery or vein. Hypertension along with diabetes greatly increases risks for vision loss. Hypertensive retinopathy causes a high risk of hypertensive damage to other end organs.

Clinical manifestations

In the early stages of hypertensive retinopathy, funduscopy will reveal constriction of the arterioles, and a decrease in the ratio of the retinal arteriole width compared to the retinal venules. If hypertension is chronic and poorly controlled, there will be permanent arterial narrowing, arteriovenous nicking (crossing abnormalities), arteriosclerosis with moderate vascular wall changes or more severe vascular wall hyperplasia and thickening. The moderate changes are referred to as *copper wiring* while the more severe changes are referred to as *silver wiring*. Total vascular occlusion may occur. Arteriovenous nicking is a major factor predisposing the patient to branch retinal vein occlusion.

If the acute disease is severe, there may be superficial flame-shaped hemorrhages, cotton-wool spots, yellow hard exudates, or papilledema. The yellow hard exudates are due to intraretinal lipid deposition because of leaking retinal vessels. The exudates may form a star-shaped macular lesion, especially with hypertension severe. The optic disc will become congested and edematous. This is papilledema that indicates a hypertensive crisis.

Diagnosis
Diagnosis of hypertensive retinopathy involves the patient history of the duration and severity of hypertension, and funduscopic examination.

Treatment
The management of hypertension is the mainstay of treatment for hypertensive retinopathy. If there are any other conditions present that can harm vision, they must also be controlled well. When vision loss occurs, retinal edema is treated using a laser or intravitreal injection of either corticosteroids or anti-VEGF drugs.

Section review
1. What are the two main forms of diabetic retinopathy?
2. What is the result of later stage nonproliferative diabetic retinopathy?
3. What are the outcomes of acute and severe hypertensive retinopathy?

Retinitis pigmentosa

Retinitis pigmentosa is a slowly progressive condition. There is bilateral degeneration of the retina and retinal pigment epithelium, due to various types of genetic mutations.

Pathophysiology
Gene transmission may be autosomal recessive or dominant, and less commonly, X-linked. This condition may be part of **Bassen-Kornzweig syndrome** or **Laurence-Moon syndrome**. Congenital hearing loss may accompany the symptoms.

Clinical manifestations
Retinitis pigmentosa affects the rods of the retina, causing defective night vision. This becomes symptomatic as early as childhood but may occur in various ages. Night vision may be totally lost over time. A peripheral ring scotoma will be detectable by visual field testing, and may slowly widen. Central vision can also be affected in advanced cases. Hyperpigmentation, in a bone-spicule configuration in the midperipheral retina, is the most obvious funduscopic finding. Others may include retinal arteriole narrowing, cystoid macular edema, a waxy yellow disk, posterior subcapsular cataracts, myopia, and less commonly, cells in the vitreous.

Diagnosis

Diagnosis of retinitis pigmentosa is based on poor night vision symptoms, or family history. Funduscopy is performed, usually along with electroretinography. Other retinopathies that can mimic this condition are excluded. These may be associated with rubella, syphilis, chloroquine or phenothiazine toxicity, and nonocular cancer. Family members must be examined and tested when necessary, or to establish a hereditary pattern. If the patient has a hereditary syndrome, he or she may want to seek genetic counseling prior to having children.

Treatment

Damage from retinitis pigmentosa is irreversible. However, vitamin A palmitate at 20,000 units orally, once per day, can help slow disease progression. Any patient taking this supplement must have regular liver function tests. Vision will decrease as the macula becomes more involved and can progress to legal blindness.

Color blindness

Human color vision is based on the cones' ability to see three basic colors: red, green, and blue. Any color in the spectrum is matched by some combination of these three primary colors. There are *long wavelength (L) cones*, *middle wavelength (M) cones*, and *short wavelength (S) cones*. The S cones are the cones for blue pigment. They only make up about 5% of the total amount of cones. The distribution of L and M cones is quite varied, and they are found in relatively random locations. In a normal color vision, the ratio of L to M cones can be as small as one-to-one, or as large as one-to-fifteen. The genes for red and green cone pigments are side by side on the X chromosome. If there is unequal crossing over during meiosis, this can cause one X chromosome to have a missing or defective red or green gene.

As a result of this error, approximately 2% of the male population is termed *red-green color blind* due to a lack of red pigment (or *protanopia*) or a lack of green pigment (or *deuteranopia*). Incidence of red-green color blindness is much less common in females since they are likely to have at least one X chromosome that has normal red and green genes. Lack of blue cone pigment (or *tritanopia*) is much rarer since the blue gene is located on chromosome 7. This condition is as uncommon in males as it is in females.

Pathophysiology

Though most cases of color blindness are inherited, there is also *acquired* color blindness. This develops later in life and affects both genders equally. It may be caused by optic nerve or retinal damage. It is important to alert a physician if color vision changes because this can indicate a serious underlying condition. Diseases that can cause color blindness include glaucoma, macular degeneration, diabetic retinopathy, cataracts, diabetes mellitus, Parkinson's disease, Alzheimer's disease, and multiple sclerosis.

Clinical manifestations

The most common color blindness involves the colors red and green. The next most common color blindness involves yellow and blue. The least common form is **achromatopsia**, in which only black and white vision exist.

Diagnosis

Testing for color blindness uses *pseudoisochromatic plates*, which contain small colored dots with numbers or symbols within them. If the vision is normal, these numbers or symbols can easily be seen. A color-blind person will either not see them at all, or see a different number or symbol. Children must be tested before beginning school since many educational materials will involve identifying colors.

Treatment

For some people, tinted glasses or contact lenses may help to distinguish colors. Otherwise, there is no treatment. Color-blind people learn to compensate for their condition by using labels that signify colors they cannot perceive, such as in clothing, in order to match various garments.

Nyctalopia

Nyctalopia, or *night blindness*, is often a symptom of another problem, such as untreated nearsightedness. It occurs because of a disorder of the rods in the retina, resulting in a reduced ability to see in dim light.

Pathophysiology

Nyctalopia may also be caused by glaucoma, medications for glaucoma, cataracts, diabetes mellitus, retinitis pigmentosa, vitamin A deficiency, and **keratoconus**.

Clinical manifestations

The only symptom of nyctalopia is difficulty seeing in the dark. The danger with nyctalopia is most serious when driving since oncoming headlights can temporarily blind a person with the condition.

Diagnosis

A complete eye examination will reveal the cause of night blindness. Specialized tests may be ordered if more information is needed.

Treatment

For nyctalopia, new eyeglasses or contacts sometimes help. Dietary supplements such as vitamin A and zinc are helpful. There are also certain vitamin formulations that contain eye-healthy ingredients. If caused by a cataract, surgery is indicated.

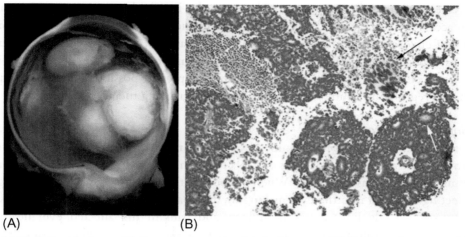

Fig. 14.11 Retinoblastoma. (A) Gross photograph of retinoblastoma. (B) Tumor cells appear viable when in proximity to blood vessels, but necrosis is seen as the distance from the vessel increases. Dystrophic calcification *(dark arrow)* is present in the zones of tumor necrosis. Flexner-Wintersteiner rosettes—arrangements of a single layer of tumor cells around an apparent "lumen"—are seen throughout the tumor, and one such rosette is indicated by the *white arrow.*

Retinoblastoma

Retinoblastoma is a rare, congenital eye tumor occurring in young children. It may originate in the retina of one or both eyes (see Fig. 14.11). This condition can be inherited or acquired and is rarely diagnosed after a child has reached the age of 5 years. Inherited retinoblastoma is usually diagnosed during the first year of life, often involves multiple tumors, and occurs in 40% of patients in both eyes. Acquired retinoblastoma is usually diagnosed in children between 2 and 3 years of age and usually occurs in only one eye. Retinoblastoma is the most common pediatric intraocular tumor, yet only about 300 cases are diagnosed annually in the United States, with one in four cases (of both forms) involving both eyes.

Pathophysiology

About 40% of retinoblastomas are due to autosomal dominant disorders, from mutations of the retinoblastoma susceptibility gene (RB1), also called a tumor-suppressor gene, and 60% of these tumors are acquired. It is believed that, in the inherited form, the first mutation occurs in a germ cell from either parent, meaning the mutation exists in every cell of the child's body. The tumor will only develop if a second, random mutation in a retinoblast cell occurs. There are multiple tumors in the inherited form since these second mutations usually occur in several of the existing one to two million-retinoblast cells. Oppositely, acquired retinoblastoma requires two independent mutations to occur in the same somatic cell—after fertilization—in order for them to transform into cancer. Initial retinoblastoma mutation occurs in the long arm of chromosome 13, band a14. Retinoblastoma tumors of the retina extend into the vitreous humor, where there may be small, free-floating tumors. These can also invade the optic nerve. The tumors can access the subarachnoid space and CNS. Retinoblastoma spreads to the choroid in 25% of children. Since the choroid is extremely vascular, metastasis via hematogenous spread is possible. Metastatic sites include the bone marrow, liver, long bones, and lymph nodes. If a tumor invades the eye orbit, lymphatic spread is possible. Infrequently, spontaneous regression occurs, possibly by tumors outgrowing their blood supply.

Clinical manifestations
Retinoblastoma is most often signified by **leukokoria**, which is a white pupillary reflex known as the *cat's eye reflex*. This is due to a tumor mass behind the lens. The other most common symptom is strabismus. When this occurs, the tumor has grown large enough to reflect light shown into the eye, making the pupil appear to be white. Other signs and symptoms include limited vision, redness and pain in the affected eye, headache, anorexia, and vomiting. If the eye manifestations exist, and the child is younger than 4 years of age, a careful ophthalmologic examination of both eyes must be performed, under general anesthesia. Any newborn with a known genetic risk for this condition must receive routine ophthalmologic examinations. The cancers will be single or multiple gray-white elevations of the retina. Cancer seeds may be visible in the vitreous as well.

Diagnosis
Orbital ultrasonography or CT scan usually confirms retinoblastoma. CT can usually detect calcification for nearly all cancers. If the optic nerve appears abnormal during ophthalmoscopy, an MRI is more accurate for finding cancer extension into the choroid or optic nerve. When the extraocular spread is suspected, testing should include a bone scan, bone marrow aspirate and biopsy, and lumbar puncture.

If the child has a parent or sibling who has had retinoblastoma, he or she must be evaluated by an ophthalmologist shortly following birth, with follow-ups every 4 months until the age of 4 years. When retinoblastoma is diagnosed, molecular genetic testing is required. When a germline mutation is identified, parents must also be tested for this mutation. If additional offspring have the germline mutation, the same testing and examination procedures are required. Recombinant DNA probing may help detect asymptomatic carriers.

Treatment
For unilateral retinoblastoma, enucleation, or removal of the eye and as much of the optic nerve as possible, is required. Vision can usually be preserved for patients with bilateral cancer. Treatments may include bilateral photocoagulation or unilateral enucleation and photocoagulation, cryotherapy, and irradiation of the opposite eye. Radiation therapy is with external beam techniques. For very small cancers, brachytherapy is used, which is the attachment of a radioactive plaque to the eye wall, near cancer. Systemic chemotherapy, using carboplatin plus etoposide, or cyclophosphamide plus vincristine, may help reduce large cancers or treat cancer that has disseminated beyond

the eye. Chemotherapy on its own usually cannot cure this cancer. Ophthalmologic reexamination of both eyes is needed. Retreatment, if it is necessary, must occur at 2- and 4-month intervals.

When retinoblastoma is treated while intraocular, more than 90% of patients can be cured. Once metastasis has occurred, the prognosis is poor. For hereditary retinoblastoma, there is an increased incidence of second cancers. Approximately 50% occur within the irradiated area. These cancers may include malignant melanoma and sarcomas. Within 30 years of diagnosis, about 70% of patients will develop second cancer.

Section review
1. What is the pathophysiology of retinitis pigmentosa?
2. When color vision becomes reduced over time, what possible underlying conditions may be present?
3. What is the pathophysiology of nyctalopia?

Clinical cases

Clinical case 1
1. How quickly must the causes of papilledema be treated?
2. Besides trauma, what are the other causes of papilledema?
3. What are the treatments for papilledema?

A 24-year-old woman was in a car accident and experienced significant head trauma. After being treated at the emergency department and allowed to go home with her boyfriend, she began to experience double vision, nausea, tinnitus, and a severe headache. She returned to the hospital to be evaluated, and after a visual examination and CT scan, was diagnosed with papilledema.

Answers:
1. Papilledema, in this case, is bilateral and is caused by major head trauma that results in increased intracranial pressure, and swelling of the optic nerve bilaterally. It must be treated as quickly as possible so that the increased pressure does not cause permanent damage and irreversible effects.
2. Hydrocephalus, brain hemorrhage, encephalitis, meningitis, hypertension, abscesses, and brain tumors may also cause papilledema. It may additionally develop as an idiopathic condition, with no apparent cause.
3. Papilledema is treated based on the underlying cause. Sometimes, a small piece of the skull must be removed in order to relieve the pressure. Mannitol may be prescribed to keep nervous system pressure at normal levels. Other measurements to decrease the intracranial pressure are transient hyperventilation (by way of reducing carbon dioxide), hypertonic saline, and in some cases, steroid medications.

Clinical case 2
1. What are the additional signs and symptoms that may accompany contralateral homonymous hemianopia?
2. Besides occipital stroke, what are the other causes of this condition?
3. Is it likely that some amount of function will return?

A 55-year-old woman was diagnosed with contralateral homonymous hemianopia after an occipital stroke. She told her physician that she sometimes experienced visual hallucinations and that it was difficult to recognize people's faces.

Answers:
1. Contralateral homonymous hemianopia may also be accompanied by other symptoms if the lesion is not in the visual cortex. For example a stroke in the thalamus can also cause contralateral homonymous hemianopia due to the involvement of the LGN however, since entire thalamus is involved patient can have other symptoms such as sensory loss in body or face (VMP, VPL nucleus), weakness (LA, VA nucleus), altered mental status (medial dorsal nucleus), etc.
2. Contralateral homonymous hemianopia may also be caused by tumors, bleeding within the brain, and other less common causes.
3. It is possible that some aspects of this form of hemianopia may resolve over time. Since an occipital stroke was the cause, other areas in the brain may be able to "take over" for the damaged areas, but this differs from patient to patient. There are certain types of visual therapies that can help patients compensate for visual deficits after an occipital stroke.

Clinical case 3
1. Which type of glaucoma is probably present in this patient's right eye?
2. What typical findings will there be when the eye is examined?
3. What is the probable treatment of this patient's condition?

A 50-year-old woman came to the emergency department with severe pain in her forehead, cheeks, and around her right eye. Her vision was blurry in the right eye, and she was seeing halos around lights, along with feeling extremely nauseous. When examined, she said that she had never had any similar symptoms before, and did not see any flashes or floaters. She denied any double vision, but her right eye was mildly reddened. The physician asked her if her family had any history of glaucoma or macular degeneration. The patient admitted to having quit smoking in the past year, but that she had smoked cigarettes for 25 years.

Answers:
1. Acute angle-closure glaucoma of the right eye is most likely present.
2. This disease causes mild conjunctival injection, a hazy cornea, a mid-point fixed pupil, a shallow angle, and an eyeball that is hard upon palpation.

3. Treatment of acute angle-closure glaucoma is either laser or surgical peripheral iridectomy. A hole is placed in the peripheral iris to restore aqueous flow from the posterior to the anterior chamber. However, the other eye should undergo a prophylactic peripheral iridectomy to prevent development of the disease there as well.

Clinical case 4

1. What type of macular degeneration is most likely present?
2. What are the chances of this patient developing the same condition in the other eye?
3. What injection treatments will be recommended and how often are they administered?

A 75-year-old man complains to his eye doctor of a loss of central vision in his left eye. He had cataract surgery in that eye 12 years before, and until now has had perfect vision. His vision loss has affected his ability to read, and as an author of books, he is worried that this will ruin his work. The patient's sister developed age-related macular degeneration when she was in her late 60s. The patient describes his left-eye vision as being all "gray" in the center, but his peripheral vision is completely normal.

Answers:

1. Wet macular degeneration is the likely diagnosis. In this form, the patient usually sees a dark spot or spots in the center of vision, due to blood or fluid under the macula. The peripheral vision usually remains normal.
2. There is a relatively high risk for the other eye developing the same condition. When there are more than five drusen of large size, pigmental clumping, and systemic hypertension, patients have an 87% risk of developing wet macular degeneration of the other eye within 5 years. If none of these risk factors is present, the risk is only 7%.
3. The first line of treatment for wet macular degeneration utilizes injections of either bevacizumab, ranibizumab, or aflibercept. They are injected into the affected eye every 4 weeks, and usually improve visual acuity to variable degrees.

Clinical case 5

Retinal detachment

1. In retinal detachment, what happens to the retina?
2. When this patient sees the ophthalmologist, what types of surgery may be performed?
3. If scleral buckling is the chosen surgical procedure, how long should the patient expect the recovery period to be?

A 49-year-old man visited his optometrist, complaining of sudden vision loss in his left eye, and seeing flashes of lights as well as floaters. He had been extremely nearsighted for most of his life, and always wore glasses. The patient reported being more sensitive to light than usual over the past week. Examination revealed that the left pupil was more dilated than the right, and the left cornea was reddened. The optometrist used a slit-lamp for the examination and diagnosed the patient with retinal detachment. The patient is referred to an ophthalmologist.

Answers:
1. In retinal detachment, the retina peels away from the inner wall of the eye. A fluid-filled space is created between the neural layer and the retinal-pigmented epithelium. This is a medical emergency.
2. Surgery for retinal detachment includes pneumatic retinopexy, scleral buckling, and if necessary, vitrectomy. Prescribing medicated eyedrops to help the eye heal will follow this.
3. Recovery time for scleral buckling is between 2 and 4 weeks. There will be some pain in the hours or days after the procedure. The swelling will usually last for a few weeks.

Key terms

- 11-cis Retinal
- Accommodation
- Achromatopsia
- All-trans-retinal
- Amacrine cells
- Amsler grid
- Angiogenic
- Aqueous humor
- Atrial myxoma
- B-scan ultrasound
- Bassen-Kornzweig syndrome
- Benedict's syndrome
- Bipolar cells
- Bitemporal hemianopia
- Blind spot
- Cataract
- Chemosis
- Choriocapillaris
- Choroid
- Circadian rhythm
- Claude's syndrome
- Complement factor H gene
- Cone pigments
- Cones
- Convergence
- Cornea
- Corneoscleral junction
- Cryopexy
- Crystallins
- Cyclic guanosine monophosphate
- Depth perception
- Drusen
- Enophthalmos
- Exophthalmos
- Fibrous layer
- Field of vision
- Fluorescein angiography
- Fovea centralis
- Funduscope
- Ganglion cells
- Glaucoma
- Graves' orbitopathy
- Hemianopia
- Homonymous
- Homonymous hemianopia
- Horizontal cells
- Inner layer
- Internuclear ophthalmoplegia
- Intraocular pressure
- Ipsilateral
- Iridotomy
- Keratoconus
- Lamina cribosa
- Laurence-Moon syndrome

Lens
Lens fibers
Leukokoria
M cells
Macula lutea
Macular degeneration
Melanopsin
Meningeal carcinomatosis
Meyer's loop
Müller cells
Myopia
Neoangiogenesis
Nyctalopia
Off-center neurons
On-center neurons
Opsin
Optic abiotrophy
Optic chiasm
Optic disc
Optic radiation
Optic tract(s)
Optical coherence tomography
Orogenital dermatitis
P cells
Papilledema
Paramedian
Perimetry test
Photoreceptors
Pinealoma
Retina
Retinal
Retinal detachment
Retinitis pigmentosa
Retinoblastoma
Rhegmatogenous retinal detachment
Rhodopsin
Rods
Rubeosis iridis
Sclera
Scleral venous sinus
Scotoma
Slit-lamp examination
Subretinal
Suprasellar
Telangiectatic microangiopathy
Trabeculoplasty
Transducin
Transtentorial
Unilateral visual loss
Vascular layer
Visual acuity
Visual axis
Visual cortex
Vitreoretinal traction
Vitreous body
Vitreous humor

Suggested readings

1. Blindness—https://www.healthline.com/symptom/blindness.
2. Brain Types, Symptoms and Radiology—https://www.lecturio.com/magazine/brain-types-symptoms-and-radiology/.
3. Chan, J. W. *Optic Nerve Disorders: Diagnosis and Management*, 2nd ed.; Springer, 2014.
4. Chang, T. C.; Thanos, A.; et al. *Clinical Decisions in Glaucoma*, 2nd ed.; Ta Chen Chang, 2016.
5. Glaucoma—https://www.mcw.edu/ophthalmology/education/ophthcstudies/case1.htm.
6. Hansen, E. *What is Color Blindness?: What to Know If You're Diagnosed With Color Blindness*. CreateSpace Independent Publishing Platform, 2013.
7. Hasan, T.; Bani, I. *Diabetic & Hypertensive Retinopathy: An Angiographic Exploration*. Lap Lambert Academic Publishing, 2016.
8. Hayreh, S.S. Ischemic Optic Neuropathies. (2012) Springer.
9. Henderson, B. A. *Essentials of Cataract Surgery*, 2nd ed.; Slack Incorporated, 2014.
10. Henry, J.; Norville, J. *Retinitis Pigmentosa: Causes, Tests, and Treatment Options*, 2nd ed.; CreateSpace Independent Publishing Platform, 2014.
11. Hunt, D. M.; Hankins, M. W.; Collin, S. P.; Marshall, N. J. *Evolution of Visual and Non-visual Pigments*. Springer, 2014.
12. Icon Group International. *Papilledema: Webster's Timeline History, 1950-2007*. Icon Group International, Inc., 2010

13. Internuclear Ophthalmoplegia—https://www.nejm.org/doi/full/10.1056/nejmicm1200499.
14. Macular Degeneration—https://www.macular.org/wet-amd.
15. Night Blindness—https://www.webmd.com/eye-health/night-blindness.
16. Papilledema—https://www.healthline.com/health/papilledema.
17. Remington, L. A.; Goodwin, D. *Clinical Anatomy and Physiology of the Visual System*, 3rd ed.; Butterworth-Heinemann, 2011.
18. Retinal Detachment—https://www.slideshare.net/cassidywendler/case-study-retina/detachment.
19. Rowe, F. *Visual Fields via the Visual Pathway*, 2nd ed.; CRC Press, 2016.
20. Salzmann, M. *The Anatomy and Histology of the Human Eyeball in the Normal State, Its Development, and Senescence*. Ulan Press, 2012.
21. Scanlon, P. H.; Sallam, A.; van Wijngaarden, P. *A Practical Manual of Diabetic Retinopathy Management*, 2nd ed.; Wiley-Blackwell, 2017.
22. Skorkovska, K. *Homonymous Visual Field Defects*. Springer, 2017.
23. Swartwout, G. *Macular Degeneration ... Macular Regeneration, Vol. 3;* CreateSpace Independent Publishing Platform, 2012.
24. Treatment of Brain Aneurysm—https://www.medicinenet.com/brain_aneurysm/article.htm#what_are_future_directions_for_the_treatment_of_brain_aneurysm.
25. Vanderah, T.; Gould, D. J. *Nolte's The Human Brain: An Introduction to Its Functional Anatomy*, 7th ed.; Elsevier, 2015.
26. Wilson, M. W. *Retinoblastoma (Pediatric Oncology)*. Springer, 2010.
27. Wray, S. H. *Eye Movement Disorders in Clinical Practice*. Oxford University Press, 2014.

CHAPTER 15

Limbic, olfactory, and gustatory systems

The limbic system is one of the most complicated structures in the brain. It is involved in homeostasis, memory, emotions, olfaction, and many other psychologic functions. This system includes the amygdala, septal nuclei, cingulate cortex, and many other structures that reach the forebrain, midbrain, lower brainstem, and the spinal cord. The limbic system is highly complex, since it connects with the neocortex and central nuclei and utilizes many different neurotransmitters. The olfactory system utilizes olfactory (receptor) cells, sustentacular (supporting) cells, and basal cells. It sends information through cranial nerve I to the brain, and handles the sense of smell. Olfactory receptors can respond to many different odorants. The sense of smell is crucial in order for the sense of taste to function properly. There are many different conditions that affect the limbic, olfactory, and gustatory systems, which will be discussed in detail in this chapter.

Limbic system

The **limbic system** is an extensive neural network. It handles emotions, memory, homeostasis, motivations, unconscious drives, and olfaction. It is highly complex, making the study of this system clinically difficult. Advances in behavioral studies, deep-brain stimulation, perfusion, and functional magnetic resonance imagings (MRIs) have allowed better understanding of the limbic system.

Anatomy of the limbic system

The **limbic lobe** contains the structures of this system. They include the following:
- *Amygdala*—also called the *amygdaloid nuclear complex*; fear and emotion center
- *Various hypothalamic nuclei*—involved in homeostasis; hunger, satiety, thermoregulation, and sleep onset
- *Olfactory cortex*
- *Septal nuclei*—below the rostrum of the corpus callosum
- *Nucleus accumbens*—a region in the basal forebrain, rostral to the preoptic area of the hypothalamus
- *Hippocampal formation*—mainly involved in memory
- *Cingulate cortex*—in the medial aspect of the cerebral cortex
- *Areas of the basal ganglia*—at the base of the forebrain and top of the midbrain

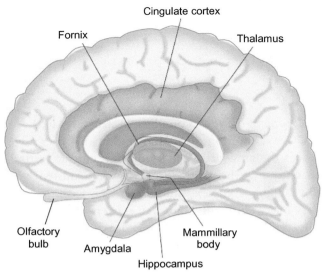

Fig. 15.1 The limbic system. Structures of the limbic system play important roles in emotion, learning, and memory. Pathophysiology in limbic structures is frequently found in mental disorders.

- *Ventral tegmental area*—close to the midline, on the floor of the midbrain
- *Limbic midbrain areas*—including the periaqueductal gray matter

The *limbic brain* includes all of these structures as well as their projections, which reach the forebrain, midbrain, lower brainstem, and the *spinal cord limbic systems* (Fig. 15.1). These spinal systems are reached mainly via the fornix, stria terminalis, **ventral amygdalofugal pathway**, and *mammillothalamic tract*.

Hippocampal formation

The **hippocampal formation** is made up of the **dentate gyrus**, **hippocampus**, and **subicular complex**. The subicular complex includes the **subiculum**, **presubiculum**, and **parasubiculum**. The neocortex of the parahippocampal gyrus passes medially, from the collateral sulcus. It joins the transitional **juxtallocortex** of the subiculum. This structure is curved superomedially, to the inferior surface of the dentate gyrus. It then curves laterally to the laminae of the hippocampus. The curve continues superiorly, then medially, above the dentate gyrus. It terminates by pointing to the center of the superior surface of the dentate gyrus. Three of the pathways in the hippocampal formation are believed to utilize glutamate, aspartate, or both as the major excitatory neurotransmitter.

The two main sources of subcortical afferents to the hippocampal formation are the medial septal complex and supramammillary area of the posterior hypothalamus. Projections come from the amygdaloid complex and claustrum, reaching the subicular complex and **entorhinal cortex**. There are monoaminergic projections from the locus ceruleus, mesencephalic raphe nuclei, and ventral tegmental area. Recall that the **locus ceruleus** is a nucleus in the **pons** that is involved in stress and panic. The main neurotransmitter is norepinephrine (NE). The locus ceruleus is a part of the reticular activating system, and is involved in arousal via NE. The ventral tegmental area is in the midbrain. The main neurotransmitter is dopamine. The raphe nucleus is an elongated nucleus stretched over the whole brainstem area, and is mainly involved in the secretion of serotonin. The raphe nucleus terminates at the dorsal horn of the spinal gray matter, where regulation of the release of **enkephalins**, to inhibit pain sensation, occurs.

Noradrenergic and serotoninergic projections reach every hippocampal field, but have highest density within the dentate gyrus. In the supramammillary area, neurons reach the hippocampal formation through the **fornix** and a ventral route. All divisions of the anterior thalamic nuclear complex, as well as the related lateral dorsal nucleus, project to the subicular complex and other parts of the hippocampal formation. Certain midline thalamic nuclei project mostly to the entorhinal cortex. These are mostly the central medial, **paratenial**, and **reuniens** nuclei.

There is a circuit within in the limbic system that was first described by the anatomist named James Papez back in 1937. Papez believed that this circuit played a major role in emotions. However, in modern neurology, the *Papez circuit* is known to play a major role in *memory* formation and processing, along with many other functions. The Papez circuit is very complex. Imagine this circuit as a connection. The structures involved in this connection are the:

- Hippocampus
- Mammillary body
- Anterior nucleus of the thalamus
- Cingulate gyrus

The Papez circuit (see Fig. 15.2) starts at the hippocampus and connects to the mammillary body through the *fornix*. The connection between the mammillary body and the anterior nucleus of the thalamus is through the **mammillothalamic fibers**. The connection between the anterior nucleus of the thalamus and the cingulate gyrus is through the *cingulate bundle, entorhinal cortex,* and **subiculum**, back to the hippocampal formation, to complete the circuit. This circuit is perhaps the most well described circuit out of the many existing circuits within the limbic system.

Subicular complex

The subicular complex is basically divided into the subiculum, presubiculum, and parasubiculum. The *subiculum* has a superficial molecular layer, a pyramidal cell layer, and a deep polymorphic layer. The superficial molecular layer has apical dendrites of the subicular pyramidal cells. The *presubiculum* lies medial to the subiculum. It has a dense

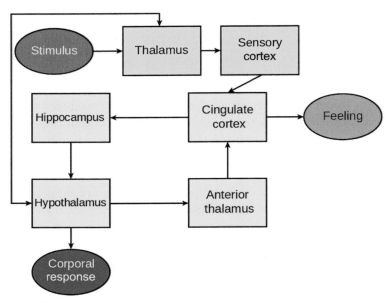

Fig. 15.2 Circuit of Papez.

superficial layer of pyramidal cells. The presubiculum creates a border between the subicular complex and entorhinal cortex. Cell layers lying deep to the parasubiculum cannot be distinguished from the deep layers of the entorhinal cortex. The subiculum is involved in memory formation.

Entorhinal cortex

The most posterior area of the piriform cortex is the *entorhinal cortex*. The entorhinal cortex extends rostrally, to the anterior edge of the amygdala. Caudally, it overlaps part of the hippocampal fields. Fibers from the olfactory bulb and the periamygdaloid and piriform cortices reach the lateral areas. Additional caudal regions do not usually receive primary olfactory inputs.

There are reciprocal connections between the entorhinal cortex and the hippocampus and neocortical regions. The entorhinal cortex is a relay center receiving afferents from the *association cortex*. It relays information back to the hippocampal formation.

Septum pellucidum

The **septum pellucidum** is a midline, paramedian structure that separates the lateral ventricles. Under this area, the septal region has four primary groups of nuclei. These are as follows:
- *Dorsal group*—basically, the dorsal septal nucleus
- *Ventral group*—the lateral septal nucleus

- *Medial group*—the medial septal nucleus and the nucleus of the **diagonal band of Broca**
- *Caudal group*—the fimbrial and triangular septal nuclei

Major afferents to this area end mostly in the lateral septal nucleus. They have fibers of the fornix that begin in the hippocampal fields CA1 and CA3, as well as the subiculum. Afferents arise from the preoptic area; lateral hypothalamic area; and the anterior, paraventricular, and ventromedial hypothalamic nuclei. The lateral septum is greatly innervated in a monoaminergic manner. It includes noradrenergic afferents from the locus ceruleus (A1) and medullary (A2) cell groups; dopaminergic afferents from the ventral tegmental region (A10); and serotoninergic afferents from the midbrain raphe nuclei.

Efferents project from the lateral septum to the medial and lateral preoptic areas; the anterior hypothalamus; and the supramammillary and midbrain ventral tegmental area. This is through the medial forebrain bundle. Efferents also project to the medial habenular nucleus and to certain midline thalamic nuclei. This is through the stria medullaris thalami. There are projections from the habenular, through the fasciculus retroflexus, reaching the interpeduncular nucleus and nearby ventral tegmental area of the midbrain. This creates a route for the forebrain limbic structures to affect the midbrain nuclear groups.

From the medial septal and vertical limb nuclei of the diagonal band, efferents travel through the dorsal fornix, fimbria, and **supracallosal striae**. They also have a ventral route, via the amygdaloid complex. These projections reach every hippocampal field. However, the largest terminations are in the dentate gyrus, the CA3 field, the entorhinal cortex, and the parasubiculum and presubiculum that many of these are cholinergic or GABAergic.

> **Focus on the limbic system and olfaction**
> The sense of smell is the only sense directly linked to the limbic system that is associated with emotional, physical, and psychological responses. "Good" smells and "bad" smells can affect a person's moods drastically. For example, the smell of chocolate increases theta brain waves, triggering relaxation.

Amygdala

The *amygdaloid nuclear complex* is also called the **amygdala**. It is made up of lateral, central, and basal nuclei. These lie in the dorsomedial temporal pole of the temporal lobe, anterior to the hippocampus. The temporal horn of the lateral ventricle lies between the hippocampus and amygdala. This is a good marking point to distinguish between these two structures when viewing a computed tomography (CT) scan or MRI image. The structure that is anterior to the temporal horn of the lateral ventricle is the amygdala, and the structure behind it is the hippocampus. These structures are near the tail of the caudate nucleus, and lie partly deep to the gyrus ambiens, gyrus semilunaris, and uncinate gyrus. Above the lateral ventricle, the amygdala is partly continuous with the inferomedial edge of the claustrum. It is partially separated from the putamen and globus pallidus by fibers of the external capsule, and the substriatal gray matter, including the cholinergic magnocellular nucleus *basalis of Meynert*. The amygdala, laterally, is close to the optic tract.

There are dorsomedial and ventrolateral subnuclei in the lateral nucleus. The basal nucleus has dorsal magnocellular and intermediate parvocellular basal nuclei. It also has a ventral band of darkly staining cells called the *paralaminar basal nucleus*, named because it borders the white matter, ventral to the amygdaloid complex. The accessory basal nucleus is medial to the basal nuclear divisions. It may have dorsal, magnocellular, and ventral, parvocellular divisions. The lateral and basal nuclei are often collectively called the basolateral area or nuclear group of the *amygdaloid complex*. The basolateral nuclear complex consists of the lateral, basal, and accessory basal portions. It may be a *quasicortical* structure, since it has similarities to the cortex. Lacking a laminar structure, it has direct and returning connections with nearby temporal and other cortical areas. It also projects to the motor or premotor cortex. Direct cholinergic and noncholinergic input from the basal forebrain's magnocellular **corticopetal system** is received. There are returning connections also with the mediodorsal thalamus. Small peptidergic neurons are distributed in the basolateral nuclear complex, such as those that contain cholecystokinin (CCK), neuropeptide Y, and somatostatin. The neurons are similar in appearance and density to those of the adjacent temporal lobe cortex. Projection neurons from this area partly use glutamate or aspartate as neurotransmitters. The neurons project to the ventral striatum, and not to the hypothalamus or brainstem. Therefore, this area of the amygdaloid complex may be a polymodal area similar to the cortex. It is separated from the cerebral cortex by fibers of the external capsule.

The central nucleus is located throughout the caudal half of the amygdaloid complex. The medial and central nuclei have an extension, across the basal forebrain, and within the stria terminalis. This extension merges with the bed nucleus of the stria terminalis. This extensive complex is sometimes called the *extended amygdala*, and is formed by a few structures. These include the centromedial amygdaloid complex, consisting of the medial nucleus, and medial and lateral portions of the central nucleus; the medial bed nucleus of the stria terminalis; and cellular columns of the sublenticular substantia innominata between them. The subnuclei of the bed nucleus are lined up, along an anterior-to-posterior gradient. Parts of the medial nucleus accumbens may be within the extended amygdala.

Other amygdala structures
Consistently, the amygdala's intrinsic connections mostly arise in the lateral and basal nuclei, and end in the central and medial nuclei. This means that information mostly flows

in one direction. The lateral nucleus projects to every division of the basal nucleus, the accessory basal nucleus, and the paralaminar and anterior cortical nuclei. Less significantly, it projects to the central nucleus, receiving only a few afferents from other nuclei. The magnocellular, parvocellular, and intermediate areas of the basal nucleus project to the accessory basal nuclei, the medial nuclei, and especially to the medial area of the central nuclei. They also project to the periamygdaloid cortex and amygdalohippocampal area. Major intra-amygdaloid afferents begin in the lateral nucleus. The medial nucleus then projects to the accessory basal nuclei, anterior cortical nuclei, and central nuclei. It also projects to the periamygdaloid cortex and amygdalohippocampal area. Afferents especially arise from the lateral nucleus. The posterior portion of the cortical nucleus projects to the medial nucleus. However, it is hard to differentiate this projection from the one arising in the amygdalohippocampal area. The central nucleus has projections to the anterior cortical nucleus, and to some cortical transitional zones. This forms important foci for afferents from many amygdaloid nuclei—primarily the basal and accessory basal nuclei. It has significant extrinsic connections.

The amygdaloid complex has extensive, rich connections with many neocortical areas, in the unimodal and polymodal areas of the cingulate, frontal, insular, and temporal neocortices. Many projections are received from the brainstem, such as the parabrachial and peripeduncular nuclei. There is a rich monoaminergic innervation received by the amygdala. The noradrenergic projection mostly arises from the locus ceruleus. Serotoninergic fibers emerge from the dorsal nuclei, and in lesser amounts, from the median, raphe nuclei. Dopaminergic innervation mostly arises in the midbrain ventral tegmental area (A10). A very dense cholinergic innervation, from the magnocellular nucleus basalis of Meynert, reaches the basal and parvocellular accessory basal nuclei, the amygdalohippocampal area, and the nucleus of the lateral olfactory tract.

There are rich cholinergic connections between the amygdaloid nuclear complex and the neocortical areas, as well as with the allocortical and juxtallocortical areas. The amygdala projects to dispersed neocortical fields, mostly from the basal nucleus. The complex projects to almost all levels of the visual cortex, in the temporal and occipital lobes. Most projects arise from the magnocellular basal nucleus. A direct pathway to the amygdala may not connect with the primary sensory cortices. There may be a crude sensory input that acts as a visual alerting system to various threats, without conscious effort. The amygdala also returns projections to the auditory cortex in the superior temporal gyrus' rostral half. Projections to the polymodal sensory areas in the temporal lobe mostly return the projections from the amygdalopetal area. Efferents from the lateral and accessory basal nuclei reach the medial perirhinal area and other parts of the temporal pole. There is heavy innervation of the insular cortex from the amygdaloid medial and anterior cortical nuclei. A heavy projection is received by the orbital cortex and medial frontal cortical areas, as well as parts of the anterior cingulate gyrus. The basal nucleus is a primary source of these projections. They also receive contributions from the lateral nuclei, and the magnocellular and parvocellular divisions of the accessory basal nuclei.

There are associated connections between every part of the primary olfactory cortex, the superficial amygdaloid structures, olfactory bulb, lateral olfactory tract nucleus, anterior cortical nucleus, and the periamygdaloid or piriform cortex. Most cortical input to the amygdala comes from the lateral nucleus of the anterior temporal lobe. Rostral area of the superior temporal gyrus may be a unimodal auditory association cortex. They project to the lateral nucleus. Projections also exist from polymodal sensory association cortices of the temporal lobe. These include the perirhinal cortex, the caudal portion of the parahippocampal gyrus, the dorsal area of the superior temporal sulcus, and the medial as well as lateral areas of the temporal pole cortex. There is heavy projection of the rostral insula to the lateral, parvocellular basal, and medial nuclei. The caudal insula is connected with the second somatosensory cortex. It additionally projects to the lateral nucleus, to provide a route for somatosensory information to reach the amygdala. The caudal orbital cortex has projections to the basal, magnocellular accessory basal, and lateral nuclei. There are also projections from the medial prefrontal cortex to the accessory and basal nuclei's magnocellular divisions.

The primary relay for amygdaloid projections to the hypothalamus is the central nucleus. Amygdaloid fibers reach the bed nucleus in the stria terminalis mostly through the stria terminalis itself, but also through the ventral amygdalofugal pathway. Central and basal nuclei mostly project to the lateral bed nucleus. Medial and posterior cortical nuclei mostly project to the medial bed nucleus. The anterior cortical and medial nuclei mostly project to the medial preoptic area and anterior medial hypothalamus, which includes the paraventricular and supraoptic nuclei, as well as to the ventromedial and premammillary nuclei. The amygdala projects to the lateral hypothalamic rostrocaudal edge. Most fibers originate in the central nucleus. They run in the ventral amygdalofugal pathway and medial forebrain bundle. Rich projections exist to the medial, magnocellular area of the thalamic mediodorsal nucleus. They are mostly from the lateral, basal, and accessory basal nuclei as well as the periamygdaloid cortex. The central and medial nuclei have projections to the midline nuclei—mostly the nucleus centralis and nucleus reuniens.

The basal nuclei's parvocellular division, along with the *magnocellular accessory* basal nucleus, and central nucleus all project to cholinergic cells of the basal forebrain—mostly the nucleus basalis of Meynert, and diagonal band horizontal lib nucleus. The nucleus accumbens and other parts of the striatum receive many projections from the amygdaloid complex—mostly the basal and accessory basal nuclei. The ventral pallidum receives many fibers from the ventral striatum. The pallidum then projects to the thalamic mediodorsal nucleus. Therefore, the ventral striatopallidal system provides another route through which the amygdala influences processes of the mediodorsal thalamus and prefrontal cortex. The largest amount of efferents to the hippocampal formation comes from the lateral, magnocellular accessory basal, and parvocellular basal nuclei. The primary projection is from the lateral nucleus, to the rostral entorhinal cortex. Many fibers also end in the hippocampus proper and subiculum. Marked polarity in amygdalohippocampal connections may exist. Therefore, the amygdala has more influence on hippocampal process than the reverse situation. Many cognitive and affective processes are linked to amygdala–cortical functional connectivity.

Development of such functional connections is related to adult emotions. There are specific, extensive changes to these connections that occur between childhood and adolescence. When functional connectivity is atypical, this may be related to bipolar disorder, depression, and schizophrenia. The major relay for amygdala projections to the brainstem is the central nucleus, which also receives many returning projections. The central nucleus projects to the periaqueductal gray matter, substantia nigra pars compacta, ventral tegmental area, mesencephalic tegmental reticular formation, peripeduncular nucleus, nucleus of the solitary tract, parabrachial nucleus, and the vagal dorsal motor nucleus.

Functions of the limbic system

The hypothalamus has **suprasegmental** integrations of the autonomic nervous system. There are efferent pathways from the hypothalamus to the neural structures involved in parasympathetic and sympathetic reflexes. The hypothalamus contains neurosecretory cells that control secretion of pituitary hormones. The hypothalamus regulates body temperature, hunger, levels of circulating electrolytes, and thirst. Perceptions of visceral activities greatly affect our emotional states.

The cingulate gyri may be involved in memory processing, exploratory behavior, and visually focused attention. This system appears to be more effective in the nondominant brain hemisphere. The cingulate gyri have functions in both cognition and emotions. Concentrations of NE are highest in the hypothalamus, and then in the medial areas of the limbic system. There is 70% or more NE concentrated in axon terminals arising in the medulla and the locus ceruleus of the rostral pons. There is a large amount of serotonin in axons of other ascending fibers, especially those that begin in the reticular formation of the midbrain, and end in the amygdala, septal nuclei, and lateral areas of the limbic lobe.

Neuronal axons in the ventral tegmental parts of the midbrain ascend in the medial forebrain bundle and the nigrostriatal pathway. They have high levels of dopamine. This may explain why a severe depressive reaction can be caused by electrical stimulation of the substantia nigra via an electrode that was previously placed to treat Parkinson's disease. Many of these structures and their connections make up a collectively functional system. Therefore, the term *limbic system* is actually a simple description, since its parts have vastly different connections with the neocortex and central nuclei. The neurotransmitters are also different, as are the effects of the parts when they become damaged. Overall, limbic system lesions usually do alter our emotions.

Section review
1. What are the primary structures of the limbic system?
2. Which structure is a good marking point to distinguish between the hippocampus and amygdala?
3. What is the role of the hypothalamus regarding the limbic system?

Olfactory system

The nerve fibers used in olfaction have cellular origins in the mucous membrane of the upper and posterior nasal cavity, the *superior turbinates* and *nasal septum*. The olfactory mucosa has a total area of about $2.5\,cm^3$.

Anatomy of the olfactory system

There are three cell types in the olfactory system. These include 6–10 million *olfactory cells* (or *receptor cells*), *sustentacular (supporting) cells*, and *basal cells*. The sustentacular cells maintain potassium and other electrolyte levels in the extracellular spaces. The basal cells are stem cells that create olfactory and sustentacular cells during regeneration. The epithelial surface has a covering layer of mucus secreted by the tubuloalveolar cells (or *Bowman's glands*). Inside are the immunoglobulins A and M, lactoferrin, lysoenzyme, and odorant-binding proteins. These molecules are believed to prevent pathogens from entering the cranium through the olfactory pathway.

The olfactory tract is formed by the axons of the mitral and tufted cells of the olfactory bulb (see Fig. 15.3). The tract continues, along the olfactory groove of the cribriform plate, to reach the cerebrum. Groups of cells making up the anterior olfactory nucleus are caudal to the olfactory bulbs. The dendrites from these cells synapse with olfactory tract fibers. Their axons project out to the olfactory nucleus and bulb on the opposite side. The neurons probably serve to reinforce olfactory impulses. The olfactory tract posteriorly divides into medial and lateral olfactory striae. The medial stria has fibers from the anterior olfactory nucleus that pass to the opposing side through the anterior commissure. In the lateral stria, the fibers begin in the olfactory bulb, with collaterals emerging that reach the anterior perforated substance. These end in the medial and cortical nuclei of the amygdaloid complex and prepiriform area. This area is also called the lateral olfactory gyrus. This represents the **primary olfactory cortex**, in an area on the anterior end of the parahippocampal gyrus and uncus. Olfactory impulses, therefore, reach the cerebral cortex without being relayed through the thalamus, which makes olfaction a unique sensory system. The fibers then project to the nearby entorhinal cortex and the thalamic

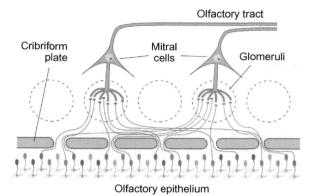

Fig. 15.3 Sorting of olfactory nerve fibers among glomeruli of the olfactory bulb. Olfactory receptors of different types—each type characterized by its receptor protein and a restricted range of odor sensitivities (represented here by different colors)—are intermingled in a given area of olfactory epithelium. The axon terminals of any given type all converge on one or two glomeruli (which in reality would contain thousands of axon terminals and the dendrites of up to dozens of mitral and tufted cells).

medial dorsal nucleus. The amygdaloid nuclei connect to the hypothalamus and septal nuclei. The nuclei may be used in reflexes related to eating and sexual function. Feedback regulation occurs throughout the afferent olfactory pathway.

Olfactory receptors

The olfactory cells are bipolar neurons. Each has a peripheral process called the **olfactory rod**. From each rod, 10–30 fine cilia project. These hair-like processes have no motility, and are the sites of the olfactory receptors. The central processes (*olfactory cilia*) are only 0.2 mm in diameter. They are unmyelinated fibers that converge, forming small fascicles surrounded by Schwann cells that pass through openings in the ethmoid bone's cribriform plate, into the olfactory bulb. The central processes of the olfactory receptor cells collectively make up cranial nerve I (the *olfactory nerve*). This is the only area in the body where neurons directly contact the external environment.

Olfactory pathways

The olfactory system's afferent pathways continue directly to the olfactory cortex, within the orbitofrontal region of the prefrontal cortex. They bypass the thalamus. The olfactory nerve is the only cranial nerve that does not relay in the thalamus. Many of these pathways' terminal fields are primitive cortical areas believed to be parts of the limbic system. This explains how some odors affect behaviors and emotions. The cortex processing olfaction is in close proximity to the *Papez circuit*. The olfactory nerves initiate from the olfactory receptor neurons within the olfactory mucosa. Axons are grouped into many small bundles. Unique glia and layers of meninges surround these. The bundles enter the anterior cranial fossa via the foramina in the cribriform plate of the ethmoid bone. They are attached to the inferior surface of the olfactory bulb. This is located at the anterior end of the olfactory sulcus, on the frontal lobe's orbital surface. They end in the olfactory bulb.

Olfactory bulb

The **olfactory bulb** is posteriorly continuous with the olfactory tract. The bulb's output passes through this tract, directly to the ipsilateral piriform cortex, the amygdala, and the rostral entorhinal cortex. The olfactory bulb has a clear laminar structure. Inwards from the surface, the laminae are organized as follows:

- *Olfactory nerve layer*—unmyelinated axons of the olfactory neurons; continuous turnover of receptor cells shows that this layer's axons are at different growth, maturity, or degeneration stages
- *Glomerular layer*—a thin sheet of glomeruli; incoming olfactory axons divide and synapse on secondary olfactory neuron terminal dendrites, such as mitral, periglomerular, and tufted cells
- *External plexiform layer*—has principal and secondary dendrites made up of mitral and tufted cells
- *Mitral cell layer*—a thin sheet of the cell bodies of mitral cells; each cell sends one principal dendrite to a glomerulus, secondary dendrites to the external plexiform layer, and one axon to the olfactory tract; it also has small amounts of granule cell bodies
- *Internal plexiform layer*—has axons, granule cell bodies, and recurrent deep collaterals of mitral and tufted cells
- *Granule cell layer*—has most of the granule cells, their superficial and deep processes, and many centripetal and centrifugal nerve fibers passing through

Various odor molecules use differing spatial activity patterns within the olfactory bulb. The main neurons are the mitral and tufted cells. Their axons create its output through the olfactory tract.

The cells are similar, morphologically. Most of them utilize glutamate, aspartate, or another excitatory amino acid as their neurotransmitter. The mitral cell receives sensory input superficially, at its glomerular tuft, and is continuous through the bulb's layers. Mitral and tufted cell axons are believed to be parallel output pathways from the olfactory bulb. The main form of interneurons within the olfactory bulb is the granule and periglomerular cells. Most periglomerular cells are dopaminergic. However, some are GABAergic. Their axons are laterally distributed. They end in the extraglomerular regions. Granule cells are almost the same size as the periglomerular cells. Their unique feature is the lack of an axon. They resemble the amacrine cells of the retinas of the eyes. Granule cells contain two primary, spined dendrites. These pass radially within the bulb and appear to be GABAergic. A granule cell is most likely a powerful inhibitor upon the olfactory bulb's output neurons.

The bulb's granule cell layer extends into the olfactory tract, appearing as scattered multipolar neurons making up the anterior olfactory nucleus. From mitral and tufted cells, many centripetal axons relay inside or send collaterals to the anterior olfactory nucleus. Axons from the nucleus continue with the other direct fibers from the bulb, into the olfactory striae. Afferent inputs to the olfactory bulb emerge from many central sites. Neurons of the anterior olfactory nucleus and pyramidal neuron collaterals within the olfactory cortex project out to the olfactory bulb granule cells. In the horizontal limb nucleus of the diagonal band of Broca, cholinergic neurons project to the granule cell and glomerular layers. They are part of the basal forebrain cholinergic system. Additional afferents to the glomeruli and granule cell layer emerge from the pontine locus ceruleus and the mesencephalic raphe nucleus.

Olfactory tract

When the olfactory tract nears the anterior perforated substance, also called the *ventral striatopallidal region*, it becomes flatter and widens as the **olfactory trigone**. The tract's fibers continue, from the caudal angles of the trigone. They appear as divergent medial and lateral olfactory striae, bordering the anterior perforated substance. The anterolateral margin of the anterior perforated substance is followed by the lateral olfactory stria to the limen insulae. Here, it is bent, posteromedially, merging with the elevated *gyrus semilunaris* at the rostral margin of the uncus within the temporal lobe. The lateral olfactory gyrus forms a thin gray layer that covers the lateral olfactory stria. Laterally, it merges with the gyrus ambiens, which is part of the limen insulae. The lateral olfactory gyrus and gyrus ambiens collectively form the prepiriform cortical region. This caudally passes into the entorhinal region of the parahippocampal gyrus. The prepiriform and periamygdaloid regions as well as the entorhinal area form the piriform cortex, which is the largest cortical olfactory area.

The medial olfactory stria medially passes along the rostral boundary of the anterior perforated substance, nearing the medial continuation of the diagonal band of Broca. Together, these curve upwards on the medial hemisphere aspect. This is anterior to the attachment of the lamina terminalis. The anterior perforated substance is laterally continuous with the peduncle of the amygdaloid complex and temporal stem. It is medially continuous with the septal region. The primary foci of the lateral olfactory tract are the piriform cortex, amygdala, and rostral entorhinal cortex. Piriform cortex neurons widely project to the orbitofrontal cortex of the neocortex, agranular insula, medial dorsal thalamic nucleus, hypothalamus, amygdala, and hippocampal formation.

Functions of the olfactory system

When we breathe quietly, only a small amount of air reaches the olfactory mucosa. We must "sniff" to bring the air into the olfactory crypt, where the olfactory receptors are located. The inhaled substance must be volatile in order to be perceived as an odor.

This means it is spread through the air as very small particles. It must also be soluble in water. Molecules that result in the same odor have related shapes, and not necessary the same chemical qualities. During sniffing, there is a slow negative potential shift or *electroolfactogram* that can be detected by placing an electrode upon the mucosa. Conductance changes underlying this receptor potential are created by molecules of odorous material being dissolved in the mucus above the receptor.

Transduction of odorant stimuli into electrical signals is partially controlled by guanosine triphosphate-dependent adenylyl cyclase, or G *protein*. Similar to other cyclic adenosine monophosphate pathways, this uses the same intracellular second messenger. The messenger opens a voltage-gated calcium channel in the receptor. Then, there are transmembrane receptor proteins, and many intracellular biochemical events, which generate axon potentials.

Olfactory sensation intensity is based on the frequency of afferent neuron firing. An odor's quality is believed to be due to cross-fiber activation and integration. The individual receptor cells can respond to many odorants. They have different types of responses to stimulants such as excitatory, inhibitory, and on-off responses. Olfactory potential can be eliminated if the olfactory receptor surface or olfactory filaments are destroyed. Loss of electroolfactogram occurs 8–16 days after the nerve is severed. The receptor cells vanish, but the sustentacular cells are not changed. Basal cells of the olfactory epithelium divide, meaning that olfactory receptor cells are always dying and being replaced by new ones. The chemoreceptors for smell and taste are one of the only examples of human neuronal regeneration.

Chemesthesis involves the trigeminal system, via undifferentiated receptors of the nasal mucosa. The receptors are highly sensitive to irritants but cannot greatly discriminate between stimuli. The trigeminal afferents release neuropeptides, resulting in hypersecretion of mucus, localized edema, and sneezing. Stimulation of the olfactory pathway, at cortical areas of the temporal lobe, may also cause olfactory experiences. The olfactory system quickly adapts to sensory stimuli. There must be repeated stimulation for sensation to be sustained. An aroma can bring back a long-ago memory of one or many experiences. Olfactory and emotional stimuli are closely linked, since they have common limbic system roots. Oppositely, the ability to recall an odor is not highly common in comparison to recalling sights and sounds. Olfactory experiences are not part of dreaming as well.

About 2% of the human genome exists to express unique odorant receptors. There are more than 500 distinct genes. These transmembrane proteins are extremely diverse, allowing for subtle differentiation of thousands of odorant molecules. This molecular specificity is neuroanatomically encoded. Different odorants activate certain olfactory receptors. Each neuron expresses just one allele of one receptor gene. Also, each olfactory glomerulus receives inputs from neurons that express just one type of odorant receptors.

Therefore, each glomerulus response to a specific type of odorant stimulus. This encoding is believed to be preserved within the olfactory cortex.

> **Focus on the sense of smell**
> Humans have evolved a higher sensitivity to odors that indicate danger or poison. Smelling a food source that is past its expiration date often provides odors that we sense as "rotten." This type of odor is due to the fact that bacteria have begun to accumulate on the food source. Therefore, our perception of a rotten smell is a method of protecting ourselves from eating something that could make us very sick.

Section review
1. Where are the locations of the olfactory receptors?
2. What are the layers of the olfactory system?
3. How are odors perceived?

Gustatory system

The gustatory system utilizes sensory receptors known as *taste buds*, which are distributed over the surface of the tongue. They are also present in smaller amounts on the soft palate, pharynx, larynx, and esophagus. The sense of taste comes from activation of the taste buds by chemical substances in solution. The receptors transmit their activity via the sensory nerves to the brainstem. The four primary taste sensations are salty, sweet, bitter, and sour. Recently, a fifth sensation called *umani* was identified. It signifies a *savory* taste of glutamate, aspartate, and some ribonucleotides. The complete range of taste sensations consists of combinations of the basic gustatory sensations. The taste receptors are extremely sensitive. A G-protein transduction system, similar to the one used in olfaction, signals taste sensations to the brain. Gustatory as well as olfactory acuity diminishes with aging.

Anatomy of taste buds

Taste buds are microscopic epithelial structures. They are shaped similarly to barrels, and contain chemosensory cells (Fig. 15.4). These cells synaptically contact the terminals of the gustatory nerves. There are many taste buds, mostly on the lateral aspects of all types of lingual papillae—except the filiform papillae. They are also scattered over nearly all dorsal and lateral tongue surfaces, and less abundantly, upon the epiglottis and lingual

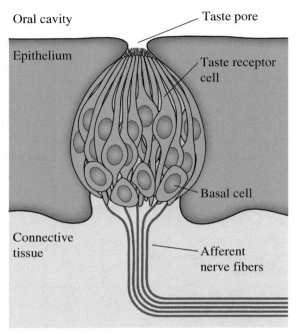

Fig. 15.4 Schematic drawing of a taste bud.

aspect of the soft palate. A taste bud is linked via synapses at its base to either the facial, glossopharyngeal, or vagus nerves—each of which carry taste sensations. Certain physiological features of taste beds resemble those of neurons, such as synaptic transmission and action potential generation. Therefore, they are sometimes called *paraneurons*.

Between individual people, taste bud distribution can be quite varied. They are most common on the posterior tongue areas, primarily around the circumvallate papillae walls as well as their surrounding sulci. Here, there are about 250 taste buds for each of the 8–12 papillae. The sides of the tongue have more than 1000 taste buds—especially over the more posterior foliate papillae folds. They are rare or absent on the fungiform papillae—usually about 53 per papilla, when present. Taste buds have been identified in fetuses on the epiglottis and soft palate. Most of these disappear during postnatal development.

Taste bud microstructure

Every taste bud contains 50–150 fusiform cells within an oval cavity of the epithelium. The cell clusters are apically converged on a gustatory pore, which is a 2 μm wide opening on the mucosal surface. The entire structure is about 40 μm wide and 70 μm high. It is separated via a basal lamina from the lamina propria below. A small afferent nerve fiber fasciculus penetrates the basal lamina and is spirally located round the sensory cells. In the oral salvia, dissolved chemical substances diffuse through the taste buds' gustatory pores. They reach taste receptor cell membranes causing membrane depolarization.

Neural pathways

There is a complicated distribution of taste bud innervation, since individual nerve fibers branch outward. Every fiber can have many terminals. These may spread, innervating widely separated taste buds, or they can innervate more than one sensory cell in each taste bud. Oppositely, individual buds can receive terminals of several different nerve fibers. Such convergent and divergent innervation may be functionally needed.

The chorda tympani is the gustatory nerve for the anterior part of the tongue, except for the circumvallate papillae. The chorda tympani travels via the lingual nerve. Taste fibers run, in most people, within the chorda tympani to cell bodies in the facial ganglion. Sometimes, they diverge to the otic ganglion, via the greater petrosal nerve. Taste buds in the inferior soft palate surface are mostly supplied by the facial nerve via the greater petrosal nerve, pterygopalatine ganglion, and lesser palatine nerve. They may additionally be supplied by the glossopharyngeal nerve. Taste buds of the circumvallate papillae, postsulcal tongue area, palatoglossal arches, and oropharynx are innervated by the glossopharyngeal nerve. The taste buds in the extreme pharyngeal part of the tongue and epiglottis receive fibers from the internal laryngeal branch of the vagus nerve.

Every taste bud receives two different fiber classes. The first branches in the taste bud's periphery, forming a perigemmal plexus. The other forms an intragemmal plexus in the taste bud itself. This innervates the bases of receptor cells. The perigemmal fibers contain different neuropeptides, including calcitonin gene-related peptide (CGRP) and substance P. They appear to represent free sensory endings. Within each taste bud, the intragemmal fibers branch, forming a series of synapses.

Central connections

Within the brainstem, the gustatory afferents make up the tractus solitaries. They end in the rostral third of the nucleus solitaries, within the medulla oblongata.

Functions of the gustatory system

As discussed above, the basic four taste sensations of salty, sour, sweet, and bitter, have been augmented by *umami*. Though most sources state that certain tongue areas are specialized for these different taste sensations, newer evidence has shown that all areas are responsive to all of them. Every afferent nerve fiber connects to largely separated taste buds, and can respond to several different types of chemical stimuli. Some respond to all of the four basic categories, while others respond to only one, or just a few. In each class of tastes, receptors may be differentially sensitive to a large range of similar chemical. Taste buds alone can detect a restricted range of chemical substances in an aqueous solution. The sensations of taste and smell are hard to separate since the oral and nasal cavities are continuous with each other. Much of perceived tastes result from airborne odorants, from the oral cavity, passing through the nasopharynx and reaching the olfactory area above. Perceive taste sensations result from (probably) central processing of complex response patterns, from various tongue areas. In newborns, oral sucrose can reduce procedural pain from single events. This nociceptive effect's mechanism is not fully understood.

Autonomic tongue innervation

The parasympathetic innervation of the lingual mucous glands occurs through the facial nerve's chorda tympani. This synapses in the submandibular ganglion. The postganglionic branches reach the lingual mucosa through the lingual nerve, or tough plexuses surrounding the lingual arteries. Postganglionic sympathetic supply to the lingual glands and vessels begins from the carotid plexus. Fibers are derived from neurons of the superior cervical ganglion. The supply enters the tongue via plexuses surrounding the lingual arteries. Isolate nerve cells, which may be postganglionic parasympathetic neurons, exist in the postsulcal region. These may innervate both glandular tissue and vascular smooth muscle.

Testing of taste sensation

Placing small amounts of salt, sugar, lemon, or quinine on various areas of the tongue, without the patient seeing each of them, can test a patient's taste sensations. The lemon proves the sour sensation, while the quinine provokes the bitter sensation. After each test, the tongue is wiped clean, and the patient is asked to describe the taste sensation. Testing of taste sensation is useful for a variety of conditions. One example is to determine the presence of Bell's palsy. Taste sensations are compared on each side of the anterior tongue.

Instead of using lemon to provoke the sour sensation, a low-voltage direct current can be administered via electrodes placed on the tongue surface. When loss of taste is bilateral, mouthwashes can be used that contain diluted sucrose, sodium chloride, citric acid, and quinine. The patient swishes the mouthwash, spits out the test fluid, rinses the mouth with water, then must indicate if a substance was tasted, and what it tasted like. There are *electrogustometers* that are used to measure taste intensity, and to determine detection and recognition thresholds of taste and olfactory stimuli. However, these are only used in extreme cases.

> **Focus on the "umami" or "savory" sense of taste**
>
> The Japanese term "umami" is equivalent to the English word "savory." It has been recognized as the fifth basic taste, in addition to the better-known tastes of sweet, sour, bitter, and salty. Savory dishes evoke pleasant emotions, and this form of taste is a signal that a food source is rich in protein. The sensory cells specifically involved in this form of taste were originally discovered in Japan in 1910.

Section review
1. Where is the distribution of taste buds?
2. What are the physiological characteristics of taste buds?
3. What is the location of the gustatory centers within the brain?

Clinical considerations

There are many conditions of the limbic, olfactory, and gustatory systems that must be understood. Limbic system conditions include emotional disturbances, which may be due to hallucinations or pain, emotional lability, spasmodic (pseudobulbar) emotional disturbances, anger, violent behaviors, rage due to temporal lobe seizures, rage without seizures, violence due to neurologic diseases, aggressiveness due to encephalopathies and drug intoxication, apathy, placidity, alterations in sexuality, anxiety, euphoria, fear, and differential limbic diagnoses. Olfactory system conditions include anosmia, parosmia, hallucinations, and agnosia. Gustatory system conditions include ageusia, hypogeusia, and dysgeusia.

Emotional disturbances of the limbic structures

Regarding the limbic system, there are a variety of neurological factors related to emotional disturbances. Illusions and hallucinations are classified as perceptual abnormalities, while delusions are classified as cognitive derangements. Emotional lability as well as pathologic laughing and crying, which is known as the *pseudobulbar state*, are examples of disinhibition of emotional expression. There are also rage reactions and aggressiveness, apathy, placidity, altered sexuality, and endogenous anxiety, depression, euphoria, and fear. Even today, not all of the limbic-related emotional disturbances are understood. Further study will allow for better psychiatric and neurologic advances.

Emotional disturbances due to hallucinations and pain

Emotional disturbances that are related to hallucinations cause patients to feel realistically threatened by imagined voices or individuals. The patient shows all the signs of fear that result from real stimuli. The patient reacts inappropriately to what he or she is perceiving as real. Also, patients having severe and acute pain may be excessively emotional and inattentive to others' words and actions. These patients are almost continuously anxious, angry, and displaying various reactions to pain. Examples of conditions causing these behaviors include explosive migraine, severe pain in many body areas, spinal subdural hemorrhage, subarachnoid hemorrhage, and trauma from multiple bone fractures.

Emotional lability

Many different cerebral diseases commonly weaken control of emotional expression. For example, vascular lesions of the cerebrum may result in uncontrolled crying or laughter. *Emotional lability* is defined as vacillating between these states, a sign of *organic brain disease*. The emotional response is excessive, but not as intense as that of pseudobulbar states. Lesions of the frontal lobes may be more often involved. Emotional lability often is present in Alzheimer's disease and other diffuse cerebral diseases, with the limbic cortex being affected by these conditions. Chronic nervous system diseases also often cause crying and laughter, as well as "joking" behaviors and lack of inhibition that can be linked to frontal lobe disease.

Spasmodic or pseudobulbar emotional disturbances

Pseudobulbar or *spasmodic* emotional disturbances have been understood for a long time. *Emotional incontinence* describes forced laughing or crying, and has a pathologic origin in the brain that can be diffuse or focal. Therefore, it is a syndrome of a variety of causes. The causes of pseudobulbar affective display include: amyotrophic lateral sclerosis with pseudobulbar palsy, bilateral strokes, bilateral traumatic lesions of the cerebral hemispheres, **progressive supranuclear palsy**, amyotrophic lateral sclerosis (ALS), stroke, **gliomatosis cerebri**, hypoxic-ischemic encephalopathy, multiple sclerosis with bilateral corticobulbar demyelinative lesions, pontine myelinolysis, progressive, and even **Wilson disease**.

Pathologic laugher and crying usually involves lesions distributed bilaterally, which involve the motor tracts—especially the corticobulbar motor system within the brainstem. Infiltrative frontal lobe gliomas and **encephalitides** may also be causative. Sudden hemiplegia may develop from a stroke linked to a preexisting, silent lesion in the opposite hemisphere. There may be imbalances between lack of voluntary movements of the muscles innervated by the motor nuclei, of the lower pons and medulla, while movement of the same muscles for other functions still occurs. The patient may be unable to elevate or retract the corners of the mouth, chew, forcefully close the eyes, open and close the mouth, speak normally, articulate, swallow, or move the tongue. However, the abilities to cough, yawn, clear the throat, or to spasmodically cry or laugh are still present. This is known as the motor syndrome of *pseudobulbar palsy*. This explains why the term *pseudobulbar affective state* is used to describe the emotional condition.

The patient may suddenly have spasms of laughter lasting for a few moments to many minutes, until he or she is exhausted. More often, however, sudden crying manifests when something that brings up an emotional memory is discussed. The emotional severity does not match the severity of the pseudobulbar paralysis or the exaggerated facial and jaw jerk tendon reflexes. Some forced crying and laughing brings with it no obvious weakness of the facial and bulbar muscles. In other patients, these muscles are severely weakened, yet there is no forced crying or laughing. In central pontine myelinolysis and progressive supranuclear palsy, in which pseudobulbar palsy often occurs, forced crying and laughing are seen in some patients.

Anger and violent behaviors

Anger, aggressiveness, and rage are important components of social behaviors. When they begin early in life, the individual establishes family and social positions. These behaviors have a relationship to the temporal lobes, and especially, the amygdala. Electrode stimulation of the medial amygdaloid nuclei causes anger, while stimulation of the lateral nuclei does not. Aggressiveness can be reduced by bilateral destruction of the amygdaloid complex. Lesions of the mediodorsal thalamic nuclei cause patients to become more calm and distant. Other involved structures include the ventromedial nuclei of the hypothalamus and the rostral cingulate gyrus.

Rage reactions that are severe in nature can occur as part of a temporal lobe seizure. They may be episodic reactions, without identified seizures, or a psychotic state after having a seizure, which is called *postictal psychosis*. This condition usually resolves on its own within a few hours. Rage reactions can be part of a recognized acute neurologic disease, and with reduced consciousness caused by metabolic or toxic encephalopathies, such as *bromo-dragonfly* (related to the *phenethylamines*), and *K4* (*Spice*).

Apathy and placidity

Humans display energized and explorative behaviors, possibly because of curiosity, controlled by *expectancy circuits* in the brain. This uses nuclear groups in the mesolimbic and mesocortical dopaminergic circuits. These are connected to the diencephalon and mesencephalon by the medial forebrain bundles. When lesions interrupt the connections, expectancy reactions disappear. A positron emission tomography (PET) study can link functional problems in initiating movements with impaired activation to the anterior cingulum, prefrontal cortex, putamen, and supplementary motor area. Perhaps, the most common psychobehavioral alteration accompanying cerebral disease is a quantitative reduction in all activity. This is especially true in patients with alterations of the anterior areas of the frontal lobes. The patient has reduced thoughts, spoken words, and movements. This is not only motor in nature, since the patient often perceives and thinks slowly, lacks associations with ideas, begins speaking less often, and shows less interest or inquisitiveness. The patient's family often notices these extreme personality changes.

The patient may appear to have an increased threshold to stimuli, inattentiveness, inability to remain attentive, apathy, impaired cognition, or lack of impulses (*abulia*). All of these may be correct, but each involved reductions in mental activity. Bilateral lesions deep within the septal region have caused the most severe lack of impulses, spontaneity, and drive (conation). There may also be impaired learning and memory. There are also lesser degrees of placidity and pathy, due to a nervous system disease, in which a patient becomes **hypobulic**.

Reduced attention is linked to lesions in the right, or nondominant, parietal lobe. The patient does not seem aware of any changes, is concerned about any other disease states, and pays little attention to family or personal problems. In severe cases of right parietal lobe stroke, the patient ignores everything to his left side, which is also called *neglect*. As mentioned in Chapter 6, neglect is a cortical sign.

Alterations in sexuality

Cerebral diseases can alter sexual behaviors in both sexes. *Hypersexuality* is well known, with orbital frontal lobe lesions removing moral and ethical restraints, leading to indiscriminate behaviors. Superior frontal lesions may cause an overall loss of initiative, reducing sexual and other types of impulses pleasure. Perforation of the dorsal septal region has caused sexually disinhibition.

Differential limbic diagnoses

Differential diagnoses of limbic system conditions involve pathogenesis and etiology. There are many neurologic diagnoses that must be considered. With uninhibited laughter or crying, and with emotional lability, there could be cerebral disease such as bilateral corticobulbar tract disease. Usually, motor and reflex changes of pseudobulbar palsy are related to increased facial and mandibular reflexes, and usually, corticospinal tract signs in the limbs. Bilateral cerebral disease is also indicated by extreme emotional lability. Signs of unilateral disease may be clinically apparent, however. Common pathologic bases include cerebrovascular lesions such as lacunar infarctions, ALS, diffuse hypoxic-hypotensive encephalopathy, and multiple sclerosis. In less common conditions such as Wilson disease and progressive supranuclear palsy, the manifestations may be extremely prominent. Abrupt onset signifies a vascular disease.

Placidity and apathy may be the earliest and clinically important signs of cerebral disease. They must be distinguished from Parkinson's disease akinesia and bradykinesia, as well as depressive illness-related reductions in mental activity. The most common pathologic states underlying apathy and placidity are Alzheimer's disease, frontal-corpus callosum tumors, and normal-pressure hydrocephalus. However, these can complicate many frontal and temporal lesions, including those that occur with demyelinating disease, or following a ruptured anterior communicating aneurysm.

Outbursts of rage and violence are often part of lifelong sociopathic behaviors. It is more significant when these occur in an individual who otherwise shows no sociopathy. If accompanying a seizure, the rage may be due to disruptive seizure activity upon temporal lobe function. Uncontrolled rage and violence is rarely due to temporal lobe epilepsy. More commonly, reduced degrees of undirected, combative behavior are due to ictal or postictal automatism. In rare cases, rage and aggressiveness are due to an acute neurologic disease, such as a glioma, involving the mediotemporal and orbitofrontal regions. The outbursts may be linked to dementia, or in stable patients, obscure encephalopathy.

Rage reactions along with continuous violence must be distinguished from mania. A manic state causes incoherent ideation, a mood that is euphoric or irritable, and uninterrupted psychomotor activity. There should also be distinguishing from *organic drivenness*, which involves continued motor activity without clear ideation. This is usually seen in children following encephalitis. Additionally, rage reactions with continued violence should be distinguished from extreme akathisias, which has restless movements and pacing along with extrapyramidal symptoms.

If a patient has extreme agitation and fright, he or she must be assessed for delirium, delusions, mania, or isolate panic attacks. Delirium includes clouded consciousness, hallucinations, and psychomotor overactivity. Delusions may be linked to schizophrenia. Mania involves overactiveness and overideation. A feeling of suffocation, palpitations, and trembling signify a panic attack. Only rarely does panic involve temporal lobe epilepsy. In adults without anxiety, an acute panic attack can signify the beginning of schizophrenia or a depressive illness.

When a patient experiences slow development of bizarre ideation, over weeks or months, the usually causes are psychosis from bipolar disease or schizophrenia. However, there should be consideration of tumors, immune or paraneoplastic encephalitis, temporal lobe lesions—especially along with temporal lobe seizures, aphasic symptoms, **quadrantic** visual field defects, and rarely, rotatory vertigo. Also to be excluded are hypothalamic disease, suggested by diabetes insipidus, somnolence, and visual field defects; and hydrocephalus.

Anosmia

Anosmia is classified as a quantitative abnormality, in which there is *loss* of the sense of smell. It is the most common olfactory abnormality in comparison to **hyposmia**, **hyperosmia**, **dysosmia**, **parosmia**, olfactory hallucinations and delusions, and **olfactory agnosia**. When anosmia is unilateral, it is usually not recognized by the patient. This form sometimes occurs with hysteria, on the same side where anesthesia, blindness, or deafness develops. *Bilateral anosmia* is common, and the patient often believes that he or she also has lost the sense of taste, a condition known as *ageusia*. The sense of taste is largely based on volatile particles in what we eat and drink. These particles reach the olfactory receptors via the nasopharynx. Therefore, the perception of flavor combines smell, taste, and tactile sensations. Patients with anosmia but without ageusia are able to distinguish sweet, sour, bitter, and salty sensations. The presence of anosmia can be verified by having the patient sniff nonirritating stimuli such as peanut butter, coffee, and vanilla, and then attempt to identify them. If the odors can be detected and described, the olfactory nerves are most likely intact or somewhat intact. Humans are able to distinguish many more odors than they are able to actually identify by name. When the odors cannot be detected, there is an olfactory defect. Irritating substances are not used in this testing since they have a primary irritating effect upon the mucosal-free trigeminal nerve endings.

There is some controversy about testing the ability to detect odors in one nostril at a time. The test may not be sensitive to the presence of a unilateral lesion such as a meningioma since air may mix within the nasopharynx. Other tests show that quick sniffing, though one nostril, does for a short time allow segregation of each nasal cavity, which can detect unilateral lesions. There is also a scratch-and-sniff test available, with 40 microencapsulated odorants. When the sense of smell is lost, it is usually classified into three categories:

- *Nasal*—odorants do not reach olfactory receptors
- *Olfactory neuroepithelial*—due to destruction of receptors or their axon filaments
- *Central*—due to olfactory pathway lesions

The most common diagnoses include viral infections of the upper respiratory tracts, nasal or paranasal sinus disease, and head injuries. Nasal diseases that cause anosmia or bilateral hyposmia usually involve hypertrophy and hyperemia of the nasal mucosa, which prevent olfactory stimuli from reaching receptor cells. The most frequent cause of hyposmia is usually heavy smoking. Other common causes include chronic atrophic rhinitis, allergic sinusitis, infective sinusitis, vasomotor sinusitis, nasal polyposis, and overuse of topical vasoconstrictors. Olfactory mucosa biopsies, in allergic rhinitis, have revealed sensory epithelial cells to still be present, but with deformed and shortened cilia that are covered by other mucosal cells. Anosmia or hyposmia can follow herpes simplex, hepatitis virus, and influenza infections. This is due to destruction of receptor cells. When the basal cells are also destroyed, the condition may be permanent.

Any type of head injury can cause anosmia, usually because of tearing of delicate filaments of the receptor cells. The damage can be unilateral or bilateral. Closed head injuries rarely cause complete anosmia (only 6% of patient), but lesser degrees of anosmia are common. Olfaction may slightly be recovered, in about 33% of patients, over several days to months. Past 6–12 months, this recovery is only very slight. Additional causes include cranial surgery, chronic meningeal inflammation, and subarachnoid hemorrhage.

Olfactory acuity changes during a woman's menstrual cycle. This may be via the **vomeronasal system**. Acuity can also be disordered during pregnancy.

Anosmia may be related to temporal lobe epilepsy, and more often in these patients who have undergone an anterior temporal lobectomy. There is impaired ability to discriminate odor qualities, and in matching odors with test objects that are observed or touched. Olfaction and taste also diminish because of aging, which is known as *presbyosmia*. Since receptor cell populations are depleted, regional losses cause the neuroepithelium to slowly be replaced with respiratory epithelium. This is normally present in the nasal cavity, and acts in the filtering, humidifying, and warming of incoming air. Neurons of the olfactory bulb may also be reduced because of aging.

An olfactory groove meningioma may involve the olfactory bulb and tract. It can posteriorly extend, involving the optic nerve, and sometimes cause optic atrophy. When combined with papilledema on the opposing side, these conditions form **Foster Kennedy syndrome**. A similar effect may be caused by a large aneurysm of the anterior cerebral or anterior communicating artery. However, these aneurysms usually cause double vision and abnormal pupils. Defects in the sense of smell are linked to lesions of the receptor cells and their axons, or of the olfactory bulbs.

Early neuronal degeneration in the olfactory area of the hippocampus occurs in Alzheimer's disease, Lewy body dementia, and Parkinson's disease. Many patients with other degenerative brain diseases have anosmia or hyposmia. The earliest neuropathologic changes of a large amount of neurodegenerative processes start in olfactory structures. There are theories that many neurodegenerative diseases are related to a pathogen entering the brain via the peripheral olfactory system, starting a reaction inside the brain leading to structural alteration of proteins and other substances. These alterations lead to a degenerative process within the brain.

Olfactory hallucinations

When an odor is reported but there is no stimulus, it is an olfactory hallucination. The perception of an odor by one person that is undetectable by others is called *phantosmia*. This is often due to a temporal lobe seizure, described as an *uncinate fit*. The hallucination is brief, and accompanied or followed by altered consciousness, or other epileptic manifestations. The hallucination becomes a *delusion* if the patient believes it originated from himself or herself. The key to diagnosis is the stereotyped nature of these episodes, meaning that these hallucinations are always about the same odor, for instance, an odor of "burned food." In contrast, combinations of olfactory hallucinations and delusions signify psychiatric illnesses. The patient often complains of many odors that are noxious, and are coming from their bodies, termed *intrinsic hallucinations*. The patient may attribute the odors as coming from others, termed *extrinsic hallucinations*. Both forms have varied intensities, and are extremely persistent phenomena. There may also be concurrent gustatory hallucinations.

Focus on olfactory hallucinations
Olfactory hallucinations are perceived abnormal smells that are usually unpleasant. They are not actually present in a person's physical environment, and may come from many different areas of the olfactory system. If an abnormal smell occurs for less than a few minutes, it is usually from the uncus, in the temporal lobe of the brain. Olfactory hallucinations lasting longer than a few minutes, up to several hours, are usually because of disturbances of the olfactory organ, nerves, or bulb. In rare cases, olfactory hallucinations can be an indication of a serious underlying disorder, so a physician should be consulted if they persist.

Section review

1. What are the various results of limbic system injuries?
2. What is the pathology of pseudobulbar emotional disturbances?
3. What are the classifications of anosmia?

Clinical cases

Clinical case 1

1. Are "phantom" odors are component of psychiatric or neurological diseases?
2. Can olfactory disorders actually enhance the likelihood of developing depression?
3. Are olfactory hallucinations possibly linked to any forms of psychosis?

A 29-year-old woman who had separated from her husband went to her physician because of palpitations, tremor, anxiety, and a preoccupation with personal odor. She took three or more showers per day because she believed that her body had an offensive odor that others would notice. She also admitted to battling depression and even had attempted suicide twice. A psychiatrist had previously treated her with sertraline, and a dermatologist had treated her with botulinum toxin to reduce sweating from her armpits. Family history revealed that her sister was diagnosed and treated for major depressive disorder and borderline personality disorder. All standard blood and urine tests were normal. She was prescribed a higher dose of sertraline, and after 2 months, since her preoccupation with body odor continued, olanzapine was reduced, which was successful.

Answers:

1. Yes, phantom odors occur in psychiatric or neurological diseases, and can also occur in isolation as a single symptom. *Phantosmia* is also referred to as *olfactory hallucination*. It is the perception of an odor that is not actually there, and can be unilateral or bilateral. Phantosmia is most likely to occur in women between the ages of 15 and 30 years, such as this patient.
2. Yes, olfactory abnormalities lead to restrictions in olfactory-related areas, which can affect quality of life and enhance the likelihood of depression. Between 25% and 33% of patients with olfactory disorders exhibit depressive symptoms, and 35% of patients with parosmia or phantosmia have profound depression.
3. Olfactory and other types of hallucinations are a hallmark symptom of schizophrenia, and are commonly observed in other psychotic disorders. However, auditory and visual hallucinations occur much more in psychotic patients.

Clinical case 2

1. Which factors in this case are linked to a metallic taste in the mouth?
2. How can taste disturbances be caused?
3. Aside from age-related changes to taste (and smell), what are other causes of taste loss or alterations?

An 84-year-old woman complained of unintentional weight loss. She also had hypertension, chronic renal insufficiency, history of chronic obstructive pulmonary disease (COPD), and depression. She had quit smoking when she was 64 years old. After recently developing an ear infection, she had been prescribed levofloxacin. Within only 2 days, the patient developed a metallic taste in her mouth, nausea, and vomiting. Though the nausea and vomiting resolved on their own, the metallic taste persisted over 2 weeks after the last dose of levofloxacin. She said that her food tasted like bile, and her severe loss of weight was linked to poor appetite. Review of her depression revealed that it was poorly controlled by fluoxetine. It was replaced with mirtazapine, in adjusted doses for her renal condition. In 1 month, she was seen with complete resolution of the metallic taste, increased appetite, and weight gain.

Answers:

1. This patient's factors related to the metallic taste include history of chronic renal insufficiency and the treatment for hypertension. Taste disturbance is a rare, but documented side effect of levofloxacin.
2. Taste disturbances can be caused by many methods. Drugs can affect peripheral receptors, chemosensory neural pathways, and even the brain. Drugs excreted in the saliva can cause taste disturbances. Drugs that cause mouth dryness cause taste disturbances due to lack of saliva. Drugs can also affect taste receptor cells by passing through the blood, altering taste cell turnover by inhibiting protein formation, inhibiting calcium channel activity, affecting sodium channels, and inhibiting receptor-coupled events. Drugs can also cause glossitis, affecting taste.
3. Other causes of taste loss or alterations include infections, head injuries, dental problems, radiation therapy, chemotherapy, heavy smoking, vitamin deficiencies, Bell's palsy, Sjögren syndrome, Alzheimer's disease, Parkinson's disease, and other neurological conditions.

Key terms

Amygdala	Dysosmia
Anosmia	Encephalitides
Basket cells	Enkephalins
Chemesthesis	Entorhinal cortex
Corticopetal system	Fornix
Diagonal band of Broca	Foster Kennedy syndrome

Gliomatosis cerebri
Hippocampal formation
Hyperosmia
Hypobulic
Hyposmia
Juxtallocortex
Limbic lobe
Limbic system
Locus ceruleus
Mammillothalamic fibers
Monoaminergic
Olfactory agnosia
Olfactory bulb
Olfactory rod
Olfactory trigone
Parasubiculum
Paratenial
Parosmia
Pons
Progressive supranuclear palsy
Presubiculum
Quadrantic
Reuniens
Septum pellucidum
Stratum lacunosum-moleculare
Stratum oriens
Stratum radiatum
Subicular complex
Subiculum
Supracallosal striae
Suprasegmental
Taste buds
Ventral amygdalofugal pathway
Verrucae hippocampi
Vomeronasal system
Wilson disease

Suggested readings

1. Aleman, A.; Laroi, F. *Hallucinations: The Science of Idiosyncratic Perception*. American Psychological Association, 2008.
2. Anderson, P.; Morris, R.; Amaral, D.; Bliss, T.; O'Keefe, J. *The Hippocampus Book. Oxford Neuroscience Series* Oxford University Press, 2006.
3. Arslan, O. E. *Neuroanatomical Basis of Clinical Neurology*, 2nd ed.; CRC Press, 2014.
4. Bevan, T. E. *The Psychobiology of Transsexualism and Transgenderism, Based on Scientific Evidence*. Praeger, 2014.
5. Brewer, G. J. *Wilson's Disease for the Patient and Family*. Xlibris, 2002.
6. Cardinali, D. P. *Autonomic Nervous System: Basic and Clinical Aspects*. Springer, 2017.
7. Doty, R. L. *Handbook of Olfaction and Gustation*, 3rd ed.; Wiley-Blackwell, 2015.
8. Geary, R. T. *The Limbic System: Anatomy, Functions and Disorders*. Nova Biomedical, 2014.
9. Golbe, L. I. *A Clinician's Guide to Progressive Supranuclear Palsy*. Rutgers University Press, 2018.
10. Hamilton, J. *Pseudobulbar Affect (PBA): Information for Caregivers & Patients*. Amazon Digital Services LLC, 2015.
11. Isaacson, R. L. *The Limbic System*. Springer, 2011.
12. Joseph, R. G. *Limbic System: Amygdala, Hypothalamus, Septal Nuclei, Cingulate, Hippocampus: Emotion, Memory, Language, Development, Evolution, Love, Attachment, … Aggression, Dreams, Hallucinations, Amnesia*. Science Publishers, 2017.
13. Jutkiewicz, E. M. *Delta Opioid Receptor Pharmacology and Therapeutic Applications. Handbook of Experimental Pharmacology, Vol. 247*; Springer, 2019.
14. Khouzam, H. R.; Tiu Tan, D.; Gill, T. S. *Handbook of Emergency Psychiatry*. Mosby, 2007.
15. Mori, K. *The Olfactory System: From Odor Molecules to Motivational Behaviors*. Springer, 2014.
16. Rice, R. *Suffering and the Search for Meaning: Contemporary Responses to the Problem of Pain*. IVP Academic, 2014.
17. Salloway, S. P.; Malloy, P. F.; Cummings, J. L. *Neuropsychiatry of Limbic and Subcortical Disorders*. American Psychiatric Association Publishing, 1998.
18. Schwartz, M. F.; Berlin, F. *Sexually Compulsive Behavior: Hypersexuality. Psychiatric Clinics of North America, Vol. 31(4).*; Saunders, 2008.

19. Sterline, P.; Laughlin, S. *Principles of Neural Design*. The MIT Press, 2015.
20. Tibbetts, T. *Identifying and Assessing Students With Emotional Disturbance*. Brookes Publishing, 2013.
21. Toko, K. *Biochemical Sensors: Mimicking Gustatory and Olfactory Senses*. Pan Stanford, 2016.
22. Trainor, P. *Neural Crest and Placodes. Current Topics in Developmental Biology, Book, Vol. 111*; Academic Press, 2015.
23. Whalen, P. J.; Phelps, E. A. *The Human Amygdala*. The Guilford Press, 2009.
24. Zucco, G. M.; Herz, R. S.; Schaal, B. *Olfactory Cognition. Advances in Consciousness Research, Vol. 85*; John Benjamins Publishing Company, 2012.
25. Zufall, F.; Munger, S. D. *Chemosensory Transduction: The Detection of Chemostimuli*. Academic Press, 2016.

CHAPTER 16

Cerebellum

The cerebellum coordinates skilled voluntary movements, posture, gait, muscular tone, and other movements. It may also be involved in modulating emotional states and cognition. Organization of the cerebellum is extremely complex, along with its afferent and efferent connections. The cerebellum's primary role is to assist in the control of conscious movements generated within the cerebral hemispheres.

Anatomy of the cerebellum

The cerebellum makes up only about 10% of the brain mass. However, it has a very large surface, similar to the cerebral cortex. It also contains more than 50% of all brain neurons, about 100 billion in total. It has tiny **granule cells** that are the most abundant neurons in the entire brain. The most distinctive neurons of the cerebellum are extremely large, round **Purkinje cells**. The Purkinje cells are in a single line, and the thick, dendritic planes are parallel to each other. The axons travel to the deep nuclei. Here, they synapse on output neurons that send fibers to the brainstem. The Purkinje cells are the main output neurons of the cerebellar cortex.

The appearance of the cerebellum is similar to that of a cauliflower. Its lobes have rough correspondences to its separate functional areas (see Fig. 16.1). The **anterior lobe** is located anterior to the **primary fissure**. This lobe receives much of its afferent information from the spinal cord, playing a great role in coordinating movements of the trunk and limbs. The **flocculonodular lobe** has three small parts. The **nodulus** is its **vermal** portion, and there are two small **flocculi** on each side, close to the vestibulocochlear nerve. The nodulus is continuous with each flocculus, but this is not easily seen except when the cerebellum is dissected. The flocculonodular lobe receives afferent input from the vestibular system. It helps to control eye movements and make postural adjustments in relation to gravity. Deep fissures separate the lobes. All parts of the cerebellum posterior to the primary fissure, except the flocculonodular lobe, make up the **posterior lobe**, which is the largest lobe in size. This lobe receives most of the afferent input from the cerebral cortex, via relays in the pontine nuclei, and transmissions through the middle cerebellar peduncle. This lobe is vital for the coordination of voluntary movements.

The **paramedian fissures** are shallow in the anterior cerebellum, but deeper posteriorly, and separate the vermis from the cerebellar hemispheres. The anterior and posterior vermis and hemisphere are subdivided into lobules named because of their shapes,

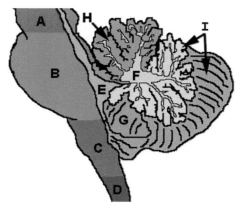

Fig. 16.1 (A-D) Anterior, dorsal, posterior, and ventral views of the human cerebellum. (E) A sagittal section of the human cerebellum. (F-G) Dorsal and ventral views of the cerebellum of *Lemur albifrons*, Bolk's (1906) prototype for his ground plan of the mammalian cerebellum. Two loops are present in the folial chain of the hemisphere: (1) as the ansiform lobule, (2) as the paraflocculus. The course of the folial chains of the vermis and hemisphere in (A-D) and (F-G) is indicated with *red lines*. (H) Bolk's stick diagram of the folial chains of the vermis and hemisphere. Key and abbreviations: 1, 2, ansiform and parafloccular loops of the folial chain of the hemisphere; *Ce*, central lobule; *Cu*, culmen; *De*, declive; *FA*, fastigium; *F/T*, folium and tuber; *Icp*, inferior cerebellar peduncle; *Li*, lingula; *Mcp*, middle cerebellar peduncle; *Nod*, nodulus; *PFLD*, dorsal paraflocculus; *PFLV*, ventral paraflocculus; *Py*, pyramis; *Scp*, superior cerebellar peduncle; *Sl*, simplex (posterior quadrangular) lobule; *Uv*, uvula; *Vma*, anterior (superior) medullary velum; *Vmp*, posterior medullary velum.

positions, or similarities to anatomical structures of other body parts. Each subdivision of the cerebellum consists of a longitudinal cortical zone, a vestibular or cerebellar target nucleus, and a supportive **olivocerebellar** climbing fiber system. The longitudinal cortical zones, except for their connections, have distinct immunohistochemical properties. Mossy fibers and climbing fibers are afferent pathways.

The cerebellum is located dorsal to the pons, medulla oblongata, and fourth ventricle. It is a protruding structure, underneath the occipital lobes of the cerebral hemispheres. It is separated from these hemispheres by the transverse cerebral fissure, or the **tentorium cerebelli**. The cerebellum is bilaterally symmetrical and is the largest portion of the hindbrain. The two-cerebellar **hemispheres** are about the same size as apples and are medially connected by the **vermis**, which resembles a worm. It is extremely convoluted on its service, with fine gyri known as **folia**, which resemble leaves, and are transversely oriented. The folia are less prominent than the folds of the cerebral cortex.

The **cerebellar peduncles** are three paired fiber tracts connecting the cerebellum to the brainstem and other parts of the brain. Nearly all fibers that enter or leave the cerebellum are **ipsilateral**, meaning *to and from the same side of the body*. These are further explained as follows:

- **Superior cerebellar peduncles**—connect the cerebellum and midbrain; they carry information from neurons in the deep cerebellar nuclei to the cerebral motor cortex, through the thalamic relays; there are no direct connections to the cerebral cortex, which is the same situation with the basal nuclei; these peduncles contain all the efferent fibers of the dentate, emboliform, and **globose nuclei** along with a small fascicle from the fastigial nucleus; fibers decussate in the caudal mesencephalon, then synapse in the contralateral red nucleus and thalamus; the ventral spinocerebellar tract reaches the upper pontine tegmentum and loops around the entrance of the trigeminal nerve, joining these peduncles, then uniting with spinocerebellar fibers that enter via the restiform body.
- **Middle cerebellar peduncles**—carry information in one direction, from the pons to the cerebellum; they "inform" the cerebellum of voluntary motor activities that are initiated by the motor cortex; they accomplish this through relays in the pontine nuclei; these are by far the largest of the three sets of peduncles, passing obliquely from the basal pons to the cerebellum; they contain a large pontocerebellar mossy fiber pathway, made almost totally of fibers arising from the contralateral basal pontine nuclei, with lesser amounts from nuclei of the pontine tegmentum. The middle peduncle is the major input to the cerebellum.
- **Inferior cerebellar peduncles**—connect the medulla oblongata and cerebellum; they carry sensory information to the cerebellum from the body's muscle proprioceptors, and from the vestibular nuclei of the brainstem; this assists with equilibrium and balance; located medial to the middle peduncles, consisting of an outer, compacted fiber tract called the *restiform (afferent) body,* and a medial, *juxtarestiform (mostly efferent) body*; the restiform body receives spinocerebellar fibers along with trigeminocerebellar, cuneocerebellar, reticulocerebellar, and olivocerebellar tracts from the medulla; the juxtarestiform body is almost totally made up of efferent Purkinje cell axons reaching the vestibular nuclei, and uncrossed fibers from the fastigial nucleus; the juxtarestiform body also has primary afferent mossy fibers from the vestibular nerve, and secondary afferent fibers from the vestibular nuclei; crossed fibers from the fastigial nucleus pass dorsally to the superior peduncles as the *uncinate tract*; they enter the brainstem at the border of the restiform and juxtarestiform bodies.

Cerebellar nuclei

The four-cerebellar nuclei include, from most medial to most lateral, the *fastigial, emboliform, globose,* and *dentate* nuclei. The emboliform nuclei are also called the *anterior interposed nuclei*, and the globose nuclei are also called the *posterior interposed nuclei*. The cerebellar nuclei form two groups that are interconnected. The *rostrolateral group* is made up of the emboliform and dentate nuclei. The *caudomedial group* is made up of the fastigial and globose nuclei.

From the flocculus to the nodulus, in the roof of the fourth ventricle, there are small, cholinergic neurons that extend into the spaces between the nuclei. These cells make up the *basal interstitial nucleus*, which has unknown connections. The most lateral, and the largest group is the dentate nucleus, which has a "crumpled" shape. From its hilus emerges the primary efferent cerebellar pathway, known as the *brachium conjunctivum*. The dentate nucleus has convolutions that are narrow in the rostromedial portion, but extremely wider in the ventrocaudal portion. In the cerebellar nuclei, there are cells of many different sizes. The main output of the nuclei is provided by *glutamatergic relay neurons*. Small GABAergic neurons provide innervation to the contralateral inferior olive. There are GABAergic as well as glycinergic interneurons. It is believed that every cell type receives inhibitory input from the Purkinje cells, as well excitatory input from the mossy and climbing fiber collaterals.

Mossy and climbing fibers

Mossy fibers of the cerebellum originate from different spinal tracts, including the Clarke nucleus (layer VII). Mossy fibers terminate in granular cells. They have myelinated axons, ending upon claw-like dendrites of granule cells, and Golgi cells. Granule cells develop an ascending axon that splits in the molecular layer, into two long **parallel fibers** that run toward the long axis of the folia. Parallel fibers end on the spines of the spiny branches of the dendritic tree of Purkinje cells, and the interneuron dendrites contacted along their courses. Mossy fibers, like climbing fibers, are highly branched in the cerebellar white matter. The parent fibers enter the cerebellum laterally. They have a transverse course, decussating within the cerebellar commissure. Along the way, they emit thin collaterals, which enter the lobule white matter and end in many, longitudinally oriented, symmetrically distributed groups of mossy fiber terminals, within the granular layer. Climbing fibers originating from the inferior olivary nucleus in the medulla provide very powerful excitatory input to the proximal dendrites and cell soma. They use aspartate to accomplish this input.

> **Focus on the roles of the cerebellum**
> The cerebellum was considered for many years to be responsible for balance and all motor functions. Today, we know that it plays an important role in other cognitive functions, including attention, and how we focus and capture images. Breathing, sleeping, blood pressure, and heart rate are also controlled by the cerebellum, because it sends information to the spinal cord and nearby sections as well.

Section review
1. What are the characteristics of the cerebellum's structure?
2. What is the difference between the anterior and posterior paramedian fissures?
3. What are the three types of cerebellar peduncles?

Cerebellar cortex

The cerebellum's outer cortex, like the cerebrum, is thin and made up of gray matter, with internal white matter, and small, deep, and paired masses of gray matter. The cerebellar cortex has three cell layers: *molecular* (basket cells, Golgi cells, and stellate cells); *Purkinje layer* (Purkinje cell bodies); and granular layers.

As a general rule, the medial areas of the cerebellum regulate motor activities of the trunk and axial muscles. The intermediate portions of each hemisphere regulate the distal areas of the limbs and skilled movements. The most lateral parts of each hemisphere manage information from the association areas of the cerebral cortex. They are believed to assist in planning movements, but not in executive movements. The flocculonodular lobes receive input from the equilibrium centers of the inner ears, adjusting posture, and maintaining balance.

Even though the cerebellum is much smaller than the cerebrum, if it were unfolded, it would be about 50% of the size of the cerebral cortex. Most cerebellar neurons are small granule cells. The cerebellar cortex is also different, in that it has a smaller number of different types of cells that are interconnected in an extremely stereotyped manner. There is an organized geometric order to the folia of the cerebellar cortex. The monolayer of large Purkinje cells separates a layer of small granule cells from the superficial molecular layer, which has few cells in comparison. The Purkinje layer has the large somata of the Purkinje cells, forming a pear shape, and the smaller Bergmann glia somata. The granular layer is made up of the somata of granule cells and initial segments of their axons, along with granule cell dendrites, afferent mossy fiber branching terminal axons, climbing fibers, and the Golgi neuronal somata, basal dendrites, and complicated axonal structures. The Purkinje cells are arranged in one layer, between the molecular and granular layers. The dendritic trees are flat, oriented perpendicular to the parallel fibers, within a plane that is transverse to the long axes of the folia. Large primary dendrites emerge from the outer pole of each Purkinje cell.

Cortical interneurons

Cerebellar cortical interneurons are divided into those of the molecular layer (stellate and basket cells) and those of the granular layer (Golgi cells). They are all inhibitory. The molecular layer interneurons use GABA as their neurotransmitter. The majority of Golgi cells are **glycinergic** (inhibitory). The stellate cells of the upper molecular layer have axons that end on Purkinje cell dendrites. The basket cells are in the deep molecular layer, with axons ending on many Purkinje cells with baskets that surround their somata and end in a feather-like plume around the initial axon. Their dendrites and axons are sagittal-oriented. The granular and molecular layers both contain Golgi cell dendrites. Their axonal plexus is in the granular layer, where it ends on the granule cell dendrites, also having the greatest dimension within the sagittal plane. Golgi cells receive innervation from collaterals of mossy fibers and Purkinje cell axons. In the molecular layer, dendrites of interneurons are contacted by parallel fibers. There are no apparent synaptic contacts between climbing fibers and dendrites, or cell bodies of cerebellar interneurons, within the molecular of granular layers. Interneurons of the molecular layer can, however, become activated by excess glutamate from the climbing fibers. Therefore, the Golgi cells give feedback inhibition to the granule cells. Molecular layer interneurons provide feedforward inhibition to the Purkinje cells. There is an electrotonic coupling between stellate and Golgi cells. This may be restricted to the sagittal planes.

The other types of interneurons are the cigar-shaped *Lugaro cells*, at the level of the Purkinje cells, and the *monopolar brush cells*. The Lugaro cells are glycinergic, innervating the stellate and basket cells. They have a long axon that is transversely oriented and ends on the Purkinje cells, receiving a strong input from an extracerebellar serotoninergic system. The monopolar brush cells are excitatory and mostly found in the cerebellum's vestibular-dominated areas. Here, they are a *booster system* for the vestibular mossy fiber input. The mossy fibers end with extremely large synapses upon the base, or *brush*, of these cells. The axons end as mossy fibers on granule cells.

Cerebellar lobules

There are 10 lobules of the cerebellum, which are identified by using Roman numerals for the vermis, and the prefix "H" for the hemisphere. These divisions are explained in Table 16.1.

Corticonuclear and olivocerebellar projections

Each modular area of the cerebellum has one or more longitudinal Purkinje cell zones projecting to one of the vestibular or cerebellar nuclei. Some of these zones are only present in certain lobules, while others are present throughout the rostrocaudal portion of the cerebellum. Climbing fibers from part of the contralateral inferior olive terminate upon Purkinje cells of a certain zone. They also send collateral innervations to related deep cerebellar nuclei. The primarily crossed nucleo-olivary pathway originating from small GABAergic neurons of the cerebellar nuclei returns this innervation. The modules can be seen based on their Purkinje cell axons and climbing fiber afferents, which are collected into compartments within the cerebellar white matter. Borders between the compartments, or modules, become visible once they are stained for acetylcholine esterase (AChE).

Table 16.1 Cerebellar lobules.

Lobule	Explanation
(H) I to V make up the anterior lobe	I—the lingual, conjoined with the superior medullary velum
VI (declive) and HVI (posterior quadrangular lobule)	Also known as Bolk's simplex lobule
VIIA (folium) and VIIB (tuber vermis) are separated, behind the primary fissure, by the deep paramedian fissure from HVIIA (superior semilunar lobule), inferior semilunar lobule, and gracile lobule	The inferior semilunar lobule and gracile lobule together correspond to the HVIIA. The gracile lobule corresponds to the rostral portion of Bolk's paramedian lobule, with the caudal portion formed by the biventral lobule (HVIII), the hemisphere from the pyramis (VIII)
Superior and inferior semilunary lobules	Correspond to the crus I and II of Bolk's ansiform lobule. Their folia spread out from the deep horizontal fissure representing the intercrural fissure
VIII (pyramis) is continuous with HVIII (biventral lobule) laterally	The biventral lobule corresponds to the caudal portion of Bolk's paramedian lobule
HIX (tonsil)	Its folial loop is directed medially
HX (flocculus)—appears as a doubled folial rosette. X is called the nodulus	The dorsal leaf is called the accessory paraflocculus of Henle. The ventral leaf is the true flocculus. Between the flocculus and nodulus, the cortex is absent. Tissue is stretched, forming the inferior medullary velum

The modular organization develops early, a long time before any transverse fissures appear. The Purkinje cells are created in the ventricular matrix of the cerebellar anlage. They move to the meningeal surface, forming many mediolaterally arranged clusters. Later, when the cerebellar surface increases and millions of granule cells are proliferating in the external granular layer, the Purkinje cell clusters lengthen to become Purkinje cell zones. The cells move to a monolayer, with the early borders between clusters becoming indistinct. In humans, the most lateral cluster is much larger than the rest, and clearly related to the anlage of the dentate nucleus. It develops into the most lateral Purkinje cell zone (called D2) and makes up the large cerebellar hemisphere's size.

Cortico-olivary system

Connections of the cerebellar nuclei and brainstem, along with the thalamus and spinal cord, are interrelated with the effects of the cerebellar modules. Of the vermis, specific zones project to certain areas, as follows:
- **A zone**—to the fastigial nucleus, which gives rise to the uncinate tract and decussates in the cerebellar commissure; it is hooked around the brachium conjunctivum, then

distributed to the vestibular nuclei and the medullary and pontine reticular formation; one branch of the uncinate tract ascends to the ipsilateral midbrain and thalamus; projections to the cerebral cortex are bilateral, since crossed ascending fibers of the uncinate fasciculus later recross within the thalamus; the uncrossed and direct fastigiobulbar tract passes, along the lateral margin of the fourth ventricle; it reaches the vestibular nuclei and reticular formation symmetrically, mirroring the uncinate tract; the direct fastigiobulbar tract is inhibitory and glycinergic; small GABAergic neurons form a nucleo-olivary pathway that ends in the contralateral caudal medial accessory olive.
   - The fastigial nucleus' caudal pole receives Purkinje cell afferents from lobule VII (the folium and tuber vermis), which is also called the *visual vermis* since it is involved in eye movements; projections of the oculomotor region of the fastigial nucleus are totally crossed, and end in the pontine paramedian reticular formation, called the *horizontal gaze center*, the superior colliculus, the rostral interstitial nucleus of the medial longitudinal fasciculus, called the *vertical gaze center,* and in the thalamic nuclei, which may include frontal and parietal eye fields as targets; the fastigial nucleus affects visceromotor systems, through projections of the vestibular nuclei, and connections with the catecholaminergic nuclei of the hypothalamus and brainstem.
- **X zone**—interstitial cell groups; projections of these groups are located between the fastigial and posterior interposed nuclei, the target nucleus of the X zone; not fully understood in humans, but in lower mammals, they provide collaterals to the superior colliculus, thalamus, and spinal cord.
- **B zone**—lateral vestibular nucleus, also called *Deiters' nucleus*; it is similar to the cerebellar nuclei and does not receive a primary input from the labyrinth; it does receive a collateral innervation from the climbing fibers that innervate the B zone, and gives rise to the lateral vestibulospinal tract; the nucleo-olivary pathway targets the caudal dorsal accessory olive.

The cerebellar vermal zones can affect neurotransmission in the reticulospinal and vestibulospinal systems. They bilaterally control vestibular and postural reflexes of the proximal and axial muscles, and in the oculomotor brainstem centers. In certain areas, the oculomotor and skeletomotor functions are located. Oculomotor functions are within lobule VII. Skeletomotor functions are in the anterior vermis and posterior lobule VIII (the pyramis). The X and B zones are restricted to these lobules. The nodulus (lobule X) is most caudal and belongs to the vestibulocerebellum.

The C1, C3, and Y zones target the anterior interposed (emboliform) nucleus. This nucleus has a similarly detailed organization to these Purkinje cell zones. They have the same climbing fiber input from a specific body area, and project to a common set of neurons. Ascending axons from the anterior interposed nucleus penetrate the brachium conjunctivum, which decussates at the edge of the pons and mesencephalon. Its ascending branch enters and surrounds the magnocellular red nucleus. It continues to the thalamus, where the anterior interposed nucleus connects with the contralateral primary motor cortex. The brachium conjunctivum's descending branch ends in the reticular tegmental nucleus of the pons. This system is also somatotopically organized and includes the magnocellular red nucleus, primary motor cortex, and their efferent tracts. A nucleo-olivary pathway from the anterior interposed nucleus ends in the rostral dorsal accessory olive.

The corticospinal (pyramidal) and rubrospinal tracts emerge from the motor cortex and magnocellular red nucleus. Both tracts cross the midline. The corticospinal tract crosses at the bulbospinal junction, while the rubrospinal tract crosses at its level of origin within the midbrain. The corticospinal tract provides collateral innervation to the magnocellular red nucleus. Both tracts affect distal limb movements. Climbing fibers that innervate the C1, C3, and Y zones, along with the anterior interposed nucleus originate from the rostral dorsal accessory olive. This receives a somatotopic cutaneous input, mostly via the dorsal column and trigeminal nuclei. It has a clear cutaneous map of the entire contralateral body surface. The tracts provide collateral innervation to the dorsal column nuclei.

Connections of the globose and dentate nuclei are similarly arranged, ascending and decussating in the brachium conjunctivum. They end in a group of nuclei located at the mesodiencephalic junction, which includes the parvocellular red nucleus and the **nucleus of Darkschewitsch** in the central gray matter, and also the thalamic nuclei projecting to motor, premotor, prefrontal, and posterior parietal cortical areas, as well as frontal and parietal eye fields. At the mesodiencephalic junction, the nuclei give rise to ipsilaterally descending tegmental tracts ending in the inferior olive to form reciprocal loops of unknown function.

Afferent mossy fibers

Mossy fiber systems originate from many sites within the brainstem and spinal cord. The major system is the pontocerebellar pathway. These systems have some common features. Individual mossy fibers are bilaterally distributed. They give off collaterals at certain mediolateral positions, ending in longitudinal groupings of mossy fiber *rosettes*. The systems end as multiple and bilaterally distributed bands of terminals. The bands are not continuous. They are usually restricted to the apices or bases of folia. Exteroceptive components terminate superficially. This differs from proprioceptive systems, which end in the folia bases. Mossy fiber aggregates are not as distinctive as climbing fiber zones. They often merge in the bases of fissures. The aggregates of various mossy fiber systems may overlap or interlock, but are not fully understood.

Spinocerebellar, reticulocerebellar, cuneocerebellar, and trigeminocerebellar tract terminations only occur in the anterior and posterior cerebellar motor regions. These include the anterior lobe, simplex lobule (VI and HVI), lobule VIII, and the paramedian lobule (combining the gracile HVIIB and biventral HVIII lobules). The lobules receive primary and secondary vestibulocerebellar inputs along with pontocerebellar mossy fibers that relay information for cortical motor areas. Many mossy fiber systems end somatotopically. A similar organization is present in the C1, C3, and Y climbing fiber zones restricted to the hemisphere of the same lobules.

Corticopontocerebellar projection

The largest source of fibers projecting to the pontine nuclei is the cerebral cortex. Fibers travel through the cerebral peduncle. The fibers from the frontal lobe are found in the medial portion of the peduncle. Corticonuclear and corticospinal fibers are within its central area. Fibers from the occipital, parietal, and temporal lobes are within its lateral portion. The cerebral peduncle fibers have a mediolateral sequence, which is basically consistent when they terminate in the pontine nuclei. The frontal eye fields and prefronto-pontine fibers project medially as well as rostrally. The motor and premotor projections end caudally and centrally. The occipital, parietal, and temporal fibers end in the lateral pontine nuclei. There is a somatotopic organization of the motor and premotor projections. This means that the "face" is represented rostrally while the "legs" are represented caudally within the nuclei. A prefrontal projection from the dorsal prefrontal cortex exists. A large number of corticopontine fibers are axon collaterals projecting to targets within the brainstem and spinal cord. There is nearly complete crossing of the pontocerebellar projection. Pontine nuclei fibers reach the cerebellum, through the middle cerebellar peduncle. They end in the entire cerebellar cortex, except for lobule X, the nodulus. There is visual cortical mossy fiber input in the paraflocculus (tonsil, HIX).

Vestibulo-ocular reflex

The vestibulo-ocular reflex stabilizes the position of the retinas, during head movements, via rotation of the eyeballs in the opposite direction. It is an open reflex. Its function is regulated by long-term adaptation of the reflex by the flocculus. Semicircular canal systems are present in the flocculus. The vestibulo-ocular reflex has various different components. One of these connects the horizontal semicircular canal, by oculomotor neurons within the vestibular nuclei, to the oculogyric muscles that move the eyes in a plane that is colinear with the lateral canal's plane. The anterior semicircular canal affects the ipsilateral superior oblique, as well as the contralateral inferior oblique muscles moving the eye in its plane.

There are five Purkinje cell zones in the follicular cortex and nearby ventral paraflocculus. Two pairs of zones are in its lateral portion, except form the medial C2 zone. Zones

F1 and F3 are connected to the oculomotor neurons in the vestibular nuclei. They subserve the anterior canal vestibulo-ocular reflex. The F2 and F4 zones connect to the oculomotor neurons of the horizontal canal vestibulo-ocular reflex. The flocculus and ventral paraflocculus receive input from the vestibular mossy fibers. These relay efferent copies of output of the vestibulo-oculomotor neurons. They additionally receive climbing fiber input, and signal a retinal slip when retinal stabilization by the vestibulo-ocular reflex is not complete. Two neuronal groups in the mesencephalon receive retinal slip. Within the horizontal plane, this is relayed by the nucleus of the optic tract. The nucleus is within the pre-tectum, and receives fibers from the contralateral optic nerve, through the optic tract, which project to the dorsal inferior olive cap. This is located dorsomedial to the caudal, medial accessory olive. The dorsal cap, to the F2 and F4 zones, provides climbing fibers. Retinal slip in the anterior plane canal occurs via lateral and medial nuclei of the accessory system. These nuclei are part of the nuclei found on the periphery of the rostral mesencephalon. This receives optic nerve fibers from part of the optic nerve, called the *accessory optic tract*.

These nuclei are projected to the ventrolateral continuance of the inferior olive, just rostral to the dorsal cap. This ventrolateral outgrowth provides innervation to the F1 and F3 zones. Continual, simultaneous activation of the vestibular mossy fiber-parallel fiber input and climbing fibers, which relay retinal slip, cause plastic changes in Purkinje cell output. These compensate for the retinal slip. Compensation of retinal slip, within all planes, occurs from combinations of the horizontal and anterior canal systems. Climbing fibers carry "error signals" that are used as part of cerebellar learning.

Functional divisions of the cerebellum

The massive activity of mossy fibers overwhelms the cerebellar activities of climbing fibers and Purkinje cells. There are several functional networks within the cerebral cortex. Connections between the cerebrum and cerebellum may be indirect, such as via cortical association systems or brainstem nuclei that exclude the pontine nuclei. The networks mirror each other in distribution within the anterior and posterior cerebellum. A central position contains the default mode network. These brain regions are active when the person is not focused on outside stimuli.

Focus on the Purkinje cells
Purkinje cells are present in the cerebellum and also in the heart. It is now known that they have some type of link with an *autistic spectrum disorder*. The Purkinje cells receive up to 10 times more connections than all other types of neurons. They have the ability to turn off the functioning of other cells that are external to the cerebellum.

Section review
1. What are the three cell layers of the cerebellar cortex?
2. What neurotransmitter does the molecular layer of the cortical interneurons use?
3. What are the names and functions of the zones of the cortico-olivary system?

Major functions of the cerebellum

The cerebellum processes inputs from the cerebral motor cortex, different brainstem nuclei, and sensory receptors. It regulates skeletal muscle contractions needed for smooth and coordinated movements, such as driving, using a computer, or playing a musical instrument. The activities of the cerebellum are subconscious. Basically, the cerebellum finely adjusts motor activities in four ways, as follows:

- Cerebral cortex motor areas, through brainstem relay nuclei, inform the cerebellum of voluntary muscle contractions that are about to happen.
- Simultaneously, the cerebellum receives information from the body's proprioceptors, concerning muscle and tendon tension and joint positions; it also receives information from the equilibrium and visual pathways, allowing for evaluation of body momentum and position.
- The cerebellar cortex determines how to coordinate muscle contraction force, direction, and extent; this ensures correct muscular action, maintenance of posture, and coordinated, smooth movements.
- Through the superior peduncles, the cerebellum sends instructions for movement coordination to the cerebral motor cortex; the cerebellar fibers also send information to the brainstem nuclei, which then regulate the spinal cord's motor neurons.

The cerebellum continuously monitors body movements with brain information, sending out control messages to make fine corrections. Therefore, injury to the cerebellum causes a loss of muscle tone and movement that are clumsy and uncertain, known as **ataxia**. When this is severe, the patient cannot sit or stand without assistance. Drugs such as alcohol also have a temporary effect. The cerebellum is also involved in emotions, language, and thinking. It is believed to analyze these events in comparison to the brain's "intentions", and make adjustments as needed. There is still a lot that is not fully understood about the cerebellum and how it affects nonmotor functions.

Clinical considerations

The cerebellum interacts with the basal ganglia, thalamus, and corticospinal tracts in order to produce voluntary movements. The corticospinal tracts pass through the medullary pyramids. They connect with the cerebral cortex, to the lower brainstem and spinal motor centers. The extrapyramidal system is formed by the basal ganglia, including the caudate nucleus, globus pallidus, putamen, subthalamic nucleus, and substantia nigra.

The basal ganglia are deep within the forebrain, directing outputs, primarily rostrally, through the thalamus and to the cerebral cortex. The majority of neural lesions causing movement disorders occur in the extrapyramidal system. This means that movement disorders are also called *extrapyramidal disorders*.

Decreased or slow purposeful movements are referred to as *hypokinesia*. Those with excessive voluntary or abnormal involuntary movements are referred to as **hyperkinesia**. Cerebellar disorders that impair gait and movement are classified separately. Most hypokinetic disorders are **Parkinsonian** in nature. These involve slow, decreased movements, muscular rigidity, postural instability, and resting tremor.

Ataxia

Ataxia is the hallmark of cerebellar disease. As mentioned before, cerebellar lesions cause ipsilateral or same side ataxia. Finger-to-nose and heel-to-shin tests can be used to reveal dysmetria. Various lesions within the CNS and PNS can cause ataxic gait. However, the cerebellar ataxia is a wide base gait, with a short stride. The best example of ataxia related to the cerebellum can be observed in individuals who have excessive alcohol. It is important to identify the onset of the symptoms. Causes of acute cerebellar ataxia include acute cerebellar stroke. Vitamin deficiencies such as B1 can lead to subacute ataxia. Hereditary forms of cerebellar ataxia are examples of chronic ataxias.

Dystonias

Dystonias are sustained and involuntary muscle contractions that often distort body posture. Classifications of dystonias include generalized, focal, and segmental. Generalized dystonia, or *dystonia musculorum deformans,* is rare and progressive. It involves movements resulting in sustained and usually bizarre postures. This is often hereditarily linked, usually as an autosomal dominant disorder with partial penetrance. A patient's sibling, who is asymptomatic, often develops a crude form of the disorder.

Pathophysiology

Dystonias can be primary (idiopathic) or secondary to degenerative or metabolic CNS disorders. These include Hallervorden-Spatz disease, Wilson's disease, multiple sclerosis, various lipidoses, cerebral palsy, stroke, and brain hypoxia. Drugs can also be causative, such as phenothiazines, butyrophenones, antiemetics, and thioxanthenes. Generalized dystonia is usually caused by the DYT1 gene.

Clinical manifestations

Symptoms of dystonias usually start in childhood. There are inversion and plantar fixation of the foot during walking. The trunk or one leg may only be affected, though the condition can involve the entire body. In the most severe form, the patient may be twisted into a grotesque fixed posture, and eventually confined to a wheelchair. When symptoms start in adulthood, they usually only affect the face or arms, with mental function usually being preserved. *Focal dystonias* affect just one body part, and usually begin in a person's 30s or 40s, with women affect more often. Initial spasms may be periodic and occur at random or in times of stress. Certain movements of an affected body part, disappearing during rest, trigger them. Spasms may progress over days, weeks, or many years. They can be triggered by movements of unaffected body parts and can continue while resting. Over time, the affected body part remains distorted, sometimes painfully, causing severe disability. The symptoms are varied based on muscles that are involved. *Occupational dystonia* involves focal dystonic spasms due to performed skilled acts. Examples include writer's cramp, typist's cramp, and golfers' "yips." The voice becomes strained, hoarse, or creaky because of abnormal involuntary contractions of the laryngeal muscles. *Torticollis*, often a genetic condition or caused by certain medications, starts with a sensation of pulling, then sustained torsion and deviation of the head and neck. Early on, the patient can voluntarily overcome it. There are sensory or tactile tricks that can stop the spasms, including touching the face on the side that is contralateral to the deviation. *Segmental dystonias* affect two or more continuous body parts. In *Meige's disease*, also called *blepharospasm-oromandibular dystonia*, there is involuntary blinking, grimacing, and jaw grinding. This usually starts in later middle age and can resemble buccal, lingual, and facial movements seen in tardive dyskinesia.

Diagnosis

Diagnoses of dystonias are clinical, usually by a neurologist. Proper diagnosis is based upon patient and family history, and complete physical and neurological examinations. Laboratory tests, imaging studies, and genetic testing may be required. It is common for dystonias to be misdiagnosed or even to remain undiagnosed when the symptoms are mild. Important factors that are assessed regarding diagnosis include the age at which symptoms began, the areas of the body affected, any possible causes, and whether there are additional neurological signs and symptoms.

Treatment

Treatment for generalized dystonia involves anticholinergics, muscle relaxants, or both. When this condition is severe or does not respond to drugs, it may require deep brain stimulation of the globus pallidus interna, which is a surgical intervention. Focal dystonia, segmental dystonia, or generalized dystonia severely affecting certain body parts requires treatment with botulinum toxin type A, injected into the affected muscles. This weakens muscular contractions, but does not change the abnormal neural stimulus. Toxin injection is extremely effective for blepharospasm and torticollis, with varied dosages. Treatments are repeated every 3–6 months.

Dysarthria

Dysarthria refers to a speech disorder characterized by poor articulation, phonation, and sometimes, respiration. The patient has speech that is slurred, slow, and difficult. Dysarthrias are characterized by weakness and often, the abnormal muscle tone of the speech musculature, which moves the lips and tongue. There are several types of dysarthria:
- **Flaccid dysarthria**—from damage of the cranial nerves or regions of the brainstem and midbrain.
- **Spastic dysarthria**—from damage to the motor regions in the cortex, on both sides of the brain.
- **Ataxic dysarthria**—from damage to pathways connecting the cerebellum with other brain regions.
- **Hypokinetic dysarthria**—from Parkinson's disease.
- **Hyperkinetic dysarthria**—from damage to the basal ganglia.

Dysarthria is related to aphasia, but unlike aphasia, it is not necessarily a language disorder. In dysarthria, the patient is still able to understand speech, find the right words to use, and use correct grammar. Instead, the problems are with the motor execution of language, such as the movement of the articulators. Sometimes, other aspects of language production are affected, such as working memory. *Tardive dystonia* is another form and is related to certain medications.

Pathophysiology
Dysarthria is caused by damage to brain areas important for the more motor aspects of speech, as opposed to the linguistic aspects. It may also be caused by many different medications, sometimes after only one exposure. Tardive dystonia may be caused by prolonged exposure to levodopa and various neuroleptics.

Clinical manifestations
Symptoms of dystonias may include intermittent spasmodic or sustained involuntary contractions of the muscles of the face, neck, trunk, extremities, and pelvis. Symptoms are usually transient, and may also include foot cramps, functional leg abnormalities, uncontrolled blinking, difficulty in speaking, and involuntary pulling of the neck muscles. *Cervical dystonia* involves twisting, pulling forward, pulling backward, or pulling sideways of the head and neck, sometimes with stiffness and pain. Aside from blinking, *blepharospasm* may involve photophobia, eye irritation, and uncontrolled closing of the eyes. *Dopa-responsive dystonia* mostly affects the legs in patients between 5 and 30 years of age, causing a stiff and unusual gait. The sole of the foot is either bent upwards, or the foot is turned outwards at the ankle.

Hemifacial spasm involves facial muscle spasms on one side that are more prominent during mental stress or fatigue. *Laryngeal dystonia* may cause the voice to be very quiet, breathy, or to sound "strangled." *Oromandibular dystonia* causes the mouth to be pulled upwards and outwards, and sometimes, problems swallowing. *Writer's cramp* involves uncontrollable cramps and movements of the arm and wrist and is similar in symptoms to *musician's cramp, typist's cramp*, and *golfer's cramp. Generalized dystonia* involves muscle spasms, an abnormal twisted posture, a limb or foot turned inwards, and sudden jerking of certain body parts. *Paroxysmal dystonia* involves sudden muscle spasms or abnormal body movements that happen only during times of mental stress, fatigue, alcohol or coffee consumption, and when the individual makes a sudden body movement.

Diagnosis

Diagnosis of dystonias is based on physical examination, patient and family history, blood and urine testing, MRI, assessment of current medications, and genetic testing. Blood and urine tests determine if there are any toxins or infections, and assess organ function. An MRI is used to rule out neurological conditions. Genetic testing evaluates gene mutations and rules out conditions such as Huntington's disease.

Treatment

Treatment of dystonias includes medications such as diphenhydramine, levodopa, botulinum toxin, anticholinergics, and muscle relaxants. Physical therapy, splints, braces, and posture-improving techniques may be helpful. If surgery is indicated, it may involve deep brain stimulation.

Pendular knee jerk

A *pendular knee jerk* is an irregular knee-jerk reflex related to a cerebellar lesion. The affected leg keeps moving regularly after an original reflex occurs. The clinical definition of pendular knee jerk is when there are more than three oscillations equidistant from the neutral position after a knee-jerk reflex is initiated. A cerebellar injury causes the patient to have a knee jerk that swings forward and backwards several times. The reflex is mediated mostly by the L4 nerve root but also by the L3 nerve root. The pendular knee jerk is characterized by marked hypotonia in relation to a cerebellar lesion. After a gentle tap to the patellar tendon, the leg keeps swinging after the knee-jerk reflex more than three times. The pendular reflexes are not quick and involve less damping of limb movement usually seen when a deep tendon reflex is initiated. Diagnosis of pendular knee jerk is very simple, based on the tapping of the patellar tendon and the resultant swinging of the leg for more than three times, with slow movements occurring. Surgical treatment is based on the underlying cause, such as a tumor or other type of cerebellar lesion.

Focus on dysarthria
All forms of dysarthria affect how we say consonants such as the letters "T," "P," and "K," causing the speech to sound slurred. In extremely severe cases, even vowels such as "A," "E," and "U" can be distorted. The extent of slurred speech depends on the amount of neurological damage. Hypernasality is often present, along with problems in respiration, phonation, and resonance.

Section review
1. What is the pathophysiology of dystonia?
2. What are the treatments for dysarthria?
3. What is a pendular knee-jerk reflex?

Clinical cases

Clinical case 1
1. What other viral infection may be linked to this condition?
2. What are other causes of this patient's condition?
3. Which symptoms are common in this disorder, and what is the likely prognosis?

A 5-year-old boy was taken to the hospital because of the rapid onset of ataxia. His parents told the attending physician that he had chickenpox recently, but had fully recovered. The child suddenly became clumsy and imbalanced. There were no seizures present, but there were diminished superficial and deep reflexes. Tests were conducted to rule out meningitis and intracranial lesions. Eventually, the boy was diagnosed with acute cerebellar ataxia.

Answers:
1. Viral infections, besides chickenpox, that may be linked to acute cerebellar ataxia include Coxsackie, Epstein-Barr, and echovirus.
2. Other causes of acute cerebellar ataxia include cerebellar abscess, alcohol use, certain medications, insecticides, cerebellar bleeding, multiple sclerosis, cerebellar stroke, and certain vaccinations.
3. Common symptoms of ataxia include dysarthria, nystagmus, uncoordinated eye movements, and unsteady gait. In patients with ataxia caused by a recent viral infection, recovery is usually complete without treatment, over a few months. However, strokes, bleeding, or other infections may cause permanent symptoms.

Clinical case 2
1. What is the major pathophysiologic finding in this condition?
2. What are the common symptoms of this form of ataxia?
3. What diagnostic methods are used for this condition?

A 22-year-old man presented with bilateral cerebellar ataxia, lost tendon jerk reflexes, kyphoscoliosis, and peripheral neuropathy with peroneal muscular atrophy. His condition started at age 12 and gradually worsened. Additional signs and symptoms included slurred speech, hearing impairment, deformity of the feet, and cardiomyopathy. A clinical diagnosis of Friedrich's ataxia was made.

Answers:
1. The major pathophysiologic finding in Friedrich's ataxia is *dying back phenomenon* of the axons. It begins in the periphery, with the ultimate loss of neurons, and a secondary gliosis. These changes usually occur in the spinal cord and spinal roots.
2. Common symptoms of Friedrich's ataxia include arm and leg muscle weakness, loss of coordination, vision and hearing impairment, slurred speech, scoliosis, high plantar arches, carbohydrate intolerance, diabetes mellitus, atrial fibrillation followed by tachycardia and hypertrophic cardiomyopathy.
3. Diagnosis of Friedrich's ataxia is based on medical history, physical examination, reflex testing, electromyogram, nerve conduction studies, electrocardiogram, echocardiogram, blood tests, MRI, CT, and genetic testing.

Clinical case 3
1. What are the possible causes of lower limb dystonia in adults?
2. Lower limb dystonia is often misdiagnosed; what are the actual common symptoms of this condition?
3. Is there such a thing as "runner's dystonia"?

A 49-year-old woman who was a successful long-distance runner began to experience symptoms that affected her feet, and then her knee and hip joints. Over time, she needed assistance walking, could no longer stand for extended periods of time and had poor circulation with a feeling of pins and needles in her feet. Her condition worsened, resulting in extreme plantar-flexion of the right foot. Her knees became unable to extend fully. The patient's left foot also became affected, and ankle-foot-orthotics were prescribed. Both feet became inverted. Her weight-bearing was on the metatarsals of the right foot and the lateral border of the left foot. Her diagnosis was *lower limb dystonia*.

Answers:
1. The causes of lower limb dystonia in adults include stroke, Parkinson's disease, trauma, and psychogenic causes. Sometimes, it appears without any apparent external cause, often primarily or only in the context of walking or running.

2. The actual common symptoms of lower limb dystonia, and not another condition, include the following: the foot turns outward with the sold facing inward; the foot or leg turns inward; the foot points downward; there is extension of the big toe or toe curling; there is a sense of tightness or stiffness in the limb; or there are changes in the walking or running gait.
3. Yes, runner's dystonia is a form of adult-onset task-specific lower limb dystonia that typically presents in long-distance runners. Dystonic spasms initially occur only when running. They often progress to interfere with walking and result in a disability that will prevent the individual from enjoying his or her pastime. It is rare for the patient to return to pre-dystonia exercise levels.

Clinical case 4

1. What are the differential diagnoses of this patient's condition?
2. How common are "mixed" dysarthrias?
3. Could this patient benefit from speech therapy?

A 72-year-old woman developed progressive dysarthria with dysphagia. Her speech became slowed, with a "nasal" sound, and she had decreased mobility in her tongue and palate. There were no other neurological signs upon examination. Multiple investigations were performed, including blood tests, brain MRI, and electrodiagnostic studies. The MRI revealed evidence of cortical cerebral atrophy with possible periventricular white matter changes. Nerve conduction studies revealed bilateral median neuropathy across the wrists, which was worse on the left side. Needle EMG revealed abnormal spontaneous activity in the left orbicularis oris, and right biceps. There was also chronic denervation in the right and left tibialis anterior.

Answers:

1. The differential diagnoses of this patient's condition include early motor neuron disease, cranial neuropathies, myopathies, neuromuscular junction disorders, and neurodegenerative movement disorders.
2. Mixed dysarthrias, where symptoms of more than one type of dysarthria are present, make up the majority of dysarthric cases. Neural damage resulting in dysarthria is rarely only within one part of the nervous system.
3. Yes, speech therapy can help the patient safely chew and swallow. It will teach various techniques, including avoiding conversations when feeling tired, repeating words or syllables so that proper mouth movements are learned, and methods to help deal with frustration while speaking. The patient may also benefit from using a computer or various cards with words on them to help aid in communication.

Key terms

Abetalipoproteinemia	Hyperkinesia
Anterior lobe	Ipsilateral
Arbor vitae	Neuroacanthocytosis syndrome
Ataxia	Nodulus
Cerebellar cortex	Nucleus of Darkschewitsch
Cerebellar hemispheres	Olivocerebellar
Cerebellar peduncles	Paramedian fissures
Cerebrotendinous xanthomatosis	Parkinsonian
Chorea	Posterior lobe
Dysarthria	Primary fissure
Dystonias	Purkinje cells
Flocculi	Sydenham's chorea
Flocculonodular lobe	Tentorium cerebelli
Folia	Vermal
Globose nuclei	Vermis
Glycinergic	Vestibulocerebellum
Granule cells	Videofluoroscopy
Hemichorea	

Suggested readings

1. Barlow, J. S. *The Cerebellum and Adaptive Control.* Cambridge University Press, 2005.
2. Berkowitz, A. *Clinical Neurology and Neuroanatomy: A Localization-Based Approach.* McGraw-Hill Education/Medical, 2016.
3. Brin, M. F.; Comella, C. L.; Jankovic, J. *Dystonia: Etiology, Clinical Features, and Treatment.* LWW, 2004.
4. Broussard, D. M. *The Cerebellum: Learning Movement, Language, and Social Skills.* Wiley-Blackwell, 2013.
5. Budd, R. *Anatomical Structures of the Human Brain, Cerebellum, and Nervous System: An Interactive Anatomy Reference Guide.* Harrison Foster Publications, 2016.
6. Dressler, D.; Altenmuller, E.; Krauss, J. K. *Treatment of Dystonia.* Cambridge University Press, 2018.
7. Gruol, D. L.; Koibuchi, N. *Essentials of Cerebellum and Cerebellar Disorders—A Primer for Graduate Students.* Springer, 2016.
8. Heck, D. *The Neuronal Codes of the Cerebellum.* Academic Press, 2015.
9. Hong, S. *Ataxia: Causes, Symptoms and Treatment (Neuroscience Research Progress).* Nova Biomedical, 2012.
10. Ito, M. *The Cerebellum: Brain for an Implicit Self.* FT Press, 2011.
11. Kaufman-Katz, B. *Dysarthria Treatment Manual.* PRO Ed, 2002.
12. Manson, A. *Medical Researches on the Effect of Iodine in Bronchocele, Paralysis, Chorea, Scrophula, Fistula Lachrymalis, Deafness, Dysphagia, White Swelling, and Distortions of the Spine.* Forgotten Books, 2017.
13. Manto, M.; Groul, D.; Schmahmann, J. D.; Koibuchi, N.; Rossi, F. *Handbook of the Cerebellum and Cerebellar Disorders, Vol. 1;* Springer, 2013.
14. Manto, M.; Huisman, T. A. G. M. The Cerebellum: Disorders and Treatment. In *Handbook of Clinical Neurology,* Vol. 15; Elsevier, 2018.
15. Marien, P.; Manto, M. *The Linguistic Cerebellum.* Academic Press, 2015.
16. Micheli, F. E.; LeWitt, P. A. *Chorea: Causes and Management.* Springer, 2014.
17. Naidich, T. P.; Duvernoy, H. M.; Delman, B. N.; Sorensen, A. G.; Kolias, S. S.; Haacke, E. M. *Duverony's Atlas of the Human Brain Stem and Cerebellum.* Springer, 2008.
18. Palay, S. L.; Chan-Palay, V. *Cerebellar Cortex: Cytology and Organization.* Springer, 2011.

19. Plaitakis, A. *Cerebellar Degenerations: Clinical Neurobiology (Foundations of Neurology)*. Springer, 2012.
20. Porter, R.; Lemon, R. *Corticospinal Function and Voluntary Movement. Monographs of the Physiological Society, Book 45* Oxford University Press, 1995.
21. Rosenbaum, D. A. *Human Motor Control*, 2nd ed.; Academic Press, 2009.
22. Rothwell, J. *Control of Human Voluntary Movement*, 2nd ed.; Chapman and Hall, 1993.
23. Schmahmann, J. D.; Doyon, J.; Toga, A. W.; Petrides, M.; Evans, A. C. *MRI Atlas of the Human Cerebellum*. Academic Press, 2000.
24. Seaman, T. *Diagnosis Dystonia: Navigating the Journey*. CreateSpace Independent Publishing Platform, 2015.
25. Swigert, N. B. *The Source for Dysarthria*; 2nd ed.; LinguiSystems, 2010.

CHAPTER 17

Autonomic nervous system

The autonomic nervous system (ANS) is the visceral part of the human nervous system. It has neurons in both the central nervous system (CNS) and peripheral nervous system (PNS). Its target tissues include the cardiac muscle, smooth muscle of the viscera and blood vessels, and the body's glands. The ANS functions to maintain homeostasis, a constant internal environment. Since it controls involuntary effectors, the ANS is considered to be the subconscious regulator of body functions under normal circumstances. To do this, the ANS uses efferent and afferent pathways, along with various neurons of the brain and spinal cord (Fig. 17.1). The ANS allows us to interact with our environment. For example, if you see a lion in a jungle, your heart rate increases immediately, making it possible for you to run away. Portions of the CNS that participate with the ANS include the *nuclei* for the cranial nerves III, VII, X, and IX as well as the ventrolateral medulla, anterior cingulate gyrus, insula, amygdala, and hypothalamus. The PNS includes some parts of the ANS, which are basically the parasympathetic and sympathetic nerves. An example is the vagus nerve (CN X).

There are two primary anatomical divisions, with some opposing actions. The **sympathetic division** is also known as the **thoracolumbar** division. The **parasympathetic division** is also known as the **craniosacral** division. These two divisions are unique in their anatomical structures. They also differ in their pharmacological responses to medications. They are often referred to as the *sympathetic nervous system* and *parasympathetic nervous system* (Fig. 17.2). However, even though these two divisions have separate pathways, many autonomic effectors are *dually innervated*, receiving input from both divisions. In this case, the effects of both the systems are often antagonistic. One division inhibits the effector while the other division stimulates it at the same time. This antagonism allows dually innervated effects to participate in complex body functions such as sexual activities where arousal is facilitated by the parasympathetic and orgasm is being facilitated by the sympathetic system. Effectors are stimulated, then quickly inhibited (or the opposite) in a specific, timed sequence. There are also autonomic effectors that are *singly innervated* and only receive input from the sympathetic division. Many common medications that treat hypertension, regulate gastrointestinal function, or maintain regularity of the heartbeat, have primary effects upon the neurons of these two systems. The autonomic effector tissues and organs are listed in Table 17.1.

The intrinsic neurons of the GI tract form the **enteric nervous system (ENS)**. This intestinal nervous system comprises complicated nerve plexuses that are in between layers

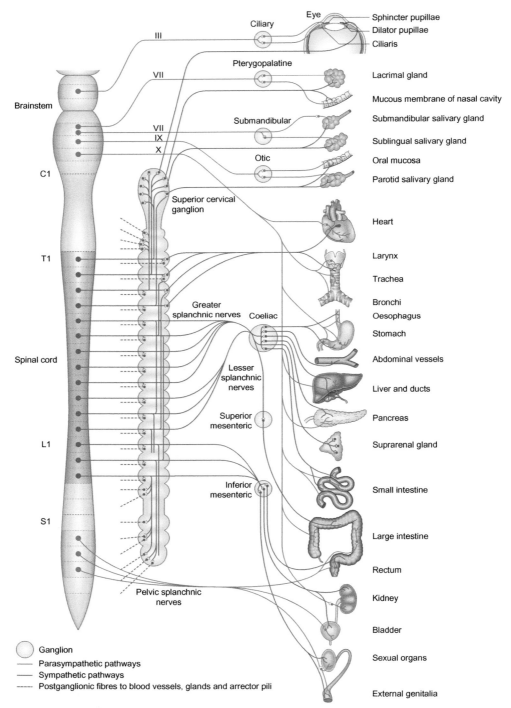

Fig. 17.1 Efferent pathways of the autonomic nervous system.

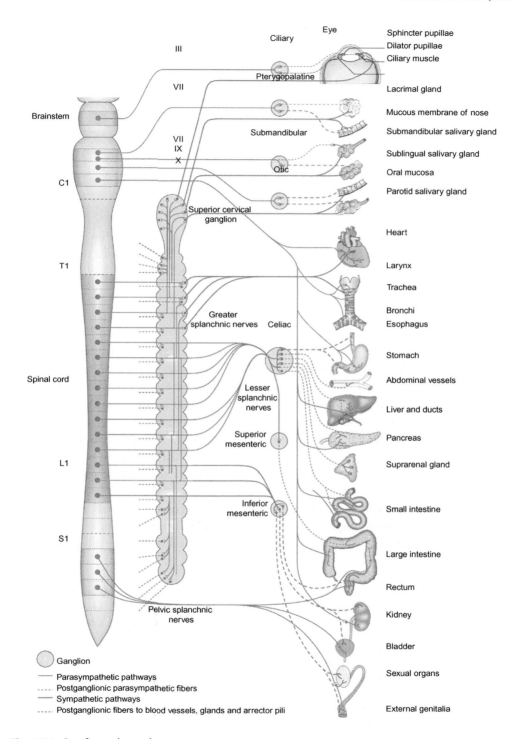

Fig. 17.2 See figure legend on next page.

Table 17.1 Tissues and organs classified as autonomic effectors.

Autonomic effector	Tissues and organs
Smooth muscle	Blood vessels, bronchial tubes, ciliary muscles and iris of the eye, gallbladder, hair follicles, intestines, spleen, stomach, urinary bladder
Cardiac muscle	Heart
Glandular epithelium	Adrenal medulla, digestive glands (gastric, liver, pancreas, salivary), lacrimal glands, sweat glands
Other tissues	Adipose tissue, kidneys, skeletal muscle (which is mostly a somatic effector, but sympathetic fibers aid in regulating contractility during intense exercise, to avoid fatigue)

of the intestinal wall. This system controls visceral effects in this wall. They include endocrine and exocrine cells as well as smooth muscles. The ENS is basically a regional part of the ANS. It controls its own area, regulated by the ANS.

Autonomic outflow

In the ANS, efferent components are organized into both sympathetic and parasympathetic divisions. These arise from the preganglionic cell bodies of different areas. Efferent autonomic neurons function in reflex arcs, depending on feedback from sensory pathways. The *autonomic outflow system* has a more diffuse organization than the *somatic motor system*, which has lower motor neurons projecting directly from the spinal cord or brain. In the somatic motor system, there are no interposed synapses, and there is innervation of a smaller group of target muscle cells. This allows individual muscles to receive separate stimulation, finely tuning motor function. The autonomic and somatic motor pathways are compared in Table 17.2.

In the autonomic outflow system, there is a slower conduction that utilizes a two-neuron chain. The **presynaptic neuron**, or **preganglionic** neuron, is the cell body of the primary neuron (Fig. 17.3). In the CNS, it is located within the intermediolateral

Fig. 17.2 Sympathetic and parasympathetic divisions of the autonomic nervous system (ANS). Parasympathetic neuron cell bodies are located in the brainstem and sacral spinal cord ("craniosacral" division). Cell bodies of sympathetic neurons are located in the thoracic and upper lumbar cord segments ("thoracolumbar" division). The axons of these neurons synapse with postganglionic neurons, which innervate smooth muscle, cardiac muscle, and glands of the body. The postganglionic neuron cell bodies may be located in distinct autonomic ganglia (represented with circles), or in or very near the wall of the innervated visceral organ. Note that sympathetic fibers provide the only innervation to peripheral effectors (sweat glands, arrector pili muscles, adipose tissue, and blood vessels). *From Cramer, D., et al. Basic and Clinical Anatomy of the Spine, Spinal Cord, and ANS, 2nd ed.; Elsevier Mosby: St Louis, 2005.*

Table 17.2 Autonomic and somatic pathways.

Features	Autonomic efferent pathways	Somatic motor pathways
Direction of information	Efferent	Efferent
Amount of neurons between CNS and effector	Two (preganglionic and postganglionic)	One (somatic motor neuron)
Is myelin sheath present?	Preganglionic: yes, Postganglionic: no	Yes
Peripheral fiber location	Most cranial nerves; all spinal nerves	Most cranial nerves; all spinal nerves
Effector that is innervated	Smooth and cardiac muscle, glands, adipose and other tissues (involuntary)	Skeletal muscle (voluntary)
Neurotransmitter	Acetylcholine or norepinephrine	Acetylcholine

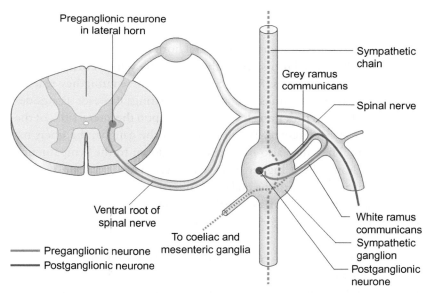

Fig. 17.3 Outflow from preganglionic sympathetic neurons in the lateral horn of the spinal cord. Preganglionic axons may synapse on a postganglionic neuron in a sympathetic ganglion at the same segmental level, or may pass into the sympathetic chain to synapse on a postganglionic neuron in a more rostral or caudal sympathetic ganglion, or may pass to a ganglion in an autonomic plexus in the abdominal cavity, e.g., the coeliac ganglion.

gray column of the brainstem or spinal cord nuclei. This neuron sends out its axon, usually a small myelinated B fiber. The axon synapses with a secondary neuron called the **postsynaptic neuron** or **postganglionic** neuron. This secondary neuron is located in one of the autonomic ganglia. From this point, the postganglionic axon continues to its terminal distribution, within a target organ. The majority of postganglionic autonomic axons is unmyelinated C fibers.

The autonomic outflow system projects outward widely to most target tissues. It is not as focused as the somatic motor system. Since postganglionic fibers are present about 32 times as much as preganglionic neurons, just one preganglionic neuron can control autonomic functions of a fairly large terminal area.

Sympathetic division

The sympathetic nervous system, or thoracolumbar division (T1 to L2) of the ANS, develops from preganglionic cell bodies within the intermediolateral cell columns of the 12 thoracic spinal cord segments, along with the upper two lumbar segments (Fig. 17.4). It consists of preganglionic sympathetic fibers, the adrenal medulla, and postganglionic sympathetic fibers. Preganglionic sympathetic neurons are located in the lateral horn of the thoracic spinal cord.

Preganglionic sympathetic fibers

The majority of preganglionic fibers in the lateral horn of the thoracic spinal cord is myelinated. Following the ventral roots, these fibers form the **white communicating rami** of the thoracic and lumbar nerves (Fig. 17.5). Through these rami, they reach the ganglia of the sympathetic chains or trunks that appear like strings of beads alongside the spinal cord on each side. These **trunk ganglia** are situated upon the lateral sides of the bodies of the thoracic and lumbar vertebrae. When they enter the ganglia, the fibers may synapse with ganglion cells. They may also pass upward or down to the sympathetic trunk and synapse with ganglion cells at higher or lower levels. They may also pass through the trunk ganglia, outward, to one of the collateral (intermediary) sympathetic ganglia. These include the **celiac ganglia** and **mesenteric ganglia**. There are usually 22 sympathetic ganglia on either side of the vertebral column: cervical (3), thoracic (11), lumbar (4), and sacral (4).

The **splanchnic nerves** emerge from the lower seven thoracic spinal cord segments. They pass through the trunk ganglia to the celiac ganglia and **superior mesenteric ganglia**. The synaptic connections in these locations occur with ganglion cells that have postganglionic axons passing to the abdominal viscera, through the **celiac plexus**. The splanchnic nerves from the lowest thoracic and upper lumbar regions send fibers to synaptic stations within the **inferior mesenteric ganglion**, and to small ganglia related to the **hypogastric plexus**. Through this plexus, postsynaptic fibers are then distributed to the viscera of the lower abdomen and pelvis. The lumbar splanchnic sympathetic nerves contribute to the superior and inferior hypogastric plexuses. This contributes to innervation of the bladder neck, ductus deferens, prostate, and other structures. If these nerves are damaged, sexual dysfunction may occur.

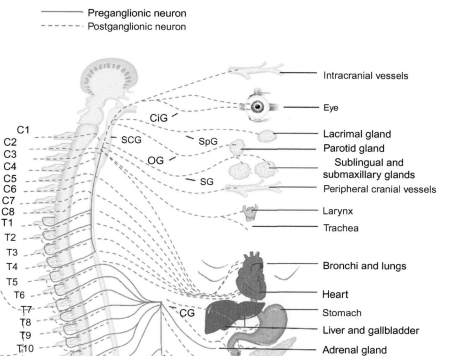

Fig. 17.4 Sympathetic division of the autonomic nervous system. *CG*, celiac ganglion; *CiG*, ciliary ganglion; *IMG*, inferior mesenteric ganglion; *OG*, otic ganglion; *PP*, pelvic plexus; *SCG*, superior cervical ganglion; *SG*, submandibular ganglion; *SMG*, superior mesenteric ganglion; *SpG*, sphenopalatine ganglion. *Redrawn from Rudy, E.B., (Ed.), Advanced Neurological and Neurosurgical Nursing, Mosby: St Louis, 1984.*

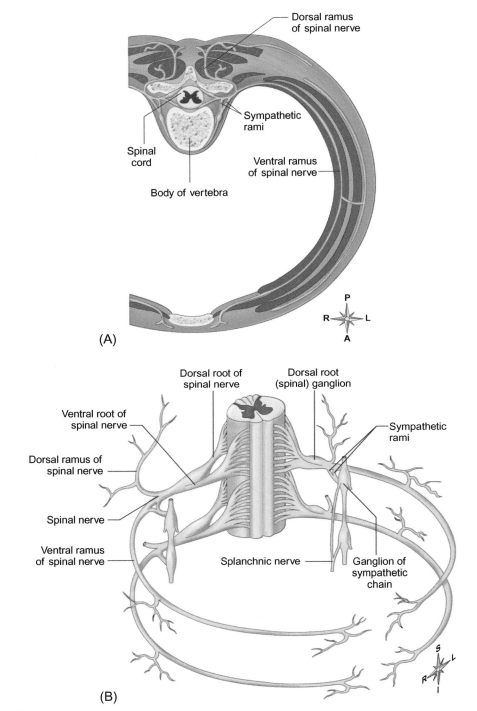

Fig. 17.5 Rami of the spinal nerves.

Focus on preganglionic fibers

By tracing the axon inside a sympathetic chain ganglion, the preganglionic fiber can branch along any of three paths. It may synapse with a sympathetic postganglionic neuron. From the preganglionic fibers, there are ascending or descending branches through the sympathetic trunk, synapsing with postganglionic neurons in other chain ganglia. It may pass through one or more ganglia, with no synapses.

Focus on collateral ganglia

Preganglionic neurons that pass-through chain ganglia with no synapses continue through the splanchnic nerves to other sympathetic ganglia. These *collateral (prevertebral) ganglia* are pairs of sympathetic ganglia very near to the spinal cord. They are named for nearby blood vessels, such as the *celiac ganglion* (or *solar plexus*), which lies next to the celiac artery.

Adrenal medulla

The preganglionic sympathetic axons of the splanchnic nerves additionally project to the adrenal glands. They synapse on the chromaffin cells of the adrenal medulla. The **adrenal chromaffin cells** receive direct synaptic input from the preganglionic sympathetic axons. These cells form from the neural crest, and therefore, are considered to be *modified postganglionic cells* without axons.

Postganglionic sympathetic fibers

The postganglionic sympathetic fibers are mostly unmyelinated (hence gray in color) and have their dendrites and cell bodies within sympathetic chain ganglia. They form the **gray communicating rami**. Their fibers may follow the spinal nerve for a long distance or course directly to their target tissues. The gray communicating rami join each spinal nerve. They distribute the vasomotor, **pilomotor**, and sweat gland innervation to all somatic areas. Branches of the **superior cervical sympathetic ganglion** enter the sympathetic carotid plexuses, surrounding the internal and external carotid arteries, for distribution of sympathetic fibers to the head.

After the carotid plexuses, the postganglionic sympathetic axons then project to the lacrimal and salivary glands, muscles that raise the eyelids, those that dilate the pupils, facial sweat glands, cranial sweat glands, and blood vessels. From the three pairs of cervical sympathetic ganglia, the superior **cardiac nerves** pass to the **cardiac plexus**, at the base of the heart. They distribute cardioacceleratory fibers to the myocardium. From the upper five thoracic ganglia, vasomotor branches pass to the thoracic aorta and posterior **pulmonary plexus**. Though this plexus, dilator fibers pass to the bronchi.

Focus on sympathetic neurons
In the sympathetic division, the preganglionic neurons are relatively short, while the postganglionic neurons are long. The axon of one sympathetic preganglionic neurons synapse with many postganglionic neurons. These often terminate in widely separated organs. Therefore, sympathetic responses are usually widespread and involve many different organs.

Focus on the "fight-or-flight reaction"
The major function of the sympathetic division is as an "emergency" system. Physical or psychological stress increases sympathetic outflow to a great degree. Related responses make the body ready to use large amounts of energy, known as the "fight-or-flight reaction." There are increases in heart rate, strength of cardiac muscle contraction, dilation of coronary vessels, dilation of blood vessels in skeletal muscles, dilation of respiratory airways, increased sweating, and many more activities.

Section review
1. What is dual innervation of autonomic effectors?
2. How are sympathetic preganglionic fibers organized?
3. What are the differences between the white and gray communicating rami?

Parasympathetic division

The parasympathetic (craniosacral) division of the ANS arises from the preganglionic cell bodies within the brainstem's gray matter and the S2 to S4 segments of the sacral portion of the spinal cord. The involved cell bodies include the medial portion of the **oculomotor nucleus**, the **Edinger-Westphal nucleus**, and the superior and inferior **salivatory nuclei**. The majority of preganglionic fibers from S2, S3, and S4 continues without interruption from their central origin in the spinal cord. They reach either the wall of the **viscus** that they supply or the location where they synapse with terminal ganglion cells. These cells are associated with the **plexus of Meissner** and **plexus of Auerbach**, within the wall of the intestinal tract. Since the parasympathetic postganglionic neurons are near to the tissues that they supply, their axons are relatively short. Parasympathetic distribution is totally to visceral structures.

As mentioned earlier, there are four cranial nerves conveying preganglionic parasympathetic fibers, which have visceral efferent functions. Three nerves *oculomotor nerve* (cranial nerve III), *facial nerve* (CN VII), and *glossopharyngeal nerve* (CN IX) distribute parasympathetic or visceral efferent fibers to the head. The parasympathetic axons of these nerves synapse with postganglionic neurons that are located in the ciliary, sphenopalatine, submaxillary, and otic ganglia, respectively. The *vagus nerve* (CN X) distributes autonomic fibers to the thoracic and abdominal viscera, through the **prevertebral plexuses**. The **pelvic nerve**, or *nervus erigentes*, distributes parasympathetic fibers to the majority of the large intestine, as well as to the pelvic viscera and genitals, through the *hypogastric plexus*.

Table 17.3 compares the pathways of the sympathetic and parasympathetic nervous systems.

> **Focus on parasympathetic preganglionic fibers**
> At least 75% of all parasympathetic preganglionic fibers travel with the vagus nerve for one foot or more prior to synapsing with postganglionic fibers in the terminal ganglia, near effectors in the chest and abdomen.

Autonomic plexuses

The large network of nerves that create a conduit for the distribution of sympathetic, parasympathetic, and afferent fibers are called **autonomic plexuses**. The *cardiac plexus* is located around the bifurcation (division point) of the trachea and roots of the great vessels of the base of the heart. This plexus is formed from the cardiac sympathetic nerves along with the cardiac branches of the vagus nerve. The plexus distributes these nerves and their branches to the myocardium and vessels that leave the heart. The right and left *pulmonary plexuses* join with the cardiac plexus. They are situated around the primary bronchi and pulmonary arteries, at the roots of the lungs. These plexuses are formed from the vagus and upper thoracic sympathetic nerves. They are distributed to the vessels and bronchi of both lungs.

The *celiac (solar) plexus* is in the epigastric region, above the abdominal aorta. It forms from vagal fibers that reach it through the esophageal plexus, the sympathetic fibers that arise from the celiac ganglia, and the sympathetic fibers that continue downward from the thoracic aortic plexus. The celiac plexus projects to the abdominal viscera through many subplexuses. These include the hepatic, phrenic, splenic, superior gastric, suprarenal, renal, ovarian or spermatic, abdominal aortic, and the superior and inferior mesenteric plexuses. The *hypogastric plexus* is in front of the fifth lumbar vertebrae and promontory of the sacrum. This plexus receives sympathetic fibers from the aortic plexus and lumbar trunk ganglia, as well as the parasympathetic fibers from the pelvic nerve. Its two lateral **pelvic plexuses** lie on either side of the rectum. The hypogastric plexus projects to the pelvic viscera and genitals, through subplexuses extending with the visceral branches of the hypogastric artery.

Table 17.3 Sympathetic and parasympathetic neurons and pathways.

Neurons	Sympathetic pathways	Parasympathetic pathways
Preganglionic		
Dendrites, cell bodies	In lateral gray columns of thoracic cord; in first 2–3 lumbar segment of cord	In brainstem nuclei, and in lateral gray columns of sacral cord
Axons	In anterior spinal nerve roots, to thoracic and first four lumbar spinal nerves, to and through white rami; terminating in sympathetic ganglia at various levels or extending through sympathetic ganglia, to and through splanchnic nerves, terminating in collateral ganglia	From brainstem nuclei—through CN III to ciliary ganglion; from pons nuclei through CN VIII—to sphenopalatine or submaxillary ganglion; from nuclei in medulla—through CN IX to otic ganglion, or through CN X and XI to cardiac and celiac ganglia, respectively
Distribution	Short fibers—from CNS to ganglion	Long fibers—from CNS to ganglion
Neurotransmitter	Acetylcholine	Acetylcholine
GANGLIA	Sympathetic chain ganglia in 22 pairs; collateral ganglia (celiac, superior, inferior mesenteric)	Terminal ganglia (in or near effector)
Postganglionic		
Dendrites, cell bodies	In sympathetic and collateral ganglia	In parasympathetic ganglia (ciliary, sphenopalatine, submaxillary, otic, cardiac, celiac)—located in or near visceral effector organs
Receptors	Cholinergic (nicotinic)	Cholinergic (nicotinic)
Axons	In autonomic nerves, plexus innervating thoracic, abdominal viscera and blood vessels in these cavities; in gray rami to spinal nerves, to smooth muscle of skin, blood vessels, hair follicles, and to sweat glands	In short nerves—to various visceral effector organs
Distribution	Long fibers from ganglion—to widespread effectors	Short fibers from ganglion—to one effector
Neurotransmitter	Norepinephrine (most); acetylcholine (less)	Acetylcholine

Autonomic innervation of the head

The visceral structures of the head are supplied by the ANS. The smooth muscle, glands, and vessels of the skin of the face and scalp receive only postsynaptic sympathetic innervation. This innervation is from the superior cervical ganglion through the carotid plexus that extends along branches of the external carotid artery. There is a dual autonomic supply from the sympathetic and parasympathetic divisions supplying the deeper structures. These structures include the intrinsic eye muscles, salivary glands, and mucous membranes of the pharynx and nose. The innervation is mediated by the internal carotid plexus and visceral efferent fibers in four pairs of cranial nerves. The internal carotid plexus provides postganglionic sympathetic innervation from the superior cervical plexus. The visceral efferent fibers supply parasympathetic innervation.

In the head, the four pairs of autonomic ganglia are the ciliary, pterygopalatine, otic, and submaxillary ganglia. Each ganglion receives one sympathetic, one parasympathetic, and one sensory root. The sensory root is a branch of the trigeminal nerve. With these ganglia, only the parasympathetic fibers have synaptic connections. The ganglia contain cell bodies of the postganglionic parasympathetic fibers. The sympathetic and sensory fibers continue through these ganglia with no interruption.

The **ciliary ganglion** is situated between the optic nerve and lateral rectus muscle, inside the posterior eye orbit. This ganglion's parasympathetic root forms from cells in or close to the Edinger-Westphal nucleus of the oculomotor nerve. Its sympathetic root is made up of postganglionic fibers, from the superior cervical sympathetic ganglion, through the carotid plexus of the internal carotid artery. Its sensory root is from the nasociliary branch of the ophthalmic nerve. The distribution is via 10–12 short ciliary nerves. These nerves supply the ciliary muscle of the lens, along with the constrictor muscle of the iris. Sympathetic nerves supply the dilator muscle of the iris and causes miosis.

The **sphenopalatine ganglion**, also called the *pterygopalatine ganglion*, is situated deep within the pterygopalatine fossa. It is related to the maxillary nerve. A parasympathetic root arises from cells of the superior salivatory nucleus, through the glossopalatine nerve and great petrosal nerve. The sphenopalatine ganglion's sympathetic root is from the internal carotid plexus via the deep petrosal nerve. This nerve joins the great superficial petrosal nerve, forming the *vidian nerve* within the pterygoid (or vidian) canal. Most sensory root fibers begin in the maxillary nerve. However, a small amount arises in cranial nerves VII and IX, through the tympanic plexus and vidian nerve. Distribution is via the **pharyngeal rami** to the mucous membranes of the pharyngeal roof. Additional distribution is through the **nasal rami** and **palatine rami**, to the mucous membranes of the nasal cavity, uvula, palatine tonsil, and the hard and soft palates. Still more distribution is via the **orbital rami**, to the periosteum of the orbit and the lacrimal glands.

Medial to the mandibular nerve, below the foramen ovale of the infratemporal fossa, is the **otic ganglion**. It has parasympathetic root fibers arising in the inferior salivatory nucleus of the medulla. These course, via cranial nerve IX, to the tympanic plexus and lesser superficial petrosal nerve. The sympathetic root is from the superior cervical sympathetic ganglion, through the plexus locate on the middle meningeal artery. The sensory root is believed to include fibers from cranial nerve IX and also from the geniculate ganglion of cranial nerve VII, through the tympanic plexus and lesser superficial petrosal nerve. The **parotid gland** receives secretory and sensory fibers from the otic ganglion.

A small amount of somatic motor fibers from the trigeminal nerve continues through the otic ganglion, supplying the **tensor tympani** and **tensor veli palatini** muscles.

The **submaxillary ganglion** is found on the medial side of the mandible, between the lingual nerve and the submaxillary duct. It has parasympathetic root fibers that arise from the superior salivatory nucleus of nerve VII. This is via the glossopalatine, chorda tympani, and lingual nerves. The sympathetic root is from the plexus of the external maxillary artery. The sensory root is from the geniculate ganglion, through the glossopalatine, chorda tympani, and lingual nerves. The submaxillary ganglion is distributed to the **submaxillary glands** and **sublingual glands**.

Visceral afferent pathways

The visceral afferent fibers have cell bodies within **sensory ganglia** of certain cranial and spinal nerves. Small amounts of these fibers are myelinated. However, most are unmyelinated, with slow conduction velocities. Table 17.4 summarizes the pain innervation of the viscera. The visceral afferent pathways include those of the brainstem and spinal cord.

Pathways to the brainstem

The visceral afferent axons within the *vagus nerve*, and to a lesser extent, the *glossopharyngeal nerve*, carry many sensations to the brainstem via the heart, great vessels, respiratory tract, and GI tract. The involved ganglia are the inferior vagus nerve ganglion and inferior glossopharyngeal nerve ganglion. Afferent fibers are additionally involved in reflexes regulating blood pressure, respiratory rate, respiratory depth, and heart rate. This requires the use of specialized receptors or receptor areas. These **baroreceptors** are stimulated by pressure and located in the aortic arch and carotid sinus (Fig. 17.6). **Chemoreceptors** are sensitive to hypoxia. They are located within the aorta and carotid bodies. In the medulla, there is a chemosensitive area that contains chemoreceptor neurons, which change their firing patterns to respond to alterations of pH and the partial pressure of carbon dioxide in the cerebrospinal fluid.

Table 17.4 Visceral pain innervation.

Nerves and segments	Nervous division	Involved structures
Vagus	Parasympathetic	Larynx, trachea, esophagus
Splanchnic (T7 to L1)	Sympathetic	Lungs, stomach, spleen, small viscera, colon, kidney, ureter, upper bladder, fundus of the uterus, ovaries
Somatic (C7 to L1)	Sympathetic	Parietal peritoneum and pleura, diaphragm
Pelvis (S2 to S4)	Parasympathetic	Upper vagina, uterine cervix, urethra, prostate, trigone of bladder, rectum

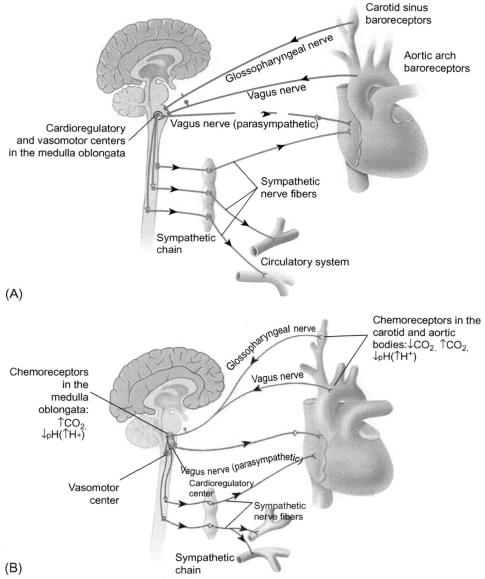

Fig. 17.6 Baroreceptor and chemoreceptor reflex control of blood pressure. (A) Baroreceptor reflexes. Baroreceptors located in the carotid sinuses and aortic arch detect changes in blood pressure. Action potentials are conducted to the cardioregulatory and vasomotor centers. The heart rate can be decreased by the parasympathetic system; the heart rate and stroke volume can be increased by the sympathetic system. The sympathetic system can also constrict or dilate blood vessels. (B) Chemoreceptor reflexes. Chemoreceptors located in the medulla oblongata and in the carotid and aortic bodies detect changes in levels of blood oxygen, carbon dioxide, or pH. Action potentials are conducted to the medulla oblongata. In response, the vasomotor center can cause vasoconstriction or dilation of blood vessels by the sympathetic system, and the cardioregulatory center can cause changes in the pumping activity of the heart through the parasympathetic and sympathetic systems. *From Seeley, R.R.; Stephens, T.D.; Tate, P., Anatomy & Physiology, 3rd ed., Mosby: St Louis, 1995.*

Pathways to the spinal cord

The visceral afferent fibers of the spinal cord enter via the **middle sacral nerve, thoracic nerve,** and **upper lumbar nerve**. The sacral nerves convey sensory stimuli from the organs of the pelvis. The nerve fibers function in reflexes of the sacral parasympathetic outflow controlling some sexual responses, micturition, and defecation. Axons that carry visceral pain impulses from the heart, upper digestive tract, kidneys, and gallbladder course along with the thoracic and upper lumbar nerves. These visceral afferent pathways are related to sensations that include hunger, nausea, and visceral pain. A pain impulse from a viscus may converge with pain impulses from a certain skin region, causing *referred pain*. This is defined as pain felt in the distribution of a nerve that is not the actual nerve innervating the source of the pain. It is usually felt deeply, across a wide area, and the patient usually describes it as *aching pain*. Examples include the shoulder pain caused by gallstones and pain in the left arm or throat due to myocardial ischemia.

Section review
1. What are the differences between axons of the sympathetic and parasympathetic pathways?
2. Which deeper structures receive dual sympathetic and parasympathetic innervation?
3. Related to spinal cord pathways, what is the description of referred pain?

Hierarchy of the autonomic nervous system

The ANS is continually influenced by the *autonomic centers*, which are neurons in various parts of the brain, with axons that conduct impulses, directly or indirectly, to the autonomic preganglionic neurons. The autonomic centers work in a hierarchy to control the ANS. The autonomic centers of the cerebral cortex have the most control, especially the frontal lobe and limbic system. In these centers, neurons send impulses to other autonomic centers in the brain—especially the hypothalamus. Neurons in the hypothalamus then send stimulating or inhibiting impulses to sympathetic and parasympathetic preganglionic neurons in the lower autonomic centers of the brainstem and spinal cord.

Functional considerations

The ANS can also be divided into **cholinergic** and **adrenergic** divisions. The cholinergic neurons include the following:
- Preganglionic and parasympathetic postganglionic neurons
- Sympathetic postganglionic neurons to the sweat glands
- Sympathetic vasodilator neurons to the blood vessels in skeletal muscle

In circulating blood, there is usually no acetylcholine (ACh). The effects of localized cholinergic discharge are usually discrete and of short duration. This is because of high concentrations of cholinesterase at the cholinergic nerve endings. Table 17.5 summarizes the responses of effector organs to ANS impulses and circulating catecholamines.

Table 17.5 Effector organ responses to ANS impulses and catecholamines.

Effector organs	Cholinergic response	Noradrenergic receptor type and response
Eyes		
Radial muscle of iris	No effect	Alpha (α)—contraction (mydriasis)
Sphincter muscle of iris	Contraction (miosis)	No effect
Ciliary muscle	Contraction for near vision	Beta (β)—relaxation for far vision
Lacrimal glands	Secretion	No effect
Nasopharyngeal glands	Secretion	No effect
Salivary glands	Profuse watery secretion	α—Thick secretion β_2—Amylase secretion
Lungs		
Bronchial muscle	Contraction	β_2—Relaxation
Bronchial glands	Stimulation	?—Inhibition (?)
Heart		
S-A node	Decreased heart rate; vagal arrest	β_1—Increased heart rate
Atria	Decreased contractility; usually an increase in conduction velocity	β_1—Increased contractility, conduction velocity
A-V node and conduction system	Decreased conduction velocity; A-V block	β_1—Increased conduction velocity
Ventricles	No effect	β_2—Increased contractility, conduction velocity
Arterioles		
Coronary, skeletal muscle, pulmonary, abdominal, viscera, renal	Dilation	α—Construction β_2—Dilation
Skin and mucosa, cerebral, salivary glands	No effect	α—Constriction
Systemic veins	No effect	α—Constriction β_2—Dilation
Stomach		
Motility and tone	Increase	α, β_2—Usually decrease
Sphincters	Usually relaxation	α—Usually contraction
Secretion	Stimulation	?—Inhibition (?)

Continued

Table 17.5 Effector organ responses to ANS impulses and catecholamines—cont'd

Effector organs	Cholinergic response	Noradrenergic receptor type and response
Intestines		
Motility and tone	Increase	α, β$_2$—Decrease
Sphincters	Usually relaxation	α—Usually contraction
Secretion	Stimulation	?—Inhibition (?)
Liver	No effect	α, β$_2$—Glycogenolysis
Gallbladder and ducts	Contraction	?—Relaxation
Pancreas		
Acini	Increased secretion	α—Decreased secretion
Islets	Increased insulin and glucagon secretion	α—Decreased insulin and glucagon secretion
		β$_2$—Increased insulin and glucagon secretion
Spleen capsule	No effect	α—Contraction
		β$_2$—Relaxation
Adrenal medulla	Secretion of epinephrine and norepinephrine	None
Ureters		
Motility and tone	Increase (?)	α—Usually increase
Urinary bladder		
Detrusor	Contraction	β—Usually relaxation
Trigon and sphincter	Relaxation	α—Contraction
Juxtaglomerular cells	No effect	β$_1$—Increased renin secretion
Uterus	Variable based on stage of menstrual cycle, amount of circulating estrogen and progesterone, pregnancy, and other factors	α, β$_2$—Variable, on palms of hands and in some other locations (adrenergic sweating)
Male sex organs	Erection	α—Ejaculation
Skin		
Pilomotor muscles	No effect	α—Contraction
Sweat glands	Generalized secretion	α—Slight localized secretion on palms of hands and in some other locations (adrenergic sweating)
Adipose tissue	No effect	β$_1$—Lipolysis
Pineal gland	No effect	β—Increased melatonin synthesis and secretion

Within the adrenal medulla, postganglionic cells no longer have axons. They are specialized for secreting the catecholamine called *epinephrine* into the blood. Cholinergic preganglionic neurons to these cells act as **secretomotor** nerve supplies to the adrenal glands. Most postganglionic neurons are considered to be adrenergic—except for the sweat gland and sympathetic vasodilator neurons. *Norepinephrine* (NE) has a longer and broader action than ACh.

> **Focus on neurotransmitters and receptors**
> Axons that release norepinephrine (NE) are called *adrenergic fibers*. Axons that release acetylcholine (ACh) are called *cholinergic fibers*. Autonomic cholinergic fibers are axons of preganglionic sympathetic neurons, and of both preganglionic and postganglionic parasympathetic neurons. The axons of postganglionic sympathetic neurons are the only autonomic adrenergic fibers except for the postganglionic sweat glands that are exceptionally cholinergic.

Clinical considerations

There are a variety of conditions related to ANS dysfunction, which affect different parts of the body. These include Horner syndrome, Adie tonic pupil syndrome, and Hirschsprung's disease.

Horner syndrome

Horner syndrome involves ptosis, miosis, and anhidrosis, due to dysfunction of cervical sympathetic output.

Pathophysiology
Horner syndrome develops when there is a disruption of the cervical sympathetic pathway. The causative lesion may be primary, which includes congenital links, or secondary to another disorder. Central lesions may involve brainstem ischemia, brain tumors, or syringomyelia. Peripheral lesions may involve cervical adenopathy, **pancoast tumor**, aortic or carotid dissection, neck and skull injuries, or thoracic aortic aneurysm. Peripheral lesions may originate as preganglionic or postganglionic lesions.

Clinical manifestations
Signs and symptoms of Horner syndrome include ptosis, miosis, anhidrosis, and hyperemia of the affected side depending on the level of the lesion.

Diagnosis
The instillation of eyedrops helps confirm and characterize Horner syndrome. A 4%–5% solution of cocaine or a 0.5% solution of apraclonidine drops are put into both eyes.

Cocaine blocks synaptic reuptake of NE. It causes the pupil of the unaffected eye to dilate. In peripheral Horner syndrome, a postganglionic lesion causes the pupil of the affected eye not to dilate. This is because the postganglionic nerve terminals are degenerated, resulting in increased anisocoria. If the lesion is below the superior cervical ganglion, known as preganglionic or central Horner syndrome, and postganglionic fibers are intact, the pupil of the affected eye will dilate in the presence of cocaine, leading to a decreased in the anisocoria or the difference between the pupil sizes.

Apraclonidine is a weak alpha-adrenergic agonist. It minimally dilates the pupil of a normal eye. In peripheral Horner syndrome, a postganglionic lesion causes the pupil of the affected eye to dilate much more than that of the unaffected eye. This is because the iris dilator muscle of the affected eye has no more sympathetic innervation. It has developed adrenergic supersensitivity. Therefore, anisocoria is decreased. Results can be falsely normal when the causative lesion is acute. If the lesion is preganglionic, or a central of Horner syndrome, the affected eye's pupil will not dilate. This is because the iris dilator muscle does not develop adrenergic supersensitivity. Therefore, anisocoria increases.

If Horner syndrome is suspected, hydroxyamphetamine (1%) is put into both eyes after 48 h to help locate the lesion. This drug works by causing NE to be released from the presynaptic terminals. It will have no effect if postganglionic lesions are present since they cause the postganglionic terminals to degenerate. This means that when hydroxyamphetamine is applied and there is a postganglionic lesion, the affected eye's pupil will not dilate, but the pupil of the unaffected eye will dilate, increasing anisocoria. If there is a central or preganglionic lesion, application of hydroxyamphetamine will cause the affected eye's pupil to dilate normally or more than normal. The pupil of the unaffected eye will dilate normally, resulting in anisocoria that is decreased or unchanged. It is important to understand that postganglionic lesions sometimes cause the same results to occur. Given the advent of technology, these tests are not done as frequently in clinical practice.

Treatment

There is no treatment for primary Horner syndrome, but if a cause is identified, it is treated.

Argyll Robertson pupil

Argyll Robertson pupil is also referred to as *AR pupils*. It refers to the pupils appearing small and bilateral, constricting when attempting to focus on a close object, but not constricting when the eyes are exposed to bright light. They react to accommodation but have no response to light. The condition develops only after a long period of untreated syphilis infection. It is a rare disease that only sometimes occurs in advanced cases of syphilis and other neurological disorders and is uncommon in developing countries.

Pathophysiology

Argyll Robertson pupil develops from mostly from syphilis, due to the *treponema pallidum* spirochete. Additional pathophysiology is linked to alcoholism and diabetes. It has been detected in various disorders with lesions in the region of the Edinger-Westphal nucleus.

Clinical manifestations

Argyll Robertson pupils are characterized by a lack of light reflexes, a small and irregular appearance, quick accommodation reflexes, sometimes an unequal pupil size, and when mydriatics are instilled, slow pupil dilation (see Fig. 17.7). When syphilis is the cause, the pupils are usually irregular and unequal. Sometimes, there is depigmentation, atrophy, and loss of the ciliospinal reflex. The condition is usually bilateral.

Diagnosis

Diagnosis is via slit-lamp examination to detect signs of iris atrophy. Other signs that help confirm diagnosis involve Charcot's joint with hyperextensibility of the joint, loss of joint and vibration sense to posterior column damage, a positive Romberg's test, and a stamping gait. Differential diagnoses include causes such as diabetes mellitus, Horner's syndrome, dorsal mid-brain syndrome, encephalitis, iritis, sarcoidosis, chronic Adie's tonic pupil, and Lyme disease.

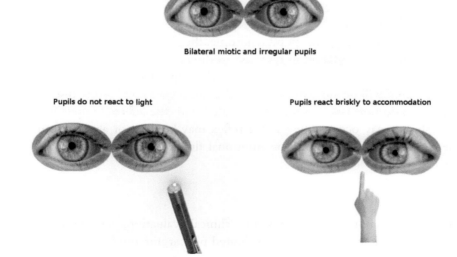

Fig. 17.7 Argyll Robertson pupils.

Treatment

Immediate treatment of underlying syphilis is required. Intravenous penicillin is the best method to cure nearly every stage of syphilis. Alternatively, a 2-to-4-week course of tetracycline or doxycycline can be used. When signs and symptoms of Argyll Robertson pupil are present, a health-care professional must be consulted immediately. There are no specific prevention guidelines except for the prevention of syphilis.

Adie tonic pupil syndrome

Adie tonic pupil syndrome is also known as *Adie syndrome, Adie's pupil, Holmes-Adie syndrome*, and **tonic pupil syndrome**. It is a rare neurological disorder that affects the pupil of the eye. In most patients, the pupil is larger than normal (dilated), and "slow to react" to direct light. Absent or poor tendon reflexes are associated with this disorder. This syndrome is most common in the third or fourth decade of life and is much more common in females. It is called Holmes-Adie syndrome when the eye abnormalities accompany knee or ankle jerks.

Pathophysiology

In most cases, the cause of Adie tonic pupil syndrome is idiopathic, but it may occur due to other conditions, including surgery, trauma, ischemia, or infection. Rarely, localized disturbance of sweat secretion is associated. There is usually nonprogressive, limited damage to the ANS. There is also a loss of deep tendon reflexes, which may not occur at the same time as the pupil abnormalities. There is a family tendency to the syndrome.

Clinical manifestations

In most patients with Adie tonic pupil syndrome, the affected pupil is dilated all the time, not constricting much or at all in response to direct light. The pupil constricts slowly when focusing on nearby objects. Eventually, the pupil that was at first larger than the unaffected pupil, will become smaller than the unaffected pupil as the disease progress. Some patients do not have symptoms associated with the affected pupil. Blurry vision or photophobia may occur. The patellar reflex may be poor or absent. Some patients experience headache, facial pain, or emotional fluctuations. The condition is usually unilateral.

Diagnosis

Adie tonic pupil syndrome is diagnosed by clinical evaluation, detailed patient history, and an ophthalmologic examination. A diluted pilocarpine mixture may be instilled as eye drops, causing the pupils to constrict. The affected pupil will constrict slowly in response to light. A slit lamp may be used to examine the eyes under high magnification.

Treatment

Treatment is usually not needed. Glasses may help correct blurred vision, and sunglasses help protect against photophobia.

Hirschsprung's disease

Hirschsprung's disease is also known as *congenital megacolon*. It is an anomaly of innervation of the lower intestine, usually only of the colon, and causes partial or total functional obstruction. It is caused by congenital lack of the Meissner and Auerbach autonomic plexus, or *aganglionosis*, in the intestinal wall. It affects about one of every 5000 live births and is usually limited to the distal colon in 75% of cases. However, it can affect all of the colon, and even all of the large and small intestines. The denervated area is always continuous. Males are affected four times as often as females, unless the entire colon is involved, which has a 1:1 distribution between the genders (Fig. 17.8).

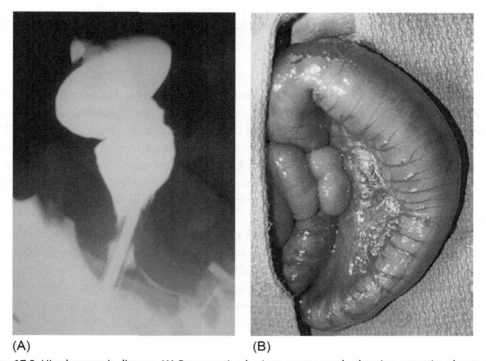

Fig. 17.8 Hirschsprung's disease. (A) Preoperative barium enema study showing constricted rectum (*bottom*) and dilated sigmoid colon. Ganglion cells were absent in the rectum, but present in the sigmoid colon. (B) Corresponding intraoperative appearance of the dilated sigmoid colon. *(Courtesy of Dr. Aliya Hussain, The University of Chicago, Chicago, Illinois.)*

Pathophysiology

Hirschsprung's disease links to lack of neuroblasts migration from the neural crest. There are at least 12 different associated genetic mutations. It is linked to congenital anomalies and genetic abnormalities, especially Down syndrome. There is absent or abnormal peristalsis in the affected GI segment. This causes continuous smooth muscle spasm, with partial or complete obstruction from the accumulation of intestinal contents. There is massive dilation of more proximal and normally innervated areas of the intestine.

Clinical manifestations

Usually, the patient with Hirschsprung's disease begins developing signs and symptoms early in life. About 98% of neonates pass meconium in the first 24h of life. However, with this condition, between 50% and 90% of affected infants do not pass meconium in the first 48h of life. There is abdominal distention, constipation, and eventually, vomiting—similar to other forms of bowel obstruction. Sometimes, patients with ultrashort segment aganglionosis have mild or intermittent constipation, usually with bouts of mild diarrhea. This causes a delay in diagnosis. Upon digital rectal examination, an empty rectum with palpable stool higher up within the colon may be seen. The infant may fail to thrive.

Diagnosis

Diagnosis of Hirschsprung's disease is needed as soon as possible. The more time without treatment, the higher the chance of developing toxic megacolon (Hirschsprung's' enterocolitis), which can be fulminant and fatal. Most infants are diagnosed early in life. The first step is usually barium enema and/or rectal suction biopsy. A barium enema may reveal a change in diameter between dilated, normally innervated colon size proximal to the narrowed distal segment, which does not have normal innervation. X-rays can help with the diagnosis in some cases. Hirschsprung's disease is the likely diagnosis if the colon is still filled with barium.

A rectal suction biopsy can reveal a lack of ganglion cells. Acetylcholinesterase staining will highlight enlarged nerve trunks. Rectal **manometrics**, performed in certain locations, may reveal a lack of relaxation of the internal anal sphincter. This is common with abnormal innervation. A definitive diagnosis requires a full-thickness biopsy of the colon or rectum, which identifies the extent of the disease and helps plan needed surgery.

Treatment

Treatment of Hirschsprung's disease involves surgical repair. Normally innervated bowel areas are brought to the anus, with the preservation of the anal sphincters. In neonate usually involves a colostomy, proximal to the aganglionic segment. This decompresses the colon, allowing for growth and development prior to the second stage of the procedure. Later, the entire aganglionic part of the colon is resected, with a pull-through procedure being performed. In certain facilities, a one-stage procedure can now be performed during the neonatal period, for short-segment disease. When a laparoscopic technique is used, results are similar to those of the open method. They cause shorter hospitalization, earlier start of feeding, and less pain. Following definitive repair, prognosis is good, but many infants have chronic dysmotility with constipation, obstructions, or both.

Section review
1. What are the basic differences between the cholinergic and adrenergic divisions of the ANS?
2. What are the clinical manifestations of Argyll Robertson pupil?
3. How is Hirschsprung's disease treated?

Clinical cases

Clinical case 1
1. Why would Horner syndrome have developed in this patient following the surgery that was performed?
2. What other nerves may have been at risk in this surgery, based on their location?
3. What is the underlying etiology of Horner syndrome?

A 5-year-old boy was brought to his pediatrician, with a swelling on the left side of his neck that had been present for 4 months. There were no other complaints. Physical examination revealed a hard, mobile mass on the anterior surface of the sternocleidomastoid muscle. A previous neck ultrasonography suggested lymphadenopathy, and an MRI revealed a lesion on the sternocleidomastoid muscle and internal carotid artery. The parotid land was displaced anteriorly. Surgery was recommended since the tumor was deep to the large neck arteries, though surgical risks could include left eyelid ptosis, respiratory changes, and loss of some vocal control. After the surgery, the boy developed an ipsilateral Horner syndrome, which subsided after 3 months.

Answers:
1. Horner syndrome likely developed because it is a group of ipsilateral signs and symptoms resulting from interruption, via surgery, of the cervical sympathetic trunk to the head.
2. The other nerves at risk were the left superior laryngeal, the left vagus, and the left glossopharyngeal nerves.
3. In Horner syndrome, injury or damage to nerve fibers may be due to carotid artery injury, lung tumors, brainstem tumors, stroke, idiopathic causes, and surgery, such as in this case study.

Clinical case 2

1. What are the likely treatments for syphilis, and for the patient's leg pain?
2. What are the characterizations of tabes dorsalis?
3. Do these treatments usually improve symptoms?

A 47-year-old man with HIV infection presented with severe paroxysmal, shooting pains in his legs, increased difficulty walking, tinnitus, and urinary incontinence. Physical examination revealed Argyll Robertson pupils that were nonreactive to bright light, but contracted quickly when focusing on a close object. Samples of blood and cerebrospinal fluid were positive for syphilis.

Answers:

1. Usually, intravenous penicillin will be used for syphilis, and gabapentin will be used for the leg pain.
2. Tabes dorsalis is a type of neurosyphilis characterized by the degeneration of the nerves in the dorsal columns of the spinal cord. Along with Argyll Robertson pupils, it is associated with ataxia and loss of proprioception.
3. Yes, treatment for this condition usually improves symptoms and mobility slowly, though often, the leg pain resists pharmacotherapy. The syphilitic condition will usually become negative when blood and cerebrospinal fluid are retested.

Clinical case 3

1. What is the likely diagnosis for this patient?
2. What would a slit-lamp examination of this patient likely reveal?
3. What are the possible differential diagnoses of this condition?

A 32-year-old woman presented with unilateral mydriasis, poor pupillary constriction in bright light, slow re-dilation, and decreased deep tendon reflexes. She described having blurred vision and photophobia. Her physical examination revealed anisocoria that was greater in light than in the dark.

Answers:

1. The likely diagnosis is Adie tonic pupil syndrome, based on the clinical manifestations, and this woman's age, since the condition has a higher occurrence in females of this age group.
2. A slit-lamp examination should reveal sectoral iris paralysis, which involves segmental contracture of the iris that is seen as worm-like movements of the pupillary margin.
3. Differential diagnoses include trauma, giant cell arteritis, varicella-zoster, Sjögren syndrome, tertiary syphilis, chronic alcoholism, diabetes mellitus, amyloidosis, cancer-related dysautonomia, Charcot-Marie-Tooth disease, and Miller-Fisher syndrome.

Clinical case 4

1. If the surgical biopsy revealed absence of ganglion cells, and acetylcholinesterase staining was abnormal, what would the likely diagnosis be?
2. What are the functions of ganglion cells in relation to normal GI function and this condition?
3. Without prompt and correct diagnosis and treatment, what are the likely outcomes?

A 1-year-old girl with a past medical history of constipation and gastroesophageal reflux was brought to the emergency department with persistent vomiting that progressed into hematemesis over a 2-day period. Physical examination evaluated hemoglobin levels, which were normal. The patient was admitted for additional evaluation and management. An upper GI study with oral contrast revealed normal anatomy. Emesis began again in 24 h. The patient's abdomen was severely distended. An abdominal X-ray revealed a large amount of retained barium contrast, with moderate dilation of the descending colon, severe dilation of the proximal sigmoid colon, and a normal-sized rectum. Surgery was scheduled.

Answers:
1. Based on the findings, this girl had Hirschsprung's disease.
2. Ganglion cells act as points in the enteric nervous system to help coordinate and facilitate bowel relaxation. When they are absent, the aganglionic areas become spastic. This causes distal intestinal obstruction.
3. Without prompt and correct diagnosis and treatment, Hirschsprung's enterocolitis (toxic megacolon) is likely. This is a life-threatening completion resulting in a grossly enlarged colon that is often followed by sepsis and shock.

Key terms

Adie tonic pupil syndrome
Adrenal chromaffin cells
Adrenergic
Argyll Robertson pupil
Autonomic plexuses
Baroreceptors
Blast sign
Cardiac nerves
Cardiac plexus
Celiac ganglia
Celiac plexus
Chemoreceptors
Cholinergic
Ciliary ganglion
Conic pupil syndrome
Craniosacral
Edinger-Westphal nucleus
Enteric nervous system
Gray communicating rami
Hirschsprung's disease
Horner syndrome
Hypogastric plexus
Inferior mesenteric ganglion
Manometrics
Mesenteric ganglia
Middle sacral nerve
Nasal rami
Oculomotor nucleus
Orbital rami
Otic ganglion
Palatine rami
Pancoast tumor
Parasympathetic division
Parotid gland

Pelvic nerve
Pelvic plexuses
Pharyngeal rami
Pilomotor
Plexus of Auerbach
Plexus of Meissner
Postganglionic
Postsynaptic neuron
Preganglionic
Presynaptic neuron
Prevertebral plexuses
Pulmonary plexus
Salivatory nuclei
Secretomotor
Sensory ganglia
Sphenopalatine ganglion
Splanchnic nerves
Sublingual glands
Submaxillary ganglion
Submaxillary glands
Superior cervical sympathetic ganglion
Superior mesenteric ganglia
Sympathetic division
Tensor tympani
Tensor veli palatini
Thoracic nerve
Thoracolumbar
Trunk ganglia
Upper lumbar nerve
Viscus
White communicating rami

Suggested readings

1. Berkowitz, A. *Lange Clinical Neurology and Neuroanatomy: A Localization-Based Approach*. McGraw-Hill Education/Medical, 2016.
2. Bujis, R. M.; Swaab, D. F. *Autonomic Nervous System. Handbook of Clinical Neurology, 117;* Elsevier, 2013.
3. Cardinali, D. P. *Autonomic Nervous System: Basic and Clinical Aspects*. Springer, 2017.
4. Carlstedt, T. *Central Nerve Plexus Injury*. Imperial College Press, 2007.
5. Colombo, J.; Arora, R.; DePace, N. L.; Vinik, A. I. *Clinical Autonomic Dysfunction: Measurement, Indications, Therapies, and Outcomes*. Springer, 2014.
6. Cramer, G. D.; Darby, S. A. *Clinical Anatomy of the Spine, Spinal Cord, and ANS*, 3rd ed.; Mosby, 2013.
7. Gabella, G. *Structure of the Autonomic Nervous System*. Springer, 2012.
8. Idiaquez, J.; Benarroch, E.; Nogues, M. *Evaluation and Management of Autonomic Disorders—A Case-Based Practical Guide*. Springer, 2018.
9. Iwase, S.; Hayano, J.; Orimo, S. *Clinical Assessment of the Autonomic Nervous System*. Springer, 2016.
10. Karczmar, A. G.; Koketsu, K.; Nishi, S. *Autonomic and Enteric Ganglia: Transmission and Its Pharmacology*. Springer, 2012.
11. Kiernan, J.; Rajakumar, R. *Barr's The Human Nervous System: An Anatomical Viewpoint*, 10th ed.; LWW, 2013.
12. Kumbhare, D.; Robinson, L.; Buschbacher, R. *Buschbacher's Manual of Nerve Conduction Studies*, 3rd ed.; DemosMedical, 2015.
13. Levine, A. C. *Adrenal Disorders: Physiology, Pathophysiology and Treatment*. Humana Press, 2018.
14. Levitan, I. B.; Kaczmarek, L. K. *The Neuron: Cell and Molecular Biology*, 4th ed.; Oxford University Press, 2015.
15. Mai, J. K.; Paxinos, G. *The Human Nervous System*, 3rd ed.; Academic Press, 2011.
16. Mathias, C. J.; Bannister, R. *Autonomic Failure: A Textbook of Clinical Disorders of the Autonomic Nervous System*, 5th ed.; Oxford University Press, 2013.
17. Murphy, N. B. *Hirschsprung's Disease—Solving the Puzzle—An Informational Resource Guide for Patients and Medical Professionals*. Hirschsprungs Help Inc, 2011.
18. Nieuwenhuys, R.; Voogd, J.; van Huijzen, C. *The Human Central Nervous System: A Synopsis and Atlas*, 4th ed.; Steinkopff, 2007.
19. Owens, P. B. *Autonomic Nervous System (ANS): Clinical Features, Disorders (Human Anatomy and Physiology)*. Nova Science Publishing Inc, 2016.
20. Pierrot-Desilligny, E. *The Circuitry of the Human Spinal Cord*. Cambridge University Press, 2012.

21. Robertson, D.; Biaggioni, I.; Burnstock, G.; Low, P. A.; Paton, J. F. R. *Primer on the Autonomic Nervous System*, 3rd ed.; Academic Press, 2011.
22. Sandroni, P.; Low, P. A. *Autonomic Disorders: A Case-Based Approach*. Cambridge University Press, 2015.
23. Struhal, W.; Lahrmann, H.; Fanciulli, A.; Wenning, G. K. *Bedside Approach to Autonomic Disorders: A Clinical Tutor*. Springer, 2017.
24. Valenza, G.; Scilingo, E. P. *Autonomic Nervous System Dynamics for Mood and Emotional-State Recognition*. Springer, 2014.
25. Wood, J. D. *Enteric Nervous System: The Brain-in-the-Gut (Integrated Systems Physiology: From Molecule to Function to Disease)*. Morgan & Claypool Life Sciences, 2011.

CHAPTER 18

Neurotransmitters

Neurotransmitters are molecules that, along with electrical signals, carry and may amplify signals between nerves and other nerves, or between nerves and other cell types. They affect many different functions, including our thoughts, emotions, memoires, movements, and sleep patterns. Synaptic transmission is primarily affected by enhancing or inhibiting neurotransmitter release, destruction, reabsorption, as well as by blocking neurotransmitter binding to receptors. Neurotransmitters are often amino acids, small peptides, or derivative chemicals. The classic neurotransmitters include acetylcholine (ACh), dopamine, gamma-aminobutyric acid (GABA), glutamine, norepinephrine (NE), serotonin, and others. More than 50 neurotransmitters, or potential neurotransmitters, have been identified.

Classifications of neurotransmitters

Neurotransmitters can be classified by their function or by their chemical components (Table 18.1).

Types of neurotransmitters

Neurotransmitters are excitatory or inhibitory, or in some cases, have both types of functions. Excitatory neurotransmitters trigger depolarization, which increases the likelihood of a response. Inhibitory neurotransmitters trigger hyperpolarization, which decreases the likelihood of a response. The major types of neurotransmitters include those primarily functioning in the fight-or-flight response, concentration, pleasure, mood, calmness and relaxation, learning, memory, and euphoria. These include the neurotransmitter groups known as ACh, the biogenic amines, amino acids, dissolved gases, neuropeptides, purines, and endocannabinoids.

Acetylcholine

ACh is a neurotransmitter used by somatic efferent neurons, preganglionic sympathetic neurons, and parasympathetic nervous system. ACh was the first neurotransmitter ever identified and is well understood since it is released at neuromuscular junctions (NMJs)—locations that are easier to study than the deeply buried CNS synapses. ACh is synthesized within the nerve terminal by **choline acetyltransferase**. This catalyzes reactions

Table 18.1 Neurotransmitters.

Neurotransmitter	Actions	Locations
Acetylcholine		
At nicotinic ACh receptors (autonomic ganglia, skeletal muscles, and in CNS)	Excitatory, direct action	CNS: through cerebral cortex, hippocampus, brain stem
At muscarinic ACh receptors (visceral effectors, and in CNS)	Excitatory or inhibitory, based on subtype of muscarinic receptor; indirect action via second messengers	PNS: all neuromuscular junctions with skeletal muscle; some autonomic motor endings (all parasympathetic postganglionic and preganglionic fibers)
Biogenic amines		
Dopamine	Excitatory or inhibitory, based on receptor type; indirect action via second messengers	CNS: substantia nigra of midbrain; hypothalamus; main neurotransmitter of indirect motor pathways; PNS: some sympathetic ganglia
Epinephrine and norepinephrine	Excitatory or inhibitory based on receptor type; indirect action via second messengers	CNS: brain stem, mostly in locus ceruleus of midbrain; limbic system; some cerebral cortex areas; PNS: main neurotransmitter of postganglionic neurons in sympathetic nervous system
Histamine	Excitatory or inhibitory based on receptor type; indirect action via second messengers	CNS: hypothalamus
Serotonin	Mostly inhibitory; indirect action via second messengers; direct action at 5-HT_3 receptors	CNS: brainstem, mostly midbrain; hypothalamus; limbic system; cerebellum; pineal gland; spinal cord
Amino acids		
Glutamate	**Mostly excitatory**; direct action	CNS: spinal cord; throughout brain, primary excitatory neurotransmitters
Gamma-aminobutyric acid (GABA, or ϒ-Aminobutyric acid)	**Mostly postsynaptic inhibition**; direct and indirect actions via second messengers	CNS: cerebral cortex, hypothalamus, Purkinje cells in cerebellum, spinal cord, granule cells in olfactory bulb, retina

Table 18.1 Neurotransmitters—cont'd

Neurotransmitter	Actions	Locations
Glycine	**Mostly inhibitory**; direct action	CNS: spinal cord, brain stem, retina
Neuropeptides		
Endorphins (including beta endorphin, dynorphin) and enkephalins	Mostly inhibitory; indirect action via second messengers	CNS: throughout brain (hypothalamus; limbic system; pituitary) and spinal cord
Tachykinins (including substance P, neurokinin A (NKA))	Excitatory; indirect action via second messengers	CNS: basal nuclei, midbrain, hypothalamus, cerebral cortex; PNS: some sensory neurons of dorsal root ganglia (pain afferents), enteric neurons
Cholecystokinin (CCK)	Mostly excitatory; indirect action via second messengers	Through CNS; second messenger actions in small intestine
Somatostatin	Mostly inhibitory; indirect action via second messengers	CNS: wide distribution (hypothalamus, basal nuclei, hippocampus, cerebral cortex); also in pancreas
Purines		
Adenosine	Mostly inhibitory; indirect action via second messengers	Throughout CNS and PNS
ATP	Excitatory or inhibitory; based on receptor; direct and indirect actions via second messengers	CNS: basal nuclei, causes calcium ion wave propagation in astrocytes; PNS: dorsal root ganglion neurons
Dissolved gases and lipids		
Carbon monoxide (CO)	Excitatory or inhibitory; indirect action via second messengers	Brain; also certain neuromuscular and neuroglandular synapses
Nitric oxide (NO)	Excitatory or inhibitory; indirect action via second messengers	CNS: brain, spinal cord; PNS: adrenal gland; nerves to penis
Endocannabinoids such as 2-arachidonoylglyerol and anandamide	Inhibitory; indirect action via second messengers	Throughout CNS

between **acetyl coenzyme A** and **choline**. Then, ACh is transported into the synaptic vesicles. When released by the presynaptic terminal, ACh is briefly bound to postsynaptic receptors. It is then released and degraded by **acetylcholinesterase** (AChE), becoming acetic acid and choline. The AChE enzyme is located within the synaptic cleft and on postsynaptic membranes. The presynaptic terminals recapture the choline that is released, reusing it to synthesize more ACh. All neurons that stimulate skeletal muscles, and many ANS neurons, release ACh, though ACh-releasing neurons are also in the CNS. The effects of ACh are prolonged when AChE is blocked by nerve gas or organophosphate insecticides such as **malathion**. This causes tetanic muscle spasms. Release of ACh can be inhibited by botulinum toxin. Levels of ACh are decreased in certain areas of the brain in patients with Alzheimer's disease.

Receptors that bind to ACh are called **cholinergic receptors**. They are additionally divided into **muscarinic** and **nicotinic** receptors. The muscarinic receptors active end-organ effector cells within the bronchial smooth muscle, the salivary glands, and the sino-atrial node. Binding of ACh to the muscarinic receptors is inhibited by atropine. The nicotinic receptors are located within the **somatic system**, on motor end plates of skeletal muscle cells, the **adrenal medulla**, and all **postganglionic neurons**. Binding of ACh to a nicotinic receptor results in an excitatory effect. The nicotinic receptors are destroyed in the condition known as *myasthenia gravis* (MG), which will be discussed later in this chapter.

Biogenic amines

Biogenic amine neurotransmitters are found throughout the brain, with roles in emotions and regulation of the body's biological clock. The biogenic amines include the **catecholamines** and **indolamines**. Dopamine, epinephrine, and NE are examples of catecholamines. NE and dopamine play a role in feeling good and euphoric. Via a common pathway, dopamine and NE are synthesized from the amino acid called *tyrosine*. The same pathway is used by the adrenal medulla and the brain's epinephrine-releasing cells. Histamine and serotonin are examples of indolamines. Histamine is synthesized from the amino acid called *histidine*. Serotonin is synthesized from *tryptophan*, another amino acid. Post ganglion of sympathetic nerves and adrenal medulla release NE.

Dopamine
Dopamine is a monoamine that is synthesized from the amino acid *tyrosine*. Many enzymatic reactions are required, producing levodopa, then dopamine, then NE, and epinephrine. Dopamine is excitatory or inhibitory—based on which receptor subtype is activated. Dopamine functions in the brain as part of behavior, cognition, learning, mood, sleep, and voluntary movements. Its actions are mostly within the substantia nigra, hypothalamic, arcuate nucleus, and the midbrain's ventral tegmental portion. Dopamine is essential in the reward system of the body. It is inactivated via reuptake, using the

dopamine transporter, followed by the enzymatic breakdown. Enzymes involved in this breakdown include *catechol-O-methyltransferase* (COMT) and monoamine oxidase (MAO). Normally, dopamine promotes smooth and coordinated muscle movements. The release of dopamine is enhanced by amphetamines and levodopa. Its reuptake is blocked by cocaine.

The D_{1-5} dopamine receptors are utilized for the increase in actions of the direct basal ganglia pathway. The receptors are further summarized as follows:
- **D_1**—help regulate growth and development of neurons, behavioral responses, and D_2-receptor signaling. These receptors activate a G_s protein, which then activates adenyl cyclase and cAMP signaling.
- **D_2**—believed to be involved in muscle tone regulation, and are associated with schizophrenia. These receptors activate a G_i protein, inhibiting cAMP production.
- **D_3**—believed to be involved in emotions, depression, cognition, drug addiction, and schizophrenia (see Fig. 18.1). These receptors also activate a G_i protein, inhibiting cAMP production.
- **D_4**—believed to be involved in thrill-seeking behaviors and schizophrenia. These receptors also activate a G_i protein, inhibiting cAMP production.
- **D_5**—believed to be involved in blood pressure regulation. These receptors also activate a G_s protein, followed by activation of adenyl cyclase and cAMP signaling.

When there is not enough biosynthesis of dopamine in the dopaminergic neurons, **Parkinson's disease** may develop. Within the frontal lobes of the brain, dopamine controls information flowing from other areas. In the frontal lobes, dopamine disorders may result in reduced neurocognitive functions. These especially include memory, attention, and the ability to solve problems. Imbalance of Dopamine is associated with mental illness. In schizophrenia, for example, there is overactive dopamine signaling. Also, some psychoactive drugs, such as mescaline and lysergic acid diethylamide (LSD) can bind to biogenic amine receptors, resulting in **hallucinations**.

> **Focus on dopamine**
> Addictions such as food, love, gambling, and drugs are all associated with dopamine. The amino acids contained in vegetables stimulate dopamine production. Though very few neurons manufacture dopamine, many of them transmit it. The regulation and stimulation of dopamine levels can be modulated by tyrosine, which is found in many types of meat and fruit.

Norepinephrine and epinephrine
NE, also called *noradrenaline*, is synthesized from tyrosine, within the cytoplasm. It is packaged into postganglionic fiber vesicles and released through exocytosis. There is the termination of NE reuptake into the postganglionic nerve ending, receptor site

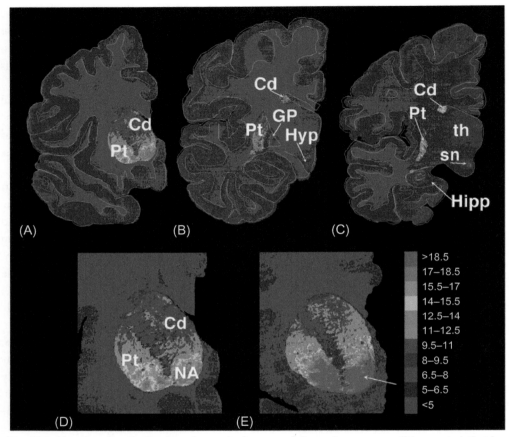

Fig. 18.1 Mapping the distribution of a particular type of dopamine receptor (D3 receptor) that has been implicated in reward circuitry, using a radioactive ligand. In coronal sections of normal brain (A–D) the receptor is preferentially localized in the ventral striatum, particularly the nucleus accumbens (NA). In the brain of a chronic cocaine user who died of a cocaine overdose (E), this preferential distribution is even more pronounced. The color bar at the lower right shows the color coding of receptor density, measured as radioligand binding (fmol/mg). *Cd*, caudate nucleus; *GP*, globus pallidus; *Hipp*, hippocampus; *Hyp*, hypothalamus; *Pt*, putamen; *sn*, substantia nigra; *th*, thalamus. *(From Staley, J. K.; Mash. D. C. J. Neurosci. 1996, 16, 6100.)*

diffusion, or enzyme metabolism. The sympathetic postganglionic neurons use NE. The release of NE is enhanced by amphetamines. Removal of NE from the synapse is blocked by cocaine and tricyclic antidepressants (TCAs). Brain levels of NE are reduced by the antihypertensive drug *reserpine*, which results in depression.

Epinephrine, also called *adrenaline*, is the final product of the tyrosine pathway, within **chromaffin cells** of the adrenal medulla. These cells also produce dopamine and NE. Epinephrine has many effects on various organ systems, along with metabolic functions. The degradation pathways of NE and epinephrine are similar. They both act as hormones and as neurotransmitters, via the adrenergic receptors. The adrenergic receptors are classified as alpha (α) and beta (β). Epinephrine and NE bind to all of the types of adrenergic receptors, but with various levels of affinity.

- **α$_1$-receptors**—postsynaptic adrenoreceptors in the eyes, smooth muscles, blood vessels, lungs, gut, and genitourinary system. Stimulation activates a G$_q$ protein, causing increased phospholipase activity. This releases diacylglycerol. It also releases inositol trisphosphate, which promotes increased calcium. The signals are excitatory, indicated by pupil dilation (mydriasis), bronchodilation, blood vessel constriction, increased heart contraction strength (positive **inotropy**), as well as decreased heart rate (negative **chronotropy**). NE binds to these receptors better than does epinephrine.
- **α$_2$-receptors**—mostly on presynaptic nerve terminals. The activation of G$_i$ proteins inhibits **adenyl cyclase** activity and cAMP production. This decreases entry of unbound calcium into the neuronal terminal, reducing amounts of NE released. Also, it causes vasoconstriction and reduced sympathetic outflow in the CNS. Epinephrine binds to these receptors better than does NE.
- **β$_1$-receptors**—stimulation of these receptors activates a G$_s$ protein. This increases the activities of adenyl cyclase, adenosine triphosphate (ATP) conversion into cAMP, and protein kinase phosphorylation of many proteins. It causes positive chronotropy (increased heart rate), inotropy (heart contraction force), and **dromotropy**, which is impulse conduction through the atrioventricular node of the heart. It is not fully understood if epinephrine and NE bind equally to this receptor, or if epinephrine binding is preferred.
- **β$_2$-receptors**—mostly postsynaptic adrenoceptors in glands and smooth muscles. They activate adenyl cyclase using a G$_s$ protein and inhibit the same enzyme via a G$_i$ protein. These opposite effects may allow different cell and tissue functions. Smooth muscle is relaxed by the B2 stimulation, causing bronchodilation, vasodilation, insulin release, and induction of **gluconeogenesis**. Epinephrine binds to these receptors with much more affinity than NE.
- **β$_3$-receptors**—primarily found in fat and adipose tissue. These receptors also activate adenyl cyclase, using a G$_s$ protein. These receptors induce the breakdown of fats and lipids via **lipolysis**, and help regulate heat production (**thermogenesis**), primarily in skeletal muscle. NE binds to these receptors better than epinephrine.

Histamine

Histamine is a vasoactive, biogenic amine that is involved in local immune responses, appetite control, learning, memory, and wakefulness. It is synthesized from the amino acid *histidine*. Histamine is involved in the inflammatory response and plays a central role in the mediation of itching. It causes temporary and rapid constriction of large blood vessel walls and dilation of postcapillary venules. These actions result in increased blood flow into the body's microcirculation. Vascular permeability is increased, from retraction of endothelial cells that line the capillaries. Pharmacologic effects of histamine are, in part, based on histamine receptors located on target cells. The two primary receptors are called *H1* and *H2*, but there are also *H3* and *H4* receptors. When histamine binds to the H1 receptor, the effect is mostly pro-inflammatory. When it binds to the H2 receptor, an antiinflammatory effect occurs due to suppression of leukocyte functions. The H1 receptor is located on smooth muscle cells—primarily in the bronchi—causing bronchoconstriction. Both receptor types are found in many different cells, often on the same cells, and are able to act antagonistically. Neutrophils often have H1 and H2 receptors, with H1 stimulation causing **chemotaxis** and H2 stimulation causing inhibition. The H2 receptors are found in significant amounts on stomach mucosa parietal cells, inducing secretion of gastric acid.

Serotonin

Serotonin is also known as *5-hydroxytryptamine* or *5-HT*. It is a monoamine neurotransmitter that is used in functions such as eating, sleep, arousal, dreaming, mood regulation, temperature regulation, and pain transmission. It is synthesized from the amino acid known as *tryptophan*, using a short metabolic pathway. This is made up of two enzymes, **tryptophan hydroxylase** and **amino acid decarboxylase**. There are seven serotonin receptors that have been identified, using G_i, G_q, or G_s proteins, which activate an intracellular second-messenger cascade. One of these receptors functions as a sodium-potassium channel, resulting in membrane depolarization. Serotonin is terminated through uptake at the synapse of the presynaptic neuron. Many different substances can inhibit serotonin reuptake and thus relieve anxiety and depression. These include

- Amphetamine
- Cocaine
- Dextromethorphan
- Ecstasy (3,4-methylenedioxymethamphetamine, MDMA)
- Selective serotonin reuptake inhibitors (SSRIs)
- Tricyclic antidepressants (TCAs)

The activity of serotonin is blocked by LSD.

Section review
1. What are the differences between excitatory and inhibitory neurotransmitters?
2. How is ACh synthesized?
3. What agents may enhance or block dopamine?
4. What are the functions of serotonin?

Amino acids

Since amino acids are present in all body cells, and are vital for many biochemical reactions, it is difficult to identify the role of an amino acid that may be acting as a neurotransmitter. The amino acids with known neurotransmitter roles include *glutamate*, *GABA*, and *glycine*. There may be other amino acids with neurotransmitter capabilities.

Glutamate, GABA, and Glycine

Glutamate, or *glutamic acid*, is the most prevalent CNS **excitatory** neurotransmitter, and perhaps the most common neurotransmitter within the brain. It is important in learning and memory among other functions. Excessive release of glutamate causes excitotoxicity, in which the neurons are stimulated continuously. Through **glutaminergic receptor** opening of nonselective ion channels, glutamate is generally excitatory. Glutamate stimulation is terminated by a membrane transport system. This is used in reabsorption of glutamate and aspartate (aspartic acid) over the presynaptic membrane. These neurotransmitters reenter cells as sodium enters via the sodium-potassium ATP pump.

GABA is an amino acid neurotransmitter synthesized from glutamic acid. It is the most prevalent **inhibitory** brain neurotransmitter of all, and is important in presynaptic inhibition at the axoaxonal synapses. It is predominantly concentrated within the substantia nigra, hypothalamus, and hippocampus. This substance reduces the excitability of neurons via hyperpolarization. The two types of GABA receptors are as follows:
- **GABA$_A$**—activation of these receptors opens chloride channels
- **GABA$_B$**—these receptors act through a second messenger, either opening potassium channels, or closing calcium channels

The inactivation of GABA is through active transport into the astrocyte glial cells, close to the synapses. The inhibitory effects of GABA are affected by alcohol, benzodiazepines, and barbiturates. Any substance that blocks GABA synthesis, release, or action causes convulsions. Many abused drugs affect either glutamate, GABA, or both, resulting in the brain being stimulated or tranquilized.

Glycine is a widely distributed neurotransmitter in the brainstem, spinal cord, and retinas of the eyes. It is actually the *main inhibitory neurotransmitter* in the spinal cord. Its receptor is a gated channel made up of five protein subunits, allowing for an influx of chloride ions. The influx acts as an **inhibitory postsynaptic potential (IPSP)**. This reduces the occurrences of postsynaptic or motor neuron action potentials in the future. Competitive inhibitors of glycine receptors include *caffeine* and the poison called **strychnine**, which causes uncontrolled convulsions and respiratory arrest. Glycine acts as an essential coagonist for the glutamate receptor.

Neuropeptides

Neuropeptides are basically chains of amino acids, produced in the brain from the cleaving of larger polypeptides. They have many different types of molecules and widely different effects. Neuropeptide receptors are classified as delta (δ), kappa (κ), mu (μ), and as nociception-type receptors. They function via many G proteins. The **endorphins** and **enkephalins** are natural opiates. They reduce pain perception in stressful conditions and producing analgesia and euphoria via inhibition of *substance P*. The endorphins include **beta-endorphin** and **dynorphin**. Endorphin release is increased when an athlete gets a *second wind* during physical activity. This is believed to be responsible for the "runner's high" experienced by a long-distance marathon participant. The *placebo effect* is believed to be caused by release of endorphins. Studies have shown that endorphins attach the same receptors that bind natural opiates. Similar and even stronger effects are produced by opiate medications such as fentanyl and morphine. The enkephalins are *pentapeptides* that are also referred to as *endogenous* ligands derived from and binding to opioid receptors. There are two forms of enkephalins. One contains *leucine* and the other contains *methionine*, which are both produced by the *proenkephalin gene*. Enkephalins are in large concentrations in the brain as well as the cells of the adrenal medulla. The activities of enkephalins are greatly increased in pregnant women during labor.

The *tachykinin* called **substance P** is an essential pain signal mediator within the PNS. It is found in the synaptic vesicle of unmyelinated C fibers. These fibers enhance pain signal transmission. The substance P receptor is a protein. In the CNS, **tachykinins** such as substance P and **neurokinin A** (NKA) are involved in the control of respirations, cardiovascular functions, and mood. Substance P acts as both a neurotransmitter and a neuromodulator and is closely related to NKA, which was formerly known as *substance K*.

Gut-brain peptides are neuropeptides produced by nonneural tissues within the body. They are located throughout the gastrointestinal tract and include **cholecystokinin** (CCK) and **somatostatin**. CCK is involved in anxiety, memory, and pain. It also inhibits appetite and is sometimes referred to as *pancreozymin*. Actual synthesis and secretion of CCK are from the enteroendocrine cells of the duodenum, causing the release of digestive enzymes from the pancreas and bile from the gallbladder. Somatostatin is often released with GABA, and inhibits the release of growth hormone. It is also known as *growth hormone-inhibiting hormone (GHIH)* and regulates the endocrine system. Somatostatin also affects neurotransmission and cell proliferation via interaction with G protein-coupled somatostatin receptors and inhibiting release of many secondary hormones. Additionally, somatostatin inhibits secretion of insulin and glucagon.

Purines

Purines are the breakdown products of nucleic acids and are chemicals that contain nitrogen. **Adenosine** acts outside of cells on adenosine receptors, along with the extracellular **ATP** neurotransmitter actions. Adenosine is a strong inhibitor within the brain. Blockage of adenosine receptors allows substances such as caffeine to have strong stimulatory effects. Adenosine may be involved in the sleep-wake cycle and control of seizures. It dilates arterioles, which increases blood flow to the heart and other tissues when needed.

ATP is the universal form of cellular energy. It is also an extremely primitive but major neurotransmitter within the CNS and PNS, and is released by sensory neurons and by injured cells. When released by injured cells, pain sensations are provoked. Similar to receptors for ACh and glutamate, some receptors create quick excitatory responses when ATP binds. Other ATP receptors have slower, second-messenger responses. When it binds to astrocyte receptors. ATP mediates unbound calcium (Ca^{2+}) influx. Other examples of purines include *adenine* and *guanine*.

Dissolved gases

Small, toxic gas molecules, of a short-lived nature, may act as neurotransmitters. These gases are referred to **gasotransmitters**, which include carbon monoxide (CO), hydrogen sulfide (H_2S), and nitric oxide (NO). All of them are different than the standard definition of a "neurotransmitter." They are not stored in vesicles or released by exocytosis in the manner of other types of neurotransmitters. Gasotransmitters do not attach to surface receptors. They travel quickly through plasma membranes of nearby cells, binding with intracellular receptors.

CO and **NO** activate the enzyme **guanylate cyclase**, which creates the second messenger known as *cyclic GMP*. Both CO and NO are present in different areas of the brain. They are believed to act via different pathways, yet have a similar mode of action. In the brain, NO participates in many different processes. These include new memory formation, via increasing strengths of specific synapses. In stroke patients, excessive release of NO is believed to contribute to brain damage. In the peripheral nervous system, NO causes relaxation of blood vessels and intestinal smooth muscle. Certain types of male impotence are treated by enhancing the action of NO with drugs such as *sildenafil*. **H_2S** is the newest gasotransmitter discovered and is not fully understood. Different from CO and NO, it is believed to act directly on proteins such as ion channels, altering their functions.

Opioid peptides

The endogenous ligands for opiate receptors are the **opioid peptides**. In the pathways of the central nervous system, these include the *enkephalins* and *endorphins*. Small interneurons that contain inhibitory interneurons help to suppress transmission by spinothalamic tract neurons at the spinal cord level. Like all peptides, the opioid peptide molecules are larger in size than many other neurotransmitter molecules. These peptides may be produced by the body itself, or absorbed from partially digested food. The effects of opioid peptides vary, but always resemble the effects of opiate medications. Brain opioid peptides are important in emotions, motivation, attachment behaviors, pain and stress responses, and control of food intake. The endogenous opioids are produced in the body, aside from enkephalins and endorphins.

Nonopioid neuropeptides

The nonopioid neuropeptides include *substance P, vasoactive intestinal peptide*, and *calcitonin gene-related peptide*. Substance P is classified as an *undecapeptide*, meaning that it is composed of a chain of 11 amino acid residues. It is part of the *tachykinin* neuropeptide family, acting as both a neurotransmitter and neuromodulator. Substance P is closely related to *NKA*, and both are produced from a polyproteins precursor, following differential splicing of the *preprotachykinin A gene*. Substance P is released from the terminals of certain sensory nerves, in the brain and spinal cord.

Vasoactive intestinal peptide is also called *vasoactive intestinal polypeptide*, or *VIP*. This peptide hormone is vasoactive inside the intestines. It has 28 amino acid residues and belongs to a glucagon/secretin superfamily, the ligand of *class II G protein-coupled receptors*. This peptide is produced in the gut, pancreas, and the hypothalamic suprachiasmatic nuclei. It stimulates heart contractility, causes vasodilation, increases glycogenolysis, and lowers arterial blood pressure. It also relaxes the smooth muscle of the gallbladder, stomach, and trachea. It is involved in the regulation of circadian rhythm, induces smooth digestive muscle relaxation, stimulates secretion of water into bile and pancreatic juice, and causes inhibition of gastric acid secretion and absorption from the intestinal lumen.

Neuromodulators

Neuromodulators are neurotransmitters that diffuse through neural tissue. They affect the slow-acting receptors of many different neurons. The major CNS neuromodulators include dopamine, ACh, serotonin, histamine, and NE. Neuromodulators have modulatory effects on target areas. Neuromodulators are not reabsorbed by presynaptic neurons or broken down into metabolites. They are present for a long time in the cerebrospinal fluid, influencing the activities of several other neurons within the brain. The actual process of *neuromodulation* is explained as a neuron using one or more chemicals to regulate many different neuron populations. This means that neuromodulation is distinct from the classical, fast synaptic transmission. In neuromodulation, the receptors are usually G-protein coupled receptors instead of the ligand-gated ion channels seen in classical neurotransmission. Neuromodulators may alter output of physiological systems by acting on associated inputs. They are believed to change transformations to produce proper effects in the neuromuscular system.

Functional considerations

There are two primary functional classifications of neurotransmitters. The function of any neurotransmitter is determined by the receptor that it binds with. A neurotransmitter may be *excitatory*, causing depolarization, or *inhibitory*, causing hyperpolarization. Some neurotransmitters have both of these effects, which is based on certain types of receptors that interact with them. Glutamate is an example of a neurotransmitter with mostly excitatory effects. The amino acids GABA and glycine have mostly inhibitory effects. ACh and NE bind to two or more receptor types, causing opposing effects. For example, ACh has excitatory effects at NMJs in skeletal muscles but has inhibitor effects in cardiac muscle.

Neurotransmitters may also be classified as having direct or indirect actions. The neurotransmitters acting directly bind to ion channels, and also open them. They cause fast responses in postsynaptic cells via alteration of membrane potential. Usually, ACh and amino acid neurotransmitters have direct actions. The indirect-acting neurotransmitters cause wider and longer-lasting effects. They act through intracellular second-messenger molecules—usually through G protein pathways. This means that their actions resemble those of hormones. Examples of indirect neurotransmitters include biogenic amines, dissolved gases, and biogenic amines.

A chemical messenger that is released by a neuron, and does not directly cause an excitatory postsynaptic potential (EPSP) or an IPSP is known as a *neuromodulator*. Instead, neuromodulators affect the strength of synaptic transmissions. They may have presynaptic actions and influence neurotransmitter synthesis, release, degradation, or reuptake. A neuromodulator can also act postsynaptically, altering sensitivity of a postsynaptic membrane to a neurotransmitter. Neuromodulator receptors may not always be located at a synapse. Neuromodulators can be released from a certain cell, and act on many nearby cells. This is similar to the *paracrines*, which are local-acting chemical messengers that are quickly destroyed. Though neuromodulators and neurotransmitters may be very similar, examples of neuromodulators generally include adenosine, chemical messengers such as NO, and many different neuropeptides.

Synthesis of neurotransmitters

Most neurotransmitters are small molecules, such as amines or amino acids, while others are larger neuropeptides or modified lipids. The amines and amino acids are synthesized within presynaptic cytoplasm. This usually requires local substrates such as acetate or choline and soluble enzymes that are carried via slow axonal transport. The synthesis of peptide transmitters is different. It requires protein synthesis within a neuronal cell body. Neuropeptides begin in this location, as larger precursor proteins. They are packaged into vesicles, then sent out via fast axonal transport to the synaptic endings. Along their route, precursors are cleaved, then processed into final and biologically active neuropeptides.

Presynaptic release of neurotransmitters into the synaptic cleft

A Ca^{2+} mediated secretory process is required to release vesicles that contain neurotransmitters. An **active zone** is a presynaptic density that contains many voltage-gated unbound calcium channels. These zones also have anchoring sites for many small vesicles kept there by an unbound calcium-sensitive protein system. Presynaptic terminal depolarization, which may occur from the propagation of an action potential down the axon, with electronic spreading into the terminal, causes the voltage-gated channels to open. The free intracellular Ca^{2+} concentration is only about 10^{-7} **molar** (M), the unbound calcium ions flow through the channels and quickly elevate calcium concentrations, near the active zone, by up to 1000 times as much. In the brief time prior to excess Ca^{2+} being diffused off or sequestered, any vesicle kept nearby can fuse with the presynaptic membrane. This causes discharge of its contents into the synaptic cleft, via **exocytosis**. The entire process takes less than 100 ms. The large-dense-cored vesicles are not near to the active zones. Therefore, their exocytosis requires repeated action potentials, more unbound calcium entry, and tens of milliseconds to occur. This exocytotic membrane addition cannot continue for long. If it did, the presynaptic endings would continuously enlarge. Vesicle membranes, via **endocytosis**, are taken back up. Those required for small-molecule transmitters are able to be recycled in under 1 min.

Since the release of neurotransmitters is based on the entry of unbound calcium, modulation of voltage-gated calcium channels often results in a significant physiological response. Certain toxins or drugs can modulate the channels, inhibiting or increasing neuronal activity. Also, botulinum toxin and other toxins, which interfere with functions of vesicles containing ACh to inhibit release, result in skeletal muscle relaxation.

Neurotransmitter binding to postsynaptic receptors

Neurotransmitters from small vesicles target postsynaptic receptor molecules, directly across the synaptic cleft (Fig. 18.2). Therefore, the contents of these vesicles need only a short time for their effects to occur. The entire synaptic delay, from presynaptic action potential to postsynaptic effect, may be less than 200 ms. The large vesicle contents take longer for release, and usually diffuse to receptors that are further away. This means that their effects occur more slowly. There is self-regulating transmitter release, and sometimes, the presynaptic neuron also contains **autoreceptors**. These often cause neurotransmitter release to be slower.

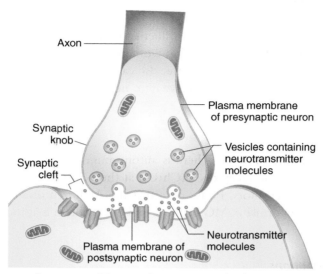

Fig. 18.2 Components of a synapse. Diagram shows synaptic knob or axon terminal of presynaptic neuron, the plasma membrane of a postsynaptic neuron, and a synaptic cleft. On the arrival of an action potential at a synaptic knob, neurotransmitter molecules are released from vesicles in the knob into the synaptic cleft. The combining of neurotransmitter and receptor molecules in the plasma membrane of the postsynaptic neuron opens ion channels and thereby initiates impulse conduction in the postsynaptic neuron.

Section review
1. What is the most prevalent CNS excitatory neurotransmitter, and what are its effects?
2. What is the name and functions of the universal form of cellular energy?
3. How do neuromodulators act?

Clinical considerations

There are a variety of conditions that may be linked to neurotransmitters. ACh is related to MG, dopamine production is related to Parkinson's disease, and may be related to schizophrenia and attention deficit hyperactivity disorder. Problems producing or using glutamate are also linked to schizophrenia, as well as to autism, obsessive-compulsive disorder (OCD), and depression. Depressive disorders are also related to many other neurotransmitter imbalances. Other neurotransmitter-related conditions include bipolar disorder and cyclothymic disorder.

Myasthenia gravis

MG is *a NMJ disease*. It is characterized by fluctuating weakness and fatigable weakness of skeletal muscles. In most cases, MG strikes women in their 30s or men in their 60s–70s. There is also **Lambert–Eaton syndrome**, also called *myasthenic syndrome*. This involves proximal muscle weakness, due to presynaptic neuromuscular transmission related to malignancies.

Pathophysiology

MG is an autoimmune disorder. There is autoimmune destruction or inactivation of postsynaptic acetylcholine receptors (AChRs), at the NMJs. This functionally reduces the amount of ACh receptors. Please note that the ACh neurotransmitter production and reabsorption are normal in MG patients, and the problem is at the level of the post-synaptic receptor.

Clinical manifestations

Symptoms include *diplopia* (double vision), *dysphagia* (difficulty swallowing), and *ptosis* (drooping of the eyelids—see Fig. 18.3), which develop toward the end of each day. The condition can be relapsing or remitting, but usually progresses slowly, and can lead to severe respiratory weakness and its complications. The symptoms may be temporarily worsened by infections, stress, menses, pregnancy, or surgery.

Diagnosis

Diagnosis of MG is clinical, via a bedside anticholinesterase test, and measurement of AChR antibody levels, with or without electromyography, which can also be diagnostic on its own. The antibodies are present in 80%–90% of patients with generalized MG, but

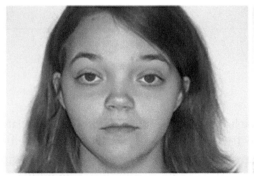 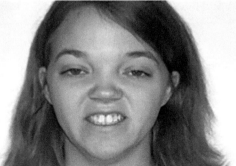

Fig. 18.3 Typical myasthenic facies. At rest (*left*), there is slight bilateral lid ptosis, which is partially compensated by asymmetric contraction of the frontalis muscle, raising the right eyebrow. During attempted smile (*right*), there is contraction of the medical portion of the upper lip and horizontal contraction of the corners of the mouth without the natural upward curling, producing a "sneer."

only in 50% with the ocular form of the disease. The antibody levels do not relate to the severity of the disease. Some patients without usual AChR antibodies (binding, blocking and modulating antibodies) will test positive for another antibody anti-*muscle-specific receptor tyrosine kinase* (MuSK). An EMG, using repetitive stimuli over 2–3 per second will show a significant decrease in amplitude of the muscle action potential response in about 60% of patients (known as a *decrementing response*). Single-fiber EMG is able to detect the decrease in more than 95% of patients. Once diagnosis is confirmed, a CT or MRI of the thorax is performed to check for the presence of a thymoma. Other tests are done to screen for vitamin B12 deficiency, hyperthyroidism, rheumatoid arthritis, systemic lupus erythematosus, and other autoimmune conditions. If the patient is in myasthenic crisis, he or she must be evaluated for an infection that triggered the condition. Myasthenic crisis or cholinergic crisis can be differentiated by administering intravenous *edrophonium*, and assessing if there is improvement, no improvement, or worsening of weakness. If there is improvement, the patient is in myasthenic crisis. If there is no improvement or worsening, the patient is in cholinergic crisis. Serial pulmonary function assessments, such as forced vital capacity (FVC) and negative inspiratory force (NIF) are needed to identify patients in acute exacerbation, who are approaching respiratory failure.

Treatment
Treatment is different for acute exacerbations, requiring steroids, intravenous immunoglobulin (IVIG) and plasmapheresis (plasma exchange). Pyridostigmine and other anticholinesterase drugs treat symptoms via increasing amounts of available ACh in the NMJs, though inhibition of AChE. Long-term immunosuppressants such as azathioprine are needed in patients with recurrent exacerbations to stop the progression of the disease. A thymoma usually requires surgical removal. The effects of this surgery take several years to appear.

> **Focus on myasthenia gravis**
> Additional symptoms of myasthenia gravis (MG) include memory loss, and inability to chew, and what patients describe as "brain fog." In most cases, there are variations in all symptoms, at some point in the disease course, to different degrees. Ongoing research about MG is trying to determine if female hormones play a part in why the disease is more prominent in women.

Parkinson's disease
Parkinson's disease is a condition of involuntary tremors, muscle weakening, stooped posture with instability, gait disturbances, and rigidity or slowness of movements. It generally begins between 45 and 70 years of age, with peak onset in the sixth decade. The motor symptoms of Parkinson's disease result from cellular death within the substantia

nigra of the midbrain. This results in a lack of dopamine in these areas. Low levels of dopamine demand greater effort in order to make voluntary movements. Dopamine depletion produces hypokinesia, an overall reduction in motor output. Parkinson's disease was discussed in greater detail in Chapter 7.

Schizophrenia

Schizophrenia is a serious neurological disorder that affects just under 1% of the worldwide population. It has an early onset, is chronic, and causes personal, social, and vocational problems on a severe level. It is characterized by disordered thinking, affect, and behaviors. Incidence is higher in social classes that have high mobility and disorganization. Schizophrenics make up about 50% of those hospitalized for mental reasons. They also make up 20%–30% of all new admissions to psychiatric facilities, which is 100,000–200,000 new cases annually in the United States. Age of admission is usually between 20 and 40 years. Peak incidence is between 28 and 34 years. Schizophrenia costs over $50 billion annually in this country for management and treatment.

Pathophysiology

Schizophrenia appears to develop from combined genetic and early development events. There may be abnormalities of synaptic connectivity that affect the hippocampus, prefrontal cortex, and cerebroventricles. Normal and schizophrenic brains are compared in Fig. 18.4. Many neurotransmitters may be implicated in schizophrenia, including dopamine, glutamate, serotonin, and GABA. Genetic factors may be implicated in 80% or more of schizophrenia cases. Risks are higher for the siblings—especially twins—of schizophrenic patients to also develop the disease, and also for children of schizophrenic parents. Genes implicated as risk factors include those that express *neuregulin, dysbindin, catechol O-methyltransferase, proline dehydrogenase,* and *disrupted in schizophrenia 1 (DISC1)*. However, environmental factors may be of higher significance, since 80% of schizophrenics have no other family members affected by the disease. Environmental factors are not proven but may include brain injury during gestation or delivery, other obstetric complications, northern-latitude or urban births during the winter, and maternal viral infections.

The brains of schizophrenics are different than in normal individuals in that the pyramidal cells are smaller and more densely packed. There is a thinning of laminae II and III, and decreased density of dendritic spines in the frontal and temporal cortices. There is an atypical distribution of interstitial neurons within the frontal lobe white matter. The number of small neurons is reduced in, usually layer II of the anterior cingulate cortex, which are the GABA-releasing neurons. Macrocolumns of the cortical neurons are smaller in the occipital lobes, with vertical axons increased in number. There is a lack of gabanergic, inhibitory interneurons in the prefrontal cortex. A reduction in the volume of the superior temporal gyrus is linked to auditory hallucinations.

Schizophrenia

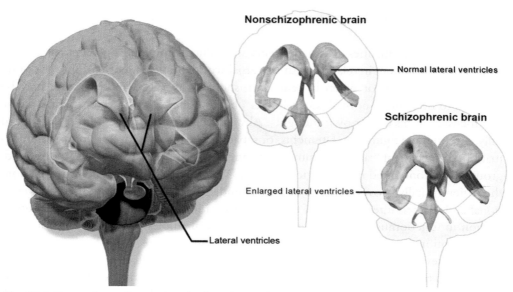

Fig. 18.4 Comparison of normal and schizophrenic brains.

Clinical manifestations
There are three "clusters" of behavior in schizophrenia, as follows:
- *Negative symptoms*—from diminished psychomotor activity; these include reduced speech patterns and spontaneous movements, and flatness of affect; the speech shows a lack of additional or unprompted content that is seen in normal speech patterns; a *reduced affect display* may also be described as *emotional blunting*, which is reduced emotional reactivity, a failure to express feelings, with rare expressive gestures, and little animation in facial expressions or vocal inflections
    - Also: latency of response, apathy, restriction in a recreational activity, inability to feel close or intimate, motor retardation—these resemble frontal lobe behavioral syndromes; they correlate with reduced frontal lobe blood flow and a poor prognosis
- *Disorganization syndrome or thought disorder*—fragmentation of ideas, loosened associations, *tangentiality*, and inappropriate emotional expressions; the patient shows tangential speech patterns, characterized by replies to questions that never approach the intended responses—this is described as digressive, irrelevant, or oblique speech
    - Also: incoherence, illogicality, bizarre behavior, aggression, and agitation—these are not related to the frontal lobes
- *Reality distortion*—including hallucinations and delusions; these are considered to be "positive" symptoms since they are considered more diagnostic of schizophrenia; these are probably related to temporal lobe function

Some patients exhibit delusions of reference, hostility, and suspicion. All of the various syndrome clusters can coexist, with various combinations. While males and females are affected with equal frequency, males appear to have more severe symptoms in general.

The focal abnormalities of schizophrenia are hallucinations and a perceptual disorder concerning the patient in relation to the external world. Some patients, prior to the onset of extreme psychosis or when in remission, show no usually symptoms, and may be believed to be entirely normal. Over longer observation, however, the patient is vague and focused on his or her own thoughts. Abstract thinking is poor, and it is difficult to fully comprehend figurative statements, or separate important from nonimportant information. The patient cannot communicate ideas clearly. Parts of ideas are confused, grouped, or condensed in an illogical manner. When analyzing a problem or situation, the patient usually is overinclusive instead of underinclusive. There are often conversational and written interruptions of thoughts or positions. There is a generalized deterioration in functioning that includes social withdrawal, aimlessness, bizarre actions, and self-absorption.

When severe, schizophrenia causes the patient to become totally preoccupied with internalized thoughts. He or she may only express groups of meaningless phrases or neologisms. The speech is often described as *word salad*—lots of different words being used, but not in understandable order. Concentration is extremely poor, and often, some intrusive idea or act gets in the way of what they were attempting to do. The patient may appear severely confused or delirious. Sometimes, schizophrenics are extremely talkative and show odd behaviors, while at other times they are quiet and remote from all others, often saying or doing nothing. In extreme cases, schizophrenics say nothing at all and may assume immobile postures, such as in catalepsy. If remission occurs, they may remember a lot of what occurred, or only have fragmented memories.

The schizophrenic patient often expresses a division between the mind and body, as if not understanding their connection, which is known as *depersonalization. Thought insertion* is an expression that someone else put an idea into the patient's mind, while *thought withdrawal* describes someone else extracting an idea from the patient. Commonly, the patient believes that all thoughts or actions are being controlled by others, via electronic devices or mental telepathy. The patient may also state that he or she can control environmental factors or even other people by simply thinking about what is desired to happen. If *derealization* is present, the patient feels that the world is altered, unnatural, or the ability to understand time passage is difficult. *Auditory hallucinations* are common, with voices perceived that accuse, threatening, or take control of the patient. There may be one or more voices that are perceived. Since the voices usually seem to come from inside the patient's body, they are not distinguishable from feelings and thoughts. Less frequent, but sometimes present, are olfactory, visual, and other types of hallucinations. These are often part of a delusion system of thinking. Visual hallucinations are a component in many neurologic processes, while auditory hallucinations are a significant part of schizophrenia.

Early in the course of the disease, normal activities may be interrupted or slowed. The patient cannot function normally in most settings. Other people begin to find the patient's ideas, actions, and complaints to be disturbing. Social withdrawal is common. When a patient goes into a panic or has a frenzy of excited behavior, this often results in a trip to the emergency department. Schizophrenics may also become *catatonic*—mute and immobile. This is not highly common, but the disease is characterized by lack of drive, lack of will, lack of assertiveness, and slowed motor activity. The individual eventually becomes untidy and malnourished, often roaming out in public, and unfortunately, living in extremely poor conditions.

Schizophrenics are more likely to commit suicide as well, most commonly in younger patients living away from their families. They become frightened and overwhelmed by the symptoms, and cannot face tasks of daily living alone. Suicide may be caused by the intensity of the auditory hallucinations. Some schizophrenics are homicidal, usually reacting to a delusion that makes them feel threatened or wronged by another person. The presence of escalating paranoia is a warning of this situation. Some schizophrenics have a progressive and potentially severe intellectual impairment. Studies have revealed frontotemporal brain changes on CT scan and single-photon emission CT.

Some schizophrenics have period exacerbations that may be in regular intervals. Remissions that provide some return of normal function are more common and longer when medication is given and extended institutionalization is avoided. About 10% of patients, following an acute schizophrenic episode, have a lengthy, nearly complete remission, prior to lapsing into a chronic disease state. At the time of acute psychosis, these patients, unfortunately, cannot be distinguished from those who will have a permanent remission. The life expectancy of a schizophrenic is slightly reduced, linked to malnutrition, neglect, infections, exposure, and crime.

Early studies of schizophrenia revealed lack of application to assigned mental and motor tasks, **astereognosis**, graphesthesia, hyperreflexia or hyporeflexia, sensory extinction, slight tendency for grasping objects and others, balance and coordination impairment, abnormal movements, abnormal motor activity, anisocoria, slight esotropia, and abnormal visual-auditory integration. Such signs were seen in 50% of patients and related to the amount of cognitive disorder. About half of patients also have slight problems with ocular tracking movements. Electroencephalographic abnormalities are seen in about 33% of patients but are usually minor and of no certain meaning. The ability to maintain attention is usually reduced. Verbal and visual pattern learning, memorizing, and problem-solving are much more difficult. Left-hemispheric functions appear to be more reduced than right-hemispheric functions. In chronic schizophrenics, however, there is usually impairment of both hemispheres.

Diagnosis

The diagnosis of schizophrenia is based on recognizing characteristic psychological disturbances that are often unsupported by physical examination and laboratory testing. Patients are often diagnosed as being *schizophrenic* when they actually have another condition resembling schizophrenia. The diagnostic criteria for schizophrenia have changed greatly over time. While there is no single objective test, the American Psychiatric Association has set forth the following criteria for schizophrenia:

- Two diagnostic criteria have to be met over much of the time, of a period of at least 1 month
- There must be a significant impact on social or occupational functioning for at least 6 months
- There must be delusions, hallucinations, or disorganized speech
- There may be negative symptoms, or severely disorganized or catatonic behavior

If signs of disturbance are present for 1–6 months, the diagnosis of *schizophreniform disorder* is applied. Psychotic symptoms for less than 1 month may be diagnosed as a *brief psychotic disorder*. Various conditions are classified as *psychotic disorder not otherwise specified*. *Schizoaffective disorder* is diagnosed if symptoms of a mood disorder are substantially coexistent. Schizophrenia is not diagnosed if symptoms of a *pervasive developmental disorder* are present unless there are also prominent delusions or hallucinations.

Schizophrenia occurs along with *OCD* much more than would be explainable simply by chance. It is difficult to distinguish obsessions in OCD from the delusions of schizophrenia. It is important to understand that the previous subtypes of schizophrenia have now been removed from today's diagnoses. These include the *paranoid*, *disorganized*, *catatonic*, *undifferentiated*, and *residual* types. Additional defined subtypes that are no longer being used include *postschizophrenic depression*, *simple schizophrenia*, and *cenesthopathic schizophrenia*.

Treatment

Treatment of schizophrenia is focused on suppression of psychotic symptoms, reducing disordered thoughts and apathy, preventing relapses, and increasing social adjustment to the best level possible. When there is a hazard of injury, suicide, or difficult home management of the patient, hospitalization is indicated. Once drug therapy is successful in managing the condition, the patient must receive supervised, planner activities, and vocational therapy that often occurs in a *halfway house* with other patients. Preferred medications include second-generation atypical nonphenothiazine antipsychotic drugs. Their extrapyramidal adverse effects are far less than those of the phenothiazines. They act by inducing calmness, blunted emotional responses, reduce hallucinations, reduced aggression and impulsive behaviors, and yet leave cognitive functions almost completely intact. With modern therapies and psychiatric support, approximately 60% of schizophrenic patient recover enough to be able to return to their homes. They usually become socially adjusted to some degree, with about

50% being able to return to work. However, about 30% remain severely handicapped by their condition, and 10% must remain hospitalized.

> **Focus on schizophrenia**
> Schizophrenia is *not* a disorder of multiple personalities. Though schizophrenics experience hallucinations and delusions, they do not have "other" personalities. Also, schizophrenia is not related to reduce intelligence. Many famous schizophrenics, including as mathematician John Nash and ballet dancer Vaslav Nijinsky, were of extremely high intelligence.

Section review
1. What is the pathophysiology of myasthenia gravis?
2. What are the three clusters of behavior in schizophrenia?
3. What are the diagnostic criteria for schizophrenia?

Obsessive-compulsive disorder

OCD involves *obsessions* such as anxiety-causing ideas, images, or impulses. It also involves *compulsions*, which are urges to do something to reduce anxiety.

Pathophysiology
This condition is of unknown cause. It occurs about equally in men and women, affecting approximately 2% of the population.

Clinical manifestations
Obsessive thoughts usually involve perceived danger, harm, contamination, doubt, loss, risk, or aggression. The patient feels "forced" to perform purposeful and repetitive rituals to balance the obsessions. Examples of rituals include
- *Checking items*—to balance doubt
- *Avoiding others who may provoke them*—to balance a fear of behaving aggressively
- *Hoarding*—to balance loss
- *Washing*—to balance contamination

The majority of rituals, such as checking door locks or repeated hand washing, can be observed. Other rituals, such as statements made very quietly, or counting items repeatedly, may not be easily observed. Usually over time, the patient understands that the obsessions do not reflect actual risks, and the compulsive behaviors are excessive and unrealistic. The preservation of insight, no matter how slight or significant, is a differentiating factor between OCD and psychotic disorders, which involve a loss of contact with reality. A person with OCD fears being embarrassed by revealing their condition, which is able to occupy several hours per day. Performance at work or school may decline, and relationships often become difficult or end altogether. A common secondary feature of OCD is depression.

Diagnosis

The diagnosis of OCD is based on criteria in the DSM-IV-TR. These criteria include the following:
- Obsessions:
  - Recurrent, persistent thoughts, urges, or impulses experienced as intrusive and unwanted, causing significant anxiety or distress in most individuals
  - The individual tries to ignore or suppress thoughts, urges, or images; or tries to neutralize them with another thought or action, such as by performing a compulsive behavior
- Compulsions:
  - Repetitive behaviors such as hand washing, putting items into specific orders, repeated checking of items and surroundings; or mental acts such as silently repeated words, counting, or praying—the individual is driven to perform these in response to an obsession, or according to rules that must be rigidly applied
  - The mental acts or behaviors are attempts at preventing or reducing anxiety or distress, or preventing some dreaded event or situation; they are not realistically connected with what they attempt to neutralize or prevent, and are clearly excessive

Additional criteria are considered. The obsessions or compulsions take more than 1 h/day, or cause significant distress or impairment in social, occupational, or other areas of functioning. The OCD symptoms are not attributable to abused drugs, medications, or other medical conditions. The disturbance is not better explained by symptoms of another mental disorder, such as the following:
- *Difficulty in discarding or parting with possessions*—as in hoarding disorder
- *Excessive worrying*—as in generalized anxiety disorder
- *Guilty ruminations*—as in major depressive disorder
- *Hair pulling*—as in trichotillomania (hair-pulling disorder)
- *Impulses*—as in disruptive, impulse-control, and conduct disorders
- *Preoccupation with appearance*—as in body dysmorphic disorder
- *Preoccupation with having an illness*—as in illness anxiety disorder
- *Preoccupation with substances or gambling*—as in substance-related and addictive disorders
- *Repetitive patterns of behavior*—as in autism spectrum disorders
- *Ritualized eating behavior*—as in eating disorders
- *Sexual urges or fantasies*—as in paraphilic disorders
- *Skin picking*—as in excoriation (skin-picking) disorder

- *Stereotypies*—as in stereotypic movement disorder
- *Thought insertion or delusional preoccupations*—as in schizophrenia spectrum and other psychotic disorders

A physical examination helps to rule out other problems that could be causing the symptoms, and also checks for related complications. Laboratory tests may include a complete blood count (CBC), thyroid function tests, and screening for alcohol and drugs. Psychological evaluation involves a discussion of the patient's thoughts, feelings, symptoms, and behavior patterns. If the patient allows, the examiner may also talk to the patient's family and friends. Also, a current or past history of a *tic disorder* must be ruled out. The patient must be evaluated as to his or her insight on the condition. With good or fair insight, the patient recognizes that OCD beliefs are definitely or probably not true, or that they may or may not be true. With poor insight, the individual thinks that OCD beliefs are probably true. With absent insight or delusional beliefs, the patient is totally convinced that the OCD beliefs are true.

Treatment

Exposure and ritual prevention therapy are often an effective treatment. The patient is exposed to people or situations that trigger that anxiety-proving obsessions and rituals. Following exposure, the patient stops the ritualized behaviors, and allows anxiety triggered by the exposure to reduce via habituation. Improvement usually continues for years, most often in patients who follow the approach and use it beyond the ending of formal treatment. The majority of patients have incomplete responses to this therapy, similar to their responses to medications. Psychotherapy and drug therapy are widely believed to be the best treatment for OCD, especially when it is severe. Medications include SSRIs such as fluoxetine, fluvoxamine, sertraline, and paroxetine. The TCA called *clomipramine* may also be effective. For highly treatment-resistant OCD, deep brain stimulation may be an effective option. It was approved under a Humanitarian Device Exemption and is a very specialized procedure. This type of deep brain stimulation requires a neurosurgeon with expertise in stereotactic and functional neurosurgery.

Depression

Depression is described as intense sadness that persists and is a mood disorder. Depressive disorders are characterized by depression that interferes with function. Often, the patient has decreased interest or pleasure in activities. The exact causes are unknown but likely involve heredity, altered neurotransmitters and neuroendocrine function, along with psychosocial factors. *Major depressive disorder* is also known as *major depression* or *unipolar disorder*. Episodes including five or more mental or physical symptoms, lasting for 2 weeks or more, are seen. **Dysthymia** involves low-level or subthreshold symptoms persisting for 2 or more years. *Depressive disorder not otherwise specified* involves clusters of symptoms that do not meet the criteria for other depressive disorders. *Depressive disorder due to a* *general physical condition* and *substance-induced depressive disorder* are two other commonly seen forms of depression.

Depressive disorders usually develop during a person's mid-teen years, 20s, or 30s, but can actually occur at any age. Patients in primary care settings report depressive symptoms 30% of the time, yet less than 10% have major depression. *Demoralization* results from disappointments or losses and resolves when events or circumstances improve. The depressive feelings of demoralization usually last for days instead of weeks or months, with suicidal thoughts and prolonged loss of function seen much less commonly.

Pathophysiology

Heredity accounts for approximately 50% of depression but is not as significantly linked to *late-onset depression*. Genetic factors likely influence the development of depressive responses to adverse events. Changes in neurotransmitter levels may include abnormal regulation of cholinergic, catecholaminergic, and serotonergic neurotransmission. Neuroendocrine dysregulation may emphasize the hypothalamic-pituitary-adrenal axis, the hypothalamic-pituitary-thyroid axis, and the effects of growth hormone. Major life stressors often precede episodes of major depression but do not usually cause lasting and severe depression except for individuals predisposed to a mood disorder.

Women are at higher risk for depression than men, and this may be related to higher daily stresses, higher levels of MAO, higher rates of thyroid dysfunction, and endocrine changes related to menstruation and menopause. In *seasonal effective disorder*, depression develops usually in the autumn or winter, mostly in more northern and colder climates. The pathophysiology of depression may be interrelated with the adrenal gland and thyroid disorders, brain tumors, AIDS, stroke, multiple sclerosis, Parkinson's disease, certain beta-blockers, corticosteroids, interferon, reserpine, alcohol, amphetamines, and because of toxicity or withdrawal from drugs.

Clinical manifestations

Depression causes cognitive dysfunction, psychomotor dysfunction, poor concentration, fatigue, loss of sexual desire, loss of pleasure, and sadness. Often, the patient also has anxiety, panic attacks, and other mental conditions. All depressed patients are more likely to abuse alcohol or other recreational drugs in an attempt to reduce their symptoms. The patient may start smoking and neglect his or her own health, resulting in worsening or development of other conditions. The immune system can be significantly weakened by depression, which also increases risks for cardiovascular disorders, heart attack, stroke, and alterations in blood clotting. Major depression is divided into subgroups of *psychotic* (involving delusions), *catatonic* (severe reduction of movements or abnormal alterations of movements), *melancholic* (loss of pleasure in nearly all activities), and *atypical* (alterations in mood that usually worsen over hours). Dysthymia usually begins insidiously in adolescence, continuing as a low-grade disorder over 2 years, up to decades, and can be intermittently complicated by episodes of major depression. A form called *anxious depression* involves concurrent mild symptoms that are common to both anxiety and depression. For patients with symptoms of mild quality or short duration, there is often spontaneous remission of the depression.

Diagnosis

Diagnosis of depression takes into account the patient's signs and symptoms and utilizes screening questionnaires with specific, close-ended questions. Depression severity is determined by the amount of pain and disability, and by the duration of the symptoms. Suicidal thoughts must be evaluated. Severe depression is indicated by psychosis and catatonia. Severity is influenced by coexisting physical conditions, anxiety disorders, and substance abuse disorders. Differential diagnoses include demoralization, anxiety disorders, bipolar disorder, dementia, hypothyroidism, and Parkinson's disease. There are no laboratory findings that are pathognomonic for depression. Though there are tests for limbic-diencephalic dysfunction, these are rarely diagnostic. Laboratory testing is used to exclude the physical causes of depression. These include CBC, thyroid-stimulating hormone (TSH) levels, and routine electrolyte, folate, and vitamin B12 levels. Sometimes, testing for illicit drug use is indicated.

Treatment

Mild depression is treated with general support and psychotherapy. Moderate to severe depression is treated with medications, psychotherapy, or both. It sometimes requires electroconvulsive therapy. Some patients need combinations of different drugs. Improvement may not be seen until 1–4 weeks of treatment. Depression is likely to recur, especially in patients who have had more than one episode. Severe cases, therefore, often require long-term maintenance drug therapy. Most patients are treated on an outpatient basis. Hospitalization is required for those with significant suicidal ideation, especially when there is a lack of family support. Those with psychotic symptoms or physical debilitation also should be hospitalized. With substance abuse, depressive symptoms often resolve within a few months after stopping substance use. While substance abuse is continuing, antidepressants are usually not successful.

The patient is usually seen initially on a weekly or bi-weekly basis. Cognitive-behavioral therapy may be done individually or in groups. Medications may include SSRIs such as citalopram, escitalopram, fluoxetine, fluvoxamine, paroxetine, and sertraline. Some patient becomes more agitated, anxious, or depressed until they adjust to these medications. Serotonin blockers primarily block the $5\text{-}HT_2$ receptors and inhibit reuptake of 5-HT and NE. They include mirtazapine, nefazodone, and trazodone. The serotonin-NE reuptake inhibitors include duloxetine and venlafaxine. Bupropion is a NE-dopamine reuptake inhibitor that may help depressed patients who also have ADHD, cocaine dependence, or if they are trying to stop smoking. Monoamine oxidase inhibitors (MAOIs) include isocarboxazid, phenelzine, selegiline, and tranylcypromine. Choice of medications is guided by past responses to certain antidepressants, and otherwise, the SSRIs are often the first drugs used. They are prescribed for administration in the morning. Continued therapy with antidepressants is usually needed for 6–12 months in order to prevent relapses, and up to 2 years in patients older than 50. Electroconvulsive therapy is indicated for severe suicidal depression, depression with agitation or psychomotor retardation, delusional depression, or depression during pregnancy. This therapy is used after drugs have been ineffective. Phototherapy may help patients with seasonal depression. Other potential therapies include psychostimulants, vagus nerve stimulation, deep brain stimulation, and transcranial magnetic stimulation.

> **Focus on depression**
> According to the Anxiety and Depression Association of America, it is not uncommon for someone with an anxiety disorder to also have depression, or vice-versa. Nearly 50% of those diagnosed with depression are also diagnosed with an anxiety disorder. A major depressive disorder is the leading cause of disability in the United States for people between ages 15 and 44, and is more prevalent in women than in men.

Bipolar disorder

Bipolar disorder involves episodes of mania and depression. These may alternate, though many patients have one of these states that predominates over the other. The exact cause is unknown, but factors may include heredity, neurotransmitter alterations, and psychosocial factors. The condition usually begins in the teens, 20s, or 30s, with a lifetime prevalence of about 4%. Bipolar disorder occurs nearly evenly in males and females.

Pathophysiology

The pathophysiology of bipolar disorder involves three different forms and may involve dysregulation of serotonin and NE. Exacerbations may be linked to the use of cocaine, amphetamines, alcohol, TCAs, and MAOIs. The three forms of bipolar disorder include the following:
- *Bipolar I disorder*—at least one manic or mixed episode, and usually depressive episodes, that disrupt the normal social and occupational function
- *Bipolar II disorder*—major depressive episodes, with at least one hypomanic episode, but no complete manic episodes
- *Bipolar disorder not otherwise specified*—clear bipolar features, but not sufficient to meet specific criteria for the other two forms

Clinical manifestations

Bipolar disorder starts with an acute phase of symptoms. This is followed by repeated remissions and relapses. Remissions are usually complete, with some patients having residual symptoms. Relapses are discrete episodes, with more intense manic, depressive, hypomanic, or mixed symptoms. Episodes last from a few weeks, up to 6 months. Cycles vary between patients. They may be infrequent, only a few over the patient's lifetime, or may be rapid cycling, with four or more episodes per year. Most patients do not alternate between mania and depression with each cycle—usually one of these features dominates the other.

A *manic episode* is 1 or more weeks of a persistently elevated, expansive, or irritable mood, plus three or more of the following symptoms:
- Decreased need for sleep
- Distractibility
- Flights of ideas, or racing of thoughts
- Greater talkativeness than normal
- Increased goal-directed activities
- Inflated grandiosity or self-esteem
- Persistently elevated mood

The manic patient is involved in activities such as dangerous sports, gambling, or promiscuous sexual activities, without seeming to be aware of possible danger. *Manic psychosis* involves psychotic symptoms that resemble schizophrenia, including delusions, hallucinations, greatly increased activity, mood swings, and full-blown delirium. *Hypomania* involves a distinct episode of 4 or more days, with brightened mood, reduced need for sleep, and accelerated psychomotor activity. A depressive episode resembles symptoms of major depression. A mixed episode blends manic or hypomanic and depressive features. In about 33% of bipolar disorder patients, entire episodes are mixed.

Diagnosis

Diagnosis of bipolar disorder is based on symptoms, and history or remissions and relapses. The patient is questioned about excessive spending, impulsive sexual activities, abuse of stimulant drugs, and any suicidal thoughts. Substance use is thoroughly reviewed. For new patients, thyroid function tests are often performed to rule out a thyroid abnormality. Bipolar disorder must be evaluated along with the presence of any anxiety disorders, which can confuse the diagnosis.

Treatment

The treatment of the bipolar disorder is usually in three phases. When acute, the focus is on stabilizing and controlling the initial and sometimes severe manifestations. Treatment continuation is focused on attaining a complete remission. Maintenance of prevention is focused on keeping the patient in remission. Treatment is usually on an outpatient basis, except for those with severe mania or depression. Medications include mood stabilizers and second-generation antipsychotics, which can also be combined. Mood stabilizers include lithium, and anticonvulsants such as valproate, carbamazepine, and lamotrigine. Second-generation antipsychotics include aripiprazole, olanzapine, quetiapine, risperidone, and ziprasidone. SSRIs and other antidepressants are sometimes added, but can trigger mania, so careful monitoring is required. Medications for pregnant women with bipolar disorder must be individually evaluated because of potential harm to the fetus. Electroconvulsive therapy and phototherapy may be helpful for some patients.

To help prevent major episodes, it is important to involve the patient's family and friends. Group therapy is often successful. Individual psychotherapy helps adjust to problems of daily living and self-identification. Those with bipolar II disorder often do not follow drug regimens because they feel less alert or creative. All bipolar disorder patients should avoid stimulant drugs and alcohol, minimize sleep deprivation, and recognize early signs of relapses. Control of finances may need to be handled by a trusted individual instead of the patient. Those with sexual excesses need counseling on possible negative outcomes and sexually transmitted infections.

Cyclothymic disorder

Cyclothymic disorder involves hypomanic and mini-depressive episodes over a few days, with an irregular course. The episodes are not as severe as those seen in bipolar disorder. Cyclothymic disorder is often a precursor of bipolar II disorder but may occur as a significant, yet not major, mood disorder. In *chronic hypomania*, the patient is excessively happy, self-assured, overenergetic, continuously planning events and activities, and often overly concerned with other people's lives. They experience restless impulses and be overly familiar with others. Though cyclothymic disorder and chronic hypomania can actually be positive in regards to successful activities and creativity, interpersonal and social relationships often deteriorate. Abuse of alcohol and drugs is a common outcome.

The patient should be instructed on ways to live with extreme symptoms. It is advised that any work schedule should be made flexible instead of rigidly structured. If the patient is artistic, a career in the arts can be ideal. Mood stabilizers may be successful for some patients and include divalproex rather than lithium. Antidepressants are only used when symptoms are severe and prolonged. Support groups can be very helpful, allowing patients to share common experiences and feelings about this condition.

Section review
1. What is the difference between the obsessions and compulsions seen in OCD?
2. What are the primary clinical manifestations of depression?
3. What are the three phases of bipolar disorder treatment?

Clinical cases

Clinical case 1
1. What is the difference between a cholinergic crisis and a myasthenic crisis?
2. What are the commonly used treatments for myasthenic crisis?
3. What may worsen the symptoms of myasthenia gravis?

A 35-year-old woman began feeling shoulder muscle fatigue when performing normal activities. Over time, the condition required her to rest frequently in order to have enough strength to continue doing what she needed to do. After testing with intravenous edrophonium, this patient was diagnosed with MG.

Answers:
1. Cholinergic crisis is when there is no improvement or worsening of weakness when edrophonium is administered. Myasthenic crisis is when there is improvement with a small dose of edrophonium.
2. Commonly, MG is treated with anticholinesterases and immunosuppressants such as steroids, cyclosporine, and azathioprine. The thymus may be surgically removed, immunoglobulins may be administered, and plasmapheresis may be performed as well.
3. The symptoms of MG may be temporarily worsened by infections, stress, menses, pregnancy, or surgery.

Clinical case 2
1. What is the likely diagnosis for this patient?
2. What are the differences between a schizophrenic brain and a normal brain?
3. Is suicide more prevalent in people with this condition?

A 20-year-old college student was suspended because of bizarre behaviors in class and on campus. He had withdrawn from social activities and his appearance became notably different, with his clothes appearing dirty and a general lack of cleanliness. In a psychological evaluation, he stated that his college was actually an organized crime operation that was spying on him. He described that the college officials were "after him" so much that he contemplated suicide because of the situation. Drug screening tests revealed nothing. He was eventually hospitalized into a facility for observation. While there, he regularly avoided eating as much as he could, stating that the healthcare staff were really spies, and were putting "truth serum" into his food. When visiting the patient, his father and sister explained that his great-grandmother had lived for 30 years in a state mental hospital. Also, the patient's mother had abandoned the family and had been also treated for mental health issues.

Answers:
1. The likely diagnosis for this patient is schizophrenia, based on his delusions, lack of drug use, and family history of mental health conditions.

2. Brain scans and microscopic tissue studies reveal many abnormalities in the schizophrenic brain. The most common structural abnormality involves the lateral brain ventricles, which appear enlarged when schizophrenia is present. There may be up to 25% loss of gray matter in certain areas of the schizophrenic brain.
3. Yes, people with schizophrenia attempt suicide more often than people in the general population. It is estimated that up to 10% of people with schizophrenia take their own lives within the first 10 years of the illness—especially young men with schizophrenia.

Clinical case 3

1. What are the differences between obsessions and compulsions as seen in obsessive-compulsive disorder (OCD)?
2. What medications are used for this condition?
3. Is neurosurgery an option for severe cases of OCD?

A 32-year-old woman was diagnosed with OCD at the age of 7. Medications had been successful for most of her life, but recently, her OCD symptoms began to recur. She had recently separated from her husband and began to have difficulty in sleeping, mood changes, increased anxiety, and new OCD behaviors. These behaviors included repeatedly calling her husband, repeatedly calling her children's babysitter, and even repeatedly calling their school to check on them. The patient was referred to receive cognitive-behavioral therapy.

Answers:

1. Obsessions are anxiety-causing ideas, images, or impulses. Compulsions are urges to do something to reduce anxiety.
2. Medications include SSRIs such as fluoxetine, fluvoxamine, sertraline, and paroxetine. The TCA called clomipramine may also be effective.
3. Yes, deep brain stimulation has been approved for OCD patients who do not respond to medications and other treatments. It requires a neurosurgeon who his experienced in stereotactic and functional neurosurgery.

Clinical case 4

1. What is the likely diagnosis of this patient?
2. What further description of this patient's condition can be made?
3. What is the difference between major depression and dysthymia?

A 27-year-old man experienced the loss of his fiancée due to an automobile accident in which she was killed by a drunk driver. He began seeing a psychiatrist for his severe depression. Since the accident, he has nightmares regularly in which he relives what happened and feels guilty that he survived. He had to quit his job since it was in a building located very close to the accident site and now drives to another town for work. The patient has become continually withdrawn, irritable, and anxious. He stopped playing the piano, going to the gym, and watching football with his friends on weekends. His detachment and emotional flatness seem to dominate his life.

Answers:
1. This patient probably has major depression, based on his symptoms. This condition is likely to last for a long time, with symptoms varying in intensity, and can become chronic without proper treatment.
2. This patient probably has the melancholic form of major depression. This is signified by detachment, emotional flatness, irritability, and guilt.
3. The symptoms of major depression are more debilitating than those of dysthymia and inhibit the ability to function throughout the day. People with major depression have a normal mood baseline when not experiencing an exacerbation of depression. Dysthymia does not come and go—it is always present and can last for years. For an adult to be diagnosed with dysthymia, symptoms must have been present for at least 2 years.

Key terms

Acetyl coenzyme A
Acetylcholine (ACh)
Acetylcholinesterase
Active zone
Adenosine
Adenosine triphosphate (ATP)
Adenyl cyclase
Adrenal medulla
Amino acid decarboxylase
Astereognosis
Autoreceptors
Beta endorphin
Biogenic amine
Bipolar disorder
Carbon monoxide (CO)
Catecholamines
Chemotaxis
Cholecystokinin
Choline
Choline acetyltransferase
Cholinergic receptors
Chromaffin cells
Chronotropy
Cyclothymic disorder
Dopamine
Dromotropy
Dynorphin
Dysthymia
Endocytosis
Endorphins
Enkephalins
Epinephrine
Excitatory
Exocytosis
Gamma-aminobutyric acid (GABA)
Gasotransmitters
Gluconeogenesis
Glutamate
Glutaminergic receptor
Glycine
Guanylate cyclase
Gut-brain peptides
Hallucinations
Histamine

Hydrogen sulfide (H$_2$S)
Indolamines
Inhibitory
Inhibitory postsynaptic potential
Inotropy
Lambert-Eaton syndrome
Lipolysis
Malathion
Molar
Muscarinic
Myasthenia gravis
Neurokinin A
Neuromodulators
Neuropeptides
Neurotransmitters
Nicotinic
Nitric oxide (NO)
Norepinephrine (NE)
Obsessive-compulsive disorder (OCD)
Opioid peptides
Parkinson's disease
Postganglionic neurons
Purines
Schizophrenia
Serotonin
Somatic system
Somatostatin
Strychnine
Substance P
Tachykinins
Thermogenesis
Tryptophan hydroxylase

Suggested readings

1. Akaike, A.; Shimohama, S.; Misu, Y. *Nicotinic Acetylcholine Receptor Signaling in Neuroprotection*. Springer, 2018.
2. Bunch, C. K. *Soft Bipolar Cyclothymia Suffering: All Three Books—Depression With Occasional Up Energy States*. Amazon Digital Services LLC, 2013.
3. De Wild-Scholten, M. *Understanding Histamine Intolerance & Mast Cell Activation*, 3rd ed.; CreateSpace Independent Publishing Platform, 2013.
4. Eiden, L. E. *Catecholamine Research in the 21st Century: Abstracts and Graphical Abstracts, 10th International Catecholamine Symposium*. Academic Press, 2013.
5. Engel, A. G. *Myasthenia Gravis and Myasthenic Disorders*, 2nd ed.; *Contemporary Neurology Series* Oxford University Press, 2012.
6. Forger, D. B. *Biological Clocks, Rhythms, and Oscillations: The Theory of Biological Timekeeping*. The MIT Press, 2017.
7. Granholm, E. L.; McQuaid, J. R.; Holden, J. L. *Cognitive-Behavioral Social Skills Training for Schizophrenia—A Practical Treatment Guide*. The Guilford Press, 2016.
8. Iversen, L.; Iversen, S.; Dunnett, S.; Bjorklund, A. *Dopamine Handbook*. Oxford University Press, 2009.
9. Kaminski, H. J.; Kusner, L. L. *Myasthenia Gravis and Related Disorders (Current Clinical Neurology)*; 3rd ed.; Humana Press, 2018.
10. Klein, J.; Loffelholz, K. *Cholinergic Mechanisms: From Molecular Biology to Clinical Significance*. Elsevier Science, 1997; Vol. 109.
11. Lack, C. W.; McKay, D. *Obsessive-Compulsive Disorder: Etiology, Phenomenology, and Treatment*. Onus Books, 2015.
12. Macaluso, M.; Preskorn, S. H. *Antidepressants: From Biogenic Amines to New Mechanisms of Action. Handbook of Experimental Pharmacology*; Springer, 2019; Vol. 250.
13. Meriney, S. D.; Fanselow, E. *Synaptic Transmission*. Academic Press, 2019.
14. Miklowitz, D. J.; Gitlin, M. J. *Clinician's Guide to Bipolar Disorder—Integrating Pharmacology and Psychotherapy*. The Guilford Press, 2015.
15. Mochida, S. *Presynaptic Terminals*. Springer, 2015.
16. Mondimore, F. M. *Bipolar Disorder: A Guide for Patients and Families*, 3rd ed.; *A Johns Hopkins Press Health Book* Johns Hopkins University Press, 2014.
17. Nestler, E. J.; Hyman, S. E.; Malenka, R. C. *Molecular Neuropharmacology: A Foundations for Clinical Neuroscience*; 3rd ed.; McGraw-Hill Education/Medical, 2015.

18. Penzel, F. *Obsessive-Compulsive Disorders: A Complete Guide to Staying Well*, 2nd ed.; Oxford University Press, 2016.
19. Schousboe, A.; Sonnewald, U. *The Glutamate/GABA-Glutamine Cycle: Amino Acid Homeostasis. Advances in Neurobiology Book, Vol. 13*; Springer, 2016.
20. Stone, T. W. *CNS Neurotransmitters and Neuromodulators: Dopamine*. CRC Press, 1996.
21. Strand, F. L. *Neuropeptides: Regulators of Physiological Processes. Cellular and Molecular Neuroscience* Bradford, 1999.
22. Swerdlow, N. R. *Behavioral Neurobiology of Schizophrenia and Its Treatment. Current Topics in Behavioral Neurosciences* Springer, 2010.
23. Types of Neurotransmitters—http://ib.bioninja.com.au/options/option-a-neurobiology-and/a5-neuropharmacology/types-of-neurotransmitters.html
24. Udenfriend, S.; Meienhofer, J. *Opioid Peptides: Biology, Chemistry, and Genetics: Analysis, Synthesis, Biology, Vol. 6;* Academic Press, 2014.
25. Wu, G. *Amino Acids: Biochemistry and Nutrition*. CRC Press, 2013.

CHAPTER 19

Spinal cord

In the study of the human central nervous system, the spinal cord is the starting point. The spinal cord has the consistency of rubber and is floating inside the cerebrospinal fluid (CSF), in a similar way that our brain is surrounded by CSF. It is between 42 and 45 cm in length, and approximately one cm in diameter at its widest portion. The spinal cord itself only weighs about 35 g. It has an overall organization that is very similar from section to section. Many functional processes occur in the same way in other nervous system levels. The spinal cord is of vital importance for subconscious activities of living, such as reflexes. It contains all motor neurons that supply the muscles, along with major autonomic efferents. The spinal cord receives all of the body's sensory input and some sensory input from the brain.

Communication to and from the brain utilizes the spinal cord tracts. The ascending tracts are sensory, delivering information to the brain. The descending tracts are motor, delivering information out to the peripheral areas of the body. If a tract's name begins with the prefix *spino-*, it is a sensory tract. If the name ends with the suffix *-spinal*, it is a motor tract. The three major sensory tracts include the posterior column tract, the spinothalamic tract, and the spinocerebellar tract. The main descending motor tracts are the corticospinal, vestibulospinal, reticulospinal, and tectospinal tracts. Damage to the spinal cord tracts may result in various conditions. The major ascending sensory tracts are the dorsal column/medial lemniscal, spinothalamic, spinoreticulothalamic, and spinocerebellar tracts.

Gross anatomy and protection

The spinal cord is enclosed within the vertebral column. It extends from the **foramen magnum** of the skull all the way to the first or second lumbar vertebra, and is slightly inferior to the ribs. The spinal cord appears as a shiny, white structure. It provides two directions of conduction, descending from the brain, and ascending to the brain. The spinal cord is protected by the vertebral bones, meninges, and CSF. The **spinal dura mater** consists of a single layer that is not connected with the bony walls of the vertebral column. This is opposed to the brain dura, which consists of two layers. An **epidural space** lies between the bony vertebrae and the spinal dura mater, which is filled with soft fat, acting as padding, and a network of veins. CSF fills up the subarachnoid space, which is situated between the *arachnoid* meninges and the *pia mater* meninges.

The dural and arachnoid membranes extend inferiorly to the level of S_2, which is far lower than the end of the spinal cord. The cord usually ends between L_1 and L_2. Therefore, the subarachnoid inside the meningeal sac, inferior to this point, is the perfect location for removal of CSF since there is no spinal cord at this level, preventing any damage. This procedure is called a *lumbar puncture*. The spinal cord is absent in this location. This means that there is not much danger of damaging the spinal cord or its nerve roots beyond the L_3 level. The spinal cord terminates inferiorly in a cone-shaped and tapered **conus medullaris**. A fibrous extension of this structure, covered by pia mater, is called the **filum terminale**. It extends inferiorly, from the conus medullaris to the coccyx. Here, the filum terminale anchors the spinal cord to provide vertical support. The spinal cord is also fastened to the dura mater meninges, throughout its length by the **denticulate ligaments**, which are saw-toothed, shelf-like structures made of pia mater that provide horizontal support to the cord. For most of its length, the spinal cord is about as wide as an adult human thumb. There are pronounced enlargements where the nerves arise that serve the arms and legs. The upper enlargements are called **cervical enlargements**, while the lower enlargements are called **lumbar enlargements.** The cervical enlargements usually extend from C_5 to T_1. The lumbar enlargements are also called *lumbosacral enlargements*, and usually, extend from L_2 to S_3.

Spinal nerves

There are 31 pairs of *spinal nerves*, which are attached to the spinal cord by paired roots (see Fig. 19.1). They function as part of the *peripheral nervous system*. Each segment of the spinal cord is designated by the paired spinal nerves arising from it. For example, the first lumbar spinal cord segment (L_1) is where the first lumbar nerves (lumbar nerve L_1) emerge from the cord. Even with these segments, the spinal cord is continuous over its entire length, with only gradual changes in its internal structure. Each nerve exits the vertebral column by passing superiorly to its related vertebra, via the intervertebral foramen, except for the eighth cervical nerve, which exits the spine below the C7 vertebrae. Each nerve then connects to the area of the body that it supplies. The 31 segments of the spinal cord are as follows:

- **Cervical**—8 segments
- **Thoracic**—12 segments
- **Lumbar**—5 segments
- **Sacral**—5 segments
- **Coccygeal**—1 segment

Since the spinal cord does not reach the end of the vertebral column and ends at the level of L1-L2 in adults, its segments are superior to where their related spinal nerves emerge, through the intervertebral foramina. Prior to reaching their intervertebral foramina, the lumbar and sacral spinal nerve roots are sharply angled downward. They travel inferiorly through the vertebral canal for a fairly large distance. At the inferior end of the vertebral

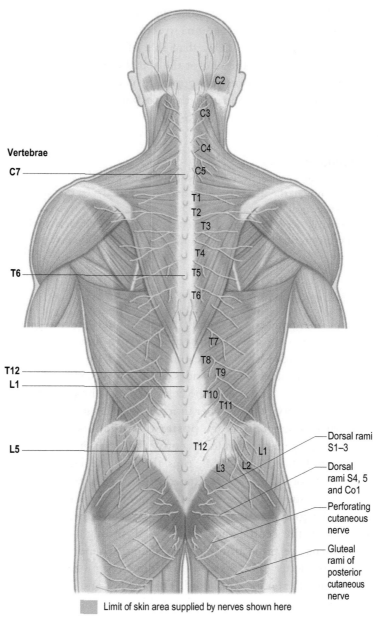

Fig. 19.1 The cutaneous distribution of the dorsal rami of the spinal nerves. The nerves are shown lying on the superficial muscles. The nerves are numbered on the right side; the spines of the seventh cervical, sixth and twelfth thoracic, and first and fifth lumbar vertebrae are labeled on the left side.

canal, there is a collection of nerve roots called the **cauda equina**, due to its resemblance to a horse's tail (Fig. 19.2). The cauda equina structure forms due to the fact that the fetal growth of the vertebral column is faster than that of the spinal cord. This leads to an extension of the nerves beyond the end of the spinal cord at the level of L1-L2. This causes the lower spinal nerve roots to "pursue" their exit points, inferiorly through the vertebral canal. The cauda equina fills the **lumbar cistern**, which is the area from L_1 or L_2 to the end of the dural sheath at S_2.

> ### Focus on spinal nerves
> Each of the first seven cervical nerves leaves the vertebral canal *above* their corresponding vertebrae. For example, the first cervical nerve emerges between the *occiput* and the *atlas*, which is the first cervical vertebra. However, since there are only seven cervical vertebrae, the eighth cervical nerve emerges between the seventh cervical and first thoracic vertebrae. This means that each subsequent nerve leaves *below* its corresponding vertebra.

Cross-sectional anatomy

From front to back, the spinal cord is slightly flattened. There are two grooves marking its surface. These are the wide **anterior median fissure** and the narrower **posterior median sulcus**. These grooves follow the entire length of the spinal cord. They partially divide the cord into two halves. The two sides of the spinal cord are able to communicate via a narrow band of neural tissue near the central canal. The **posterior intermediate sulcus** is located at the cervical and upper thoracic levels. A **glial septum** projects from this sulcus, and partially subdivides each **posterior funiculus**.

The spinal cord's gray matter is inside its core, while the white matter is in the periphery. A **central canal**, filled with CSF, runs through the entire cord. When cross-sectioned, the spinal cord's gray matter appears like the letter "H," or somewhat like a butterfly. It has mirrored-image lateral gray masses that are connected by a "crossbar" called the **gray commissure**, which encloses the central canal. Through the gray commissure, the central canal carries CSF throughout the spinal cord. The two **posterior horns** are projections of gray matter. There are also similar **anterior horns** resembling the letter "H". Seen in three dimensions, these horns form columns of gray matter through the entire spinal cord. The thoracic and superior lumbar segments have other gray matter columns called the **lateral horns**, which are smaller in size. The posterior horns are continuous cell columns and not a series of discrete nuclei. They interact with cells from many other levels.

Neurons with cell bodies in the spinal cord's gray matter are all *multipolar*, and the posterior horns are entirely made up of **interneurons**. The anterior horns contain mostly cell bodies of **somatic motor neurons**, as well as some interneurons. The motor neurons send axons out to the skeletal muscles, which are their effector organs, through the *antral rootlets*.

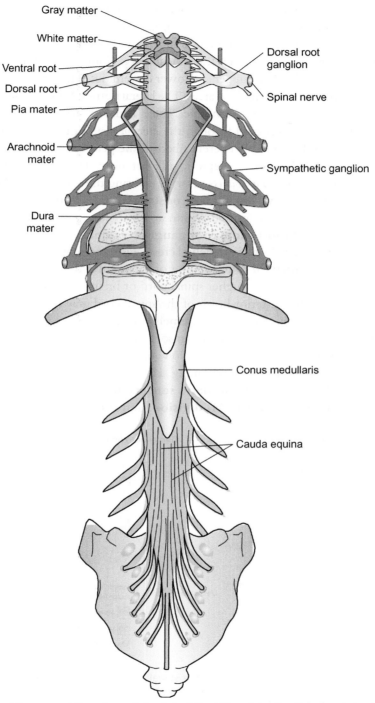

Fig. 19.2 Spinal Nerves and Coverings of the Spinal Cord. Note how the dura mater extends to cover the spinal nerve roots and nerves. *Modified from Thibodeau, G. A.; Patton, K. T.* Structure and function of the body, *14th ed.; Mosby: St Louis, 2012*

They emerge from a poorly defined **anterolateral sulcus**. These rootlets fuse together, becoming the **anterior roots** of the spinal cord. Anterior gray matter size, at any level of the spinal cord, reflects how much skeletal muscle is innervated at that level. Therefore, the anterior horns are largest in the cervical and lumbar regions that are limb-innervating, as they innervate the muscles of the upper and lower extremities. They are responsible for the enlargements of the spinal cord in these areas. The lateral horns mostly contain cell bodies of the sympathetic division of the autonomic motor neurons serving the visceral organs. Motor axons leave the spinal cord through the *anterior root*. Note that the anterior roots have both somatic and autonomic efferent fibers of the peripheral nervous system.

From the peripheral sensory receptors, afferent fibers form the **posterior roots**. These spread out as the *posterior rootlets* prior to entering the spinal cord. These rootlets enter the cord in the shallow, longitudinal **posterolateral sulcus**. Cell bodies of their associated sensory neurons are located within an enlarged region of the posterior root. This is called the **dorsal (posterior) root ganglion** or simply called DRG. Recall that dorsal root ganglia cells originate from neural crest cells. Once entering the cord, axons of these neurons take a variety of routes. Some of them enter the posterior white matter directly, eventually synapsing at higher spinal cord or brain levels. Other axons synapse with interneurons in the posterior horns of the spinal cord gray matter, at their level of entry. Posterior and anterior roots are extremely short, fusing laterally, and forming the **spinal nerves** (which contain motor and sensory nerves). The spinal gray matter is subdivided based on the neurons' relative involvement in the innervation of the body's somatic and visceral regions. As shown in Fig. 19.3, the spinal gray matter has four divisions or zones. These include the **somatic sensory (SS) division, visceral sensory (VS) division, visceral motor (VM) division, and somatic motor (SM) division**.

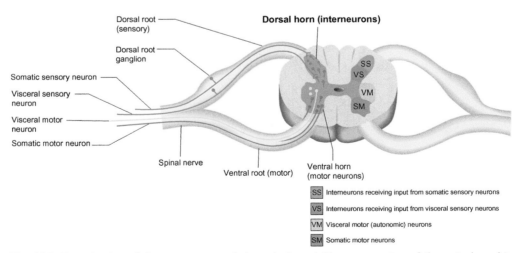

Fig. 19.3 Organization of the gray matter of the spinal cord. The gray matter of the spinal cord is divided into a sensory half dorsally and a motor half ventrally. Note that the dorsal and ventral roots are part of the PNS, not of the spinal cord.

White matter

The spinal cord's **white matter** is made up mostly of myelinated nerve fibers, and to a lesser degree, nonmyelinated nerve fibers. These allow communication between different areas of the cord, as well as between the spinal cord and the brain. The fibers run in three different directions, as follows:
- **Ascending**—running upward to higher centers; these are sensory inputs
- **Descending**—running downward to the spinal cord from the brain, or within the cord, to lower levels; these are motor outputs
- **Transverse**—these run across the midline, from one side of the spinal cord to the other, and are also known as *commissural fibers*

These myelinated ascending and descending tracts make up the majority of the white matter of the spinal cord.

On each side of the spinal cord, the white matter is divided into three *white columns*, called **funiculi**. They are named by their position, as follows:
- Dorsal (posterior) funiculi
- Lateral funiculi
- Ventral (anterior) funiculi

Each funiculus has several fiber tracts. Each tract consists of axons with similar functions and destinations. The names of the spinal tracts, in most cases, identify their origins and destinations.

Focus on white matter
Several columns of myelinated axons around the spinal cord's gray matter. The lipid in the myelin causes the columns to appear nearly white in color. It is sometimes called *superficial tissue* since it is located in the outer regions of the brain and spinal cord.

Section review
1. What is the description of the spinal dura mater and epidural space?
2. What is the cross-sectional description of the spinal cord?
3. What are the directions of fibers within the spinal cord, and their descriptions?

Innervation of specific body regions

All of the body's somatic region from the neck downward—the skin and skeletal muscles—are supplied by the spinal nerve rami and their primary branches. The posterior body trunk is supplied by the posterior rami. The thicker anterior rami supply the remainder of the trunk and limbs. *Roots* are medial to, and form, the spinal nerves, with each root being only sensory or motor. *Rami* are distal to and are lateral spinal nerve branches. Like these nerves, rami carry sensory and motor fibers.

Except for T_2 to T_{12}, all anterior rami branch. They join each other lateral to the vertebral column. The rami form complex, interlaced networks of nerves called **nerve plexuses**. These occur in the cervical, brachial, lumbar, and sacral regions. They mostly service the limbs. However, only *anterior* rami form plexuses. In a plexus, fibers from different anterior rami are crossed. They are then redistributed. Each resulting branch contains fibers from several of the spinal nerves. Fibers for each anterior ramus travel to the peripheral parts of the body via different routes. This means that each limb muscle receives its nerve supply from more than one spinal nerve. Therefore, damage to a single spinal segment or root does not totally paralyze any limb muscle.

Cervical plexus and the neck

Deep inside the neck, under the sternocleidomastoid muscle, a lopped **cervical plexus** is formed by the anterior rami of the first four cervical nerves. The majority of its branches are **cutaneous nerves**, only supplying the skin. They carry sensory impulses from the neck skin, the ears, the posterior of the head, and the shoulder. Other branches innervate the anterior neck muscles. The most important of these nerves is the **phrenic nerve**, receiving fibers from C_3, C_4, and C_5. This nerve runs inferiorly, through the thorax. It supplies the motor fibers of the diaphragm—the main muscle needed for breathing.

Brachial plexus and upper limbs

As mentioned before, nerve roots from spinal nerves C5 to T1 form a network to innervate all the muscles of the upper limbs. This network is also called the **brachial plexus** (see Fig. 19.4).

Lumbosacral plexus and lower limbs

There is much overlapping of the sacral and lumbar plexuses (see Fig. 19.5). Since many lumbar plexus fibers contribute to the sacral plexus, through the **lumbosacral trunk**, the two plexuses are often collectively called the **lumbosacral plexus**. Though each lumbosacral plexus mainly serves a lower limb, it sends certain branches to the abdomen, pelvis, and buttock.

Anterolateral thorax and abdominal wall

In the thorax alone, there are anterior rami, in a simple and segmented pattern. This corresponds to the pattern of the posterior rami. The anterior rami of T_1 to T_{12} primarily course anteriorly, deep to each of the ribs, as the **intercostal nerves.** They supply the intercostal muscles, skin, and muscle of the anterolateral thorax, and the majority of the abdominal wall. Along this course, these nerves have *cutaneous branches*

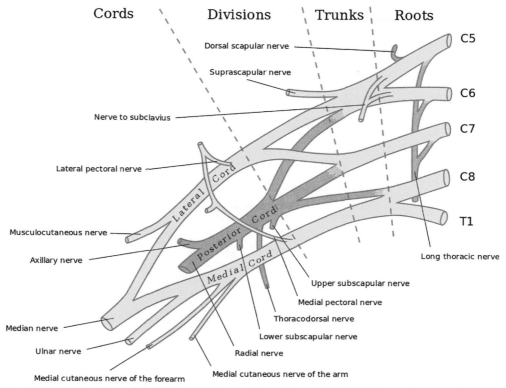

Fig. 19.4 Anterior view of right brachial plexus. *From Figure 15-8 of McCance, K. L.; Huether, S. E.; Felver, L.* Pathophysiology: The Biologic Basis for Disease in Adults and Children, *7th ed.; Mosby, 2014; p. 456.*

to the skin. Two of the thoracic nerves are unique. The very small T_1 nerve has most fibers entering the brachial plexus. The T_{12} nerve, lying inferior to the twelfth rib, is called a **subcostal nerve**.

Innervation of joints

Hilton's law describes the nerves that serve synovial joints. It states that "Any nerve serving a muscle that produces movement at a joint also innervates the joint and the skin over the joint." Therefore, when we learn which nerves service the different major muscles and muscle groups, we fully understand this concept. An example concerns the muscles that cross the knee. These include the quadriceps, gracilis, and hamstring. The nerves for these muscles are the *femoral nerve* anteriorly, and branches of the *sciatic* and *obturator nerves* posteriorly. This means that these nerves also innervate the knee joint.

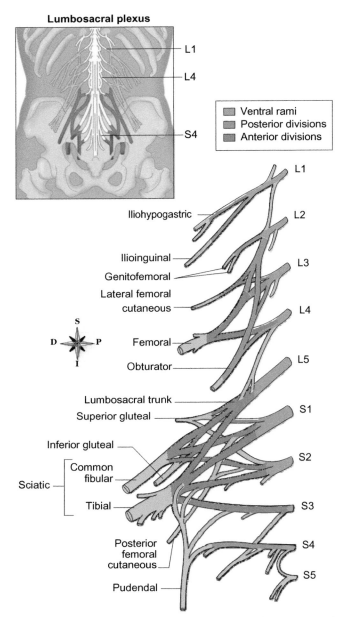

Fig. 19.5 Lumbosacral plexus, formed by the combination of the lumbar plexus with the sacral plexus, as shown in the inset. Note that the ventral rami split into anterior and posterior "divisions" before reorganizing into the various individual nerves that exist this plexus.

Ascending and descending pathways

The white matter of the spinal cord includes ascending and descending pathways in specific locations. There are three general types of nerve fibers in the white matter. These include the following:

- *Long ascending fibers*—these project to the thalamus, cerebellum, cortex, and various brainstem nuclei
- *Long descending fibers*—these project from the cerebral cortex or from several brainstem nuclei, toward the spinal gray matter
- *Shorter propriospinal fibers*—these interconnect different levels of the cord, and include fibers used in coordinating flexor reflexes

Fibers with similar connections usually are banded together. They form the various spinal cord tracts. Propriospinal fibers are mostly within a thin shell that surrounds the gray matter. This is known as the **propriospinal tract** and is also called the **fasciculus proprius**. There are many ascending and descending tracts that are primarily described by where they originate and terminate. Some are of unknown function.

The actual appearance of the fibers is not as simple as how they are drawn in most books. Each primary afferent is believed to be part of one or more reflex arcs, and also in one or more ascending tracts. Similarly, most sensory information reaches the thalamus and cerebellum by several routes. Therefore, loss of a single tract can often be compensated for by other tracts.

Ascending pathways

In humans and other primates, the ascending tracts are found in all three of the funiculi. Information reaching the thalamus is relayed to the cerebral cortex, and consciously perceived. Information reaching the cerebellum helps to regulate movements, which is unconsciously performed. Each *posterior column* consists of all parts of a posterior funiculus, except for its area of interaction with the propriospinal tract. The posterior columns are mostly made up of ascending collaterals of large, myelinated primary afferents. These carry data from various types of mechanoreceptors. However, large numbers of second-order and unmyelinated fibers are also present. This is the primary pathway by which information reaches the cerebral cortex, from the low-threshold receptors of the skin, joints, and muscles (see Fig. 19.6).

The cell bodies of spinal primary afferent fibers are in the ipsilateral dorsal root ganglia. These fibers are of many different diameters and amounts of myelination. At the point, each posterior rootlet enters the spinal cord, the fibers are separated into a **medial division** and a **lateral division**. The medial division has heavily myelinated afferents of a large diameter. The lateral division has finely myelinated or unmyelinated afferents that are smaller in diameter. The medial division's fibers enter the posterior column. They ascend to the brainstem, with many collaterals to the deeper laminae of gray matter in

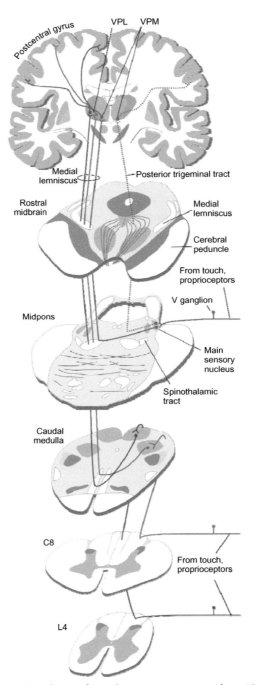

Fig. 19.6 Ascending trigeminal pathways from the main sensory nucleus. *VPL*, ventral posterolateral nucleus; *VPM*, ventral posteromedial nucleus.

the spine. Many of these reach the caudal medulla prior to their final synapse. Each posterior column, caudal to the T_6 level, is an undivided bundle known as a **fasciculus gracilis**. Rostral to T_6, the fibers may leave this bundle, but not many fibers are added. The afferents that enter rostral to T_6 form another bundle. This is the almost triangular **fasciculus cuneatus**, and is lateral to the fasciculus gracilis. The two bundles are partially separated by a glial partition that extends inward. This partition is called the **posterior intermediate septum**.

> ### Focus on ascending pathways
> When sensory nerve fibers reach the spinal cord, they are bundled based on their function. They are known as *nerve tracts* or *fasciculi*, and are found in the spinal white matter. Each ascending nerve tract is named based on its origin and termination.

Somatotopic organization

At each level of the spine, fibers that enter the posterior columns are added, laterally, to the existing fibers. Lamination occurs as layers of fibers from the sacral levels are mostly medial, and layers from the cervical levels are mostly lateral. This arrangement illustrates the **somatotopic organization**, a characteristic of most motor and sensory pathways. In this organization, certain parts of the body are represented in specific regions of a pathway or nucleus.

The posterior column fibers reaching the brainstem synapse in the **nucleus gracilis** or the **nucleus cuneatus**, which are collectively known as the **posterior column nuclei**, located in the *caudal medulla*. The second-order fibers beginning in these nuclei cross the midline. They form the flattened and ribbon-shaped **medial lemniscus**. This fiber bundle continues rostrally, through the brainstem, terminating in the thalamus. Third-order fibers beginning in the thalamus—especially in the **ventral posterolateral nucleus (VPL) and ventral posteromedial nucleus (VPM)**, ascend through the internal capsule, mostly synapsing in the primary somatosensory cortex of the postcentral gyrus.

The somatotopic organization is maintained when primary afferents of the posterior columns end in the posterior column nuclei. Sacral level fibers end in the most medial parts of the nucleus gracilis. Fibers from the cervical levels end in the most lateral parts of the nucleus cuneatus. The somatotopic organization exists in the remainder of this pathway. Therefore, information from the sacral segments travels through a certain part of the medial lemniscus, projects to a specific part of the anterior posterolateral nucleus, and continues to a certain region of the postcentral gyrus. The sacral-to-cervical sequence does not continue along a medial-to-lateral line everywhere in the pathway. However, sacral information is near lumbar information but separated from cervical information

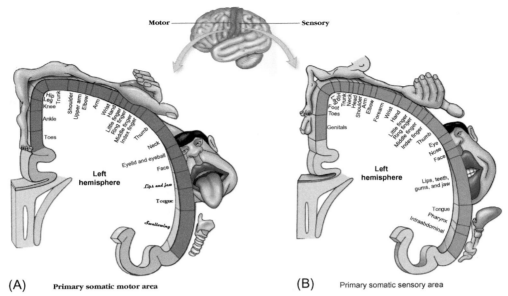

Fig. 19.7 Primary Somatic Sensory (SS) (A) and motor (B) areas of the cortex. *From Patton, K. T.; Thibodeau, G. A.* Anatomy & Physiology, *8th ed.; St Louis: Mosby.*

everywhere in the pathway, up to the somatosensory cortex. The **homunculus** is a "body map" used to understand the somatotopic arrangement of the pathway at different levels (Fig. 19.7). The sacral and lumbar levels correspond to the legs. The thoracic levels correspond to the trunk. The cervical areas correspond to the arms and neck. When viewed in this manner, the homunculus is lying down, with the "feet" toward the midline, up to the level of the posterior column nuclei. As mentioned before, there are similar body maps or homunculi existing at the level of cortex as well.

The posterior column carries information used to perceive pressure, light touch, vibration, and the position and movement of the joints. Since input from cutaneous receptors also reaches the cortex via other routes, when the posterior column is damaged, there will be impairment, but not total inability to have tactile perceptions. Complicated discrimination tasks are more greatly affected than simple stimuli detection. After posterior column destruction, **proprioception** and vibration sensations are usually lost. This results in a distinctive form of **ataxia**. The brain cannot control motor activity properly without sensory feedback concerning current positions of body parts. This form of ataxia is at its worst when the patient closes his or her eyes since visual compensation cannot occur.

The posterior columns exemplify how sensory information travels in multiple pathways. A tumor or anything else impinging on the posterior columns would likely affect the adjacent posterior horns and roots, along with the ascending pathways in the posterior area of the lateral funiculus. Common conditions that specifically cause damage to the posterior column are tabes dorsalis, and vitamin B12 and vitamin E deficiencies.

Tabes dorsalis is a condition occurring in the late stages of neurosyphilis. It was believed to be a typical outcome of posterior column damage. The patient would have impaired vibratory sensations as well as two-point discrimination. Movement and position are affected, and walking is very hard unless the patient can actually view his or her own limbs. When trying to stand up with the eyes closed and feet together, the patient sways and can fall without support. This is known as **Romberg's sign**. There is pronounced degeneration in the posterior columns with this condition. Yet, there is *also* degeneration of the posterior root fibers. This is most evident in the heavily myelinated fibers of the medial division. Mechanoreceptive input to all spinal pathways is affected.

Another example is when the posterior columns are transected, resulting in severe initial impairment. There is extreme difficulty in coordination of the affected limbs. However, if the patient is encouraged to use the limbs, the recovery is impressive over a few months. Eventually, movement and coordination become nearly normal. Tactile threshold is normal, and there is only a slight impairment of position sense and two-point discrimination. The only thing that is permanently impaired is the ability to use somatosensory information in more complicated tasks. This includes judging the speed or direction of a stimulus that moves across the skin, or *stereogenesis*, which is the judgment of the shape of an object pressed against the skin.

Focus on somatotopic organization
Somatotopy involves the direct correspondence of a body area to a specific point on the CNS. Most motor and sensory pathways are organized in this manner.

Section review
1. What are the three general types of nerve fibers in the white matter?
2. What is the definition of the somatotopic organization?
3. When there is damage to the posterior column, would the type of disability will occur?

Spinothalamic tract

The **spinothalamic tract** is, like the posterior column-medial lemniscus, an ascending pathway. This tract is one of many that carry **nociceptive** information, rostrally, from the spinal cord. Pain is a complicated sensation. A noxious stimulus creates a perception of where it occurs, and also, perception of a rapidly increased attention level, autonomic responses, and emotional reactions. There is a much higher chance that the event and its surrounding circumstances will not be forgotten. The spinothalamic tract reaches the VPL nucleus and nearby nuclei of the thalamus. It is involved in awareness and the localization of nociceptive stimuli. Other tracts carry pain information to many sites in the thalamus, limbic system, and reticular formation. The spinothalamic and spinoreticulothalamic tracts are together inside the spinal cord. They are collectively called the **anterolateral system.** This reflects their location in the anterior half of the lateral funiculus.

In the lateral part of the posterior root, nociceptive, thermoreceptive, and certain mechanoreceptive fibers enter the spinal cord. These project their branches into the posterior horn, where they synapse in its superficial laminae. The synapses occur on neurons of *lamina I*, on neurons of deeper laminae with dendrites projecting dorsally into the **substantia gelatinosa**, and on small interneurons of the substantia gelatinosa. These carry the information to the neurons of other laminae. The anterolateral system or pathway is formed as these second- and third-order cells of the pain and temperature pathways send axons across the midline, with a slight rostral inclination.

Anterolateral pathway

The anterolateral pathway makes up most of the anterior area of the lateral funiculus. At its anteromedial edge, new fibers join the pathway, also in somatotopic organization. The fibers from the most caudal segments make up its most posterolateral portion. The fibers from the more rostral segments fill the more anteromedial portions. There are subdivisions based on the origin, destination, and likely functions of the fibers. Most direct spinothalamic fibers arise from laminae I and V. They project to their own areas of the VPL nucleus and nearby nuclei of the thalamus, somatotopically, like that of the medial lemniscus. The fibers are believed to be involved in conscious awareness of different types of pain, such as aching, burning, or stinging. They are also involved in the location of the pain and make up an alternative pathway through which mechanoreceptive input travels to the thalamus and cerebral cortex.

Cortical locations of termination are wider spread than those from the posterior column-medial lemniscus pathway. They reach the postcentral gyrus as well as the insula and other areas, showing multiple levels of conscious awareness of painful stimuli. Another set of spinothalamic fibers that mostly arise from the intermediate gray matter projects to the intralaminar and other nuclei of the thalamus, with no somatotopic arrangement. Many of the final projections are indirect. They have a polysynaptic course through the reticular formation, which is most of the brainstem core. Therefore, they are referred to as **spinoreticular fibers**, likely involved in changes in attention levels as a response to pain. A collection of **spinomesencephalic fibers** also arises mostly from laminae I and V. These are important for intrinsic mechanisms of pain control. There are also nociceptive projections to autonomic control areas of the brainstem and spinal cord, as well as **spinohypothalamic fibers** directly reaching the hypothalamus. All of these various types of fibers are mixed with, or close to each other in the spinal cord.

The substantia gelatinosa is not fully understood concerning its transmission of pain signals. It has many small cells in which unmyelinated afferents terminate in large numbers. Several different types of these neurons are understood, based on their neurotransmitters. For instance, it is believed that the substantia gelatinosa secretes natural opiates, thereby regulating pain by blocking the release of cytokines such as *substance P*. These cytokines are involved in the generation of pain.

> **Focus on pain perception**
> Pain signals can excite ANS pathways as they pass through the medulla oblongata. This causes increased heart rate and blood pressure, followed by rapid breathing and sweating. These reactions range in severity based on the intensity of pain. They can be depressed by brain centers in the cortex via various descending pathways. Recall that we have spinoreticulothalamic pathways. These are pain fibers that synapse at the level of the reticular activating system in the brainstem before reaching the thalamus. Since the reticular activating system is involved in arousal, painful stimuli cause arousal of the patient when the brainstem is intact. This is a common test in patients who are in a coma, to assess their response to pain.

Descending pathways

The descending pathways are mostly found in the lateral and anterior funiculi. They influence activities of the **lower motor neurons**. The alpha and gamma motor neurons of the anterior horn are controlled in many ways by the supraspinal centers. These centers are partially located in the brainstem. The primary descending outflow is from the cerebral cortex—especially from the precentral gyrus. These structures form the **corticospinal system**. The **lateral corticospinal tract** is also called the **pyramidal tract** because it runs through a pyramid of the medulla (Fig. 19.8). This tract is large in size and crossed in structure. It has about 85% of the fibers from the contralateral pyramid, crossing in the pyramidal decussation. This tract is located in the lateral funiculus of the spinal cord. It is medial to the posterior spinocerebellar tract, with fibers originating in the precentral gyrus and nearby areas of the cerebral cortex. They descend through the internal capsule, cerebral peduncle, basal pons, and medullary pyramid. The fibers decussate at the spinomedullary junction, ending in the anterior horn or intermediate gray matter. They terminate on the motor neurons of the anterior horn, but in large numbers on the smaller interneurons that eventually synapse on these motor neurons. There is no anatomical evidence of a somatotopic arrangement of these fibers below the midbrain level.

The generation of movement involves the interrelated yet different fibers of the corticospinal tract and the motor axons of the anterior root. The alpha neurons and anterior root fibers that these neurons form make direct contact with striated muscles. They are called the lower motor neurons, and sometimes, the *final common pathway* of the motor

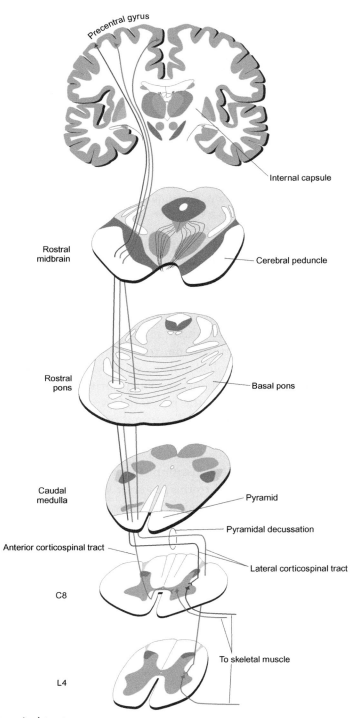

Fig. 19.8 Corticospinal tracts.

system. This is because they are the only structures that give the nervous system control of body movements. Interruption of the lower motor neurons that supply a muscle will cause **flaccid paralysis** and eventual muscle atrophy.

The **upper motor neurons** are those with axons descending from the cerebral cortex or brainstem, ending on the lower motor neurons, either directly or via an interneuron. If there is corticospinal damage, and upper motor neuron lesion will have extremely different effects than a lower motor neuron lesion. These effects are compared in Table 19.1. The muscles involved usually show hyperactive reflexes. They are **hypertonic**, with increased resting tension. There is **paresis**, which is paralysis or weakness of fine voluntary movements and involuntary movements. These symptoms are collectively called **spastic paralysis**. Many pathologic reflexes are related to upper motor neuron lesions. *Babinski's sign* is most well known, which involves dorsiflexion of the big toe and fanning of the other toes after the sole of the foot is firmly stroked.

The 15% of fibers in each pyramid that do not cross in the pyramidal decussation eventually reach the anterior funiculus. This is located adjacent to the anterior median fissure. When they reach it, these fibers are described as the **anterior corticospinal tract**. The fibers end on motor neurons or interneurons in medial parts of the anterior horn or intermediate gray matter. They are able to affect the activity of motor neurons for the axial muscles. Many fibers cross in the anterior white commissure prior to synapsing, but not all of them. Most of the fibers end in the cervical and thoracic segments, probably aiding in control of the neck and shoulder muscles. When this tract is damaged, there is no obvious weakness, possibly because of the bilateral fiber distribution from the contralateral tract. Therefore, the *pyramidal tract* refers to both the lateral and anterior corticospinal tracts.

Focus on neuron damage and its effects

Upper motor neuron damage may result in *clonus*, a condition of repetitive, rhythmic contractions of a muscle when attempting to hold it in a stretched state. Lower motor neuron damage may cause *fasciculations*, which are involuntary twitchings of groups of muscle fibers, and *fibrillations*, which are rapid, irregular, and unsynchronized contractions of muscle fibers.

Table 19.1 Upper and lower motor neuron damage and related effects.

Effect	Upper motor neuron damage	Lower motor neuron damage
Atrophy	Mild	Severe
Muscle tone	Increased	Decreased
Strength	Decreased	Decreased
Stretch reflexes	Increased	Decreased
Other signs	Clonus; pathologic reflexes such as Babinski's sign	Fasciculations; fibrillations

Section review
1. What is the description of the anterolateral pathway?
2. What structures form the corticospinal system?
3. Which neurons are the only structures used by the nervous system to control the body?

Integrative pathways

The process of *integration* involves incoming information being combined with other incoming information as well as previously stored information. Integration occurs in many areas of the spinal cord, brainstem, cerebellum, basal nuclei, and cerebral cortex. Sensations result, evoking conscious perception as well as subconscious awareness of changes that have occurred in the external or internal body environment. Integrative functions in the brain include wakefulness, sleep, learning, and memory. Each individual type of sensation is called a **sensory modality**, with sensory neurons carrying only information, which may be somatic, visceral, or "special". There are many types of specialized neurons. One example is the *Purkinje cells* found in the cerebellum. They are able to receive and integrate an extremely large number of synaptic inputs. The *interneurons* have integrative functions and are found in large numbers in the integrative areas. They are also known as *internuncial, relay, association, connector, intermediate,* or *local circuit* neurons.

The events that occur in a sensation cause stimulation of the receptor. There is transduction and conversion of the stimulus into a graded potential, which may have a large variance in amplitude, and diminishes over distance. There is a generation of impulses when graded potential reaches the threshold. Then, the integration of sensory input by the CNS occurs. In the SM pathways, control of body movement involves motor portions of the cerebral cortex, initiating and controlling precise movement. The basal ganglia help in establishing muscle tone. They integrate semi-voluntary automatic movements. The cerebellum helps to make movements smooth and aids in posture and balance. The direct motor pathways, or pyramidal tracts, convey impulses from the cerebral cortex, resulting in precise muscular movements. They also integrate excitatory and inhibitory input, to become the final common pathway.

The autonomic nervous system carries signals from the CNS to all organs and tissues, except the skeletal muscle fibers. It consists of preganglionic and postganglionic neurons that are linked in functionally unique pathways. The ANS can also integrate reflex interactions between different parts of the peripheral nervous system, even without utilizing the spinal cord in its activities. The ANS is discussed in **Chapter** 17.

Reflex activity

Reflexes comprise many types of control systems in the body. They can be *inborn (intrinsic)* or *learned (acquired)*. Inborn reflexes are rapid and predictable motor responses to stimuli. They are performed unconsciously, not learned, and involuntary. Such reflexes help maintain posture, control visceral activities, and avoid pain. Touching a hot surface causes an instantaneous inborn spinal reflex that withdraws the hand from the surface. Other similar reflexes continue without any awareness, such as visceral reflexes regulate by the subconscious lower CNS regions—primarily the brain and spinal cord.

A *learned* reflex is developed through repetition or practice, such as all of the many activities required to drive an automobile. Though the process is mostly automatic, it only develops because a lot of time was needed to develop the related skills. The difference between inborn and learned reflexes is not clear. Most inborn reflexes are modified by learning and by conscious effort. Regarding the hot surface, pain signals picked by the spinal cord's interneurons are quickly transmitted to the brain. This makes us aware of pain as well as the cause of the pain. The **withdrawal reflex** is serial processing controlled by the spinal cord. Pain awareness is based on the simultaneous and parallel process of the sensory stimuli.

Components of a reflex arc

Reflexes occur over extremely specific neural paths. These *reflex arcs* consist of five components, as follows:
- **Receptor**—the site of stimulus action
- **Sensory neuron**—which transmits afferent impulses to the CNS
- **Integration center**—in a simple reflex arc, this center can be a single synapse, between sensory and motor neurons, such as the **monosynaptic reflex**; in a more complex reflex arc, multiple synapses and chains of interneurons are involved, such as the **polysynaptic reflex**
- **Motor neuron**—conducts efferent impulses from an integration center to an effector organ
- **Effector**—a gland cell or muscle fiber that responses, via contraction or secretion, to efferent impulses

The functional classes of reflexes include **somatic reflexes** and **autonomic (visceral) reflexes**. The somatic reflexes activate skeletal muscle. The autonomic reflexes activate the smooth or cardiac muscle, or glands, which are all described as *visceral effectors*.

Spinal reflexes

The spinal cord is involved in reflexes, sensory processing, and motor outflow. **Spinal reflexes** are stereotyped motor outputs caused by certain afferent inputs. They often involve neural circuitry that is only found in the spinal cord. Except for axon reflexes, all reflex pathways involve at least one receptor structure and an associated afferent neuron. The cell body of this afferent neuron must lie in a posterior root ganglion or another sensory ganglion. Reflex pathways also involve efferent neurons, which has its cell body with the CNS. Except for the **stretch reflex**, all reflexes involve one or more interneurons. Reflexes can be simple to highly complex.

All skeletal muscles, except for some of the head muscles, contract to various extents, as responses to stretching. The responsible reflex arc uses the simplest possible route through the CNS. This is because it only uses two neurons and one intervening synapse. This is sometimes called a *monosynaptic reflex* or *myotatic reflex*. The afferent limb of the arc is called a *Ia afferent*, with its related muscle spindle primary ending. Central processes of the *Ia afferent* form synapses in the spinal cord directly on alpha motor neurons innervating muscle containing the stimulated spindle.

Stretch reflexes can be used for clinical testing, such as by tapping the patellar tendon, to stimulate the **knee-jerk reflex**, which slightly stretches the quadriceps muscle. The *Ia* endings in this muscle's spindles become excited. Then, they excite the quadriceps alpha motor neurons, resulting in contraction of the quadriceps, which completes the reflex. Testing many stretch reflexes helps determine the integrity of peripheral nerves and predictable spinal cord segments (Table 19.2). Since stretch reflexes are usually initiated by tapping a tendon, they are often called **deep tendon reflexes** (DTRs). The actually responsible receptors in the muscles are attached to the tapped tendons. Stretch reflexes may be important for continual automatic corrections of movements and postures, though other reflexes may be even more important. As we stand, there is a slight swaying of the body, with some muscles being stretched, causing a reflex contraction that helps return the individual to the desired position.

Focus on stretch reflexes
Stretch reflexes are usually hypoactive or absent in cases of peripheral nerve damage or ventral horn injury related to a certain area of the body. The reflexes are absent in patients with chronic diabetes mellitus, neurosyphilis, and when in patients who are comatose. It is important to understand that they are hyperactive when corticospinal tract lesions reduce the inhibitory effect of the brain upon the spinal cord, as in stroke patients.

Table 19.2 Common clinically-tested deep tendon reflexes.

Reflex	Primary cord segment	Involved muscle	Peripheral nerve
Biceps	C5	Biceps brachii	Musculocutaneous
Brachioradialis	C6	Brachioradialis	Radial
Triceps	C7	Triceps brachii	Radial
Knee-jerk (patellar)	L4	Quadriceps femoris	Femoral
Ankle-jerk (Achilles)	S1	Gastrocnemius, soleus	Tibial

Flexor and crossed-extensor reflexes

Painful stimuli initiate the *withdrawal reflex*, which is also called the *flexor reflex*. This causes the automatic withdrawal of the potentially affected body part from the stimulus. Flexor reflexes are ipsilateral as well as polysynaptic, which is required when several muscles are needed to withdraw the body part. Flexor reflexes are crucial to survival. Their actions are above those of the spinal pathways. They prevent all other reflexes from utilizing them at the same time. Like other spinal reflexes, however, descending signals from the brain are able to override the flexor reflexes, such as when we are anticipating a needle being inserted into the skin to give an injection.

A **crossed-extensor reflex** may occur with a flexor reflex in the weight-bearing limbs. It is very important for maintaining balance. This complex spinal reflex consists of an ipsilateral withdrawal reflex and a contralateral extensor reflex. The incoming afferent fibers are synapsed with interneurons controlling the flexor withdrawal response, on the same side of the body, with other interneurons controlling the extensor muscles on the opposing side. The crossed-extensor reflex is obvious when stepping on a sharp object. The ipsilateral response results in quick lifting of the injured foot, as the contralateral response affects the extensor muscle of the opposing leg, supporting the weight that was suddenly shifted upon it. Another example is when someone grabs one of your arms, which is quickly withdrawn, and the other arm pushes the person away.

Superficial reflexes

Superficial reflexes are caused by gentle stimulation of the skin, such as by stroking. The superficial reflexes are important clinically, depending on functional upper motor pathways and upon spinal cord-level reflex arcs. The most common examples are plantar and abdominal reflexes.

Plantar reflex

The **plantar reflex** helps to test spinal cord integrity, from L_4 to S_2. It determines, in a direct manner, whether the corticospinal tracts have proper function. This reflex is elicited by drawing a blunt object down along the lateral aspect of the plantar surface, or sole, of the foot. When the primary motor cortex or corticospinal tract has been damaged, the plantar reflex is replaced by an abnormal reflex, known as **Babinski's sign**. In this reflex, the great toe dorsiflexes while the smaller toes fan laterally. Infants usually show this sign until about one year of age, since their nervous systems are not completely myelinated. The physiological mechanism of Babinski's' sign is not fully understood.

Abdominal reflexes

The **abdominal reflexes** are used to determine the integrity of the spinal cord and anterior rami, from T_8 to T_{12}. These reflexes are elicited by stroking the skin of the lateral abdomen above or below the umbilicus, or to its side. There will be a reflex contraction of the abdominal muscles. The umbilicus will move toward the stimulated site. These reflexes are very different among various people. When they are absent, this indicates corticospinal tract lesions.

Muscle tension and motor neuron inhibition

When a *Ib fiber* from a **Golgi tendon organ** is stimulated, effects are different based on the limb's position and activity during stimulation. The effect may oppose the stimulation of an Ia fiber, such as when alpha motor neurons innervating the muscle that are connected to the tendon organ are inhibited. This is a type of **autogenic inhibition**. It involves an inhibitory interneuron between the afferent and efferent fibers. Excitation of the motor neurons, through an interneuron, can also occur from stimulation of a tendon organ that is attached to a weight-supporting muscle. While not fully understood, it is known that this reflex is activated before any hazardous levels can be reached. The Golgi tendon organs are believed to contribute to fine adjustments in muscle contraction force during ordinary motor activities. Other receptors initiate various forms of autogenic inhibition at higher levels of tension.

A clinical example of autogenic inhibition is the **clasp-knife response**. Following damage to descending motor pathways, there is a greatly increased resistance of muscles to manipulation. The leg, for example, may be very difficult to flex, but if enough force is applied, the leg slowly flexes—until all resistance suddenly disappears. The leg then collapses in flexion, like the snapping shut of a clap knife. Other receptors are primarily involved, instead of the Golgi tendon organs.

Withdrawal reflexes coordinated by painful stimuli

The **flexor reflex** is initiated by cutaneous receptors, involving an entire limb. This is exemplified by pulling the hand back from a hot object, via flexing of the arm. Spinal flexor reflex pathways are slightly inhibited from descending influences of the brainstem. Therefore, only noxious stimuli cause a strong reflex. When the descending influences are removed for any reason, flexion can occur because of harmless **tactile stimulation**. Therefore, most or all of the cutaneous receptors are linked to the pathway. Ordinarily, only nociceptors are strong enough to cause reflex withdrawal. Since a flexor reflex involves the entire limb, the pathway is spread over several segments of the spinal cord. This includes the motor neurons that innervate all of the flexor muscles of the limb. All primary afferent fibers bifurcate upon entering the spinal cord. Their processes extend, one or more segments, in rostral as well as caudal directions. The flexor pathway includes one or more interneurons, which may have processes extending over several segments. The term *withdrawal reflex* may be more appropriate to use than *flexor reflex*. It is not an all-or-none phenomenon. It has different patterns based on the stimulated portion of the limb. Modification of the reflex response to reflect the area being stimulated is referred to as a **local sign**.

Section review
1. How does the autonomic nervous system function as an integrative pathway?
2. What are the descriptions of the various types of spinal reflexes?
3. What is the definition of the term "withdrawal reflex"?

Reciprocal and crossed effects of reflexes
It is easier to shorten a stretched muscle when motor neurons to its synergists are exited, and motor neurons to its antagonists are inhibited. This is known as **reciprocal inhibition**. Reflex activity in a muscle produces similar activity in its **ipsilateral synergists**, as well as opposite activity in its **ipsilateral antagonists**. Therefore, a tap on the patellar tendon causes excitation of quadriceps motor neurons, and via an interneuron, excitation of motor neurons to the hamstring muscles. The stretching of an extensor muscle of the thigh monosynaptically excites its motor neurons, as well as those of the other thigh extensors. Stimulation of a Golgi tendon organ may have a reverse pattern. Tension applied to the patellar tendon in certain phases of a movement will inhibit the quadriceps while exciting the hamstring muscles. Both of these actions occur via interneurons. The flexor reflex is accompanied by inhibition of that limb's extensors.

The **crossed effects** of reflex actions refer to the flexor reflex (Fig. 19.9). If pain is sensed by the left foot, there are simultaneous, opposing activity patterns in the right leg. When the left leg is flexed and withdrawn, the right leg is extended to aid in body support. Similarly, stimulation of muscle spindles and Golgi tendon organs results in related effects, but not as pronounced. Crossed effects may be components of more complicated activities. Individual interneurons receive many inputs. They may participate in withdrawal reflexes, tendon organ-mediated reflexes, and more complicated movements.

Sensory processing
Afferent fibers enter the spinal cord via the posterior roots. They end almost completely on the ipsilateral side of the CNS. They reach their termination sites by either synapsing on neurons in the ipsilateral gray matter or by directly ascending in an uncrossed form to the relay nuclei in the medulla oblongata. The relay cells project axons through certain sensory pathways to structures that are more **rostral**. Each primary afferent fibers branches often, to various ascending sensory pathways and local reflex circuits. One *Ia afferent* from a muscle spindle may have 500 or more branches in the spinal cord. Motor neurons innervating skeletal muscles are within the anterior horns. Many preganglionic autonomic neurons are found in the intermediate gray matter of certain spinal segments.

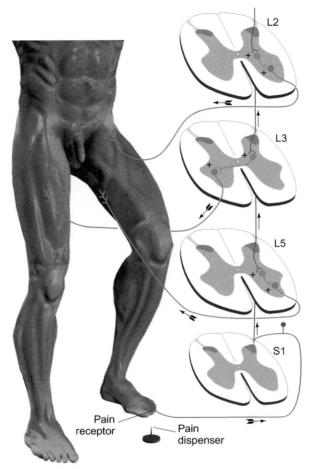

Fig. 19.9 Crossed extension of reflex actions.

Their axons leave the cord in the anterior roots. Neuronal activity is regulated by local reflex circuits as well as by pathways descending through the spinal white matter form the cerebral cortex and from diencephalic and brainstem nuclei.

Regional specializations of gray matter

The posterior horn is made up mostly of interneurons with processes inside the spinal cord, as well as projection neurons with axons collected into long and ascending sensory pathways. This area contains the *substantia gelatinosa* and *body* of the posterior horn. These are present at all levels of the cord. The substantia gelatinosa "caps" the posterior horn. When stained and microscopically examined, this region appears paler than the remainder of the gray matter. This is because it has finely myelinated and unmyelinated sensory fibers that carry pain and temperature data. A paler-staining area of white matter known as **Lissauer's tract** lies between the substantia gelatinosa and surface of the spinal cord. It contains finely myelinated and unmyelinated fibers utilized by the substantia gelatinosa. The posterior horn's body is mostly made up of interneurons along with projection neurons transmitting various somatic and VS information. It functionally overlaps some of the intermediate gray matter.

Motor neurons of the anterior horn

The anterior horn has cell bodies of large motor neurons, which supply skeletal muscles. These are alpha motor neurons, also called lower motor neurons. They are the only neurons involved in the control of body movements, no matter if they are voluntary or involuntary. Many different components can influence these neurons, but only they can cause muscle contractions. There would be complete paralysis of a muscle when the lower motor neurons supplying occurs, or if there is an interruption of its axons. Lesions of the lower motor neuron cause flaccid paralysis, and the muscle is a limb, lacking contraction. Reflex contractions can no longer occur. The muscle eventually atrophies due to lack of **trophic factors** delivered normally by the motor axons. An example of this condition is in *poliomyelitis* in which a viral disease damages the anterior horn motor neurons. Another example involves damage to the anterior roots.

Alpha motor neurons are longitudinally oriented in groups shaped like cigars, with each group innervating on muscle. In cross sections, they appear to be arranged in clusters separated by areas of interneurons. Clusters innervating axial muscles are medial to those innervating limb muscles. In the cervical and lumbar enlargements, the anterior horns are laterally enlarged, accommodating additional motor neurons. Smaller gamma motor neurons are mixed with alpha motor neurons. They innervate intrafusal muscle fibers of the muscle spindles and are also called **fusimotor neurons**.

In the anterior horn of the cervical area, the **spinal accessory nucleus** extends from the caudal medulla to approximately C_5. The axons emerge from the lateral surface of the spinal cord, slightly posterior to the denticular ligament, as separate series of rootlets forming the accessory nerve. The **phrenic nucleus** contains the motor neurons innervating the diaphragm. It is located in the medial area of the anterior horn, in segments C_3 to C_5. As a result, injuries to the upper cervical spinal cord are serious. Destruction of descending pathways controlling the phrenic nucleus and respiratory motor neurons makes the patient unable to breathe.

Autonomic neurons of the intermediate gray matter

The gray matter intermediate to the anterior and posterior horns has some of the characteristics of both of these. It also contains the spinal preganglionic autonomic neurons, and at some points, a unique region called **Clarke's nucleus**. Preganglionic sympathetic neurons for all of the body are found in segments T_1 through L_3, primarily located in the **intermediolateral cell column**. This forms a pointed lateral horn on the spinal gray matter. Axons leave through the anterior roots. In segments S_2 to S_4, cells in a related location make up the **sacral parasympathetic nucleus**, while not forming a distinct lateral horn. Axons leave via the anterior roots, synapsing on the postganglionic parasympathetic neurons of the pelvic viscera.

Clarke's nucleus, also called the *nucleus dorsalis of Clarke*, or **posterior thoracic nucleus**, is a rounded grouping of large cells on the medial surface of the posterior horn's base, from approximately T1 to L2. It is very prominent at lower thoracic levels. An important relay nucleus for information to the cerebellum, this nucleus may also help forward proprioceptive information from the legs to the thalamus. It is considered to be part of the posterior horn due to its significant role in sensory processing. The rest of the intermediate gray matter contains various projection neurons, sensory interneurons, and interneurons synapsing on motor neurons.

The layered arrangement of spinal cord gray matter

There are a variety of *lamina*, or layers, of gray matter in the spinal cord, as follows:
- **Lamina I (marginal zone)**—thin layer covering the substantia gelatinosa
- **Lamina II**—the substantia gelatinosa itself
- **Laminae III to VI**—the body of the posterior horn
- **Lamina VII**—basically related to the intermediate gray matter, including Clarke's nucleus; also includes large extensions into the anterior horn
- **Lamina VIII**—makes up some of the interneuronal zones of the anterior horn
- **Lamina IX**—clusters of motor neurons embedded in the anterior horn
- **Lamina X**—zone of gray matter surrounding the central canal

Histological differences between the laminae are related to functional differences (see Table 19.3). One example is the functional difference between large- and small-diameter peripheral nerve fibers, based on fiber termination patterns in the gray matter.

Table 19.3 Significant spinal cord gray matter subdivisions.

Nucleus	Lamina	Levels	Functions
Marginal zone	I	All	Certain spinothalamic tract cells
Substantia gelatinosa	II	All	Pain and temperature information
Body of posterior horn	III to VI	All	Sensory processing
Clarke's nucleus	VII	T1 to L2	Posterior spinocerebellar tract cells
Intermediolateral column	VII	T1 to L3	Preganglionic sympathetic neurons
Sacral parasympathetic nucleus	VII	S2 to S4	Preganglionic parasympathetic neurons to pelvic viscera
Accessory nucleus	IX	Medulla to C5	Motor neurons to sternocleidomastoid and trapezius
Phrenic nucleus	IX	C3 to C5	Motor neurons to diaphragm

Section review
1. What are crossed effects of reflex actions?
2. What are the effects of the lower motor neurons?
3. What do layers II, VII, and IX of the spinal cord laminae consist of?

Blood supply of the spinal cord

The most important blood supply branches of the subclavian arteries are the vertebral arteries. Smaller branches develop the rostral origin of the anterior spinal artery and the smaller posterolateral spinal arteries. These are the major blood supplies of the cervical cord. The thoracic and lumbar cord areas are supplied by arteries from the aorta and internal iliac arteries. The sacral cord is supplied by segmental branches of the lateral sacral arteries.

A segmental artery divides into an anterior and a posterior ramus. Every posterior ramus develops into a spinal artery that enters the vertebral foramen, pierces into the dura, then supplying the spinal ganglion and roots through its anterior and posterior radicular branches. The majority of anterior radicular arteries are small. Some of them do not reach the spinal cord. Between four and nine of these arteries arise irregularly, but are larger, supplying the majority of the spinal cord's blood. The radicular arteries help supply blood to the surrounding ligaments and vertebral bodies. Venous drainage is into the posterior veins that form the spinal plexus.

The radiculomedullary arteries are divided into three groups, as follows:
- **Upper (cervicothoracic)**—from the anterior spinal arteries and branches of the costovertebral and thyrocervical arteries
- **Intermediate (middle thoracic)**—the T3 to T8 cord segments, usually from one T7 radicular artery
- **Lower (thoracolumbar)**—from a large T10 or L1 anterior radicular artery (*of Adamkiewicz*), which supplies the lower 66% of the cord; there is much variation in the areas supplied between individual people; there is no way to predict which area or amount of the spinal cord that will be infarcted when one of the vessels is occluded; the junction in between vertebral spinal and aortic circulations is usually at the T2 to T3 spinal segment, yet most ischemic lesions are very much below this area

The single anterior spinal artery, running the entire length of the cord within its anterior sulcus, is formed by the anterior medullary arteries. There are penetrating branches via the **sulcocommissural** arteries. The branches supply most of the anterior gray columns and anterior parts of the posterior gray neuronal columns. A pial radial network supplies the peripheral rim of the white matter of the anterior two-thirds of the spinal cord. This originates from the anterior median spinal artery as well. Therefore, the anterior median spinal artery branches supply approximately the anterior two-thirds of the spinal cord. Infarction of this region results in an anterior spinal cord syndrome, with loss of pain and temperature sensation as well as paralysis below the lesion's level. There is no effect upon proprioception and vibration sense corresponding to the transaction of the spinothalamic and corticospinal tracts, but not that of the posterior columns.

The posterior medullary arteries form the paired posterior spinal arteries. These supply the posterior one-third of the spinal cord via direct penetrating vessels and a pial vessel plexus. In the cord, there is an area of capillaries where penetrating branches of the anterior spinal artery meet the penetrating branches of the posterior spinal arteries as well as the circumferential pial network branches. Since collateral arteries have varied sizes, spinal segments do not have the same amount of circulatory protection.

There are usually eight to 12 anterior medullary veins, and even more posterior medullary veins near each other, at all segmental levels. These drain in the radicular veins. Also, a valve-less vein network extends along the vertebral column. It runs from the pelvic venous plexuses to the intracranial venous sinuses, without passing through the *Batson plexus* of the lungs. This may be a route for metastatic disease from the pelvis.

> **Focus on clinical significance of spinal cord blood supply**
> Since the **artery of Adamkiewicz** supplies most of the cord, any disruption of its flow can result in cord ischemia. This artery receives most of its blood from the aorta. This is the reason that during most surgeries involving the aorta, a drain is placed to remove cerebrospinal fluid (CSF) and control the pressure within the spinal cord system. This release of pressure helps with the flow of blood within the aorta and the Adamkiewicz artery, preventing spinal cord ischemia during these operations.

Clinical considerations

A variety of spinal cord conditions and disorders occur, which include inflammatory myelopathies, vascular disorders, spina bifida, syringomyelia, trauma, compression, and tumors of the spinal cord. These conditions and disorders will be discussed in detail in Chapter 20.

Key terms

Abdominal reflexes
Anterior corticospinal tract
Anterior horns
Anterior median fissure
Anterior roots
Anterolateral sulcus
Anterolateral system

Artery of Adamkiewicz
Ataxia
Autogenic inhibition
Autonomic (visceral) reflexes
Babinski's sign
Brachial plexus
Cauda equina

Central canal
Cervical enlargements
Cervical plexus
Clarke's nucleus
Clasp-knife response
Conus medullaris
Corticospinal system
Crossed effects
Crossed-extensor reflex
Cutaneous nerves
Deep tendon reflexes
Denticulate ligaments
Dorsal (posterior) root ganglion
Epidural space
Fasciculus cuneatus
Fasciculus gracilis
Fasciculus proprius
Filum terminale
Flaccid paralysis
Flexor reflex
Foramen magnum
Funiculi
Fusimotor neurons
Glial septum
Golgi tendon organ
Gray commissure
Homunculus
Hypertonic
Intercostal nerves
Interneurons
Intermediolateral cell column
Ipsilateral antagonists
Ipsilateral synergists
Knee-jerk reflex
Lateral corticospinal tract
Lateral division
Lateral horns
Lissauer's tract
Local sign
Lower motor neurons
Lumbar cistern
Lumbar enlargements
Lumbosacral plexus
Lumbosacral trunk
Medial division
Medial lemniscus
Monosynaptic reflex
Nerve plexuses
Nociceptive
Nucleus cuneatus
Nucleus gracilis
Paresis
Phrenic nerve
Phrenic nucleus
Plantar reflex
Polysynaptic reflex
Posterior column nuclei
Posterior funiculus
Posterior horns
Posterior intermediate septum
Posterior intermediate sulcus
Posterior median sulcus
Posterior roots
Posterior thoracic nucleus
Posterolateral sulcus
Proprioception
Propriospinal tract
Pyramidal tract
Reciprocal inhibition
Romberg's sign
Rostral
Sacral parasympathetic nucleus
Sensory modality
Somatic motor neurons
Somatic motor (SM) division
Somatic reflexes
Somatic sensory (SS) division
Somatotopic organization
Spastic paralysis
Spinal accessory nucleus
Spinal dura mater
Spinal nerves
Spinal reflexes
Spinohypothalamic fibers
Spinomesencephalic fibers
Spinoreticular fibers
Spinothalamic tract
Stretch reflex
Subcostal nerve
Substantia gelatinosa
Sulcocommissural
Superficial reflexes

Tabes dorsalis
Tactile stimulation
Trophic factors
Upper motor neurons
Ventral posterolateral nucleus (VPL)
Ventral posteromedial nucleus (VPM)
Visceral motor (VM) division
Visceral sensory (VS) division
White matter
Withdrawal reflex

Suggested readings

1. Althaus, J. *On Sclerosis of the Spinal Cord: Including Locomotor Ataxy, Spastic Spinal Paralysis, and Other System-Diseases of the Spinal Cord, their Pathology, Symptoms, Diagnosis, and Treatment.* Ulan Press, 2012.
2. Campagnolo, D. I.; Kirshblum, S. *Spinal Cord Medicine*, 2nd ed.; LWW, 2011.
3. Chung, K. C.; Yang, L. J. S.; McGillicuddy, J. E. *Practical management of Pediatric and Adult Brachial Plexus Palsies.* Saunders, 2011.
4. Cohen-Adad, J.; Wheeler-Kingshott, C. *Quantitative MRI of the Spinal Cord.* Academic Press, 2014.
5. Cramer, G. D.; Darby, S. A. *Clinical Anatomy of the Spine, Spinal Cord, and ANS*, 3rd ed.; Mosby, 2013.
6. Crock, H. V.; Yoshizawa, H. *The Blood Supply of the Vertebral Column and Spinal Cord in Man.* Springer, 2013.
7. Feldman, A. G. G. *Referent Control of Action and Perception: Challenging Conventional Theories in Behavioral Neuroscience.* Springer, 2016.
8. Fix, J. *Atlas of the Human Brain and Spinal Cord*, 2nd ed.; Jones & Bartlett Learning, 2008.
9. Freeman Somers, M. *Spinal Cord Injury: Functional Rehabilitation*, 3rd ed.; Pearson, 2009.
10. Hadzic, A. *Hadzic's Peripheral Nerve Blocks and Anatomy for Ultrasound-Guided Regional Anesthesia*, 2nd ed.; McGraw-Hill Education, 2011.
11. Kaya, D.; Yosmaoglu, B.; Doral, M. N. *Proprioception in Orthopaedics, Sports Medicine and Rehabilitation.* Springer, 2018.
12. Kirshblum, S.; Lin, V. W. *Spinal Cord Medicine*, 3rd ed.; DemosMedical, 2018.
13. Latash, M. L.; Zatsiorsky, V. *Biomechanics and Motor Control: Defining Central Concepts.* Academic Press, 2015.
14. Light, A. R.; Reichmann, H. The Initial Processing of Pain and its Descending Control: Spinal and Trigeminal Systems. In *Pain and Headache*, Vol. 12; S. Karger, 2012.
15. Marcano-Reik, A. J. *Brain and Spinal Cord Plasticity: An Interdisciplinary and Integrative Approach for Behavior, Cognition, and Health.* Nova Science Publishers, Inc, 2016.
16. Minassian, K.; Hofstotter, U. *The Human Spinal Cord Circuitry: From the Generation of Simple Reflexes to Rhythmic Activity.* Lap Lambert Academic Publishing, 2012.
17. Patterson, M. M.; Grau, J. W. *Spinal Cord Plasticity: Alterations in Reflex Function.* Springer, 2011.
18. Pierrot-Desilligny, E. *The Circuitry of the Human Spinal Cord.* Cambridge University Press, 2012.
19. Sabharwal, S. *Essentials of Spinal Cord Medicine.* DemosMedical, 2013.
20. Sengul, G.; Watson, C. *Atlas of the Spinal Cord.* Academic Press, 2012.
21. Thron, A. K. *Vascular Anatomy of the Spinal Cord: Radioanatomy as the Key to Diagnosis and Treatment*, 2nd ed.; Springer, 2016.
22. Tubbs, R. S.; Loukas, M.; Hanna, A.; Oskouian, R. *Surgical Anatomy of the Lumbar Plexus.* Thieme, 2018.
23. Vaccaro, A. R.; Fehlings, M. G.; Boakye, M. *Essentials of Spinal Cord Injury: Basic Research to Clinical Practice.* Thieme, 2012.
24. Villiger, E.; Piersol, G. A. *Brain and Spinal Cord: A Manual for the Study of the Morphology and Fibre-tracts of the Central Nervous System.* Ulan Press, 2012.
25. Willis, W. D.; Coggeshall, R. E. *Sensory Mechanisms of the Spinal Cord: Volume 2—Ascending Tracts and Their Descending Control*, 3rd ed.; Springer, 2004.

CHAPTER 20

Spinal cord lesions and disorders

Spinal cord disorders, including various types of lesions, can cause permanent and severe neurologic disabilities. Quick evaluation and treatment may help prevent this occurrence for some patients. Spinal cord disorders usually occur because of conditions that are extrinsic, such as compression from spinal stenosis, herniated disks, abscesses, tumors, or hematomas. Less often, spinal cord disorders are intrinsic to the spinal cord itself. These include hemorrhage, infarction, arteriovenous malformation, transverse myelitis, HIV infection, poliovirus infection, syphilis, trauma, vitamin B12 deficiency, decompression sickness, radiation therapy, and others. Sometimes, the nerve roots outside of the actual spinal cord are also damaged.

Myelopathies

Myelopathy is an umbrella term for any lesion involving the spinal cord. If there is any sign of inflammation, then it is called **myelitis**. *Transverse myelitis* is often used to describe the entire cross-sectional area of the spinal cord being affected, at one or multiple levels. *Diffuse* or *disseminated myelitis* describes multiple, widespread lesions. *Longitudinally extensive transverse myelitis (LETM)* describes a unique form of lesion that is extended longitudinal over more than three vertebrae levels, mostly related to certain circulating autoantibodies. *Meningomyelitis* describes inflammation of both the meninges and spinal cord. *Meningoradiculitis* refers to the disease of the meninges and spinal roots. *Pachymeningitis* is an inflammatory disease process of only the spinal dura. *Epidural* or *subdural spinal abscess* or *granuloma* describes infection within the epidural or subdural spaces. When spinal disease is *acute*, it occurs usually within a few days. When *subacute*, it occurs within 2–6 weeks. When *chronic*, it occurs over more than 6 weeks.

Transverse myelitis

Transverse myelitis is most often due to an immunologic process. It happens mostly after viral illnesses, such as enteroviruses, polio, herpes zoster, mycoplasma infection, and even **arboviruses** that include the equine encephalitis virus and West Nile virus. The enteroviruses have a high affinity for the neurons of the anterior spinal cord horns and the motor nuclei in the brainstem. They are **neuronotropic**, causing the generically named *poliomyelitis*. The herpes zoster virus has a high affinity for the dorsal root ganglia. The resulting functional abnormalities are related to the motor and sensory neurons, and not of the spinal tracts. West Nile virus also commonly damages the anterior horn cells. Less common forms of transverse myelitis may be linked to herpes simplex virus, cytomegalovirus, varicella-zoster virus, hepatitis viruses, Epstein-Barr virus, and the SV70 virus, which causes epidemic conjunctivitis. These are apparent in AIDS patients and others with impaired immunity. These complicated disease processes are made more confusing because of the possibility of a postinfectious type of myelitis.

Clinical manifestations
Symptoms are usually sensory loss, motor weakness, plus bowel and bladder dysfunction, depending on the level of the lesion. Back pain is common, and some patients may have a fever.

Diagnosis
A rapid evaluation can be done by cerebrospinal fluid (CSF) sampling. Magnetic resonance imaging (MRI) with contrast is needed for all patients. The spinal cord may appear normal, or have cord lesions over multiple levels. Brain imaging is usually done to exclude lesions in the brain. The viral myelitis diagnosis can also be confirmed by detection of virus-specific immunoglobulin M (IgM) in the CSF. This is especially helpful in the setting of West Nile virus.

Treatment
Treatment is usually aimed at decreasing inflammation with steroids and immune-modulating medications.

Spinal epidural abscess

Spinal epidural abscess may affect people of all ages and is often missed or mistaken for other diseases. The most common causative organisms include *Staphylococcus aureus*, streptococci, gram-negative bacilli, and anaerobic organisms. Back injuries, various skin or wound infections, and bacteremia may allow seeding of the spinal epidural spaces or vertebral bodies. This results in osteomyelitis, with purulent process extension to the epidural space. Sometimes, spreading is due to an infected disc. Another cause is bacteremia from the use of nonsterile needles or injection of contaminated drugs.

Pathophysiology
Microorganisms can be introduced into the epidural space during spinal surgery, or less commonly, via lumbar puncture needles during epidural or spinal anesthesia. They can also be transmitted via other epidural infections, with localization over the lumbar and sacral roots. Some cases, even fulminant in nature, have no clear source within the body for the bacterial abscess.

Clinical manifestations

In *cauda equina epidural abscess*, there may be severe back pain, but minimal neurologic symptoms until the infection extends to the upper lumbar and thoracic spinal cord segments. In the cervical or thoracic region, the purulent process is only accompanied by a low-grade fever and usually intense local backache. This is usually followed within several days by radicular pain. There may be a headache and nuchal rigidity, but usually just continued pain and the patient's inability to move the back. Within days, there may be quickly progressing paraparesis, with sensory loss in the lower body areas, and paralysis of the **sphincteric** muscles. The paralysis can develop rapidly in a matter of hours. Over the site of the infection, spinal percussion causes extreme tenderness. Examination shows signs of transverse cord lesion, and sometimes spinal shock when paralysis has quickly developed.

Diagnosis

The diagnosis is clinical with the help of imaging. MRI with contrast is the imaging method of choice (see Fig. 20.1).

Treatment

Ideally, laminectomy and drainage should be performed before the onset of paralysis. If this does not occur, the spinal cord lesion, probably caused by venous ischemia, will become irreversible. Initially, broad-spectrum antibiotics are administered. Treatment is then based on cultures, if available, which usually reveal staphylococcus. If vertebral

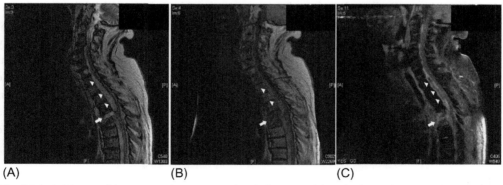

Fig. 20.1 A 66-year-old woman who presented with fevers and four-limb weakness. (A) T2-weighted sagittal image showing discitis (*arrow*) and extensive ventral epidural abscess (*arrowheads*) with spinal cord displacement and compression. (B) T1-weighted sagittal image without contrast also shows discitis and abscess, as well as abnormal vertebral body signal secondary to osteomyelitis. (C) T1-weighted sagittal image postcontrast showing the rim-enhancing epidural abscess (*arrowheads*) and discitis and vertebral osteomyelitis (*arrow*). (Courtesy of Dr. Nancy Fischbein, Stanford University.)

body osteomyelitis is the primary manifestation, there may be only a few spinal motor and sensory roots involved. Sometimes, therefore, antibiotics are sufficient without surgery. Surgical drainage is required for all cases of osteomyelitis. Antibiotics are continued over several weeks. The patient is reexamined at regular intervals, with sequential MRI scans. If partial spinal cord compression develops due to a fibrous, granulomatous reaction after surgery, there may have been incomplete surgical drainage of the abscess. Inadequate treatment of spinal subdural abscesses can result into a localized, chronic *adhesive meningomyelitis*.

Guillain-Barré syndrome

Guillain-Barré syndrome is also known as *acute idiopathic polyneuritis* and *acute inflammatory demyelinating polyradiculoneuropathy*. It is usually a quickly progressive, inflammatory polyneuropathy involving muscular weakness and slight distal sensory abnormalities. It is the most common acquired inflammatory neuropathy.

Pathophysiology

Guillain-Barré syndrome is not fully understood and believed to be an autoimmune condition with several variants. Demyelination of the peripheral nervous system is the primary feature in some of these, while in others, axons are affected. The syndrome usually begins after viral or bacterial illness, and rarely after vaccination. Common pathogens include *Campylobacter jejuni* and more recently *Zika virus*.

Clinical manifestations

In most patients, Guillain-Barré syndrome causes flaccid weakness, which is more prominent than sensory abnormalities. The relative symmetric weakness starts distally in most patients and then proceeds to involve the more proximal muscles. Paresthesias usually start in the legs, progressing to the arms. Less often, it begins in the head or arms. About 90% of patients have a maximal weakness at 4 weeks. Deep tendon reflexes are lost while sphincters are usually normal. Respiratory paralysis is the most severe and dangerous complication. It may require endotracheal intubation. The *Fisher variant* of this syndrome may only cause ataxia, areflexia, and ophthalmoparesis with no obvious weakness in the limbs.

Diagnosis

Diagnosis of Guillain-Barré syndrome is clinical. In this syndrome, symptoms usually appear on both sides of the body, with quick worsening. Deep tendon reflexes such as knee jerks are usually absent. The CSF will contain more protein than usual but very few immune cells. Abnormal sensations in the feet accompany or occur before weakness; absent or reduced deep tendon reflexes in weak limbs following a recent viral infection or diarrhea are diagnostic.

Treatment

Guillain-Barré syndrome usually requires admission to the hospital. Plasma exchange (**plasmapheresis**) and/or immunoglobulin therapy can speed recovery and reduce the severity of the illness. The patient will usually be given pain medications and other medications to prevent blood clots. The patient requires physical assistance and therapy, before and during recovery. There must be movement of the arms and legs to strengthen muscles, physical therapy to help deal with fatigue and regain normal functions, and training with adaptive devices such as braces or a wheelchair.

Multiple sclerosis

Multiple sclerosis (MS) involves patches of demyelination within the brain, spinal cord, or both. It is believed to involve an immunologic mechanism and may be related to infection by a latent virus. Genetic susceptibility for MS is suggested by an increased incidence in certain families as well as the presence of human leukocyte antigen (HLA) allotypes such as HLA-DR2. There may be a relation to lower levels of vitamin D, due to lack of sufficient sun exposure. However, none of these theories alone can explain the pathophysiology of this disease. MS usually affects people between ages 20 and 40, though it has developed in people from 15 to 60 years of age. Woman are affected slightly more than men. *Neuromyelitis optica*, also known as *Devic disease*, is a separate disorder that was previously believed to have been a variant of MS.

Pathophysiology

The pathophysiology of MS involves localized areas of demyelination (plaques). There is destruction of the oligodendroglia, perivascular inflammation, and chemical alterations of lipid and protein constituents of myelin, within and surrounding the plaques. Usually, the cell bodies and axons are mostly preserved, but axonal damage can occur. In plaques throughout the central nervous system (CNS), but mostly in the white matter, fibrous gliosis develops. This is mostly in the lateral and posterior columns, especially in the cervical regions, but also in the optic nerves and periventricular areas. There may be damage in the midbrain, pons, and cerebellum. The gray matter of the cerebrum and spinal cord are affected to a much lower degree. Both immune B cells and T cells are involved in the pathogenesis of this disease.

Clinical manifestations

MS is characterized by various CNS deficits. In most cases, there are remissions and recurrences of exacerbation. However, their frequency is highly varied. Overall, the common patterns of MS progression are as follows:
- **Relapse/remittance**—exacerbations are alternated with remissions; partial or nearly complete recovery occurs, with stable symptoms; disability develops over time.

- **Primary progression**—this is the most aggressive form, in which the patient continues to get worse after the diagnosis.
- **Secondary progression**—starts with relapses that alternate with remissions; it then becomes a progressive form, with no remissions.

Initially, symptoms can include paresthesias in one or more extremities, the trunk, or one side of the face. There is weakness or clumsiness of a hand or leg. Visual disturbances and partial vision loss with pain are common.

Additional early symptoms include unusual fatigability or slight stiffness of one limb, bladder control problems, slight gait disturbances, mild affective disturbances, and vertigo. All of these usually indicate scattered CNS involvement and may be very subtle. Signs and symptoms may be temporarily worsened by fever, hot baths, or warm weather. There is often mild cognitive impairment. The patient may show apathy, inattention, or poor judgment. The affective disturbances of depression and emotional lability are common.

Asymmetric or unilateral optic neuritis and bilateral internuclear ophthalmoplegia are common. In severe forms, optic neuritis will cause loss of vision that ranges from scotomas to near total blindness. They also cause eye pain, abnormal visual fields, swelling of the optic disk, and partial or complete afferent pupillary defects. A lesion in the medial longitudinal fasciculus connecting the third, and sixth cranial nerve nuclei will cause internuclear ophthalmoplegia. In horizontal gaze, there is decreased adduction of one eye, with nystagmus of the abducting eye, yet convergence is normal.

Uncommon yet characteristic signs include quick, small-amplitude eye oscillations in straight-head (primary) gaze, known as *Pendular nystagmus*. There may be intermittent unilateral facial numbness, pain that resembles trigeminal neuralgia, palsy, or spasms. Mild dysarthria may be caused by cerebellar damage, bulbar weakness, or disturbances of cortical control. Brainstem injury can cause other unusual cranial nerve deficits.

Motor deficits usually include weakness that reflects corticospinal tract damage in the spinal cord. This affects the lower extremities usually, being bilateral and spastic. Knee and ankle jerks, and other deep tendon reflexes are usually increased since this is an upper motor neuron disease. Babinski's sign and clonus are usually present. The gait becomes stiff and imbalanced because of spastic paraparesis. When advanced, the patient may be confined to a wheelchair. Late in the disease course, painful flexor spasms may result from touching of bedclothes or other sensory stimuli. Hemiparesis may be caused by cerebral or cervical spinal cord lesions. This is sometimes the presenting symptom.

In advanced disease, cerebellar ataxia with spasticity may cause severe disability. Additional cerebellar symptoms include slurred speech, scanning speech, and Charcot's triad. Scanning speech is defined as slow enunciation, usually with hesitation at the beginning of a word or syllable. Charcot's triad combines scanning speech with intention tremor and nystagmus.

Paresthesias and partial loss of sensations are common, usually being localized to one or both legs or hands. Painful sensory disturbances, such as electric-shock-like pain or burning pain, can be spontaneous, or due to touching—especially with spinal cord damage. *Lhermitte's sign* is an example of an electric-shock-like pain, radiating downward through the spine or into the legs when the neck is flexed. Objective sensory changes are usually transient and hard to evoke.

Spinal cord involvement usually results in bladder dysfunction, including urinary hesitancy or urgency, mild urinary incontinence, and partial retention of urine. In men, constipation or erectile dysfunction may occur. In women, genital anesthesia is possible. In advanced MS, severe urinary and fecal incontinence may occur.

Diagnosis

Diagnosis of MS is clinical. There must be a history of exacerbations and remissions, with objective demonstration via examination or testing of two or more separate neurologic abnormalities. The symptoms should last for at least 24 h. Brain and spinal MRI is required. With clinical findings, MRI may be diagnostic. If inconclusive, more tests may be needed to demonstrate the separate neurologic abnormalities. This usually begins with CNS analysis, and sometimes includes evoked potentials. An MRI is able to exclude other treatable and mimicking disorders. These may include nondemyelinating lesions where the spinal cord and medulla oblongata meet, such as foramen magnum tumors or subarachnoid cysts. Gadolinium-contrast enhancement helps distinguish active inflammation from older plaques. The sensitivity of MRI and computed tomography (CT) is increased by administering twice the dose of the contrast agent and delaying the scan. This is known as *double-dose delayed scanning*.

Examination of CSF includes opening pressure, cell count with differential, glucose, protein, immunoglobulin, oligoclonal bands, and usually, myelin basic protein and albumin. The IgG is often increased as a percentage of components of CSF. Oligoclonal bands are often detected. Myelin basic protein can be elevated during active demyelination. There may be slight increases in CSF lymphocytes and proteins.

Additional tests include evoked potentials, which are delays in electrical responses to sensory stimulation. This is often more sensitive for MS than the signs and symptoms. Somatosensory evoked potentials and brainstem auditory evoked potentials are sometimes measured. Systemic disorders such as lupus and infections such as Lyme disease can mimic MS. They are excluded with specific blood tests. There can also be blood tests to measure an antibody that is specific for neuromyelitis optica, known as neuromyelitis optica (NMO)-IgG, to differentiate this from MS.

Treatment

The goals of treatment for MS include shortening acute exacerbations and decreasing their frequency, relieving symptoms, and especially, maintaining the ability to walk.

To help treat loss of vision, coordination, or strength, brief courses of corticosteroids are used, including prednisone and methylprednisolone. Intravenous corticosteroids may shorten acute exacerbations. However, this does not have an overall effect on disease progression. Immunomodulators decrease exacerbations and delay eventual disability. Oral medications are now available in addition to the injectable and infusion therapies. Injectable medications include interferons and glatiramer. Natalizumab is an example of an infusion form that reduces exacerbations and new brain lesions but has increased risks for a viral infection in the brain called progressive multifocal leukoencephalopathy or PML.

Immunosuppressants such as azathioprine, methotrexate, cyclophosphamide, mycophenolate, and cladribine have been used for more severe cases, but these are controversial. Other treatments that help to control certain symptoms are as follows:
- *Baclofen or tizanidine*—for spasticity; gait training and range-of-motion exercises are often also done
- *Gabapentin, amitriptyline, desipramine, carbamazepine*—for painful paresthesias
- *Antidepressants plus counseling*—for depression
- *Amantadine or modafinil*—for fatigue

Bladder dysfunction is treated by focusing on the underlying mechanism. Supportive care involves reassurance and encouragement. Regular exercise is advised. It may include treadmills, stationary biking, stretching, and swimming. Patients can continue normal activities as tolerated but must avoid overwork and excessive heat exposure. If the patient smokes, he or she must stop.

> **Focus on multiple sclerosis**
> MS is the most common immune-mediated disorder that affects the CNS. In 2015, about 2.3 million people were affected globally, and it caused 18,900 deaths in that year. The name of the disease refers to the numerous scars (actually plaques or lesions), which develop on the white matter of the brain and spinal cord.

Amyotrophic lateral sclerosis

Amyotrophic lateral sclerosis (ALS) is also known as *Lou Gehrig's disease*. It is the most common form of motor neuron disorder (MND).

Pathophysiology

This disease is progressive, with symptoms worsening over time. The alpha motor neurons located in the anterior horn of the gray matter degenerate. This degeneration is responsible for the paralysis of the motor muscles, which include the muscles of respiration. The C3 to C5 spinal cord levels innervate the diaphragm and therefore, degeneration of alpha motor neurons at this level leads to respiratory paralysis and inability to breathe.

Clinical manifestations

The patient usually presents with unilateral weakness in the arms or legs, and usually, a wrist or a foot drop. The hallmark is the presence of both upper and lower motor neuron signs and symptoms. Fasciculations, hyperactive deep tendon reflexes, spasticity, clumsiness, extensor plantar reflexes, stiffness of movement, fatigue, and weight loss are common. Additional symptoms include dysphagia, hoarseness, increased saliva production, slurred speech, and choking when attempting to drink liquids. Later in the disorder, a **pseudobulbar affect** occurs. There are inappropriate, involuntary, and uncontrollable excesses of crying or laughter in some patients. Death results from failure of the respiratory muscles. About half of patients die within 3 years of onset. About 20% will live for 5 years, and 10% for 10 years. Survival for 30 or more years is rare. In the bulbar form, deterioration and death occur more quickly.

In *progressive bulbar palsy*, the muscles innervated by the cranial nerves and corticobulbar tracts are mostly affected. There is increased difficulty with chewing, swallowing, and talking. The voice sounds more "nasal." There is a reduced gag reflex, fasciculations and weak movements of the tongue and facial muscle, and weak movement of the palate. A pseudobulbar affect with emotional lability can occur when the corticobulbar tract is affected. Usually, the disorder spreads, and the extrabulbar segments are affected. This is called *bulbar-variant ALS*. The patient with dysphagia has very poor prognosis. Respiratory complications from aspiration usually cause death within 1–3 years.

In *progressive muscular atrophy*, inheritance is often autosomal recessive, especially when developing in childhood. Other cases are sporadic, and the disorder can develop at any age. Anterior horn cell involvement occurs on its own, or is more significant than corticospinal involvement. Progression is usually more benign compared to other MNDs. The earliest manifestation may be fasciculations. Muscle wasting and significant weakness starts in the hands, progressing to eh arms, shoulders, and legs. This eventually becomes generalized. Patients may survive for 25 years or more.

In *primary lateral sclerosis and progressive pseudobulbar palsy*, signs of distal motor weakness and muscle stiffness increase gradually. These affect the limbs in primary lateral sclerosis. In progressive pseudobulbar palsy, the lower cranial nerves are affected. Fasciculations and muscle atrophy may follow in many years. These disorders usually require several years to cause total disability.

Diagnosis

Diagnosis of ALS is suggested by progressive and generalized motor weakness, with no clear explanation. Disorders that must be ruled out include neuromuscular disorders, myopathies, spinal lesions, thyroid disorders, electrolyte abnormalities, and infections such as hepatitis C, Lyme disease, and syphilis.

Upper and lower motor neuron deficits, plus weakness in the facial muscles, are strongly suggestive for ALS.

Electrodiagnostic tests check for disorders of neuromuscular transmission or demyelination. This evidence is not present in motor neurons disorders. Nerve conduction velocity is usually normal until late in the disease course. The most useful test is needle electromyography (EMG), which shows fibrillations, fasciculations, positive waves, and sometimes, giant motor units. A brain MRI is required. An MRI of the cervical spine is also required in most cases to rule out structural lesions.

Laboratory tests to determine other treatable causes are varied. These include complete blood count, creatine kinase, electrolytes, and thyroid function tests. To check for a **paraprotein** that is rarely related to MNDs, serum and urine protein electrophoresis with **immunofixation** for monoclonal antibodies is performed. If an underlying **paraproteinemia** is discovered, this may indicate **paraneoplasia** of the disorder. Treatment of this may relieve the disorder. Demyelination motor neuropathies that can mimic ALS are related to antimyelin-associated glycoprotein (MAG) antibodies. A 24-h urine collection will check for heavy metal exposure if suspected. Lumbar puncture is advised since elevated white blood cells (WBCs) or protein in the CSF strongly suggests another condition. If risk factors or history suggest them, other tests may include the venereal disease research laboratory (VDRL) tests, erythrocyte sedimentation rate (ESR), and measurement of antibodies. These include Lyme titer, rheumatoid factor, hepatitis C virus, HIV, antinuclear, and anti-Hu antibodies. Genetic counseling may be requested, though disorders detected by genetic testing have no specific treatments.

Treatment

While there are no specific treatments for ALS, for progressive bulbar palsy, an **antiglutamate** drug called *riluzole* can prolong life for a few weeks. Recently, a new medication called edaravone (Radicava®) was approved, which can prolong the disease for a few months. Progressive neurologic disabilities require a multidisciplinary team approach. Drugs that may reduce symptoms include baclofen, quinine, phenytoin, glycopyrrolate, amitriptyline, benztropine, trihexyphenidyl, transdermal Hyoscine, atropine, and fluvoxamine.

Section review
1. What are the causes of transverse myelitis?
2. What is the pathophysiology of spinal epidural abscess?
3. What are the treatments for Guillain-Barré syndrome?
4. What are the common patterns of multiple sclerosis progression?
5. How is amyotrophic lateral sclerosis diagnosed?

Radiculopathies

Radiculopathies are also known as *nerve root disorders*. They cause segmental radicular deficits (see Fig. 20.2). These include pain or paresthesias, in a dermatomal distribution; and weakness of muscles that are innervated by the nerve root. These disorders are precipitated by chronic pressure upon a root within or adjacent to the spinal column—usually from a herniated intervertebral disk. Bone changes, caused by rheumatoid arthritis or osteoarthritis—especially in the cervical and lumbar regions—may also compress isolated nerve roots. Less often, carcinomatous meningitis causes multiple root dysfunctions in a patchy pattern. Rare causes include mass spinal lesions such as epidural abscesses and tumors, neurofibromas, and spinal meningiomas. These may develop with radicular symptoms, instead of the common spinal cord dysfunction. Diabetes mellitus may cause a painful radiculopathy of the thorax or extremities, due to ischemia of the nerve root. The nerve roots are sometimes affected by infections caused by fungi or spirochetes. Herpes zoster infection

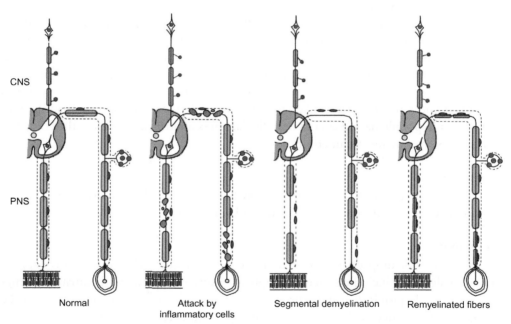

Fig. 20.2 Diagram of the main pathological events of primary segmental demyelination in immune-mediated inflammatory polyradiculoneuropathies. Attack by inflammatory cells causes patchy multifocal demyelination along nerve fibers but spares their axons. Recovery occurs by remyelination. Demyelinated segments become invested by several Schwann cells, resulting in a decrease in internodal length of those areas. *CNS*, central nervous system; *PNS*, peripheral nervous system. *(Adapted from Herskovitz, S., Scelsa, S., Schaumburg, H. H., 2008. Disorders of Peripheral Nerves. Oxford University Press: New York.)*

usually causes a painful radiculopathy. There are dermatomal sensory loss and a definitive rash, but sometimes there may be a motor radiculopathy with myotonic weakness and reflex loss. A condition called *cytomegalovirus-induced polyradiculitis* is a complication of AIDS.

Pathophysiology
Usually, radiculopathies are located in the lower spine. However, injuries that can lead to radiculopathies include lifting heavy objects improperly, trauma, tumors, and diabetes. Radiculopathies usually cause characteristic radicular syndromes of pain and segmental neurologic deficits that radiate below the knees. Muscles innervated by the affected motor root become weak and atrophied. Sensory root involvement results in a dermatomal distribution of impairment.

Clinical manifestations
Clinical manifestations of radiculopathies include segmental deep tendon reflexes being diminished or absent. Electric shock-like pains can radiate along the distribution of the affected nerve root. Pain can be worsened by movements transmitting pressure to the nerve root, through the subarachnoid space. Cauda equina lesions affect multiple lumbar and sacral roots, mostly due to cancer. They cause radicular symptoms in both legs and can impair the function of sphincter muscle and sexual organs. Findings showing spinal cord compression include a sensory level. This is an abrupt change in sensation below a horizontal line across the spine. They may also show flaccid paraparesis or quadriparesis, early-onset hyporeflexia and later hyperreflexia, reflex abnormalities below the compression site, and sphincter dysfunction.

Diagnosis
The diagnosis is clinical. MRI or CT scan may be needed to help with diagnosis. Electrodiagnostic studies are the gold standard and should be attempted prior to MRI in most cases.

Treatment
Treatment of radiculopathies is focused on the specific causes. They may involve appropriate analgesics, including acetaminophen, nonsteroidal antiinflammatory drugs (NSAIDs), and opioids. If pain is chronic, pain management is more difficult. Tricyclic antidepressants, anticonvulsants, physical therapy, and even mental health consultation should be considered. Some patients benefit from alternative medical treatments. These include spinal manipulation, transdermal electrical nerve stimulation, acupuncture, and medicinal herbs.

Herniated nucleus pulposus
Herniated nucleus pulposus is also known as a herniated, ruptured, or prolapsed intervertebral disk (see Fig. 20.3). The prolapse occurs through a tear in the surrounding

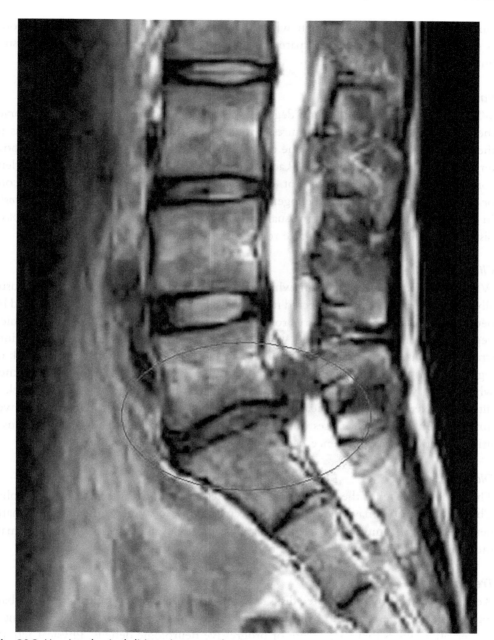

Fig. 20.3 Herniated spinal disk and narrowed nerve root.

annulus fibrous. This causes pain when the disk impinges on a nearby nerve root. A segmental radiculopathy, with paresthesias and weakness in the affected root's distribution, occurs.

Pathophysiology

When traumatic or nontraumatic degenerative changes occur in the spinal vertebrae, they can cause protrusion or rupture of the nucleus through the annulus fibrosus in the cervical or lumbosacral area. The nucleus will be displaced, posterolaterally or posteriorly, into the extradural space. Radiculopathy develops when the herniated nucleus compresses or irritates the nerve root. Posterior protrusion can compress the spinal cord or cauda equina—especially in a congenitally narrow spinal canal, known as *spinal stenosis*. In the cervical area, C6 and C7 are usually affected. In the lumbar area, more than 80% of disruptures affect the L5 or S1 nerve roots. Herniated disks are very common.

Clinical manifestations

When a herniated disk causes pain, which is not always, the signs and symptoms are usually in the distribution of the affected nerve roots. Pain is usually acute, often relieved by bed rest. This is different from nerve root pain due to an abscess or epidural tumor, which begins more slowly and is made worse by bed rest. With lumbosacral herniation, straight-leg raises cause stretching of the lower lumbar roots and worsening of back or leg pain. This will be bilateral if the disk herniation is centralized. Straitening the knee while sitting will also cause pain. Cervical herniation results in pain when the neck is tilted or flexed. Chronic cervical cord compression causes spastic paresis of the lower limbs. When acute, this causes quadriparesis. Cauda equina compression often causes urine retention or incontinence because of loss of sphincter function.

Diagnosis

The cause and level of the lesion can be identified via CT, MRI, or CT myelography. The involved roots may be revealed by electrodiagnostic studies. Since asymptomatic herniated disk is common, symptoms must be carefully evaluated with MRI abnormalities prior to any invasive procedures being done.

Treatment

Regardless of treatment, herniated disks desiccate over time, with symptoms usually abating. Without surgery, within 3 months, as many as 95% patients recover. Treatment is conservative unless there are severe or progressive neurologic deficits. Physical activity is restricted, with walking and light activity allowed as long as these can be tolerated. Prolonged bed rest and traction are contraindicated. For pain relief, acetaminophen, NSAIDs, and other analgesics are given. Physical therapy and exercise can improve posture and strengthen back muscles.

Invasive procedures should be considered for persistent or worsening neurologic deficits. The procedures of choice include microscopic diskectomy and laminectomy, with surgical removal of the herniated material. Immediate surgical evaluation is required for lesions acutely compressing the spinal cord or cauda equina. Surgical decompression is required immediately for cervical radiculopathies causing signs of spinal cord compression.

Spinal muscular atrophies

Spinal muscular atrophies (SMA) include several forms of hereditary disorders involving skeletal muscle wasting because of progressive degeneration of anterior horn cells in the spinal cord, as well as those of the motor nuclei in the brainstem. They are usually due to autosomal recessive mutations of a single gene locus, on the short arm of chromosome-5. This causes a homozygous deletion. SMA are not only peripheral nervous system disorders but also involve the CNS.

Pathophysiology

The four main forms of SMA include the following:
- **Type I (Werdnig-Hoffman disease)**—present in utero with symptoms beginning by about 6 months of age
- **Type II (intermediate)**—symptoms usually being between 3 and 15 months of age; less than one in four patients learn to sit, and none of them learns to crawl or walk
- **Type III (Wohlfart-Kugelberg-Welander disease)**—usually develops between 15 months and 19 years of age, with similarities to type I; progression is slower and life expectancy is longer; this form may be survivable
- **Type IV (adult-onset)**—this form may be recessive, dominant, or X-linked; adult onset is between 30 and 60 years of age

Clinical manifestations

Symptoms of type I include hypotonia, hyporeflexia, tongue fasciculations, and great difficulty in swallowing, sucking, and over time, breathing; death is usually due to respiratory failure, mostly within the first year (95%), and by the age of 4 years overall. Symptoms of type II include flaccid muscle weakness, fasciculations, dysphagia, and lack of deep tendon reflexes; this form is often fatal early in life, often because of respiratory complications; progression can stop spontaneously, resulting in permanent, nonprogressive weakness, and high risks for severe scoliosis and its complications.

Symptoms of type III in certain families may be secondary to hexosaminidase deficiency and other enzyme defects; there is symmetric weakness and wasting, progressing from the proximal to distal areas—mostly in the legs, but beginning in the quadriceps and hip flexors; the arms are affected later; life expectancy is based on whether respiratory complications occur. Symptoms of type IV include a slow progression of primarily proximal muscle weakness and wasting; this form may be hard to differentiate from ALS.

Diagnosis

Diagnosis of SMA is suspected when there is unexplained muscle wasting and flaccid weakness—especially in infants and children. EMG and nerve conduction velocity studies are required, including muscles innervated by the cranial nerves. Conduction will be normal, but the affected muscles are denervated. A definitive diagnosis is obtained by genetic testing, which is effective in about 95% of patients. Muscle biopsy is sometimes performed. Serum levels of aldolase and creatine kinase remain normal or near-normal in most cases.

Treatment

Recently, the Food and Drug Administration (FDA) approved the first drug for the infantile form of SMA. It is called nusinersen (Spinraza®). There is no treatment for the other forms, but bracing, physical therapy, and special appliances may be beneficial. Adaptive devices from therapists may improve independence and self-care by allowing for most daily living activities.

Thoracic outlet compression syndromes

Thoracic outlet compression syndromes are poorly defined disorders, with pain and paresthesias in one hand, the neck, a shoulder, or an arm. They are believed to involve compression of the brachial plexus, and maybe, the subclavian vessels. These structures go through the thoracic outlet.

Pathophysiology

The pathophysiology of thoracic outlet compression syndromes is often unknown. Sometimes, there is compression of the lower trunk of the brachial plexus or subclavian vessels, below the scalene muscles, above the first rib, and prior to these structures entering the axilla. Compression can be caused by an abnormal first thoracic rib, a cervical rib, abnormal insertion or position of the scalene muscles, or an abnormally joined clavicle fracture. Thoracic outlet syndromes are most common in women, developing between the ages of 35 and 55.

Clinical manifestations

Pain and paresthesias usually start in the neck or shoulder. They extend to the medial aspect of the arm and hand, and occasionally, to the adjacent anterior chest wall. There is often mild to moderate sensory impairment in the C8 to T1 distribution on the side with pain. Less often, there are significant vascular-autonomic changes in the hand such as inflammation and cyanosis. In even fewer cases, all of the affected hand becomes weak. Rare complications include distal gangrene and Raynaud's syndrome.

Diagnosis
Diagnosis of these syndromes is based on symptom distribution. Diagnosis may be aided by auscultating bruits at the clavicle or apex of the axilla, or the finding of a cervical rib via X-ray. Other tests are debated, but evaluation for brachial plexopathy may be justified.

Treatment
Without any objective neurologic deficits, most patients respond well to NSAIDs, physical therapy, and low doses of tricyclic antidepressants. Surgery is reserved for those with extreme or progressive neurovascular deficits, and for those who do not respond to the initial treatments.

Cervical spondylosis and spondylotic cervical myelopathy

Cervical spondylosis is osteoarthritis of the cervical spine. It causes stenosis of the canal, and sometimes, cervical myelopathy because of encroachment of bone osteoarthritis growth known as *osteophytes* upon the lower cervical spinal cord. There may be involvement of the lower cervical nerve roots, known as *radiculomyelopathy*. Cervical spondylosis is common.

Pathophysiology
Sometimes, when the spinal canal is congenitally narrow, osteoarthritis causes stenosis of the canal and bone impingement upon the spinal cord. This causes compression and myelopathy of the spinal cord. This may be aggravated by hypertrophy of the ligamentum flavum. Radiculopathy is usually caused by osteophytes in the neural foramina between C5 and C6 or between C6 and C7. Pain is common, but manifestations are varied based on the neural structures involved.

Clinical manifestations
Spinal cord compression usually causes gradual spastic paresis, paresthesias, or both in the hands and feet. It may cause hyperreflexia. The deficits may be asymmetric, nonsegmental, and aggravated by coughing or the Valsalva maneuver. Patients with cervical spondylosis due to trauma may develop central cord syndrome. Over time, muscle atrophy and flaccid paresis may occur in the upper extremities at the lesion level, with spasticity below this level. Nerve root compression usually causes early radicular pain. Later on, there may be hyporeflexia, weakness, and muscle atrophy.

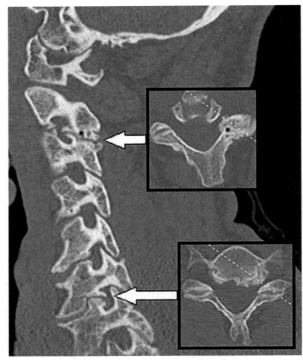

Fig. 20.4 CT of spondylosis causing radiculopathy.

Diagnosis

With the characteristic neurologic deficits, in patients who have osteoarthritis, are elderly, or have radicular pain at the C5 or C6 levels, cervical spondylosis is suspected. Diagnosis is via MRI or CT (see Fig. 20.4).

Treatment

When the spinal cord is involved, a cervical laminectomy may be required. A posterior technique may relieve compression, but leaves the anterior compressive osteophytes, possibly resulting in spinal instability and kyphosis. If the patient only has radiculopathy, nonsurgical treatments with NSAIDs and a soft cervical collar may be affective. If not, surgical decompression is likely required.

> ### Focus on cervical spondylosis
> The risk factors for cervical spondylosis include aging, occupations that put repetitive stress upon the neck, previous neck injuries, smoking, and genetic factors. It is very important to seek medical attention if there is a sudden onset of numbness, weakness, or loss of bladder or bowel control.

> **Section review**
> 1. What are radiculopathies?
> 2. How is herniated nucleus pulposus diagnosed?
> 3. What are the four main forms of spinal muscular atrophies?
> 4. How do you describe thoracic outlet compression syndromes?

Spina bifida

Spina bifida is defective closure of the vertebral column (Fig. 20.5). It is of unknown cause, but risks for this condition are increased by low folate levels during pregnancy. Spina bifida is one of the most serious neural tube defects that is still compatible with a prolonged lifespan. Overall, it is relatively uncommon, occurring in about 0.7 of every 1000 births in the United States. Most commonly, it occurs in the lower thoracic, lumbar, or sacral region, extending for three to six vertebral segments. Severity may range from occult with no apparent abnormalities, to protruding sacs, to a completely open spine. This final form is accompanied by severe neurologic disabilities, and usually, death.

Pathophysiology

In *occult spinal dysraphism*, the skin overlying the lower back, usually in the lumbosacral area, is abnormal. There may be hyperpigmented areas and even tufts of hair over the affected area. In *spina bifida cystica*, the protruding sac can contain meninges, the spinal cord, or both. When both occur, this is called *myelomeningocele*. The sac usually has meninges and a central neural plaque. When not covered sufficiently with skin, this sac can rupture easily, with increased risk for meningitis. Spina bifida is associated with *Chiari II type malformation* in some patients. Other congenital anomalies may include syringomyelia and soft-tissue masses surrounding the spinal cord.

Clinical manifestations

Though many affected children are asymptomatic, manifestations may be neurologic, orthopedic, or urologic. If the spinal cord or lumbosacral nerve roots are affected, which is common, there are various degrees of paralysis and sensory deficits below the lesion. The rectal tone is affected if the lesion involves the sacral roots. Reduced muscle innervation leads to leg atrophy. Since paralysis occurs in the fetus, orthopedic problems at birth may include arthrogryposis of the legs, clubfoot, or dislocated hip. Later, scoliosis may develop. This is more common when the lesions are above the L3 level. Paralysis may impair bladder function when the lesion is in the sacral area. This sometimes causes a neurogenic bladder, then urinary reflux. Other outcomes may include frequent urinary tract infections, hydronephrosis, and kidney damage.

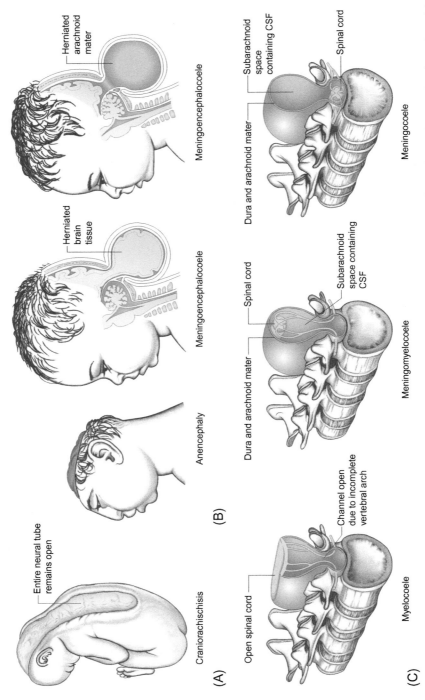

Fig. 20.5 Defects caused by failure of neural tube formation. (A) Total failure of neurulation. (B) Failure of rostral neurulation. (C) Failure of caudal neurulation. Abbreviations: *CSF*, cerebrospinal fluid.

Diagnosis

Ultrasound or MRI of the spinal cord is essential for children with spina bifida occulta. There may be underlying spinal abnormalities. Normal X-rays of the spine, hips, and if malformed, lower extremities should be performed. After diagnosis, urinary tract evaluation is required, including urine culture, urinalysis, blood urea nitrogen (BUN) and creatinine assays, and ultrasound. Prognosis and interventions are determined by measuring bladder capacity and pressure at which the urine exits into the urethra. Sometimes, urodynamic and voiding cystourethrogram tests are needed. Prenatal screening involves fetal ultrasound, and measurement of maternal serum levels of alpha-fetoprotein, best between 16 and 18 weeks of gestation. The levels can be assessed via amniotic fluid sampling if the previous testing suggests increased risks. Elevated levels suggest higher risk for spina bifida cystica, but the occult form rarely elevates these levels.

Treatment

The neurologic damage can progress in the occult form without early surgical treatment. Several disciplines will be needed for all cases of spina bifida treatment. The form, vertebral segments affected, and lesion extent must be assessed along with the health status of the infant and related abnormalities. At birth, a present meningomyelocele is immediately covered with a sterile dressing. If it is leaking CSF, antibiotics are begun to prevent meningitis. Neurosurgical repair is usually done within 72h after birth. Plastic surgeons may be needed to ensure adequate closure of the lesion.

Kidney functions are closely monitored. Urinary tract infections are promptly treated. To prevent infection, obstructive uropathy at the bladder outlet or ureteral level requires prompt treatment. Sterile intermittent catheterization is required to regularly empty the bladder in some patients, making them susceptible to infections.

Syringomyelia

A **syringomyelia** is a fluid-filled cavity, known as a *syrinx*, within the spinal cord (see Fig. 20.6). It usually results from lesions partially obstructing CSF flow. About 50% of syrinxes occur in patients with congenital abnormalities of the craniocervical junction. This may involve herniation of cerebellar tissue into the spinal canal—known as a Chiari malformation. There may also be other abnormalities such as encephalocele and myelomeningocele. The abnormalities often expand during adolescence. A syrinx may also develop after spinal cord tumors, scarring due to previous spinal trauma, or from no revealed predisposing factor.

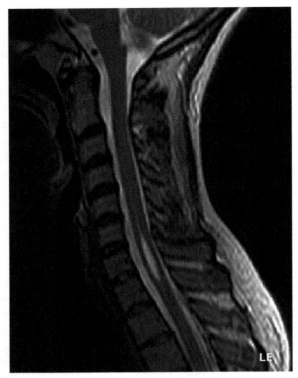

Fig. 20.6 Syringomyelia.

Pathophysiology

Syringomyelia involves a paramedian and usually irregular longitudinal cavity. It often begins in the cervical area but can extend down the entire spinal cord. A related but rarer condition, *syringobulbia*, affects the brainstem.

Clinical manifestations

Signs and symptoms are of insidious onset, with the lesion developing in the center of the spinal cord, resulting in a central cord syndrome. Pain and temperature sensory fibers decussate (cross) in front of the spinal canal. Therefore, a lesion in this area causes pain and temperature deficits early in the course of the disease. The first abnormality seen may be a burn or cut that is not painful. Usually, manifestations include atrophy, weakness, and often, hand and arm fasciculations and hyporeflexia. There may be a reduction in pain and temperature sensation over the shoulders, arms, and back in a "cape-like" fashion. Light touch and position-vibration sensation are normal in most cases.

Diagnosis

A syrinx is suggested by unexplained central cord syndrome or other significant neurologic deficits. These especially include pain and temperature sensory deficits over the shoulders, arm, and back. An MRI of the spinal cord and brain is performed. Gadolinium enhancement is helpful in detecting any related tumors.

Treatment

If possible, any underlying craniocervical junction abnormalities, spinal tumors, or postoperative scarring is first corrected. The surgical treatment of syringomyelia includes decompression of the foramen magnum and upper cervical cord. Surgery usually is not able to reverse severe neurologic deterioration and should be avoided if possible.

Spinal cord trauma and compression

Every year in the United States, there are about 12,000 new spinal cord injuries, usually in young adults. The most common causes of spinal cord trauma include car accidents, falls, violence, and sports injuries. This can lead to complete or incomplete spinal cord damage. Hyperextension and hyperflexion injuries of the spine are illustrated in Fig. 20.7.

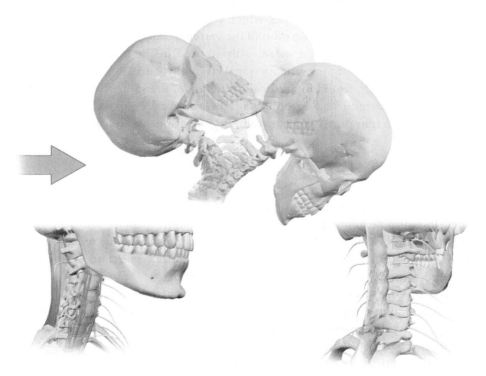

Fig. 20.7 Hyperextension and hyperflexion injuries of the spine.

Pathophysiology

Primary spinal cord injury occurs with an initial mechanical trauma and immediate tissue destruction. Nerve fibers can be compressed or damaged by vertebral dislocations or fractures, due to bone fragment or connective tissues. Acceleration, deceleration, or deformation forces upon impact cause vertebral injuries. There may be compression, traction, or tissue shearing. Bones, ligaments, neural tissues, and joints of the vertebral column may be damaged by hyperextension, hyperflexion, vertical compression, or rotation. In the vertebrae, dislocation and fracture may occur separately or together. The basic classifications of spinal injuries include the following:

- Simple fractures—single breaks, usually affecting the transverse or spinous processes
- Compressed (wedged) fractures—one of the vertebral bodies is compressed anteriorly
- Comminuted (burst) fractures—one of the vertebral bodies is shattered into multiple fragments
- Dislocations—in which there is a displacement of vertebral structures

The vertebrae may fracture easily from direct or indirect trauma. With tearing of supporting ligaments, the vertebrae are moved out of alignment, and dislocations occur. Horizontal forces move vertebrae directly forward. If the patient is in a relaxed position when injured, the vertebrae are in an angulated position. Both flexion and extension injuries may cause dislocations. Most vertebral injuries occur at the C1 to C2, C4 to C7, and T1 to L2 vertebrae, which are the most mobile areas of the vertebral column. The spinal cord makes up most of the vertebral canal in the cervical and lumbar regions. The size here makes it more easily injured in these areas. Noncontinuous vertebral injuries are common. Primary spinal cord injury may occur without vertebral fracture or dislocation, from longitudinal stretching of the spinal cord. This can occur with or without flexion and extension of the vertebral column. Stretching causes alterations in axon transport, edema, myelin degeneration, and retrograde or Wallerian degeneration.

Clinical manifestations

Immediately after injury, **spinal shock** can occur in some patients. This is characterized by total loss of reflex function, flaccid paralysis, sensory deficits, and loss of bladder and rectal control in all areas below the lesion. There is a transient reduction in blood pressure, poor venous circulation, and disturbed thermoregulation. The individual's body temperature approaches that of the surrounding air. Spinal shock can last for 24h. **Neurogenic shock** occurs with cervical or upper thoracic injury. It is caused by the disruption of sympathetic activity via loss of supraspinal control, and unopposed parasympathetic tone controlled by the vagus nerve. Its symptoms include hypotension, vasodilation, bradycardia, and hypothermia.

Continued deficits are based on the extent and level of the injury. All functions stop below any transected area, or below areas of concussion, contusion, compression, or ischemia. Return of spinal neuron excitability is slow. There may be a return to motor, sensory, reflex, and autonomic functions, or autonomic neural activity in the isolated segment will develop. Hyperactivity phases are varied. They may include, slight reflex activity, flexor spasms, alterations between flexor and extensor spasms, and predominant extensor spasms. Initial manifestations of acute spinal cord injury include rapid loss of voluntary movements in body parts below the injury level, sensations in the lower extremities and sometimes lower trunk, based on the injury level, and spinal and autonomic reflexes below the injury level. In most individuals, reflex activity returns in 1–2 weeks.

Flexion reemerges first, then reflex voiding and bowel elimination appear. Flexor spasms occur along with intense sweating, piloerection, and automatic bladder emptying—these are collectively called a **mass reflex**. Extensor spasms may develop, but usually, after all flexor spasms are developed. Autonomic hyperreflexia can occur after several months. This involves sudden, massive reflex sympathetic discharge, related to injuries at T5 to T6, or above. There is an imbalance between the sympathetic and parasympathetic nervous systems. There is vasoconstriction of the muscular, splanchnic, and cutaneous vascular beds. Blood pressure increases, but heart rate decreases. There is paroxysmal hypertension, bradycardia, a pounding headache, blurred vision, sweating with skin flushing above the level of the lesion, nasal congestion, nausea, and piloerection. There may be distension of the bladder or rectum and sensory stimulation. Immediate treatment is required due to possible cerebrovascular accidents occurring. Emptying of the bladder or bowel usually relieves the syndrome.

Diagnosis

Diagnosis of spinal cord trauma is based on physical examination, and MRI or CT scan. Diagnosis should be quick since impairments are permanent when endogenous repair events do not restore the damaged axonal circuits.

Treatment

The immediate intervention is immobilization o the spine to prevent additional injury. Decompression and surgical fixation may be required, and usually, are performed early. A new treatment with spinal electrical stimulation is now being investigated in academic centers for patients with chronic spinal cord injury. The preliminary results are reported to be positive.

Focus on spinal cord injury
It is of vital importance, for a possible spinal cord injury, to immobilize the spine as gently and quickly as possible. The use of a rigid neck collar and rigid carrying board is required to move the patient. Once at the hospital, treatment focuses on maintaining the patient's airway and further immobilizing the neck to prevent addition damage to the spinal cord.

Spinal cord tumors

Spinal cord tumors can develop within the spinal cord parenchyma, known as *intramedullary tumors*. These tumors also develop outside of the cord parenchyma, known as *extramedullary tumors*, and often compress the cord or nerve roots. Of the intramedullary tumors, the most common forms are gliomas, which include ependymomas and low-grade astrocytomas. They may infiltrate and destroy the parenchyma, and extend over many segments, or cause a syrinx (see Fig. 20.8). Extramedullary tumors may be *intradural* or *extradural*.

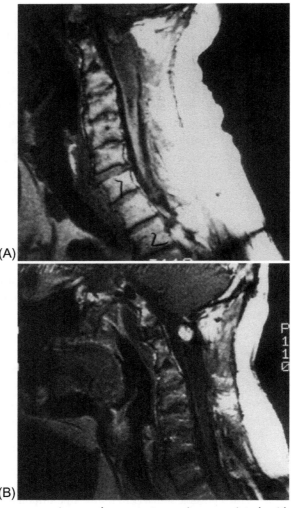

Fig. 20.8 Magnetic resonance image demonstrates syrinx associated with spinal cord tumor (hemangioblastoma). (A) T1-weighted image shows a nodule in upper cervical cord and a low-signal central mass suggestive of a cyst. (B) Postgadolinium image shows that nodule intensely enhances, which is classic for hemangioblastoma. *(Courtesy of Erik Gaensler.)*

Pathophysiology

Intradural and extradural tumors cause neurologic damage via compressing the spinal cord or the nerve roots. Most extradural tumors, before compressing the spinal cord, will invade and destroy bones.

Clinical manifestations

Especially for extradural tumors, early pain is common. It will be progressive and unrelated to physical activity but worsened by lying down or leaning against an object. There may be back pain that radiates down the sensory distribution of a certain dermatome. Neurologic deficits are usually referable to the spinal cord, developing over time. These commonly include incontinence, spastic weakness, and dysfunction of sensory tracts in a certain cord region and below. Deficits are commonly bilateral. Extramedullary tumors often cause pain. They sometimes cause distal lower extremity sensory deficits, segmental neurologic deficits, or symptoms of spinal cord compression. Symptoms may worsen quickly, resulting in paraplegia and bowel and bladder incontinence. Symptoms of nerve root compression are common. These include pain and paresthesias, then sensory loss, and muscular weakness. If the compression is chronic, there will be muscle wasting along the distribution of the affected roots.

Diagnosis

For segmental neurologic deficits or possible spinal cord compression, emergency diagnosis and treatment are needed. Spinal tumors are suggested by back or radicular pain that is progressive, nocturnal, or unexplained. Segmental neurologic deficits and unexplained neurologic deficits referred to the spinal cord or nerve roots are also diagnostic. Unexplained back pain in patients who have primary lung, breast, prostate, kidney, thyroid, or lymphoma-related tumors may also indicate a spinal cord tumor. Diagnosis is via MRI of the affected area. CT with myelography is less accurate.

Treatment

With neurologic deficits, corticosteroids are started immediately to reduce cord edema and preserve functionality, especially in the setting of cauda equina syndrome. Tumors compressing the cord are treated quickly. Some tumors, if well-localized and primary, can be excised surgically. Radiation therapy may be used with or without surgical decompression.

Section review
1. What are the causes of spina bifida?
2. How do you describe syringomyelia?
3. What is spinal shock?
4. What are the most common forms of intramedullary tumors?

Clinical cases

Clinical case 1
1. What is the definition of clonus, as seen in this case of spasticity?
2. What type of spinal cord tract damage is related to this condition?
3. What other symptoms may be present in this condition?

A 38-year-old man presented with spinal cord damage that had resulted in spasticity. Previous medications had been partially effective, but due to many adverse effects, had been stopped. Electroencephalogram and EMG revealed myoclonic jerks of spinal origin that initiated nonsustained clonus. Each episode lasted between 4 and 6s. Eventually, treatment with valproic acid greatly reduced the frequency of the myoclonic jerks with minimal side effects.

Answers:
1. Clonus is a self-sustained, oscillating stretch reflex. It is caused when a hyperreflexic muscle is stretched with continued force. This results in a rhythmic series of muscle contractions. Clonus most often occurs in the foot, ankle, quadriceps, finger flexors, jaw muscles, and others.
2. Spasticity is usually caused by an upper motor neuron lesion in the spinal cord or brain. It can occur in relation to other conditions as well, including MS, cerebral palsy, stroke, brain or head trauma, ALS, hereditary spastic paraplegias, and various metabolic diseases.
3. Other symptoms include hypertonicity, which is increased muscle tone; exaggerated deep tendon reflexes; muscle spasms; scissoring, which is an involuntary crossing of the legs; and fixed joints (contractures).

Clinical case 2
1. How did physicians arrive at this diagnosis?
2. What are other common clinical manifestations of this disease?
3. What type of MS is likely present?

A 28-year-old woman presented with irregular fatigue, pins and needles in her arms and legs, vision and balance problems, and depression lasting for months. Recently, she reports that she had the flu, and her neurologic symptoms became much worse. After several appointments and testing, she was diagnosed with MS. The patient stated that in between these episodes she had minimal symptoms.

Answers:
1. Diagnosis of MS involves an MRI and spinal tap in most cases. The neurologic examination includes funduscopic procedures. The motor examination tests the patient's ability to move the arms and legs.
2. Other common symptoms include cognitive difficulties and pain, at least early in the disease course. Certain colors may be perceived as "dull," and there may be eye pain.

Facial numbness is sometimes reported, as is an electric shock-like feeling when moving the head or neck, which travels down to the arms and legs. Over time, many patients develop bladder and bowel problems, clumsiness or lack of coordination, emotional changes, nystagmus, and diplopia. There may be problems related to heat generated by exercise, muscle spasms, sexual problems, and unusual sensations.

3. Based on the clinical manifestations, this patient probably has relapsing-remitting MS. It usually starts in the patient's 20s or 30s. It attacks, then relapses for weeks, months, or longer, and during these periods, the disease does not become progressively worse.

Clinical case 3

1. What is the likely diagnosis for this patient?
2. What are the goals of treatment for this condition?
3. Have stem cell treatments been successful for this condition?

A 51-year-old man was admitted to the hospital with a progressive limb movement disorder that he says has persisted for about 1 year. He has also had difficulty speaking for about 3 months. His mental status was poor, and it was difficult to understand what he was saying. The patient's tongue muscles showed alight atrophy, with inflexible movements, and the pharyngeal reflex was decreased. His muscle strength was decreased in all four limbs, along with muscle tone. The knee reflex was active in his left leg, but reflexes in the other three limbs were decreased or absent.

Answers:

1. The likely diagnosis, based on the signs and symptoms, is ALS.
2. Treatment is focused on improving neural function, promoting cerebral blood circulation, and improving muscle function.
3. Stem cell treatments have been shown to improve sleep and emotions, reduce speaking difficulties and swallowing problems, improve muscle activity and flexibility, along with increased muscle volume.

Key terms

Amyotrophic lateral sclerosis	Mass reflex
Antiglutamate	Multiple sclerosis
Arboviruses	Myelitis
Cervical spondylosis	Myelopathy
Guillain-Barré syndrome	Neurogenic shock
Herniated nucleus pulposus	Neuronotropic
Immunofixation	Paraneoplasia

Paraprotein
Paraproteinemia
Plasmapheresis
Pseudobulbar affect
Radiculopathies
Sphincteric

Spina bifida
Spinal shock
Syringomyelia
Werdnig-Hoffman disease
Wohlfart-Kugelberg-Welander disease

Suggested readings

1. Al Mosawi, A. *Wohlfart Kugelberg Welander Syndrome: Pediatric Neurology*. Lap Lambert Academic Publishing, 2018.
2. Anson, J. A.; Benzel, E. C.; Awad, I. A. *Syringomyelia and The Chiari Malformation*. AANS, 1997.
3. Ariza-Jemenez, A. B.; Jacinto, M. A. *Transverse Myelitis: A Review and Update*. Lap-Lambert Academic Publishing, 2015.
4. Bedlack, R. S.; Mitsumoto, H. *Amyotrophic Lateral Sclerosis: A Patient Care Guide for Clinicians*. Demos Medical, 2012.
5. Berenstein, A.; Lasjaunias, P. *Surgical Neuroangiography: 5 Endovascular Treatment of Spine and Spinal Cord Lesions*. Springer, 2012.
6. Birnbaum, G. L. *Multiple Sclerosis: Clinician's Guide to Diagnosis and Treatment*, 2nd ed.; Oxford University Press, 2013.
7. Bowen, B. C.; Rivera, A.; Saraf-Lavi, E. *Spine Imaging: Case Review Series*, 2nd ed.; Mosby, 2008.
8. Brochet, B. *Neuropsychiatric Symptoms of Inflammatory Demyelinating Diseases*. Springer, 2015.
9. Czervionke, L. F.; Fenton, D. S. *Imaging Painful Spine Disorders*. Saunders, 2011.
10. Durrant, D. H.; True, J. M.; Dishman, D.; Leahy, M.; Patrick, E.; Shooker, C. *Myelopathy, Radiculopathy, and Peripheral Entrapment Syndromes*. Scholars Consortium, LLC, 2012.
11. Ferner, R. E.; Huson, S.; Evans, D. G. R. *Neurofibromatoses in Clinical Practice*. Springer, 2011.
12. Hong, S. *Ataxia: Causes, Symptoms and Treatment*. Nova Biomedical, 2012.
13. Kaiser, M. G.; Haid, R. W.; Shaffrey, C. I.; Fehlings, M. G. *Degenerative Cervical Myelopathy and Radiculopathy: Treatment Approaches and Options*. Springer, 2019.
14. Kalb, R. *Multiple Sclerosis: The Questions You Have, The Answers You Need*, 5th ed.; Demos Health, 2011.
15. Keren, D. F. *High-Resolution Electrophoresis and Immunofixation: Techniques and Interpretation*, 2nd ed.; Butterworth-Heinemann, 2017.
16. Kirshblum, S.; Campagnolo, D. I.; DeLisa, J. A. *Spinal Cord Medicine*. LWW, 2001.
17. Love, L.; Nelson, A.; Zejdlik, C. *Nursing Practice Related to Spinal Cord Injury and Disorders: A Core Curriculum*. Demos Medical, 2001.
18. Noiri, E.; Hanafusa, N. *The Concise Manual of Apheresis Therapy*. Springer, 2014.
19. Ozek, M. M.; Cinalli, G.; Maixner, W. *Spina Bifida: Management and Outcome*. Springer, 2008.
20. Page, H. W. *Injuries of the Spine and Spinal Cord Without Apparent Mechanical Lesion and Nervous Shock*. Ulan Press, 2012.
21. Pierrot-Desilligny, E.; Burke, D. *The Circuitry of the Human Spinal Cord: Spinal and Corticospinal Mechanisms of Movement*. Cambridge University Press, 2012.
22. Rhodes, M. A.; Haacke, E. M.; Moore, E. A. *CCSVI as the Cause of Multiple Sclerosis—The science Behind the Controversial Theory*. McFarland, 2011.
23. Taylor, J. A. *The Physiology of Exercise in Spinal Cord Injury (Physiology and Disease)*. Springer, 2016.
24. Vasilakis, N.; Gubler, D. J. *Arboviruses: Molecular Biology, Evolution and Control*. Caister Academic Press, 2016.
25. Vodusek, D. B.; Boller, F. *Neurology of Sexual and Bladder Disorders (Handbook of Clinical Neurology)*. Number 130, Elsevier, 2015.

CHAPTER 21

Complete neurological exam

For patients who have neurologic symptoms, there is a series of steps known as the **neurologic method** that must be followed. The lesions' anatomy and pathophysiology must be identified via careful history and accurate neurologic examination. A differential diagnosis is generated, and specific and appropriate tests are performed. Neurology is perhaps the only specialty in medicine that is, able to diagnose most, if not all, neurological diseases simply by understanding the anatomy of the nervous system. The brain, spinal cord, and peripheral nerves form a neuroaxis, which starts at the cerebral cortex and ends at the level of the muscles. In neurology, the lesion must be located along the neuroaxis in order to come up with the correct differential diagnosis. A weakness in the hand can be due to any lesion along the neuroaxis; cortex, subcortical white matter, basal ganglia, thalamus, cerebellum, brainstem, spinal cord, brachial or pelvic plexus, peripheral nerves, neuromuscular junctions, and muscles. After the lesion's location is identified, a variety of pathophysiologic causes should be evaluated. These include vascular, infectious, neoplastic, degenerative, traumatic, toxic-metabolic, and immune-mediated causes. Following the neurologic method helps to prevent incorrect diagnoses due to symptoms that mimic those of other conditions. In order to identify a lesion along the neuroaxis, a few rules need to be followed. A thorough history must be obtained, followed by a complete examination to locate the lesion responsible for any signs and symptoms. For example, unsteadiness is probably from a lesion in the cerebellum, whereas bilateral vision loss is most likely due to a lesion in the occipital visual cortex.

History

Patient history is the single most important portion of the neurologic evaluation. The patient should be allowed to explain the signs and symptoms in his or her own words. It is usually easy for the clinician to determine if the patient's history is accurate, and to assess if a family member should also be interviewed. Specific questions are used to understand the intensity, quality, distribution, duration, and frequency of symptoms. Anything that worsens and attenuates each symptom should be evaluated, as well as any past treatments that were effective. Specific disabilities must be described quantitatively, such as how long the patient is able to walk before needing rest, and how these disabilities affect daily routines.

It is essential to document past medical history and a thorough review of body systems. This is because neurologic complications are common in many other conditions such as alcoholism, cancer, diabetes, HIV infection, and vascular disorders. Family history is significantly important since many disorders are inherited. These include migraine headache, metabolic conditions, muscle disorders, nerve abnormalities, and neurodegenerative disorders. Unusual infections and exposure to parasites and toxins can be revealed by complete occupational, social, and travel histories. Occasionally, neurologic manifestations are functional or hysterical and are linked to a psychiatric disorder. Usually, these manifestations do not follow anatomical or physiological rules, resulting in the patient often being depressed or unexplainably frightened. Distinguishing between functional and physical disorders may be difficult since these sometimes coexist.

Neurologic examination

A patient's neurologic examination begins when he or she enters the examination area and requires careful observation of all movements and other activities. The examination continues while the history is taken. As the patient moves to the examination table, the coordination, speed, symmetry, gait, and posture should be noted. Mood and social adaptation is assessed based on the patient's responses, demeanor, and even clothing. Often, prior to formal tests, movement disorders, unusual postures, neglect of space, and abnormalities of language can be noted. The skilled practitioner can add or exclude various types of information based on the preliminary hypothetical location of the lesion. Complete neurologic screening must be performed on all patients. A neurological exam is not complete unless the patient's ability to walk is assessed.

Mental status

To explain in simple words, in order for a human to be fully conscious, he or she needs to be both "awake" and "aware." A coma is defined as the patient being neither awake nor aware of surroundings. Wakefulness can be assessed by the patient's attention. Inattentive patients are unable to fully cooperate, which makes testing more difficult. Examination of mental status searches for cognitive declines. This requires testing many areas of cognitive function, including attention and concentration; orientation to persons, times, and places; memory; mathematical and verbal abilities; judgment; and reasoning. Loss of orientation to self or other people usually occurs only in severe delirium, dementia, or obtundation. The patient's awareness of the possible illness, knowledge level, educational history, affect, and mood are all assessed.

To test higher cortical function (see Chapter 6), the patient is requested to perform complex movements involving three body parts that discriminate between the right and left sides. These movements may include touching the left ear with the right hand while sticking out the tongue, or other suggested activities (testing complex task performance). The patient must name simple objects as well as body parts, then read, write, and repeat simple phrases (testing language). Five main language aspects need to be tested in all patients: fluency, comprehension (understanding of what they are being told), repetition ability, reading, and writing. Spatial perception is assessed by having the patient perform simple, then complex finger movements as demonstrated by the examiner. Then, he or she is asked to draw objects such as clocks, houses, cubes, or interlocking pentagons. Cognitive ability is another high cortical function that needs to be assessed.

Examination of mental status assesses current capacity by evaluating overall appearance, behavior, unusual or strange beliefs and perception including delusions and hallucinations, mood, and all cognitive aspects, including attention, memory, and orientation. This type of examination is done for any patient with altered mental status, or with acute or chronic cognitive impairment. The **Mini-Mental State Examination** is a commonly used screening tool. Baseline results are documented, with the examination repeated every year, and when a mental status change is suspected. The patient must be instructed that recording of mental status is a routine medical procedure and there is no embarrassment attached to this. The examination room must be quiet so that the patient hears all questions clearly. The patient must be questioned in the language that he or she speaks fluently. Mental status examination begins with an assessment of attention, by having the patient repeat three words. The parameters of mental status examination are summarized in Table 21.1.

Cranial nerves

The 12 cranial nerves include the following:
- 1st—Olfactory (smell)
- 2nd—Optic (vision)
- 3rd—Oculomotor (general eye movements)
- 4th—Trochlear (finer eye movements)
- 5th—Trigeminal (primary sensory nerve of the face, chewing, biting)
- 6th—Abducens (outward eye movements)
- 7th—Facial (motion of the face)
- 8th—Vestibulocochlear (acoustic, auditory, balance)
- 9th—Glossopharyngeal (sensory information of mouth, nose, throat, ear, tongue, tonsils, carotid bodies)
- 10th—Vagus (interfaces with the GI tract, heart, lungs, many organs, skeletal muscles)
- 11th—Spinal accessory (head, neck, shoulders, pharynx, larynx movements)
- 12th—Hypoglossal (tongue muscle movements)

Table 21.1 Mental status examination parameters.

Parameter	Description
Orientation	Person—what is your name? Time—what is today's date? Place—what is the name of this place or facility?
Short-term memory	The patient must recall three objects after a 3-min delay.
Long-term memory	The patient must answer a question about the past, such as his or her wedding, first car they owned, where their first house was located, etc.
Mathematics	Simple, serial tests are used to assess basic math abilities. The patient may be asked to subtract the number seven from 100, then subtract it from 93, etc. Also, the amount of a certain type of coin, such as a quarter, that makes up a sum such as $5.75, should be calculated by the patient.
Word finding	Over 1 min, the patient is asked to name as many objects as possible within a certain category, such as animals, clothing, or household objects.
Concentration	The patient is asked to spell a five-letter word both forward and backward, such as "World" or "Globe".
Object naming	The patient is asked to name several objects from the examination room, such as a pen, book, tongue depressor, etc.
Following commands	A one-step command is requested first, such as "touch the tip of your nose with your right index finger." Then a three-step command is requested, such as "take this piece of paper in your left hand, fold it in half, and put it on the floor."
Writing	Ask the patient to write a sentence that contains a subject and an object, which makes sense. Ignore any spelling errors.
Spatial orientation	Have the patient draw a clock, house, or other simple object. If the clock is drawn, have the patient mark it with a certain time. Another option is to ask the patient to draw two intersecting pentagons (five-sided, five-angled shapes).
Abstract reasoning	Name three or four related objects and ask the patient to identify a "unifying" theme between them. (These can be fruit, musical instruments, type of vehicles, etc.) State a slightly challenging proverb such as "People who live in glass houses should not throw stones," and ask the patient to explain what it means.
Judgment	Ask the patient about a hypothetical situation that requires a good judgment to be made, such as "What would you do if you found a stamped letter on the sidewalk?" The correct answer is put the letter into a mailbox. If the patient says that he or she would open the letter, a personality disorder is suggested.

Olfactory nerve (I)

If the patient reports head trauma or abnormal smells or tastes, the *olfactory nerve* is evaluated. This is also usually performed when lesions of the anterior fossa, such as meningioma, are suspected. The patient is requested to identify odors such as coffee, cloves, or soap, which are presented to each nostril separately. Normal function is assumed when the patient can detect each smell, even if he or she cannot identify it. Irritants such as alcohols are not used for testing since they can be detected as *noxious stimuli* without using the olfactory receptors.

Optic nerve (II)

The optic nerve transmits visual data from the retina. This is through the optic chiasm, and then the optic tracts, to the lateral geniculate nuclei of the thalami. In the optic chiasm, there are crossed fibers from the nasal (medial) sides of both retinas that convey data from the temporal (lateral) halves of the visual fields. To assess the **optic nerve**, pupillary light reflex should be tested by shining a light into each pupil. Each optic nerve (*afferent* arm) carries the light signal into a parasympathetic preganglionic nucleus in the midbrain called the *Edinger-Westphal nucleus*, which in turn constricts the pupils through the parasympathetic fibers of cranial nerve three (*efferent* arm). Each reflex has an afferent arm as well as an efferent arm. When normal, the *optic disk* is yellowish in color, and oval in shape. It is located nasally, at the posterior pole of the eye. There should be a sharp demarcation of the disk margins and its crossing blood vessels. The veins should have spontaneous pulsations. The *macula* is paler than the other parts of the retina. It is situated approximately two disk diameters temporal to the optic disk's temporal margin. The macula can be seen by having the patient look at the light emitted by the ophthalmoscope.

For neurologic patients, the most important condition to identify is optic disk swelling caused by increased intracranial pressure (*papilledema*). Early in this condition, there is engorgement of the retinal veins, and lack of spontaneous venous pulsations. The disk may appear hyperemic and have linear border hemorrhages. Its margins are blurred, first at the nasal edge. Once papilledema is fully developed, the optic disk will be elevated above the retinal plane. Blood vessels that cross the disk border will be obscured. Nearly always, papilledema is bilateral and does not usually impair vision or cause pain. The only vision impairment may be an enlarged blind spot. *Optic disk pallor* is another abnormality and is caused by optic nerve atrophy. It can be visualized in patients who have multiple sclerosis or other optic nerve disorders, and it is related to defects of visual acuity, the visual fields, or the reactive ability of the pupils.

Visual acuity testing uses a **Snellen chart** (about 20 ft away) for distance vision and a handheld chart (about 14 in. away) for near vision. Each eye is separately assessed, with the other eye being covered. The smallest line of print that the patient can read is recorded. Acuity is expressed as a fraction. The numerator is the distance at which the line of print can be read by someone with normal vision. The denominator is the distance that it can be read by the individual patient. Therefore, 20/20 indicates normal acuity, and the denominator increases as vision is worsened. More severe abnormalities can be graded based on the distance at which the patient can count the examiner's fingers, see hand movements, or perceive light.

The optic fundus is inspected using direct ophthalmoscopy for **funduscopic** examination. A darkened room is used, with the patient's pupils dilated, allowing for easier visualization of the fundus. Mydriatic eye drops may be used to enhance dilation, but not until visual acuity and pupillary reflexes have been tested—or in patients with untreated closed-angle glaucoma, or an intracranial mass lesion, due to possible transtentorial herniation. Consensual and direct pupillary responses are tested. It is important that visual acuity testing occurs with refractive errors corrected. This means that patients who wear glasses should be examined while wearing them.

Visual fields are tested via directed confrontation in all four visual quadrants. Each eye is tested separately. The examiner stands at about one arm's length from the patient. The patient's eye not being tested, and the examiner's eye opposite it is closed or covered. The patient is asked to stare at the examiner's open eye, which superimposes the monocular fields of both individuals. The examiner uses his or her index finger, of either hand, to determine the peripheral limits of the patient's visual field. The finger is moved slowly inward, in all directions, until the patient can see it. The size of the patient's blind spot (*central scotoma*), which is in the temporal half of the visual field, is also measured in relation to the examiner's blind spot. Confrontation testing determines if the patient's visual field is similar to or more restricted than the examiner's. Another method is to use a pin as the visual target. Slight field abnormalities can be detected by having the patient compare the brightness of colored objects that are positioned at different sites in the visual field, or by measuring the fields using a pin with a red-colored head as the target.

In the patient who is not fully alert, significant abnormalities can be detected by determining if the patient blinks when the examiner's finger is moved toward the patient's eye from different directions. For progressive or resolving abnormalities, the visual fields can be better mapped by using perimetry, with *tangent screen* or automated perimetry testing. In tangent screen testing, the patient is tested with a 9 mm white stimulus at 1 m away. The areas of the patient's response are recorded, and the patient is moved backward to 2 m away. The tangent test is repeated using an 18 mm white stimulus. The field should expand to twice the original size. When it does not, it is called a *tubular field* or *gun-barrel field*. This is nonphysiologic, indicating a nonorganic component to the field constriction.

Perception of color is tested using the standard *pseudoisochromatic Ishihara* or *Hardy-Rand-Ritter* plates, which have figures or numbers embedded in a series of specially colored dots. Red-green color vision is often impaired out of proportion to other colors, due to optic nerve lesions. The visual and oculomotor systems are involved with the optic, oculomotor, trochlear, and abducens nerves.

Oculomotor (III), trochlear (IV), and abducens (VI) nerves

The oculomotor, trochlear, and abducens nerves control actions of the intraocular (pupillary sphincter) and extraocular muscles. These nerves are observed for symmetry of eye movement, globe position, asymmetry or drooping of the eyelid (ptosis), and twitching or fluttering of the lids or globes. With the lights dimmed, *anisocoria* or differences in the sizes of the pupils should be ruled out. Pupillary light response is evaluated for quickness and symmetry. Normally, the pupils average about 3 mm in diameter in a room with adequate light. They can range, however, from about 6 mm in children to <2 mm in elderly patients. Also, one patient's pupils may be different from each other by about 1 mm in size. This is called *physiologic anisocoria*. The pupils are normally round and regularly shaped, constricting quickly from direct illumination, and slightly less to illumination of the pupils on the opposite side. This is called the *consensual response*. They dilate again quickly when illumination is removed. Normal pupils constrict when the eyes converge to focus on a closer object, such as the tip of the nose. This is known as *accommodation*.

Miosis (pupil constriction) is controlled by parasympathetic fibers originating in the midbrain. They travel along with the oculomotor nerve to the eye. If this pathway is interrupted, such as by a hemispheric mass lesion, with compression of the oculomotor nerve as it leaves the brainstem, it causes a dilated, unreactive pupil, about 7 mm in size. Pupil dilation requires a three-neuron sympathetic relay. This is from the hypothalamus, through the brainstem, to the T1 spinal level, then to the superior cervical ganglion and the eye. Lesions in this pathway cause constricted, unreactive pupils, of about 1 mm or less in size. *Adie (tonic) pupil* is unilateral and large, with sluggish reactivity to light, normal accommodation, and caused by a ciliary ganglion lesion. *Argyll Robertson pupils* are bilateral, small, and irregular, with no light reactivity but normal accommodation, due to a midbrain lesion. *Horner syndrome* causes a unilateral, small pupil, ptosis, normal light reactivity and accommodation, and is due to sympathetic innervation of the eye. *Marcus Gunn pupil* appears normal, with consensual light reactivity being more than direct light reactivity, normal accommodation, due to an optic nerve lesion.

Examination of the patient's eyelids (*palpebrae*) should be done when the patient's eyes are open. The interpalpebral fissure between the upper and lower lids is usually about 10 mm, and nearly equal for both eyes. The upper lid usually covers 1–2 mm of the iris. This is increased by ptosis (drooping) of the lid caused by lesions of the levator palpebrae muscle, or its oculomotor or sympathetic nerve supply. In Horner syndrome, ptosis occurs along with miosis, and sometimes, defective sweating (anhidrosis) of the forehead. Exophthalmos (proptosis) is an abnormal protrusion of the eye from its orbit. This is best detected by the examiner standing behind the seated patient and looking down at the patient's eyes.

Movement of the eyes is due to the six muscles attached to each globe, moving the eye into the six cardinal gaze positions. These include the superior and inferior rectus, superior and inferior oblique, and the medial and lateral rectus positions. In the resting state, equal and opposed muscle actions cause the eye to be in the mid- or primary position, which is looking directly forward. When an extraocular muscle's function is disrupted, the eye cannot move in the direction of action of the affected muscle, known as *ophthalmoplegia*. It may deviate in the opposing direction due to the unopposed action of other extraocular muscles. In this misalignment, visual images of perceived objects are on different places on each retina. This causes *diplopia* (double vision). The extraocular movements controlled by cranial nerves III, IV, and VI are tested by having the patient follow a moving target such as a penlight or the examiner's finger, to all four quadrants, as well as across the midline (in an H-shaped pattern). This may detect palsies or paresis of the ocular nerves and can demonstrate nystagmus. The oculomotor nerve innervates all of the extraocular muscles, except for the superior oblique. This muscle is innervated by the trochlear nerve. The lateral rectus is innervated by the abducens nerve. Therefore, patterns of ocular muscle involvement help distinguish ocular muscle disorders from those affecting a cranial nerve.

During examination, the patient looks at the light source held in each cardinal gaze position. The examiner observes if the eyes move completely, and in a *conjugate* or *yoked* fashion in each direction. This means that the eyes move in the same direction to maintain binocular gaze. If normal, light from the flashlight will fall at the same spot on both corneas. Any limited eye movements or inconsistencies off movement are recorded. If the patient has diplopia, the weak muscle should be identified by asking the patient to gaze in the direction in which the image separation is most intense. Each eye is then covered separately. The patient is asked which of the two (near or far) images then disappears. The image that is, displaced father in the direction of gaze always refers to the weak eye. Another method is to cover one eye with translucent red plastic, glass, or cellophane. This allows the eye responsible for each image to be identified. If the left lateral rectus muscle is weak, for example, diplopia will be worse on leftward gaze. The leftmost of the images will disappear when the left eye is covered.

Nystagmus is a condition of rhythmic eye oscillation. It can occur at the extremes of voluntary gaze in normal patients. However, it may be caused by anticonvulsant or sedative drugs, or be from disease that affects the extraocular muscles, their innervation, or the cerebellar or vestibular pathways. Most commonly, a condition called *jerk nystagmus* is seen, with a slow phase of movement that is, followed by a fast phase, but in the opposite direction. The eyes are observed in their primary position, and then in each of the cardinal gaze positions. If nystagmus is indicated, it is described by the position of gaze in which it occurs, the direction, fine or coarse amplitude, any related factors such as head position changes, and related symptoms such as vertigo. The direction of jerk nystagmus is usually the direction of the fast phase. This condition often increases in amplitude with gaze in the direction of the fast phase, known as the *Alexander law*. *Pendular nystagmus* is less common, usually staring in infancy, with equal velocity in both directions.

Trigeminal nerve (V)
The trigeminal nerve carries sensory fibers from the face and motor fibers to the muscles used for mastication. The three sensory divisions (mandibular, maxillary, and

ophthalmic) of the *trigeminal nerve* are evaluated via pinprick, touching, or tuning fork tests. The mandibular division is tested by touching the jaw; the maxillary division is tested by touching the cheek; and the ophthalmic division is tested by touching the forehead. These tests measure facial sensation, as does brushing a piece of cotton against the lateral or lower cornea to evaluate corneal reflexes. The corneal reflex *afferent* arm is a branch of the trigeminal nerve, which leads to closure of the eye through the facial nerve (*efferent* arm). The blink reflex is mediated by the nasociliary branch of the ophthalmic branch of the trigeminal nerve, sensing stimuli on the cornea only (via the afferent fiber). Weak blinking, caused by facial weakness such as seventh cranial nerve paralysis, must be distinguished from depressed or absent corneal sensation. This is common in people who wear contact lenses. If there is facial weakness present, the patient will feel the cotton normally on both sides, even though blinking is decreased. If there is impaired trigeminal function, neither eye will blink. Unilateral blinking indicates a facial nerve lesion on the unblinking side.

Trigeminal motor function is assessed by palpation of the masseter muscles as the patient clenches the teeth, then having the patient open the mouth against resistance. When one of the pterygoid muscles is weak, the jaw will deviate to that side as the mouth is opened. Normal jaw strength will not be able to be overcome by the examiner.

Facial nerve (VII)

The *facial nerve* supplies the muscles of the face. It also mediates taste sensation from approximately the anterior two-thirds of the tongue. The facial nerve is evaluated by checking for any hemifacial weakness. Asymmetrical facial movements are often more pronounced during spontaneous conversation, especially when smiling. If obtunded, these occur if the patient grimaces at something that smells noxious. On the side of weakness, the nasolabial fold is depressed, and the palpebral fissure is wider. When the patient only has lower facial weakness, meaning that eye closure or wrinkling of the forehead is preserved, the etiology of facial nerve weakness is central, not peripheral. If there is a peripheral lesion, an entire side of the face will be weak, and the eye cannot fully close. With a hemispheric (central) lesion, the forehead is normal, and closure of the eye is partial. This may be due to dual cortical motor input to the upper part of the face. Bilateral facial weakness is detected by having the patient squeeze the eyes shut, press the lips together tightly, and puff the cheeks outward. If normal, the examiner should be unable to open the patient's eyelids, force the lids apart, or force air out of the mouth by pushing on the cheeks. Facial weakness with dysarthria is most obvious when making "m" sounds. In a patient who can whistle, this ability may be lost due to facial weakness. Taste sensation in the anterior two-thirds of the tongue is tested with bitter, salty, sour, and sweet solutions applied with a cotton swab, first on one side of the tongue, and then on the other. Hyperacusis is detected via a vibrating tuning fork held next to the ear.

Vestibulocochlear nerve (VIII)

The *vestibulocochlear nerve* has auditory (hearing) and vestibular (equilibrium) divisions. Evaluation of the vestibulocochlear nerve, therefore, involves testing of the hearing and balance. Otoscopic inspection of the auditory canals and tympanic membranes is done, as well as assessment of each ear's auditory acuity. Weber and Rinne tests are performed using a 512-Hz tuning fork. Auditory acuity is tested in a rough form by the examiner rubbing his or her thumb and forefinger together, about 2 in. from each ear. When the patient cannot hear this or otherwise complains of hearing loss, further testing is performed.

In the Weber test, the handle of a vibrating tuning fork is placed in the middle of the forehead. With conductive hearing loss, the tone will sound louder in the affected ear. With sensorineural hearing loss, the tone will be louder in the normal ear. In the Rinne test, the base of a lightly vibrating tuning fork is placed upon the mastoid process of the temporal bone, until the sound can no longer be perceived. Then, the tuning fork is moved near the opening of the external auditory canal. If there is normal hearing or sensorineural hearing loss, air inside the auditory canal conducts sound more effectively than bone. Therefore, the tone can still be heard. IF there is conductive hearing loss, the patient hears the bone-conducted tone while the tuning fork is on the mastoid process for a longer time than he or she can hear the air-conducted tone.

For patients with positional vertigo, a test known as either the *Nylen-Bárány* or *Dix-Hallpike maneuver* is used to attempt to reproduce the condition. The patient is seated on a table. The head and eyes are directed forward. The examiner then quickly lowers the patient to a supine position. The head positioned over the edge of the table, 45 degrees below horizontal. Then, the test is repeated with the patient's head and eyes turned 45 degrees to the right, and once again, with the head and eye turned 45 degrees to the left. The eyes are checked for any present nystagmus. The patient is asked to describe the development, intensity, and ending of vertigo, should it occur.

Glossopharyngeal (IX) and vagus (X) nerves

The *glossopharyngeal* and vagus nerves innervate the muscles of the pharynx and larynx. These are involved in swallowing and speaking. The glossopharyngeal nerve also carries touch sensation from the posterior one-third of the tongue, tonsils, tympanic membrane, and Eustachian tube. It also carries taste sensation from the posterior one-third of the tongue. The vagus nerve has sensory fibers from the pharynx, larynx, external auditory canal, tympanic membrane, and posterior fossa meninges.

The glossopharyngeal and vagus nerves are usually evaluated at the same time. When the palate elevates symmetrically, this must be noted. A tongue depressor is used to touch one side of the posterior pharynx and then the other side. The patient is asked to say "ah." Symmetry of the gag reflex is then observed. If the gag reflex is bilaterally absent, this is a common occurrence in healthy people and may be insignificant. For unresponsive, intubated patients, suctioning of an endotracheal tube usually triggers coughing. If the patient has hoarseness, the vocal cords must be inspected. If hoarseness is isolated, but there is normal palatal elevation and gag reflex, a search should be initiated for lesions such as aortic aneurysm or mediastinal lymphoma, both of which may compress the recurrent laryngeal nerve. When the palate is weak, it is difficult for the patient to make the "k" sound.

Spinal accessory nerve (XI)

The *spinal accessory nerve* innervates the trapezius and sternocleidomastoid muscles. To test the upper trapezius, the patient must elevate the shoulders against resistance supplied by the examiner. To assess the sternocleidomastoid muscle, the patient must turn the head against resistance supplied by the examiner's hand, while the active muscle, opposite the turned head, is being palpated. Weakness of the sternocleidomastoid causes a decreased ability to rotate the head away from the weak side.

Hypoglossal nerve (XII)

The *hypoglossal nerve* innervates the muscles of the tongue. It is evaluated by having the patient extend the tongue and inspecting it for weakness (with deviation toward the side of a lesion), atrophy, and **fasciculations**. If weakness is unilateral, the patient has a reduced ability to press the tongue against the opposite cheek. Facial weakness can cause false-positive test results. Weakness of the tongue may produce dysarthria, with significant slurring of labial or "l" sounds. Denervation is linked to atrophy and fasciculations (twitches).

Motor system

The patient's limbs and shoulder girdle are totally exposed, then inspected for asymmetric development, atrophy, hypertrophy, fasciculations, myotonia, tremor, and other involuntary movements. These include **chorea**, **athetosis**, and **myoclonus**. In a relaxed patient, passive flexion and extension of the limbs give information on muscle tone. Decreased muscle bulk indicates atrophy. However, bilateral atrophy, or atrophy of large or concealed muscles, unless it is advanced, may not be noticed. Loss of certain amounts of muscle mass in the elderly is common. Hypertrophy develops when a muscle must work harder in order to compensate for another muscle's weakness. *Pseudohypertrophy* is due to muscle tissue being replaced by excessive amounts of connective tissue or storage materials.

Fasciculations are generally common in lower motor neuron diseases and may occur rarely in normal individuals—especially in the calf muscles of the elderly. However, they usually indicate lesions of the lower motor neuron, such as from nerve degeneration, or injury with regeneration. **Myotonia** reveals myotonic dystrophy. It may be shown by an inability to quickly open the hand after it has been clenched. The **clasp-knife phenomenon** and spasticity, as well as **Lead-pipe rigidity**, often present with **cogwheel rigidity**, usually due to a basal ganglia disorder.

Muscle strength

Often, a patient describes the feeling of muscle "weakness," which may actually be clumsiness, fatigue, or true muscle weakness. The exact character of the symptom must be correctly documented. This includes the time of occurrence, exact location, any ameliorating or precipitating factors, and all associated clinical manifestations. The limbs are then inspected for weakness, tremor, and other involuntary movements. When a weak limb is extended, it will drift downward. Specific muscle groups are tested for strength against resistance. Each side of the body is compared against the other. Pain often precludes a full effort in strength testing. If weakness is nonorganic, resistance to movement may at first be normal, then suddenly dissipate.

Slight weakness can be indicated by decreased arm swinging during walking, decreased spontaneous use of a limb, a pronator drift in an outstretched arm, a leg that is, externally rotated, slowed alternating movements, or fine dexterity impairment. The patient, for example, may be unable to remove a small object from a container or to fasten a shirt button. *Tiller* and *mini-Tiller testing* can detect subtle motor weaknesses. In the Tiller test, the patient makes a fist with each hand and then rotates these two fists around each other. In the mini-Tiller test, the patient makes a fist with each hand and then extends the index finger of each hand, followed by the same rotating movements. In either test, the weaker limb will become fixed in space as the stronger limb revolves around it.

Strength must be graded, using the following scale, which was originally developed by the Medical Research Council of the United Kingdom:
- 0—No visible muscle contraction
- 1—Visible muscle contraction; no movement or only trace movement
- 2—Limb movement occurs when gravity is eliminated
- 3—There is movement against gravity, but not against resistance
- 4—There is movement against resistance supplied by the examiner
- 5—Full muscle strength

Difficulty in using this scale and other similar scales is because there is a large range between grades four and five. Distal strength may be semiquantitatively measured using a handgrip ergometer, or an inflated blood pressure cuff squeezed by the patient.

Functional testing often reveals a better assessment of strength and disability. With various movements, the patient's deficiencies are documented and quantified to the highest possible degree, such as the number of steps that can be climbed. If the patient rises from a squatting position, or steps onto a chair, proximal leg strength is tested. Distal strength is tested by walking on the heels and up on tiptoe. Weakness of the quadriceps is indicated if the patient has to push with the arms in order to rise out of a chair. Shoulder girdle weakness is indicated by swinging the body in order to move the arms. **Gowers' sign** suggests pelvic girdle weakness (see Fig. 21.1). It involves rising from the supine position by turning prone, then kneeling, and using the hands to climb up the thighs to slowly push erect.

Fig. 21.1 Gower's sign.

Gait, stance, and coordination

Integrity of the motor, vestibular, and proprioceptive pathways is required for normal gait, stance, and coordination. A lesion in any of these pathways causes specific deficits. When assessing gait, neurologists must observe the stride length, number of steps, and the width between the legs while walking. A wide gait is required for stability if there is cerebellar ataxia. A steppage gait is caused by **dropfoot**, such as lifting the leg higher than is normal, in order to avoid catching the foot on irregularities of the walking surface. Waddling is caused by pelvic muscle weakness. A **magnetic gait** can be due to Parkinson's or normal pressure hydrocephalus. If the patient has impaired proprioception, he or she must constantly watch how the feet are placed in order to avoid falling or tripping. Coordination is tested with knee-to-shin maneuvers, or finger-to-nose maneuvers to assess **dysmetria**.

Sensation

Loss of sensation is best tested by using a pin to lightly prick areas such as the face, torso, and limbs. The patient is asked if the pinprick feels the same on both sides of the body and if the sensation is sharp or dull. The pin is then discarded in order to avoid possible transmission of hepatitis, HIV infection, or other bloodborne disorders. To assess cortical sensory function, the patient holds a coin, key, or other familiar object in the palm of the hand (**stereognosis**). Then, the patient is asked to distinguish numbers written on his or her palm (**graphesthesia**) and to distinguish between several simultaneous, closely placed pinpricks to the fingertips (**2-point discrimination**).

Temperature sense is tested by using a cold tuning fork, with one of its prongs made warm by the examiner, either using body temperature or test tubes holding warm and cold water. Joint position sense is assessed by moving the patient's terminal phalanges of the fingers, then the toes, up or down by a few degrees. When the patient's eyes are closed, if he or she cannot identify these small movements, larger up-and-down movements are made. Then the next proximal joints are tested, such as the ankles or wrists. **Pseudoathetosis** is the involuntary writhing in a snake-like pattern, of a limb. It results from a severe loss of sense of position. However, motor pathways, including those in the basal ganglia, are preserved. Since the brain cannot sense where the limb is in space, the limb moves on its own, and the patient has to use vision to control the movements of the limb. Usually, when the eyes are closed, the patient cannot locate the limb in space. If there is an inability to stand with the feet placed together as the eyes are closed (**Romberg test**), it indicates impaired positioning sensing in the lower extremities. Cerebellar disease causes the patient to stand with the feet apart but as close together as possible without falling. Only then can the eyes be closed. A rare positive result is caused by severe bilateral loss of vestibular function, such as aminoglycoside toxicity.

Vibration sensing is tested by having the examiner place a finger under the patient's distal interphalangeal joint, then pressing a lightly tapped 128-cycle tuning fork on top of the joint. The patient should sense the end of the vibration at almost the same time as the examiner, who feels it *through* the patient's joint. Light touch is tested by using strips of cotton. Impaired sensation, and the location of a lesion can be anatomically suggested as follows:

- *Crossed face-body pattern*—brainstem
- *Hemisensory loss*—brain
- *Midline hemisensory loss*—thalamus or functional (psychiatric)
- *Saddle area sensory loss*—cauda equina
- *Sensation reduced below a certain dermatomal level*—spinal cord
- *Single dermatomal or nerve ranch distribution*—isolated nerves (mononeuritis multiplex) or nerve roots (radiculopathy)
- *Stocking-glove distribution*—distal small peripheral nerves

Location of the lesion will be confirmed by determination of whether motor weakness and reflex changes are similarly patterned. Lesions of the brachial or pelvic plexus are suggested by "patchy" limb sensory, motor, and reflex deficits.

Position

To test for the ability to sense joint position, the examiner grasps the sides of the distal phalanx of a finger or toe, slightly displacing the joint in an upward or downward direction. The patient, while the eyes or closed, is asked to report the position change that is, perceived. Normally, the sensing of joint positions is extremely sensitive. The patient is able to detect even a slight movement. When joint position sensing is diminished distally, the more proximal limb joints are tested, until normal position sensing occurs. Another test is to have the patient close the eyes and extend the arms, then attempt to touch the tips of the index fingers together.

Pain

To assess for pain, a disposable pin is used to prick the skin, but not puncture it, hard enough to cause a mildly unpleasant sensation. The patient is then asked if he or she feels the sharp stimulus. If using a safety pin, the rounded end is used to show the patient the difference between sharp and dull stimuli. Based on the individual patient, the examiner can compare pain sensation. This is done from side to side, proximal to distal, dermatome to dermatome, and from an area of deficit toward areas of the body that are normal.

Coordination

There are various testing methods to assess coordination. When coordination is impaired, it is known as *ataxia*. This usually occurs due to lesions that affect the cerebellum or its connections and may affect eye movements, the limbs, trunk, and even speech. *Distal limb ataxia* is assessed by having the patient perform fast, alternating movements. These may involve tapping of the palm and dorsum of the hand upon the other hand or tapping the sole of the foot upon the examiner's hand. Any irregularities of movement rate, rhythm, amplitude, or force must be noted. The *finger-to-nose test* requires the patient to move one index finger back and forth between his or her nose and the examiner's finger. Ataxia can be linked to *intention tremor*, which is worse at the beginning and end of every movement.

Often, the force of muscular contractions is impaired and can be identified. The patient is asked to raise the arms quickly to a certain height, or the arms are extended and outstretched in front of the patient's body. When displaced by a sudden force, there may be *rebound* (overshooting). This can be shown by asking the patient to forcefully flex the arm at the elbow against resistance. Then, the resistance is suddenly removed. When a limb is ataxic, continued contraction without resistance can cause the hand to actually strike the patient. The *heel-knee-shin test* reveals ataxia of the lower limbs. The patient lies in the supine position and is asked to run the heel of the foot smoothly, up, then down the opposite shin, from ankle to knee. If ataxia is present, there will be jerky, inaccurate movements, and the patient will be unable to keep the heel contacting the shin. In *truncal ataxia*, the patient is required to sit on the side of a bed or in a chair, with no lateral support. Any tendency to tilt to one side is recorded.

Reflexes

Muscle stretch (deep tendon) reflex testing assesses afferent nerves, motor nerves, synaptic connections in the spinal cord, and descending motor pathways. Reflexes are depressed by lower motor neuron lesions, such as those affecting the anterior horn cell, peripheral nerve, or spinal root. Reflexes are increased by upper motor neuron lesions, such as in nonbasal ganglia disorders, anywhere above the anterior horn cell.

The noncranial reflexes that are tested include the following:
- *Biceps*—innervated by C5 and C6
- *Radial brachialis*—innervated by C6
- *Triceps*—innervated by C7
- *Quadriceps knee jerk*—innervated by L4
- *Ankle jerk*—innervated by S1

Asymmetric increases or depressions are documented. **Jendrassik's maneuver** is used to augment hypoactive reflexes. This involves the patient locking the hands together, then pulling them vigorously apart as a tendon in the lower extremity is tapped by the examiner (see Fig. 21.2). Another method is to have the patient push the knees together against each other as the upper limb tendon is tested. A superficial abdominal reflex can be caused

Fig. 21.2 Jendrassik's maneuver.

by lightly stroking the four quadrants of the abdomen. If this reflex is depressed, it may be due to a central lesion, obesity, or lax skeletal muscles, such as following pregnancy. The absence of this reflex can indicate spinal cord injury.

Babinski's, Chaddock's, and Oppenheim reflexes

Babinski's reflex, **Chaddock's reflex**, and **Oppenheim reflex** are pathologic in nature and are reversions to primitive responses that indicate loss of cortical inhibition. All of these evaluate plantar response. The normal reflex response involves flexion of the great toe. Abnormal responses are slower. They consist of great toe extension with spreading of the other toes, and often, flexion of the knees and hips. This reaction is from the spinal reflexes, indicating spinal disinhibition from an upper motor neuron lesion. In Babinski's reflex, the foot's lateral sole is firmly stroked, from the heel to the ball of the foot, with a tongue depressor or the end of a reflex hammer (Fig. 21.3). This stimulus must be noxious but should not cause injury. Stroking is not too close to the medial aspect, which can accidentally cause a primitive grasp reflex. For sensitive patients, reflex response may be hidden by a quick, voluntary withdrawal of the foot. This is not a component of Chaddock's or Oppenheim reflex testing.

For Chaddock's reflex, the lateral foot, from the lateral malleolus to the small toe, is stroked with a blunt instrument. For the Oppenheim reflex, the anterior tibia, from just

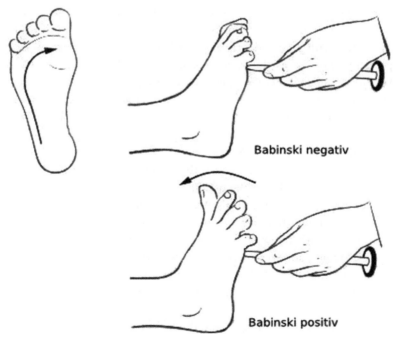

Fig. 21.3 Babinski's (plantar) reflex.

below the patella to the foot, is firmly stroked with the examiner's knuckle. Additional signs and reflexes include the following:

- **Bulbospongiosus reflex**—the dorsum of the penis is tapped to test the S2 to S4 levels; normal response is contraction of the bulbospongiosus muscle
- **Cremasteric reflex**—the medial thigh (3 in.) below the inguinal crease is stroked upward; normal response is elevation of the ipsilateral testis; this tests the L2 level
- **Glabellar sign**—present if tapping of the patient's forehead causes blinking; normally, each of the first five taps results in a single blink, and then the reflex becomes fatigued; blinking continues in patients with diffuse cerebral dysfunction
- **Grasp reflex**—present if gentle stroking of the palm of the patient's hand results in the fingers flexing and grasping the examiner's finer
- **Hoffman's sign**—present if flicking the nail on the patient's third or fourth finger causes involuntary flexion of the distal phalanx of the thumb and index finger
- **Palmomental reflex**—present if stroking the palm of the patient's hand causes contraction of the ipsilateral mentalis muscle of the lower lip
- **Rooting reflex**—present if stroking the patient's lateral upper lip causes movement of the mouth toward the stimulus
- **Snout reflex**—present it tapping a tongue depressor across the patient's lips causing pursing of the lips
- **Sphincteric reflexes**—tested during a rectal examination; to test tone at the S2 to S4 nerve root levels, the examiner inserts a gloved finger into the patient's rectum and asks him or her to squeeze it; another method involves the perianal region being lightly touched with a cotton strip—normal response is contraction of the external anal sphincter, known as the *anal wink reflex*; rectal tone usually becomes lax in patients with acute spinal cord injury or cauda equina syndrome

Testing for *clonus* is done by rapid dorsiflexion of the foot at the ankle. Sustained clonus reveals an upper motor neuron disorder.

Stance and gait

To detect instability from cerebellar ataxia, the patient is asked to stand, keeping the feet together and the eyes open. Then, the patient is asked to close the eyes. The *Romberg sign* that signifies sensory ataxia results in instability happening with the eyes closed, but not when they are open. The patient is asked to walk normally, first on the heels, then on the toes, then in *tandem* (with 1 foot placed directly in front of the other). These procedures help identify classic gait abnormalities, which include the following:

- *Hemiplegic gait*—the affected leg is extended and internally rotated with the foot inverted and plantar flexed; the leg moves in a circular direction at the hip, known as *circumduction*
- *Paraplegic gait*—a slow and stiff gait, and the legs cross in front of each other, which is called *scissoring*

- *Cerebellar ataxic gait*—a wide-based gait that may be related to staggering or reeling; it makes the patient appear as if he or she is intoxicated
- *Sensory ataxic gait*—a wide-based gait in which the feet are slapped down upon the floor; the patient may watch his or her own feet
- *Steppage gait*—often due to a peroneal (fibular) nerve lesion, the patient cannot dorsiflex the foot; there is exaggerated hip and knee elevation, allowing the foot to clear the floor during walking
- *Dystrophic gait*—a lordotic, waddling gait caused by pelvic muscle weakness
- *Parkinsonian gait*—a flexed posture with slow starting of gait, steps that are shuffling and small, with reduced arm swinging; involuntary acceleration called *festination* may be present
- *Choreic gait*—a jerky and lurching gait; however, falls are not common
- *Apraxic gait*—affected by *apraxia*, which is the loss of a previously learned act; a frontal lobe disease may cause loss of walking ability; initiating walking may be difficult, and the patient may appear as if he or she is stuck to the floor; once walking begins, the gait is slow and shuffling, but there is no difficulty in performing identical leg movements when lying down, with the legs not bearing any weight
- *Antalgic gait*—the patient favors one leg over the other, attempting to avoid putting weight upon the injured leg, thereby avoiding pain

Screening without neurologic complaints

When a patient has no neurologic complaints, there are six screening methods performed. The screening procedures focus on mental status, cranial nerve conditions, assessments of motor and sensory functions, reflexes, and problems with coordination, stance, and gait. The patient is observed to be alert and awake, confused, or unable to be aroused. He or she is tested for orientation to person, place, and time. Aphasia is assessed by asking the patient to repeat the words "no ifs, ands, or buts." To assess the cranial nerves, the optic disks are examined for papilledema and the visual fields are tested using confrontation. The six cardinal gaze directions are used to confirm the ability to move the eyes conjugately. Facial strength is assessed by having the patient close the eyes tightly and then showing the teeth.

Motor function is checked by comparing the two sides of the body in relation to the speed of fine finger movements, strength of extensor muscles of the upper limbs, and strength of lower limb flexor muscles. This helps detect corticospinal tract lesions. The patient is asked to describe any perceived sensory deficits. Light tough and vibration sense is tested in the feet. If this is impaired, the examiner determines upper limits of impairment in both the arms and legs. The two sides of the body are compared regarding the activity of the biceps, triceps, quadriceps, and Achilles tendon reflexes. The plantar responses are also assessed. The patient is observed standing and walking. Any asymmetries or instabilities of stance or gait are recorded.

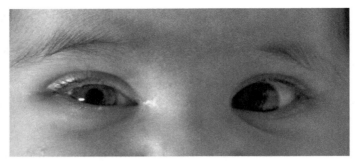

Fig. 21.4 Horner's syndrome.

Autonomic nervous system

Assessment of the autonomic nervous system involves evaluation of heart rate changes in response to the Valsalva maneuver, postural hypotension, decreased or absent sweating, and evidence of **Horner's syndrome** (see Fig. 21.4), which includes pupillary constriction, facial anhidrosis, and unilateral ptosis; disturbances of bladder, bowel, hypothalamic, and sexual function should be documented. As a rule, any lesion in the lateral brainstem can cause involvement of the spinosympathetic pathway, resulting in Horner's syndrome on the same side.

Key terms

2-Point discrimination	Horner's syndrome
Athetosis	Jendrassik's maneuver
Babinski's reflex	Lead-pipe rigidity
Chaddock's reflex	Magnetic gait
Chorea	Micrographia
Clasp-knife phenomenon	Mini-Mental State Examination
Cogwheel rigidity	Myoclonus
Dropfoot	Myotonia
Dysmetria	Neurologic method
Fasciculations	Oppenheim reflex
Funduscopic	Pseudoathetosis
Gowers' sign	Romberg test
Graphesthesia	Snellen chart
Hemispatial neglect	Stereognosis

Suggested readings

1. Armitage, A. *Advanced Practice Nursing Guide to the Neurological Exam.* Springer, 2015.
2. Banich, M. T.; Compton, R. J. *Cognitive Neuroscience*, 4th ed.; Cambridge University Press, 2018.

3. Berkowitz, A. *Lange Clinical Neurology and Neuroanatomy: A Localization-Based Approach*. McGraw-Hill Education/Medical, 2016.
4. Binder, D. K.; Sonne, D. C.; Fischbein, N. J. *Cranial Nerves: Anatomy, Pathology, Imaging*. Thieme, 2010.
5. Campbell, W. W. *DeJong's The Neurologic Examination*, 7th ed.; LWW, 2012.
6. Cardinali, D. P. *Autonomic Nervous System: Basic and Clinical Aspects*. Springer, 2018.
7. Colombo, J.; Arora, R.; DePace, N. L.; Vinik, A. I. *Clinical Autonomic Dysfunction: Measurement, Indications, Therapies, and Outcomes*. Springer, 2015.
8. Dadio, G.; Nolan, J. *Clinical Pathways: An Occupational Therapy Assessment for Range of Motion & Manual Muscle Strength*. LWW, 2018.
9. Elwell, V. A.; Kirollos, R.; Al-Haddad, S. *Neurosurgery: The Essential Guide to the Oral and Clinical Neurosurgical Examination*. CRC Press, 2014.
10. Fuller, G. *Neurological Examination Made Easy*, 5th ed.; Churchill Livingstone, 2013.
11. Futrell, N.; Jamieson, D. G. *Vascular Neurology Board Review: Questions and Answers*, 2nd ed.; Demos Medical, 2018.
12. Goodfellow, J. A. *Pocket Tutor—Neurological Examination*, 2nd ed.; JP Medical Publishing, 2018.
13. Harrigan, M. R.; Deveikis, J. P. *Handbook of Cerebrovascular Disease and Neurointerventional Technique*, 3rd ed.; *Contemporary Medical Imaging* Humana Press, 2018.
14. Jankovic, J.; Mazziotta, J. C.; Pomeroy, S. L.; Daroff, R. B. *Bradley's Neurology in Clinical Practice, 2-Volume Set*, 7th ed.; Elsevier, 2015.
15. Kumar, N. *Handbook of Neurological Examination*. PHI, 2011.
16. Mesulam, M. M. *Principles of Behavioral and Cognitive Neurology*, 2nd ed.; Oxford University Press, 2000.
17. Norrving, B. *Oxford Textbook of Stroke and Cerebrovascular Disease*. Oxford Textbooks in Clinical Neurology Oxford University Press, 2014.
18. Robertson, D.; Biaggioni, I.; Burnstock, G.; Low, P. A.; Paton, J. F. R. *Primer on the Autonomic Nervous System*, 3rd ed.; Academic Press, 2011.
19. Ropper, A. H.; Samuels, M. A.; Klein, J. *Adams and Victor's Principles of Neurology*, 10th ed.; McGraw-Hill Education/Medical, 2014.
20. Sandroni, P.; Low, P. A. *Autonomic Disorders: A Case-Based Approach*. Cambridge University Press, 2015.
21. Seidel, H. M.; Stewart, R. W.; Ball, J. W.; Dains, J. E.; Flynn, J. A.; Solomon, B. S. *Mosby's Guide to Physical Examination*, 7th ed.; Mosby, 2010.
22. Shibasaki, H.; Hallett, M. *The Neurologic Examination: Scientific Basis for Clinical Diagnosis*. Oxford University Press, 2016.
23. Strub, R. L.; Black, F. W. *The Mental Status Examination in Neurology*, 4th ed.; F. A. Davis Company, 2000.
24. Weiner, W. J.; Goetz, C. G.; Shin, R. K.; Lewis, S. L. *Neurology for the Non-Neurologist*, 6th ed.; LWW, 2010.
25. Wilson-Pauwels, L.; Stewart, P.; et al. *Cranial Nerves: Function and Dysfunction*, 3rd ed.; PMPH-USA, Ltd, 2016.

CHAPTER 22

Neurologic diagnostic procedures

Neurologic diagnostic procedures are usually not used for preliminary screening of patients, except in emergency situations. This may be when a complete neurological evaluation is justified to rule out severe neurological disease, such as subarachnoid hemorrhage or stroke. Testing is guided by the evidence revealed during patient history and physical examination. Neurologic diagnostic procedures include, but are not limited to lumbar puncture, computed tomography (CT), magnetic resonance imaging (MRI), encephalography, cerebral catheter angiography, duplex Doppler ultrasonography, myelography, electroencephalography (EEG), and electromyography. Historically, neurologists believed that diagnosis almost always could be made by localizing a lesion, using patient history and examination. Authors still believe that localization of the lesion through examination and history is the most important step toward arriving on a correct diagnosis. With the advent of technology, however, neurologists are now more dependent on various diagnostic tests. New imagining techniques proved that localization through the neurological exam and history was not always accurate.

Lumbar puncture

A *lumbar puncture* is commonly referred to as a *spinal tap*. It evaluates both intracranial pressure and cerebrospinal fluid (CSF) composition. It allows for the therapeutic reduction of intracranial pressure and also the administration of radiopaque agents for myelography, and intrathecal drugs. The general contraindications for lumbar puncture include infection at the puncture site, bleeding diathesis, and increased intracranial pressure due to various lesions, obstructions, or blockages, which can lead to herniation with spinal tap. These include intracranial mass lesions, obstructions of CSF outflow due to Chiari I malformations or aqueductal stenosis, or spinal cord CSF blockage, such as from tumor cord compression.

CT or MRI should be done if focal neurologic deficits or papilledema are present, prior to lumbar puncture. This helps rule out the presence of masses that may result in cerebellar or transtentorial herniation. For immunosuppressed patients (such as with HIV or transplantation), performing a spinal tap without obtaining imaging first is contraindicated. Abnormalities of the CSF are seen in many different conditions. These include acute bacterial meningitis, subacute meningitis, acute syphilitic meningitis, paretic neurosyphilis, brain abscesses or tumors, spinal cord tumors, cerebral hemorrhage or thrombosis, Guillain-Barré syndrome, idiopathic cranial hypertension, lead encephalopathy, Lyme disease of the CNS, multiple sclerosis, and viral infections.

For lumbar puncture, the patient is usually placed in the *lateral decubitus position*, hugging the knees, and curling the body up as tightly as possible (fetal position). For a patient who cannot maintain this position, assistants may have to help hold the patient. Sometimes, the spine is better flexed if the patient (especially when obese) sits on the side of the bed and leans over a tray table. A puncture site of about 20 cm in diameter is cleaned with iodine and then wiped with sterile gauze. This removes the iodine and prevents it from entering the subarachnoid space. A lumbar puncture needle, with a *stylet*, is inserted into the L3–L4 or L4–L5 interspace. The L4 spinous process is usually on a line that lies between the posterior and superior iliac crests. The needle is pointed rostrally, toward the patient's umbilicus, and kept exactly parallel to the floor. As the needle enters into the subarachnoid space, there is usually an audible popping sound. The stylet is then removed, allowing CSF to flow outward. The opening pressure is measured with a *manometer*. Four tubes are filled with approximately 2–10 mL of CSF for testing. The puncture site is covered with a sterile adhesive strip. Up to 10% of patients will experience a postlumbar puncture headache. Historically, it was thought that lying in bed for an hour after the spinal tap would prevent this headache. However, recent studies showed that this maneuver does not prevent postlumbar puncture headache, and hydration is the mainstay of its treatment and prevention.

When the CSF is normal, it is clear and lacks color. It will be cloudy or turbid if there are 300 or more cells per microliter of CSF, which may suggest bacterial meningitis. If the fluid is bloody, this may indicate a traumatic puncture or subarachnoid hemorrhage. A traumatic puncture occurs from pushing the needle in too far, into the venous plexus that runs along the anterior spinal canal. This is signified by gradual clearing o the CSF between the first and fourth tubes, confirmed by decreasing red blood cells counts. There will also be the absence of xanthochromia, a yellowish CSF caused by lysed RBCs in a centrifuged sample, as well as fresh, **uncrenated** RBCs. If there is intrinsic subarachnoid hemorrhage, the CSF will be blood throughout all collection tubes. Xanthochromia is often present within 6–12 h following ictus. RBCs are usually crenated and older. The CSF may be faintly yellow if there are senile **chromogens**, severe jaundice, or increased protein, of over 100 mg/dL.

The diagnosis of many neurologic disorders is aided by cell count and differential, as well as levels of glucose and protein. If there is a suspected infection, the centrifuged sediment in the CSF is stained for bacteria via a Gram stain. It may also be stained for tuberculosis (TB) with an acid-fast stain or immunofluorescence, as well as for *Cryptococcus* species, using *India ink*. Fluid amounts of 10 mL or more will improve the likelihood of detecting the pathogen, especially acid-fast bacilli and some types of fungi, in cultures and stains. For early meningococcal meningitis or severe leukopenia, protein in the CSF may be insufficient for bacterial adherence to the microscope slide during Gram staining. This can produce a false-negative result but can be prevented by mixing one drop of aseptic serum with the CSF sediment. For suspected hemorrhagic meningoencephalitis, a wet mount is used, to examine for amoebas. Rapid bacterial identification may be via latex particle agglutination and co-agglutination tests. This is especially true when cultures and stains are negative, such as in partially treated meningitis. The CSF must be cultured aerobically as well as anaerobically, and for fungi and acid-fast bacilli. Viruses, except enteroviruses, are usually not isolated from the CSF. Viral antibody panels can be used. Often, regular testing includes Venereal Disease Research Laboratories (VDRLs) and cryptococcal antigen testing. Today, polymerase chain reaction tests for herpes simplex virus and other CNS pathogens are easily obtained.

In normal CSF, the blood glucose ratio is approximately 0.6, except in cases of severe hypoglycemia, in which it is usually >50 mg/dL (2.78 mmol/L). Disease may be nonspecifically indicated by increased CSF protein, over 50 mg/dL. Advanced TB meningitis, complete block by a spinal cord tumor, bloody spinal puncture, or purulent meningitis will cause protein in the CSF to increase to over 500 mg/dL. The diagnosis of demyelinating disorders is aided by examinations for globulin, myelin basic protein, IgG index, and oligoclonal banding. Normally, globulin levels are <15%. It has recently been discovered that CSF antibodies can be checked to diagnose autoimmune neurological diseases. One of the main concerns prior to performing lumbar puncture is to make sure that the patient does not have any bleeding diathesis, either due to a disease or medication. In the authors' personal experience, the spinal tap must be avoided if the platelet level is below 50,000 or if the patient received a medication called clopidogrel (Plavix) in the previous 48 h. Heparin drip can be withheld for 4 h prior to lumbar puncture. If warfarin was administered, the international normalized ratio should be normalized before attempting to tap. For *novel oral anticoagulant*, you should wait 24 h from the last dose before attempting a spinal tap. Table 22.1 shows normal and abnormal findings for CSF obtained via lumbar puncture.

CT scan

CT scan allows for quick and noninvasive imaging of the skull and brain (Fig. 22.1). It is superior to MRI for the visualization of fine bone detail in the posterior fossa, base of the skull, and spinal canal. However, CT is not superior to MRI for visualizing the contents of these structures. A radiopaque contrast agent will help to detect abscesses and brain tumors. Noncontrast CT is used to quickly reveal acute hemorrhage and many gross structure alterations, with no danger of contrast allergy or kidney failure. Using an intrathecal agent, CT can show abnormalities of the brainstem, spinal cord, or spinal nerve roots, and can detect a spinal cord syrinx. Abnormalities revealed by CT include herniated disks and meningeal carcinomas. If CT angiography is used, which requires a contrast agent, the cerebral blood vessels can be visualized. This means that an MRI or other angiography will not be needed. The adverse effects of contrast agents generally include contrast nephropathy and allergic reactions. Contrast-induced nephropathy with

Table 22.1 Normal and abnormal cerebrospinal fluid findings.

Parameters	Normal	Abnormal	Possible cause
Pressure (initial readings)	120–180 mm H$_2$O (9–14 mmHg)	<60 mm H$_2$O	Faulty needle placement Dehydration Spinal block along subarachnoid space Block of foramen magnum
		200 mm H$_2$O	Muscle tension Abdominal compression Brain tumor Subdural hematoma Brain abscess Brain cyst Cerebral edema (any cause) Hydrocephalus
Color (turbidity)	Clear, colorless	Cloudy	Increased cell count Increased microorganisms
		Yellow	Xanthochromic (caused by red blood cell [RBC] pigments) High protein content Presence of RBCs
Red blood cells	None	Smoky	Traumatic tap
		Blood-tinged	Traumatic tap
		Grossly bloody	Subarachnoid hemorrhage
White blood cells	0–6/mm^3	10/mm^3 (cell counts range from <100/mm^3 to many thousands depending on causative factor; all are abnormal findings)	Occurs in many conditions: Bacterial infections of meninges Viral infections of meninges Neurosyphilis Tuberculous meningitis Metastatic neoplastic lesions Parasitic infections

Protein[a]	15–45 mg/dL (1% of serum protein)	
		<10 mg/dL	Acute demyelinating diseases
			Following introduction of air or blood into subarachnoid space
		60 mg/dL	Little clinical significance
			Occurs in many conditions: Complete spinal block; Guillain-Barré syndrome; Carcinomatosis of meninges; Tumors close to pial or ependymal surfaces or in cerebellopontine angle; Acute and chronic meningitis; Meningeal hemorrhage; Demyelinating disorders; Degenerative diseases
Glucose (CSF: Serum ratio)	0.6: approximately 60% of blood glucose level—(50–55 mg/dL)	<0.4: <40 mg/dL >0.6: >60 mg/dL	Acute bacterial meningitis; Tuberculous meningitis; Meningeal carcinomatosis; Acute viral meningitis; Diabetes
Chloride	700–750 mg/dL 116–130 mEq/L	100 mg/dL <625 mg/dL <110 mEq/L 800 mg/dL	Hypochloremia; Tuberculous meningitis; Not of neurologic significance; correlates with blood levels of chloride

[a]Note: If CSF contains blood, this will raise the protein level.
Courtesy: McCance, K., S. Huether. Pathophysiology: The Biologic Basis for Disease in Adults and Children, 7th edn., 122013. Mosby VitalBook file, 2014.

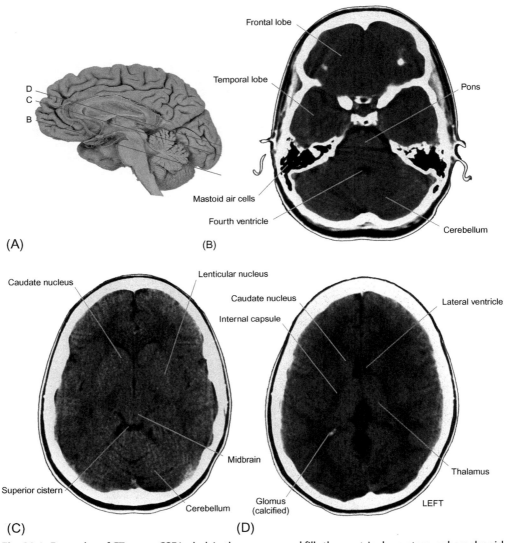

Fig. 22.1 Examples of CT scans. CSF is dark in these scans and fills the ventricular system, subarachnoid cisterns, and cerebral sulci around the edge of the brain. Bone is white, air is black, and gray matter is slightly lighter than white matter. (A) Planes of section produced by the computer and displayed as (B), (C), and (D). The streaks cutting across the brainstem and cerebellum in (B) are artifacts reflecting the dense bone surrounding these regions. *Courtesy: Dr. Raymond F. Carmody, University of Arizona College of Medicine.*

permanent irreversible kidney damage is extremely rare, even in patients with mild to moderate kidney failure. CT angiography became less common because of the development of MR angiography. This has changed in the past few years as CT angiography is the main imaging modality in the setting of acute stroke.

Magnetic resonance imaging

MRI allows for a better resolution of neural structures than CT scan (see Fig. 22.2). Primarily, this is significant in the visualization of brainstem lesions, the cranial nerves, the spinal cord, and abnormalities of the posterior fossa. When CT is used, the images are often affected by bony streak artifacts. MRI is also preferred for detecting early infarction, demyelinating plaques, cerebral contusions, subclinical brain edema, abnormalities of the **craniocervical junction**, incipient transtentorial herniation, and syringomyelia. It is highly effective in revealing spinal abscesses or tumors that may compress the spinal cord and cause emergency interventions.

Contraindications of MRI include patients who have had cardiac or carotid stents or a pacemaker that are not MRI compatible, or any other metallic objects in the body that may move in the magnetic field of the MRI machine. This can overheat or be displaced inside the body by the intense magnetization used in the procedure. Enhancement, using intravenous paramagnetic contrast agents such as gadolinium may be needed to visualize demyelinated, inflammatory, and neoplastic lesions. Gadolinium is overall much safer than the contrast agents that are used with CT scans. However, nephrogenic systemic fibrosis, also called *nephrogenic fibrosing dermopathy*, has been reported in patients with impaired kidney function and acidosis.

The choice of which type of MRI technique to use is based on the patient's suspected disorder, and the tissues and locations of the body to be viewed. The various MRI techniques include the following:

- **Diffusion-weighted imaging (DWI)**—allows quick, early detection of ischemic stroke
- **Diffusion tensor imaging**—an extension of DWI; it can reveal white matter tracts in three dimensions, known as **tractography**, and can be used for monitoring integrity of CNS tracts that are affected by aging and disease.

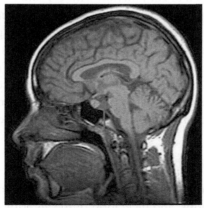

Fig. 22.2 Normal MRI of the brain, with arrow pointing to location of the hypothalamus. *Source: https://commons.wikimedia.org/wiki/File:Hypothalamus.jpg (public domain, free to use).*

- **Functional MRI**—reveals brain regions activated by certain cognitive or motor tasks, though its use is still being developed; activated brain regions show an increased flow of oxygenated blood; it is now used to localize the epileptiform lesion in patient with focal seizures
- **Perfusion-weighted imaging**—may detect areas of hypoperfusion in early ischemic stroke, or to diagnose metastatic diseases in the brain
- **Magnetic resonance angiography (MRA)**—used with or without a contrast agent, it shows cerebral vessels and major arteries as well as their branches, in the head and neck; it is used when cerebral angiography cannot be done, such as when the patient has increased risks or refuses the procedure; when checking for stroke, MRA usually exaggerates severity of arterial narrowing, and therefore, usually reveals occlusive disease of large arteries. MRA is shown in Fig. 22.3
- **Magnetic resonance venography**—it shows major veins and dural sinuses of the cranium; it replaces cerebral angiography for diagnosing cerebral venous thrombosis; it is also useful for monitoring the resolution of thrombi, and for guiding the duration of anticoagulation
- **Magnetic resonance spectroscopy (MRS)**—measures brain metabolites, per region, distinguishing tumors from abscesses or strokes

The most important advantage of MR studies over CT is the lack of radiation side effects with MRI modalities.

Cerebral catheter angiography

Cerebral catheter angiography uses X-rays, which are taken following injection of a radiopaque agent through an intra-arterial catheter. It reveals individual cerebral arteries and venous brain structures. Using digital data processing known as *digital subtraction angiography*, small amounts of the radiopaque agent can produce high-resolution images. Cerebral angiography is used in conjunction with CT and MRI for delineating locations and vascularity of intracranial lesions. This procedure has been the preferred method for the diagnosis of occluded or stenotic arteries, aneurysms, congenitally absent vessels, and arteriovenous malformations. It allows visualization of vessels as small as 0.1 mm. Today, however, because of MRA and CT angiography, the use of cerebral catheter angiography has decreased greatly. It is still commonly used when cerebral vasculitis is suspected, or when angiographic interventions may be required. These interventions include angioplasty, aneurysm obliteration, intra-arterial thrombolysis, and stent placement.

Duplex Doppler ultrasonography

Duplex Doppler ultrasonography is a noninvasive procedure that is able to evaluate dissection, occlusion, stenosis, and ulceration of the anterior circulation. It is performed

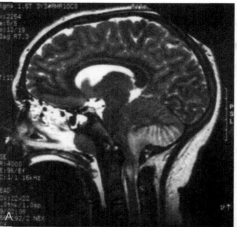
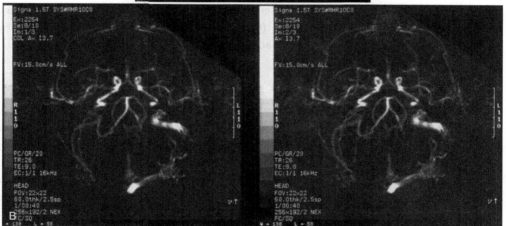

Fig. 22.3 A 10-year-old girl with otomastoiditis was evaluated because of unresponsiveness. Magnetic resonance imaging shows areas of increased signal in the right cerebellum greater than the left cerebellum, consistent with infarctions. The cerebellar tonsils are herniated. (A) Associated edema occurs in the superior cervical cord and inferior medulla. Phase-contrast magnetic resonance angiography (MRA) images demonstrate lack of flow in the straight sinus and the right transverse sinus. Only a small amount of signal in the region of the right sigmoid and internal jugular vein is seen. (B) Some arterial flow is represented in the examination. *Source: Figure 65.34 from Page 966. e1 of Bradley's Neurology in Clinical Practice 7th Edition.*

quickly and safely but does not give the same amount of detail as angiography. This method is preferred over periorbital Doppler ultrasonography and oculoplethysmography in the evaluation of patients with carotid artery transient ischemic attacks. It is useful for following various abnormalities over time. **Transcranial** Doppler ultrasonography (TCD) aids in the evaluation of residual blood flow following the vasospasm of the middle cerebral artery due to a subarachnoid hemorrhage. It can be used to assess the brain vessels. Another important indication of TCD is the detection of emboli from the heart.

Myelography

Myelography uses X-rays, which are taken after injection of a radiopaque agent into the subarachnoid space, via lumbar puncture. It has been replaced by MRI for the evaluation of intraspinal abnormalities. However, CT myelography is still performed when MRI is not available. The contraindications are the same as those for lumbar puncture. Myelography can worsen the effects of spinal cord compression, primarily when excess fluid is removed too quickly. CSF leak is a condition where CSF leaks out of the subarachnoid space due to a pore, which can be caused by lumbar puncture. This leak can lead to a decrease in the intracranial pressure, which ultimately causes a headache such as the one you see after the lumbar puncture (post-LP headache). CT or MR myelography are still the best ways to identify the exact location of the CSF leak.

Electroencephalography

EEG involves electrodes that are distributed over the brain. They detect electrical changes, such as those related to seizure or sleep disorders, and metabolic or structure encephalopathies. A total of 20 electrodes are symmetrically distributed over the patient's scalp. Normally, when a patient is awake, the EEG shows 8–12-Hz, 50-mV sinusoidal alpha waves. These wax and wane over the brain's occipital and parietal lobes. There are also 12-Hz or higher, 10- to 20-mV beta waves frontally, mixed with 4- to 7-Hz theta waves. The EEG is monitored, checking for asymmetries between the two brain hemispheres, which suggests a structural disorder. It is also checked for excessive slowing and for abnormal brain wave patterns. Excessive slowing, such as the appearance of 1- to 4-Hz, 50- to 350-mV delta waves, for example, occurs in dementia, depressed consciousness, and encephalopathy. Fig. 22.4 shows an example of electroencephalogram results.

Abnormal brain wave patterns may be nonspecific or diagnostic. Nonspecific patterns include *epileptiform sharp waves*. Diagnostic patterns include the three-Hertz spike and wave discharges seen in absence seizures, and the 1-Hz periodic sharp waves seen in Creutzfeldt-Jakob disease. For the evaluation of episodic altered consciousness, of uncertain cause, EEGs are highly useful. When a seizure disorder is suspected, but a routine EEG is normal, activities that cause electrical activation of the cortex may occasionally result in evidence of a seizure disorder. These activities include hyperventilation, photic stimulation, and sleep or sleep deprivation. When an EEG is otherwise not informative, nasopharyngeal leads may detect temporal lobe seizure foci that would not normally be detected with surface electrodes. A link between seizure activity and

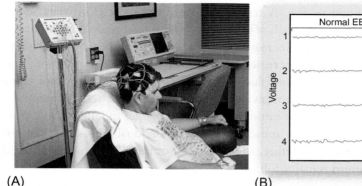
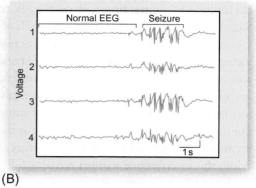

Fig. 22.4 Electroencephalography. (A) Photograph of a person with voltage-sensitive electrodes attached to his skull. Information from these electrodes is used to produce a graphic recording of brain activity—an electroencephalogram (EEG). (B) An EEG tracing showing activity in four different places in the brain (obtained from four sets of electrodes). Compare the moderate chaotic activity identified as normal with the explosive activity that occurs during a seizure.

symptoms such as fleeting memory lapses, unusual episodic motor behaviors, and subjective auras may be shown via continuous ambulatory monitoring of the EEG over 24 h. This can be used with or without video monitoring. For more details on EEG, please see Chapter 6.

Focal cortical electrical activity results when auditory, tactile, or visual stimuli are used to activate the related areas of the cerebral cortex. Normally, such small potentials are mixed in with the EEG *background noise*. However, computer processing is able to cancel out this noise, revealing waveforms. The amplitude, duration, and latency of the evoked responses indicates if the tested sensory pathway is functionally intact. Evoked responses are very useful in the detection of clinically hidden deficits of demyelinating disorders, assessing the sensory systems of infants, confirming deficits believed to be histrionic, and in the following subclinical disease courses. Visual evoked responses can show unsuspected optic nerve damage, for example, by multiple sclerosis. If the brainstem integrity is questioned, auditory evoked responses are objectively tested. A somatosensory evoked response can identify a physiologic disturbance occurring when a structural disorder affects many levels of the neuroaxis. This could be a metastatic carcinoma invading the plexus and spinal cord.

Electromyography

The use of **electromyography** and *nerve conduction velocity studies* can identify neuromuscular junctions, nerves, and muscles affected by weakness. In electromyography, a needle with a microphone is inserted into a muscle. The electrical activity is recorded both while the muscle is contracting and resting. A resting muscle is normally electrically silent. Slight contraction causes action potentials of single motor units to appear. With increased contraction, the amount of potentials increases, to form an interference pattern. **Denervated** muscle fibers are shown via increased activity with the needle's insertion, and abnormal, spontaneous fibrillations and fasciculations. There are fewer motor units recruited during contraction. This produces a reduced interference pattern. The surviving axons branch to innervate adjacent muscle fibers. This enlarges the motor unit, producing giant action potentials. In a muscle disorder, individual fibers are affected regardless of their motor units. Therefore, the amplitude of their potentials is reduced while the interference pattern remains full.

Nerve conduction velocity studies involve stimulation of a peripheral nerve with electrical shocks, at several points along the nerve's course to a muscle. The time to initiation of a contraction is recorded. The time that impulse needs to move down a measured length of a nerve determines the *conduction velocity*. The time needed to move over the segment closest to the muscle is called the *distal latency*. For sensory nerves, measurements are similar. If weakness is caused by a muscle disorder, nerve conduction will be normal. Conduction is often slowed in neuropathy. Response patterns may show dispersed potentials, due to unequal involvement of myelinated and unmyelinated axons. A nerve may be repeatedly stimulated, as part of a process of evaluating a neuromuscular junction for *fatigability*. One example is the progressive decremental response seen in myasthenia gravis.

Key terms

Cerebral catheter angiography	Electromyography
Chromogens	Myelography
Craniocervical junction	Tractography
Denervated	Transcranial
Duplex Doppler ultrasonography	Uncrenated
Electroencephalography	

Suggested readings

1. Alpert, J. N. *The Neurologic Diagnosis: A Practical Bedside Approach*. Springer, 2011.
2. Anderson, M. W.; Fox, M. G. *Sectional Anatomy by MRI and CT*, 4th ed.; Elsevier, 2016.
3. Berkowitz, A. *Lange Clinical Neurology and Neuroanatomy: A Localization-Based Approach*. McGraw-Hill Education/Medical, 2016.
4. Bermudes, R. A.; Lanocha, K.; Janicak, P. G. *Transcranial Magnetic Stimulation: Clinical Applications for Psychiatric Practice*. American Psychiatric Publications Inc., 2017
5. Biller, J.; Gruener, G.; Brazis, P. *DeMyer's The Neurologic Examination: A Programmed Text*, 7th ed.; McGraw-Hill Education/Medical, 2016.

6. Borden, N. M. *3D Angiographic Atlas of Neurovascular Anatomy and Pathology*. Cambridge University Press, 2006.
7. Boriani, S.; Presutti, L. *Atlas of Craniocervical Junction and Cervical Spine Surgery*. Springer, 2017.
8. Chernecky, C. C.; Berger, B. J. *Laboratory Tests and Diagnostic Procedures*, 6th ed.; Saunders, 2012.
9. Corbett, J. V.; Banks, A. *Laboratory Tests and Diagnostic Procedures With Nursing Diagnoses*; 8th ed.; Pearson, 2012.
10. Deisenhammer, F.; Sellbjerg, F.; Teunissen, C. E.; Tumani, H. *Cerebrospinal Fluid in Clinical Neurology*. Springer, 2015.
11. Hofer, M. *CT Teaching Manual: A Systematic Approach to CT Reading*, 4th ed.; Thieme Medidak Publishing GmbH, 2010.
12. Koch, B. L.; Hamilton, B. E.; Hudgins, P. A.; Harnsberger, H. R. *Diagnostic Imaging: Head and Neck*, 3rd ed.; Elsevier, 2016.
13. Lee, T. C.; Mukundan, S. *Netter's Correlative Imaging*. Neuroanatomy: Saunders, 2014.
14. Leite, C.; Castillo, M. *Diffusion Weighted and Diffusion Tensor Imaging: A Clinical Guide*. Thieme, 2016.
15. McKinney, A. M. *Atlas of Normal Imaging Variations of the Brain, Skull, and Craniocervical Vasculature*. Springer, 2017.
16. Mori, S.; Tournier, J. D. *Introduction to Diffusion Tensor Imaging: And Higher Order Models*. Academic Press, 2013.
17. Osborn, A. G.; Salzman, K. L.; Jhaveri, M. D.; Barkovich, A. J. *Diagnostic Imaging: Brain*, 3rd ed.; Elsevier, 2015.
18. Pozniak, M. A.; Allan, P. L. *Clinical Doppler Ultrasound: Expert Consult*, 3rd ed.; Churchill Livingstone, 2013.
19. Preston, D. C.; Shapiro, B. E. *Electromyography and Neuromuscular Disorders: Clinical-Electrophysiologic Correlations*. Saunders, 2012.
20. Rea, P. *Essential Clinical Anatomy of the Nervous System*. Academic Press, 2015.
21. Romans, L. E. *Computed Tomography for Technologists: A Comprehensive Text*. Wolters Kluwer Health/LWW, 2010.
22. Ross, J. S.; Moore, K. R. *Diagnostic Imaging: Spine*, 3rd ed.; Elsevier, 2015.
23. Schomer, D. L.; Lopes da Silva, F. H. *Niedermeyer's Electroencephalography: Basic Principles, Clinical Applications, and Related Fields*, 7th ed.; Oxford University Press, 2018.
24. Smith, F. W.; Dworkin, J. S. *The Craniocervical Syndrome and MRI*. S. Karger, 2015.
25. Wible, B. C. *Diagnostic Imaging: Interventional Procedures*, 2nd ed.; Elsevier, 2017.
26. Yumoto, E. *Pathophysiology and Surgical Treatment of Unilateral Vocal Fold Paralysis: Denervation and Reinnervation*. Springer, 2016.

Glossary

Abducens nerves (VI) Small motor nerves supplying the lateral rectus muscles of the eyes; originating in the facial colliculus of the tegmentum of the pons below the surface of the rhomboid fossa.

Aberrant corticospinal fibers Abnormal fibers pertaining to or connecting the cerebral cortex and spinal cord.

Absolute refractory period The period following stimulation, during which no additional action potential can be evoked.

Accelerometry The quantitative determination of acceleration and deceleration in the human body, or part of the body, in performance of a task.

Accessory nerves (XI) The pair of nerves that convey motor impulses to the pharynx and muscles of the upper thorax, back, and shoulders; originating in the medulla oblongata.

Achromatopsia Also spelled achromatopsia; color blindness.

Acoustic neuroma A benign tumor involving cells of the myelin sheath surrounding the vestibulocochlear nerve.

Action potentials Sequences of electrical changes in the axon of a neuron exposed to a stimulus that exceeds threshold.

Acute vestibular neuronitis A sudden, severe attack of vertigo, not accompanied by deafness or tinnitus, affecting young to middle-aged adults, often after an upper respiratory infection, caused by unilateral vestibular dysfunction.

Adenohypophyseal placode The ectodermal thickening that plays a role in the development of the hypophysis (pituitary gland).

Adherens junctions Protein complexes at the junction between epithelial cells, at which actin filaments from inside the cells pass across adjacent cell membranes.

Ageusia Loss or absence of the sense of taste.

Agnosia Inability to recognize the import of sensory impressions, including auditory, gustatory, olfactory, tactile, and visual forms.

Agraphia Loss of ability to express thoughts in writing.

Akinesia Absence or loss of the power of voluntary movement; or temporary paralysis of a muscle by injection of procaine.

Alar plate The neural structure in the embryonic nervous system that involves the communication of general somatic and visceral sensory pulses.

Alexander disease A rare, fatal degenerative disease of the CNS of infants, characterized by psychomotor retardation, seizures, and paralysis.

Alexia A form of receptive aphasia in which there is an inability to understand written language.

All-*trans*-retinal The orange retinaldehyde resulting from the action of light on the rhodopsin of the retina, which converts the 11-*cis*-retinal component of rhodopsin to all-*trans*-retinal plus opsin.

Alpha-fetoprotein A major plasma protein produced by the yolk sac and fetal liver during development.

Alpha-synuclein A polypeptide protein found primarily in brain neurons, but also in the fibrils of Parkinson's disease, the amyloid plaques of Alzheimer's disease, and in brain tissue affected by other neurodegenerative diseases.

Amacrine cells Nerve cells with short branching dendrites that are believed to lack axons; they are present in the retinas of the eyes.

Ambient cistern A structure located on the lateral aspect of the midbrain, dorsally continuous with the quadrigeminal cistern.

Amblyopia An uncorrectable decrease in vision in one or both eyes with no apparent structural abnormality; commonly referred to as "lazy eye."

Ampulla A small dilatation in a canal or duct, such as in the semicircular canals of the ears.

Ampullary crest The most prominent part of a localized thickening of the membrane lining the ampullae of the semicircular ducts, covered with neuroepithelium containing endings of the vestibular nerve.

Amsler grid Also called an Amsler chart; a grid of lines with a center black dot used by patients with macular degeneration to detect early worsening of their disease.

Amygdaloid body A small mass of subcortical gray matter in the tip of the temporal lobe, anterior to the inferior horn of the lateral brain ventricle; part of the limbic system.

Amyloid angiography The X-ray study of the blood vessels to check for amyloids, which are aggregates of proteins that become folded, allowing protein copies to stick together and form fibrils.

Anaxonic neurons The neurons that cannot be differentiated from dendrites.

Angiogenic Of vascular origin.

Anosmia Lack of the sense of smell.

Anterior (ventral) horn In the lateral ventricle of the brain, the horn that passes forward, laterally, and slightly downward from the interventricular foramen into the frontal lobe; in the spinal cord, it is the ventral gray matter section that contains motor neurons affecting the skeletal muscles.

Anterior cerebral artery One of two terminal branches of the internal carotid artery; it passes anteriorly, around the genu of the corpus callosum, then posteriorly in the longitudinal fissure, to be joined by the anterior communicating artery.

Anterior choroidal artery The artery that originates from the internal carotid or (rarely) the middle cerebral artery; it distributes to the lateral and third ventricles, optic chiasm and tract, internal capsule, globus pallidus, and many other structures.

Anterior commissure Also called the precommissure; a nerve fibers bundle (white matter) connecting the temporal lobes of the cerebral hemispheres across the midline, in front of the fornix; it plays a key role in pain and pain sensation.

Anterior communicating artery A short vessel joining the two anterior cerebral arteries across the midline, completing the cerebral arterial circle of Willis anteriorly.

Anterior corticospinal tracts Also called ventral corticospinal tracts; small bundles of descending fibers connecting the cerebral cortex to the spinal cord.

Anterior cranial fossa The portion of the internal base of the skull, anterior to the sphenoidal ridges and limbus, in which the frontal lobes of the brain rest.

Anterior inferior cerebellar artery A branch of the basilar artery, running dorsally along the posterior edge of the pons, supplying blood to the hindbrain, superior and middle cerebellar peduncles, and portions of the ventral cerebellum.

Anterior perforated substances Regions at the base of the brain through which many small branches of the anterior and middle cerebral arteries enter the depth of the cerebral hemisphere.

Anterograde flow The flow of freshly oxygenated blood away from the heart, toward the brain.

Aphasia A condition characterized by partial or total loss of the ability to communicate verbally, or using written words.

Apraxia A neurological condition characterized by loss of ability to perform activities that the individual is physically able and willing to do.

Aptamers Oligonucleotide or peptide molecules that bind to a specific target molecule.

Aqueductal stenosis Narrowing of the aqueduct of Sylvius, which blocks the flow of CSF in the ventricular system.

Aqueous humor The watery fluid that fills the anterior and posterior chambers of the eye, secreted by the ciliary processes in the posterior chambers.

Arachnoid granulations Tufted prolongations of pia-arachnoid, composed of many arachnoid villi that penetrate dural venous sinuses and allow the transfer of cerebrospinal fluid to the venous system.

Arachnoid mater A delicate fibrous, web-like membrane forming the middle of the three coverings of the central nervous system.

Arachnoid trabeculae Fine filaments that pass from the arachnoid to the pia mater, which are embryological remnants.

Arbor vitae Fern-like patterns that branch from the white matter in the sagittal section of the cerebellum.

Archicortex The part of the cerebral cortex, that with the paleocortex, develops in association with the olfactory system; it lacks the layered structure of the neocortex, having less than six layers of cells.

Arcuate fibers Nervous or tendinous fibers passing as an arch, from one structure, usually to an immediately adjacent structure, such as the gyri in the cerebral cortex.

Arnold-Chiari phenomenon Also called a Type II Chiari malformation, consisting of a downward displacement of the cerebellar tonsil through the foramen magnum, sometimes causing noncommunicating hydrocephalus.

Arterio-venous malformation (AVM) A potentially fatal congenital intracranial anomaly with large arteries feeding in a mass of communicating vessels that empty into large draining veins filled with arterialized blood.

Artery of Adamkiewicz Also called the arteria radicularis magna; the largest anterior segmental medullary artery.

Arytoenoepiglottic folds Also called aryepiglottic folds; triangular folds of mucous membrane enclosing ligamentous and muscular fibers at the entrance of the larynx.

Asplenia Congenital or acquired absence of the spleen or impaired reticuloendothelial function of the spleen.

Association areas Large areas of the cerebral cortex that are not sensory or motor, but are involved in advanced stages of sensory information processing, multisensory integration, or sensorimotor integration.

Astereognosis Loss or lack of ability to understand the form and nature of objects that are touched.

Astrocytes Types of neuroglia that connect neurons to blood vessels and provide growth factors.

Astrocytomas Primary brain tumors composed of astrocytes.

Astrocytosis An increase in the number of neuroglial cells due to the destruction of nearby neurons from CNS trauma, infection, ischemia, stroke, autoimmune responses, and neurodegenerative disease.

Asynchronous discharge Electrical discharging of nearby motor units in an irregular pattern.

Athetosis Repetitive involuntary, slow, sinuous, writhing movements.

Atonia Lack of normal tone or strength; flaccidity.

Atrial myxoma A benign, pedunculated, gelatinous tumor that originates in the interatrial septum of the heart.

Auditory association area An area in the temporal lobe, within Wernicke's area, near the lateral cerebral sulcus; it is critical for processing acoustic signals so that they can be interpreted as speech, music, or other sounds.

Auditory cortex The region of the cerebral cortex that receives auditory radiation from the medial geniculate body, a thalamic cell group receiving input from the cochlear nuclei in the rhombencephalon.

Auditory ossicles The articulated, small bones of the middle ear; the malleus, incus, and stapes.

Aura A subjective symptom at the onset of a migraine headache, or an epileptic ictal phenomenon perceived only by the patient; it may involve visual disturbances, dizziness, numbness, and many other sensations.

Auricle Also called the pinna; the projecting part of the ear lying outside the head.

Auriculotemporal nerve roots The nerve roots of the ear and temporal region.

Autogenic inhibition A sudden relaxation of muscle in response to high-magnitude tension, which protects muscles against tearing.

Axolemma The plasma membrane of an axon.

Axon hillock A specialized part of the cell body (soma) of a neuron that connects to the axon.

Axon terminals Distal terminations of the telodendria (branches) of an axon.

Axons Neuron processes that carry impulses away from the nerve cell body (efferent processes), the conducting portion of a nerve cell.

Axoplasm The cytoplasm within the axon of a neuron.

Azygos system Veins on each side of the vertebral column that drain the back, thoracic, and abdominal walls; there is much variation to the origins, courses, tributaries, anastomoses, and terminations of this vein system.

Babinski's sign A reflex movement of the big toe upward instead of downward when the plantar aspect of the foot is stroked; used to test for injury or disease of the upper motor neurons.

Ballismus A type of involuntary movement affecting the proximal limb musculature, manifested in jerking, flinging movements of the extremity, caused by a lesion of or near the contralateral subthalamic nucleus.

Basal ganglia The specific gray matter areas located deep within the white matter of the cerebral hemispheres.

Basal nuclei Another name for *basal ganglia*.

Basal plate The region, during development, of the neural tube that is ventral to the sulcus limitans; it contains mostly motor neurons.

Basal vein of Rosenthal A vein along the base of the brain running from the cavernous sinus, around the cerebral peduncle, into the vein of Galen; its tributaries include the anterior cerebral, deep middle cerebral, and striate veins.

Basilar artery An artery in the base of the brain formed by the union of the right and left vertebral arteries, which courses along the clivius from the lower to upper borders of the pons, bifurcating into the posterior cerebral arteries.

Basilar membrane The lower border of the scala media of the ear.

Basilar venous plexus A venous plexus on the clivius, connected with the cavernous and petrosal sinuses and internal vertebral venous plexus.

Basket cells In the cerebellum, cells with axis-cylinder processes terminating in a basket-like network around the Purkinje cells.

Bassen-Kornzweig syndrome Also called abetalipoproteinemia; a rare, autosomal recessive syndrome marked by lack of low-density lipoproteins in the blood, acanthocytosis, hypercholesterolemia, retinitis pigmentosa, and neuropathy.

Behçet's syndrome A group of symptoms including ulceration of the mouth or genital area, skin lesions, and inflammation of the uvea of the eye.

Bell's palsy Neuropathy of the facial nerve, resulting in paralysis of the muscles on one side of the face.

Benedict's syndrome See Benedikt syndrome below.

Benedikt syndrome Hemiplegia with clonic spasm or tremor and oculomotor paralysis on the opposite side.

Beta waves Also called beta rhythms; in an electroencephalogram, they are smaller than those of alpha rhythms, with an average frequency of 25 per second, usually during periods of intense nervous system activity.

Bill's bar A bony anatomical landmark that divides the superior compartment of the internal acoustic meatus into an anterior and posterior compartment.

Bipolar neurons: Also called Bipolar cells Nerve cells with only two cell processes associated with their cell body.

Bitemporal hemianopia Blindness in the temporal field of vision of both eyes.

Blepharospasm Spasm of the orbicular muscle of the eyelid.

Blind spot The small, circular, optically insensitive region in the retina where fibers of the optic nerve emerge from the eyeball; it lacks rods or cones, and is also called the optic disk.

Blood-brain barrier (BBB) The mechanism that inhibits passage of materials from the blood into the brain tissues; it reflects relative impermeability of the brain capillaries.

Brachium conjunctivum Also simply called the brachium; any anatomic structure resembling an arm.

Bradykinesia Abnormal slowness of movement.

Bradyzoites Slow-growing microorganisms such as *Toxoplasma gondii* and others responsible for parasitic infections.

Brainstem The structure consisting of the superior midbrain, middle pons, and inferior medulla oblongata.

Broca's area An area in the frontal lobe usually of the left cerebral hemisphere, associated with the motor control of speech.

Brudzinski signs Physical signs of meningitis, evoke by either passive flexion of one leg, resulting in a similar movement on the opposite side, or if the neck is passively flexed and flexion occurs in the legs.

B-scan ultrasound Also called B-mode ultrasound; a display that uses dots of differing intensities to represent echoes received from tissues that more strongly or weakly reflect sound waves.

Callosomarginal artery The second branch of the pericallosal artery, running in the cingulate sulcus, sending branches to supply part of the medial and superolateral surfaces of the cerebral hemisphere.

Calvarial dura The strongest, outermost membrane of the calvaria of the skull.

Calvarium A term used to describe the calvaria, which is the dome-like superior portion of the cranium, including the superior parts of the frontal, parietal, and occipital bones.

Canaliculi perforantes Micropores widely distributed in the bony surfaces that line the perilymphatic space, most numerous in the scala tympani, and in the peripheral and modiolar parts of the osseous spiral lamina.

Canavan disease A rare, autosomal recessive form of leukodystrophy, involving early onset demyelination and vascuolation of the cerebral white matter, severe mental retardation, head enlargement, neck muscle atony, spasticity, and blindness.

Caroticotympanic nerves The nerves related to the carotid canal and tympanum.

Carotidynia Pain caused by pressure upon the carotid artery.

Cataplexy A condition characterized by sudden, brief attacks of muscle weakness that sometimes cause falling, usually triggered by strong emotions; often associated with narcolepsy.

Cauda equina The collection of spinal roots descending from the lower spinal cord, occupying the vertebral canal below the cord.

Caudate nucleus The "C"-shaped nucleus arched superiorly over the diencephalon, which along with the putamen, forms the striatum.

Cavernoma A cavernous vascular tumor.

Cavernous angioma A vascular malformation composed of sinusoidal vessels without a large feeding artery; can be multiple, especially if inherited as an autosomal dominant trait.

Central sulcus The sulcus separating the frontal and parietal lobes of the cerebrum.

Centrioles Cellular structures built of microtubules that organize the mitotic spindle.

Centrum semiovale The central area of white matter underneath the cerebral cortex.

Cerebellar tonsils Paired cerebellar lobules on either side of the uvula, projecting from the inferior surface of the cerebellum.

Cerebellar vermis A worm-shaped structure in the medial, cortico-nuclear zone of the cerebellum, within the posterior fossa of the cranium; it is functionally associated with body posture and locomotion.

Cerebellomedullary cistern The largest subarachnoid cistern between the cerebellum and medulla oblongata, divided into posterior and lateral portions.

Cerebellopontine angle The recess at the junction of the cerebellum, pons, and medulla oblongata.

Cerebellum The portion of the brain, with two hemispheres like the cerebrum, with a cortex of gray matter and a layer of deeper white matter; all input travels to the cortex and all output comes from the deep nuclei.

Cerebral angiography A radiographic procedure used to visualize the vascular system of the brain after injection of a radiopaque contrast medium.

Cerebral aqueduct (of Sylvius) The slender cavity of the midbrain that connects the third and fourth ventricles, and contains CSF.

Cerebral arterial circle The roughly pentagonal-shaped circle of vessels on the ventral aspect of the brain, in the area of the optic chiasm, hypothalamus, and interpeduncular fossa.

Cerebral cortex The brain location of the conscious mind, allowing for awareness, communication, memory, understanding, vision, language, and voluntary movements; it makes up about 40% of the total brain mass.

Cerebral hemispheres The two large, identical masses of the cerebrum, connected by the corpus callosum, and marked with many gyri.

Cerebral ischemia Reduction or loss of oxygen to the cerebrum; if prolonged, this may lead to cerebral infarction.

Cerebral peduncles Anterior halves of the midbrain, divisible into tegmenti and crus cerebri, separated by the substantia nigra.

Cerebral white matter The internal cerebral tissue, deep to the cortical gray matter; it communicates between areas of the cerebrum, and between the cortex and lower centers of the central nervous system.

Cerebrospinal fluid (CSF) The fluid within the subarachnoid space, the central canal of the spinal cord, and the four ventricles of the brain; it protects the CNS.

Cerebrotendinous xanthomatosis A metabolic disorder associated with deposition of cholestanol and cholesterol in the brain and other tissues.

Cerebrum The anterior and largest part of the brain, consisting of two hemispheres; it controls voluntary movements and coordinates mental actions.

Cerulea dolens Also called phlegmasia cerulea dolens; an acute, fulminating form of deep venous thrombosis, with pronounced edema and severe cyanosis of the limb.

Cervical spondylotic myelopathy A functional disturbance or pathological change secondary to encroachment of cervical spondylosis upon a congenitally small cervical spinal canal.

Chaddock's reflex Also called Chaddock's sign; in lesions of the pyramidal tract, stimulation below the external malleolus causes extension of the great toe.

Chemesthesis The chemical sensibility of the skin and mucous membranes.

Chemical synapse Biological junctions through which neuronal signals can be exchanged with each other and with nonneuronal cells.

Chemically gated ion channels Ion channels that open when a molecule (ligand) binds to it. Also called ligand-gated channels.

Chemosis Edema of the conjunctiva of the eye.

Chemotaxis Movement in response to the influence of chemical stimulation.

Cheyne-Stokes respirations Breathing characterized by rhythmic waxing and waning of the depth of the respirations.

Cholesteatomas Cyst-like masses with linings of stratified squamous epithelium, filled with desquamating debris that often includes cholesterol; they are most common in the middle ear and mastoid region.

Cholinergic synapses Gaps where a neuron that produces acetylcholine sends messages to other neurons or to skeletal muscle cells.

Chorda tympani A nerve given off from the facial nerve that conveys taste sensation from the anterior two-thirds of the tongue, and aids in the innervation of the submandibular and sublingual salivary glands.

Chordomas Malignant tumors arising in the axial skeleton from embryonic remains of the notochord.

Chorea Ceaseless occurrence of rapid, jerky involuntary movements.

Choreoathetoid movements The movements related to or characterized by choreoathetosis, which involves choreic and athetoid movements.

Choreoathetosis Abnormal movements of combined choreic and athetoid patterns.

Choriocapillaris Also called the capillary lamina of choroid; the internal or deep portion of the choroidea of the eye, composed of a close capillary network.

Choroid epithelium The cellular covering of the middle, vascular coat of the eye, between the sclera and retina.

Choroid fissure Also called the retinal fissure; a ventral groove formed by invagination of the optic cup and its stalk by vascular mesenchyme, from which hyaloid vessels develop.

Choroid plexus The plexus of cells that produces cerebrospinal fluid in the ventricles of the brain; it consists of modified ependymal cells.

Choroid The middle, vascular coat of the eye, between the sclera and retina.

Choroidal vein The vein that follows and drains the choroid plexuses in the cerebral ventricles, later merging with the thalamostriate vein, to form the internal cerebral vein.

Chromaffin cells Endocrine cells that elaborate many catecholamines, secreting these substances into the blood.

Chromogens Substances that lack color, but may be transformed into pigments.

Chronic bacterial meningoencephalitis Long-term bacterial inflammation of the brain and its meninges; also called encephalomeningitis.

Chronotropy Affecting a time or rate, such as heart rate.

Chvostek sign Unilateral spasm due to muscle tetany, induced by tapping over the facial nerve, which occurs in severe hypocalcemia.

Ciliary ganglion A parasympathetic ganglion in the posterior part of the eye orbit.

Circadian rhythm The regular recurrent in cycles of about 24 h from one point to another, including biological activities that cycle regardless of long periods of darkness or other environment condition changes.

Circumferential blindness Lack of vision affecting the circumference or perimeter of the visual fields.

Circumventricular organs Structures in the brain characterized by extensive vasculature and lack of a normal blood-brain barrier, allowing for a link between the CNS and peripheral blood flow.

Clarke's nucleus Also called the posterior thoracic nucleus.

Claude's syndrome Paralysis of the oculomotor nerve on one side, and asynergia (lack of coordination) on the other side, together with dysarthria.

Claustrum A thin layer of gray matter lateral to the external capsule of the lentiform nucleus, separating it from the white matter of the insula.

Clivius A sloped structure in the base of the skull, between the occipital and sphenoid bones, which provides support for the pons.

Coccygeal ligament The thread-like termination of the spinal dura mater, surrounding and fused to the filum terminale of the cord, and attached to the deep dorsal sacrococcygeal ligament.

Cochlear nerve One of two divisions of the vestibulocochlear nerve that facilitate hearing by conducting stimuli to the brain.

Cochlear nuclei The nuclei on the posterior and lateral surface of the inferior cerebellar peduncle in the floor of the lateral recess of the rhomboid fossa are the major source of origin of the lateral lemniscus or central auditory pathway.

Cochlear recess A small concavity between the two limbs of the vestibular crest in the vestibule of the ear near the beginning of the cochlear duct.

Coenurosis A condition caused by tapeworm larvae, resulting in grape-like cysts obstructing the outflow of cerebrospinal fluid in the fourth ventricle of the brain.

Cogan's syndrome Also called oculovestibuloauditory syndrome; an autoimmune condition characterized by interstitial keratitis and bilateral, rapidly progressive audiovestibular dysfunction.

Cogwheel rigidity Muscle tension that subsides in little jerking movements when the muscle is passively stretched.

Collaterals Branches of an axon or blood vessel.

Commissures Fibers that connect the gray matter areas of the cerebral hemispheres, helping to coordinate functions; the largest of the commissures is the corpus callosum.

Communicating hydrocephalus Buildup of cerebrospinal fluid and intracranial pressure due to an abnormality of CSF absorption; there is no obstruction causing this type of hydrocephalus.

Complement factor H gene The gene that provides instructions for making a protein called complement factor H, which helps regulate some of the body's immune responses (known as the complement system).

Conductive hearing loss A loss of hearing associated with the impaired transmission of sound waves through the external ear canal to the bones of the middle ear; it may be caused by blockage or dysfunction.

Cone pigment Photosensitive pigment in the outer segment of the cones of the eyes; the three pigments of the cones form the basis of normal trichromatic color vision.

Cones Receptor cells in the eyes that allow for the perception of colors.

Confluence of sinuses A meeting place, at the internal occipital protuberance, of the superior sagittal, straight, and occipital sinuses, drained by the two transverse sinuses of the dura mater.

Congenital oculosympathetic paralysis An inherited form of paralysis of the sympathetic pathway of the eye.

Connexins Complex protein assemblies that, in groups of six, form connexons.

Connexons Assemblies of six connexin proteins that form pores for gap junctions between the cytoplasm of two adjacent cells.

Continuous propagation The process during which an action potential in an axon spreads to a neighboring region of its membrane by a series of small steps.

Contralateral homonymous hemianopia A visual field loss on the left or right side of the vertical midline, which usually affects both eyes; the contralateral form may be caused by lesions from the optic tract to the visual cortex.

Contralateral Anatomically opposite to another structure; the left eye is contralateral to the right eye.

Conus medullaris The anterosuperior portion of the right ventricle of the heart, at the entrance to the pulmonary trunk.

Convergence The coordinated inclination of the two lines of sight toward their common point of fixation, or the point itself.

Convulsions Contortions of the body caused by violent, involuntary muscular contractions of the extremities, trunk, and head.

Cornea The clear, transparent anterior covering of the eye.

Corneal reflex A reflex action of the eye, resulting in automatic closing of the eyelid when the cornea is stimulated.

Corneoscleral junction The margin of the cornea overlapped by the sclera.

Corona radiata The radiating crown between the cortex and basal ganglia, through which fibers radiate through the cerebral white matter to the cortex.

Corpora arenacea Also called brain sand; calcified structure in the pineal gland and other areas, such as the choroid plexus.

Corpora quadrigemina The two inferior and two superior colliculi on the tectum of the dorsal aspect of the midbrain.

Corpus callosum The wide, flat, and heavily myelinated axon bundle that connects the cerebral hemispheres.

Corticobulbar tracts Two-neuron white matter motor pathways connecting the motor cortex in the cerebral cortex to the medullary pyramids.

Corticonuclear Another term that means "corticobulbar," related to the motor pathway connecting the motor cortex of the cerebral cortex to the medullary pyramids.

Corticopetal system Related structures to nerve fibers or impulses originating outside and passing toward the cerebral cortex.

Corticopontine fibers Projections from the cerebral cortex to the pontine nuclei.

Corticospinal pathway The white matter motor pathway beginning at the cerebral cortex, terminating on lower motor neurons and interneurons of the spinal cord, controlling movements of the limbs and trunk.

Corticosubthalamic fibers Projects from the cerebral cortex, below the thalamus.

Cortilymph The fluid filling the intercellular spaces of the organ of Corti, similar in composition to perilymph.

Cranial fossae The anterior, medial, or posterior hollows or depressed areas of the cranium.
Cranial meninges The three membranes covering the brain: the dura mater, arachnoid, and pia mater.
Cranial nerves The nerves that emerge from or enter the cranium or skull; they include the olfactory, optic, oculomotor, trochlear, trigeminal, abducens, facial, vestibulocochlear, glossopharyngeal, vagal, accessory, and hypoglossal nerves.
Cranial placodes Areas of thickening of the epithelium in the embryonic head ectoderm layer, which gives rise to structures of the sensory nervous system.
Cranial root Any of the roots of the accessory nerve that arise from the medulla.
Cranioarchischisis The most severe type of neural tube defect, in which both the brain and spinal cord remain open; both anencephaly and spina bifida are present.
Craniopharyngioma A type of brain tumor derived from pituitary gland embryonic tissue, most common in children.
Craniosynostosis One or more fibrous sutures in an infant's skull prematurely fuses by ossifying, changing the growth pattern of the skull.
Creutzfeldt-Jacob disease A transmissible, rapidly progressing, neurodegenerative disorder called a spongiform degeneration; related to *mad cow disease.*
Crista ampullaris Also called the ampullary crest.
Crista galli The triangular midline process of the ethmoid bone, extending superiorly from the cribriform plate; it gives anterior attachment to the falx cerebri.
Crocodile tears A flow of tears, usually unilateral, when eating or in anticipation of eating; due to damage and abnormal regrowth of nerve fibers of the salivary gland into the lacrimal gland.
Crouzon syndrome A genetic disorder characterized by premature closure of one or more cranial sutures, resulting in craniofacial abnormalities such as oxycephaly, wide-spaced protruding eyes, and hypoplasia of the maxilla.
Crus commune A structure formed by the joining of the opposite end of the superior semicircular canal and upper part of the posterior canal; this structure opens into the upper and medial part of the vestibule of the ear.
Cryopexy In retinal detachment surgery, sealing the sensory retina to the pigmented epithelium and choroid by a freezing probe applied to the sclera.
Crystallins Water-soluble proteins in the lenses of the eyes.
Cuneate nucleus A wedge-shaped nucleus in the closed part of the medulla oblongata.
Cupula A small, inverted cup or dome-shaped cap over a structure.
Cushing disease A disorder caused by excess production of the hormone ACTH by a pituitary tumor.
Cushing syndrome Swelling of the face, fat accumulation around the torso, fat loss from the extremities, and easy bruising; due to elevated cortisol caused by excess production of ACTH or administration of excessive corticosteroids.
Cyclic guanosine monophosphate A cyclic nucleotide or intracellular second messenger, similar in action to cyclic adenosine monophosphate; it activates protein kinases, usually producing opposite effects upon cell function.
Decussation of pyramids The crossing of the fibers at the lower limit of the medullary pyramids of the medulla oblongata.
Deep middle cerebral vein The vein that accompanies the middle cerebral artery in the depths of the lateral sulcus, emptying into the basal vein of Rosenthal.
Deep nuclei The four gray matter masses, within the white matter of the cerebellum, from which all cerebellar output occurs.
Deep vein thrombosis (DVT) A blood clot in a major vein, which usually develops in the legs or pelvis.
Deep veins Many systemic veins accompanying arteries, usually enclosed in sheaths wrapping the veins and associated arteries.

Deiters' cells Also called phalangeal cells; the supporting cells of the spiral organ, attached to the basement membrane, receiving the hair cells between the free extremities.

Delirium A state of mental confusion that develops quickly, usually fluctuating in intensity, due to a disturbance of normal brain function.

Delta waves Also called delta rhythms; electrical oscillations in the brain at a frequency of less than approximately 4 Hz, associated with a state of deep sleep or unconsciousness.

Dementia A loss of mental ability that interferes with normal activities of daily living, for more than 6 months, not present since birth, and not associated with a loss or alteration of consciousness.

Dendrites Thin, sensitive, branched extensions from a cellular body.

Denticulate ligaments In the pia mater of the spinal cord, a pair of ligaments, one on each side of the spinal cord, with 21 attachments per side, attaching them to the arachnoid and dura mater.

Depolarization The membrane of a neuron becoming less negative (more positive) than the resting potential.

Depth perception The ability to recognize depth or relative distances to different objects in space.

Diabetes insipidus A chronic disorder of excessive urination and usually intense thirst and dehydration; caused by insufficient vasopressin from the pituitary, or inability of the kidneys to respond to vasopressin.

Diagonal band of Broca A diagonal tissue mass involved in speech articulation.

Diaphragm sellae A fold of dura mater extending transversely across the sella turcica, roofing over the hypophyseal fossa; it is perforated in its center for the passage of the infundibulum.

Diencephalon The part of the forebrain between the cerebral hemispheres and midbrain, including the thalamus, epithalamus, and hypothalamus.

Diploic Double or "twin"; also, related to the spongy layer between the inner and outer compact layers of the flat bones of the skull.

Diplopia The perception of two images of a single object; also called double vision.

Dix-Hallpike maneuver A test for assessing paroxysmal vertigo and nystagmus, in which the patient is brought from a sitting to a supine position; the symptoms develop when the head is rotated toward the affected ear.

Dorsal median septum An often poorly defined septum formed by the mantle and marginal layers of the alar plate; when they grow dorsally, the marginal layers become fused on the median plane.

Dorsal median sulcus A narrow grove that exists only in the closed part of the medulla oblongata, ending at about the middle of the medulla oblongata.

Dorsomedial Located on the back, toward the midline.

Downregulating Inhibiting or suppressing normal responses of an organ or system.

Dretrecogin alfa A recombinant form of human activated protein C, with antithrombotic, antiinflammatory, and pro-fibrinolytic properties.

Dromotropy Affecting conductivity of a nerve fiber; either positive or negative.

Drusen Small, bright structures seen in the retina and in the optic disk.

Duplex Doppler ultrasonography The combination of real-time and Doppler ultrasonography.

Dura mater The outermost, toughest, and most fibrous of the three meninges covering the brain and spinal cord.

Dural folds The folds of the dura mater.

Dural venous sinuses Endothelium-lined venous channels in the dura mater.

Duret hemorrhage Small brainstem bleeding, resulting from brainstem distortion secondary to transtentorial herniations.

Dynamic equilibrium The ability to adjust to displacements of the body's center of gravity by changing the base of support.

Dynamic stretch reflex The reflex elicited by the potent dynamic signal transmitted from the primary sensory endings of the muscle spindles, caused by rapid stretch or unstretch.

Dysarthria A disturbance of speech due to emotional stress, brain injury, or from paralysis, incoordination, or spasticity of muscles used for speaking.

Dysconjugate gaze Failure of the eyes to turn together in the same direction.
Dysmetria Disturbance of the power to control the range of movement in muscular action.
Dysosmia Any disorder of the sense of smell.
Dysprosody Impairment in the ability to apply normal speech intonation patterns.
Dystonic movements Those that show impairment of muscular tonus.
Dystonic postures Those caused by impairment of muscular tonus.
Echinococcal cysts Lesions caused by *Echinococcus granulosus*, with the outer layer composed of host-derived inflamed fibrous tissue, an intermediate acellular layer, and an innermost layer from a developed infection.
Ectatic basilar artery The distention or stretching of the basilar artery.
Edinger-Westphal nucleus A small group of preganglionic parasympathetic motor neurons in the oculomotor nucleus of the midbrain; involved in innervation of the pupils and ciliary muscles.
Ehrlichiosis A bacterial infection spread by the insects known as ticks; symptoms include fever, chills, headache, muscle aches, and tiredness.
Electrochemical gradient The combined difference in concentration and charge; influences the distribution and direction of diffusion of ions.
Electronystagmography Graphic recordings of eye movements that provide objective documentation of induced and spontaneous nystagmus.
Elliptical recess A small concavity lying superiorly and posteriorly on the medial wall of the vestibule that lodges the utricle of the ear.
Embolism An obstruction in a blood vessel due to a blood clot or other foreign matter that becomes lodged while traveling through the bloodstream.
Embolus A clot or other plus, usually part or all of a thrombus, brought by the blood from another vessel, then forced into a smaller one, obstructing circulation.
Emissary veins Small vessels in the skull connecting sinuses of the dura mater with veins on the exterior of the skull, via a series of anastomoses.
Encephalocele A herniation or protrusion of parts of the brain and meninges through a skull defect, causing a sac-like structure to develop.
Endolymph The fluid in the membranous labyrinth of the ear; it is entirely separate from the perilymph.
Endolymphatic potential The standing direct current potential in the endolymph relative to the perilymph, measuring positive 80 mV.
Endoscopic third ventriculostomy Via use of an endoscope, an operation to establish an opening from the third ventricle to the prechiasmal and interpeduncular cisterns, or to the interpeduncular cistern alone.
Enkephalins Pentapeptide endorphins, found in many parts of the brain.
Enophthalmos A backward displacement of the eyeball into the orbit.
Ependymal cells Types of CNS supporting cells; they line the ventricular system of the brain and the central canal of the spinal cord.
Ependymomas Tumors that arise from the ependymal tissues of the CNS.
Ephaptic transmission The passage of a neural impulse from one nerve fiber, axon, or dendrite to another, through membranes; it may be a factor in epileptic seizures.
Epicanthal folds The folds of skin extending from the root of the nose to the medial termination of the eyebrows, overlapping the medial angles of the eyes.
Epidural space The space between the dura mater and the lining of the spinal canal.
Epidural venous plexus Also called the internal vertebral venous plexus; a plexus of unvalved veins in the fat of the epidural space within the vertebral canal.
Epilepsy Paroxysmal transient disturbances of nervous system function, resulting from the abnormal electrical activity of the brain.
Epileptiform Resembling epilepsy; denoting certain convulsions, especially of a functional nature.
Epithalamus A division of the diencephalon, including the habenular, pineal body, and posterior commissure.

Epley maneuver Also called the canalith repositioning maneuver; a technique used to manage benign paroxysmal positional vertigo (BPPV), which involves the sequential movement of the head into four positions.

Epworth Sleepiness Scale A scale based on a questionnaire that rates an individual's probability of falling asleep, with increasing probability from zero to three, for eight different situations of daily living.

Equilibrium potential Also called reversal potential; the membrane potential at which there is no overall flow of a particular ion from one side of the membrane to the other.

Equilibrium The condition of being evenly balanced.

Erdheim-Chester disease A rare disease of abnormal multiplication of histiocytes, or tissue macrophages; usually begins in middle age.

Esotropia Also called "cross-eye"; deviation of the visual axis of one eye toward that of the other.

Excessive daytime sleepiness A subjective difficulty in maintaining an awake state, and an increased ease of falling asleep when an individual is sedentary.

Excitatory postsynaptic potential (EPSP) The depolarizing graded potential in a postsynaptic neuron.

Exophthalmos The bulging of the eyes outward due to an increase in the volume of the tissues behind the eyes; usually related to Graves' disease.

Exotropia Also called "wall-eye"; deviation of the visual axis of one eye away from that of the other.

External acoustic meatus The passage leading inward through the tympanic part of the temporal bone, from the auricle to the tympanic membrane; also called the external auditory canal.

External branch The branch of the accessory nerves that controls the sternocleidomastoid and trapezius muscles.

Exteroceptors Sensory receptors that respond to stimuli from the external world.

Extradural hemorrhage Extravasation of blood between the skull and dura mater.

Extraventricular drains Devices placed outside of a brain ventricle to allow for drainage of CSF.

Eye stroke Also called retinal artery occlusion; it occurs when blood flow to the retina is blocked; this should be treated or it may result in blindness.

Fabry disease A condition caused by the deficiency of alpha-galactosidase, characterized by abnormal accumulations of neutral glycolipids in endothelial cells in blood vessel walls.

Facial hemiatrophy Atrophy that is usually progressive, affects the tissues on one side of the face.

Facial myokymia A disorder of the facial muscles that causes narrowing of the palpebral fissure and continuous undulation of the facial skin surface, with a "bag of worms" appearance.

Facial nerves (VII) The cranial nerves that control the facial muscles and relay sensation from the taste buds of the front part of the tongue.

Fasciculus cuneatus The more lateral of the two large ascending axon tracts that fill the dorsal funiculus of the spinal cord.

Fasciculus gracilis The smaller medial subdivision of the posterior funiculus.

Fasciculus lenticularis A tract connecting the globus pallidus to the thalamic fasciculus.

Fasciculus proprius One of the ascending and descending spinal association fiber systems.

Fasciculus subthalamicus A tract that connects the subthalamic nucleus and globus pallidus.

Fasciculus thalamicus Part of the subthalamus, with fibers from the ansa lenticularis and the lenticular fasciculus.

Fenestra vestibuli Also called the oval window; a membrane-covered opening on the medial wall of the tympanic cavity leading into the vestibule; it is closed off by the foot of the stapes bone.

Festination Related to a person's gait, it involves short, accelerating steps, often on tiptoe, with the trunk flexed forward and the legs flexed stiffly at the hips and knees.

Field of vision The area of space in which objects are visible at the same time when the eye is fixed and the face is turned, so as to exclude the limiting effects of the orbital margins and nose.

Fields of Forel Areas in the diencephalon that are also known as H fields; these include the thalamic fasciculus, lenticular fasciculus, and pallidothalamic tracts.

Filum terminale The slender, thread-like termination of the spinal cord.
Fissure A groove on the surface of the cerebrum that is very deep.
Flocculi One of the small, paired lateral lobules continuous with the nodule of the cerebellum, forming part of the flocculonodular lobe.
Fluorescein angiography A technique used to diagnose chorioretinal disease, based on the enhancement of anatomic and vascular details in the retina after intravenous injection of a dye, known as fluorescein.
Focal dystonias Musculoskeletal problems resulting from overuse or repetitive stress.
Folia Thin, transverse, parallel folds of the cerebellar hemispheres, separated by shallow sulci.
Fontanelles Fibrous membranes at the angles of the cranial bones that accommodate brain growth in the fetus and infant.
Foramen cecum The small foramen formed by the frontal crest of the frontal bone ending at an articulation with the ethmoid bone.
Foramen ovale The septal opening in the fetal heart providing communication between the atria; it normally closes at birth; also, an aperture in the great wing of the sphenoid for vessels and nerves.
Foramen rotundum A round opening in the great wing of the sphenoid for the maxillary branch of the trigeminal nerve.
Foramen singular Also called the foramen singulare; it is located in the internal acoustic meatus, posterior to the cochlear area, which transmits nerves to the ampulla of the posterior semicircular duct.
Fornix An anatomical arch-like structure.
Foster Kennedy syndrome Pallor and loss of nerve fibers in the optic nerve of one eye, with swelling of the head of the optic nerve in the other.
Fourth ventricle One of four connected, fluid-filled cavities in the brain; it extends from the cerebral aqueduct to the obex and is filled with CSF.
Fovea centralis A small depression near the center of the retina, constituting the area of most acute vision.
Fulminant Occurring suddenly, with great intensity or severity.
Funduscope Also called an ophthalmoscope; a device used for studying the interior of the eyeball, through the pupil.
Funiculi Cord-like structures, especially the large nerve tract bundles in the white matter of the spinal cord.
Fusimotor neurons Those related to efferent innervation of intrafusal muscle fibers.
Gait unsteadiness A condition of unstableness in the manner or style of walking.
Ganglion cells Neurons with cell bodies located outside the limits of the brain and spinal cord, forming part of the peripheral nervous system.
Ganglion A mass of neuron cell bodies, located outside the central nervous system.
Ganglionopathy A term that refers to any disease of ganglia.
Gap junctions Passageways between two adjacent cells; formed by transmembrane proteins called connexons.
Gasserian ganglion The trigeminal or semilunar ganglion.
Gated ion channels Pore-forming membrane proteins that control which ions are able to pass through.
Gaze palsies Also called supranuclear gaze disturbance; an inability to direct the eyes to the side contralateral to a lesion in the frontal lobe.
Gelastic epilepsy A rare type of seizure involving a sudden burst of energy, usually as laughing or crying.
Geniculate ganglia The ganglia that are bent abruptly, similar to a knee joint; this refers to the facial nerve.
Giemsa A staining substance used to add color to protozoan parasites, blood smears, and viral inclusion bodies, allowing for microscopic study and identification.
Glabellar reflex Seen in geriatric patients, it involves blinking induced by tapping over the glabella; with repeated tapping it is normally suppressed, but may remain active in frontal lobe lesions and Parkinsonism.
Glasgow Coma Scale A standardized system for assessing response to stimuli in a neurologically impaired patient.

Gliomas Types of tumors that form in the glial cells of the brain or spine.

Gliomatosis cerebri A diffuse intracranial neoplasm of astrocytic origin.

Globus pallidus The output nuclei of the basal nuclei; along with the putamen, this forms a mass shaped like lens and called the lentiform nucleus.

Glomus cells Modified smooth muscle cells that act as peripheral chemoreceptors; mainly located in the carotid and aortic bodies, they help regulate breathing.

Glomus jugulare A microscopic collection of chemoreceptor tissue in the adventitia of the jugular bulb; a tumor here may cause paralysis of the vocal cords, attacks of dizziness, blackouts, and nystagmus.

Glomus A small body composed mostly of fine arterioles connecting directly with veins, and having a rich nerve supply.

Glossectomy Excision of all or a portion of the tongue.

Glossopharyngeal nerves (IX) The cranial nerves arising directly from the brain, providing taste sensation to the back of the tongue, and in the lining of the throat; they also supply the carotid body and sinus, plus muscles used in swallowing.

Glossopharyngeal neuralgia Severe paroxysmal pain originating on the side of the throat, extending to the ear, and affecting the petrosal and jugular ganglion of the glossopharyngeal nerve.

Glycinergic An agent that functions to directly modulate the glycine system in the body or brain.

Gnathostomiasis A rare infection that causes necrotic tracts, surrounded by inflammation along nerve roots, the spinal cord, or brain; it can also cause subarachnoid hemorrhage.

Golgi apparatus The organelle that prepares cellular products for secretion.

Gowers' sign When a patient has to use the hands and arms to "walk" the body from a squatting position to standing, due to lack of hip and thigh muscle strength.

Gracile nucleus The medial of the three nuclei of the dorsal column.

Graded potential A change in membrane potential, of variable size, as opposed to being all-or-none.

Gradenigo syndrome Suppurative otitis media, pain in the trigeminal nerve distribution, and abducens palsy, usually associated with infection or cancer at the base of the skull.

Granulomatosis Any condition involving the formation of multiple granulomas, which are either mononuclear inflammatory cells or modified macrophages that resemble epithelial cells, usually surrounded by lymphocytes.

Graphesthesia The ability to recognize writing on the skin purely by the sensation of touch.

Graves' orbitopathy The dysthyroid orbitopathy (disease of the eye orbits and their contents) seen in Graves' disease.

Great cerebral vein (of Galen) A cerebral vein formed by the two internal cerebral veins, continuing into the sinus rectus.

Guillain-Barré syndrome A disorder in which the immune system attacks part of the peripheral nervous system, resulting in eventual paralysis.

Gustatory cortex The brain structure mostly responsible for the perception of taste, made up of the anterior insula on the insular lobe, and the frontal operculum on the inferior frontal gyrus of the frontal lobe.

Gyri The many convolutions and ridges, separated by grooves, which mark the surfaces of the cerebral hemispheres.

Habenular nuclei Those that regulate key nervous system neurotransmitters, connecting the forebrain and midbrain within the epithalamus.

Habenular perforata Many small canals that radiate through the osseous lamina, from Rosenthal's canal, to the rim of the lamina, carrying fascicles of the cochlear nerve to the organ of Corti.

Hair cells Sensory receptors in the inner ear that transform sound vibrations into messages that travel to the brain.

Hallervorden-Spatz disease Also called pantothenate kinase-associated neurodegeneration; a degenerative brain disease that can lead to Parkinsonism, dystonia, dementia, and death.

Hand-Schüller-Christian disease A triad of exophthalmos, lytic bone lesions (often in the skull) and diabetes insipidus.

Helicotrema The passage that connects the scala tympani and scala vestibuli at the apex of the cochlea.

Helminthic brain infections Those that involve parasites and infect the central nervous system; they may cause encephalitis, meningitis, cerebral masses, hydrocephalus, myelopathy, stroke, and neurologic disorders.

Hematoxylin and eosin A staining method used for tissue sections in the microscopic examination; hematoxylin is obtained from certain trees and eosin is derived from the fluorescent dye called fluorescein.

Hemianesthesia Loss of sensation on one side of the body.

Hemianopia Defective vision or blindness in half of the visual field, usually with bilateral defects caused by a single lesion.

Hemiataxia Ataxia on one side of the body.

Hemiballismus Violent motor restlessness of half of the body, most severe in the upper limbs.

Hemidystonia Dystonia involving only one side of the body, usually associated with a structural lesion in the contralateral basal ganglia.

Hemifacial spasm A disorder of the facial nerve characterized by unilateral involuntary paroxysmal facial muscle contractions, due to high-frequency bursts of motor units, lasting from a few milliseconds to several seconds.

Hemisensory Referring to the loss of sensation on one side of the body.

Hemispatial neglect Attention deficit involving one side of the field of vision, due to damage to one brain hemisphere.

Hemitremor Tremor affecting one side of the body.

Hensen's stripe A band on the undersurface of the membrana tectoria of the cochlear duct.

Herpes zoster An acute viral disease caused by a herpesvirus, which also causes chickenpox; commonly known as shingles, involving inflammation of spinal ganglia and a vesicular eruption along the sensory nerve's distribution.

Hirschsprung's disease Congenital or aganglionic megacolon, in which nerve fibers are partially absent in the bowel, causing severe obstruction.

Holoprosencephaly A congenital defect caused by the failure of the prosencephalon to divide into hemispheres during embryonic development.

Homonymous hemianopia A condition affecting the right halves or the left halves of the visual fields of both eyes, often from damage to the optic tract or occipital lobe; the patient must turn the head from side to side to compensate.

Homonymous Pertaining to the corresponding vertical halves of the visual fields of both eyes.

Horizontal cells Interneurons located in the outer plexiform layer of the retinas, influencing signal processing in response to visual stimuli, at the level of contact between photoreceptor and bipolar cells.

Horner syndrome Ipsilateral myosis, ptosis, and facial anhidrosis, due to a lesion of the cervical sympathetic chain or central pathway.

Huntingtin A cytoplasmic protein in peripheral tissues of unclear function; it is elongated by polyglutamate residues of various lengths in Huntington disease and other neuropathies.

Huntington's chorea Also called Huntington's disease; a rare hereditary condition of quick involuntary movements, speech disturbances, and mental deterioration due to degenerative changes in the cerebral cortex and basal ganglia.

Hydrocephalus A condition of accumulated cerebrospinal fluid within the brain, usually causing increased pressure inside the skull.

Hygroma An accumulation of fluid in a sac, cyst, or bursa.

Hyperacusis Abnormal acuteness of the sense of hearing.

Hyperkinetic disorders Brain-based motor system disorders of excessive involuntary movements and hypotonia; the most common form is Huntington's chorea.

Hyperphosphorylation A signaling mechanism used by cells to regulate mitosis; it occurs when a biochemical with multiple phosphorylation sites is fully saturated.

Hyperpolarization An increase in membrane potential in which the membrane becomes more negative than the resting membrane potential.

Hypersomnolent Being excessively sleepy, or actually sleeping excessively, related to a variety of causative disorders.

Hyperthermia Excessively high body temperature.

Hypertonia Commonly known as spasticity from CNS damage, especially upper motor neuron lesions.

Hypertropia Strabismus in which there is a permanent upward deviation of the visual axis of the eye.

Hypobulic Reduced ability to make decisions or to act.

Hypodensities Areas of X-ray images that are less dense than normal, or than the surrounding areas.

Hypoglossal nerves (XII) The cranial nerves arising directly from the brain that supply the muscles of the tongue, and are necessary for talking and swallowing.

Hypokinesia Decreased body movement such as in Parkinson's disease.

Hypokinetic disorders Those involving decreased body movements, such as Parkinson's disease.

Hypomimia A reduced degree of facial expression, as in Parkinson's disease.

Hypophonic Having a weak voice, due to incoordination of the vocal muscles.

Hyposmia Diminished sensitivity of the sense of smell.

Hypothalamic sulcus A groove in the lateral wall of the third ventricle that marks the boundary between the thalamus and hypothalamus.

Hypothalamic-releasing hormones The hormones produced in the hypothalamus and carried by a vein to the anterior pituitary gland, where they stimulate release of the anterior pituitary hormones.

Hypothalamopituitary axis A feedback system that coordinates the activity of major peptide hormones; the hypothalamus synthesizes releasing hormones that act on the pituitary and evoke end-organ responses.

Hypothalamus The region of the diencephalon that forms the floor of the third brain ventricle.

Hypothermia Excessively low body temperature.

Hypotropia Strabismus with permanent downward deviation of the visual axis of one eye.

Immunofixation Separation of antigens within a protein mixture upon an electrophoretic gel, for identification by application of labeled antibodies.

Immunohistochemical Related to application of antigen-antibody interactions to histochemical techniques, such as in immunofluorescence.

Indolamine An indole or indole-derivative containing a primary, secondary, or tertiary amine group, such as serotonin.

Inferior colliculi Two small rounded elevations on the dorsal aspect of the midbrain, just below the two superior colliculi; they are relay centers for auditory fibers.

Inferior ganglion The lower sensory ganglion in the glossopharyngeal nerve, or on the vagus nerve.

Inferior olivary complex The three nuclei that form the inferior olivary nucleus; including the principal, medial accessory, and posterior accessory olivary nuclei.

Inferomedial Related to a location situated below and toward the center.

Infratentorial Beneath the tentorium of the cerebellum.

Infundibulum A hollow funnel-shaped mass in front of the tube cinereum that extends to the posterior lobe of the pituitary gland.

Inhibitory postsynaptic potential (IPSP) A graded potential in a postsynaptic neuron that inhibits action potential generation, usually hyperpolarizing.

Inotropy The force of muscle contraction.

Insomnia The inability to obtain an adequate amount or quality of sleep.

Insula The lobe of the cerebral cortex that is buried in the lateral sulcus, beneath portions of the parietal, frontal, and temporal lobes.

Internal acoustic meatus A canal beginning at the opening of the internal acoustic meatus in the posterior cranial fossa; it gives passage to the facial and vestibulocochlear nerves, with the labyrinthine artery and veins.

Internal capsule A compact band, formed by projection fibers at the superior brain stem, which passes between the thalamus and parts of the basal nuclei.

Internal carotid arteries Primary divisions of the common carotid artery, distributing blood through three sets of branches to the cerebrum, eyes, forehead, nose, internal ears, trigeminal nerves, dura mater, and hypophysis.

Internal cerebral vein One of two paired veins passing caudally near the midline, uniting to form the great cerebral vein.

Internal jugular veins Continuations of the sigmoid sinus of the dura mater, they join the subclavian vein to form the brachiocephalic vein.

Interneurons Neurons between a sensory neuron and a motor neuron, internuncial, or association neuron.

Internuclear ophthalmoplegia A disorder of conjugate lateral gaze, in which the affected eye shows impairment of adduction; when gazing contralaterally, the affected adducts minimally but the other eye abducts with nystagmus.

Interoceptors Sensory receptors in viscera that are sensitive to changes and stimuli within the body's internal environment; also called visceroceptors.

Interpeduncular cistern A dilation of the subarachnoid space, rostral to the basilar pons, and ventral and caudal to the mammillary bodies, where the arachnoid membrane stretches across over the base of the diencephalon.

Interpolaris Related to an interpolar position, meaning situated between or connecting poles.

Interthalamic adhesion A variable connection between the two thalamic masses, across the third ventricle; it is absent in about 20% of human brains.

Interventricular foramen One of the channels that connect the lateral ventricles with the third ventricle, at the midline of the brain.

Interventricular foramina (of Monro) A passage from the third to the lateral ventricle of the brain.

Intracerebral hemorrhage Also known as a cerebral bleed; a type of intracranial bleed that occurs within the brain tissue or ventricles, causing headache, one-sided weakness, vomiting, seizures, decreased consciousness, and neck stiffness.

Intralaminar nuclei A diffuse group of nuclei in the internal medullary lamina of the thalamus.

Intrapontine Within the pons of the brainstem.

Intumescences Enlargements at specific areas of the spinal cord, which are related to innervation of the limbs.

Ipsilateral Also called homolateral; situated on the same side.

Iridotomy Surgical incision into the iris, especially to create an artificial pupil by transverse division of iris fibers.

Isodense Denoting a tissue having a radiopacity similar to that of another or adjacent tissue.

Jaw jerk reflex Also called the masseter reflex; a stretch reflex used to test status of the trigeminal nerve and to help distinguish an upper cervical cord compression from lesions above the foramen magnum.

Jendrassik's maneuver A distraction maneuver used in neurological testing; the patient is asked to hook the fingers of each hand together and try to pull them apart, while the lower extremity reflexes are being tested.

Juvenile Parkinsonism A form of Parkinson's disease presenting in younger patients with atypical features, mostly dystonia and pain without resting tremor.

Juxtallocortex Collective term for regions of the cerebral cortex between the isocortex and allocortex.
Keratoconus An abnormal cone-shaped protrusion of the cornea of the eye; also referred to as astigmatism.
Kernig signs An inability, while laying supine with the thighs and knees flexed, to passively extend the legs at the knees; these are signs of meningeal irritation.
Kinesin A protein that carries the flow of materials from the cell body to the axon terminal as part of anterograde flow.
Kinociliary bundles Nerve fibers related to kinocilia, which are motile, protoplasmic filaments on the free surfaces of cells.
Kyphoscoliosis Backward and lateral curvature of the spine, in vertebral osteochondrosis (Scheuermann's disease).
Labyrinth The inner ear, consisting of the vestibule, cochlea, and semicircular canals.
Labyrinthine artery An artery that is a branch of the basilar artery, supplying the labyrinth.
Lacunes Deficiencies or gaps between structures or tissues.
Lambert-Eaton syndrome A neuromuscular transmission disorder due to a defect in release of acetylcholine from presynaptic nerve terminals, often linked to a long history of cigarette smoking.
Lamina cribosa Also called the cribriform plate; the horizontal plate of the ethmoid bone, part of the floor of the anterior cranial fossa.
Lamina terminalis The thin lamina at the median portion of the forebrain wall; stretching from the interventricular foramen to the base of the optic stalk, and contains the vascular organ that regulates the osmotic concentration of the blood.
Lateral corticospinal tracts Fibers that cross to the opposite side in the corticospinal decussation, descending in the posterior half of the lateral funiculus of the spinal cord.
Lateral preoptic nucleus The part of the hypothalamus related to thermoregulation.
Lateral sulcus The deepest, most prominent cortical sulcus, extending from the anterior perforated - substance; composed of a large posterior ramus, a short anterior ramus, and a short ascending ramus.
Lateral ventricle The cavity in each cerebral hemisphere, derived from the cavity of the embryonic tube, containing CSF.
Laurence-Moon syndrome A rare, autosomal recessive genetic disorder associated with retinitis pigmentosa, spastic paraplegia, and mental disabilities.
Leak channels Ion channels that are always open (nongated channels).
Lens fibers The fibers that make up most of the lens of the eye; they are long, thin, transparent, and firmly packed cells, with diameters of 4–7 μm, and lengths of up to 12 mm; also called laminae.
Lens placode A thickened portion of ectoderm that serves as the precursor to the lens of the eye.
Lenticulostriate arteries Branches of the middle or anterior cerebral arteries, supplying blood of the basal ganglia and much of the internal capsule.
Lentiform nucleus The large, cone-shaped mass of gray matter forming the central core of the cerebral hemisphere.
Letterer-Siwe disease A syndrome of Langerhans cell histiocytosis characterized by skin lesions, ear drainage, lymphadenopathy, osteolytic lesions, and hepatosplenomegaly.
Leukokoria Any condition marked by the appearance of a whitish reflex or mass in the pupillary area behind the lens.
Lewy bodies Intracytoplasmic neuronal inclusion; especially noted in pigmented brainstem neurons and seen in Parkinson's disease, diffuse Lewy body disease, and sometimes in Alzheimer's disease.
Lewy body dementia A common neurodegenerative disease, with gradual, progressive loss of intellectual abilities and a movement disorder resembling Parkinson's disease.
Lipofuscin A yellow-brown pigment consisting of granules composed of lipid-containing residues of lysosomal digestion.
Lipohyalinosis Degenerative changes in small blood vessels, marked by the accumulation of a glassy- or waxy-appearing lipid within the vessel wall.

Lissauer's tract Also called the dorsolateral fasciculus, a long narrow axon tract between the dorsolateral tip of the dorsal horn and outer edge of the spinal cord.
Local potentials Graded potentials.
Locus ceruleus A pigmented eminence in the superior angle of the floor of the fourth brain ventricle.
Longitudinal fasciculi The dorsal, inferior, and medial longitudinal bundles of fibers within the brain.
Lyme disease Also called Lyme borreliosis; an infectious disease signified by initial redness after a tick bite, then fever, headache, fatigue, loss of facial movement, joint pain, headache, palpitations, and inflammation.
Lymphocytic hypophysitis An autoimmune condition involving inflammation of the pituitary gland.
Lymphoepithelioma A type of nasopharyngeal carcinoma, with prominent infiltration of lymphocytes in the tumor area; classified as a class III nasopharyngeal carcinoma; usually associated with the Epstein-Barr virus.
M cells Also called magnocellular cells; neurons within the adina magnocellular layer of the lateral geniculate nucleus of the thalamus that are part of the visual system.
Macroglossia Enlargement of the tongue, either developmental in origin or secondary to a neoplasm or vascular hamartoma.
Macula lutea An irregular yellowish depression on the retina, lateral to and slightly below the optic disk, that receives and analyzes light only from the center of the visual field.
Macula A small strip of epithelium; its movements provide information related to head position or acceleration, and it lies within the utricle and saccule.
Macular degeneration Progressive deterioration of the macula of the retina, which is responsible for central vision; this leads to irreversible loss of central vision, though peripheral vision is retained.
Magnetic gait Also called Bruns ataxia; difficulty initiating forward movement of the feet while they are in contact with the ground; due to frontal lobe disease.
Magnetic resonance angiography (MRA) Imaging of blood vessels using special magnetic resonance sequences to enhance the signal of flowing blood and suppress signals from other tissues.
Magnetic resonance venography (MRV) Imaging of veins to visualize and assess blood flow, toward the diagnosis of vascular disease, and to monitor and evaluate therapeutic interventions.
Magnocellular Characterized by a relatively large size compared to related structures, such as a magnocellular cell compared to a parvocellular cell.
Malathion An organophosphorus insecticide used in topical applications for lice.
Mammillary bodies Small rounded structures on the undersurface of the brain, that as part of the diencephalon, form part of the limbic system.
Mandibulomaxillary advancement Moving of the maxilla and mandible forward, effectively enlarging the airway in the palate and tongue regions; it is successful in treating obstructive sleep apnea.
Manometrics Related to a manometer, the instrument for ascertaining pressures of liquids or gases.
Marantic endocarditis Nonbacterial, thrombotic infection of the endocardium, associated with cancer and other debilitating diseases.
Mastoid aditus The opening within the mastoid bone that leads back from the epitympanic recess into an air space called the tympanic or mastoid antrum.
Meckel's cave The trigeminal cave, an arachnoidal pouch containing cerebrospinal fluid.
Medial group of nuclei One of the two lateral types of nuclei along the length of the brainstem, along with the small group of nuclei.
Medial lemniscus A band of white fibers originating from the gracile and cuneate nuclei and decussating in the lower medulla oblongata.
Medulla oblongata The most inferior part of the brain stem.
Medulla spinalis The spinal cord.
Medullary pyramids Paired white matter structures of the medulla oblongata that contain motor fibers of the pyramidal tracts.

Melanopsin An opsin-like protein, sensitive to light, found in a small number of retinal ganglion cells that are photosensitive; it is believed to synchronize the circadian day-night cycles, control pupil size, and assist in the release of melatonin.

Membrane potential Stored electrical energy across a cell membrane due to the unequal distribution of ions on the two sides of the membrane.

Memories Retained and recalled previously experienced sensations, impressions, information, and ideas.

Memory consolidation A process whereby a memory becomes increasingly resistant, due to physical and psychological changes, to interference from competing or disrupting factors with the passage of time.

Memory engram A lasting memory, due to experiences or repetition of stimuli.

Meniere's disease A condition characterized by recurrent vertigo, hearing loss, and tinnitus.

Meningeal arteries The anterior meningeal branch of the anterior ethmoidal artery, the meningeal branches of the vertebral artery, the middle meningeal artery, and the posterior meningeal artery.

Meningeal carcinomatosis The spread of cancer to the meninges, most common in small cell lung cancer.

Meninges The three membranes covering the brain and spinal cord: the dura mater, arachnoid, and pia mater.

Meningismus A condition in which the patient shows signs of meningitis, but there are no pathological changes in the meninges; it is associated with pneumonia in younger children.

Mesencephalic nucleus A long, thin plate of unipolar neurons through the length of the midbrain; the only known example of primary sensory neurons enclosed in the CNS instead of in a peripheral sensory ganglion.

Mesencephalon One of the three primary vesicles of the developing brain; it becomes the midbrain.

Mesoderm One of the three primary germ layers during early embryonic growth.

Metencephalon The embryonic part of the hindbrain that differentiates into the pons and cerebellum.

Meyer's loop Also called the optic radiation; a nerve pathway from the lateral geniculate body to the visual cortex.

Microatheroma A lipid-containing lesion that forms on the innermost layer of the wall of an artery in atherosclerosis; a plaque, commonly referred to as an atheroma.

Microfilaments Rods of the protein called actin; part of the cytoskeleton.

Microglia Neuroglia that support neurons and phagocytize bacteria and debris.

Micrognathia Abnormal smallness of the jaws, especially the lower jaw.

Microtubules Hollow rods of the protein called tubulin; part of the cytoskeleton.

Midbrain Also known as the mesencephalon; the short part of the brainstem just above the pons and containing the center for visual reflexes.

Middle cerebral artery One of two branches of the internal carotid artery, dividing into three branches.

Middle cranial fossa A butterfly-shaped portion of the internal base of the skull, posterior to the sphenoidal ridges and limbus, and anterior to the crests of the petrous part of the temporal bones and dorsum sellae.

Middle meningeal artery The third branch of the first segment of the maxillary artery; entering the cranial cavity through the foramen ovale, running dorsally in the dura, and branching widely along the side of the skull.

Mitophagy Selective degradation of mitochondria by autophagy, which is natural cell degradation; the condition often occurs following mitochondrial damage or stress.

Mixed hearing loss A form of hearing loss that mixes factors from the other forms, which include conductive, sensorineural, and central hearing loss.

Molar In relation to mass, something that is very large, not molecular.

Monoaminergic Referring to nerve cells or fibers that transmit nervous impulses via a catecholamine or indolamine.

Motor neurons The neurons that conduct impulses from the central nervous system to an effector.

Müller cells Supporting cells that extend through the retina to form its inner and outer limiting membranes.

Multipolar neurons Nerve cells that have many processes associated with their cell bodies.

Myasthenia gravis A condition of episodic muscle weakness and an ability to become easily fatigued, caused by the destruction of the acetylcholine receptors.

Myectomy Surgical excision of a muscle.

Myelencephalon The secondary brain vesicle, at the most posterior region of the embryonic hindbrain, from which the medulla oblongata develops.

Myelomeningocele Hernial protrusion of the spinal cord and its meninges through a defect in the vertebral arch (spina bifida or neural tube defect).

Myerson's sign Also called a glabellar tap sign; a clinical examination finding in which the patient cannot resist blinking when tapped repetitively above the nose and between the eyebrows.

Myokymic Related to myokymia, a benign condition in which there is persistent quivering of the muscles.

Myopia Nearsightedness; objects are seen more closely when close to the eye, while distant objects appear blurred or fuzzy.

Myotome In embryos, the part of the somite that develops into skeletal muscle; also refers to all muscles derived from one somite and innervated by one segmental spinal nerve.

Myringoplasty Surgical closure of the perforation of the pars tensa of the tympanic membrane.

Narcolepsy A disorder marked by excessive daytime sleepiness, uncontrollable sleep attacks, and cataplexy.

Nasal hemianopia Vision loss affecting the medial half of the visual field, which is the half closer to the nose.

Neoangiogenesis The development of new blood vessels; when this occurs in abnormal tissues or positions, it is called neovascularization.

Neocortex The newer, six-layered portion of the cerebral cortex, with the most highly evolved stratification and organization.

Neologisms Newly created words of a patient, often seen in schizophrenia, or the use of an existing word in a new sense, often having meanings only to the individual and no other person.

Nervous spinosus A branch of the mandibular nerve, passing superiorly through the foramen spinosum; distributed to the meninges of the posterior portion of the middle cranial fossa.

Nervus conarii Single or paired areas of the tentorium cerebelli where the postganglionic adrenergic sympathetic axons enter the dorsolateral part of the pineal gland.

Neural crest cells Ectodermally derived cells along the outer surface of each side of the neural tube in early embryonic development; they migrate laterally throughout the embryo and give rise to spinal, cranial, enteric, and sympathetic ganglia as well as pigment cells, Schwann cells, and the adrenal medulla.

Neural fold Either of the paired longitudinal elevations resulting from the invagination of the neural plate in the early developing embryo; they unit to enclose the neural groove, and form the neural tube.

Neural groove The shallow median groove of the neural plate, between the neural folds of an embryo.

Neural plate A thick sheet of ectoderm surrounded by the neural folds, two longitudinal ridges in front of the primitive streak of the developing embryo.

Neural tube The fetal structure that gives rise to the brain, spinal cord, and associated neural structures; formed from ectoderm by day 23 of development.

Neurilemma The part of a Schwann cell outside the myelin sheath.

Neuroablative Destroying or inactivating nerve tissue, using surgery, cauterization, sclerosing agents, lasers, or cryotherapy.

Neuroacanthocytosis syndrome Several neurological conditions in which the blood contains misshapen, speculated red blood cells (acanthocytes).

Neuroblastoma A type of cancer that forms in certain types of nerve tissue; most frequently beginning in one of the adrenal glands.

Neurocysticercosis A disorder caused by the pork tapeworm *Taenia solium*; larvae migrate to the CNS and other tissues, causing cysts, gliosis, inflammation, edema, seizures, neurologic deficits, and personality changes.

Neurofibrils Fine cytoplasmic threads that extend from the cell bodies into the axon of a neuron.

Neurofilaments Intermediate filaments found in the cytoplasm of neurons that help to form the cytoskeleton.

Neuroglandular junction The area of contact between a motor neuron and a gland.

Neuroglia Specialized cells of the nervous system that produce myelin; they maintain the ionic environment and provide growth factors that support neurons. They also provide structural support and play a role in cell-to-cell communication.

Neurohypophysis The neural portion or posterior lobe of the hypophysis.

Neuromuscular junction The synapse between a motor neuron and a skeletal muscle fiber; a myoneural junction.

Neuromyotonia Also called Isaac syndrome; myotonia caused by electrical activity of a peripheral nerve; characterized by stiffness, delayed relaxation, fasciculations, and myokymia.

Neuronotropic Having an affinity for neurons.

Neuropores Cranial and caudal openings, in early embryonic development. The cranial neuropore closes on day 24, and the caudal neuropore closes on day 28. Failure of these to close results in anencephaly and spina bifida, respectively.

Neurosyphilis CNS manifestations of tertiary syphilis, either symptomatic or asymptomatic; due to infection with *Treponema pallidum*.

Neurotransmitters Chemicals that an axon end secretes into a synapse.

Neurotubules Elongated microtubules in the cell body, dendrites, axon, and certain synaptic endings of neurons.

Neurulation The folding process that includes transformation of the neural plate into the neural tube.

Nigrostriatal Projecting from the substantia nigra to the corpus striatum.

Nigrosubthalamic fibers The fibers of the substantia nigra that project below the thalamus.

Nodes of Ranvier Any of the many gaps in the myelin sheath along axons of neurons of the peripheral nervous system.

Notochord A flexible rod made out of a material similar to cartilage, lying along the anteroposterior axis, composed of cells derived from the mesoderm.

Nuchal rigidity A form of rigidness that affects the upper part of the neck.

Nuclear pores Protein-lined channels in the nuclear envelopes.

Nuclei of Meynert Also called the nucleus basalis; a group of neurons located mainly in the substantia innominata of the basal forebrain, rich in acetylcholine, with widespread projections to the neocortex and other brain structures.

Nucleus of Darkschewitsch An ovoid cell group in the anterior central gray substance rostral to the oculomotor nucleus.

Numb-chin sign A neurological condition causing numbness in the mental nerve distribution; this may be a sign of primary breast cancer, lymphoma that has spread to the jaw, a tumor at the base of the skull, or tooth abnormalities.

Nyctalopia Night blindness; the decreased ability to see in reduced illumination.

Nystagmus Rhythmic, oscillating motion of the eyes, usually involuntary; it can be a normal physiological response or a result of a pathologic problem.

Obex The point at which the fourth ventricle narrows to become the central canal of the spinal cord; it occurs in the caudal medulla.

Objective tinnitus A noise produced in the ear that can be heard by another person, particularly when a stethoscope is used.

Obstructive hydrocephalus Also called noncommunicating hydrocephalus; interference with flow of CSF results in enlarged brain ventricles; it is caused by a blockage of the ventricular system and/or spinal canal.

Oculomotor nerves (III) The cranial nerves that provide motor and parasympathetic innervation to some structures within the eye orbits.

Off-center neurons Nerve cells that are stimulated by activation of surrounding areas and inhibited by stimulation of centered areas.

Olfactory bulbs The bulbous distal ends of the olfactory tract, beneath the anterior lobes of the cerebrum.

Olfactory cortex The part of the cerebral cortex, including the piriform lobe and hippocampus formation, concerned with the sense of smell.

Olfactory nerves (I) The cranial nerves for the sense of smell, lying on the floor of the frontal section of the cranial cavity, running forward over the roof of the nose.

Olfactory tracts Nerve-like white bands composed primarily of nerve fibers, originating from the mitral and tufted cells of the olfactory bulb, but also containing scattered cells of the anterior olfactory nucleus.

Oligoclonal bands Immunoglobulins in the cerebrospinal fluid of patients with multiple sclerosis, and sometimes in other neurological conditions.

Oligodendrocytes Types of neuroglia in the CNS that produce myelin.

Olives Prominent bulges along the ventrolateral surface of the medulla oblongata created by most of the olivary nuclei.

Olivocochlear system Part of the auditory system that is involved in descending control of the cochlea.

On-center neurons Nerve cells that are stimulated by activation of centered areas and inhibited by stimulation of surrounding areas.

Operculum The folds of pallium from the frontal, parietal, and temporal lobes of the cerebrum overlying the insula.

Ophthalmic artery A branch of the internal carotid artery, which supplies the eyes and neighboring structures.

Oppenheim reflex Extension of the toes due to scratching the inner leg, or after sudden flexion of the thigh on the abdomen and leg on the thigh; a sign of cerebral irritation.

Opsin The protein portion of the rhodopsin molecules of the retinal rods and cones.

Optic abiotrophy Progressive loss of vitality of the eyes, leading to disorders or loss of function; usually linked to increased aging.

Optic chiasm A structure in the forebrain formed by the decussation of fibers of the optic nerve from each half of each retina.

Optic disc Also spelled optic disk; the intraocular part of the optic nerve, formed by fibers converging from the retina; there are no sensory receptors, and so, no response to stimuli; also called the blind spot.

Optic nerves (II) The cranial nerves originating in the retinas that connect to the brain via a channel in the bone at the back of the eye orbits; they partially cross over in the optic chiasma.

Optic neuritis A vision disorder characterized by inflammation of the optic nerve.

Optic radiation Either of two large, fan-shaped fiber tracts in the brain extending from the lateral geniculate body on either side, to the striate cortex.

Optic tracts Nerve tracts proceeding backward from the optic chiasm, around the cerebral peduncle, dividing into lateral and medial roots that end in the superior colliculus and lateral geniculate body, respectively.

Optical coherence tomography A radiographical method used to obtain high-resolution cross-sectional images of tissues and their defects, such as the structures of the eye.

Oralis An area of the spinal trigeminal nucleus; it is associated with transmission of fine tactile senses from the orofacial region, and is continuous with the principal sensory nucleus of the trigeminal nerve.

Organ of Corti Also called the spiral organ; a ridge of specialized epithelium in the floor of the cochlear duct that contains row of hair cells (the actual auditory receptor cells, innervated by the cochlear nerve).

Orogenital dermatitis Skin inflammation that is caused by contact between the mouth of one person with the genitals of another; this is a component of Strachan's syndrome.

Otic ganglion A parasympathetic ganglion next to the medial surface of the mandibular division of the trigeminal nerve, just inferior to the foramen ovale; its postganglionic fibers supply the parotid gland.

Otoconial crystals Also called otoliths; crystalline particles of calcium carbonate and a protein adhering to the gelatinous membrane of the maculae of the utricle and saccule.
Otolithic membrane A layer of gelatinous substance containing otoconia or otoliths, found on the surface of maculae in the inner ear.
Otoliths Also called the otoconial crystals.
Oval window Also called the vestibular window or fenestra vestibuli.
Oximetry Also called pulse oximetry; a noninvasive method for monitoring a person's oxygen saturation.
P cells Small parvocellular ganglion cells that make up about 90% of the total amounts of ganglion cells; they are more sensitive than M cells to shapes and details of visual images.
Paleocortex The portion of the cerebral cortex that, with the archeocortex, develops in association with the olfactory system.
Pallidal division The portion of the subthalamic nucleus that sends out pallidosubthalamic fibers.
Pallidosubthalamic fibers The fibers of the lateral pallidal division of the subthalamic nucleus.
Pancoast tumor Any carcinoma of the lung apex causing Pancoast syndrome, by invasion or compression of the brachial plexus and stellate ganglion.
Papez circuit A long, circuitous conduction chain in the forebrain, leading from the hippocampus via the fornix to the mammillary and then returning via the anterior thalamic nuclei, cingulate gyrus, and parahippocampal gyrus.
Papilledema A swelling of the optic nerve, at the point where it joins the eye, due to increased intracranial pressure.
Papilloma A circumscribed, benign tumor derived from epithelium.
Parabrachial pontine reticular formation Interconnected, surrounding nuclei of the pons, with ascending and descending pathways reaching the spinal cord.
Paramedian Close to the midline of a body structure.
Paraneoplasia Clinical and biochemical disturbances associated with malignant neoplasms, but not direct relation to invasion by a primary tumor or its metastases.
Parasomnias Primary sleep disorders, in which abnormal events occur during sleep, due to inappropriately timed activation of physiological systems.
Paratenial Referring to a nucleus of the midline nuclear group of the thalamus.
Paratonia Involuntary resistance to passive movement, which increases with the velocity of movement, continuing through the full arc of motion.
Parenchyma The essential or functional elements of an organ.
Parieto-occipital sulcus An oblique groove on the medial surface of the brain, marking the boundary between the parietal and occipital lobes.
Parkinson's disease A slowly progressive disease, usually in later life, with degeneration of the extrapyramidal system, a mask-like face, resting tremor, slowed movements, a festinating gate, abnormal posture, and muscular weakness.
Parkinsonian Pertaining to Parkinsonism.
Parkinsonism Any disorder with the symptoms of Parkinson's disease, or any such symptom complex occurring secondarily to encephalitis, cerebral arteriosclerosis, certain toxin exposure, and neurosyphilis.
Pars externa One of the two parts of the globus pallidus, along with the pars interna; it acquires input from the putamen and caudate, and communicates with the subthalamic nucleus.
Pars interna Part of the globus pallidus, with some shared functions of the pars externa; it sends GABAergic inhibitory output to the thalamus, and sends projections to the midbrain, affecting control of posture.
Pars reticulata Also called the substantia nigra; a dark layer of gray matter separating the tegmentum of the midbrain from the crus cerebri.
Perforating arteries The arteries, usually three in number, named because they perforate the tendon of the adductor magnus to reach the back of the thigh.

Pericallosal artery The continuation of the anterior cerebral artery after the anterior communicating artery; it supplies branches to the cerebral cortex as it passes along the corpus callosum.

Perikaryon The cytoplasm of a cell body, exclusive of the nucleus and any processes.

Perilymph The fluid in the space separating the membranous labyrinth from the osseous labyrinth; it is entirely separate from the endolymph.

Perilymphatic compartment The reservoir for perilymph within the ear.

Perimetry test The part of visual field testing that assesses peripheral vision.

Perineural infiltration Also called paraneural infiltration; it is infiltration adjacent to or along a nerve.

Perivascular space An immunological space between an artery and a vein, and the pia mater, which can be expanded by leukocytes.

Perivenous demyelination Loss of the myelin sheaths surrounding veins.

Pharyngotympanic tube The narrow channel that connects the middle ear with the nasopharynx, equalizing pressure on either side of the eardrum; also called the auditory or eustachian tube.

Phlegmasia alba Also called phlegmasia alba dolens; phlebitis of the femoral vein, with swelling of the lower limb, sometimes following childbirth or an acute febrile illness.

Phorias Tendencies of the eyes to deviate from the normal position when fusional stimuli are absent, or fusion is prevented; a latent or unusually unmanifested form of tropia.

Photoreceptors Specialized cells, including the rods and cones of the retina, which originate nerve impulses when stimulated by light.

Pia mater Also called simply the pia; the delicate, innermost layer of the meninges of the brain and spinal cord.

Pillar cells The cells forming the outer and inner walls of the tunnel in the spiral organ.

Piloerection Commonly called goose bumps, caused by cold, fear, euphoria, or sexual arousal.

Pineal gland A small, cone-shaped organ in the brain that secretes the hormone melatonin.

Pinealoma A tumor of the pineal body, composed of neoplastic nests of large epithelial cells.

Planum semilunatum The area of epithelium bounding the sensory area of the crista ampullaris.

Plasmapheresis A specialized form of dialysis used in treating refractory viral or bacterial encephalitis.

Pleocytosis An increased cell count, especially an increase in white blood cell count, in a body fluid such as the CSF.

Plethysmography The determination of changes in volume by using a plethysmography, which measures and records variations in organs, limbs, and other body parts.

Plexus of Auerbach Unmyelinated fibers and postganglionic autonomic cell bodies in the muscular coat of the esophagus, stomach, and intestines.

Plexus of Meissner An autonomic plexus in the submucosa of the alimentary tube, which regulates mucosal secretions.

Polyglutamine Such as in the *polyglutamine tract*, a protein consisting of a sequence of several glutamine units.

Polymerase chain reaction (PCR) A technique used in molecular biology to amplify one copy or a few copies of a DNA segment across several orders of magnitude, generating thousands to millions of copies of a DNA sequence.

Pons A band of nerve fibers on the ventral surface of the brainstem linking the medulla oblongata and cerebellum with upper brain portions.

Pontine arteries Branches of the basilar artery that serve the pons, divided into medial branches or paramedian pontine branches, and lateral branches or circumferential pontine branches.

Pontine cistern A large space on the ventral aspect of the pons.

Pontine flexure A bend in the axis of the embryological central nervous system; it marks the junction between the metencephalon and myelencephalon.

Pontine hemorrhage Bleeding in the pons, usually seen in hypertensive patients.

Pontomedullary junction An area between the pons and medulla oblongata.

Porus acusticus Also called the internal acoustic opening; it is the opening to the meatus, located inside the posterior cranial fossa of the skull, near the center of the posterior surface of the petrous part of the temporal bone.

Posterior (dorsal) horn In the lateral brain ventricle, the horn that passes forward, laterally and slightly downward, from the corpus callosum into the occipital lobe; in the spinal cord, the dorsal gray matter section that receives light touch, proprioception, and vibrational sensory information.

Posterior cerebral artery The artery formed by the bifurcation of the basilar artery, passing around the cerebral peduncle to reach the medial aspect of the hemisphere.

Posterior choroidal arteries Two branches of the P_2 segment of the posterior cerebral artery, supplying the choroid plexus of the third ventricle and parts of the choroid plexus of the lateral ventricle.

Posterior commissure An axon tract running transversely through the gray matter forming the roof of the cerebral aqueduct in the midbrain.

Posterior communicating artery An artery interconnecting the posterior cerebral and middle cerebral arteries at the base of the brain.

Posterior cranial fossa The part of the cranial cavity located between the foramen magnum and tentorium cerebelli; it contains the brainstem and cerebellum.

Posterior perforated substances Triangular areas forming the floor of the interpeduncular fossa, immediately behind the corpora mammillaria, containing many openings for blood vessels.

Postsynaptic neuron A neuron on the "receiving" side of a synapse.

Postsynaptic potentials Decreased (excitatory) or increased (inhibitory) membrane polarization in the postsynaptic neuron, with repeated stimulation over an excitatory or inhibitor pathway; the neuron will either fire or not respond.

Postural instability A tendency to fall, or the inability to keep oneself from falling; imbalance.

Potential spaces Those that occur between two adjacent structures that are normally pressed together, such as the subdural space.

Prefrontal cortex The anterior part of the brain's frontal lobe.

Prefrontal lobotomy Severing of the white fibers connecting the thalamus to the prefrontal and frontal lobes of the brain; this was formerly performed to treat abnormal behaviors.

Preganglionic sympathetic nerve Related to the nerve fibers supplying a ganglion, especially one of the autonomic nervous system, those that function as part of the sympathetic nervous system.

Pregeniculate nucleus Also called the ventral lateral geniculate nucleus; a small cell group located rostral to the dorsal lateral geniculate nucleus.

Premotor cortex An area of the motor cortex within the frontal lobe, just anterior to the primary motor cortex.

Prerubral field Also called one of the fields of Forel; the three regions of the subthalamus.

Presbycusis Progressive, bilaterally symmetrical perceptive hearing loss occurring usually after age 50, caused by structural changes in the organs of hearing.

Presynaptic neuron The neuron releasing neurotransmitter at a synapse on the "sending" side of a synapse.

Pretectal region A narrow, transversely oriented rostral area of the mesencephalic tectum, bounded caudally by the superior colliculus, rostrally by the habenular trigone, and laterally by the pulvinar thalami.

Primary motor cortex Also called Brodmann area 4; located in the dorsal portion of the frontal lobe, working in association with other motor areas.

Primary somatosensory cortex Also called the postcentral gyrus; the convolution of parietal lobe bounded in front by the central sulcus.

Primary vesicles The three earliest subdivisions of the embryonic neural tube, including the prosencephalon, mesencephalon, and rhombencephalon.

Principal sensory nucleus The portion of the trigeminal nerve that carries touch, two-point discrimination, and pressure signals from the face and mouth.

Prodrome An early symptom indicating the onset of a disease.

Propagation The process of increasing or causing to increase.
Proprioceptive The ability to sense the relative position of the body; it is provided by proprioceptors in skeletal muscles, tendons, and the fibrous membrane of joint capsules.
Proprioceptors Sensory receptors that sense changes in muscles, tendons, and body position.
Proptosis A bulging protrusion of a body organ or area.
Prosody Speech variables, including speed, pitch, and relative emphasis, which distinguish vocal patterns.
Pseudobulbar affect Expression of emotions that are out of proportion to feelings experienced by the patient.
Pseudoxanthoma elasticum An inherited connective tissue disorder in which elastic fibers of the skin, eyes, and cardiovascular system gradually become calcified and inelastic.
Pterion The region where the frontal, parietal, temporal, and sphenoid join together on the side of the skull, just behind the temple.
Pterygopalatine ganglia The ganglia related to the pterygoid process and palatine bone.
Purkinje cells Nerve cells with large, pear-shaped bodies, and many dendrites, found in the middle layer of the cerebellar cortex.
Purulent labyrinthitis Also called acute suppurative labyrinthitis; a form in which pus enters the labyrinth, usually through a fistula after a middle ear infection, or through temporal bone erosion due to meningitis.
Putamen The larger, more lateral part of the lentiform nucleus.
Pyramidal cells Neurons of the cerebral cortex of a triangular shape, with a long apical dendrite directed toward the cortex surface, lateral dendrites, and a basal axon descending to deeper layer.
Pyramids A pair of thick bands along the anterior surface of the medulla oblongata, where many axons decussate and enter the descending lateral corticospinal tracts on the opposite side of the spinal cord.
Radiculomedullary veins The veins of the nerve roots and spinal cord.
Radiculopathies Diseases of the nerve roots.
Ramsay-Hunt syndrome Also called geniculate neuralgia; facial paralysis with otalgia, a vesicular eruption in the external canal of the ear that may extend to the auricle, due to herpes zoster virus infection of the geniculate ganglion.
Raphe nuclei A variety of nerve cell groups in and along the median plane of the medulla oblongata, nucleus raphes pallidus, and caudal portions of the nucleus raphes magnus.
Recesses Small cavities or depressions in a body organ, part, or structure.
Red nucleus An oval mass of gray matter that appears pink in fresh specimens, in the anterior part of the tegmentum, extending into the posterior part of the hypothalamus.
Refractory period The period following stimulation during which a neuron or muscle fiber will not respond to another stimulus.
Reissner's membrane Also called the vestibular membrane; located inside the cochlear of the inner ear, it separates the cochlear duct and vestibular duct; with the basilar membrane, it creates the endolymph compartment.
Relative refractory period Follows the absolute refractory period; it is the interval when a threshold for action potential stimulation is markedly elevated.
Restiform body One of two large, cord-like bundles of nerve fibers in each cerebellar hemisphere, connecting the cerebellum with the medulla oblongata; also called the inferior cerebellar peduncle.
Resting membrane potential The voltage that exists across the plasma membrane during the resting state of an excitable cell; ranges from -90 to -20 mV depending on cell type.
Resting tremor A tremor occurring in a relaxed and supported limb, such as in Parkinson's disease.
Reticular formation A large, vaguely defined structure made up of intermingled gray and white matter, extending through the central core of the brainstem, into the diencephalon.
Reticular lamina Connective tissue with reticular fibers, formed by the apical ends of hair cells and apical processes of supporting cells; it is blanketed by the tectorial membrane, and maintains specific electrochemical gradients of the ear.

Retinitis pigmentosa A group of inherited disorders that slowly lead to blindness due to abnormalities of the photoreceptors, primarily the rods, in the retina.

Retinoblastoma A malignant tumor of the retina that occurs mostly in young children.

Retrognathia A condition in which one or both jaws recede with respect to the frontal plane of the forehead.

Retrograde flow The flow of fluid in a direction opposite of the normal direction.

Rett syndrome A genetic condition, mostly affecting females, involving small hands and feet, and deceleration of the rate of head growth, plus stereotyped hand repetitive movements and seizures.

Reuniens Referring to a nucleus of the thalamic midline nuclear group.

Reversible cerebral vasoconstriction syndrome A condition of narrowing of the cerebral blood vessels, which can be reversed with treatment.

Rhodopsin Visual purple.

Rhombencephalon The caudal portion of the developing brain, which constricts to form the metencephalon and myelencephalon; it includes the pons, cerebellum, and medulla oblongata, and is also called the hindbrain.

Rhombic lips Structures formed by thickening of the lateral areas of the alar plate in the rostral metencephalon; parts of these lips form the cerebellum-related nuclei, a transverse ridge, and eventually, the cerebellum.

Righting reflexes Also called labyrinthine righting reflexes, which correct body orientation when it is taken out of the normal upright position.

Rinne test A tuning fork test that compares duration of perception by bone and by air condition; normally, the fork is heard twice as long by air conduction as by bone conduction.

Romberg's sign A tendency to sway or fall while standing upright with the feet together, the arms stretched out, and the eyes closed.

Rosenthal's canal Also called the spiral canal of the cochlea; the winding tube of the bony labyrinth that is incompletely divided into two compartments by a winding shelf of bone called the bony spiral lamina.

Rostral Toward the rostrum, meaning beak or a similar anatomical structure.

Rostrocaudal Along the long (head-to-tail) axis of the body.

Round window Also called the cochlear window or fenestra cochlea; a round opening in the middle ear covered by the secondary tympanic membrane.

Rubeosis iridis A condition of new formation of vessels and connective tissue on the surface of the iris.

Saccule The smaller of the two divisions of the membranous labyrinth, which communicates with the cochlear duct via the ductus reuniens.

Saltatory propagation Impulse conduction along an axon that seems to jump from one node to the next.

Sarcopenia Degenerative loss of skeletal muscle mass, quality, and strength due to aging; it is a component of the frailty syndrome, and often of cachexia.

Satellite cells Glia in the peripheral nervous system that support neurons in ganglia.

Schistosomiasis A disease in which necrotizing eosinophilic granulomas develop within the brain, causing seizures, increased intracranial pressure, and diffuse, focal neurologic deficits.

Schwann cells A type of neuroglia that surrounds an axon of a peripheral nerve, forming the neurilemma and myelin.

Scotoma Area of lost or depressed vision in the visual field.

Secondary neurulation A process of the formation of the sacral spinal cord, when the neural tube has closed, and a second cavity extends into the solid cell mass at its caudal end; it occurs during weeks 5 and 6 of development.

Secondary vesicles The five brain vesicles formed by specialization of the prosencephalon (telencephalon and diencephalon), the mesencephalon, and the rhombencephalon (metencephalon and myelencephalon) in later embryonic development.

Segmental dystonias Impairments of muscular tonus, but in segmented areas of a body part or area.

Semicircular canals The passages in the body labyrinth that control the sense of equilibrium.

Semicircular duct One of the three ducts that make up the membranous labyrinth of the inner ear.

Semilunar ganglion Also called the trigeminal ganglion; on the sensory root of the CN V, in a cleft of the dura mater on the anterior surface of the temporal bone; the ophthalmic, maxillary, and part of the mandibular nerves emerge from it.

Sensorineural hearing loss The form associated with some pathological change in structures within the inner ear, or in the acoustic nerve.

Septal vein A vein related to a septum; in the cerebrum, a septal vein runs posteriorly across the septum pellucidum.

Septum pellucidum A thin membrane of nervous tissue forming the medial wall of the lateral brain ventricles.

Septum transversum A thick mass of cranial mesenchyme, formed in the embryo, that gives rise to parts of the thoracic diaphragm and ventral mesentery of the foregut.

Seroconversion Development of antibodies in blood serum due to infection or immunization.

Sjaastad syndrome Also called chronic paroxysmal hemicrania, a severe and debilitating unilateral headache usually affecting the area around the eye, with no neurological symptoms.

Small group of nuclei One of the two lateral types of nuclei along the length of the brainstem, along with the medial group of nuclei.

Solitariothalamic tract Also called the solitary tract; a compact fiber bundle extending longitudinally through the posterolateral region of the medulla; made up of fibers of the vagus, glossopharyngeal, and facial nerves.

Solitary nuclei Various nuclei of termination of visceral afferent fibers of the facial, glossopharyngeal, and vagus nerves.

Somatosensory association cortex The association cortex involved in sensations received in the skin and deep tissues.

Somatotopic organization Organized in relation to specific areas of the body; relates to the homunculus used for mapping CNS control of body areas.

Somnambulism Sleepwalking.

Spasmodic dystonias Impairment of muscular tonus combined with muscle spasms.

Spatial summation Increasing the number of fibers stimulated.

Sphenopalatine ganglion Also called the pterygopalatine ganglion, within a fossa of the sphenoid bone.

Spina bifida aperta Spina bifida associated with a meningocele, meningomyelocele, or myelocele; also called spina bifida cystica or manifesta.

Spinal cord ischemia An insufficient supply of blood to the spinal cord.

Spinal root Any of the roots of the accessory nerve that arise from the ventrolateral part of the first five segments of the spinal cord.

Spinal trigeminal nucleus A nucleus in the medulla that receives information about deep touch, pain, and temperature from the ipsilateral face.

Spiral ligament The periosteum, forming the outer wall of the ductus cochlearis; it is greatly thickened.

Splanchnic nerves Mostly preganglionic fibers in a network, with filaments innervating the penis and clitoris, along with other abdominal cavity structures.

Split-brain syndrome Disconnection of the two hemispheres, either by congenital absence or the corpus callosum, or by surgery as a treatment for epilepsy.

Static equilibrium The type of equilibrium characterized by constant properties, such as lack of movement, temperature regulation, and pressure regulation.

Stellate cells Star-shaped cells such as astrocytes or Kupffer cells, with many filaments extending radially.

Stenting Insertion or application of a stent, which is an appliance or material intended to support a graft or keep a passage open.

Stereociliary bundles The nerve fibers of the homogeneous cilia in simple membrane coverings, found on the free-surface hair cells.

Stereognosis The sense by which the form of objects is perceived through touch.

Stereotaxically In relation to stereotaxis, which refers to precise positioning, especially of discrete brain areas controlling specific functions.

Strabismus A condition in which the eyes do not point in the same direction; also referred to as a tropia or squint.

Stratum lacunosum-moleculare A layer of the hippocampus that, along with the stratum radiatum, contains the perforant pathway.

Stratum oriens A layer of the hippocampus that contains basal dendrites and basket cells.

Stria vascularis The upper portion of the spiral ligament, which forms the outer wall of the cochlear duct, containing many capillary loops and small blood vessels; it produces endolymph for the scala media.

Striatum A basal nuclei structure formed by the caudate nucleus and putamen; it is named because fibers of the internal capsule pass through, creating a striped appearance.

Striola The narrow central area of the utricular macula, where orientations of the tallest stereocilia and kinocilia change.

Sturge-Weber syndrome A congenital syndrome of the face, commonly called port-wine stains; angiomas that lead to anoxia; late glaucoma; and often, intracranial calcification, mental retardation, and epilepsy.

Stylopharyngeus The muscle connecting the styloid process and pharynx, which elevates and dilates the pharynx.

Subarachnoid cisterns Small subarachnoid spaces serving as reservoirs for CSF.

Subarachnoid hemorrhage An abnormal, dangerous condition in which blood collects beneath the arachnoid mater, within the subarachnoid space; this may lead to stroke, seizures, brain damage, harmful biochemical events, and death.

Subarachnoid space The space between the arachnoid and pia mater, containing CSF.

Subclavian steal syndrome Cerebral or brainstem ischemia resulting from the diversion of blood flow from the basilar artery to the subclavian artery, in the presence of occlusive disease of the proximal part of the subclavian artery.

Subdural space A narrow fluid-containing space, often only a potential space, between the dura mater and arachnoid.

Subjective tinnitus Noise in the ears, including ringing, buzzing, or roaring that is only heard by the patient.

Submandibular ganglia The ganglia situated below the mandible.

Substance P A peptide neurotransmitter made of 11 amino acyl residues, normally in minute quantities in the nervous system and intestines; primarily involved in pain transmission.

Substantia gelatinosa The substance sheathing the posterior horn of the spinal cord, and lining its central canal.

Substantia nigra The dark layer of gray matter separating the tegmentum of the midbrain from the crus cerebri.

Subthalamic nucleus A small, lens-shaped nucleus in the brain that is functionally part of the basal ganglia system; located ventral to the thalamus.

Subthalamonigral fibers The fibers of the subthalamic nucleus that project to the substantia nigra.

Subthalamopallidal fibers The fibers of the subthalamic nucleus that project to the medial globus pallidus.

Subthalamus The part of the diencephalon wedged between the thalamus on the posterior side, and the cerebral peduncle anteriorly, lateral to the posterior half of the hypothalamus, from which it cannot be sharply delineated.

Sulcocommissural Related to the sulci of the commissural structures of the spinal cord.

Sulcus limitans In the floor of the fourth ventricle, the structure that separates the cranial nerve motor nuclei from the sensory nuclei.

Sulcus A shallow to slightly deep groove on the cerebral hemispheres.

Summation Increased force of contraction by a skeletal muscle fiber when a twitch occurs before the previous twitch relaxes.

Sundowning Appearance of confusion, agitation, and other severely disruptive behavior coupled with inability to remain asleep, occurring solely or markedly worsening at night; often seen with dementia or other mental disorders.

Superchiasmatic nucleus Also called the suprachiasmatic nucleus; a tiny region in the hypothalamus, directly above the optic chiasm, responsible for controlling the circadian rhythms.

Superficial veins Those that course in the subcutaneous tissue, emptying in to deep vein, forming prominent systems of vessels in the limbs; they are usually not accompanied by arteries.

Superior cerebellar artery A branch of the basilar artery, sending branches to the midbrain, pons, medial cerebellum, and deep cerebellar nuclei.

Superior cistern Also called the quadrigeminal cistern; an enlarged portion of the subarachnoid space immediately superior to the tectum of the mesencephalon; it contains portions of the internal cerebral veins and other blood vessels.

Superior colliculi Two small, round elevations on the dorsal aspects of the midbrain, just below the thalamus.

Superior ganglion The ganglion above the glossopharyngeal or vagus nerves; also refers to the superior cervical ganglion.

Superior orbital fissure The upper portion of the space between the floor and lateral wall of the orbit, which serves as a conduit for nerves and blood vessels.

Supernatant Located above or on top of something; also, a clear upper part of a mixture after it has been centrifuged.

Supracallosal striae Narrow ridges located above the corpus callosum.

Suprasellar Above or over the sella turcica.

Supratentorial Related to a location above a tentorium; also, often used to describe functional symptoms.

Supratrochlear Above a trochlea, denoting a nerve.

Synapse A functional connection between the axon of one neuron and the dendrite or cell body of another neuron, or the membrane of another cell type.

Synaptic cleft A narrow extracellular space between the cells at a synapse.

Synaptic delay The time required for a signal to cross a synapse between two neurons.

Synaptic fatigue Short-term synaptic depression or plasticity; it causes the temporary inability of neurons to fire and transmit input signals.

Synkineses Associated movements that are unintentional and accompany voluntary movements.

Syringomyelia A slowly progressive syndrome of cavitation in the central (usually cervical) spinal cord segments.

Tabes dorsalis A slowly progressive nervous disorder from degeneration of the dorsal columns of the spinal cord and sensory nerve trunks.

Tachyzoites Rapidly multiplying states in the development of the tissue phase of certain coccidial infections, such as *Toxoplasma gondii*.

Tardive dyskinesia A mostly irreversible neurologic disorder or involuntary movements, caused by long-term use of antipsychotic or neuroleptic drugs.

Tectohabenular fibers Crossed fibers that accompany the interconnections between the hippocampal cortices and amygdaloid complexes.

Tectorial membrane The thin, jelly-like membrane projecting from the vestibular lip of the osseous spiral lamina, overlying the spiral organ of Corti.

Tectorins Proteins that are present in the tectorial membrane.

Tectotegmentospinal tracts Also called the tectospinal tracts; those that coordinate head and eye movements as part of the extrapyramidal system.

Tectum The roof-like covering of any body structure.
Tegmentum A covering of a body structure.
Telangiectatic microangiopathy A disorder of the small blood vessels involving dilatation.
Telencephalon The anterior subdivision of the primary forebrain that develops into the olfactory lobes, cerebral cortex, and basal nuclei.
Telodendria The ends of dendrites.
Temporal summation Increasing the frequency of nerve impulses in each fiber.
Tensor veli palatini A muscle originating in the scaphoid fossa of the pterygoid process, wall of the auditory tube, and spine of the sphenoid; it innervates the mandible, tensing the soft palate and opening the auditory tube.
Tentorial nerve The meningeal branch arising in a recurrent fashion from the intracranial portion of the ophthalmic nerve, supplying the tentorium cerebelli and supratentorial falx cerebri.
Terminomas Types of germ-cell tumors.
Tetraparesis Muscular weakness affecting all four extremities.
Thalamic nuclei The nuclei of the thalamus.
Thalamostriate vein The major vein draining the caudate nucleus and thalamus; it merges with the choroid vein, forming the internal cerebral vein.
Thalamus A mass of gray matter in the diencephalon of the brain.
Theta waves Brain oscillation patterns of 4–7 Hz, normally present in minimal amounts in the temporal lobes; they are common in young children, but in older children and adults, appear in light sleep, drowsiness, and meditation.
Third ventricle A narrow, vertically oriented, irregularly quadrilateral cavity in the midplane, extending from the lamina terminalis to the rostral opening of the mesencephalic aqueduct.
Threshold The point at which a stimulus is great enough to produce an effect.
Thrombosis Formation, development, or presence of a thrombus whenever flow of blood in arteries or veins is impeded.
Thrombus A stationary blood clot along the wall of a blood vessel, often causing vascular obstruction.
Thunderclap headache Severe, sudden nonlocalizing head pain not associated with any abnormal neurologic findings; it occurs days to weeks before an episode of intracranial bleeding.
Tight junctions Joined areas between astrocytes in the central nervous system.
Tinnitus Ringing, buzzing, or roaring in the ears of a patient; it may be subjective (only heard by the patient) or objective (able to be perceived by another individual).
Todd's paralysis Weakness, usually on one side of the body, after a seizure; it lasts usually for a few minutes or hours, but sometimes for several days.
Torticollis A type of movement disorder in which the muscles controlling the neck cause sustained twisting or frequent jerking.
Trabeculoplasty Photocoagulation of the trabecular meshwork of the eye, in the treatment of glaucoma.
Tractography A 3D modeling technique used to visually represent neural tracts using diffusion-weighted images, MRI, and computer image analysis.
Tractus solitarius nucleus The nucleus that receives taste sensations carried along the facial nerve; the tractus solitarius is a thin tract of visceral sensory axons from cranial nerves VII, IX, and X.
Transducin A G protein of the disk membrane of the retinal rods that assists in triggering visual nerve impulses.
Transependymal Across the ependyma, which is the outer covering of a body structure.
Transient ischemic attack (TIA) Often described as a mini-stroke, but with symptoms usually disappearing within a few minutes; like a stroke, it is caused by disruption of blood flow to the brain.
Transient monocular blindness An episode of total or partial loss of vision in one eye, due to ischemia of the eye, lasting several minutes or longer; also called amaurosis fugax or transient monocular vision loss.
Transtentorial Passing across or through the tentorial notch or tentorium cerebelli.

Transverse cerebral fissure The triangular space between the corpus callosum and fornix above, and the dorsal surface of the thalamus below; it is bounded laterally by the choroid fissure of the lateral ventricle and lined by pia mater.

Transverse pontine fibers The fibers of the pons that extend from side to side.

Transverse sinus(es) A paired, dural venous sinus that drains the confluence of sinuses, running along the occipital attachment of the tentorium cerebelli, and terminating in the sigmoid sinus.

Trigeminal autonomic cephalgia A primary headache occurring with pain on one side of the head, in the trigeminal nerve area, and same-sided symptoms in autonomic systems such as eye watering and redness or drooping eyelids.

Trigeminal nerves (V) The cranial nerves arising in the pons, composed of sensory and motor fibers, having ophthalmic, maxillary, and mandibular regions.

Trigeminal neuralgia A disorder of the trigeminal nerve, causing sharp, stabbing pain in the cheek, lips, gums, or chin on one side of the face.

Trigeminocerebellar fibers Located in the inferior cerebellar peduncles, they transmit proprioceptive information from the face to the cerebellum.

Trigeminovascular system Small pseudounipolar sensory neurons originating from the trigeminal ganglion and upper cervical dorsal nerve roots; they innervate cerebral vessels, including those of the pia, dura mater, and large venous sinuses.

Trisomy 13 A common chromosomal anomaly in which there are three copies of chromosome 13, resulting in severe congenital malformations, structural brain abnormalities, and polydactyly.

Trochlear nerves (IV) The cranial nerves originating in the midbrain that are essential for eye movement and eye muscle sensibility.

Tropias Abnormal deviations of the eyes.

Truncal hypotonia Lack of muscle tone in the trunk of the body.

Tuberoinfundibular Related to a population of dopamine neurons projecting from the arcuate nucleus in the tuberal region of the hypothalamus.

Tubular vision Also called tunnel vision; a constriction of the visual field as if looking through a hollow cylinder or tube.

Tunnel of Corti The spiral passage in the organ of Corti.

Tympanic cavity The major portion of the middle ear; consisting of a narrow air-filled cavity in the temporal bone containing the auditory ossicles.

Tympanic duct Also called the scala tympani.

Tympanic membrane Also called the eardrum or tympanum.

Tympanogram A graphic representation of the compliance and impedance of the eardrum and ossicles of the middle ear.

Tympanometry Indirect measurement of the compliance and impedance of the eardrum and ossicles.

Uncal herniation Also called transtentorial herniation; a downward displacement of medial brain structures through the tentorial notch by a supratentorial mass, causing pressure upon the underlying structures, including the brainstem.

Uncrenated Not containing a structural notch.

Unipolar neuron A neuron that has a single process associated with its cell body.

Upregulating To increase the responsiveness of a cell or organ to a stimulus; or to increase the number of receptors on a cell membrane.

Utricle The larger of the two divisions of the membranous labyrinth of the inner ear.

Vagus nerves (X) The cranial nerves arising from the medulla oblongata that supply and control the throat, larynx, bronchi, lungs, esophagus, heart, stomach, and upper intestines.

Valsalva maneuver Attempting to forcibly exhale while keeping the mouth and nose closed; it is used to equalize pressure on either side of the eardrum.

Vasculitides Various forms of vasculitis, disorders that destroy blood vessels by inflammation.

Vasculopathy Any damage to vessels, such as blood vessels, within the body.
Venous angle An angle formed by the junction of the internal jugular vein and subclavian vein, at either side of the neck; in neuroradiology, the angle of union of the superior thalamostriate vein with the internal cerebral vein.
Venous dural sinuses The large endothelia-lined collecting channels into which veins of the brain and inner skull empty, which then empty into the internal jugular vein.
Ventral intermediate nucleus The middle third of the ventral nucleus receiving projections from the contralateral half of the cerebellum and ipsilateral globus pallidus.
Ventral median fissure Also called the anterior median fissure; it contains a fold of pia mater and extends along the length of the medulla oblongata.
Vermis A worm-like structure, such as the *vermis cerebelli*.
Verrucae hippocampi Wart-like elevations on the hippocampus.
Vertebral arteries The first branches of the subclavian artery, divided into the prevertebral, cervical, suboccipital, and intracranial portions.
Vertigo A sensation of rotation or movement of one's self (subjective vertigo) or of one' surroundings (objective vertigo) in any plane; it is not identical to dizziness, and has a variety of causes.
Vestibular crest A ridge on the inner wall of the vestibule of the labyrinth; also called the crista vestibuli.
Vestibular nerve The branch of the vestibulocochlear nerve that innervates the vestibule and semicircular canals, maintaining balance.
Vestibular nuclei A group of four main nuclei in the lateral region of the hindbrain, including the inferior vestibular nucleus, medial vestibular nucleus, lateral vestibular nucleus, and superior vestibular nucleus.
Vestibular pyramid The anterior end of the crista vestibuli.
Vestibule The central part of the bony labyrinth of the inner ear.
Vestibulocochlear nerves (VIII) The cranial nerves that each divide to form the cochlear nerve and vestibular nerve; also called the acoustic or auditory nerves.
Visceral motor neurons Those that stimulate all peripheral effectors except skeletal muscles; they innervate cardiac and smooth muscle, adipose tissue, and glands.
Visceral sensory neurons Those that monitor the internal environment and organ systems.
Visceromotor neurons The neurons concerned in essential movements of the viscera.
Viscus Any large interior organ in any of the great body cavities, especially those of the abdomen.
Visual association area Either the second or the third visual area of the visual cortex in the occipital lobe of the cerebral cortex.
Vitreoretinal traction Friction on the internal limiting membrane of the retina by adherent vitreous fibrils in vitreous humor detachment.
Waldenström's macroglobulinemia A rare, chronic cancer of the immune system, characterized by hyperviscosity (thickening) of the blood.
Wallerian degeneration A process of disintegration of an axon that occurs when it is crushed or severed, and cannot receive nutrients from the cell body.
Wartenberg sign Flexion of the thumb when attempting to flex the four fingers against resistance; in the cases of cerebral tumor, there may be intense itching of the tip of the nose and nostrils.
Waterhouse-Friderichsen syndrome The malignant or fulminating form of meningococcal meningitis, with sudden onset and short course, fever, collapse, cyanosis, hemorrhaging, and coma.
Weber syndrome A midbrain tegmentum lesion characterized by ipsilateral oculomotor nerve paresis and contralateral paralysis of the extremities, face, and tongue.
Weber's test A procedure for differentiating conductive hearing impairment from sensorineural impairment; a vibrating tuning fork is applied to several points in the midline of the forehead to determine hearing in both ears.
Werdnig-Hoffman disease Autosomal recessive disease of degeneration of the CNS cells, leading to atrophy of skeletal muscles and flaccid paralysis.

Wernicke's area An area in the posterior left hemisphere of the brain cortex at the Sylvian fissure; it is important for language comprehension and speech.

Wernicke-Korsakoff syndrome A disorder caused by a lack of thiamine, due to long-term alcohol abuse, in which the patient has difficulty walking, seeing, or thinking clearly.

Wilson disease A rare genetic disease caused by a defect in copper metabolism, with the accumulation of copper in the liver, brain, kidneys, and cornea, resulting in neurological symptoms and liver disease.

Wohlfart-Kugelberg-Welander disease Juvenile spinal muscular atrophy.

Xanthochromia Yellowish discoloration of the skin or spinal fluid; in the CSF, this usually indicates hemorrhage into the CNS and is due to xanthematin, a yellow pigment derived from hematin.

Zona arcuate Also called the zona arcuata or inner tunnel; the triangular canal between the inner and out pillars of Corti in the organ of Corti.

Zona incerta A horizontally elongated region of gray matter in the subthalamus below the thalamus.

Zona pectinata Also called the pectinate zone; the outer two-thirds of the basilar membrane of the cochlear duct.

Index

Note: Page numbers followed by *f* indicate figures, *t* indicate tables, and *b* indicate boxes.

A

Abdominal reflexes, 607–608
Abdominal wall, anterolateral thorax and, 592–593
Abducens nerves (VI), 652–654
   extrinsic eye muscles, 325
   eyes movement via, 325
   function of, 324–325
   intrinsic eye muscles, 325
Aberrant corticospinal fibers, 346–347
Accessory meningeal artery, 100
Accessory nerves (XI), 330
   external branch, 330
   internal branch, 330
Accessory optic tract, 506–507
Accommodation, 428, 652–653
Acetylcholine (ACh), 377, 549–552, 550–551*t*
Acetylcholinesterase (AChE), 549–552
   staining, 542, 545*b*
Acetyl coenzyme A, 549–552
Aching pain, 534
Achromatopsia, 458
Acoustic, auditory, and stato-acoustic nerves. *See* Vestibulocochlear nerves (VIII)
Acoustic neuroma, 409
   clinical manifestations, 410
   diagnosis, 411
   pathophysiology, 409
   treatment, 411, 414–415*b*
Actinomyces infections, 118
Action potentials, 21–22
   generation of, 22–24
   propagation, 24–25
   rate of action potential generation, 33–34
Active zone, neurotransmitter, 562
Acute angle-closure glaucoma, 443–444
Acute confusional state. *See* Delirium
Acute vestibular neuronitis, 408
   stroke, 393
Adenoid cystic carcinoma, of parotid gland, 358*b*
Adenosine, 558–559
Adenosine triphosphate (ATP), 558–559
Adenyl cyclase, 555
Adherens junctions, 370–371
Adhesive meningomyelitis, 619–620
Adiculomedullary veins, 142
Adie pupil, 653
Adie tonic pupil syndrome, 540–541, 544*b*
Adrenal chromaffin cells, 527
Adrenaline, 555
Adrenal medulla, 527, 552
Adrenergic divisions, 534
Adrenergic fibers, 537*b*
Adult-onset spinal muscular atrophies, 631
Afferent mossy fibers, 505–506
Aflibercept, 448
Aganglionosis, 541
Age-related macular degeneration, 446–448, 448*b*
Ageusia, 331, 489–490
Aggressiveness, 487
Agnosia, 201–202
Agraphia, 191
Akinesia, 246–247
Alexander disease, 123
Alexander law, 654
Alexia, 191
All-trans retinal, 422, 432
Alpha motor neurons, 611
α_1-receptors, 555
Alpha-synuclein, 245–246
α_2-receptors, 555
Alpha waves, 185
Altered mental status, 84–85
Alzheimer's disease
   causes of, 193
   clinical manifestations, 195–196
   diagnosis, 196–197
   nonmedication interventions, 197*t*
   pathophysiology, 194–195
   risk factors, 193
   treatment, 197–199
Amacrine cells, 419–420
Amantadine, 624
Amaurosis fugax, 157–158
Amblyopia, 334–335
   See also Strabismus
Amino acid, 550–551*t*, 557

Amino acid decarboxylase, 556
Amitriptyline, 624
Amplitude, 382
Ampulla, 374
Ampullary crest, 374, 388
Amsler grid, 447–448
Amygdala, 467, 471–472
Amygdaloid body, 242
Amygdaloid complex, 472–473
Amygdaloid fibers, 474
Amygdaloid nuclear complex, 467, 471–472
Amyloid angiopathy, 194
Amyotrophic lateral sclerosis (ALS), 624–626
Anal wink reflex, 664
Anastomosis of Oort, 381–382
Anencephaly, 67–68
Aneurysm
  cerebral, 168–170
Anger, 487
Angiogenic vessels, 447
Angle-closure glaucoma, 443–444
Ankle jerk, 662
Anosmia, 331–332, 489–491
ANS. See Autonomic nervous system (ANS)
Antalgic gait, 665
Anterior horns, 588
  motor neurons of, 611
Anterior inferior cerebellar artery (AICA), 134, 350–351
Anterior ischemic optic neuropathy (AION), 438–439, 441f
Anterior limb, 277
Anterior lobe, 497
Anterior median fissure, 588
Anterior nuclei (AN), 271–272, 279
Anterior roots, 588–590
Anterolateral pathway, 600–601
Anterolateral sulcus, 588–590
Anterolateral system, 599–600
Anterolateral thorax
  and abdominal wall, 592–593
Anticoagulants, 144–145
Antidepressants plus counseling, 624
Anti-glutamate, 626
Anti-vascular endothelial growth factors (anti-VEGFs), 448
Antrum (trigone), 105, 363–364
Anxious depression, 574–575
Apathy, 487–488

Aphasia, 191–192
Apnea, sleep, 225–227
Applanation tonometry, 426
Apraxia, 190–191
Apraxic gait, 665
Aqueductal stenosis, 308
Aqueduct atresia and stenosis, 71
Aqueduct of Sylvius, 122
Aqueous humor, 425–426, 443–444
Arachnoid granulations, 97
Arachnoid mater
  barrier function, 98
  functions, 97
  meninges of spinal cord, 103–104
Arachnoid trabeculae, 103–104
Arborvitae, 83–84
Arboviruses, 617–618
Archicortex, 179
Arcuate nucleus, 279
Areolar tissue, 103
Argyll Robertson pupil, 538–540, 653
Arnold-Chiari phenomenon, 302–304, 303f
  asymptomatic, 302
  cerebellar tonsils, 303
  clinical manifestations, 303–304
  diagnosis, 304
  myelomeningocele, 302–303
  symptomatic, 302
  treatment, 304
  type II, 71
Arterial vasocorona, 141
Arteries
  of brain, 133f
    carotid arteries, 131–133
    cerebral arterial circle, 135–136
    meningeal arteries, 136
    vertebral and basilar arteries, 134–135
  of spinal cord, 140–141
Arteriovenous malformation, 306
Artery of Adamkiewicz, 141, 614
Arytenoepiglottic folds, 227
Ascending pathways, 591, 595–601
Ascending pharyngeal artery, 100
Asplenia, 112
Association cortex, 470
Astereognosis, 569
Astrocytes, 4–6
Astrocytomas, 40–43
Astrocytosis, 40

Asynchronous discharge, 356
Ataxia, 508–509, 513–514*b*, 598, 661
Ataxic dysarthria, 511
Atherosclerosis, 142–143
Athetosis, 244, 657
Audiograms, 397–398
Audiometer, 385, 397–398, 398*f*
Auditory association area, 181–182
Auditory cortex, 181
Auditory hallucinations, 568–569
Auditory ossicles, 364–365
Auditory system
    anatomical structures
        external ear, 363–364, 364*f*
        inner ear, 364*f*, 366–367, 366*f*
        middle ear, 364–366, 364*f*
    cochlea and cochlear duct, 374–382
    equilibrium, 386–390
    hearing process, 367–371
    hearing tests, 385–386
    neural pathway of hearing, 383–385
    pathway of sound waves, 383
    perceiving sound, 382–383
    semicircular canals, 373–374
    vestibule, 371–373
Auditory tube, 364–365
Aura and seizures, 211–212
Auricle, 363
    lobule/lobe of, 364
    tragus of, 364
Auriculotemporal nerve roots, 100
Autistic spectrum disorder, 507*b*
Autogenic inhibition, 608
Automated auditory brainstem response, 385
Autonomic centers, 534
Autonomic effectors, 519, 522*t*
Autonomic innervation, of head, 530–532
Autonomic nervous system (ANS), 1
    Adie tonic pupil syndrome, 540–541
    Argyll Robertson pupil, 538–540
    autonomic innervation of head, 530–532
    autonomic outflow, 522–524
    autonomic plexuses, 529
    functional considerations, 534–537
    hierarchy of, 534
    Hirschsprung's disease, 541–543
    Horner syndrome, 537–538
    impulses, 535–536*t*
    neurologic examination, 666
    parasympathetic division, 519, 521–522*f*, 528–529
    regulation, 280–281
    sympathetic division, 521–522*f*, 524–528
    visceral afferent pathways, 532–534
Autonomic neurons, of intermediate gray matter, 611–612
Autonomic outflow, 522–524
Autonomic pathways, 522, 523*t*
Autonomic plexuses, 529
Autonomic reflexes, 605
Autonomic tongue innervation, 484
Autoreceptors, 562–563
Axolemma, 12
Axon, 11
    diameter and propagation speed, 25–26
    hillock, 10
    terminals, 12
Axoplasm, 12
A zone, cortico-olivary system, 503
Azygos system, 141–142

B

Babinski's reflex, 663–665
Babinski's sign, 603, 607
Baclofen, 624
Ballismus, 244
Baroreceptors, 532, 533*f*
Basal cells, 476
Basal ganglia areas, 467
Basal interstitial nucleus, 500
Basal nuclei
    clinical considerations
        dystonia, 259–261
        essential tremor, 256–259
        Hemiballismus, 256
        Huntington's chorea, 252–255
        Parkinson's disease, 244–251
    functional considerations, 242–244
    structure, 241–242
Basal vein (of Rosenthal), 137
Basilar artery, 134
    of circulus arteriosus, 140–141
Basilar fibers, 377
Basilar membrane, 376, 378
Basilar venous plexus, 102
Bassen-Kornzweig syndrome, 456
Batson plexus, 614
Behçet's syndrome, 118

Bell's palsy, 337, 355, 358b
Benedikt syndrome, 311–312
Benign paroxysmal positional vertigo
    (BPPV), 405–406
  clinical manifestations, 407
  diagnosis, 407
  pathophysiology, 407
  treatment, 407
Beta-amyloids, 200–201
Beta-endorphin, 558
β_1-receptors, 555
β_2-receptors, 555
β_3-receptors, 555
Beta waves, 185
Bevacizumab, 448
Biceps, 662
Biconvex, 426
Bilateral anosmia, 489–490
Bill's bar, 379
Biogenic amines, 550–551t, 552–556
Bipolar cells, 419–420
Bipolar disorder, 576–578
Bitemporal hemianopia, 435–436
Bizarre behaviors, 579–580b
Blepharospasm, 338, 357, 511
Blepharospasm-oromandibular dystonia, 260,
    509–510
Blind spot, 424
Blink reflex, 347–348
Blood pressure, baroreceptor and chemoreceptor
    reflex control, 533f
Blood-retina barrier system, 419
Blood supply
  of brain *see* Arteries; Ventricles
  of spinal cord, 613–614
Blurred spot, 447
Bone conduction hearing aids, 400–401
Bony labyrinth microstructure, 378–379
Bony orbit, 417
Bovine spongiform encephalopathy.
    See Creutzfeldt-Jakob disease
BPPV. *See* Benign paroxysmal positional vertigo
    (BPPV)
Brachial plexus and upper limbs, 592
Brachium conjunctivum, 311, 500
Bradykinesia, 243, 246–247
Bradyzoites, 90
Brain
  abscess, 86–87
  anatomy
    basal nuclei, 82–83
    brainstem, 83
    cerebellum, 83
    cerebral cortex, 80–81
    cerebral hemispheres, 78
    cerebral white matter, 81–82
    clinical considerations, 84–90
    diencephalon, 83
    divisions, 77–83
    lobes, 78–80
  arteries of, 133f
    carotid arteries, 131–133
    cerebral arterial circle, 135–136
    meningeal arteries, 136
    vertebral and basilar arteries, 134–135
  veins of
    deep veins, 137–139
    superficial veins, 137
Brainstem
  clinical considerations
    Arnold-Chiari phenomenon, 302–304, 303f
    Benedikt syndrome, 311–312
    cerebral aqueduct blockage, 308–309
    infarctions of pons, 307
    pontine hemorrhage, 305–306
    Weber syndrome, 310–311
  corticobulbar tracts, 301
  functional considerations, 301–302
  implants, 401
  medulla oblongata
    cardiovascular centers, 298
    components and functions, 297, 300t
    location and shape, 297
    reflex centers, 297–298
    relay stations, 298–301
    respiratory rhythmicity centers, 298
    sensory and motor nuclei, 298, 299f
  midbrain
    cerebral peduncles, 294
    periaqueductal gray matter, 295
    reticular activating system (RAS), 295
    trauma to, 307–308
  pathways to, 532–533
  pons
    components and functions, 296, 297t
    location, 296
  rule of 4, 293–294
  visual processing in, 431

Brain waves, 184–185
Branch retinal artery occlusion (BRAO), 451
Broca's aphasia, 191–192
Broca's area, 182–183
Bruch membrane, 447
Brudzinski signs, 204
B-scan ultrasound, 450
Bulbar-variant ALS, 625
Bulbospongiosus reflex, 664
B zone, cortico-olivary system, 504

C

Calcitonin gene-related peptide (CGRP), 483
Callosal disconnection syndrome, 186
Callosomarginal artery, 132
Calvarial dura, 98–99
Calvarium, 100
Canal hearing aids, 400
Canaliculi perforantes, 379
Canaliculus, 375–376
Canal of Schlemm, 426
Canavan disease, 123
Carbamazepine, 624
Carbohydrates, 131
Carbon monoxide (CO), 559
Cardiac nerves, 527–528
Cardiac plexus, 527–529
Caroticotympanic nerves, 380
Carotid arteries, 131–133
Carotidynia, 339
Cataplexy, 228
Cataract, 426–427, 446
Catecholamines, 535–536t, 552
Catechol-O-methyltransferase (COMT), 552–553
Cat's eye reflex, 460
Cauda equina, 104, 586–588
Cauda equina epidural abscess, 619
Cauda equina syndrome, 163
Caudal group, 471
Caudal insula, 474
Caudal medulla, 597
Caudal orbital cortex, 474
Caudate nucleus, 82, 241–242
Caudomedial group, 499–500
Cauliflower ear, 364
Cavernous angioma, 169
Cavernous malformations, 306
Cavernous sinus, 102, 140
Celiac ganglia, 524, 527

Celiac plexus, 524–527, 529
Cell morphology, 378–379
Central canal, 588
Central nervous system (CNS), 346, 519
  malformation, 61–63
Central nervous system (CNS) blood vessels, 131
  brain arteries, 133f
    carotid arteries, 131–133
    cerebral arterial circle, 135–136
    meningeal arteries, 136
    vertebral and basilar arteries, 134–135
  brain veins
    deep veins, 137–139
    superficial veins, 137
  clinical considerations
    cerebral aneurysm, 168–170
    embolism, 145–147
    epidural hematomas, 159–160
    focal cerebral ischemia, 149
    global cerebral ischemia, 150–157
    intracerebral hemorrhage, 166–168
    spinal cord ischemia, 170–171
    spinal epidural/subdural hematomas, 162–163
    stroke, 148–149
    subarachnoid hemorrhage, 163–166
    subdural hematomas, 160–166
    thrombosis, 142–145
    transient ischemic attack, 157–159
  spinal cord
    arteries of, 140–141
    veins of, 140–141
Central nucleus, 472
Central processing, 429–431
Central retinal artery (CRA), 424
Central retinal artery occlusion (CRAO), 451–452
Central retinal vein (CRV), 424
Central retinal vein occlusion (CRVO), 452–453
Central scotoma, 652
Central sulcus, 178
Centrioles, 11
Centrum semiovale, 277
Cerebellar ataxic gait, 665
Cerebellar lobules, 502, 503t
Cerebellar nuclei, 499–500
Cerebellar peduncles, 498–499
Cerebellar tonsils, 303
Cerebellomedullary cistern, 97
Cerebellopontine angle tumor, 395

Cerebellum
    afferent mossy fibers, 505–506
    anatomy of, 497–501
    ataxia, 509
    cerebellar cortex, 501–508
    cerebellar lobules, 502
    cerebellar nuclei, 499–500
    cortical interneurons, 501–502
    corticonuclear projections, 502–503
    cortico-olivary system, 503–505
    corticopontocerebellar projection, 506
    dysarthria, 511–512
    dystonias, 509–510
    functional divisions of, 507–508
    functions, 508
    mossy and climbing fibers, 500–501
    olivocerebellar projections, 502–503
    pendular knee jerk, 512–513
    roles of, 500b
    vestibulo-ocular reflex, 506–507
Cerebral aneurysm, 168–170
Cerebral aqueduct (of Sylvius), 106
Cerebral aqueduct blockage, 308–309
Cerebral arterial circle, 135–136
Cerebral artery, anterior, 131–132
Cerebral blood flow (CBF), 142
Cerebral catheter angiography, 676
Cerebral cortex, 177
    association areas, 181–182
    clinical considerations
        Alzheimer's disease, 193–199
        delirium, 199–207
        dementia, 199–207
        headache, 232–234
        prion diseases, 208–211
        sleep disorders, 221–232
    electroencephalogram (EEG), 184–186
    functional areas, 179–184
    functions, 181t
    higher cortical functions, 187–193
    integrative centers and higher mental
        functions, 182–184
    primary motor cortex, 180
    pyramidal cells, 179
    sensory areas, 180–181
    stellate cells, 179
Cerebral ischemia, 148
    focal, 149
    global, 150–157

Cerebral peduncles, 294, 301
Cerebrospinal fluid (CSF), 95, 366–367, 669–671,
    672–673t
    formation and circulation, 107–112, 108f
    functions, 106–112
    indicators in, 109–110
Cerebrum
    anterior commissure, 186
    arcuate fibers, 186
    association fibers, 186
    commissural fibers, 186
    functions of, 79
    gross anatomy, 177–179
    internal capsule, 186
    longitudinal fasciculi, 186
    projection fibers, 186
    white matter of, 186–187
Cerumen, 364
Cervical dystonia, 511
Cervical enlargements, 586
Cervical nerves, 586
Cervical plexus
    and neck, 592
Cervical spondylosis, 633–635
Cervical spondylotic myelopathy, 348
Cervicothoracic arteries, 613
Chaddock's reflex, 663–665
Chemesthesis, 480
Chemoreceptors, 532, 533f
Chemotaxis, 555–556
Cherry-red spot, 452
Cheyne-Stokes respirations, 305
Chiari II type malformation, 122, 635
Cholecystokinin (CCK), 558
Cholesteatomas, 353
Choline, 549–552
Choline acetyltransferase, 549–552
Cholinergic divisions, 534
Cholinergic receptors, 552
Cholinergic synapses, 28–29
Chorda tympani, 348–349, 483
Chordomas, 353
Chorea, 244, 657
Choreic gait, 665
Choreoathetoid movements, 259
Choreoathetosis, 254
Choriocapillaris, 447
Choroid, 419
Choroidal arteries

anterior, 131–132
  posterior, 134–135
Choroidal neovascularization, 447
Choroidal vein, 137
Choroid epithelium, 107
Choroid fissure, 107
Choroid plexus, 107
Chromaffin cells, 555
Chromogens, 670
Chronic angle-closure glaucoma, 443–444
Chronic bacterial meningoencephalitis, 119
Chronic hypomania, 578
Chronic labyrinthitis, 412*b*
Chronic paroxysmal hemicranias, 352
Chronotropy, 555
Chvostek sign, 357
Ciliary epithelium prevents, 419
Ciliary ganglion, 322, 531
Cingulate bundle, 469
Cingulate cortex, 467
Cingulate gyri, 475
Circadian rhythm, 431
  24-h, 422
Circle of Willis, 140–141
Circumduction, 664
Circumferential blindness, 331
Circumventricular organs, 275
11-Cis retinal, 422, 432
Clarke's nucleus, 611–612
Clasp-knife phenomenon, 657
Clasp-knife response, 608
Claudius' cells, 370–371
Claustrum, 242
Climbing fibers, 500–501
Clivius, 102
Clonus, 356, 603*b*, 644*b*
  testing for, 664
Cluster headache, 233
Coccygeal ligament, 103
Coccygeal nerves, 586
Cochlea, 374–382
Cochlear aqueduct, 375–376
Cochlear damage, 378*b*
Cochlear duct, 366, 374–382
Cochlear implants, 401
Cochlear nerve, 327, 363, 376–378
See also Vestibulocochlear nerves (VIII)
Cochlear recess, 371–372
Coenurosis, 89

Cogan's syndrome, 403
Cog-wheel rigidity, 246, 657
Collateral ganglia, 527*b*
Color blindness, 457–458
Color vision, 429*b*
Commissural fibers, 81, 591
Communicating artery
  anterior, 132
  posterior, 131–132
Communicating hydrocephalus, 122
Complement factor H gene, 447
Computed tomography (CT), 671–674
Conduction velocity, 680
Conductive hearing loss, 395
Cone pigments, 422
Cones, 420
Confluence of sinuses, 101
Confusion assessment method (CAM), 203
Congenital glaucoma, 443–444
Congenital hydrocephalus, 70–71
Congenital megacolon. *See* Hirschsprung's disease
Congenital oculosympathetic paralysis, 355–356
Connexins, 366–367
Connexons, 27
Consciousness, 190
  altered level of, 190
Consensual response, 652–653
Constructional apraxia, 190
Continuous positive airway pressure (CPAP), 227
Contracture, 356
Contralateral routing of signals (CROS) hearing aid, 400
Conus medullaris, 586
Convulsions, 211
See also Seizure disorders
Coordination, 659, 661
Copper wiring, 455
Cornea, 417
Corneal limbus, 417
Corneal reflex, 347–348
Corneoscleral junction, 417
Corona radiata, 81
Corpora arenacea, 269
Corpora quadrigemina, 294–295
Corpus callosotomy, 186
Corpus callosum, 78, 177
Cortical interneurons, 501–502
Corticobulbar tracts, 301
Corticonuclear fibers, 307–308

Corticonuclear projections, 502–503
Cortico-olivary system, 503–505
Corticopetal system, 472
Corticopontine fibers, 277
Corticopontocerebellar projection, 506
Corticospinal pathway, 301
Corticospinal system, 601
Corticospinal tract, anterior, 301, 603–604
Corticosteroids, 440
Corticosubthalamic fibers, 272
Cortilymph, 368
Coryza, 116–117
Cranial fossa, 136
  anterior, 100
  posterior, 100
Cranial nerves, 649–657
  abducens nerves (VI), 324–325
  accessory nerves (XI), 330
  classification, 319, 320f, 321t
  clinical considerations
    anosmia, 331–332
    Bell's palsy, 337
    fourth cranial nerve palsy, 332–333
    glossopharyngeal neuralgia, 339
    hemifacial spasm, 338–339
    hyposmia, 331–332
    sixth cranial nerve palsy, 333–334
    strabismus, 334–337
  facial nerves (VII), 326
  glossopharyngeal nerves (IX), 327–329
  hypoglossal nerves (XII), 331
  motor, 319
  oculomotor nerves (III), 322
  olfactory nerves (I), 319–320
  optic nerves (II), 321–322
  sensory, 319
  sensory and motor fibers, 319
  trigeminal nerves (V), 323
  trochlear nerves (IV), 322–323
  vagus nerves (X), 329–330
  vestibulocochlear nerves (VIII), 326–327
Cranial placodes, 56
Cranial root, 330
Craniocervical junction, 675
Craniopharyngioma, 281–282
Craniosacral division, 519
Cremasteric reflex, 664
Creutzfeldt-Jakob disease (CJD), 200–201, 208–209
Crista ampullaris, 374

Crista galli, 96
Crocodile tears, 356
Crossed effects of reflexes, 609
Crossed-extensor reflexes, 607
Crouzon syndrome, 335
Crus commune, 374
Cryopexy, 450–451
Crystallins, 426–427
Cuneate nucleus, 298
Cupula, 388
Cushing disease and Cushing syndrome, 287–288
Cutaneous nerves, 592
Cuticular plate, 369–370
Cyclic guanosine monophosphate
    (cGMP), 422
Cyclophotocoagulation, 445–446
Cyclothymic disorder, 578
Cytomegalovirus-induced polyradiculitis, 627–628

D

Dandy-Walker malformation, 71, 122
D-Dimer, 144
Decibels, 383
Decussation of pyramids, 298
Deep middle cerebral vein, 138–139
Deep tendon reflexes (DTRs), 606–607
Deep veins, 137
Deep vein thrombosis (DVT), 143
Degeneration, nerve cells, 34–36
Deiters' cells, 368, 370–371
Delirium, 199–207
  clinical manifestations, 201–203
  vs. dementia, 199t
  pathophysiology, 200–201
  treatment, 205–207
Delta waves, 185–186
Delusion, 489–492
Dementia, 199–207
  characteristics, 200
  classifications, 202t
  clinical manifestations, 201–203
  delirium vs., 199t
  diagnosis, 203–205
  intermediate stage, 202–203
  pathophysiology, 200–201
  treatment, 205–207
Demoralization, 574
Demyelination, 7, 84
Dendrites, 11

Denervated muscle fibers, 679–680
Dentate gyrus, 468
Dentatothalamic tract, 272
Denticulate ligament, 98, 104, 586
Depersonalization, 568–569
Depolarization, 17–18
Depression, 492b, 573–576, 580–581b
Depth perception, 431
Derealization, 568–569
Descending pathways, 591, 595, 601–604
Desipramine, 624
Desmosomes, 370–371
Deuteranopia, 457
Developmental venous anomaly, 306
Devic disease. *See* Neuromyelitis optica
Diabetes insipidus, 283–284
Diabetic retinopathy, 453–455
Diagnostic and Statistical Manual of Mental Disorders (DSM), 203
Diagonal band of Broca, 471
Diaphragm sellae, 101
Diatheses, 162
Diencephalon
    autonomic nervous system regulation, 280–281
    blood supply to, 276
    clinical considerations
        Cushing disease and Cushing syndrome, 287–288
        diabetes insipidus, 283–284
        hyperthermia, 285–287
        hypothalamic syndromes, 281–282
        hypothermia, 285–287
        pituitary insufficiency, 284–285
    development of, 60
    epithalamus, 267–270
    hypothalamus, 273–276
        functional considerations, 279–280
    internal capsule, 277–278
    interventricular foramen, 267
    subthalamus, 272–273
    telencephalon, 267
    thalamus, 267, 270–272
        thalamic nuclei functions, 278–279
Diffuse ischemic/hypoxic encephalopathy, 150
Diffuse myelitis, 617
Diffusion tensor imaging, 675
Diffusion-weighted imaging (DWI), 675
Digital subtraction angiography, 676
Diphtheria, 8
Diploic vein, 102
Diplopia, 322, 653–654
Disconnection syndromes, 186
Disseminated myelitis, 617
Dissolved gases, 550–551t, 559
Distal latency, 680
Distal limb ataxia, 661
Dix-Hallpike maneuver, 407, 656
Dizziness, 404
Dominant optic atrophy, 436
Dopamine, 552–553
Dopamine system, 244
Dorsal group, 470
Dorsal/posterior root ganglion, 590
Dorsolateral, 279
Dorsolateral (DL) nuclei, 271–272
Dorsomedial (DM) nucleus, 271–272
Dorsomedial (DM), thalamus, 269
Double-dose delayed scanning, 623
Dropfoot, 659
Drug-induced ototoxicity, 408–409
Drusen, 447
Dry macular degeneration, 446–448
Duplex Doppler ultrasonography, 676–678
Duplex ultrasonography, 144
Dura mater, 96
    arterial supply, 100–101
    dural folds, 96
    dural venous sinuses, 96
    falx cerebelli, 96
    falx cerebri, 96
    nerve supply, 98–100
    pain sensation in, 99–100
    tentorium cerebelli, 96
    venous sinuses, 101–103
Duret hemorrhage, 305
Dynamic equilibrium, 386, 388–390
Dynamic stretch reflex, 348
Dynorphin, 558
Dysarthria, 191, 511–512, 513b, 515b
Dysconjugate gaze, 308
Dysmetria, 659
Dysosmia, 489–490
Dysprosodies, 192–193
Dysthymia, 573–574, 580–581b
Dystonia, 259–261, 509–510
    acquired, 261
    focal, 260
    generalized, 260, 511–512

Dystonia *(Continued)*
  hemidystonia, 260
  movements, 259
  musculorum deformans, 260
  postures, 259
  segmental, 260
  spasmodic, 260
Dystrophic gait, 665

E

Ear
  external, 363–364, 364f
  inner, 364f, 366–367
  middle, 364–366, 364f
Eardrum, 363
Echinococcal cysts, 89
Ectatic basilar artery, 350–351
Edinger-Westphal nucleus, 528, 651
Effector, 605
Electrocochleography, 399
Electroencephalography (EEG), 678–679
Electromyography (EMG), 626, 679–680
Electronystagmography, 403
Electroolfactogram, 479–480
Elliptical recess, 372–373
Embolism, 145–147
  air, 146
  brain, 146
  diagnosis, 147
  fat, 146
  pulmonary, 146
  retinal, 146
  septic, 146
Embolus, 145
Emissary veins, 136–137
Emotional disturbances
  alterations in sexuality, 488
  anger, 487
  apathy, 487–488
  due to hallucinations and pain, 485
  emotional lability, 486
  of limbic structures, 485
  placidity, 487–488
  spasmodic/pseudobulbar, 486–487
  violent behaviors, 487
Emotional lability, 486
Encephalitides, 486
Encephalitis, 84–86
Encephalocele, 68

Endocytosis, 562
Endogenous ligands, 558
Endolymph, 366–368, 376
Endolymphatic duct/sac, 373
Endolymphatic hydrops. *See* Meniere's disease
Endolymphatic potential, 366
Endorphins, 558–560
Endoscopic third ventriculostomy (EVD), 309–310
Endotoxin, 115
Enkephalins, 468–469, 558–560
Enteric nervous system (ENS), 519–522
Enteroviruses, 116
Entorhinal cortex, 468–470
Enzyme-linked immunosorbent assay (ELISA), 144
Ependymal cells, 7
Ependymomas, 43–44
Ephaptic transmission, 381
Epicanthal folds, 336
Epidural hematomas, 159–160
Epidural space, 95, 585
Epidural spinal abscess, 617
Epilepsy, 211
  See also Seizure disorders
Epileptiform sharp waves, 678–679
Epinephrine, 537, 553–555
Epithalamus, 267
  epiphysis cerebri, 267–268
  habenular nuclei, 267, 269
  paraventricular nuclei, 267
  pineal gland, 267–269
  pineal recess, 267–268
  posterior commissure, 267
  stria medullaris thalami, 267
Epley maneuver, 405
Epworth Sleepiness Scale, 222–223
Equilibrium
  clinical considerations, 390
  dynamic equilibrium, 388–390
  static equilibrium, 387–388
  vestibule and semicircular canals, 386–387
Erdheim-Chester disease, 281–282
Esotropia, 335
Essential tremor, 256–259
Ethmoidal artery
  anterior, 100
  posterior, 100

E3 ubiquitin ligase, 246
Eustachian tube, 364–365
Excessive daytime sleepiness (EDS), 221
See also Sleep disorders
Excitatory neurotransmitter, 557
Exocytosis, 562
Exophthalmos, 653
Exotropia, 335
Extended amygdala, 472
External acoustic meatus, 363
External ear, 363–364
External plexiform layer, 478
Exteroceptors, 15
Extradural hemorrhage, 100–101
Extradural veins, 141
Extramedullary tumors, 642
Extrapyramidal disorders, 508–509
Extraventricular drains, 306–307
Extrinsic hallucinations, 491–492
Eyeball
    chambers of, 425–426
    fibrous layer, 417–418
    inner layer, 419–425
    lens, 426–427
    structure of, 418*f*, 421*f*
    vascular layer, 419

F

Fabry disease, 151
Facial hemiatrophy, 355–356
Facial motor nuclei, 349–350
Facial myokymia, 356
Facial nerve, 348–350, 528–529, 655
Facial nerve palsy
    problems with recovery from, 356
Facial nerves (VII), 326
Facial palsy
    additional causes of, 355
Facial paresis, 355–356
Falx cerebelli, 96
Falx cerebri, 139–140
Familial fatal insomnia. See Prion diseases
Fasciculations, 603*b*, 657
Fasciculus cuneatus, 595–597
Fasciculus gracilis, 595–597
Fasciculus lenticularis, 272
Fasciculus proprius, 595
Fasciculus subthalamicus, 272
Fasciculus thalamicus, 272

Fatal insomnia, 210
Fatigability, 680
Femoral nerve, 593
Fenestrae cochleae, 376
Fenestra vestibuli, 371
Festination, 247
Fibrillations, 603*b*
Fibrin, 144
Fibrous layer, 417–418
Field of vision, 431
Fields of Forel, 272
Fight-or-flight reaction, 528*b*
Filum terminale, 98, 586
Fissure, 78
Flaccid dysarthria, 511
Flaccid paralysis, 601–603
Flexor reflex, 608–609
Flexor reflexes, 607
Flocculi, 497
Flocculonodular lobe, 497
Fluid attenuating inversion recovery (FLAIR), 440
Fluorescein angiography, 447–448
Focal cerebral ischemia, 149
Focal dystonias, 260, 509–510
Folia, 498
Fontanelles, 121
Foramen magnum, 585
Foramen ovale, 345
Foramen rotundum, 345
Foramen singular, 379
Foramina of Luschka, 122
Foramina of Magendie, 122
Forebrain. See Diencephalon
Fornix, 469
Foster Kennedy syndrome, 491
Fourth cranial nerve palsy, 332–333
Fourth ventricles, 106
Fovea, 422–423
Fovea centralis, 422–423
Frontal lobe, 78, 178
Fuchs spot, 447
Fukada stepping test, 405–406
Functional magnetic resonance imaging (fMRI), 184, 676
Funduscope, 447–448, 652
Fungal meningoencephalitis, 87–88
Funiculi, 591
Fusimotor neurons, 611

G

Gabapentin, 624
Gait, 659, 664–665
Gait unsteadiness, 405
Gamma-aminobutyric acid (GABA), 377–378, 557
Ganglion, 14–15
Ganglion cell, 419–420
Ganglion cell layer, 424
Ganglionic arteries, 133
Ganglionopathy, 353–354
Gasotransmitters, 559
Gasserian ganglion, 345
Gated ion channels, 20–21
Gelastic epilepsy, 281–282
Geniculate ganglia, 325
Gentamicin, 409, 414b
Genu, 277
Gerstmann-Sträussler-Scheinker disease, 210
See also Prion diseases
Glabellar reflex, 247
Glabellar sign, 664
Glasgow Coma Scale, 306–307
Glaucoma, 443–445, 445b, 462–463b
Glial cells, 423
Glial septum, 588
Gliomas, 40
Gliomatosis cerebri, 486
Global cerebral ischemia, 150–157
Globose nuclei, 499
Globus pallidus, 82, 241–242
Glomerular layer, 478
Glomus, 107
Glossectomy, 227
Glossopharyngeal nerves (IX), 328f, 528–529, 532, 656–657
   function of, 327–329
   inferior ganglion, 327–329
   otic ganglion, 327–329
   superior ganglion, 327–329
Glossopharyngeal neuralgia, 339
Glucocorticoids, 403
Gluconeogenesis, 555
Glutamate/glutamic acid, 557
Glutamatergic relay neurons, 500
Glutaminergic receptor, 557
Glycine, 557
Glycinergic, 501–502
Gnathostomiasis, 89
Golfer's cramp, 511–512
Golgi apparatus, 10
Golgi cells, 501–502
Golgi tendon organ, 608
Gowers' sign, 658
G-protein, 422, 480
Gracile nucleus, 298
Graded potentials, 17–20
Gradenigo syndrome, 353
Granule cell, 497
Granule cell layer, 478
Granuloma, 617
Granulomatosis, 355
Graphesthesia, 659–660
Grasp reflex, 664
Gray commissure, 588
Gray communicating rami, 527
Gray matter
   intermediate, 611–612
   layered arrangement of, 612–613
   regional specializations of, 610–611
   of spinal cord, 590f
   subdivisions, 612t
Great cerebral vein, 138–139
Growth hormone-inhibiting hormone (GHIH), 558
Guillain-Barre syndrome (GBS), 8, 620–621
Gun-barrel field, 652
Gustatory cortex, 181
Gustatory system, 481–485
   functions of, 483–484
   neural pathways, 483
   taste buds, 481–482
   taste sensation, 484–485
Gut-brain peptides, 558
Gyrus, 78, 177
Gyrus semilunaris, 479

H

Habenular nuclei, 269
Habenular perforata, 376
Hair cells, 363
Hallervorden-Spatz disease, 205
Hallucinations, 553
   emotional disturbances due to, 485
Hand-Schüller-Christian disease, 281–282
Hardy-Rand-Ritter plates, 652
Head
   autonomic innervation of, 530–532
Headache, 232–234

Hearing
    neural pathway of, 383–385
Hearing aids, 400
Hearing loss, 386, 386b, 394–403
    causes of, 395f, 396t, 413b
    clinical manifestations, 396
    diagnosis of, 413b
    pathophysiology, 395–396
    treatment, 399–403, 413b
Hearing process, 367–371
Hearing tests, 385–386
Heel-knee-shin test, 661
Helicotrema, 376
Helminthic brain infections, 88–89
Hematomas
    epidural, 159–160
        spinal epidural/subdural, 162–163
    subdural, 160–166
Hemiataxia, 311
Hemiballismus, 256
Hemidystonia, 260
Hemifacial spasm, 338–339, 354, 511–512
Hemiplegic gait, 664
Hemisphere dominance, 79–80
Hemispheres, 498
Hemispheric lateralization, 183–184
Hemitremor, 314
Hensen's cells, 370–371
Hensen's stripe, 371
Hereditary optic neuropathies, 436–438
Herniated nucleus pulposus, 628–631
Herpeszoster, 353
Herpes zoster oticus. *See* Ramsay-Hunt syndrome
Hertz, 382
Higher cortical functions, 187–193
Hippocampal formation, 467–469
Hippocampus, 468
Hirschsprung's disease, 541–543
Histamine, 552, 555–556
Histidine, 552
Hodgkin lymphoma, 118
Hoffman's sign, 664
Holmes-Adie syndrome, 540
Holoprosencephaly, 69–70
Homonymous hemianopia, 434–435, 462b
Homunculus, 132, 597–598
Horizontal cells, 419–420
Horizontal gaze center, 504

Horner syndrome, 537–538, 543b, 653, 666
Huntingtin, 254
Huntington's chorea, 252–255
Hydrocephalus, 121–123
    communicating, 122
    noncommunicating, 122
    obstructive, 122
Hydrogen sulfide (H_2S), 559
Hydrops, 373
5-Hydroxytryptamine (5-HT). *See* Serotonin
Hygroma, 161–162
Hyperacusis, 337
Hyperbaric chamber, 147–148
Hyperkinesia, 509
Hyperkinetic disorders, 244
Hyperkinetic dysarthria, 511
Hyperosmia, 489–490
Hyperphosphorylation, 246
Hyperpigmentation, 456
Hyperpolarization, 387
Hypersexuality, 488
Hypertensive retinopathy, 455–456
Hyperthermia, 285–287
Hypertonia, 259, 603
Hypertrophy, 657
Hypobulic, 487–488
Hypogastric plexus, 524–529
Hypoglossal nerves (XII), 331, 657
Hypokinesia, 246–247, 509
Hypokinetic disorders, 245
Hypokinetic dysarthria, 511
Hypomania, 577
Hypomimia, 246–247
Hypophonic, 247
Hyposmia, 331–332, 489–490
Hypothalamopituitary axis, 281–282
Hypothalamus
    functional considerations, 279–280
    homeostatic activities, 274t
    hypothalamic-releasing hormones, 275–276
    location, 273
    neurohypophysis, 276
    syndromes, 281–282
Hypothermia, 285–287
Hypotropia, 335

I

Ia afferent, 606, 609–610
Idiopathic hypersomnia, 230

Immunofixation, 626
Inborn reflexes, 604–605
Indolamine, 268–269
Indolamines, 552
Infarct, 148
Inferior anastomotic vein (of Labbé), 137
Inferior cerebellar artery
    posterior, 134
Inferior cerebellar peduncles, 499
Inferior colliculi, 294–295
Inferior fibers, 436
Inferior ganglion, 327–329
Inferior mesenteric ganglion, 524–527
Inferior olivary complex, 300
Inferior vena cava filter (IVCF), 145
Inferomedial area, 345
Infraorbital branch, 353
Infratentorial dura mater, 99
Infundibular recess, 106
Infundibulum, 273
Inger-to-nose test, 661
Inhibitory brain neurotransmitter, 557
Inhibitory postsynaptic potential (IPSP), 557
Inner ear, 366–367, 366f
Inner layer, of eye, 419–425
Inner limiting membrane, 425
Inner nuclear layer, 424
Inner plexiform layer, 424
Innervation, 380
Inotropy, 555
Insomnia, 221, 224–225
    See also Sleep disorders
Insula, 78, 178
Integration center, 182, 605
Integrative pathways, 604
Intention tremor, 661
Intercavernous sinus, 101
Intercostal nerves, 592–593
Intermediate arteries, 613
Intermediate gray matter
    autonomic neurons of, 611–612
Intermediate spinal muscular atrophies, 631
Intermediolateral cell column, 611–612
Internal acoustic canal, 379
Internal acoustic meatus, 372–373, 379
Internal auditory canal, 379
Internal capsule, 241–242, 277–278
Internal carotid arteries, 131

Internal cerebral vein, 137
Internal jugular veins, 136–137
Internal plexiform layer, 478
International normalized ratio (INR), 163
Interneurons, 15–16, 588–590
Interoceptors, 15
Interpeduncular cistern, 97
Interpolaris, 347
Interthalamic adhesion, 105, 271
Interventricular foramina (of Monro), 105, 267
Intracerebral hemorrhage, 166–168
Intradural veins, 141
Intralaminar nuclei, 243
Intramedullary tumors, 642
Intraocular pressure, 426
Intraosseous veins, 141
Intraparenchymal hemorrhage. *See* Intracerebral hemorrhage
Intrapontine area, of facial nerve, 356
Intraretinal capillaries, 419
Intratemporal vestibular nerve, 381–382
Intrathecal drugs, 118
Intrinsically photosensitive retinal ganglion cell (ipRGC), 422
Intrinsic hallucinations, 491–492
Intrinsic muscles, 419
Inverse Marcus-Gunn sign, 356
Ipsilateral antagonists, 609
Ipsilateral cerebellum, 498–499
Ipsilateral ear, 339
Ipsilateral eye, 435
Ipsilateral synergists, 609
Iridotomy, 445–446
Ischemic penumbra, 142
Ischemic stroke, 153
    See also Stroke
Isodense, 306
Isolated trigeminal neuropathy, 353–354

J

Jaw jerk, 346
Jaw jerk reflex, 348, 348b
Jaw-winking phenomenon, 356
Jendrassik's maneuver, 662–663
Jerk nystagmus, 654
Joint position sensing, 660–661
Joints
    innervation of, 593–594

Jugular ganglion. *See* superior ganglion
Juxtallocortex, 468
Juxtarestiform body, 499

K

Keratoconus, 458
Kernig signs, 204
Kilohertz, 382
Kinociliary bundles, 386–387
Kinocilium, 369–370, 389
Knee-jerk reflex, 606–607
Kuru, 210–211
See also Prion diseases
Kyphoscoliosis, 223

L

Labyrinth, 366
Labyrinthine artery, 134, 379–380
Labyrinthitis, 411–412
Lacunes, 153
Lambert-Eaton syndrome, 564
Lamina, 612–613
Lamina cribos, 440–442
Lamina terminalis, 105
Lancinating, 350–351
Laryngeal dystonia, 511–512
Late-onset depression, 574
Lateral corticospinal tract, 301, 601
Lateral division, 595–597
Lateral efferents, 377
Lateral geniculate nucleus (LGN), 272, 436
Lateral horns, 588
Lateral hypothalamic nucleus, 279
Lateral posterior (LP) nuclei, 271–272
Lateral sulcus, 178
Lateral ventricles, 104–105
Laurence-Moon syndrome, 456
Lead-pipe rigidity, 657
Leak channels, 20
Learned reflex, 605
Leber's hereditary optic neuropathy, 437
Lens, 426–427
Lens fibers, 426–427
Lenticulostriate arteries, 133
Lentiform nucleus, 241–242
Lesion, in visual pathway, 436
Letterer-Siwe disease, 281–282
Leucine, 558

Leucine-rich repeat kinase 2 (LRRK2), 246
Leukokoria, 460
Lewy bodies, 200–201, 245–246
Lewy body dementia, 200–201, 245–246
Lhermitte's sign, 623
Limbic brain, 468
Limbic lobe, 467–468
Limbic midbrain areas, 468
Limbic system, 467
  anatomy of, 467–475
  differential diagnoses, 488–489
  emotional disturbances of, 485
  functions of, 475–476
  and olfaction, 471*b*
  structures of, 468*f*
Limulus amebocyte lysate test, 115
Lipids, 550–551*t*
Lipodystrophy, 355–356
Lipofuscin, 11
Lipohyalinosis, 305
Lipolysis, 555
Lissauer's tract, 610–611
Listeria, 113
Local sign, 608–609
Locus ceruleus, 194, 468–469
Long ascending fibers, 595
Long descending fibers, 595
Longitudinal fasciculi, 186
Longitudinally extensive transverse myelitis (LETM), 617
Long wavelength (L) cones, 457
Lou Gehrig's disease. *See* Amyotrophic lateral sclerosis (ALS)
Lower arteries, 613
Lower limb dystonia, 514–515*b*
Lower limbs
  lumbosacral plexus and, 592
Lower motor neuron, 601
  damage, 603*t*
Low-tension glaucoma, 443–444
Lugaro cells, 502
Lumbar cistern, 586–588
Lumbar enlargements, 586
Lumbar nerves, 586
Lumbar puncture, 110–111*b*, 586, 669–671
Lumbosacral enlargements, 586
Lumbosacral plexus
  and lower limbs, 592

Lumbosacral trunk, 592
Lyme disease, 121, 355, 359b
Lymphocytic hypophysitis, 284
Lymphoepithelioma, 353
Lysergic acid diethylamide (LSD), 553

M

Macroglossia, 227
Macula, 372–373, 387–388, 651
Macula cribrosa media, 371–372
Macula cribrosa superior, 372–373
Macula lutea, 422–423
Macular degeneration
    age-related, 446–448, 448b
    dry, 446–448
    neovascular, 447
    treatments, 463b
Mad cow disease. *See* Creutzfeldt-Jakob disease
Magnetic gait, 659
Magnetic resonance angiography (MRA), 144, 676
Magnetic resonance imaging (MRI), 675–676
Magnetic resonance spectroscopy (MRS), 676
Magnetic resonance venography (MRV), 144, 676
Magnocellular accessory basal nucleus, 474–475
Magnocellular neurosecretory extensions, 274–275
Magnocellular neurosecretory system, 276
Major depressive disorder, 573–574
Malathion, 549–552
Mammillary bodies, 273
Mammillary nucleus lesion, 279
Mammillothalamic fibers, 469
Mammillothalamic tract, 468
Mandibular nerve (V1), 323
Mandibular reflex, 348
Mandibulomaxillary advancement, 227
Manic episode, 577
Manic psychosis, 577
Manometer, 670
Manometrics, 542
Marantic endocarditis, 146
Marcus Gunn pupil, 653
Masking sound, 397–398
Masseter reflex, 348
Mass reflex, 641
Masticatory nucleus, 348
Mastoid aditus, 363–364
Mastoiditis, 112, 365
Maxillary nerve (V1), 323

M cells, 429
Meckel's cave, 345
Medial division, 595–597
Medial efferents, 377
Medial geniculate nucleus (MGN), 272
Medial group, 471
Medial lemniscus, 294, 307, 597
Medial longitudinal fasciculus (MLF), 294
Medulla oblongata
    cardiovascular centers, 298
    components and functions, 297, 300t
    location and shape, 297
    reflex centers, 297–298
    relay stations, 298–301
    respiratory rhythmicity centers, 298
    sensory and motor nuclei, 298, 299f
Medulla spinalis (spinal cord)
    development of, 56–57, 57f
Meige's disease, 509–510
Melanopsin, 422
Melatonin, 268
Membrane potential, 16–26
Memory
    brain regions used in, 188–189
    cellular memory formation and storage, 189–190
    engram, 189
    fact memories, 188
    long-term, 188
    memory consolidation, 188
    Papez circuit, 190
    short-term, 188
    skill memories, 188
Meniere's disease, 387, 405
    clinical manifestations, 405
    complications, 413–414b
    diagnosis, 405–406
    pathophysiology, 405
    testing, 413–414b
    treatment, 406
Meningeal arteries, 136
Meninges, 95
    clinical considerations
        chronic bacterial meningoencephalitis, 119–121
        meningioma, 124–125
        meningitis, 112–119
    cranial, 95
    epidural space, 95

potential spaces, 95
  of spinal cord, 103–104
  subdural space, 95
Meningioma, 124–125
Meningismus, 115, 164
Meningitis
  acute bacterial, 112–116, 113t
  chronic, 118–119
  viral, 116–118
Meningococci, 112
Meningoencephalitis, 112
Meningomyelitis, 617
Meningomyelocele, 113
Meningoradiculitis, 617
Mental status examination, 648–649, 650t
Mescaline, 553
Mesencephalic nucleus, 346
Mesencephalon (midbrain), 53
  development of, 59–60
Mesenteric ganglia, 524
Metencephalon
  development of, 59
Methionine, 558
Meyer's loop, 436
Microatheroma, 143
Microcephaly, 69
Microfilaments, 11
Microglia, 3–4
Micrognathia, 227
Microorganisms, 618
Microtubules, 11
Midbrain, 294–296
  cerebral peduncles, 294
  periaqueductal gray matter, 295
  reticular activating system (RAS), 295
  trauma to, 307–308
Middle cerebellar peduncles, 499
Middle cerebral artery, 131–132
Middle cranial fossa, 100
Middle ear, 364–366
Middle meningeal artery, 100
Middle sacral nerve, 534
Middle thoracic arteries, 613
Middle wavelength (M) cones, 457
Migraine headache, 233
Mild cognitive impairment (MCI), 197
Mild hearing loss, 386
Mini-Mental State Examination (MMSE), 196–197, 649

Mini-Tiller testing, 658
Miosis, 653
Mitophagy, 246
Mitral cell layer, 478
Mixed hearing loss, 395
Modafinil, 624
Moderate hearing loss, 386
Modified postganglionic cells, 527
Modiolus, 374
Molar, 562
Mollaret's meningitis, 116
Monoamine oxidase (MAO), 552–553
Monopolar brush cells, 502
Monosynaptic reflex, 605–606
Mossy fibers, 500–501
Motor impersistence, 254
Motor neuron, 15, 605
  of anterior horn, 611
Motor neuron damage
  upper and lower, 603t
Motor neuron inhibition, 608
Motor nucleus and nerve, 294
Motor pathway, 294
Motor system, 657
Muller cells, 423
Multiple sclerosis (MS), 8, 621–624, 644–645b
Muscarinic receptors, 552
Muscle strength, 658
Muscle tension, 608
Musician's cramp, 511–512
Myasthenia gravis (MG), 36–39, 552, 564–565
Myasthenic crisis, 564–565, 579b
Myectomy, 357
Myelencephalon (medulla oblongata)
  development of, 58–59
Myelitis, 617
Myelography, 678
Myelomeningocele, 302–303, 635
Myelopathies, 617–635
Myerson's sign, 247
Myoclonus, 657
Myokymia, 356
Myokymic activity, 356
Myopia, 444–445
Myotatic reflex, 606
Myotome, 355–356
Myotonia, 657
Myringoplasty, 363

N

Narcolepsy, 228–230
Narrow-angle glaucoma, 443–444
Narrow band noise, 397–398
Nasal, 490
Nasal rami, 531
Nasal septum, 476
Near reflex, 428
Neck
  cervical plexus and, 592
Neoangiogenesis, 446–448
Neocortex, 179
Neologisms, 201
Neovascularization, 447
Neovascular macular degeneration, 447
Nephrogenic fibrosing dermopathy, 675
Nerve conduction velocity studies, 679–680
Nerve edema, 438
Nerve fiber layer, 424
Nerve plexuses, 592
Nerve root disorders. *See* Radiculopathies
Nervous system
  anatomical divisions of, 1
  functional divisions, 2–3
Nervus conarii, 268
Nervus erigentes, 528–529
Nervus intermedius of Wrisberg, 348–349, 349f
Nervus spinosus, 99
Neural crest, 55
Neural hearing loss, 395, 396t
Neural layer, 420–423
Neural tube
  bulges and flexures of, 52–54
  clinical considerations
    anencephaly, 67–68
    congenital hydrocephalus, 70–71
    encephalocele, 68
    holoprosencephaly, 69–70
    microcephaly, 69
    spina bifida, 63–65
  derivatives, 53t
  development of, 51–54
Neurilemma, 7
Neuroablative treatments, 352–353
Neuroblastoma, 39–40
Neuroborreliosis (lyme disease), 121
Neurocysticercosis, 88
Neurofilaments, 11

Neurogenic shock, 640
Neuroglia
  of CNS, 3–7
  of PNS, 7–9
  types, 4t
Neurohypophysis, 276
Neurokinin A (NKA), 558
Neurologic complaints
  screening without, 665
Neurologic diagnostic procedures
  cerebral catheter angiography, 676
  computed tomography, 671–674
  duplex Doppler ultrasonography, 676–678
  electroencephalography, 678–679
  electromyography, 679–680
  lumbar puncture, 669–671
  magnetic resonance imaging, 675–676
  myelography, 678
Neurologic examination
  autonomic nervous system, 666
  Babinski's reflex, 663–665
  Chaddock's reflex, 663–665
  coordination, 659
  cranial nerves, 649–657
  gait, 659
  mental status, 648–649
  motor system, 657
  muscle strength, 658
  Oppenheim reflex, 663–665
  patient history, 647–648
  reflexes, 661–663
  sensation, 659–661
  stance, 659
Neurologic method, 647
Neuromodulators, 560–561
Neuromuscular junctions (NMJs), 549–552
Neuromyelitis optica, 621
Neuromyotonia, 356
Neuronal axons, 475–476
Neuronal integration, excitatory and inhibitory stimuli, 30–34
Neuronotropic, 617–618
Neurons
  anaxonic, 13
  bipolar, 13
  characteristics, 9
  classification, 13–16
  functional characteristics, 9–10
  multipolar, 13

sensory, 14–15
structure of, 10–13
unipolar, 14
Neuropeptides, 550–551t, 558
Neurosyphilis, 120, 200–201
Neurotransmitters, 549
acetylcholine, 549–552
amino acids, 557
binding to postsynaptic receptors, 562–563
biogenic amines, 552–556
bipolar disorder, 576–578
classifications of, 549
cyclothymic disorder, 578
depression, 571
dissolved gases, 559
functional considerations, 561–563
myasthenia gravis, 564–565
neuromodulators, 560–561
neuropeptides, 558
nonopioid neuropeptides, 560
obsessive-compulsive disorder, 571–573
opioid peptides, 559–560
Parkinson's disease, 565–566
presynaptic release of, 562
purines, 558–559
schizophrenia, 566–571
synthesis of, 561–562
types of, 549–559, 550–551t
Neurotransmitters and receptors, 537b
Neurotubules, 11
Nicotinic receptors, 552
Night blindness, 458
Nigrostriatal pathway, 245
Nigrosubthalamic fibers, 272
Nitric oxide (NO), 559
N-methyl-*D*-aspartate (NMDA) receptors, 190
Nociceptive information, 599–600
Nodes of Ranvier, 6
Nodulus, 497
Nonarteritic anterior ischemic optic neuropathy, 439
Noncommunicating hydrocephalus, 122
Nonopioid neuropeptides, 560
Non-vitamin K oral anticoagulants (NOACs), 145
Noradrenergic projections, 469
Norepinephrine (NE), 468–469, 537, 537b, 553–555

Normal glaucoma, 443–444
Normal hearing, 386
Noxious stimuli, 650–651
Nuchal rigidity, 112
Nuclear pores, 10
Nuclei
anterior interposed, 499–500
Nuclei of Mynert, 194
Nucleus accumbens, 467
Nucleus basalis, 188–189
Nucleus cuneatus, 597
Nucleus gracilis, 597
Nucleus of Darkschewitsch, 505
Numb-chin sign, 353
Nusinersen (Spinraza®), 632
Nutritional amblyopia, 442
Nyctalopia, 458
Nylen-Bárány maneuver, 656
Nystagmus, 405–406, 654

O

Objective tinnitus, 393
Obsessive-compulsive disorder (OCD), 571–573, 580b
Obstructive hydrocephalus, 122
Occipital arteries, 100–101
Occipital lobe, 78, 178
Occipital sinus, 101
Occult spinal dysraphism, 635
Occupational dystonias, 509–510
Oculomotor nerves (III), 322, 528–529, 652–654
Oculomotor nucleus, 528
Off-center neurons, 429
Olfaction
limbic system and, 471b
Olfactory agnosia, 489–490
Olfactory bulb, 478–479
Olfactory cells, 476
Olfactory cortex, 181, 467
Olfactory hallucinations, 489–492, 491b
Olfactory nerves (I), 319–320, 321f, 650–651
layer, 478
olfactory afferents, 319–320
olfactory bulbs, 319–320
olfactory tracts, 319–320
Olfactory neuroepithelial, 490
Olfactory rod, 477
Olfactory system
anatomy of, 476–477

Olfactory system *(Continued)*
   functions of, 479–481
      olfactory pathways, 477–479
      olfactory receptors, 477
Olfactory tract, 479
Olfactory trigone, 479
Oligodendrocytes, 6
Olives, 300
Olivocerebellar climbing fiber system, 497–498
Olivocerebellar projections, 502–503
Olivocochlear system, 377
On-center neurons, 429
Open-angle glaucoma, 443–444
Ophthalmic artery, 100, 131
Ophthalmic nerve (V1), 323
Ophthalmoplegia, 653–654
Opioid peptides, 559–560
Oppenheim reflex, 663–665
Opsin, 422
Optic abiotrophy, 436
Optical coherence tomography, 447–448
Optical reflex, 347–348
Optic chiasm, 430
Optic disc, 424, 651
Optic disk pallor, 651
Optic nerves (II), 651–652
   optic chiasm, 321
   optic tracts, 322
   superior optic fissure, 321
Optic neuritis, 439–440
Optic papilla, 424
Optic radiation, 431
Optic recess, 106
Optic tract, 430
Oralis, 347
Orbital rami, 531
Organic brain disease, 486
Organ of Corti, 367–371
Orogenital dermatitis, 442
Oromandibular dystonia, 511–512
Oscillopsia, 404, 408–409
Osmophobia, 233
Osteophytes, 633
Otic ganglion, 327–329, 531–532
Otitis media, 363–364
Otoacoustic emission screening, 385
Otoacoustic emissions testing, 399
Otoconial crystals, 387–388

Otolithic membrane, 371–372, 390
Otoliths, 372
Otoscope, 363
Ototoxic drugs, 399–400
Outer limiting membrane, 423
Outer nuclear layer, 423
Outer phalangeal cells, 368
Outer plexiform layer, 423
Oval window, 364–365, 371
Oximetry, 204

P

Pacemaker mechanism, 184
Pachymeningitis, 617
Pain, 661
   emotional disturbances due to, 485
Painful stimuli, 608–609
Palatine rami, 531
Paleocortex, 179
Pallidal division, 272
Pallidosubthalamic fibers, 272
Palmomental reflex, 664
Palpebrae, 653
Pancoast tumor, 537
Pancreozymin, 558
Papez circuit, 190, 469, 470f, 477–478
Papilledema, 117, 440–442, 441f, 461b, 651
Papillitis, 439–440
Papilloma, 121–122
Parabrachial pontine reticular formation, 272
Paradoxical emboli, 145
Paralaminar basal nucleus, 472
Parallel fibers, 500–501
Paramedian fissures, 497–498
Paramedian midbrain syndrome. *See* Benedikt syndrome
Paraneoplasia, 626
Paraneurons, 481–482
Paraplegic gait, 664
Paraprotein, 626
Paraproteinemia, 626
Parasomnias, 230–231
Paraspinal veins, 141
Parasubiculum, 468
Parasympathetic nervous system, 519, 521–522f, 528–529, 530t
Parasympathetic preganglionic fibers, 528–529, 529b
Parasympathetic tone, downregulating, 200

Paratenial nuclei, 469
Paratonia, 259
Paraventricular nucleus, 278
Paraventricular/supraoptic lesion, 278
Paresis, 603
Parietal lobe, 78, 178, 436
Parieto-occipital sulcus, 178
Parkinsonian, 509
Parkinsonian gait, 665
Parkinsonism, 243, 248, 252
    primary and secondary causes, 253*t*
Parkinson's disease (PD), 200–201, 243, 553, 565–566
    clinical manifestations, 246–247
    diagnosis, 247–248
    extrapyramidal disorders, 244–245
    hypokinetic disorders, 245
    pathophysiology, 245–246
    symptoms, 244–245
    treatment, 248–251
    typical posture, 245*f*
Parosmia, 489–490
Parotid gland, 531–532
    adenoid cystic carcinoma of, 358*b*
Paroxysmal dystonia, 511–512
Parry-Romberg syndrome. *See* Facial hemiatrophy
Pars externa, 371–372
Pars interna, 371–372
Pars reticulata, 256
Parvocellular neurosecretory system, 276
Pathologic myopia, 447
P cells, 429–430
Pelvic nerve, 528–529
Pelvic plexuses, 529
Pendular knee jerk, 512–513
Pendular nystagmus, 622, 654
Pentapeptides, 558
Perceiving sound, 382–383
Perforating arteries, 133
Perfusion-weighted imaging, 676
Periaqueductal gray matter, 295
Pericallosal artery, 132
Perikaryon, 10
Perilymph, 367, 376
Perilymphatic compartment, 366
Perineural infiltration, 353
Periodic limb movement disorder, 231
Peripheral nervous system (PNS), 1, 519, 586

Perivascular space, 98
Pesticide toxicity, 44–45
Petrosal ganglion. *See* inferior ganglion
Petrosal sinus
    inferior, 102
    superior, 102
Phantosmia, 491–492
Pharyngeal rami, 531
Pharyngotympanic tube, 364–365
Phonemes, 398–399
Phoria, 335–336
Phosphodiesterase, 422
Photocoagulation, 450–451
Photoreceptors, 419–420, 431–432
photoreceptors, 420
Phrenic nerve, 592
Phrenic nucleus, 611
Physiologic anisocoria, 652–653
Physiologic cup, 440–442
Pia mater, 98
Pigmented layer, 420
Pillar cells, 368
Piloerection, 275
Pilomotor, 527
Pineal gland, 267–269, 431
Pinealoma, 332
Pineal recess, 106
Pinna, 363
Pitch, 382
Pituitary gland, 273
Pituitary insufficiency, 284–285
Placidity, 487–488
Plantar reflex, 607
Planum semilunatum, 374
Plasmapheresis, 621
Plastic/lead-pipe rigidity, 246
Pleocytosis, 116
Plethysmography, 226–227
Plexus of Auerbach, 528
Plexus of Meissner, 528
Pneumatic retinopexy, 451
Pneumococci, 112
Polarization, 389–390
Poliomyelitis, 611
Polyglutamine, 254
Polymerase chain reaction (PCR), 115, 355
Polysynaptic reflex, 605
Pons, 468–469
    components and functions, 296, 297*t*

Pons *(Continued)*
　infarctions of, 307
　location, 296
Pontine arteries, 134
Pontine cistern, 97
Pontine hemorrhage
　causes, 305
　clinical manifestations, 305
　diagnosis, 306
　downward herniation, 305
　lipohyalinosis, 305
　pathophysiology, 305
　treatment, 306
　tumors, 305
　vascular malformations, 305
Pontomedullary junction, 306
Porencephalic cysts, 123
Porus acusticus, 380
Positron emission tomography (PET), 184, 487
Posterior cerebral artery (PCA), 132
Posterior column, 595
Posterior column nuclei, 597
Posterior fossa craniectomy, 352–353
Posterior funiculus, 588
Posterior horns, 588
Posterior intermediate septum, 595–597
Posterior intermediate sulcus, 588
Posterior interposed nuclei, 499–500
Posterior limb, 277
Posterior lobe, 497
Posterior median sulcus, 588
Posterior nucleus, 279
Posterior roots, 590
Posterior thoracic nucleus, 612
Posterolateral, 279
Posterolateral sulcus, 590
Postganglionic neuron, 522–523, 552
Postganglionic sympathetic fibers, 527–528
Postherpetic pain, 352
Postictal psychosis, 487
Postsynaptic neuron, 522–523
Postsynaptic potentials, 31–32
Postural instability, 247
Prefrontal cortex, 182–183
Prefrontal lobotomy, 183
Preganglionic fibers, 527*b*
Preganglionic neuron, 522–523
Preganglionic sympathetic fibers, 524–527
Preganglionic sympathetic nerve, 278

Premotor cortex, 182
Preoptic area, 276
Preoptic (median/lateral) nucleus, 279
Preprotachykinin A gene, 560
Prerubral field, 272
Presbycusis, 394
Presbyopia, 428
Presbyosmia, 490
Presubiculum, 468–470
Presumed ocular histoplasmosis syndrome, 447
Presynaptic neuron, 522–523
Presynaptic regulation, 32
Pretectal region, 273
Prevertebral plexuses, 528–529
Primary fissure, 497
Primary lateral sclerosis, 625
Primary motor cortex, 180
Primary olfactory cortex, 476–477
Primary progression, 622
Primary somatosensory cortex, 180
Primary spiral lamina, 376
Principal sensory nucleus, 345–346
Prion diseases
　Creutzfeldt-Jakob disease, 208–209
　prion protein (PrP), 208
Proenkephalin gene, 558
Profound hearing loss, 386
Progressive bulbar palsy, 625
Progressive muscular atrophy, 625
Progressive pseudobulbar palsy, 625
Progressive supranuclear palsy, 486
Proprioception, 598
Proprioceptors, 15
Propriospinal tract, 595
Prosencephalon, 53, 267
Prosody, 191
Proteoglycans, 426
Protozoal brain diseases, 89–90
Pseudoathetosis, 660
Pseudobulbar affect, 625
Pseudobulbar affective state, 486
Pseudobulbar emotional
　disturbances, 486–487
Pseudobulbar palsy, 486
Pseudobulbar state, 485
Pseudohypertrophy, 657
Pseudoisochromatic Ishihara, 652
Pseudoisochromatic plates, 458
Pseudostrabismus, 336

See also Strabismus
Pseudoxanthoma elasticum, 169
Pterion, 100
Pterygopalatine ganglia, 326
Pterygopalatine ganglion, 531
Pulmonary plexus, 527–528
Pulmonary plexuses, 529
Pulvinar nucleus, 272
Pupillary light reflexes/pathway, 428
Pure-tone audiometry, 385, 397–398
Purines, 550–551*t*, 558–559
Purkinje cells, 497, 502–503, 507*b*, 604
Purulent exudate, 114
Purulent labyrinthitis, 411
Putamen, 82, 241–242
Pyramidal cells, 179
Pyramidal system, 301
Pyramidal tract, 601, 603–604
Pyramids, 301

Q

Quadrantic visual field
 defects, 489
Quadriceps knee jerk, 662
Quasicortical structure, 472

R

Radial brachialis, 662
Radiculomedullary arteries, 613
Radiculomyelopathy, 633
Radiculopathies, 627–628
Ramsay-Hunt syndrome, 355
 clinical manifestations, 408
 diagnosis, 408
 pathophysiology, 407
 treatment, 408
Ranibizumab, 448
Raphe nuclei, 295
Rapid eye movement (REM) sleep behavior
 disorder, 247
Receptive field, 429
Receptor, 605
Receptor potential, 422
Reciprocal inhibition, 609
Red-green color blind, 457
Red nucleus, 295
Referred pain, 534
Reflex activity, 604–605
Reflex arcs
 components of, 605
Reflexes
 crossed effects of, 609
 neurologic examination, 661–663
Reflexive saccade, 431
Regeneration, nerve cells, 34–36
Reissner's membrane, 376
Relapse, 621
Relay stations, 383–385
Remittance, 621
Responsive dystonia, 511
Restiform body, 349–350, 499
Resting tremor, 246
Restless legs syndrome, 231–232
Reticular activating system (RAS), 295
Reticular formation, 297–298
Reticular lamina, 368
Reticular nucleus, 272
Retina, 419
 cell layers of, 418*f*, 421*f*
Retinal, 431–432
Retinal detachment, 448–451, 449*f*, 463–464*b*
Retinal layers, 420
Retinal pigment epithelial cells, 419
Retinal processing, 429
Retinal scanning, 425*b*
Retinitis pigmentosa, 456–457
Retinoblastoma, 459–461, 459*f*
Retrobulbar neuritis, 439–440
Retrognathia, 223
Retrolenticular part, 277
Reuniens nuclei, 469
Rhegmatogenous retinal detachment, 448–450
Rhodopsin, 422
Rhombencephalon, 53
Righting reflexes, 388
Rinne test, 397, 656
Rippling effect, 383
Rods, 420
Romberg's sign, 599, 664–665
Romberg test, 660
Rooting reflex, 664
Rosenthal's canal, 374–375
Rostral, 609–610
Rostrocaudal direction, 306
Rostrolateral group, 499–500
Round window, 364–365
Rubeosis iridis, 452–453
Rubrothalamic fibers, 272

S

Saccular macula, 371–372
Saccule, 366–367
Sacral nerves, 586
Sacral parasympathetic nucleus, 611–612
Sagittal sinus
    inferior, 96, 101, 139
    superior, 96, 101, 139
Salivatory nuclei, 528
Sarcopenia, 152
Satellite cells, 7–8
Savory, 484b
Savory taste, 481
Scala media, 376
Scala tympani, 376
Scala vestibuli, 376
Schistosomiasis, 89
Schizophrenia, 566–571
Schizophrenics, 569
Schwann cells, 7
Sclera, 417
Scleral buckling, 451
Scleral venous sinus, 426
Scleroderma, 353–354
Scotoma, 428, 447
Seasonal effective disorder, 574
Secondary glaucoma, 443–444
Secondary progression, 622
Secretomotor, 537
Segmental dystonias, 509–510
Seizure disorders, 184
    causes of, 212t
    characteristics, 211
    classifications of generalized and partial, 213–215t
    clinical manifestations, 211–215
    diagnosis, 216–219
    epileptiform, 217–218
    pathophysiology, 211
    symptomatic seizures and possible causes, 216t
    Todd's paralysis, 211
    treatment, 219–221
Semicircular canals, 373–374, 386–387
Semicircular duct, 366, 374
Semilunar ganglion, 323–324
Senile cataracts, 426–427
Sensation, 659–661
Sense of smell, 471, 481b
Sense of taste, 484b, 493b
Sensorineural hearing loss, 395
Sensory ataxic gait, 665
Sensory ganglia, 532
Sensory hearing loss, 395, 396t
Sensory modality, 604
Sensory neuron, 605
Sensory nucleus, CN V, 294
Sensory processing, 609–610
Septal nuclei, 467
Septal vein, 137
Septum pellucidum, 104–105, 470–471
Septum transversum, 374
Seroconversion, 116
Serotonin, 268–269, 552, 556
Serotoninergic projections, 469
Serous detachment, 448–450
Severe hearing loss, 386
Sexuality, alterations in, 488
Shorter propriospinal fibers, 595
Short wavelength (S) cones, 457
Sigmoid sinus, 102
Sildenafil, 559
Silver wiring, 455
Sixth cranial nerve palsy, 333–334
Sjögren's disease, 353–354
Sleep apnea, 225–227
Sleep disorders, 247
    characteristics, 221
    diagnosis, 222–223
    Epworth Sleepiness Scale, 222–223
    excessive daytime sleepiness, 221
    idiopathic hypersomnia, 230
    insomnia, 221, 224–225
    narcolepsy, 228–230
    nonrapid eye movement (NREM), 221
    parasomnias, 230–231
    pathophysiology, 221–222
    rapid eye movement (REM), 222
    restless legs syndrome, 231–232
    sleep apnea, 225–227
    treatment, 223–224
Slit-lamp examination, 446
Snellen chart, 651
Snout reflex, 664
Sodium-potassium exchange pump, 24
Solar plexus, 529
Solitariothalamic tract, 272
Solitary nuclei, 298
Somatic motor association area, 182
Somatic motor (SM) division, 590

Somatic motor neurons, 588–590
Somatic motor system, 522
Somatic nervous system (SNS), 1
Somatic pathways, 522, 523t
Somatic reflexes, 605
Somatic sensory (SS), 598f
   division, 590
Somatic system, 552
Somatosensory association cortex, 181
Somatotopic organization, 597–599
Somatotopy, 599b
Somnambulism, 230
Sonophobia, 233
Sound
   perceiving, 382–383
Sound waves
   pathway of, 383
Spasmodic emotional disturbances, 486–487
Spastic dysarthria, 511
Spastic paralysis, 603
Speech audiometry, 385, 398–399
Sphenopalatine ganglion, 531
Sphenoparietal sinus, 102
Sphincteric muscles, 619
Sphincteric reflexes, 664
Spillane's trigeminal neuritis, 354
Spina bifida, 635–637
   complications, 63–65
   with meningocele, 65
   with myelomeningocele, 65–67
Spina bifida cystica, 635
Spinal accessory nerve, 657
Spinal accessory nucleus, 611
Spinal accessory/spino-accessory nerves, 330
Spinal cord
   arteries of, 140–141
      anterior, 134
      posterior, 134
   ascending pathways, 595–601
   blood supply of, 613–614
   body regions innervation, 591–594
   coverings of, 589f
   cross-sectional anatomy, 588–591
   descending pathways, 601–604
   development of, 56–57
   gray matter of, 590f
   gross anatomy and protection, 585–588
   integrative pathways, 604
   meninges of

      arachnoid mater, 103–104
      dura mater, 103
      pia mater, 104
   pathways to, 534
   reflex activity, 604–605
   spinal nerves, 586–588
   spinal reflexes, 605–613
   trauma and compression, 639–641
   veins of, 140–141
   white matter, 591
Spinal cord ischemia, 170–171
Spinal cord lesions/disorders
   amyotrophic lateral sclerosis, 624–626
   cervical spondylosis, 633–635
   Guillain-Barré syndrome, 620–621
   herniated nucleus pulposus, 628–631
   multiple sclerosis, 621–624
   myelopathies, 617–635
   radiculopathies, 627–628
   spina bifida, 635–637
   spinal epidural abscess, 618–620
   spinal muscular atrophies, 631–632
   spondylotic cervical myelopathy, 633–635
   syringomyelia, 637–639
   thoracic outlet compression syndromes, 632–633
   transverse myelitis, 617–618
Spinal cord limbic systems, 468
Spinal cord tumors, 642–643
Spinal dura mater, 585
Spinal epidural abscess, 618–620
Spinal epidural/subdural hematomas, 162–163
Spinal meninges, 103
Spinal muscular atrophies (SMA), 631–632
Spinal nerves, 586–588, 589f, 590
Spinal reflexes, 605–613
Spinal root, 330
Spinal shock, 640
Spinal trigeminal nucleus, 347
Spinocerebellar pathways, 294
Spinohypothalamic fibers, 600
Spinomesencephalic fibers, 600
Spinoreticular fibers, 600
Spinothalamic pathway, 294
Spinothalamic tract, 599–600
Spiral ligament, 366–367
Spiral organ, 367–368
Spiral vessel, 378
Splanchnic nerves, 524–527
Split-brain syndrome, 186

Spondees, 398–399
Spondylotic cervical myelopathy, 633–635
Stance, 659, 664–665
Stapedius reflex, 347
Staphylococci, 113
Startle reflex, 294–295
Static equilibrium, 386–388
Stellate cells, 179
Stem cell treatments, 645b
Stenting, 144–145
Steppage gait, 665
Stereocilia, 369–370
Stereociliary bundles, 386–387
Stereogenesis, 599
Stereognosis, 659–660
Stereotactic radiosurgery, 125
Stereotaxically positioned needle, 352–353
Strabismus, 334f
    characteristics, 334–335
    clinical manifestations, 336
    comitant, 335
    diagnosis, 336
    epicanthal folds, 336
    esotropia, 335
    exotropia, 335
    hypertropia, 335
    hypotropia, 335
    incomitant, 335
    pathophysiology, 335
    phoria, 335–336
    prefixes, 334–335
    treatment, 336–337
    tropia, 335–336
Strachan's syndrome, 442
Straight sinus, 101
Stretch reflex, 605–606, 606b
Striate cortex, 429
Stria vascularis, 366–367
Striola, 387–388
Stroke, 148–149, 393
Strychnine, 557
Stylopharyngeus, 100
Subarachnoid cisterns, 97
Subarachnoid hemorrhage, 97, 163–166, 169
Subarachnoid space, 97
Subclavian arteries, 100–101
Subcostal nerve, 592–593
Subdural hematomas, 160–166
Subdural space, 95

Subdural spinal abscess, 617
Subicular complex, 468–470
Subiculum, 468–470
Subjective tinnitus, 393
Sublenticular part, 278
Sublingual glands, 532
Submandibular ganglia, 326
Submaxillary ganglion, 532
Submaxillary glands, 532
Subretinal space, 447
Substance K, 558
Substance P, 558, 560
Substantia gelatinosa, 600
Substantia nigra, 242, 295
Subthalamonigral fibers, 272–273
Subthalamopallidal fibers, 272–273
Subthalamus
    subthalamic nucleus, 272–273
    zona incerta, 273
Sudden deafness, 403
Sulcocommissural arteries, 613–614
Sulcus, 78
Sulcus limitans, 52
Sundowning, 203–204
Superficial middle cerebral vein, 137
Superficial reflexes, 607–613
Superficial veins, 136–137
Superior anastomotic vein (of Trolard), 137
Superior cerebellar artery (SCA), 134
Superior cerebellar peduncles, 499
Superior cervical sympathetic ganglion, 527
Superior cistern, 97
Superior colliculi, 294–295
Superior ganglion, 327–329
Superior mesenteric ganglia, 524–527
Superior orbital fissure, 345
Superior petrosal sinus, 102
Superior turbinates, 476
Suppurative labyrinthitis, 411
Supracallosal striae, 471
Suprachiasmatic nucleus, 279
Supraclinoid process, 143
Supranuclear facial palsy, 350
Supraorbital branch, 353
Suprapineal recess, 106
Suprasegmental integrations, 475
Suprasellar tumors, 435–436
Supratentorial dura mater membrane, 98–99
Supratrochlear branch, 353

Sustentacular cells, 476
Sympathetic nervous system, 519, 521–522f, 524–528, 525f, 528b, 530t
Sympathetic pathway, 294
Sympathetic tone, upregulating, 200
Synapses
  chemical
    function, 28–30
    types of, 26–28
Synkineses, 356
Syphilis, 538–539, 544b
Syringomyelia, 122, 637–639
Systemic histoplasmosis, 447

T

Tabes dorsalis, 121, 599
Tachykinin, 558
Tachykinin neuropeptide, 560
Tachyzoites, 90
Tactile stimulation, 608–609
Tandem, 664–665
Tangent screen, 652
Tardive dyskinesia, 356
Tardive dystonia, 511
Taste buds, 481–482
  anatomy of, 481–482
  microstructure, 482
Taste sensation
  testing of, 484–485
Tectohabenular fibers, 269
Tectorial membrane, 367–368, 371
Tectorins, 371
Tectotegmentospinal tracts, 269
Tectum, 294–295
Tegmentum, 295
Telangiectatic microangiopathy, 437
Telencephalon, 267
  development of, 60–61
Telodendria, 11
Temporal lobe, 78, 178
Tension, 426
Tension-type headache, 233
Tensor tympani, 347
Tensor tympani muscles, 531–532
Tensor veli palatini muscles, 531–532
Tentorial nerve, 98–99
Tentorium cerebelli, 96, 498
Terminal vein, 137
Terminomas, 281–282

Tetraparesis, 305
Thalamostriate vein, 137
Thalamus, 267
  external medullary lamina, 270–271
  functions of, 271–272
  hypothalamic sulcus, 271
  internal medullary lamina, 271
  interthalamic adhesion, 271
  lentiform complex, 270–271
  reticular nucleus, 270–271
  shape and location, 270, 270f
  stria medullaris, 271
  stria terminalis, 270–271
  thalamic nuclei functions, 278–279
  thalamostriate vein, 270–271
Thermogenesis, 555
Theta waves, 185
Third ventricles, 105–106
Thoracic nerve, 534, 586
Thoracic outlet compression syndromes, 632–633
Thoracolumbar arteries, 613
Thoracolumbar division, 519, 524
Thought insertion, 568–569
Thought withdrawal, 568–569
Thrombectomy, 156
Thrombosis, 142–145
Thrombus, 146
Thunderclap headache, 169, 234
Tic convulsif. See Hemifacial spasm
Tic douloureux. See Trigeminal neuralgia
Tics, 356
Tight junction, 370–371
Tiller testing, 658
Tinnitus
  clinical manifestations, 393
  diagnosis, 394
  pathophysiology, 393
  treatment, 394
Tip links, 369–370
Tizanidine, 624
Tobramycin, 409
Todd's paralysis, 211
See also Seizure disorders
Tone, 382
Tonic pupil, 653
Tonic pupil syndrome, 540
Torticollis, 260, 509–510
Total blindness, 432–434
Toxic amblyopia, 442–443

Toxoplasmosis, 90
Trabeculectomy, 445–446
Trabeculoplasty, 445–446
Traction retinal detachment, 448–450
Tractography, 675
Tractus solitarius nucleus, 348–349
Tractus spiralis foraminosus, 374–375, 379
Transcranial Doppler ultrasonography (TCD), 676–678
Transducin, 422
Transependymal, 313
Transient ischemic attack, 157–159
Transient monocular blindness, 157–158
Transmissible spongiform encephalopathies. *See* Prion diseases
Transverse cerebral fissure, 97, 107
Transverse myelitis, 617–618
Transverse pontine fibers, 296
Transverse sinus, 96, 102, 137
Transverse tracts, 591
Triamcinolone, 448
Triceps, 662
Trigeminal herpes zoster, 358–359b
Trigeminal motor neuropathy, 354
Trigeminal motor nucleus, 346–347
Trigeminal nerves (V), 324f, 345–348, 654–655
    mandibular nerve (V1), 323
    maxillary nerve (V1), 323
    ophthalmic nerve (V1), 323
    semilunar ganglion, 323–324
    trigeminal autonomic cephalgia (TAC), 323–324
Trigeminal neuralgia, 350–351, 357b
    clinical manifestations, 351–352
    diagnosis, 352
    pathophysiology, 351
    treatment, 352–353
Trigeminal neuropathies, 353–354
Trigeminal reflexes, 347–348
Trigeminal sensory nuclei, 345–347
Trigeminocerebellar fibers, 347
Trigeminovascular system, 99
Tritanopia, 457
Trochlear nerves (IV), 322–323, 652–654
Trophic factors, 611
Tropias, 335–336
Truncal ataxia, 661
Trunk ganglia, 524
Tryptophan, 552

Tryptophan hydroxylase, 556
Tuberculosis meningoencephalitis, 119–120
Tuberoinfundibular, 275
Tubular field, 652
Tumor-suppressor gene, 459–460
Tunnel of Corti, 368
2-point discrimination, 659–660
Tympanic cavity, 364–365
Tympanic duct, 376
Tympanic membrane, 363
Tympanogram, 385
Tympanometry, 385, 398, 403
Typist's cramp, 511–512
Tyrosine, 552–553

U

Umami, 483–484, 484b
Umani, 481
Uncal herniation, 167
Uncinate fit, 491–492
Uncinate tract, 499
Unconscious, 190
Uncrenated red blood cell, 670
Unilateral vision loss/abnormalities, 435
Upper arteries, 613
Upper limbs
    brachial plexus and, 592
Upper lumbar nerve, 534
Upper motor neuron damage, 603t
Upper motor neurons, 603
Usher syndrome, 389
Utricle, 366, 372
Utricular macula, 371–372
Utriculus, 367, 372
Uvea, 419

V

Vagus nerves (X), 329–330, 329f, 528–529, 532, 656–657
Valsalva maneuver, 365–366
Vancomycin, 409
Variant CJD. *See* Prion diseases
Various hypothalamic nuclei, 467
Vascular dementia, 200–201
Vascular layer, 419
Vascular supply, 379–380
Vasculitides, 149
Vasoactive intestinal peptide, 560
Vas spirale, 378

Vein of Galen malformation, 169
Veins
  of brain
    deep veins, 137–139
    superficial veins, 137
  of spinal cord, 140–141
Venous angle, 138–139
Venous dural sinuses, 139–140
Ventilation perfusion (V/Q) scan, 147
Ventral amygdalofugal pathway, 468
Ventral group, 470
Ventral median fissure, 56
Ventral posterolateral nucleus (VPL), 597
Ventral posteromedial nucleus (VPM), 597
Ventral posteromedial (VPM) thalamic nucleus, 346
Ventral striatopallidal region, 479
Ventral tegmental area, 468
Ventral thalamus, 272
Ventricles
  fourth, 106
  hydrocephalus, 121–123
  lateral, 104–105
  third, 105–106
Ventricular system, 61
Ventriculoatrial shunts, 123
Ventriculoperitoneal shunts, 123
Ventroanterior (VA) nuclei, 272
Ventrolateral, 279
Ventrolateral (VL) nuclei, 272
Ventro posterolateral (VPL) nuclei, 272
Ventro posteromedial (VPM) nuclei, 272
Vermal portion, 497
Vermis, 83–84, 498
Vertebral and basilar arteries, 134–135
Vertebral arteries, 131, 140–141
Vertical gaze center, 504
Vertigo, 404
  cause, 412b
  clinical manifestations, 404
  complications, 412b
  diagnosis, 404
  pathophysiology, 404
  treatment, 405
Vestibular crest, 371–372
Vestibular disorders, 382b
Vestibular duct, 376
Vestibular ganglion, 381
Vestibular nerve, 363, 380–381

Vestibular pyramid, 371–372
Vestibular rehabilitation therapy (VRT), 412b
Vestibular system microstructure, 389–390
Vestibule, 371–373
Vestibule canals, 386–387
Vestibulocochlear nerves (VIII), 656
  acoustic, auditory, and stato-acoustic nerves, 327
  cochlear nerve, 327
  cochlear nuclei, 327
  function of, 326–327
  vestibular nerve, 327
  vestibular nuclei, 327
Vestibulofacial anastomosis, 381–382
Vestibulo-ocular reflex, 506–507
Vidian nerve, 531
Violent behaviors, 487
Viomycin, 409
Visceral afferent pathways, 532–534
Visceral motor (VM) division, 590
Visceral motor neurons, 15
Visceral pain innervation, 532t
Visceral reflexes, 605
Visceral sensory (VS) division, 590
Visceral sensory neurons, 14–15
Visceromotor neurons, 350
Viscus, 528
Visual acuity, 427–428
Visual association area, 181–182
Visual axis, 423
Visual cortex, 181, 428–430
Visual field, 431
Visual hallucinations, 568–569
Visual pathways, 428–431
Visual pigments, 431–432
Visual processing, in brainstem, 431
Visual reflexes, in equilibrium, 387
Visual reinforcement audiometry, 385
Visual system
  age-related macular degeneration, 446–448
  anterior ischemic optic neuropathy, 438–439
  bitemporal hemianopia, 435–436
  cataract, 446
  central retinal artery occlusion, 451–452
  central retinal vein occlusion, 452–453
  color blindness, 457–458
  diabetic retinopathy, 453–455
  eyeball anatomy, 417–427
  glaucoma, 443–445

Visual system *(Continued)*
   hereditary optic neuropathies, 436–438
   homonymous hemianopia, 434–435
   hypertensive retinopathy, 455–456
   nyctalopia, 458
   optic neuritis, 439–440
   papilledema, 440–442
   pupillary light reflexes and pathway, 428
   retinal detachment, 448–451
   retinitis pigmentosa, 456–457
   retinoblastoma, 459–461
   total blindness, 432–434
   toxic amblyopia, 442–443
   unilateral vision loss and abnormalities, 435
   visual acuity, 427–428
   visual field, 431
   visual pathways, 428–431
   visual pigments, 431–432
Visual vermis, 504
Vitrectomy, 451
Vitreoretinal traction, 448–450
Vitreous body, 426
Vitreous humor, 426
Vogt-Koyanagi-Harada syndrome, 450
Volume, 382
Vomeronasal system, 490

W

Waldenström's macroglobulinemia, 403
Wallerian degeneration, 9
Wartenberg sign, 356
Waterhouse-Friderichsen syndrome, 115
Weber's test, 397
Weber syndrome, 310–311
Weber test, 656
Werdnig-Hoffman disease, 631
Wernicke-Korsakoff syndrome, 200
Wernicke's aphasia, 191–192
Wernicke's area, 182
West Nile virus, 85b, 617–618
White communicating rami, 524
White matter, 591
Willis-Ekbom Disease, 231
Wilson disease, 486
Wilson's disease, 209
Withdrawal reflex, 605, 607
Withdrawal reflexes, 608–609
Wohlfart-Kugelberg-Welander disease, 631
Writer's cramp, 511–512

X

Xanthochromia, 165
X zone, cortico-olivary system, 504

Z

Zona arcuate, 378
Zona incerta, 273
Zona pectinata, 378

9780128174241